Lecture Notes in Computer Science 16595

Founding Editors

Gerhard Goos
Juris Hartmanis

The series Lecture Notes in Computer Science (LNCS), including its subseries Lecture Notes in Artificial Intelligence (LNAI) and Lecture Notes in Bioinformatics (LNBI), has established itself as a medium for the publication of new developments in computer science and information technology research, teaching, and education.

LNCS enjoys close cooperation with the computer science R & D community, the series counts many renowned academics among its volume editors and paper authors, and collaborates with prestigious societies. Its mission is to serve this international community by providing an invaluable service, mainly focused on the publication of conference and workshop proceedings and postproceedings. LNCS commenced publication in 1973.

Tias Guns

Editor

Integration of Constraint Programming, Artificial Intelligence, and Operations Research

23rd International Conference, CPAIOR 2026
Rabat, Morocco, May 26–29, 2026
Proceedings

 Springer

Editor
Tias Guns
KU Leuven
Leuven, Belgium

ISSN 0302-9743 ISSN 1611-3349 (electronic)
Lecture Notes in Computer Science
ISBN 978-3-032-27241-6 ISBN 978-3-032-27242-3 (eBook)
https://doi.org/10.1007/978-3-032-27242-3

This Springer imprint is published by the registered company Springer Nature Switzerland AG
The registered company address is: Gewerbestrasse 11, 6330 Cham, Switzerland

If disposing of this product, please recycle the paper.

Preface

This book constitutes the proceedings of the 23nd International Conference on the Integration of Constraint Programming, Artificial Intelligence, and Operations Research (CPAIOR 2025). The conference was held in Rabat, Morocco at the ESSEC campus from May 26–29, 2026.

The conference received a total of 100 submissions of original unpublished research papers. After an initial screening, 92 papers were sent out to the Program Committee for single-blind peer review, with each paper receiving at least three reviews. The reviewing phase was followed by an author response period. The Program Committee then discussed each paper and made a recommendation for acceptance or rejection. At the end of this process, 37 papers were accepted for presentation at the conference and publication in these proceedings.

The conference included three inspiring invited talks, by Ambros Gleixner (HTW Berlin, Germany) on "On correctness of numerical solvers in mixed-integer optimization", Axel Parmentier (École Nationale des Ponts et Chaussées, France) on "Recent trends in Combinatorial Optimization Augmented Machine Learning" and Yingqian Zhang (Eindhoven University of Technology, Netherlands) on "Learning to Solve Combinatorial Optimization from Data and Language". The conference also featured a masterclass, on Tuesday 26th, on the topic of "Large Language Models for CP/OR", chaired by Serdar Kadioglu. In addition, a number of extended abstracts were accepted for presentation at the conference, but not included in these proceedings.

Of the accepted papers, the paper "Transit Network Design with Two-Level Demand Uncertainties: A Machine Learning and Contextual Stochastic Optimization Framework" by Hongzhao Guan, Beste Basciftci and Pascal Van Hentenryck was selected for the Best Paper Award. The paper "Multi-objective Maximum Satisfiability by Single-Objective Implicit Hitting Set Optimization" by Christoph Jabs, Jeremias Berg and Matti Järvisalo was selected for the Best Student Paper Award. The awards were selected by a sub-committee consisting of the program and conference chairs.

The organization of this conference would not have been possible without the help of many individuals. We would like to thank the Program Committee members and external reviewers for their hard work. We are also very grateful to the Masterclass chair, Serdar Kadioglu. Special thanks go to the conference chairs, Emiliano Traversi and Diego Delle Donne, as well as the entire local organizing team at ESSEC Rabat for their contributions to making this conference a success.

Finally, I would like to thank our sponsors for their generous financial support. At the timing of writing, these include Hexaly, the Association for Constraint Programming (ACP), The Optimization Firm and Cosling.

May 2026   Tias Guns

Organization

Program Chair

Tias Guns — KU Leuven, Belgium

Conference Chairs

Emiliano Traversi — ESSEC Business School, France
Diego Delle Donne — ESSEC Business School, France

Masterclass Chair

Serdar Kadioglu — Fidelity Investments, USA

Program Committee

Chris Beck — University of Toronto, Canada
Senne Berden — KU Leuven, Belgium
Timo Berthold — Technische Universität Berlin, Germany
Armin Biere — University of Freiburg, Germany
Henk Bierlee — KU Leuven, Belgium
Ignace Bleukx — Katholieke Universiteit Leuven, Belgium
Merve Bodur — University of Edinburgh, UK
Víctor Bucarey — Universidad de O'Higgins, Chile
Quentin Cappart — École Polytechnique de Montréal, Université de Montréal, Canada
Mats Carlsson — RISE Research Institutes of Sweden AB, Sweden
Andre Augusto Cire — University of Toronto, Canada
Sophie Demassey — Mines Paris - PSL, France
Guillaume Derval — University of Liège, Belgium
Bistra Dilkina — University of Southern California, USA
Pierre Flener — Uppsala University, Sweden
Luca Di Gaspero — University of Udine, Italy
Simon de Givry — INRAE, National Research Institute for Agriculture and Environment, France

Ambros Gleixner	HTW Berlin, Germany
Paul Grigas	University of California, Berkeley, USA
Stefano Gualandi	University of Pavia, Italy
Pascal Van Hentenryck	Georgia Institute of Technology, USA
Willem-Jan van Hoeve	Carnegie Mellon University, USA
Matti Järvisalo	University of Helsinki, Finland
Serdar Kadioglu	Fidelity Investments, USA
George Katsirelos	INRAE, France
Zeynep Kiziltan	University of Bologna, Italy
Lucas Kletzander	Technische Universität Wien, Austria
Thorsten Koch	Zuse Institute Berlin, Germany
Lars Kotthoff	University of Wyoming, USA
Jimmy H. M. Lee	Chinese University of Hong Kong, China
Michele Lombardi	University of Bologna, Italy
Orestis Lomis	KU Leuven, Belgium
Pierre Lopez	CNRS, France
Sven Löffler	Brandenburgische Technische Universität Cottbus–Senftenberg, Germany
Irfan Mahmutogullari	KU Leuven, Belgium
Arnaud Malapert	Université Côte d'Azure, France
Jayanta Mandi	KU Leuven, Belgium
Ciaran McCreesh	University of Glasgow, UK
Kostis Michailidis	KU Leuven, Belgium
Laurent Michel	University of Connecticut, USA
Andrea Micheli	Fondazione Bruno Kessler, Italy
Ioannis Mourtos	Athens University of Economics and Business, Greece
Nysret Musliu	Technische Universität Wien, Austria
Mohsen Nafar	Brandenburgische Technische Universität Cottbus–Senftenberg, Germany
Giacomo Nannicini	University of Southern California, USA
Nina Narodytska	VMware, USA
Peter Nightingale	University of York, UK
Sophie N. Parragh	Johannes Kepler Universität Linz, Austria
Justin Pearson	Uppsala University, Sweden
Laurent Perron	Google, France
Gilles Pesant	École Polytechnique de Montréal, Université de Montréal, Canada
Thierry Petit	Emotia (Politics), France
Wout Piessens	KU Leuven, Belgium
Claude-Guy Quimper	Université Laval, Canada
Günther Raidl	Technische Universität Wien, Austria

Jean-Charles Régin	Université Côte d'Azur, France
Louis-Martin Rousseau	École Polytechnique de Montréal, Université de Montréal, Canada
Michael Römer	Universität Bielefeld, Germany
Elina Rönnberg	Linköping University, Sweden
Khadija Hadj Salem	KU Leuven, Belgium
Domenico Salvagnin	University of Padua, Italy
Scott Sanner	University of Toronto, Canada
Andrea Schaerf	University of Udine, Italy
Pierre Schaus	KU Leuven, Belgium
Andreas Schutt	CSIRO, Australia
Thomas Sergeys	KU Leuven, Belgium
Thiago Serra	University of Iowa, USA
Paul Shaw	International Business Machines, France
Mohamed Siala	LAAS/CNRS, France
Angelo Sifaleras	University of Macedonia, Greece
Kostas Stergiou	University of Western Macedonia, Greece
Kevin Tierney	Universität Vienna, Austria
Michael Trick	Carnegie Mellon University, Qatar
Dimos Tsouros	KU Leuven, Belgium
Petr Vilím	Charles University, Czechia
Mathijs de Weerdt	Delft University of Technology, Netherlands
Neil Yorke-Smith	Delft University of Technology, Netherlands
Yingqian Zhang	Eindhoven University of Technology, Netherlands

Contents

Backbone-Based Predict and Search for Pseudo-Boolean Optimization

Bryan Alvarado-Ulloa[1], Bistra Dilkina[2], Dorit S. Hochbaum[3],
Ricardo Ñanculef[1], and Roberto Asín-Achá[1]([✉])

[1] Universidad Técnica Federico Santa María, Valparaíso, Chile
`{bryan.alvarado,ricardo.nanculef,roberto.asin}@usm.cl`
[2] University of Southern California, Los Angeles, CA, USA
`dilkina@usc.edu`
[3] University of California, Berkeley, USA
`dhochbaum@berkeley.edu`

Abstract. Diverse combinatorial optimization problems can be modeled as instances of the Pseudo-boolean Optimization (PBO) problem. The Predict-and-Search (PaS) framework is a powerful technique applied to Mixed Integer Linear Programming (MIP) that uses Graph Neural Networks (GNNs) to predict candidate variable values for guiding the search process of an optimization solver. Current PaS implementations rely on heuristically chosen labels during training and produce fast, good solutions, sacrificing optimality guarantees.

We present *Backbone-based Predict and Search* (BACKPAS), a specialized PaS framework for PBO. Our main contribution is redefining the GNN prediction task to identify *backbones*—literals fixed across all optimal solutions. Predicting these critical variables accelerates the search toward optimality if the network's prediction is correct. BACKPAS uses a specialized GNN architecture built on a literal-based bipartite graph to predict backbone membership and polarity, formulated as a multiclass classification problem. Then, a parameterized adaptive trust region incorporates these GNN predictions to adjust the solver's search space.

Empirically, we demonstrate that BACKPAS effectively learns backbone patterns across PBO benchmarks, including Maximum Independent Set (MIS), Minimum Vertex Cover (MVC), and Combinatorial Auctions (CA). We trained the model on small instances for which we computed their backbones. When tested on much larger instances (up to 6× the training size), our method achieves significant performance gains compared to the commercial solver GUROBI and the state-of-the-art PaS implementation CONPAS, substantially improving its anytime behavior and demonstrating strong generalization capabilities.

Keywords: Backbone · PBO · GNN · Predict-and-Search

1 Introduction

Pseudo-Boolean Optimization (PBO) [10] extends Boolean satisfiability (SAT) by adding an objective function and allowing inequalities over weighted Boolean

© The Author(s), under exclusive license to Springer Nature Switzerland AG 2026
T. Guns (Ed.): CPAIOR 2026, LNCS 16595, pp. 1–18, 2026.
https://doi.org/10.1007/978-3-032-27242-3_1

polynomials-the product of Boolean variables and their negations. PBO is an important framework for solving fundamental combinatorial optimization problems [3, 12], including Minimum Vertex Cover (MVC), Maximum Independent Set (MIS), and Set Cover (SC). Despite significant advances in PBO-solving techniques [1, 13, 14, 18, 32], many practical instances remain challenging.

Recent developments in machine-learning-guided search have shown promising results in improving traditional solving approaches [4–6, 11, 15, 21, 23, 26, 27, 29, 30, 33, 34, 37]. For SAT problems, NeuroBack [38] demonstrated how predicting *backbone* variables—those with fixed values across all solutions—can effectively guide solvers. For Mixed Integer Linear Programming (MILP) problems, the Predict-and-Search [19] (PaS) framework introduced a different approach based on learning from heuristically picked labels (high/low-quality solutions) and constrained neighborhoods around predicted values (*trust regions*), where the solver searches for high-quality feasible solutions. Within these trust regions, the original problem is solved with additional constraints that limit the distance between the final solution and the machine-learning model's prediction, achieving both solution quality and feasibility. The rationale behind PaS is that, in typical applications, new instances have the same structure as previously solved ones and, hence, it is practical to specialize a machine learning model to predict variable values for a specific distribution of instances of a given problem. While previous approaches have proven effective, the concept of backbones has not been exploited in machine-learning-guided search for optimization problems like PBO or Mixed Integer Linear Programming (MILP).

Here we present BackPaS, a machine-learning-guided framework that uses GuroBack, a newly developed PBO backbone extractor, for training data generation and combines backbone prediction with trust region search to solve PBO instances. Our key insight is that by using backbones, we learn structural information about the entire instance (all optimal solutions) instead of relying on the quality of specific assignments, improving PaS's sample efficiency. In addition, we found that a GNN trained on small instances where GuroBack efficiently computes backbones can be successfully transferred to larger instances. By using the model's predictions, we propose a PBO-specific learning approach that adapts the selection of variables and trust regions of the MILP PaS framework to the prediction of Backbones and uses a GNN specialized for PBO formulas and backbone prediction.

The key contributions of our work are:

1. A scalable training approach that uses small instances (for MIS, MVC, and CA, respectively) with GuroBack-computed backbones to learn patterns that generalize to much larger problems, up to a factor of 6X, as demonstrated empirically.
2. A novel GNN architecture, specialized for PBO, which captures the binary nature of the variables, the logical constraints, and optimization objectives, while maintaining generalization capabilities.
3. Significant performance improvements over previous PaS implementations in the primal integral metric [2], which captures the anytime performance

compared to the best-known-solution as a function of the running time. This is empirically demonstrated here, particularly for the hardest large-scale instances.

2 Background and Related Work

2.1 Pseudo-Boolean Optimization

The Pseudo-Boolean Optimization (PBO) problem extends SAT by allowing PBO constraints and adding an objective function to optimize. Comprehensive reviews can be found in [9, 12].

Let $\mathbf{x} = \{x_1, \ldots, x_n\}$ be Boolean variables with literals $\ell_{2i} = x_i$ and $\ell_{2i+1} = \bar{x}_i$ for all $i \in \{1, \ldots n\}$. A general pseudo-Boolean constraint (C_s) is defined as $\sum_{i=1}^{k} a_i \prod_{j \in T_i} \ell_j \geq b$ where $a_i \in \mathbb{R}^+$, $b \in \mathbb{R}^+$, and term T_i indexes groups of literals multiplied by a_i. A PBO instance consists of constraints $\phi = \{C_1, \ldots, C_m\}$ and objective function $\min f(\mathbf{x}) = \sum_{i=1}^{p} c_i \prod_{j \in T_i} \ell_j$ with $c_i \in \mathbb{R}^+$ and sets T_i as above. The goal is to assign Boolean values to the variables to satisfy all constraints while minimizing the objective.

Although general PBO involves products of literals, any multi-linear polynomial PBO instance reduces to a quadratic form in polynomial time. Furthermore, quadratic PBO instances can be linearized in polynomial time, yielding linear PBO instances equivalent to Binary Integer Linear Programming (BILP) [10]. Most current state-of-the-art PBO solvers only accept linearized PBO instances that are equivalent to BILP.

Extensive research has developed Machine Learning (ML) models to predict optimal binary variable values for enhancing Mixed Integer Linear Programming (MILP) optimization [6, 11, 15, 17, 21, 25–27, 29, 30]. One particularly successful approach is the Predict and Search (PaS) framework [19], reviewed in the next subsection.

2.2 The Predict and Search (PaS) Framework

The Predict and Search (PaS) framework [19] enhances MILP solver performance by using ML to predict binary variable values. PaS operates in two stages:

1. **Prediction Step:** An ML model is trained on instances with known quality assignments (e.g., low-quality, high-quality, optimal) to predict binary variables for unseen MILP instances, guiding the subsequent search process.
2. **Search Step:** The ML prediction constructs a partial assignment by selecting a subset S of binary variables and assigning predicted values $v_i \in \{0, 1\}$ to variable x_i ($i \in S$). The original MILP is then modified to include only solutions within a *trust region* that allows controlled deviations from this partial assignment, ensuring robustness to prediction inaccuracies.

The trust region is defined by a user-specified non-negative integer parameter Δ called *tolerance*. For binary indicators $\delta_i \in \{0, 1\}$, $\delta_i = 1$ allows x_i to deviate from v_i, and $\delta_i = 0$ enforces $x_i = v_i$. The tolerance permits up to Δ violations of the predicted assignments as follows:

$$x_i \leq \delta_i \qquad \forall i \in \{j \in S \mid v_j = 0\} \tag{1}$$

$$1 - x_i \leq \delta_i \qquad \forall i \in \{j \in S \mid v_j = 1\} \tag{2}$$

$$\sum_{i \in S} \delta_i \leq \Delta. \tag{3}$$

We now explain how this framework has been implemented in recent works.

Predict and Search (PaS) by [19]. This is the implementation provided by the authors that first proposed the framework. Their **Prediction Step** uses a Graph Neural Network (GNN) to learn solution distributions for binary variables of MILP problems from high-quality solutions.

The approach transforms MILP instances into bipartite graphs where variables and constraints become nodes, connected by edges when variables appear in constraints. Node features encode relevant variable and constraint metrics. The GNN uses two Bipartite Graph Convolution layers with message passing, followed by a final layer mapping each binary variable to a probability score between 0 and 1. The GNN is trained using energy-weighted cross-entropy loss.

For the **Search Step**, the partial assigment is constructed using parameters k_0 and k_1 to select variables: the k_1 highest-scoring variables are set to $v_i = 1$, and the k_0 lowest-scoring variables to $v_i = 0$. This partial assignment defines the trust region constraints with radius Δ.

Contrastive Predict and Search (ConPaS) by [20]. The CONPAS framework [20] advances the PaS methodology by incorporating contrastive learning into its **Prediction Step**. While it uses the same GNN architecture as [19] and transforms MILP instances into bipartite graphs in the same way, its training methodology differs. Unlike PaS, which trains solely on high-quality solutions, CONPAS trains its GNN using both high-quality solutions (as positive samples) and infeasible or low-quality solutions (as negative samples). [20] provides the algorithms to construct these high and low quality solutions.

The **Search Step** in CONPAS is executed identically to that in original PaS, utilizing the same approach for constructing the partial assignment based on the ML predictions (via k_0 and k_1) and defining the trust region with parameter Δ. The primary distinction between the two works lies in their distinct training methodologies for the prediction.

2.3   Backbones

The *backbone* [31] of an optimization instance is the set of pairs (variable, value) present in all the optimal solutions of the instance. Formally, for an optimization instance P with a set of constraints ϕ and objective function f, let $\mathrm{OPT}(P)$ be the optimal value of f. A pair (x_i, v_i) is in the backbone if $x_i = v_i$ for every assignment vector $\mathbf{v}$ such that $f(\mathbf{v}) = \mathrm{OPT}(P)$. Determining the backbone of a MIP or PBO instance is co-NP-complete [24].

2.4 NeuroSAT and NeuroBack

NeuroSat is among the earliest methods to incorporate neural networks into SAT solving. In our work here, we use concepts that originate in NeuroSat for Boolean problems. NeuroBack is the first method to use backbones as training signals for machine-learning-guided search. We extend this idea to the PBO domain and the PaS framework.

NeuroSAT [35] relies on a message-passing neural network trained to classify whether an instance is satisfiable or not. It represents formulas as graphs connecting literals and clauses, and iteratively propagates messages to find satisfying assignments. Notably, when NeuroSAT predicts a formula to be satisfiable, it also provides a certificate, the corresponding variables' assignment.

Based on NeuroSAT's ideas, NeuroBack [38] integrates GNN predictions within Conflict-Driven Clause Learning (CDCL) SAT solvers [9], rather than adopting an end-to-end strategy. It changes the learning task from feasible assignment prediction to identifying backbone variables, those that maintain a constant value in all solutions, and performs this prediction once at the start of the tree search to avoid frequent GNN calls during solving. The extraction of backbone variables is done with CadiBack [8], a solver extension for SAT backbone identification. By combining CadiBack's efficient backbone extraction with neural phase prediction, NeuroBack is reported to achieve significant gains, solving up to 5.2% and 7.4% more problems than baseline solvers on recent SAT competition tests.

3 BackPaS: Backbone-based Predict and Search for PBO

We introduce BackPaS[1], Backbone-based Predict and Search, an adaptation and extension of the PaS framework that uses backbone information to enhance performance on PBO problems, among other innovations. Our focus on PBO instances aligns with the original predict-and-search framework's primary emphasis on binary variable predictions and capitalizes on the natural extensibility of backbones to this domain. We note that an example in BackPaS conveys information about the entire instance (all optimal solutions) while an example in previous PaS implementations conveys information only about the quality of specific assignments. This difference allows us to hypothesize that BackPaS captures global structural properties, providing richer guidance for the search process. Our modifications to the PaS framework are in both the Prediction and Search stages, as follows:

3.1 Prediction Step

Backbone Extraction: To extract the ground-truth backbone labels needed for training, we used GuroBack[2], a general-purpose backbone extractor

---

[1] https://github.com/bryan-alvarado-ulloa/backpas.
[2] https://github.com/bryan-alvarado-ulloa/guroback.

for optimization problems with binary variables we implemented [16]. For Pseudo-Boolean Optimization (PBO) instances, where all variables are binary, GuroBack computes the complete backbone. The algorithm is an adaptation of CadiBack [8], a SAT backbone extractor, and employs an iterative refinement strategy. In each step, GuroBack classifies a subset of the binary variables as being part of the backbone or not. This classification is performed by calling Gurobi, which checks if it's possible to change the assignments of these variables while maintaining the optimum objective value.

Prediction of the Backbone: To predict the backbones of PBO instances, we propose a new GNN architecture that receives as input a modified version of the standard bipartite graph representation previously proposed for MILP instances. In this modified representation, each variable node is replaced by two nodes representing its positive and negative literals. To ensure a consistent encoding, we normalize the instances to strictly use non-negative coefficients. Specifically, any weighted sum of variables $\sum_{i=1}^{n} a_i x_i$ where a_i denotes the coefficient of variable x_i, is transformed into:

$$\sum_{i \in I^+} a_i x_i + \sum_{i \in I^-} (-a_i)\bar{x}_i + \sum_{i \in I^-} a_i, \tag{4}$$

where a_i is the coefficient associated with the variable x_i, $I^+ = \{i \in \{1, \ldots, n\} \mid a_i \geq 0\}$, and $I^- = \{i \in \{1, \ldots, n\} \mid a_i < 0\}$. Here, I^+ and I^- represent the indices of variables with non-negative and negative coefficients, respectively. The resulting constant term $\sum_{i \in I^-} a_i$ is used to adjust the constraint's right-hand side or the objective function's offset. Edges are then established between literal nodes and constraint nodes based on this normalized formulation, with node embeddings initialized using literal-based metrics. The details of these metrics are provided in thesupplementary material[3].

We further enhance the GNN architecture by replacing the original Bipartite Graph Convolutional layers with *Bipartite Graph Transformer* layers (adapting the graph transformer from [36] for the bipartite case), and increasing the network depth from 2 to 8 layers. The output layer is reconfigured for 3-class classification, predicting the probability distribution $\mathbf{p}_i$ for each variable x_i over the classes: B0 (negative literal in backbone), B1 (positive literal in backbone), and NB (non-backbone).The network is trained to minimize the Mean Cross-Entropy (MCE). To prevent bias toward larger instances, we calculate MCE by averaging the negative log-likelihood first over the variables V_G in each graph G, and subsequently across the dataset $\mathcal{G}$:

$$\text{MCE} = \frac{1}{|\mathcal{G}|} \sum_{G \in \mathcal{G}} \left[\frac{1}{|V_G|} \sum_{i \in V_G} -\log(p_{i,y_i}) \right], \tag{5}$$

where y_i denotes the true class label of variable x_i.

---

[3] https://github.com/bryan-alvarado-ulloa/backpas/blob/main/appendix/A.pdf.

3.2   Search Step

We propose a new method to construct the trust region that automatically chooses which variables are selected (S), which values should be assigned (v_i), and how many mistakes will be allowed in those assignments (Δ). These decisions are guided by the model's predicted probabilities, adapting the trust region characteristics for each instance. The configuration is controlled by two parameters: $\theta \in [0, 1]$, which defines the minimum confidence required for selecting variables, and $\alpha \in [-1, 1]$, which allows for the calibration of the model's confidence on unseen distributions of instances, as explained below.

Given predicted probabilities $p_{i,\mathrm{B0}}$, $p_{i,\mathrm{B1}}$, and $p_{i,\mathrm{NB}}$ for each variable x_i, the trust region construction proceeds as follows:

Variable Selection (S_θ): For a threshold parameter $\theta \in [0, 1]$, we select all variables whose predicted backbone class has confidence at least θ:

$$S_\theta = \{i \mid \max(p_{i,\mathrm{B0}}, p_{i,\mathrm{B1}}) \geq \theta\}.$$

Value Assignment (v_i): Each selected variable x_i for $i \in S_\theta$ is assigned the value corresponding to its most probable backbone class:

$$v_i = \begin{cases} 0 & \text{if } p_{i,\mathrm{B0}} \geq p_{i,\mathrm{B1}}, \\ 1 & \text{otherwise.} \end{cases}$$

Adaptive Tolerance (Δ): The parameter Δ corresponds to the number of incorrect predictions we allow when enforcing the assignments. This directly affects the size of the trust region: setting Δ too small risks excluding the optimal solutions, while a large Δ will not significantly reduce the search space. Therefore, selecting Δ appropriately for each instance is key to balancing efficiency and solution quality.

If we want a trust region that reduces the search space as much as possible without losing the optimal solutions, we would ideally set Δ equal to the actual number of incorrect predictions in S_θ. However, since this is not possible, we estimate the expected number of errors based on the model's confidence. We define:

$$E = \sum_{i \in S_\theta} (1 - \max(p_{i,\mathrm{B0}}, p_{i,\mathrm{B1}})),$$

which aggregates the uncertainty across all selected variables. A larger E suggests greater uncertainty and, consequently, a higher expected number of errors.

When inference-time instances differ from those seen during training, E may systematically (under/over) estimate the true error count. Hence, we do not directly set $\Delta = E$. Instead, we introduce a hyperparameter $\alpha \in [-1, 1]$ (to be tuned in validation) to flexibly adjust Δ based on E and the size of the selected set $|S_\theta|$ consistently:

$$\Delta = \begin{cases} \lceil E \cdot (1 + \alpha) \rceil, & \text{if } \alpha \leq 0, \\ \lceil (|S_\theta| - E) \cdot \alpha + E \rceil, & \text{if } \alpha > 0. \end{cases}$$

This formulation interpolates between different strategies:

- $\alpha = -1$: $\Delta = 0$, trusting all predicted assignments with no room for errors.
- $\alpha = 0$: $\Delta = \lceil E \rceil$, relying directly on the estimated number of errors.
- $\alpha = 1$: $\Delta = |S_\theta|$, meaning all assignments can be violated simultaneously, so the trust region does not restrict the search space.

We note here that this construction of the trust region, in contrast with the one used for PAS and CONPAS, is *instance-size-independent* and can be used for a testing on sets consisting of instances of different sizes.

3.3 BACKPAS Variants

To isolate and evaluate the impact of the specific design choices underpinning BACKPAS, we introduce intermediate variants that incrementally incorporate the proposed modifications. This methodical approach allows us to attribute performance changes to individual components: the prediction task redesign, the proposed GNN architecture, and the adaptive trust region parameterization.

- **BackPaS-V0: Backbone Prediction Only.** This foundational variant maintains the original GNN architecture and bipartite graph representation used by CONPAS and earlier PAS implementations, along with their fixed trust region parameters (k_0, k_1, Δ). The sole modification is the replacement of the original regression output with the proposed *three-class classification* scheme. Each variable x_i is classified into one of three backbone classes: B0 (fixed to 0), B1 (fixed to 1), or NB (not in the backbone). Consequently, the trust regions are constructed by selecting the top k_0 variables with the highest $p_{i,\text{B0}}$ scores and the top k_1 variables with the highest $p_{i,\text{B1}}$ scores.
- **BackPaS-Net: Incorporating the GNN Architecture.** This variant builds upon BACKPAS-V0 by integrating our novel *specialized GNN architecture*. This architecture features eight Bipartite Graph Transformer layers and operates on the literal-based bipartite graph representation of the PBO instances. All other components remain identical to BACKPAS-V0.
- **BackPaS-Param: Incorporating Adaptive Parameters.** This variant also builds upon BACKPAS-V0, but instead of the network change, it incorporates the proposed *adaptive trust region parametrization* defined by the parameters (θ, α). This allows the search space adjustment to dynamically respond to the GNN's prediction confidence, replacing the fixed parameters (k_0, k_1, Δ) used in BACKPAS-V0.
- **BackPaS: Full Model.** This is the complete and final model, incorporating *all proposed modifications*: the backbone classification task, the specialized GNN architecture (BACKPAS-NET change), and the adaptive trust region parametrization (BACKPAS-PARAM change).

4   Experimental Setup

We present a comprehensive empirical evaluation of our (BACKPAS) method and its variants. A series of experiments was conducted to assess the impact of our core innovations: backbone-based predictions, literal-based bipartite graph representation, specialized GNN architecture, and the novel parameterization of the search step. Our evaluation primarily uses the primal integral metric (defined below) to compare the different methods, besides standard ML metrics (loss, F1 score) for analyzing the prediction model's performance.

4.1   Computational Environment

Hardware: For CPU-intensive tasks, such as CONPAS solutions extraction (both good- and bad-quality), backbone extraction with GUROBACK, and instance solving with Gurobi, we used a computing cluster with nodes equipped with Intel Xeon E5-2670 v3 processor and 64 GB of RAM. For GPU-intensive tasks, such as network training and network inference for instance modification, we used an NVIDIA A40 GPU with 48 GB of VRAM.
Software: We used Gurobi v10.0.0rc2, PyTorch v2.7.1 (built with CUDA 12.8), and PyTorch Geometric v2.6.1. The code for GUROBACK and BACKPAS are publicly available and the repositories are provided below. Both hardware configurations described before run under Rocky Linux 8.
Resource Allocation: For backbone extraction with GUROBACK and instance solving with Gurobi, we used up to 12 concurrent executions. Each instance was allocated 5 GB RAM and one CPU core. For CONPAS solution extraction (good- and bad-quality solutions), up to 12 instances were processed in parallel using single cores without setting RAM limits.

4.2   Datasets/Benchmarks

Table 1. Benchmarks and partitions used, including their main characteristics and origins. All MIS and MVC instances are BarabasiAlbert graphs, and all CA instances follow the Arbitrary distribution [28]. Note that the partitions used for training BACKPAS are considerably smaller than those used for CONPAS and for `test`.

Benchmark	Partitions	Characteristics	Origin
MIS	`train/val-backpas`	1000 nodes, 4 avg. degree	D-MIPLIB
MIS	`train/val-conpas, val, test`	6000 nodes, 5 avg. degree	Generated
CA	`train/val-backpas`	300 items, 600 bids	Generated
CA	`train/val-conpas, val, test`	2000 items, 4000 bids	D-MIPLIB
MVC	`train/val-backpas`	1200 nodes, 5 avg. degree	D-MIPLIB
MVC	`train/val-conpas, val, test`	6000 nodes, 5 avg. degree	Generated

Table 1 summarizes the details of each benchmark and partition used in this work. We use three benchmarks: Maximum Independent Set (MIS), Minimum Vertex Cover (MVC), and Combinatorial Auctions (CA), matching CONPAS [20] benchmarks except for the Item Placement benchmark, which is not PBO.

Due to backbone extraction costs, BACKPAS requires smaller training instances than CONPAS, which trains on test-sized instances without computing backbones. We define the following partitions for MIS, CA, and MVC:

- `train-backpas`/`val-backpas`: The BACKPAS GNN was trained on 800 instances and validated on 100 instances, all utilizing computed backbones. The model checkpoint corresponding to the lowest validation Cross-Entropy Loss was selected for inference.
- `train-conpas`/`val-conpas`: The CONPAS GNN was trained on 400 instances and validated on 100 instances, using the positive and negative labels obtained using CONPAS. The model checkpoint corresponding to the lowest validation Contrastive Loss was selected for inference.
- `val`: 100 instances for hyperparameter tuning in BACKPAS (α, θ) and in BACKPAS-V0 (k_0, k_1, Δ).
- `test`: 100 instances for final primal integral evaluation.

Partitions `val-conpas` and `val` are the same and distinguished only by their use. We used D-MIPLIB [22] datasets when available, otherwise generating instances using D-MIPLIB generators (MVC, CA) and GeCo library (MIS).

For BACKPAS, extracting backbone labels from the `train-backpas` and `val-backpas` partitions required approximately 55 CPU hours for CA, 279 for MIS, and 12 for MVC. In contrast, CONPAS obtains positive training samples by solving instances from the `train-conpas` and `val-conpas` partitions using Gurobi with a 1-hour timeout per instance. Since none of these instances are solved within this limit, the extraction of the training signal for CONPAS required at least 500 CPU hours per benchmark.

4.3   Evaluation Metrics

Pipeline Performance (Primal Integral): The Primal Integral [2] quantifies how quickly the best solution $\mathbf{v_t}$ at time t approaches the best known solution $\mathbf{v_{opt}}$ (over a time frame of $1000\,\text{s}$). It integrates the Primal Gap function $p(t)$, which measures the relative quality of the incumbent solution $\mathbf{v}$, found by t:

$$p(t) = \begin{cases} 1 & \text{if no incumbent at } t \\ 1 & \text{if } f(\mathbf{v_t}) \cdot f(\mathbf{v}_{opt}) < 0 \\ 0 & \text{if } f(\mathbf{v_t}) = f(\mathbf{v}_{opt}) = 0 \\ \frac{|f(\mathbf{v_t}) - f(\mathbf{v}_{opt})|}{\max\{|f(\mathbf{v_t})|, |f(\mathbf{v}_{opt})|\}} & \text{otherwise} \end{cases}$$

where $f(\mathbf{v})$ denotes the objective function value of solution $\mathbf{v}$. We report the primal integral (PI) at $1000\,\text{s}$ $PI(1000) = \int_0^{1000} p(t)\,dt$, with lower values indicating faster convergence. When reporting this metric, we exclude the time required to run the GNN and construct the trust regions for all methods. A detailed report of the overheads on each pipeline step is provided in thesupplementary material[4].

As a summary of the three tasks contributing to overhead, graph construction is the dominant cost (5075% of the overhead in BackPaS-V0/ConPaS; 6785% in BackPaS). The literals-based graph in BackPaS is up to 3x slower than BackPaS-V0/ConPaS, yet in the worst-case (1.79 s for MVC) it consumes only 0.18% of the 1000 s budget, a negligible cost for 1540% primal integral improvements on MIS/MVC.

GNN Prediction Performance (Cross-Entropy): We evaluate the divergence between the predicted class distribution and the true labels using the Mean Cross-Entropy (MCE) defined in Eq. 5. This metric ensures that performance is weighted equally across instances regardless of their size.

4.4   Hyperparameter Tuning

The trust region parameters, k_0, k_1, and Δ for CONPAS and BACKPAS-V0, or θ and α for BACKPAS, were fine-tuned. For BACKPAS and BACKPAS-V0, this was performed using Bayesian Optimization with Tree-structured Parzen Estimators (TPE) [7], optimizing over the `val` partitions to minimize the primal integral at 1000 s. The optimization process began with a random sample of 30 points in the hyperparameter space, evaluated in parallel. Subsequently, the TPE method was used to suggest three sequential batches of 10 points each, which were also evaluated in parallel. For CONPAS, since we used the same validation sets, we utilized the same hyperparameter values as reported in their original paper to ensure consistency with their findings. The final parameter values used for each method and benchmark are provided in the appendix[5].

5   Experimental Results

We evaluate the performance of BACKPAS through a series of experiments. Subsection 5.1 benchmarks our method against previous PaS implementations and ablation variants. Subsection 5.2 details the effect of architectural choices on the GNN. Finally, although the PaS framework was not originally designed for varying instance distributions, in Subsects. 5.3 we investigate the robustness of BACKPAS when applied to the MIS problem over different graph distributions.

5.1   Performance Evaluation of BACKPAS and Variants

Table 2 presents our experimental results by comparing the anytime performance of four methods across three domains—Combinatorial Auctions (CA),

---

[4] https://github.com/bryan-alvarado-ulloa/backpas/blob/main/appendix/B.pdf.
[5] https://github.com/bryan-alvarado-ulloa/backpas/blob/main/appendix/C.pdf.

Table 2. Primal Integral comparison on the test sets. Mean (PI) values, with standard deviations (in parentheses), are reported for the CA, MIS, and MVC test partitions. *Lower PI values indicate better anytime performance.* The ranks provided (e.g., 1st) are determined by pairwise Wilcoxon signed-rank tests ($p < 0.05$) corrected for multiple tests using BenjaminiHochberg correction. Methods with the same rank are not statistically significantly different. Rank n methods perform statistically better than all methods at ranks $> n$, and statistically worse than all methods at ranks $< n$.

Method	CA	MIS	MVC
GUROBI	20.74 (6.08) 2nd	44.09 (7.73) 6th	7.79 (1.15) 6th
CONPAS	19.50 (7.36) 2nd	8.93 (1.94) 5th	6.83 (2.53) 5th
BACKPAS	**16.09** (6.79) 1st	**0.82** (0.53) 1st	**0.49** (0.50) 1st
BACKPAS-PARAM	30.13 (9.54) 4th	4.09 (1.44) 4th	1.50 (0.45) 2nd
BACKPAS-NET	24.05 (10.22) 3rd	1.13 (0.17) 2nd	1.74 (0.49) 3rd
BACKPAS-V0	17.11 (6.79) 1st	2.31 (0.61) 3rd	2.20 (0.44) 4th

Maximum Independent Set (MIS), and Minimum Vertex Cover (MVC)—using the primal integral at 1000 s.

The compared methods are: GUROBI, CONPAS, BACKPAS-V0, BACKPAS-PARAM, BACKPAS-NET, and BACKPAS. GUROBI represents the performance of directly solving the original PBO instances using the [18] commercial solver without any PaS augmentation. CONPAS [20] corresponds to the Contrastive PaS method explained above. For fair comparison, we utilize the CONPAS-LQ variant, which demonstrated superior performance among their reported configurations. CONPAS uses the same hyperparameters (k_0, k_1, Δ) as fine-tuned and reported in the original paper.

A systematic evaluation of the proposed components—the BACKBONES, the Network Architecture (NET), and the Search Parameters (PARAM)—shows that they are strongly complementary. The fully integrated BACKPAS model attains the lowest Primal Integral (PI) across all benchmarks and is statistically superior to every other variant and baseline. Notably, the BACKPAS-V0 variant, which uses only the extracted backbones, already provides a solid performance by outperforming both GUROBI and CONPAS on the test partitions of all benchmarks.

Adding the proposed architecture yields substantial gains over this backbone-only foundation. On MIS benchmark, the PI decreases from 2.31 (BACKPAS-V0) to 1.13 (BACKPAS-NET), corresponding to a proportional reduction of 51%. A similar effect appears on MVC, where the PI drops from 2.20 to 1.74 (21%). These improvements confirm that the architecture directly enhances anytime performance by providing better guidance signals.

Although the search parameters alone can be detrimental on CA and MIS, the largest overall gains arise from their combination with the new architecture. For instance, on MIS, the PI for BACKPAS-PARAM is 4.09, which decreases to 0.82 in the full BACKPAS model—a proportional improvement of 80%. On MVC, the reduction from 1.50 to 0.49 (67%) illustrates the same synergy. These

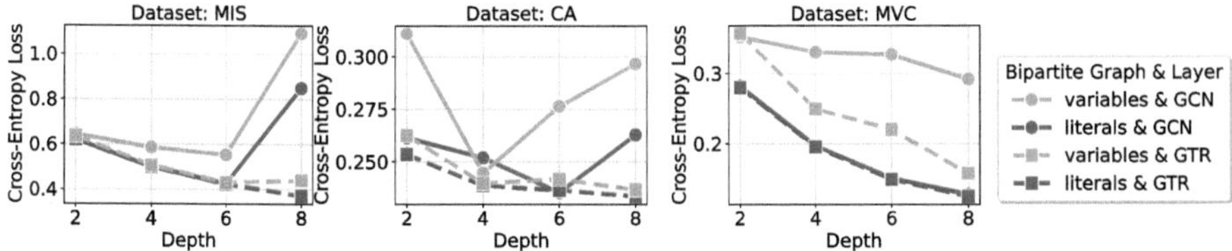

Fig. 1. Evaluation of alternative BACKPAS architectures: Cross-Entropy Loss on the validation set (`val-backpas`) for MIS, CA, and MVC benchmarks. Performance is shown by network depth (number of layers), comparing Graph Convolutional (GCN) vs. Graph Transformer (GTR) layers, and variable-based vs. literal-based graph representations. Lower values are better.

dramatic improvements show that the specialized parameters realize their full potential only when coupled with accurate predictions from the architecture.

Due to a lack of available space, a similar table including the overhead of the BACKPAS pipeline is included in thesupplementary material[6]. Even after accounting for overheads, the $PI(1000)$ scores and rankings remain largely stable. BackPaS outperforms ConPaS and Gurobi even with overhead. Moreover, significant code optimizations remain available, though previously deprioritized due to their minimal runtime impact.

5.2   Alternative GNN Architectures Performance

We perform a comparative study to better understand the effect of different architectural design choices. In this analysis, we consider the following variations:

Graph Layer Type: We compare the Graph Convolutional Layers (GCN) used in CONPAS/PAS with our Graph Transformer Layers (GTR).

Network Depth: We study the effect of different layer counts (2, 4, 6, 8) on backbone prediction performance.

Bipartite Graph Representation: We contrast the classic variable-based bipartite graph with our proposed literal-based representation.

Figure 1 shows the Cross-Entropy Loss for each configuration on the MIS, CA, and MVC datasets. Models using GTR layers consistently outperform those using GCN layers. Similarly, the literal-based bipartite graph yields lower loss than the variable-based alternative.

In terms of depth, deeper networks generally lead to lower loss, though GCN-based architectures hit a minimum earlier on MIS and CA. The best-performing configuration for all datasets combines the literal-based bipartite graph with an 8-layer GTR architecture; that is the configuration adopted in BACKPAS.

---

[6] https://github.com/bryan-alvarado-ulloa/backpas/blob/main/appendix/D.pdf.

5.3   Generalization on MIS

So far, we have tested BACKPAS on instances larger than those used for training. Despite the usage of larger instances, the overall graph density of the test instances was significantly smaller than that of the training instances. This lower density usually implies computationally easier instances (if not for the increased node counts), potentially masking generalization difficulties. To test if BACKPAS can generalize to larger instances while maintaining or increasing the graph density, we conducted a new experiment.

We generated the instances using the GeCo library. Each benchmark comprises 1000 BarabasiAlbert graphs, strictly split into 800 training instances, 100 validation instances, and 100 test instances. The specific partitions utilized for this experiment and their characteristics are detailed in Table 3.

Table 3. Partitions used for MIS generalization experiments. All instances are BarabasiAlbert graphs.

Benchmark	Partitions	Characteristics
MIS-666	train, val	666 nodes, 2 avg. degree
MIS-1000	train, val	1000 nodes, 3 avg. degree
MIS-1333	train, val	1333 nodes, 4 avg. degree
MIS-1666	val, test	1666 nodes, 5 avg. degree
MIS-2000-6	val, test	2000 nodes, 6 avg. degree
MIS-2000-10	val, test	2000 nodes, 10 avg. degree

For this experiment, the model was trained on the available **train** partitions of MIS-666, MIS-1000, and MIS-1333. The best training epoch was selected using the corresponding **val** partitions. To ensure robust hyperparameter selection, the trust region parameters were tuned using the **val** partitions of all benchmarks (from MIS-666 up to MIS-2000-10). Finally, we report the performance on the **test** partitions of the larger instances: MIS-1666, MIS-2000-6, and MIS-2000-10.

The results, measured by the Primal Integral, are summarized in Table 4.

Table 4. Primal Integral mean and standard deviations among the **test** partition are reported for MIS-1666, MIS-2000-6, and MIS-2000-10 benchmarks. Standard deviations are given in parentheses. Lower values are better.

Method	MIS-1666	MIS-2000-6	MIS-2000-10
GUROBI	3.81 (1.58)	14.94 (3.57)	32.64 (6.56)
BACKPAS	**0.04** (0.17)	**0.19** (0.25)	-

BACKPAS demonstrates strong generalization capabilities on instances where the density is preserved relative to the training set. On both MIS-1666 and MIS-

2000-6, our method significantly outperforms GUROBI, achieving Primal Integrals of 0.04 and 0.19 respectively, compared to 3.81 and 14.94 for the solver. This indicates that BACKPAS can successfully scale to larger graphs provided the density remains consistent with the training distribution or smaller.

However, the method failed to generalize to the MIS-2000-10 benchmark. As indicated by the missing entry in Table 4, the model did not generate any predictions for these instances. This failure stems from the model's inability to identify a backbone; the predicted probabilities for variable inclusion were not certain enough to meet the selection threshold. We attempted to mitigate this by re-selecting the trust region parameters using strictly the MIS-2000-10 `val` instances, but the model remained too uncertain to select any variable. This suggests a limitation in BACKPAS's ability to handle simultaneous increases in both graph size and graph density significantly beyond the training regime.

6   Conclusion

This work introduced BACKPAS, a new variant of the Predict-and-Search (PaS) framework designed to improve the solving process of Pseudo-Boolean Optimization problems. BACKPAS's main contribution lies in its multi-part approach, which includes: 1) using *backbone-based supervision* for learning; 2) a new Graph Neural Network architecture built on *Graph Transformer layers*; 3) a *literal-based bipartite graph representation*; and 4) an *adaptive trust region construction* that adjusts based on prediction confidence and *is not dependent on the instance size*.

To test these changes, we ran a systematic ablation study. This evaluation showed that the proposed components are *highly complementary* and work together effectively. The fully integrated BACKPAS model achieved the lowest Primal Integral (PI) values across all benchmarks and was statistically superior to every other method and baseline. Notably, the foundational BACKPAS-V0 variant, which uses only the backbones, already provides a strong starting point by statistically outperforming both the commercial solver GUROBI and the leading PaS implementation CONPAS. Furthermore, architectural components like the literal-based graph representations and Graph Transformer layers were shown to significantly lower prediction loss, establishing them as a *critical requirement* for the accurate predictions needed in the search process.

In summary, the final BACKPAS implementation, which combines all proposed modifications, established competitive performance across MIS, CA, and MVC benchmarks. This work not only provides a powerful new tool but also confirms the design principle that linking high-quality machine learning predictions with nuanced, adaptive search guidance leads to superior runtime performance.

Future work will focus on expanding the BACKPAS approach to other types of combinatorial problems and developing ways to weigh backbone variables based on their importance to the solver's success.

Acknowledgments. The research of the 1st, 2nd, 3rd, and 5th authors is supported in part by grant #2112533: "NSF Artificial Intelligence Research Institute for

Advances in Optimization (AI4OPT)". The 4th author is supported by ANID CCTVal (CIA250027) grant. The 1st author is also supported by ANID-Subdirección de Capital Humano/Magíster Nacional/2025 - 22252175. The second author is also supported by grant #2346058: "NRT-AI: Integrating Artificial Intelligence and Operations Research Technologies".

Disclosure of Interests. The authors have no competing interests to declare that are relevant to the content of this article.

References

1. Achterberg, T.: Scip: solving constraint integer programs. Math. Program. Comput. **1**, 1–41 (2009)
2. Achterberg, T., Berthold, T., Hendel, G.: Rounding and propagation heuristics for mixed integer programming. In: Klatte, D., Lüthi, HJ., Schmedders, K. (eds.) Operations Research Proceedings 2011: Selected Papers of the International Conference on Operations Research (OR 2011), August 30–September 2, 2011, Zurich, Switzerland, pp. 71–76. Springer, Cham (2012). https://doi.org/10.1007/978-3-642-29210-1_12
3. Aloul, F.A., Ramani, A., Markov, I.L., Sakallah, K.A.: Generic ilp versus specialized 0–1 ilp: An update. In: Proceedings of the 2002 IEEE/ACM International Conference on Computer-Aided Design, pp. 450–457 (2002)
4. Asín-Achá, R., Espinoza, A., Goldschmidt, O., Hochbaum, D.S., Huerta, I.I.: Selecting fast algorithms for the capacitated vehicle routing problem with machine learning techniques. Networks **84**(4), 465–480 (2024)
5. Bello, I., Pham, H., Le, Q.V., Norouzi, M., Bengio, S.: Neural combinatorial optimization with reinforcement learning, arXiv preprint (2016)
6. Bengio, Y., Lodi, A., Prouvost, A.: Machine learning for combinatorial optimization: a methodological tour d'horizon. Eur. J. Oper. Res. **290**(2), 405–421 (2021)
7. Bergstra, J., Bardenet, R., Bengio, Y., Kégl, B.: Algorithms for hyper-parameter optimization. Adv. Neural. Inf. Process. Syst. **24** (2011)
8. Biere, A., Froleyks, N., Wang, W.: Cadiback: extracting backbones with cadical. In: 26th International Conference on Theory and Applications of Satisfiability Testing (SAT 2023). Schloss-Dagstuhl-Leibniz Zentrum für Informatik, (2023)
9. Biere, A., Heule, M., van Maaren, H., Walsh, T. (eds.): Handbook of Satisfiability - Second Edition, Frontiers in Artificial Intelligence and Applications, vol. 336. IOS Press (2021). https://doi.org/10.3233/FAIA336
10. Boros, E., Hammer, P.L.: Pseudo-boolean optimization. Discrete Appl. Math. **123**(1–3), 155–225 (2002)
11. Cai, J., Huang, T., Dilkina, B.: Learning backdoors for mixed integer programs with contrastive learning, arXiv preprint (2024)
12. Crama, Y., Hammer, P.L.: Boolean functions: Theory, algorithms, and applications, Cambridge University Press (2011)
13. Devriendt, J., Gocht, S., Demirović, E., Nordström, J., Stuckey, P.J.: Cutting to the core of pseudo-boolean optimization: Combining core-guided search with cutting planes reasoning. In: Proceedings of the AAAI Conference on Artificial Intelligence, vol. 35, pp. 3750–3758 (2021)
14. Divide and conquer: Elffers, J., Nordström. J. Towards faster pseudo-boolean solving. IJCAI **18**, 1291–1299 (2018)

15. Ferber, A., Wilder, B., Dilkina, B., Tambe, M.: Mipaal: Mixed integer program as a layer. In: Proceedings of the AAAI Conference on Artificial Intelligence, vol. 34, pp. 1504–1511 (2020)
16. Francia-Carramiñana, M., Alvarado-Ulloa, B., Dilkina, B., Hochbaum, D.S., anculef, R., Asín-Achá, R.J.: Backbones in pseudo-boolean optimization: Extraction and analysis. In: 18th International Conference on Agents and Artificial Intelligence (ICAART) (2026). in Press
17. Gasse, M., Chetelat, D., Ferroni, N., Charlin, L., Lodi, A.: Exact combinatorial optimization with graph convolutional neural networks. In: Advances in Neural Information Processing Systems, vol. 32 (2019)
18. Gurobi Optimization, LLC: Gurobi Optimizer Reference Manual (2024). https://www.gurobi.com
19. Han, Q., et al.: A gnn-guided predict-and-search framework for mixed-integer linear programming. In: The Eleventh International Conference on Learning Representations (2023)
20. Huang, T., Ferber, A.M., Zharmagambetov, A., Tian, Y., Dilkina, B.: Contrastive predict-and-search for mixed integer linear programs. In: Forty-First International Conference on Machine Learning (2024)
21. Huang, T., Li, J., Koenig, S., Dilkina, B.: Anytime multi-agent path finding via machine learning-guided large neighborhood search. In: Proceedings of the AAAI Conference on Artificial Intelligence, vol. 36, pp. 9368–9376 (2022)
22. Huang, W., Huang, T., Ferber, A.M., Dilkina, B.: Distributional miplib: A Multi-domain Library for Advancing ml-Guided MILP Methods (2024). https://arxiv.org/abs/2406.06954
23. Huerta, I.I., Neira, D.A., Ortega, D.A., Varas, V., Godoy, J., As'in Ach'a, R.J.: Improving the state-of-the-art in the traveling salesman problem: an anytime automatic algorithm selection. Expert Syst. Appl. **187**, 115948 (2022). https://doi.org/10.1016/J.ESWA.2021.115948
24. Janota, M., Lynce, I., Marques-Silva, J.: Algorithms for computing backbones of propositional formulae. AI Commun. **28**(2), 161–177 (2015)
25. Khalil, E., Dai, H., Zhang, Y., Dilkina, B., Song, L.: Learning combinatorial optimization algorithms over graphs. Adv. Neural Inf. Process. Syst. 30 (2017)
26. Khalil, E., Le Bodic, P., Song, L., Nemhauser, G., Dilkina, B.: Learning to branch in mixed integer programming. In: Proceedings of the AAAI Conference on Artificial Intelligence, vol. 30 (2016)
27. Lederman, G., Rabe, M.N., Seshia, S.A.: Learning heuristics for automated reasoning through deep reinforcement learning, vol. 57 (2018). arXiv preprint
28. Leyton-Brown, K., Pearson, M., Shoham, Y.: Towards a universal test suite for combinatorial auction algorithms. In: Proceedings of the 2nd ACM conference on Electronic commerce, pp. 66–76 (2000)
29. Li, Z., Chen, Q., Koltun, V.: Combinatorial optimization with graph convolutional networks and guided tree search. In: Advances in Neural Information Processing Systems, p. 31 (2018)
30. Liang, J.H., Ganesh, V., Poupart, P., Czarnecki, K.: Learning Rate Based Branching Heuristic for SAT Solvers. In: Creignou, N., Le Berre, D. (eds.) SAT 2016. LNCS, vol. 9710, pp. 123–140. Springer, Cham (2016). https://doi.org/10.1007/978-3-319-40970-2_9
31. Monasson, R., Zecchina, R., Kirkpatrick, S., Selman, B., Troyansky, L.: Determining computational complexity from characteristic 'phase transitions'. Nature **400**(6740), 133–137 (1999)

32. Nieuwenhuis, R., Oliveras, A., Rodríguez-Carbonell, E., Zhao, R.: Speeding up pseudo-boolean propagation. In: 27th International Conference on Theory and Applications of Satisfiability Testing (SAT 2024). Schloss Dagstuhl-Leibniz-Zentrum für Informatik, (2024)
33. Pezo, C., Hochbaum, D.S., Godoy, J., As'in Ach'a, R.J.: Automatic algorithm selection for pseudo-boolean optimization with given computational time limits. Comput. Oper. Res. **173**, 106836 (2025). https://doi.org/10.1016/J.COR.2024.106836
34. Selsam, D., Bjørner, N.: Guiding High-Performance SAT Solvers with Unsat-Core Predictions. In: Janota, M., Lynce, I. (eds.) SAT 2019. LNCS, vol. 11628, pp. 336–353. Springer, Cham (2019). https://doi.org/10.1007/978-3-030-24258-9_24
35. Selsam, D., Lamm, M., Bünz, B., Liang, P., de Moura, L., Dill, D.L.: Learning a SAT solver from single-bit supervision. In: 7th International Conference on Learning Representations, ICLR 2019, New Orleans, LA, USA, May 6–9, 2019. OpenReview.net (2019). https://openreview.net/forum?id=HJMC_iA5tm
36. Shi, Y., Huang, Z., Feng, S., Zhong, H., Wang, W., Sun, Y.: Masked label prediction: Unified message passing model for semi-supervised classification, arXiv preprint (2020) (2020)
37. Vinyals, O., Fortunato, M., Jaitly, N.: Pointer networks. In: Advances in neural information processing systems, p. 28 (2015)
38. Wang, W., Hu, Y., Tiwari, M., Khurshid, S., McMillan, K.L., Miikkulainen, R.: Neuroback: Improving CDCL SAT solving using graph neural networks. In: The Twelfth International Conference on Learning Representations, ICLR 2024, Vienna, Austria, May 7–11, 2024. OpenReview.net (2024). https://openreview.net/forum?id=samyfu6G93

CORL: Reinforcement Learning of MILP Policies Solved via Branch-and-Bound

Akhil S. Anand[(✉)], Elias Aarekol, Martin Mziray Dalseg, Magnus Stålhane,
and Sebastien Gros

Norwegian University of Science and Technology (NTNU), Trondheim, Norway
`akhil.s.anand@ntnu.no`

Abstract. Combinatorial sequential decision-making problems are typically
modeled as mixed-integer linear programs (MILPs) and solved via branch-and-
bound (B&B) algorithms. The inherent difficulty of modeling MILPs that accurately represent stochastic real-world problems leads to suboptimal performance
in the real world. Recently, machine-learning methods have been applied to learn
MILP models for decision quality rather than how accurately they model the real-
world problem. However, these approaches typically rely on supervised learning,
assume access to optimal decisions, and use surrogates for the MILP gradients. In
this work, we introduce a proof-of-concept CORL framework that end-to-end fine-
tunes an MILP scheme using reinforcement learning (RL) on real-world data to
maximize its operational performance. We enable this by casting an MILP solved
by B&B as a differentiable stochastic policy compatible with RL. We validate
the CORL method in two illustrative combinatorial sequential decision-making
examples.

1 Introduction

Combinatorial Optimization (CO) is widely used for decision-making in Operations
Research (OR), with applications ranging from production scheduling and supply-chain
design to vehicle routing and portfolio management [18]. A broad class of CO problems can be formulated as Mixed-Integer Linear Programs (MILPs) and solved using
the Branch and Bound (B&B) algorithm. The standard model-based decision-making
workflow constructs an MILP model that closely approximates the real-world problem
and then solves it. Recently, machine learning has been integrated to improve the predictive accuracy of MILP models as a proxy for improving their real-world operational
performance, referred to as Prediction Focused Learning (PFL) [9,23,27]. However, in
model-based decision-making for stochastic systems, as in the case of MILPs, maximizing the operational performance of a decision scheme is generally not aligned with
maximizing the predictive accuracy of the underlying model [2,5]. This has been identified as the *objective mismatch* issue within Reinforcement Learning (MBRL) [13,38]
and has been observed empirically [11,13,36].

In principle, CO problems can also be solved using Reinforcement Learning (RL)
directly, without modeling them as MILPs [7]. However, such model-free RL policies
often lack guarantees on constraint satisfaction and explainability, struggle to generalize beyond the training distribution, and require large amounts of real-world interaction,

T. Guns (Ed.): CPAIOR 2026, LNCS 16595, pp. 19–35, 2026.
https://doi.org/10.1007/978-3-032-27242-3_2

limiting their practical applicability. In contrast, MILPs provide these properties by construction. Alternatively, the Decision Focused Learning (DFL) framework uses machine learning to directly fine-tune MILPs for decision quality, thereby preserving the MILP structure while improving performance through learning [24]. DFL methods prioritize decision performance over predictive accuracy of the underlying MILP models. RL naturally supports this paradigm, offering a way to learn or fine-tune model-based decision policies (e.g., Model Predictive Control (MPC) or MBRL) directly for real-world performance, rather than relying on black-box neural network policies [1,3,32]. However, existing DFL methods do not exploit this, as they rely on supervised learning assuming access to an oracle providing the true value of the objective function [23].

An MILP scheme inherently defines a deterministic, non-differentiable model-based decision policy by mapping states to the integer decision that minimizes a given objective. By using this MILP policy as a parameterized policy within the RL framework, one can make use of policy-gradient RL methods to adapt the MILP policy for improved real-world performance. Policy-gradient algorithms require the policy to be both stochastic and differentiable, properties that a standard MILP-based decision scheme lacks. We propose the CORL framework by converting a discrete MILP+B&B policy into a differentiable stochastic policy that is compatible with RL. CORL provides the necessary tools to integrate this MILP policy with the stochastic policy gradient RL methods. CORL enables end-to-end fine-tuning of MILP parameters directly from data to maximize its real-world decision performance while preserving its structure.

The CORL approach is related to emerging DFL methodologies in combinatorial decision-making, but differs fundamentally in that it does not rely on supervised learning or oracle access to true objective values. Unlike existing learning-based approaches for CO that focus on improving solver components (e.g., branching or node selection) [35], CORL instead treats a parametric MILP solved via B&B as a structured decision policy and uses RL to adapt its parameters based on real-world performance.

Organization: The rest of the paper is organized as follows: Sect. 2 reviews related work; Sect. 3 provides the necessary background; Sect. 4 introduces the CORL framework; Sect. 5 presents illustrative examples; Sect. 6 discusses the findings, limitations, and future work and Sect. 7 provides the conclusions.

2 Related Works

DFL is a major class of learning-based CO methods that seek to improve the performance of decisions by training the CO in an end-to-end fashion. Like PFL, DFL first predicts parameters for a CO instance whose solution defines the decision. Instead of minimizing parameter error, DFL minimizes a task loss, most commonly the *regret* [19,24,34]. DFL depends on proxy gradients of the MILP layer, which existing methods obtain via three main strategies [24]. The first class, analytical smoothing approaches, relaxes the MILP to a continuous Linear Programming (LP) and smooths the solution map to obtain approximate gradients [12,39]. The second class of methods models the decision variable as a stochastic variable and estimates the sensitivity by perturbing its parameters [8,28]. The third class optimizes a surrogate objective (e.g.

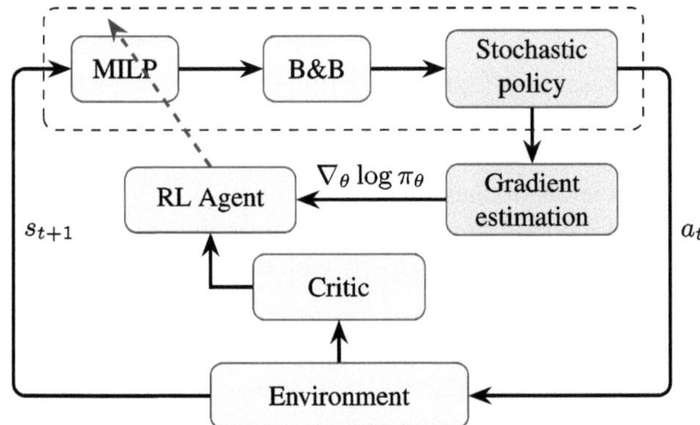

Fig. 1. Overview of the CORL framework

SPO+), avoiding differentiation through the solver [10]. Unlike DFL methods that rely on proxy gradients through either of these methods, the CORL approach uses the B&B tree to define a softmax-based stochastic policy that is directly differentiable, avoiding the need to find a suitable proxy. Although DFL methods enable regret minimization, they typically assume access to the true MILP parameters and their gradients. Therefore, DFL methods may fail in settings with noisy data or unknown true parameters, limiting their applicability in real-world scenarios [24]. They also incur additional complexity due to backpropagation through the argmin operator [24]. CORL avoids these issues by representing a general MILP-based combinatorial sequential decision scheme as a differentiable stochastic policy, enabling end-to-end RL to directly optimize its real-world performance.

RL has been used in CO primarily to enhance solver performance or solution quality under a given model [25,31]. For MILPs, early work focused on learning branching rules within B&B [17], while others trained RL agents to construct feasible solutions directly [22]. However, tuning an existing MILP model, i.e., adapting its parameters or heuristics to improve real-world decision performance via RL, remains an open challenge. In a closely related line of work, mixed-integer MPC has been used as a policy for RL [14]. The proposed CORL approach is more general, as it exploits the structure of the B&B solver and allows all leaf nodes to enter the policy assuming a uniform sampling strategy. Xu *et al.* embed a modified DQN within an MILP to select feasible integer decisions; however, this approach is tied to the black-box nature of DQN and cannot fine-tune a pre-trained MILP model [41]. In another related work [21], a Q-function estimate is incorporated into an MILP model to improve decision performance, similar to our setting. However, CORL is more general, does not require a Q-function estimate, and can be applied to any MILP model.

3 Background

This section introduces the necessary background on sequential decision-making, RL, CO, MILP, and B&B.

3.1 Modeling Sequential Decision-Making Using MDPs

Markov Decision Process (MDP) provides a framework to formulate and solve sequential decision-making problems for stochastic systems with the Markovian property [30]. MDPS consider dynamic systems with underlying states $s \in \mathbb{S}$, actions (decisions) $a \in \mathbb{A}$, and the associated stochastic state transition: $s_+ \sim \mathcal{P}(.\,|\,s,a)$. Solving an MDP involves finding a policy $\pi(a|s)$ that minimizes the expected sum of discounted costs $\ell \colon \mathbb{S} \times \mathbb{A} \to \mathbb{R}$ under stochastic closed-loop trajectories, given as the MDP cost:

$$J(\pi) = \mathbb{E}_{\mathcal{P}^\pi}\left[\sum_{k=0}^{\infty} \gamma^k \ell\,(s_k, a_k)\,\middle|\, a_k \sim \pi(.|s_k)\right], \tag{1}$$

for a discount factor $\gamma \in (0,1)$. The expectation is taken over the distribution of s_k and a_k in the Markov chain induced by $\pi(a|s)$. The solution to MDP provides an optimal policy $\pi^\star$ from the set Π of all admissible policies by minimizing $J(\pi)$, defined as:

$$\pi^\star = \arg\min_{\pi \in \Pi} J\,(\pi)\,. \tag{2}$$

The solution of an MDP is described by Bellman equations [6]:

$$Q^\star(s,a) = \ell(s,a) + \gamma\,\mathbb{E}_\rho\big[V^\star(s_+)\mid s,a\big], \tag{3a}$$

$$V^\star(s) = \min_a Q^\star(s,a), \tag{3b}$$

$$\pi^\star(a|s) = \arg\min_a Q^\star(s,a)\,, \tag{3c}$$

where $V^\star$ and $Q^\star$ are the optimal value and action-value functions, respectively. $\mathbb{E}_\rho$ is the expectation over the distribution of $\mathcal{P}$.

RL approximately solves the MDP using data. Stochastic policy gradient is a major class of methods that can be applied to both discrete and continuous state–action space problems. Soft ActorCritic (SAC) and Proximal Policy Optimization (PPO) are two widely used stochastic policy gradient algorithms in RL [16,37]. For a differentiable policy $\pi_\theta(a|s)$ parametrized by θ, the stochastic policy-gradient is:

$$\nabla_\theta J(\pi_\theta) = \mathbb{E}_{s,a\sim\pi_\theta}\big[\nabla_\theta \log \pi_\theta(a|s)\, A^{\pi_\theta}(s,a)\big]\,, \tag{4}$$

where $A^\pi(s,a) = Q^\pi(s,a) - V^\pi(s)$ and Q^π, V^π satisfy the Bellman equations with π in place of $\pi^\star$.

3.2 Combinatorial Optimization, MILP and B&B

A broad class of CO problems can be modeled using MILP. Typical examples of MILPs include knapsack, facility location, network design, and scheduling [40]. MILPs are non-convex and NP-hard to solve. The canonical exact algorithm for MILP is B&B [40], which solves a series of LP relaxations of the MILP model. In a B&B tree for solving an MILP (see Fig. 2a), the *root* node represents the original problem. Each node corresponds to a subproblem obtained by fixing or bounding some variables. Solving its LP relaxation produces a lower bound on the objective value Q of all integer-feasible

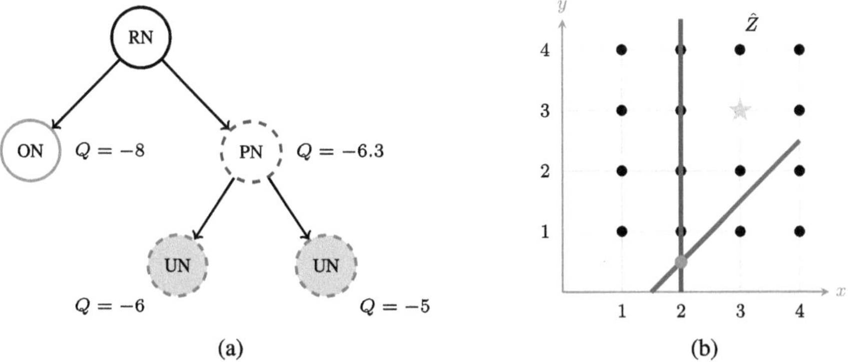

Fig. 2. (a) B&B tree: A B&B tree may contain unexplored but feasible decisions corresponding to the unexplored leaf nodes (UN). Since their true cost is unknown, we approximate them using their parent, which is a pruned node (PN). (b) Sampling child leaves of a pruned node: For a PN, black dots are unexplored child leaves, the red line represents a constraint, where solutions over it are feasible, and the blue line represents the local branching cut. $\hat{Z}$ represents the feasible solution space, defined by the blue and red constraints. The green dot is the fractional LP solution at the PN, and let the yellow star be the unknown true global optimum for RL. The stochastic policy should sample decisions from this space $\hat{Z}$ for RL. (Color figure online)

solutions (unexplored leaf nodes) in the subtree defined by that node. A node is a *leaf* if it is integer-feasible, in which case the bound equals its exact objective value. A processed node that will not be branched further is *fathomed*; this occurs if (i) its relaxation is infeasible, (ii) its bound is no better than the current optimum, or (iii) it is an integer-feasible leaf. A *pruned* node is a fathomed, non-leaf node; pruned nodes provide lower bounds for all their (still unexplored) feasible child leaves, whereas leaf nodes provide exact objective values.

4 CORL

In this section, we introduce the CORL framework: an actorcritic RL approach for learning high-performance combinatorial sequential decision-making policies, as illustrated in Fig. 1. We first introduce MILP-based policies for sequential decision-making, then show how RL can be used to fine-tune them for real-world performance, and finally derive a differentiable stochastic MILP policy compatible with RL.

4.1 MILP Policy for Combinatorial Sequential Decision-Making

When a combinatorial sequential decision-making problem is subject to uncertainty that resolves over stages, it can be formulated as a multi-stage Stochastic Dynamic Programming (SDP), wherein iteratively solving a CO subproblem at each stage t forms a policy that approximates the optimal solution of the underlying MDP (1). While there are many ways to form such CO-based policies [29], we use a two-stage *parametric* SDP formulated as an MILP and solved using B&B:

$$\min_{a_t,\, y_t} \quad L_\theta(s_t, a_t, y_t) \; + \; V_\theta(\hat{s}_{t+1}, y_t) \tag{5a}$$

$$\text{s.t.} \quad g_\theta(s_t, a_t, y_t) \le 0, \tag{5b}$$

$$\hat{s}_{t+1} = h_\theta(s_t, a_t), \quad a_t^i \in \mathbb{Z}, \quad i \in \mathcal{I}. \tag{5c}$$

where θ represents the MILP parameters, a_t represents the decision at stage t, L_θ is the instantaneous cost of the state-decision pair. It may be identical to ℓ in the real-world MDP (1), but it can be different. Function g_θ represents the inequality constraints. Function h_θ models the transition dynamics of the system, predicting the next state $\hat{s}_{t+1}$ given the current state s_t and decision a_t. For simplicity, we assume $h_\theta(s_t, a_t)$ provides a single next state $\hat{s}_{t+1}$ deterministically. Ideally, in (5) one would solve a problem along (possibly infinite) future stages; accordingly, the term $V_\theta(\hat{s}_{t+1})$ serves as a surrogate for the cumulative cost from stage $t+1$ onward. Ideally, it would equal the optimal value $V^\star(s_{t+1})$ [33]; in practice, we approximate this future cost with V_θ. y_t represents the set of auxiliary variables used to enforce the piecewise linear structure to the functions L_θ, V_θ, and g_θ.

Additionally, we assume that the feasibility of decisions is ensured by constraint relaxation, i.e., by penalizing constraint violations in the objective function. In a sequential decision-making context, (5) defines a policy: at any given stage t, it solves the two-stage optimization given the current state s_t to derive model-based decisions a_t that minimize the given objective. This two-stage formulation (5) simplifies solving the multi-stage problem: instead of optimizing across all the stages, we solve a single-stage MILP augmented with the value approximation V_θ. However, this formulation naturally extends to the full multi-stage setting.

Assuming the MILP scheme (5) has a unique minimizer, we can view the MILP scheme (5) as a model of the action-value function Q in Bellman equations (3) by fixing the decision variable a_t in (5):

$$Q_\theta(s_t, \hat{a}_t) := (5a) \\ \text{s.t. } (5b) - (5c), \quad a_t = \hat{a}_t \,. \tag{6}$$

We use the term objective value or action-value for Q_θ interchangeably. This definition of Q_θ is valid in the sense of fundamental Bellman relationships (3) between optimal action-value functions, value functions, and policies, i.e.

$$V_\theta(s) = \min_a Q_\theta(s, a), \tag{7a}$$

$$\pi_\theta(s) \in \arg\min_a Q_\theta(s, a) \,. \tag{7b}$$

In this sense, the MILP model (5) can be seen as a model of the real-world MDP. This connection helps to explain the theoretical basis of the CORL framework developed in the following sections.

While the CO-based policy is presented in the context of a two-stage SDP approximation, *the* CORL *framework is not limited to this setting and extends more generally to any parametric MILP-based policy for sequential decision-making solved via B&B.*

4.2 CORL Problem Formulation

The overarching goal of any CO-based decision-making scheme is to derive optimal decisions for the real-world system. Although accurately modeling a stochastic real-world system with an MILP is nearly impossible, it has been shown that such inexact decision models can nonetheless recover the true optimal policy $\pi^\star$ [5]. This demands the MILP policy (7b) to match the true optimal policy for the real system, $\pi^\star$ (2). Assuming the model of Q_θ given by (6) is bounded and the MILP parameterization is rich enough, this (necessary and sufficient) optimality condition can be represented as:

$$\arg\min_a Q_\theta\left(s, a\right) = \arg\min_a Q^\star\left(s, a\right), \quad \forall s. \tag{8}$$

The identification of the optimal MILP parameters $\theta^\star$ that satisfy the optimality condition (8) can be framed as the following optimization problem:

$$\theta^\star \in \arg\min_\theta J(\pi_\theta), \tag{9}$$

where J is given by (1) and the policy π_θ is given by the solution to (5). The core principle of our CORL framework is to *employ RL to approximately solve this optimization problem (9) directly from real-world data.* In CORL, the policy consists of the MILP model (5) and the B&B solver, where the RL agent adapts the MILP parameters θ to maximize real-world performance (see Fig. 1). Within RL, stochastic policy-gradient methods provide a robust framework for learning stochastic policies [16]. We adopt stochastic policy-gradient methods in an actorcritic architecture to solve the CORL problem (see Fig. 1): the actor implements the MILP-based policy, while the critic can be implemented using a generic function approximator such as neural networks. The stochastic policy-gradient methods require the policy to be both stochastic and differentiable, properties that the MILP policy (5) lacks. To address this, in the next two sections, we derive a differentiable stochastic policy from this discrete MILP policy.

4.3 Deriving a Stochastic Policy from B&B

We can derive a differentiable stochastic policy by applying a softmax distribution over the Q_θ-values of the MILP scheme (6):

$$\pi_\theta(a_t | s_t) = \frac{e^{-\beta Q_\theta(s_t, a_t)}}{\sum_{\tilde{a}_t \in \mathcal{A}} e^{-\beta Q_\theta(s_t, \tilde{a}_t)}}, \tag{10}$$

where $\beta > 0$ represents the exploration parameter. However, directly applying (10) is infeasible because B&B algorithms do not explore every feasible decision $a_t \in \mathcal{A}$ (leaf), so their true objective values $Q_\theta(s_t, a_t)$ are not known. Commercial solvers, e.g. [15], can return a subset of feasible decisions with their objective values, which can be used to form a policy. However, that would exclude a possibly large set of feasible decisions, which would be beneficial to explore (see Fig. 2b). Therefore, we exploit the structure of the B&B tree to derive an alternative form to the policy (10).

Let $\mathcal{K}_t$ be the set of all explored nodes k (leaf or pruned) in the B&B tree at t. All such (leaf or pruned) nodes $k \in \mathcal{K}_t$ are assigned a scalar objective value that we denote by $Q_\theta^{k,t}$. For a pruned node, $Q_\theta^{k,t}$ is a lower bound on every (still unexplored) child node, and for a leaf, it is the exact objective of its integer-feasible solution. Formally,

$$Q_\theta(s_t, a) \geq Q_\theta^{k,t} \quad \forall a \in \mathcal{A}_k, \tag{11}$$

with $Q_\theta(s_t, a_{t,k}) = Q_\theta^{k,t}$ if k is a leaf. $\mathcal{A}_k$ denotes the set of feasible decisions at node k, i.e. the unexplored child nodes belonging to k, and let $a_{t,k}$ denote the integer solution stored at an already-explored leaf.

We can use $Q_\theta^{k,t}$ to sample a node k by a softmax over the finite set $\mathcal{K}_t$ as:

$$P(k \mid s_t) = \frac{e^{-\beta Q_\theta^{k,t}}}{\sum_{i \in \mathcal{K}_t} e^{-\beta Q_\theta^{i,t}}}. \tag{12}$$

Sampling a node k from (12) allocates probability mass both to explored leaves and to promising pruned nodes through their $Q_\theta^{k,t}$ values. This allows us to score unexplored regions of the tree and bias exploration toward branches with a better lower bound. Since for a pruned node, $Q_\theta^{k,t}$ is only a lower bound, their $P(k|s_t)$ in (12) can be *optimistic*, as a pruned branch may receive higher probability than it would if exact values of its children were available (see (10) and Fig. 2a). At the same time, a pruned node with a poor $Q_\theta^{k,t}$ may contain a large number of child leaves, which may not be reflected in the distribution (12). However, for the proof-of-concept, we limit our methodology to (12).

Now, in order to approximate the intractable softmax over $\mathcal{A}$ in (10), a decision a_t ought to be selected randomly among the child leaf nodes of a node k sampled according to (12). To that end, let us define the optimal solution (integer or fractional) at a (pruned or leaf) node k corresponding to $Q_\theta^{k,t}$ as $a_{k,t}^\star$. A decision a_t can be sampled from a selected node k as follows:

- If k is a leaf node: we take its integer decision $a_t = a_{k,t}^\star$ deterministically.
- If k is a pruned node: we sample an (unexplored) child leaf node using a heuristic (see Fig. 2b). For the proof-of-concept, we propose two simple heuristics:

  (1) *Uniform sampling*: treat all integer-feasible decisions $a_t \in \mathcal{A}_t$ corresponding to the (unexplored) child leaves of the node k as equally likely, assuming it has a finite set of decisions and they are known.

  (2) *Nearest-Neighbor Sampling (NNS)*: sample a decision a_t from the neighbourhood of the fractional solution $a_{k,t}^\star$ of the pruned node. This can be represented using a softmax over (negative) Manhattan distances with β_d as the temperature parameter:

$$p(a_t \mid k, s_t) \propto e^{-\beta_d \|a_t - a_{t,k}^\star\|}. \tag{13}$$

Note that by construction, the B&B tree induces a partition in which each integerfeasible decision a_t belongs to a unique (leaf or pruned) node $k^\star \in \mathcal{K}_t$. Therefore, only that particular node $k^\star$ contributes to the probability of its associated decision a_t in the policy.

Putting these together forms an approximation to the MILP policy as:

$$\pi_\theta(a_t|s_t) = P(k^\star \mid s_t)\, p(a_t \mid k^\star, s_t), \tag{14}$$

where $p(a_t \mid k^\star, s_t) = 1$ if $k^\star$ is a leaf. For a pruned node, $p(\cdot \mid k^\star, s_t)$ is given by a sampling strategy (e.g., uniform sampling or NNS heuristic).

4.4 Policy Gradient Estimation

The term $p(a_t|k^\star, s_t)$ in (14) depends on θ via the cardinality of node $k^\star$, i.e., the number of its child leaf nodes. The cardinality can change when θ alters the pruning pattern. However, the cardinality of a node $k^\star$ is piecewise constant in θ. Therefore, the gradient built by assuming it as a constant is correct almost everywhere except on the measure-zero set where the cardinality changes. Therefore, we assume $\nabla_\theta \log p(a_t \mid k^\star, s_t)$ is zero when it is defined. As a result, the gradient of the log policy needed for stochastic policy gradient (4) follows from (14):

$$\nabla_\theta \log \pi_\theta(a_t|s_t) = \nabla_\theta \log P(k^\star \mid s_t)$$
$$= -\beta \nabla_\theta Q_\theta^{k^\star,t} - \nabla_\theta \log \sum_{i \in \mathcal{K}_t} e^{-\beta Q_\theta^{i,t}} . \tag{15}$$

In order to calculate $\nabla_\theta Q_\theta^{k,t}$ at any node k, we consider the partial Lagrangian of the MILP scheme (5) given by:

$$\mathcal{L}(a_t, y, \lambda, \mu; \theta) = L_\theta(s_t, a_t, y_t) + V_\theta(\hat{s}_{t+1}, y_t)$$
$$+ \lambda^\top g_\theta(s_t, a_t, y_t) \tag{16}$$
$$+ \mu^\top \left(\hat{s}_{t+1} - h_\theta(s_t, a_t)\right) .$$

where λ and μ represent the vector of Lagrange multipliers for inequality and equality constraints, respectively. At a pruned node, the Karush–Kuhn–Tucker (KKT) stationarity condition $\nabla_{a_t^\star} \mathcal{L} = 0$ holds, so the total derivative of $\mathcal{L}$ with respect to θ reduces to its partial derivative. At a leaf node, the solution $a_t^\star$ is piecewise constant in θ, hence $\nabla_\theta a_t^\star = 0$ almost everywhere. In both cases (at a leaf or pruned node), $\nabla_\theta Q_\theta$ is obtained by differentiating the Lagrangian with respect to θ while holding the optimal primal/dual variables fixed. Therefore, under standard regularity conditions, the envelope theorem [26] implies that, at the optimal solution $(a_{t,k}^\star, y^\star, \lambda^\star, \mu^\star)$:

$$\nabla_\theta Q_\theta(s_t, a_{t,k}^\star) = \nabla_\theta \mathcal{L}\big|_{(a_{t,k}^\star, y^\star, \lambda^\star, \mu^\star)} . \tag{17}$$

Since both L_θ and V_θ are piecewise-linear in θ, the auxiliary variables y act as linear coefficients that remain constant within each region, so $\nabla_\theta y^\star = 0$ almost everywhere; resulting in:

$$\nabla_\theta Q_\theta(s_t, a_{t,k}^\star) = \nabla_\theta L_\theta(s_t, a_{t,k}^\star) + \nabla_\theta V_\theta(\hat{s}_{t+1})$$
$$+ \left(\lambda^\star\right)^\top \nabla_\theta g_\theta(s_t, a_{t,k}^\star) \tag{18}$$
$$- \left(\mu^\star\right)^\top \nabla_\theta h_\theta(s_t, a_{t,k}^\star) .$$

Note that $a_{k,t}^\star$ is the optimal solution (integer or fractional) at a (pruned or leaf) node k corresponding to $Q_\theta^{k,t}$. Since each B&B subproblem is solved to optimality, the optimal primal solution $(a_{t,k}^\star)$ and dual solution $(\lambda^\star \, \mu^\star)$ of the MILP subproblem at node k satisfy the KKT conditions [20] and ensure that (18) holds exactly. Note that (18) assumes that any discontinuity caused by a change in the active integer solution arises only at a measure-zero set and therefore does not affect the gradient-based update in practice.

An illustrative pseudocode for a basic on-policy RL implementation of the CORL methodology is provided in Algorithm 1.

Algorithm 1

1: **Input:** Parametric MILP-actor θ, critic ϕ
2: **for** iter = 1 to N **do**
3:     Initialize buffers: $\mathcal{Q} \leftarrow \{\}, \quad \mathcal{D}_{QG} \leftarrow \{\}, \quad \mathcal{D}_{PG} \leftarrow \{\}, \quad \mathcal{D} \leftarrow \{\}$
4:     **for** $t = 1$ to T **do**
5:         Solve MILP with B&B, append $(Q_\theta^{k,t}, a_{k,t}^\star)$ to $\mathcal{Q}$
6:         Form $P(k \mid s_t)$ via (12) over $\{Q_\theta^k\}$
7:         Sample node $k \sim P(\cdot \mid s_t)$
8:         **if** k is leaf **then**
9:             $a_t \leftarrow a_{k,t}^\star$
10:        **else**
11:            $a_t \sim p(a \mid k, s_t)$ {uniform or NNS}
12:        **end if**
13:        Compute $\nabla_\theta Q_\theta(s_t, a_{k,t}^\star)$ via (18) and append to $\mathcal{D}_{QG}$
14:        Compute $\nabla_\theta \log \pi_\theta$ via (15) and append to $\mathcal{D}_{PG}$
15:        Apply a_t to the environment
16:        Append (s_t, a_t, r_t, s_{t+1}) to $\mathcal{D}$
17:    **end for**
18:    Update critic ϕ on $\mathcal{D}$
19:    Evaluate policy on $\mathcal{D}$, compute advantages A
20:    Update actor using $\mathcal{D}, \mathcal{D}_{PG}, A$ via (4)
21: **end for**

---

5   Illustrative Examples

5.1   Example 1

We consider a CO problem governed by a linear stochastic transition function formulated as:

$$\min_{a_t} \quad \sum_{t=1}^{T} \ell^\mathsf{T} a_t + p\zeta_t$$

$$\text{s. t.,} \quad Ds_t + Ea_t \leq F + \zeta_t, \tag{19}$$

$$s_{t+1} = Ms_t + Ba_t + w_t,$$

$$lb \leq a_t \leq ub, \quad \zeta_t \geq 0, \quad a_t^i \in \mathbb{Z}, \quad \forall a_t^i \in a_t \quad t = 1, \ldots, T,$$

where ℓ denotes the cost vector, s_t the continuous system state, and a_t the integer decision at stage t. The matrices D, E, F define linear constraints, while M, B define the linear system dynamics, and $w_t \sim \mathcal{N}(0, \sigma^2)$ represents the stochastic disturbance. The slack variable $\zeta_t \geq 0$ penalizes constraint violations with weight p, and lb, ub denote bounds on a_t.

We formulate an MILP model of (19) that represents a single stage t and summarizes the remaining stages in a value function, forming a two-stage MILP:

$$\min_{a_t, v, \zeta_t} \quad L^\intercal a_t + v + p\zeta_t$$

$$\text{s. t.,} \quad Ds_t + Ea_t \leq F + \zeta_t,$$

$$\psi_j(\hat{M}s_t + \hat{B}a_t) + b_j \leq v, \tag{20}$$

$$lb \leq a_t \leq ub, \quad \zeta_t \geq 0, \quad a_t^i \in \mathbb{Z}, \quad \forall a_t^i \in a_t.$$

Here v denotes the value function estimate, defined as a piecewise linear function of the next state s_{t+1}. Since the value function of a stochastic program is a convex polyhedral function, we approximate it using a set of affine value function pieces ψ_j, b_j. D, E, F, lb and ub are assumed to be known. The state transition and immediate cost are assumed to be unknown, and $L, \hat{M}, \hat{B}$ denote their estimates. The value function pieces ψ_j, b_j are to be learned. However, to simplify gradient calculation, the products of $\psi_j \hat{M}$ and $\psi_j \hat{B}$ are learned instead of learning $\psi_j, \hat{M}, \hat{B}$ separately. We collect all the parameters of the model that need to be learned into $\theta = \{L, \psi_j \hat{M}, \psi_j \hat{B}, b_j\}$.

5.2 Example 2

Using the model in (20) as a template, we define a simplified portfolio optimization problem where an investor allocates wealth across n assets over time. At each step $t \in \{1, \ldots, T\}$, an integer allocation $a_t \in \mathbb{Z}_+^n$ is chosen under budget and risk constraints, after which stochastic returns $r_t \in \mathbb{R}^n$ are realized, and wealth is updated. We consider $n = 10$ assets with daily historical returns from January 2023 to January 2026, primarily Exchange-Traded Funds (ETFs) along with Bitcoin. The asset set spans low-risk instruments such as short-term US Treasury bonds (SHV) and the S&P 500 (SPY), to more volatile assets such as Bitcoin (BTC) and the Nasdaq-100 (QQQ) [42]. The resulting portfolio investment problem can be modeled as follows:

$$\min_{a_t, v, \zeta_t} \quad \sum_{i \in \mathcal{I}} (\tau - r_{i,t}) \, c_i \, a_{i,t} + v + p\zeta_t$$

$$\text{s.t.} \quad \sum_{i \in \mathcal{I}} c_i a_{i,t} \leq W_t + \zeta_t,$$

$$\sum_{i \in \mathcal{I}} \sigma_i a_{i,t} \leq \rho W_t + \zeta_t, \tag{21}$$

$$\psi_j\left(\hat{M}s_t + \hat{B}a_t\right) + b_j \leq v, \quad \forall j \in \mathcal{J},$$

$$0 \leq a_{i,t} \leq U_i, \quad \forall i \in \mathcal{I},$$

$$a_{i,t} \in \mathbb{Z}, \quad \forall i \in \mathcal{I}, \quad \zeta_t \geq 0, \quad v \in \mathbb{R}.$$

The objective models immediate costs as transaction costs, with rate τ, offset by realized returns $r_{i,t}$, scaled by asset cost c_i. A slack variable $\zeta_t \geq 0$ is penalized by p to softly enforce feasibility. The first constraint enforces the budget based on current wealth W_t, while the second limits portfolio risk using a linear proxy with asset-wise standard deviations $\sigma \in \mathbb{R}_+^n$ and risk tolerance ρ.

The system state is given by the current wealth, $s_t = w_t$, with initial wealth $w_0 \in \mathbb{R}_+$. After applying action a_t, the next state evolves as

$$s_{t+1} = w_t + \sum_i r_{i,t} a_{i,t} - \tau \sum_i c_i a_{i,t}.$$

Each asset allocation is bounded by $0 \leq a_{i,t} \leq U_i$, where U_i denotes the maximum allowable units per asset. Note that the problem formulation in (21) shares the same structure as (20). To ensure notational consistency for gradient estimation in the next section, we introduce the following definitions to treat both examples equivalently:

$$L = \left[(\tau - r_{1,t})c_1 \cdots (\tau - r_{n,t})c_n \right]^\top, \quad D = \left[-1 \; -\rho \right], \quad E = \left[\mathbf{1}^\top \; \sigma^\top \right], \quad F = 0.$$

5.3 Gradient Estimation

The Lagrangian for a state-decision pair is computed as:

$$\begin{aligned}
\mathcal{L}(s_t, a_t, \lambda_t) = {} & \lambda_{ineq}(F + \zeta_t - Ds_t - Ea_t) \\
& + \sum_j \lambda_j(v - b_j - \psi_j \hat{M} s_t - \psi_j \hat{B} a_t) \\
& + \lambda_{lb}(a_t - lb) + \lambda_{ub}(ub - a_t) + L^\mathsf{T} a_t + v + p\zeta_t,
\end{aligned} \tag{22}$$

where $\lambda_t = [\lambda_{\text{ineq}}, \lambda_1, \ldots, \lambda_J, \lambda_{lb}, \lambda_{ub}]^\top$ collects the dual variables, and the bounds lb and ub are specific to the subproblem from which a_t was sampled. We can then compute the following:

$$\left[\frac{\partial \mathcal{L}}{\partial L}, \frac{\partial \mathcal{L}}{\partial \psi_j \hat{M}}, \frac{\partial \mathcal{L}}{\partial \psi_j \hat{B}}, \frac{\partial \mathcal{L}}{\partial b_j} \right] = \left[a_t, \; -\lambda_j s_t, \; -\lambda_j a_t, \; -\lambda_j \right]. \tag{23}$$

The gradient of $Q_\theta(s_t, a_t)$ is given by (17) as:

$$\nabla_\theta Q_\theta(s_t, a_t) = \left[\frac{\partial \mathcal{L}}{\partial L}, \frac{\partial \mathcal{L}}{\partial \psi_j \hat{M}}, \frac{\partial \mathcal{L}}{\partial \psi_j \hat{B}}, \frac{\partial \mathcal{L}}{\partial b_j} \right]. \tag{24}$$

5.4 RL Experiments

RL experiments are conducted on both examples in (19) (Sect. 5.1) and (21) (Sect. 5.2). The parameter values for both the experiment are shown in Table 1. The reward is defined as the negative cost. The critic and actor learning rates are set to 0.01 and 0.001, respectively, with discount factor $\gamma = 0.9$ and exploration noise $\sigma = 1$. The penalty factor p is set to 1000 for Example 1 and 400 for Example 2. The MILP model together with the B&B solver forms the actor, while the critic is implemented as a neural network and trained using Generalized Advantage Estimation (GAE). The policy gradient is computed using (15). In all experiments, both L and v are learned using RL, and all experiments are repeated five times with different random seeds.

Table 1. Summary of upper and lower bounds for all problem parameters used in Experiments 1 and 2. Fixed parameters in Example 1: $D_2 = 0$, $a_{lb} = 0$, $a_{ub} = 10$. $D = [D_1; D_2]$ and $F = [F_1; F_2]$. Fixed parameters in Example 2: $D = [-1, -0.15]$, $E_1 = 1$, while E_2, B, L, l are unconstrained.

	Exp 1		Exp 2	
Variable	ub	lb	ub	lb
ℓ, L	10	0	x	x
D_1, E, B	1	0	x	x
F_1	15	5	0	0
F_2	10	1	0	0
M, $\psi_j \hat{M}$, $\psi_j \hat{B}$, b_j	0.1	0	0.1	0

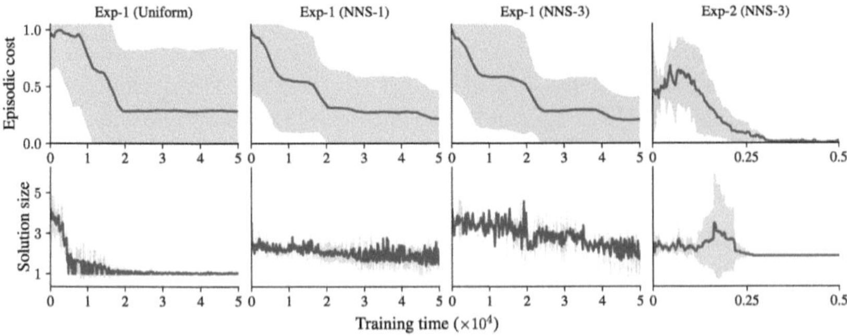

Fig. 3. The first three columns show the results for Experiment 1, with each subplot corresponding to a different sampling method: (i) uniform sampling, (ii) using 1 nearest neighbour (NNS-1), and (iii) using 3 nearest neighbours (NNS-3). The last column shows the results for Experiment 2. (**top row**): Mean and variance of the training progress over 5 random seeds for Experiments 1 and 2. The episodic cost values are normalized between 0 and 1. (**bottom row**): Comparison of how the size of the solution set changes during training. The episodic costs and solution sizes are smoothed using a rolling window of size W (Experiment 1: $W = 35$; Experiment 2: $W = 50$).

For both examples, the RL agent interacts with a simulated environment during training, while the initial MILP formulations in (20) and (21) act as imperfect approximations of that environment. In Example 1, each episode begins from a randomly sampled initial state. In Example 2, each episode begins from a randomly sampled starting day and proceeds over a fixed horizon. The initial wealth is uncertain and is sampled uniformly from $W \pm 0.1W$ at the beginning of each episode. During training, the RL agent uses Algorithm 1 to update the MILP parameters to improve performance.

The objective of all experiments is to demonstrate that the CORL algorithm can both learn from data and converge. In the experiment with Example 1 (Sect. 5.1), we additionally compare the effect of the two sampling methods. For each seed, three training runs are performed: one with uniform sampling and two with NNS. The latter includes NNS-1 (using a single nearest neighbor) and NNS-3 (using three nearest neighbors), where candidates are selected based on the shortest Manhattan distance to the fractional solution at the pruned node. In the experiment with Example 2 (Sect. 5.2), we use only the NNS-3 sampling strategy.

5.5 Results

The results for all experiments are shown in Fig. 3, illustrating both how training improves performance and how the size of the solution set in the B&B solver evolves throughout training. We observe that the cost curves decrease and converge, demonstrating the effectiveness of the algorithm. The sudden jumps in the learning curves are expected due to the discontinuous nature of MILPs. We further observe that in Experiment 1, both the NNS and uniform sampling approaches tend to converge to similar episodic costs. However, the NNS approach appears to perform better, is more robust across different seeds, and learns faster in the early training stage. Additionally, we observe that the size of the solution set consistently shrinks across all sampling methods, and learning converges as the solution set converges. Since the solution set size is directly linked to the B&B tree size, the B&B tree also shrinks during convergence. Among the variants, NNS-3 appears to be the most explorative, as its solution set decreases more gradually compared to the others.

6 Discussion

Results in Sect. 5 validate the CORL proof-of-concept, demonstrating that a general MILP model can be fine-tuned via RL on real-world data to directly optimize decision performance. Importantly, rather than trying to accurately model the real-world problem in the traditional sense, the CORL approach uses RL to adapt a CO decision scheme directly for maximizing its real-world performance while preserving its MILP structure [4,5]. While this work provides the mathematical foundations and a proof-of-concept, several extensions are needed for real-world deployment, including scalable CORL algorithms based on SAC and PPO. Additionally, integrating offline RL for data-efficient learning and extending the framework to commercial B&B solvers with improved sampling strategies, and further exploiting the structure of the CO problem are important directions. Solving an MILP at every RL step can be computationally prohibitive for large problems. In practice, this motivates using solver acceleration such as warm-starting, tree reuse, heuristics, or applying multiple policy-gradient updates using local or learned policy approximations. Additionally, CORL needs to be extended beyond the specific MILP+B&B combination to a broader class of CO methods.

Regarding sampling the nodes, using (12) is effective as a proof-of-concept; however, it suffers from two biases. On one hand, it can be overly optimistic, assigning high probability to pruned branches; and on the other hand, overly pessimistic, since a low $Q_\theta^{k,t}$ may mask a large number of potential child leaves. Addressing these biases is an important direction for future work. We find that the decision sampling strategy strongly affects RL convergence. We observe that NNS outperforms uniform sampling in Example 1; however, the current experiments are too simple to support strong conclusions. Unlike uniform sampling, NNS leverages pruned-node solutions and branch bounds to focus exploration on high-value regions, reducing suboptimal action selection and smoothing the learning. We therefore hypothesize that it will significantly outperform uniform sampling in larger-scale settings. However, NNS may reduce exploration when the B&B solution set shrinks, since each pruned node contributes only a few representative decisions. Therefore, the choice of how many neighbors to include, more broadly, how to balance exploration versus exploitation, remains problem-specific. As training progresses, RL could lower the objective values of suboptimal branches, prompting their early pruning and shrinking the B&B tree. As the B&B solution set collapses toward convergence, the policy gradient approaches zero, as observed in the examples.

In our analysis, we assume all decisions are feasible in order to restrict the scope of this work. However, for most real-world problems, relaxing a constraint by penalizing the cost might not be applicable. Therefore, a key direction for future work is needed to address this limitation by deriving sampling schemes that guarantee feasibility. Additionally, calculating the policy gradient over a discontinuous policy function can be problematic, since the true gradient is undefined at the discontinuities. The Lagrangian of (5) changes instantaneously as constraints activate or deactivate, and the approximate gradient in (17) is computed without accounting for these discontinuities. Similarly, we ignore the dependence of the cardinality of the feasible solution set on policy parameters θ. Although this may produce non-zero gradients at discontinuities, such events are rare, and their errors tend to average out; nonetheless, further work is needed to understand the implications of these simplifications and to devise strategies for explicitly handling such discontinuities. Results from the illustrative examples demonstrate the effectiveness of the CORL approach despite these approximations; however, evaluation on real-world problems is necessary to confirm its robustness, and we will pursue this in future work.

On an interesting side note, the proposed CORL framework demonstrates how an MILP scheme can serve as a compatible function approximator for RL, extending mathematical programmingbased structured policies for RL [32]. By embedding an MILP policy, the agent can exploit the problem structure more effectively and reduce the sample complexity compared to using model-free RL. We further hypothesize that CORL could enable the development of safe and explainable RL policies for combinatorial decision-making problems.

7   Conclusions

We presented CORL, a proof-of-concept RL framework for combinatorial sequential decision-making specific to MILPs solved via B&B. We formulate the MILP+B&B pipeline as a stochastic policy and derive its policy gradient to make the MILP scheme a compatible policy for RL. Compared to the traditional approach of improving the modeling accuracy of MILP to enhance decision quality, the CORL framework directly fine-tunes an existing MILP scheme to maximize its real-world performance through RL. The CORL framework establishes the foundations for building a scalable CORL algorithm for learning high-performance combinatorial policies for real-world processes. However, further work is required to scale the approach to real-world applications.

References

1. Amos, B., Yarats, D.: The differentiable cross-entropy method. In: Proceedings of the 37th International Conference on Machine Learning, pp. 291–302. PMLR (2020) (2020)
2. Anand, A.S., Kordabad, A.B., Zanon, M., Gros, S.: Optimality conditions for model predictive control: Rethinking predictive model design. arXiv:2412.18268 (2024)
3. Anand, A.S., Sawant, S., Hoffmann, J., Reinhardt, D., Gros, S.: Closing the sim2real performance gap in rl. arXiv:2510.17709 (2025)
4. Anand, A.S., Sawant, S., Reinhardt, D., Gros, S.: Data-driven predictive control and mpc: Do we achieve optimality? IFAC-Papers OnLine **58**(15), 73–78 (2024)
5. Anand, A.S., Sawant, S., Reinhardt, D.P., Gros, S.: Predicting what matters: Training ai models for better decisions. IEEE Trans. Neural Netw. Learn. Syst. (2025)
6. Bellman, R.: Dynamic Programming Princeton University Press Princeton, pp. 24–73. New Jersey Google Scholar (1957)
7. Bello, I., Pham, H., Le, Q.V., Norouzi, M., Bengio, S.: Neural combinatorial optimization with reinforcement learning, arXiv:1611.09940 (2016)
8. Berthet, Q., Blondel, M., Teboul, O., Cuturi, M., Vert, J.P., Bach, F.: Learning with differentiable perturbed optimizers. Adv. Neural. Inf. Process. Syst. **33**, 9508–9519 (2020)
9. Eiselt, H.A., Sandblom, C.L.: Operations research: A model-based approach, Springer Nature (2022)
10. Elmachtoub, A.N., Grigas, P.: Smart "predict, then optimize". Manag. Sci. **68**(1), 9–26 (2022)
11. Farahmand, A., Barreto, A., Nikovski, D.: Value-aware loss function for model-based reinforcement learning. In: Artificial Intelligence and Statistics. pp. 1486–1494. PMLR (2017)
12. Ferber, A., Wilder, B., Dilkina, B., Tambe, M.: Mipaal: Mixed integer program as a layer. In: Proceedings of the AAAI Conference on Artificial Intelligence, vol. 34, pp. 1504–1511 (2020)
13. Grimm, C., Barreto, A., Farquhar, G., Silver, D., Singh, S.: Proper value equivalence. Adv. Neural. Inf. Process. Syst. **34**, 7773–7786 (2021)
14. Gros, S., Zanon, M.: Reinforcement learning for mixed-integer problems based on mpc. IFAC-PapersOnLine **53**(2), 5219–5224 (2020)
15. Gurobi Optimization, LLC: Gurobi Optimizer Reference Manual (2024). https://www.gurobi.com
16. Haarnoja, T., Zhou, A., Abbeel, P., Levine, S.: Soft actor-critic: Off-policy maximum entropy deep reinforcement learning with a stochastic actor. In: International Conference on Machine Learning, pp. 1861–1870. PMLR (2018)
17. Khalil, E., Le Bodic, P., Song, L., Nemhauser, G., Dilkina, B.: Learning to branch in mixed integer programming. In: Proceedings of the AAAI Conference on Artificial Intelligence, vol. 30 (2016)
18. Korte, B., Vygen, J.: Combinatorial optimization: theory and algorithms, Springer (2008)
19. Kotary, J., Fioretto, F., Van Hentenryck, P., Wilder, B.: End-to-end constrained optimization learning: A survey, arXiv:2103.16378 (2021)
20. Kuhn, H.W., Tucker, A.W.: Nonlinear programming. In: Traces and emergence of nonlinear programming, pp. 247–258. Springer, Cham (2013)
21. Lee, T.H., Kim, M.S.: Rl-milp solver: A reinforcement learning approach for solving mixed-integer linear programs with graph neural networks, arXiv:2411.19517 (2024)
22. Ma, Q., Ge, S., He, D., Thaker, D., Drori, I.: Combinatorial optimization by graph pointer networks and hierarchical reinforcement learning, arXiv:1911.04936 (2019)
23. Mandi, J., Bucarey, V., Tchomba, M.M.K., Guns, T.: Decision-focused learning: Through the lens of learning to rank. In: International Conference on Machine Learning, pp. 14935–14947. PMLR (2022)

24. Mandi, J., et al.: Decision-focused learning: Foundations, state of the art, benchmark and future opportunities. J. Artif. Intell. Res. **80**, 1623–1701 (2024)
25. Mazyavkina, N., Sviridov, S., Ivanov, S., Burnaev, E.: Reinforcement learning for combinatorial optimization: a survey. Comput. Oper. Res. **134**, 105400 (2021)
26. Milgrom, P., Segal, I.: Envelope theorems for arbitrary choice sets. Econometrica **70**(2), 583–601 (2002)
27. Petropoulos, F., Laporte, G., Aktas, E., Alumur, S.A., Archetti, C., Ayhan, H., Battarra, M., Bennell, J.A., Bourjolly, J.M., Boylan, J.E., et al.: Operational research: methods and applications. J. Oper. Res. Soc. **75**(3), 423–617 (2024)
28. Pogančić, M.V., Paulus, A., Musil, V., Martius, G., Rolinek, M.: Differentiation of blackbox combinatorial solvers. In: International Conference on Learning Representations (2019)
29. Powell, W.B.: Clearing the jungle of stochastic optimization. In: Bridging Data and Decisions, pp. 109–137. Informs (2014)
30. Puterman, M.L.: Markov decision processes: discrete stochastic dynamic programming. John Wiley & Sons (2014)
31. Qi, M., Wang, M., Shen, Z.J.: Smart feasibility pump: Reinforcement learning for (mixed) integer programming, arXiv preprint arXiv:2102.09663 (2021)
32. Reiter, R., et al.: Synthesis of model predictive control and reinforcement learning: Survey and classification. arXiv preprint arXiv:2502.02133 (2025)
33. Ruszczyński, A., Shapiro, A.: Stochastic programming models. Handbooks Oper. Res. Manag. Sci. **10**, 1–64 (2003)
34. Sadana, U., Chenreddy, A., Delage, E., Forel, A., Frejinger, E., Vidal, T.: A survey of contextual optimization methods for decision-making under uncertainty. Eur. J. Oper. Res. **320**(2), 271–289 (2025)
35. Scavuzzo, L., Aardal, K., Lodi, A., Yorke-Smith, N.: Machine learning augmented branch and bound for mixed integer linear programming. Math. Program. 1–44 (2024)
36. Schrittwieser, J., et al.: Mastering atari, go, chess and shogi by planning with a learned model. Nature **588**(7839), 604–609 (2020)
37. Schulman, J., Wolski, F., Dhariwal, P., Radford, A., Klimov, O.: Proximal policy optimization algorithms, arXiv preprint arXiv:1707.06347 (2017)
38. Wei, R., Lambert, N., McDonald, A.D., Garcia, A., Calandra, R.: A unified view on solving objective mismatch in model-based reinforcement learning. Trans. Mach. Learn. Res. (2024), survey Certification
39. Wilder, B., Dilkina, B., Tambe, M.: Melding the data-decisions pipeline: decision-focused learning for combinatorial optimization. In: Proceedings of the AAAI Conference on Artificial Intelligence, vol. 33, pp. 1658–1665 (2019)
40. Wolsey, L.A., Nemhauser, G.L.: Integer and combinatorial optimization. John Wiley & Sons (1999)
41. Xu, L., Wilder, B., Khalil, E.B., Tambe, M.: Reinforcement learning with combinatorial actions for coupled restless bandits, arXiv preprint arXiv:2503.01919 (2025)
42. Yahoo Finance: Yahoo finance (nd). https://finance.yahoo.com/. Accessed 17 Mar 2026

Enhancing Scalability in Distributed Flexible Flowshop Scheduling: A Hybrid RL-CP Approach

Ioannis Avgerinos[1], Christos Katrinakis[1], Andreas Ktenidis[1]([✉]),
Aggelos Ioannis Lagos[1], Ioannis Mourtos[1], and Georgios Zois[2]

[1] ELTRUN Research Lab, Department of Management Science and Technology,
Athens University of Economics and Business, Athens 10434, Greece
`{iavgerinos,chr.katrinakis,and.ktenidis,t8210079,mourtos}@aueb.gr`
[2] Optiscale, Athens 11472, Greece
`georzois@aueb.gr`

Abstract. Manufacturing-as-a-Service (MaaS) leverages decentralised resources provided by a network of manufacturers, to deliver on-demand production services to consumers. The core challenge lies in optimally allocating resources to service requests, balancing resource utilisation with consumer satisfaction. This context introduces large-scale, multi-objective optimisation problems that traditional exact methods, such as Constraint Programming (CP), struggle to solve efficiently. Motivated by a MaaS setting of two competing producers of electronic boards for white appliances, we propose a hybrid Reinforcement Learning (RL)-CP approach for the distributed hybrid flexible flowshop scheduling problem. We aim at minimising total earliness and tardiness, while ensuring fairness among the providers by minimising load imbalance. An RL agent divides the set of requests into smaller subsets to minimise load imbalance, while yielding smaller job sets solvable through a CP model to minimise total earliness-tardiness. To evaluate our hybrid RLCP framework, we benchmark it against a relaxed CP model and a multi-phase constructive metaheuristic. As we show, the hybrid RL-CP outperforms both the CP model and the metaheuristic baselines in most instances, within practical computation times.

Keywords: distributed flowshop · constraint programming · reinforcement learning · manufacturing-as-a-service · large-scale

1 Introduction

Overview. MaaS enables the access to manufacturing capabilities on demand, often via digital platforms, and leverages distributed supplier networks, applying the "as-a-service" model to production. Production must be jointly orchestrated across multiple facilities in an efficient way that suits both the **providers** (who offer their manufacturing capacities to the network) and **consumers** (who submit manufacturing requests). Production-related constraints, particularly those

T. Guns (Ed.): CPAIOR 2026, LNCS 16595, pp. 36–53, 2026.
https://doi.org/10.1007/978-3-032-27242-3_3

tied to the sequence of processes, create challenges for typical optimisation methods in scheduling. Although Constraint Programming (CP) has opened the door to richer models of these industry-specific conditions through interval variables, real-life instances involving thousands of jobs per week remains beyond the reach of exact CP methods. A common workaround is to decompose weekly workload into shorter planning horizons, hence obtaining subproblems manageable by CP. Yet these split-horizon decisions often lead to inefficient schedules, particularly in the presence of tight deadlines.

Problem Description. We examine a MaaS paradigm motivated by a real industrial setting in which multiple competing manufacturers (*providers*) produce Electronic Boards (EB) for white appliances; each provider makes its production lines available for specific intervals. The resulting problem resembles a hybrid flexible flowshop scheduling (HFFS) setting: production lines follow a fixed sequence of stages that each job must traverse, stages may contain multiple parallel identical machines, and EB types may skip certain stages. Moreover, since no inter-industry transfers are allowed, the problem becomes a distributed HFFS (DHFFS). Each product is associated with a due-date, and deviations, whether early or late, incur penalties. This Just-In-Time (JIT) setting therefore calls for minimising the total earliness and tardiness across all products. The viability of MaaS depends on its trustworthiness from the providers' perspective, i.e., demand must be allocated as fairly as possible. This requirement introduces the additional objective of minimising the maximum load difference across providers.

Since no work-in-process storage is allowed in the examined problem, a product must move directly to the next stage as soon as it finishes the previous one. This restriction can introduce significant idle time: if no machine is available at the next stage, the product blocks its current machine until it can proceed. Transportation times between consecutive stages are assumed to be negligible.

Literature Review. The DHFFS can be viewed as the culmination of a long line of scheduling models that progressively incorporate more realistic manufacturing conditions. Researchers introduced hybrid flowshops (HFS) to capture parallel identical machines at each stage [16]. The flexible extension of HFS captures that jobs may not need to be processed at all stages, but can skip some [12]. These conditions are further compounded by the need to assign jobs to distinct factories, each with its own hybrid flowshop structure. The DHFFS therefore requires three main decisions: assignment of jobs to factories, sequencing across stages, and machine selection per stage.

Scheduling problems in distributed hybrid flowshop settings are NP-hard, given that the hybrid flow shop is strongly NP-hard [13], and traditional exact methods (e.g., Integer Programming or CP) often struggle to scale [17]. Mixed-Integer Linear Programs (MILPs) for variations of the DHFFS have been proposed by [9,22]. Also, [11] formulated three MILPs and a CP model for DHFFS with setup times to minimise the makespan. A state-of-the-art CP model for the HFFS with transportation times was proposed by [1], and it has largely inspired our own adaptation for the DHFFS.

Surveys of hybrid flowshops emphasise that high-performing metaheuristics combine constructive schedules based on dispatching rules with local search, iterated local search or iterated greedy [16]. For earliness/tardiness criteria, iterated local search and iterated greedy algorithms tailored to due-datedriven objectives have shown strong performance [14]. For distributed and blocking settings, [19] propose two problem-specific constructive heuristics, demonstrating that strong initial construction combined with targeted neighbourhood search substantially improves scalability.

Our standalone constructive metaheuristic follows this line of work: it generates diverse high-quality initial schedules using a portfolio of dispatching rules, and then applies large-scale neighbourhood search and simulated annealing tailored to the distributed, blocking DHFFS with fairness considerations. This provides a competitive, scalable non-learning baseline for large instances where standalone CP becomes impractical.

A recent systematic review by [15] highlights a rapid rise of RL methods for HFS, mostly in classical settings and predominantly using Q-learning or Deep Q-Network (DQN), with policy-gradient methods such as Proximal-Policy Optimisation (PPO) showing better scalability. In distributed settings, RL has predominantly been used to enhance metaheuristics for DHFFS variants. [21] integrate Q-learning into Teaching Learning-based Optimisation (TLBO) for a two-stage fuzzy DHFFS, while [4] employ Q-learningguided local search within a conditional Markov Chain search. Energy- and priority-aware extensions also embed RL for operator control: [8] apply double DQN in a co-evolutionary algorithm, [25] incorporate Q-learning into metaheuristic hybrids with tailored local search, and [24] combine Particle Swarm Optimisation (PSO) with Q-learning-driven local search for bi-objective energy-efficient scheduling.

Recent studies have integrated RL with CP through several complementary paradigms. Some embed RL directly into CP solvers, where learned branching or value-selection heuristics operate alongside propagation and backtracking, as in SeaPearl and DRL-guided branch-and-bound for 3-D bin packing [3,6]. A further line jointly formulates problems as RL environments and CP models, allowing learned policies to be integrated into CP search [2]. Finally, end-to-end scheduling approaches employ RL to learn dispatching rules through CP-backed simulators, achieving competitive job-shop performance [20]. To date, such RLCP hybrids have not been explored for the DHFFS.

Contribution. We start by proposing a relaxed CP model for DHFFS, with the goal only to minimise total earliness and tardiness. Our main contribution is a RL-guided decomposition of the problem into two sequential decisions: (a) partitioning the dataset into smaller sequential instances, and (b) assigning each EB to one of the available providers. These decisions, made by a RL agent, decompose the original instance into a series of smaller and easier subproblems that can be efficiently solved by a restricted version of our CP within short time limits, ultimately producing complete high-quality solutions. To evaluate our hybrid RL-CP mechanism, we generate instances (from 500-10000 jobs) grounded in real-world data and expert knowledge, and establish baseline values from our relaxed CP as well as from a state-of-the-art constructive metaheuristic, which builds full hybrid flow-shop schedules, combining dispatching rules, large-scale neighbourhood search and simulated annealing. Our hybrid RL-CP surpasses both CP and heuristic baselines for workloads of up to 2,000 jobs. As instance sizes increase, its performance remains stable, delivering solutions that are systematically superior in terms of earliness and tardiness while simultaneously achieving low load imbalance, all achieved within practical computational times.

Section 2 introduces the CP model for the problem under study. Section 3 presents the hybrid RLCP framework, which is the main contribution of this work, followed by a summary of the constructive metaheuristic, which serves as a baseline for large-scale datasets. Section 4 provides numerical evidence of the framework's effectiveness compared with the standalone CP and metaheuristic components. Finally, Sect. 5 summarises the key findings of the study.

2   CP Modelling

Problem Formulation. Let J denote the set of jobs representing EB demands. Each job $j \in J$ has a due-date d_j; completing j one time unit early or late incurs an equal penalty. Let F denote the set of service providers, each consisting of a production line with stages S. A job j requires processing on a subset of stages $S_j \subseteq S$, which must be visited sequentially. To explicitly represent the processing order of stages, we denote each subset S_j as $\{s_j^1, s_j^2, ..., s_j^{|S_j|}\}$, where the superscript indicates the position of the stage in the sequence required by job j. Let M be the set of all machines, and let $M_{s,f} \subset M$ denote the machines in provider f dedicated to stage s. Processing job j at stage s in provider f requires at least $p_{j,s,f}$ minutes. H stands for the planning horizon.

CP Model for JIT Scheduling. We combine modelling components of state-of-art CP formulations for the HFFS problem to construct a model which captures all constraints of the problem as described above. For this model, we use a set of interval variables $\mathtt{x}_{j,s}$, indicating the processing of job j at stage s. Variables $\mathtt{providerOf}_j$ denote the provider which handles job j, thus their domain is set F. Optional interval variables $\mathtt{y}_{j,s,m}$ are synchronised with the respective variables $\mathtt{x}_{j,s}$ for exactly one machine $m \in M_{s,f}$, f being the value of variable $\mathtt{providerOf}_j$. Last, variables $\mathtt{E}_j$ and $\mathtt{T}_j$ represent the earliness and tardiness of job j.

$$\min \sum_{j \in J} \mathtt{E}_j + \mathtt{T}_j \tag{1}$$

$$\mathtt{alternative}(\mathtt{x}_{j,s}, [\mathtt{y}_{j,s,m} | f \in F, m \in M_{s,f}])) \qquad \forall j \in J, s \in S_j \tag{2}$$

$$\mathtt{startAtEnd}(\mathtt{x}_{j,s_j^i}, \mathtt{x}_{j,s_j^{i-1}}) \qquad \forall j \in J, i \in [2, |S_j|] \tag{3}$$

$$\mathtt{noOverlap}([\mathtt{y}_{j,s,m} | j \in J, s \in S_j \text{ if } m \in M_{s,f}]) \qquad \forall f \in F, m \in M \tag{4}$$

$$\mathtt{E}_j = \max(0, d_j - \mathtt{endOf}(\mathtt{x}_{j,s_j|S_j|_j})) \qquad \forall j \in J \tag{5}$$

$$\mathtt{T}_j = \max(0, \mathtt{endOf}(\mathtt{x}_{j,s_j|S_j|_j}) - d_j) \qquad \forall j \in J \tag{6}$$

$$\text{if } \mathtt{providerOf}_j = f \rightarrow \sum_{m \in M_{s,f}} \mathtt{presenceOf}(\mathtt{y}_{j,s,m}) = 1 \qquad \forall j \in J, s \in S_j, f \in F \tag{7}$$

$$\mathtt{x}_{j,s} : \text{interval variable} \qquad \forall j \in J, s \in S_j$$
$$\mathtt{y}_{j,s,m} : \text{interval variable}$$
$$\qquad \mathtt{size} \in [p_{j,s,f}, H], \text{optional} \qquad \forall j \in J, s \in S_j, f \in F, m \in M_{s,f}$$
$$\mathtt{providerOf}_j \in F \qquad \forall j \in J$$

The objective function (1) minimises the total earliness and tardiness. Constraints (2) synchronise the time intervals of $\mathtt{x}_{j,s}$ and $\mathtt{y}_{j,s,m}$ only for one eligible machine, which is also the one which processes stage s of job j. The succession of stages is ensured by (3): each stage of a job starts at the end of the previous one. The $\mathtt{noOverlap}$ predicate of (4) prevents any processes from being simultaneously scheduled at the same machine. The values of earliness and tardiness are set by (5) and (6), given that $\mathtt{endOf}(\mathtt{x}_{j,s_j|S_j|})$ denotes the end of the last stage of job j (i.e., the completion time of j). Finally, (7) ensure that each stage s of a job j is assigned to one of the machines of the selected provider, as indicated by the value of variable $\mathtt{providerOf}_j$.

Notably, the CP model addresses only the JIT objective. It is possible to incorporate a load-imbalance objective, defined as the maximum difference between the total processing time of the demand allocated to any two providers. However, incorporating this objective has been observed to significantly degrade the model's performance, limiting its scalability to only a few hundred jobs. For this reason, the experiments in this work consider minimisation of total earliness and tardiness as the sole objective of the CP model.

Reduced CP Model. Datasets containing thousands of EBs cannot be directly handled efficiently by the CP model described above. To address this, we assume that our learning mechanism provides subsets of jobs $\bar{J}^i \subset J$, where i indicates the scheduling order. Additionally, for each job j, the assigned provider $f_j \in F$ is predetermined. These prior decisions simplify the CP model significantly.

Specifically, all variables and constraints that previously involved the entire set of jobs J are now restricted to the subset $\bar{J}^i$. Furthermore, the variables $\mathtt{providerOf}_j$ are no longer necessary, as the provider assignments are already determined by the learning mechanism. Consequently, constraints (7) are eliminated, and constraints (2) are replaced with:

$$\texttt{alternative}(\mathtt{x}_{j,s}, [\mathtt{y}_{j,s,m} | m \in M_{s,f^j}])) \qquad \forall j \in \bar{J}, s \in S_j \qquad (8)$$

Additionally, the scheduling of $\bar{J}^i$ depends on the decisions made during the scheduling of $\bar{J}^{i-1}$. Since some machines are partially occupied by previously scheduled jobs, the minimum start times for these machines must be updated. Let $u_m^{i-1} = \max(\texttt{endOf}(\mathtt{y}_{j,s,m}) | j \in \bar{J}^{i-1}, s \in S_j)$ denote the maximum completion time of machine $m \in M$ after scheduling $\bar{J}^{i-1}$. Then, for the successive subset $\bar{J}^i$, the start times of all present variables $\mathtt{y}_{j,s,m}$ must be greater/equal than u_m^{i-1}. Clearly, the quality of the final schedule, that is, the combined schedule across all subsets $\bar{J}^i$, depends heavily on the appropriate selection of job subsets and their assignments to providers.

3   A Hybrid RL-CP Approach

The decomposition of J into EB subsets $\bar{J}^i$ and the assignment of each job j to a provider f_j are produced by a learning mechanism that interacts repeatedly with the environment. In line with recent work on deep reinforcement learning and neural combinatorial optimisation for complex shop scheduling problems [10,18,23], we use a RL policy to guide the higher-level orchestration, while leaving the exact scheduling to a CP engine.

At a high level, our approach intertwines a RL decision layer with the CP scheduling layer. At first, the RL policy observes an aggregated state of the network, including information on backlog jobs, current provider workloads and projected completion times. Subsequently, based on this state, the policy selects a subset of jobs $\bar{J}^i$ and assigns each of them to one of the providers in F. We refer to these subsets as *baskets* hereinafter. Conditioned on these decisions, the reduced CP model is instantiated and solved, producing detailed schedules that respect precedence constraints of the flowshop environment. The resulting schedules induce a cost in terms of total earliness/tardiness and load imbalance across providers. This cost is converted into a scalar reward signal, which is used to update the RL policy. Figure 1 depicts the interaction between the employed components.

The figure illustrates the interaction loop of the hybrid RL–CP. At each decision step, the current weekly instance (jobs and candidate providers) is embedded as a heterogeneous job–provider graph, which forms the state s_i. This graph is processed by the PPO actor–critic network: the actor outputs basket-level assignments of jobs to providers (e.g., baskets of size B for Provider 1 and Provider 2), while the critic estimates the state value $V_\theta(s_i)$. The selected baskets are then passed to the environment, where a CP Optimizer model constructs detailed hybrid flowshop schedules for each provider. A heuristic layer maintains a lightweight approximate global schedule between CP calls. Its role is to complement the basket-level reward from the CP solver by estimating how the agent's latest assignment influences the estimated final global earliness/tardiness objective and the resulting load imbalance. After each basket is

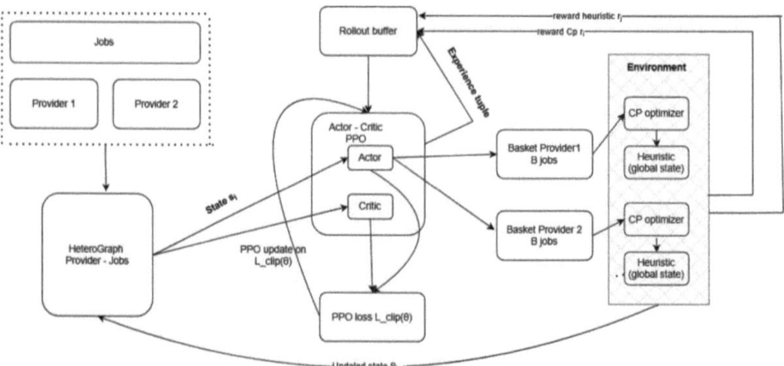

Fig. 1. A graph-based PPO allocates jobs to providers in baskets, CP Optimizer or the heuristic compute schedules and rewards, and PPO updates its policy via Lclip(θ).

scheduled by CP, the heuristic updates these quantities by summing the processing times allocated to each provider, extending projected completion times accordingly, and estimating earliness/tardiness for unscheduled jobs via a simple earliest-finish prediction. This provides a global reward component that the CP solver alone cannot supply. We also experimented with more sophisticated constructive heuristics, such as the Phase-1 scheduler used in our standalone baseline, but these richer approximations led to worse policies. The resulting experience tuples (s_i, a_i, r_i, s_{i+1}) are stored in a rollout buffer and used to compute the PPO loss $L_{\text{clip}}(\theta)$, which updates the policy parameters. The updated state s_{i+1} is then re-encoded as a new job–provider graph, closing the loop for the next decision step.

In contrast to classical approaches, where batching and provider assignment are governed by hand-crafted heuristics, the RL agent *learns* over time which patterns of decomposition and allocation lead to subproblems which can be handled more easily by the CP optimizer, thus implying better performance overall. Importantly, the CP solver remains fully responsible for detailed scheduling: the learning component never replaces, but rather *steers*, the exact optimisation engine. Unlike prior RL-based methods that directly construct machine-level schedules for (flexible) job shops [10,23], our agent operates at a higher level of abstraction, learning only the decomposition and provider-assignment policy while delegating fine-grained sequencing to CP.

3.1   MDP Formulation and Environment

We cast the sequential selection and assignment of subsets $\bar{J}^i$ as a Markov Decision Process (MDP) $\langle \mathcal{S}, \mathcal{A}, \mathcal{P}, r, \gamma \rangle$. One episode corresponds to an instance with job set J and time horizon H, decomposed into decision steps $i = 1, \ldots, I$, where each step schedules one basket (up to size B), and the episode terminates when all jobs are scheduled.

Operationally, the RL layer interacts with a simulation environment that emulates a planning horizon. Jobs are known upfront, but are revealed to the agent through a sequence of decision steps. At each step i, the environment provides a compact representation of (a) the set of unscheduled jobs, including due dates, processing times $p_{j,s,f}$, and stage masks S_j, (b) the current and projected workloads of each provider (accumulated processing time and estimated completion horizon), and (c) heuristic estimates of global earliness/tardiness.

State. We aggregate this information into a bipartite graph between jobs and providers. Job nodes carry features such as processing times, due-date, and stage-visit pattern, while provider nodes encode current load and projected completion time. This graph constitutes the state $\mathbf{s}_i \in \mathcal{S}$ fed to the policy, in a spirit similar to recent graph-based architectures for flexible job-shop scheduling [18,23].

Action. Given state $\mathbf{s}_i$, the RL agent selects an action $a_i \in \mathcal{A}$ that (i) chooses a subset of jobs $\bar{J}^i \subset J$ to be scheduled next, and (ii) assigns each $j \in \bar{J}^i$ to a provider $f_j \in F$. To make this combinatorial decision tractable, we fix a maximum basket size B and parameterise the action via a matrix of scores $q_{j,f}$ over job-provider pairs. For each remaining job j, we identify the provider f with the highest score $q_{j,f}$, and use this maximum score to rank jobs. The top-B jobs under this ranking form the basket $\bar{J}^i$, and their corresponding best providers define the assignment $\{f_j : j \in \bar{J}^i\}$.

CP-in-the-Loop Transition. Once $(\bar{J}^i, f_j, j \in \bar{J}^i)$ are chosen, they are passed to the reduced CP model. CP schedules $\bar{J}^i$ on the machines and stages of the selected providers, considering the precedence constraints between stages, and the inherited machine occupancy from previously scheduled subsets $\bar{J}^1, \ldots, \bar{J}^{i-1}$ through updated release times. The CP solver returns completion times and a detailed machine-level schedule, which are summarised by a heuristic layer into updated provider loads, projected completion times, load imbalance and global earliness and tardiness estimates, thus defining the next state $\mathbf{s}_{i+1} \sim \mathcal{P}(\cdot \mid s_i, a_i)$.

Reward. The environment computes a scalar reward $r_i = r(s_i, a_i)$ that captures the quality of the CP schedule. The reward aggregates the total earliness/tardiness of $\bar{J}^i$ and a load-imbalance term reflecting the spread between provider workloads. A supplementary heuristic component (see Fig. 1) adjusts this reward by estimating the impact of the current assignment on the final global objective across all remaining jobs. Crucially, the RL agent does not modify the internal CP formulation: it influences only the sequence of baskets and their provider assignments.

3.2   Policy Architecture and Training

The policy is implemented using a graph-based actor-critic architecture trained with PPO. Formally, the *actor* parameterises a stochastic policy $\pi_\theta(a \mid s)$, while the *critic* approximates the state-value function $V_\phi(s)$ used for variance reduction in policy-gradient updates. At each decision step, the bipartite job-provider graph is processed by a Graph Neural Network (GNN), which produces embeddings for both job and provider nodes. These embeddings are fed to an *actor* head, which outputs the job-provider score matrix $q_{j,f}$ from which baskets and assignments are derived, and a *critic* head, which estimates the value function $V(s_i)$. This design is inspired by recent advances in GNN-based and attention-based architectures for job-shop and flexible job-shop scheduling [18,23], but adapted here to operate at the level of provider assignment and instance decomposition rather than direct machine-level dispatching.

Training proceeds by simulating episodes on a set of instances of varying sizes. For each episode, the agent iteratively observes the current graph-encoded state s_i, samples an action (basket and assignments) from the stochastic policy, and receives a reward obtained from the CP schedule and transitions to s_{i+1}. The resulting on-policy trajectories are stored in a *rollout buffer*, i.e., a dataset of tuples $(s_i, a_i, r_i, s_{i+1}, \log \pi_{\theta_{\mathrm{old}}}(a_i \mid s_i), V_\phi(s_i))$, from which advantage estimates $\hat{A}_i$ (e.g., via GAE) and value targets are computed. PPO (Proximal Policy Optimisation) then updates the actor by maximising the clipped surrogate objective

$$L^{\mathrm{PPO}}(\theta) = \mathbb{E}_i \Big[\min \big(\rho_i(\theta)\hat{A}_i, \ \mathrm{clip}(\rho_i(\theta), 1 - \epsilon, 1 + \epsilon)\hat{A}_i \big) \Big], \tag{9}$$

where $\rho_i(\theta) = \exp(\log \pi_\theta(a_i \mid s_i) - \log \pi_{\theta_{\mathrm{old}}}(a_i \mid s_i))$ is the probability ratio and ϵ is the clipping parameter. The overall loss additionally includes a value regression term for the critic and an entropy bonus to encourage exploration. In our experiments, we train on a small, fixed collection of instances, cycling through one representative instance per job-size category at each training epoch, and performing a limited number of PPO epochs on the collected trajectories. Despite this modest training budget, the learned policy is able to discover allocation patterns that systematically improve downstream CP schedules compared to random or purely rule-based provider-assignment strategies, in line with empirical findings from related RL-based scheduling studies [10,23].

3.3 (Meta)heuristic Alternatives

A natural question is whether the decomposition and provider assignment decisions could be handled using alternatives methods. A first class of approaches would treat the assignment of jobs to providers, and possibly the partition of J into subsets $\bar{J}^i$, as a combinatorial optimisation problem on its own. Meta-heuristics such as genetic algorithms, tabu search or simulated annealing could be designed to explore the space of assignment vectors f_j, or even full sequences of baskets $\{\bar{J}^1, \bar{J}^2, \dots\}$. Each candidate solution would then be evaluated by instantiating and solving the corresponding CP model, thus forming a classic *matheuristic* where CP acts as an embedded evaluation oracle.

While this strategy is conceptually appealing, it faces several practical challenges in the present MaaS context. At first, the number of possible provider assignments grows exponentially with $|J|$, and the number of ways to partition jobs into ordered subsets $\bar{J}^i$ is even larger. For instances containing thousands of jobs, exploring this space directly with a metaheuristic becomes extremely costly, as each evaluation requires solving multiple CP subproblems. Also, metaheuristics would typically treat each weekly planning instance as a new optimisation problem. Even if the underlying demand patterns and provider characteristics are similar from instance to instance, there is no built-in mechanism to reuse information learned on the past; the search restarts essentially from scratch. Crucially, the quality of the final global schedule depends not only on which jobs are assigned to each provider, but also on when they are scheduled within the planning horizon, and how earlier decisions affect later CP subproblems. Encoding this long-term, sequential coupling within a static metaheuristic objective is non-trivial.

In principle, more sophisticated matheuristics could be designed to address some of these issues; for example, by iteratively refining the assignment and re-optimising batches via CP, or by employing large neighbourhood search moves that reassign subsets of jobs between providers. However, these approaches still rely on manually crafted neighbourhoods and hand-tuned parameters, and they do not explicitly exploit the repetitive nature of the planning process (e.g., weekly rolling horizons with similar characteristics).

By contrast, the proposed RL layer is explicitly designed to treat successive instances as episodes drawn from a common distribution, allowing the policy to generalise across similar demand profiles and provider configurations, optimise decisions in a multi-step setting, where the consequences of allocating a job to a provider at time i are evaluated over the entire trajectory of subsequent CP schedules, thus reducing the amount of problem-specific algorithm design. In this sense, the RL-CP framework can be viewed as a *data-driven matheuristic*: CP remains the core optimisation engine for each subproblem, while the role traditionally played by a metaheuristic (navigating the space of high-level decisions) is taken over by a learned policy.

A Standalone Metaheuristic. Given the challenges described above, we implement a metaheuristic to provide a strong non-learning baseline for large instances where the CP model may time out. This metaheuristic constructs a complete flow-shop schedule without relying on the CP model and optimises the same objectives as the RLCP framework; namely, the total earliness/tardiness of all jobs and the load imbalance between providers. The latter is considered only during the initial construction phase, since the standalone metaheuristic already achieves stronger fairness than RLCP while underperforming in total earliness/tardiness; the subsequent phases therefore focus exclusively on improving earliness/tardiness.

The metaheuristic operates in three phases. **Phase 1** generates multiple complete schedules using a single constructive scheduler. Jobs are first ordered by non-decreasing due date and assigned to the provider that minimises a score combining total processing time and a small load-balancing penalty. For each provider, we vary the *priority order* in which jobs are released to the constructive scheduler by applying classic dispatching rules (EDD, slack, critical ratio, weighted EDD, ATC) solely for ranking. The multi-stage scheduler then builds full blocking-aware schedules respecting machine availability and component capacities. Additional initial solutions are obtained by combining random provider assignments with the best sequencing rule. The best schedule found in this phase, measured by total earliness/tardiness, seeds the next phase.

Phase 2 performs iterative improvement using several operators: (i) job reassignment moves, which move a high-cost job to a different provider and rebuild the global schedule; (ii) job-swap moves, which exchange the providers of two jobs and reschedule; (iii) critical-job rescheduling, which removes a small set of jobs with the highest earliness/tardiness and reinserts them; and (iv) destruction–reconstruction steps, which temporarily remove a random fraction of scheduled jobs and reconstruct their assignments and sequences. Each move is evaluated by rebuilding the schedule with the same constructive scheduler and recomputing the earliness/tardiness objective. Improving moves are accepted greedily; if no improvement is observed for a given number of iterations, the search is restarted from the best solution found so far.

Phase 3 applies a simulated-annealing procedure to escape local optima. The same neighbourhoods as in Phase 2 (job reassignment, job swaps and provider-specific reordering) are sampled at random, but worsening moves may be accepted with probability $\exp(-\Delta/T)$, where Δ is the increase in total earliness/tardiness and T is a temperature parameter that is gradually cooled.

The best schedule found across all phases is returned as the final solution.

4   Experimental Evaluation

All experiments run on a server equipped with 32 AMD Ryzen Threadripper PRO 5955WX 16-core processors and 32 GB of RAM, running Ubuntu 22.04.5 LTS. The CP model is solved by *CPLEX 22.1.1* optimiser, using the DOCplex module. All coding for the CP, learning and metaheuristic components are implemented in *Python 3.10.12*.

4.1   Generation of Instances

The two providers in F resemble two major EB manufacturers. For our study, we focus on 10 EB types that typically account for a substantial share of the weekly demand and require a production line consisting of 10 stages, denoted by $\mathbf{s}_i$, $i \in \{1, ..., 10\}$.

To generate instances with diverse characteristics, we assume that each EB type skips a randomly selected set of 02 stages, ensuring that at least 8 stages are required for processing. Processing times do not exceed 3.7 min. To ensure that the domains of start/end times for the CP interval variables contain strictly integer values, we measure time in deciminutes; consequently, the maximum processing time becomes 37 deciminutes.

A nominal processing time for each EB type is then drawn uniformlly between 0 and 37. To capture realistic but not excessive variation across providers, we introduce a noise parameter drawn uniformly between -2 and 2. For each provider, this noise value is added to the nominal processing time to obtain the final processing time used in the experiments. For every stage at each provider, we define between 1 and 9 identical parallel machines.

We generate 720 training instances, with 8 instances for each problem size (100, 200, 500, 800, 1,000, 2,000, 5,000, and 10,000 jobs). For evaluation, we use 18 distinct instances with 500, 800, 1,000, 2,000, 5,000, and 10,000 jobs. Each job is randomly assigned to one of the ten EB types, thereby inheriting its corresponding parameter values (e.g., processing times, component requirements). The due-date of each job is selected uniformly at random between 0 and H (i.e., a large value used as planning horizon).

4.2   RL Training and Learning Protocols

Training. The provider-assignment policy is trained with PPO on the synthetic instances described above. Each training epoch consists of 8 different episodes, one per instance size (100, 200, 500, 800, 1,000, 2,000, 5,000 and 10,000 jobs). We train for 90 epochs in total, so the agent observes 720 distinct training instances and executes 141,120 decision steps (baskets of 100 jobs). After each epoch, we perform three PPO passes over the collected trajectories using a minibatch size of 16. Unless otherwise stated, we use a learning rate of 10^{-4}, discount factor $\gamma = 0.99$, GAE parameter $\lambda = 0.95$, entropy regularisation weight 0.01, value-loss weight 0.5, and gradient clipping at 0.2. Training is carried out on a single GPU-equipped machine and requires approximately five days of wall-clock time.

Learning and Evaluation Protocol. We adopt an on-policy RL scheme in which each episode corresponds to the full scheduling of a single instance over the weekly horizon. At every decision step, the agent observes the aggregated state (backlog jobs, provider/factory loads, projected completion times, component-related indicators), selects a basket of 100 jobs and a provider assignment, and then receives a scalar reward computed from the CP-derived schedule (earliness/tardiness, load imbalance and idle time). After each epoch, the current parameters of the actor–critic network are stored as a checkpoint and evaluated in a strictly offline manner on a disjoint pool of 18 evaluation instances that covers the same size profile (500, 800, 1,000, 2,000, 5,000 and 10,000 jobs). During evaluation, the learned policy is run in deterministic (greedy) mode, and its performance (measured primarily by the total earliness/tardiness and secondarily by load imbalance) is compared against CP baselines that use non-learning provider-assignment rules. The best model is selected as the checkpoint that achieves the lowest mean (ET) per job on this evaluation pool. In addition, we monitor a normalised load-imbalance metric that quantifies the fairness of the workload distribution across factories; this metric is reported for the selected checkpoint but is not used as a model-selection criterion.

During evaluation, we run the learned providerassignment policy in *deterministic* (greedy) mode: at each decision step we select the factory with the highest probability under the trained actor, without exploration noise. The downstream CP solver and the heuristic component are also deterministic for a fixed instance. As a result, repeated evaluations of the same instance with the same checkpoint always produce identical schedules (identical total earliness/tardiness, load imbalance, and solution times).

For completeness, we run three repetitions per test instance and checkpoint, but these repetitions coincide up to numerical noise. The empirical standard deviation of the performance metrics across repeats is therefore (almost) zero, and we report only the mean values, which are equal to the outcome of a single deterministic run. The repetitions are used solely as a sanity check for determinism, and do not affect the reported results.

4.3   Results

Baselines from the CP Model. To establish baseline objective values for evaluating the hybrid RLCP mechanism, the first set of experiments applies our CP model directly to the full problem. Each instance is given a 3,600-second time limit. Because the CP model struggles to provide strong bounds, the time limit is reached for all but the smallest instances, and no meaningful lower bounds can be reported. The CP model managed to compute solutions for datasets of 1,000 jobs at most; the larger scales could not be handled within the time limit of one hour.

To obtain baseline objective values for datasets of 2,000 jobs, we relax our CP model, replacing constraints (3) with:

$$\texttt{endBeforeStart}(\texttt{x}_{j,s_j^{i-1}}, \texttt{x}_{j,s_j}) \qquad \forall j \in J, i \in [2, |S_j|] \tag{10}$$

In practice, substituting `startAtEnd` with `endBeforeStart` permits more flexible scheduling among the production flows of each job. This modification allows a machine to be released as soon as a stage is completed, even if the job is not immediately processed at the subsequent stage. Although the resulting schedules are not feasible within the actual industrial environment, the relaxed model is sufficiently accurate to serve as a reliable baseline for evaluating the hybrid RLCP framework. For the 5,000 and 10,000 jobs instances, neither the initial model nor its relaxation succeeded in generating feasible solutions within the given time.

As noted earlier, load imbalance cannot be incorporated directly into the CP model without significantly reducing its scalability. Nevertheless, the corresponding metric is still reported, using the workload distribution observed in the CP solutions, to illustrate that the two objectives of the problem are not necessarily aligned. In practice, a schedule that performs well in terms of earliness and tardiness may still exhibit substantial disparities in the allocated workload.

Baselines from the Metaheuristic. The second set of experiments evaluates the standalone constructive metaheuristic, which generates complete DHFFS schedules without invoking any CP models. The metaheuristic is applied to the same set of evaluation instances. In contrast to the CP model, it simultaneously considers the minimisation of both total earliness/tardiness and load imbalance.

Table 1 summarises the results of both sets of experiments. For each method, we report the total earliness and tardiness (in $\times 10^6$ minutes), shown in the column 'total E/T ($\times 10^6$)', as well as the Normalised Load Imbalance index $\mathrm{LI^{norm}} = \frac{(\max_f L_f - \min_f L_f)}{\sum_f L_f}$, where L_f denotes the total scheduled processing load assigned to provider f. A value of $\mathrm{LI^{norm}} = 0$ corresponds to a perfectly balanced allocation (all providers carry the same load), whereas values close to 1 indicate extreme imbalance, with almost all work concentrated in a single provider.

For the CP model, which was optimised only for the first objective, we report the best objective value found after 1,200, 2,400, and 3,600 s, as shown in the corresponding columns. Since the metaheuristic completed within a few minutes for all datasets, we report its final objective values directly, along with the total elapsed time in the 'Time' column (in seconds).

Table 1. Results of the baseline methods

Instance	CP model				Standalone metaheuristic		
	total E/T (minutes)($\times 10^6$)			$\mathrm{LI^{norm}}$	total E/T ($\times 10^6$)	$\mathrm{LI^{norm}}$	Time
	1,200	2,400	3,600				
500_1	0.362	0.273	0.087	0.14	1.154	0.08	206
500_2	0.361	0.284	0.090	0.02	1.126	0.19	204
500_3	0.952	0.318	0.255	0.25	1.070	0.09	205
800_1	-	1.559	0.267	0.30	1.716	0.08	223
800_2	0.858	0.223	0.222	0.08	1.662	0.25	220
800_3	1.513	0.465	0.465	0.96	1.482	0.11	225
1000_1	-	1.816	1.546	0.42	1.996	0.08	236
1000_2	1.896	1.317	0.420	0.10	1.938	0.24	232
1000_3	1.798	1.798	1.798	0.88	1.685	0.13	240
2000_1*	3.479	3.479	3.479	0.52	3.161	0.05	240
2000_2*	3.645	3.645	3.399	0.40	1.989	0.43	240
2000_3*	3.382	3.382	3.382	0.91	2.006	0.09	240
5000_1	-	-	-	-	1.343	0.02	240
5000_2	-	-	-	-	2.556	0.04	240
5000_3	-	-	-	-	2.246	0.05	240
10000_1	-	-	-	-	17.804	0.03	427
10000_2	-	-	-	-	14.960	0.00	240
10000_3	-	-	-	-	29.972	0.00	387

*For instances of 2,000 jobs, a relaxation of the CP model has been solved

Results of Hybrid RL-CP. Table 2 reports the detailed performance of the hybrid RL-CP policy for each evaluation instance. For small and medium instances (500, 800, 1,000 and 2,000 jobs) the method attains very low total earliness/tardiness, ranging from 0.003×10^6 to 0.010×10^6 minutes, with solve times between approximately 86 and 342 seconds. The normalised load-imbalance index remains below 0.025 in all these cases, indicating nearly perfectly balanced schedules across the two factories.

For the larger 5000- and 10000-job instances, the RL-CP policy continues to scale nicely. Total E/T lies between 0.226×10^6 and 0.536×10^6 minutes, while solve times range from approximately 860 to 1,780 s (well under 30 min). The corresponding normalised load-imbalance values remain modest, in the interval $[0.049, 0.058]$. Overall, the learned policy maintains low E/T and fair distribution, even as the instance size increases by an order of magnitude, demonstrating robust scalability of our RL-CP approach.

Table 2. Hybrid RL-CP performance per evaluation instance

Instance	Objective values		Time
	total E/T (minutes)($\times 10^6$)	LI^{norm}	
500_1	0.003	0.00	86
500_2	0.003	0.00	87
500_3	0.003	0.00	86
800_1	0.004	0.01	139
800_2	0.004	0.01	139
800_3	0.004	0.01	136
1000_1	1.038	0.02	170
1000_2	0.005	0.01	173
1000_3	0.005	0.01	170
2000_1	0.010	0.02	341
2000_2	0.010	0.02	340
2000_3	0.010	0.02	342
5000_1	0.304	0.05	865
5000_2	0.304	0.05	863
5000_3	0.226	0.06	870
10000_1	0.536	0.05	1777
10000_2	0.536	0.05	1781
10000_3	0.536	0.05	1781

Comparing the results of Tables 1 and 2, for instances with 500-2,000 jobs, the best CP solutions (over all time checkpoints) obtain total E/T values between 0.09×10^6 and 3.5×10^6 min, while the metaheuristic baseline lies in a similar range (1-3×10^6 minutes). In contrast, the hybrid RL-CP policy in achieves E/T values between 0.003×10^6 and 0.010×10^6 minutes on the corresponding instances, i.e., one to two orders of magnitude smaller than both CP and the metaheuristic. For 5000- and 10000-job instances, the CP model cannot be applied to the original problem, whereas RL-CP still delivers high-quality schedules with E/T in the range $0.226\text{-}0.536 \times 10^6$, compared to $1.343\text{-}29.972 \times 10^6$ for the metaheuristic.

The hybrid approach also markedly improves fairness. The solutions of the CP model show that the load-imbalance objective is not always aligned with the minimisation of earliness/tardiness, as seen by the large deviations between the reported values, and the metaheuristic baseline exhibits normalised load-imbalance values between 0.02 and 0.43. By contrast, RL-CP attains $\text{LI}^{\text{norm}} = 0$ on all 500-job instances and stays below 0.025 for most 800, 1,000 and 2,000 job cases. Only for the largest 10,000-job instances we observe a trade-off: the metaheuristic achieves almost perfectly balanced solutions with very high E/T, whereas RL-CP accepts a modest imbalance ($\text{LI}^{\text{norm}} \approx 0.05$) in order to reduce E/T by more than an order of magnitude.

Regarding computation times, the metaheuristic remains the fastest method (about 200–430 s per instance), while RL-CP requires between 85 s on small instances and about 30 min on the largest 10,000 job cases. Given the substantial gains in both E/T and load imbalance, this extra CPU time is a reasonable cost in large-scale industrial settings. Overall, the experiments demonstrate that the proposed hybrid RL-CP policy not only exploits CP to scale to large instances, but also outperforms both the CP model and the metaheuristic baselines in most instances, within practical computation times.

To verify that the performance gains stem from the learned policy rather than from the CP decomposition itself, we also evaluated an ablated variant in which the RL agent was replaced by a simple greedy heuristic. In this variant, baskets are constructed using an EDD-based rule and jobs are assigned to providers by minimising projected completion time, while the same CP subsolver is used for scheduling. While this heuristic performs competitively on small instances, its performance deteriorates rapidly as instance size increases, becoming several-to-orders-of-magnitude worse than RLCP on the largest workloads. This confirms that the RL policy plays a critical role in capturing long-horizon interactions between successive baskets that simple rule-based strategies fail to model.

5 Concluding Remarks

This work presents an effective framework for integrating RL approaches, hybridised with standard optimisation models, for large-scale scheduling problems. As manufacturing demand in a MaaS environment grows to levels that cannot efficiently be handled by CP models or metaheuristics, delegating critical decomposition decisions to an RL agent offers faster and higher-quality solutions for both of our objectives.

Future work of this study strengthen this hybrid scheme along several directions. First, a more explicit fairness-aware training regime (e.g., stronger regularisation on the normalised load-imbalance, or hard balance constraints enforced by CP) could reduce the remaining imbalance for the largest instances without dramatically worsening E/T. Second, richer training curricula and longer training runs may further improve the learned policy, as the current model was trained for a limited number of epochs and is unlikely to have fully saturated its performance. Finally, learning to adapt CP search parameters dynamically during training could combine the strong optimality guarantees of CP with the scalability and flexibility of RL, leading to even better solutions on industrial-scale scheduling problems.

Acknowledgment. This research has been motivated and financially supported by Horizon Europe, as part of the Tec4MaaSEs research and innovation action, Grant Agreement No. 101138517.

References

1. Armstrong, E., Garraffa, M., O'Sullivan, B., Simonis, H.: The hybrid flexible flow-shop with transportation times. In: 27th International Conference on Principles and Practice of Constraint Programming (CP 2021) 210, pp. 1–18 (2021)
2. Cappart, Q., Moisan, T., Rousseau, L.-M., Prémont-Schwarz, I., Cire, A.A.: Combining reinforcement learning and constraint programming for combinatorial optimization. In: Proceedings of the AAAI Conference on Artificial Intelligence Vol. 35, no. (5), pp. 3677–3687 (2021)
3. Chalumeau, F., Coulon, I., Cappart, Q., Rousseau, L-M.: SeaPearl: a constraint programming solver guided by reinforcement learning. In: International Conference on Integration of Constraint Programming, Artificial Intelligence, and Operations Research (CPAIOR 2021), pp. 392–409 (2021)
4. Gholami, H., Sun, H.: Toward automated algorithm configuration for distributed hybrid flow shop scheduling with multiprocessor tasks. Knowl.-Based Syst. **264**, 110309 (2023)
5. IBM ILOG CPLEX Optimization Studio Documentation: State functions. https://www.ibm.com/docs/en/icos/22.1.2?topic=scheduling-state-functions. Last updated 19 Aug 2025
6. Jiang, Y., Cao, Z., Zhang, J.: Learning to solve 3-D bin packing problem via deep reinforcement learning and constraint programming. IEEE Trans. Cybern. **53**(5), 2864–2875 (2023)
7. Lee, J., Kao, H.-A., Yang, S.: Service innovation and smart analytics for industry 4.0 and big data environment. Procedia CIRP **16**, 3–8 (2014)
8. Li, R., Gong, W., Wang, L., Lu, C., Pan, Z., Zhang, X.: Double DQN-based coevolution for green distributed heterogeneous hybrid flowshop scheduling with multiple priorities of jobs. IEEE Trans. Autom. Sci. Eng. **21**(4), 6550–6562 (2023)
9. Li, Y., et al.: A discrete artificial bee colony algorithm for distributed hybrid flow-shop scheduling problem with sequence-dependent setup times. Int. J. Prod. Res. **59**(13), 3880–3899 (2021)
10. Li, Y., Yu, C.: Flexible job shop scheduling with job precedence constraints: a deep reinforcement learning approach. J. Manufact. Mater. Process. **9**(7), 216 (2025)
11. Meng, L., Gao, K., Ren, Y., Zhang, B., Sang, H., Chaoyong, Z.: Novel MILP and CP models for distributed hybrid flowshop scheduling problem with sequence-dependent setup times. Swarm Evol. Comput. **71**, 101058 (2022)
12. Naderi, B., Gohari, S., Yazdani, M.: Hybrid flexible flowshop problems: models and solution methods. Appl. Math. Model. **38**(24), 5767–5780 (2014)

13. Pan, Q.-K., Gao, L., Li, X.-Y., Gao, K.-Z.: Effective metaheuristics for scheduling a hybrid flowshop with sequence-dependent setup times. Appl. Math. Comput. **303**, 89–112 (2017)
14. Pan, Q.-K., Ruiz, R., Alfaro-Fernández, P.: Iterated search methods for earliness and tardiness minimization in hybrid flowshops with due windows. Comput. Oper. Res. **80**, 50–60 (2017)
15. Pugliese, V., Ferreira, O., Faria, F.: Hybrid flow shop scheduling through reinforcement learning: A systematic literature review. In: Proceedings of the 40th ACM/SIGAPP Symposium on Applied Computing, pp. 1240–1249 (2025)
16. Ruiz, R., Vázquez-Rodríguez, J.A.: The hybrid flow shop scheduling problem. Eur. J. Oper. Res. **205**(1), 1–18 (2010)
17. Shao, W., Shao, Z., Pi, D.: Modeling and multi-neighborhood iterated greedy algorithm for distributed hybrid flow shop scheduling problem. Knowl.-Based Syst. **194**, 105527 (2020)
18. Smit, I.G., Wu, Y., Troubil, P., Zhang, Y., Nuijten, W.P.M.: Neural combinatorial optimization for stochastic flexible job shop scheduling problems. In: Proceedings of the 39th AAAI Conference on Artificial Intelligence, Vol. 39, no. 25, pp. 26678–26687 (2025)
19. Sun, X., Shen, W., Fan, J., Vogel-Heuser, B., Zhang, C.: An improved non-dominated sorting genetic algorithm II for distributed heterogeneous hybrid flowshop scheduling with blocking constraints. J. Manuf. Syst. **77**, 990–1008 (2024)
20. Tassel, P., Gebser, M., Schekotihin, K.: An end-to-end reinforcement learning approach for job-shop scheduling problems based on constraint programming. In: Proceedings of the International Conference on Automated Planning and Scheduling, Vol. 33, no. (1), pp. 614–622 (2023)
21. Xi, B., Lei, D.: Q-learning-based teaching-learning optimization for distributed two-stage hybrid flow shop scheduling with fuzzy processing time. Complex Syst. Model. Simul. **2**(2), 113–129 (2022)
22. Ying, K.-C., Lin, S.-W.: Minimizing makespan for the distributed hybrid flowshop scheduling problem with multiprocessor tasks. Expert Syst. Appl. **92**, 132–141 (2018)
23. Zhang, W., Bao, X., Geng, H., Zhang, G., Gen, M.: Graph neural network and expert-guided deep reinforcement learning for solving flexible job-shop scheduling problem. Comput. Oper. Res. **183**, 107155 (2025)
24. Zhang, W., Li, C., Gen, M., Yang, W., Zhang, G.: A multiobjective memetic algorithm with particle swarm optimization and Q-learning-based local search for energy-efficient distributed heterogeneous hybrid flow-shop scheduling problem. Expert Syst. Appl. **237**(C), 121570 (2024)
25. Zhu, Q., Gao, K., Huang, W., Ma, Z., Slowik, A.: Q-learning-assisted metaheuristics for scheduling distributed hybrid flow shop problems. Comput. Mater. Continua **80**(3), 3573 (2024)

The Theory of Lagrangian Filtering Zones: A Case Study on the Knapsack Constraint

Frédéric Berthiaume[ID] and Claude-Guy Quimper[✉][ID]

Université Laval, Québec, Canada
frederic.berthiaume.1@ulaval.ca, claude-guy.quimper@ift.ulaval.ca

Abstract. CP-based Lagrangian filtering has proven to be a reliable method for enhancing constraint programming solvers. The general theory lacks a uniform framework to explain many known observations. This paper introduces such a framework, along with theoretical results and illustrations. We show why optimal Lagrange multipliers are not always the most effective for filtering. We propose a strategy to move the Lagrange multipliers away from the optimal one to launch a gradient descent that leads to filtering. This paper presents an application of this framework to the versatile Multidimensional Knapsack constraint. The algorithm is tested on two well-known problems : the multidimensional knapsack problem and the uncapacitated facility location problem. The results show a significant speed up compared to the traditional CP-based Lagrangian filtering method on both problems.

Keywords: Constraint programming · Lagrangian relaxation · Knapsack constraint

1 Introduction

Constraint programming is a sub-branch of Artificial Intelligence interested in combinatorial problems. Constraint solvers get their strength from their filtering algorithms that prune a search tree. Some filtering algorithms encode constraints with a linear program formulation. The reduced cost of each variable can be computed and used for filtering. The Lagrangian relaxation of the constraint's linear program formulation helps to compute efficiently a bound on the objective. This combination of cost-based filtering and Lagrangian relaxation has proven to be a reliable technique in constraint programming [2,4,6].

Sellmann [19] shows that optimal Lagrange multipliers, with respect to the objective value, do not provide the best filtering. Suboptimal Lagrange multipliers are more suitable for filtering. Some authors have explored the search for suboptimal Lagrange multipliers that provide more filtering [1,3,5]. The general procedure performs Lagrangian optimization and locally changes the Lagrange multipliers. The alteration phase allows for the exploration of new regions in the Lagrange multipliers space, and the discovery of missed filtering.

T. Guns (Ed.): CPAIOR 2026, LNCS 16595, pp. 54–70, 2026.
https://doi.org/10.1007/978-3-032-27242-3_4

We have 6 contributions: (1) we propose a unifying framework that explains many known observations of Lagrangian cost-based filtering, (2) we formally define *Lagrangian filtering zones*, (3) we show in which situation optimal Lagrange multipliers do not provide filtering, (4) we present a complete RELAXED($\boldsymbol{\lambda}$) filtering algorithm for the multidimensional KNAPSACK constraint, (5) we present empirical evidence supporting the theory, (6) we improve the solving times of the knapsack and uncapacitated facility location problems.

2   Background

Column and row vectors are written $\boldsymbol{v}$ and $\boldsymbol{v}^\top$. The dot product of $\boldsymbol{v}$ and $\boldsymbol{w}$ is written $\langle \boldsymbol{v}, \boldsymbol{w} \rangle$. The norm of a vector is written $\|\boldsymbol{v}\| = \sqrt{\langle \boldsymbol{v}, \boldsymbol{v} \rangle}$. A matrix is written M, with the j-row $\boldsymbol{M}_{j*}$ and the i-th column $\boldsymbol{M}_{*i}$, and the components m_{ji}. $\boldsymbol{\nabla} f(\boldsymbol{x})$ is the gradient of f at $\boldsymbol{x}$. The cardinality of a set A is $|A|$.

2.1   0–1 Multidimensional Knapsack Problem and Constraint

The multidimensional knapsack problem is defined as

$$V = \max\{\langle \boldsymbol{v}, \boldsymbol{x} \rangle : W\boldsymbol{x} \le \boldsymbol{c}, \boldsymbol{x} \in \{0,1\}^n\}, \tag{1}$$

where $\boldsymbol{v} \in \mathbb{Z}^n$, $\boldsymbol{c} \in \mathbb{Z}^m$ and $W \in \mathbb{Z}^{m\times n}$. In (1), there are n items and each consumes m resources. The value V is the optimal value.

The multidimensional knapsack constraint $\mathrm{MK}(\boldsymbol{X}, P; \boldsymbol{v}, W, \boldsymbol{c})$ is a constraint on n binary variables X_i and an integer variable P that is satisfied when:

$$\mathrm{MK}(\boldsymbol{X}, P; \boldsymbol{v}, W, \boldsymbol{c}) \iff \langle \boldsymbol{v}, \boldsymbol{X} \rangle \ge P \wedge W\boldsymbol{X} \le \boldsymbol{c}. \tag{2}$$

Enforcing domain consistency on MK is NP-hard, since deciding if there exists a $\boldsymbol{X}$ for any value in the domain of P that satisfies (2) is NP-complete; [8,12].

2.2   Cost-Based Filtering

Before 1999, the filtering for (2) was performed by incremental updates of the lower bound of P. This filtering provides a poor relation between the values of the binary variables $\boldsymbol{X}$ and the value of P. Focacci et al. [9] present a new approach to design filtering algorithms to overcome this problem. They introduce the linear relaxation of (1) within the filtering algorithm. Any V coming out of (1) and the linear relaxation of (1) is a valid upper bound of P. The value

$$V[x_i = a] = \max\{\langle \boldsymbol{v}, \boldsymbol{x} \rangle : W\boldsymbol{x} \le \boldsymbol{c}, x_i = a, \boldsymbol{x} \in \{0,1\}^n\} \tag{3}$$

and the linear relaxation of (3) are also valid upper bounds of P for any $a \in \mathrm{dom}(X_i)$. Focacci et al. [9] propose to use the bounds $V[x_i = a]$ to detect filtering. Cost-based filtering removes a value a from a domain $\mathrm{dom}(X_i)$ whenever:

$$V[x_i = a] < \min(\mathrm{dom}(P)). \tag{4}$$

Drawback 1 *The time needed to compute the bounds (3), each time the filtering algorithm is called, can hinder the overall performance.*

2.3 Cost-Based-Lagrangian-Filtering

Sellman and Fahle [20] adapt the pioneer work of Held and Karp [10] to Focacci et al. [9] method, and solved a Lagrangian relaxation of (1) like this one

$$V'(\boldsymbol{\lambda}) = \max\{\langle \boldsymbol{v}, \boldsymbol{x}\rangle + \langle \boldsymbol{\lambda}, \boldsymbol{c} - \mathrm{W}\boldsymbol{x}\rangle : \forall\, i,\ x_i \in \mathrm{dom}(X_i)\}. \tag{5}$$

$V'(\boldsymbol{\lambda})$ gives, for any Lagrange multipliers $\boldsymbol{\lambda} \geq \boldsymbol{0}$, a valid upper bound on P. With this Lagrangian relaxation, the optimization problem is *separable*:

$$V'(\boldsymbol{\lambda}) = \textstyle\sum_{i=1}^{n} \max\{(v_i - \langle \boldsymbol{W}_{*i}, \boldsymbol{\lambda}\rangle)x_i : x_i \in \mathrm{dom}(X_i)\} + \langle \boldsymbol{\lambda}, \boldsymbol{c}\rangle, \tag{6}$$

making the computation of $V'(\boldsymbol{\lambda})$ trivial; resolving the Drawback 1. Equation (6) means that any x_i, associated with a non-fixed variable X_i, is determined by

$$x_i = \begin{cases} 1, & \text{if } v_i - \langle \boldsymbol{W}_{*i}, \boldsymbol{\lambda}\rangle > 0 \\ 0, & \text{otherwise} \end{cases}. \tag{7}$$

Let $\bar{\boldsymbol{x}}$ be the solution of (6) with components given by (7). With the Lagrangian relaxation (5), $V'[x_i = 1 - \bar{x}_i](\boldsymbol{\lambda})$ is easier to compute, because the reduced cost of x_i is $V'(\boldsymbol{\lambda}) - V'[x_i = 1 - \bar{x}_i](\boldsymbol{\lambda}) = |v_i - \langle \boldsymbol{W}_{*i}, \boldsymbol{\lambda}\rangle|$, which gives this equation

$$V'[x_i = 1 - \bar{x}_i](\boldsymbol{\lambda}) = V'(\boldsymbol{\lambda}) - |v_i - \langle \boldsymbol{W}_{*i}, \boldsymbol{\lambda}\rangle|. \tag{8}$$

Not every valid Lagrange multipliers vector provides either a good estimation of the upper bound of P or filtering. The bound is tightened by minimizing V' :

$$\min\{V'(\boldsymbol{\lambda}) : \boldsymbol{\lambda} \geq \boldsymbol{0}\}. \tag{9}$$

Drawback 2 *The function V' is not smooth (there are points where the gradient of the function is not defined), making it challenging to optimize.*

2.4 Subgradient Descent, a More General Gradient Descent

The next definition provides an alternative to the classical gradient when a function is not differentiable.

Definition 1 (Subgradient). *Let $I \subset \mathbb{R}^n$ be an open convex subset of $\mathbb{R}^n$, let $f : I \to \mathbb{R}$ be a convex piecewise linear function, and let $\boldsymbol{x}, \boldsymbol{x_0} \in I$. A subgradient of f at $\boldsymbol{x_0}$ is a vector $\mathrm{f}(\boldsymbol{x_0}) \in \mathbb{R}^n$ that satisfies*

$$f(\boldsymbol{x}) - f(\boldsymbol{x_0}) \geq \langle \mathrm{f}(\boldsymbol{x_0}), \boldsymbol{x} - \boldsymbol{x_0}\rangle, \forall \boldsymbol{x} \in I$$

i.e., $\mathrm{f}(\boldsymbol{x_0})$ generates a plane $p(\boldsymbol{x}) = \langle \mathrm{f}(\boldsymbol{x_0}), \boldsymbol{x} - \boldsymbol{x_0}\rangle + f(\boldsymbol{x_0})$ tangent to f at $\boldsymbol{x_0}$.

Sellmann and Fahle [20] propose a filtering algorithm based on a *subgradient descent* to solve (9). The algorithm builds a sequence $(\boldsymbol{\lambda}^{(k)})_{k\geq 0}$ with

$$\boldsymbol{\lambda}^{(k+1)} = \max\left(\mathbf{0}, \boldsymbol{\lambda}^{(k)} - \alpha_k \mathbb{v}(\boldsymbol{\lambda}^{(k)})\right), \tag{10}$$

where $\max(\cdot)$ acts as the projector on $\mathbb{R}^m_+$ (the valid Lagrange multipliers), α_k is a step size, and $\mathbb{v}(\boldsymbol{\lambda}^{(k)})$ is a subgradient. To ensure that $(\boldsymbol{\lambda}^{(k)})_{k\geq 0}$ converges to a minimizer $\boldsymbol{\lambda}^\star$ of V', one can use Polyak [17] step size's

$$\alpha_k = \frac{\beta_k\left(V'(\boldsymbol{\lambda}^{(k)}) - \min(\mathrm{dom}(P))\right)}{\left\|\mathbb{v}(\boldsymbol{\lambda}^{(k)})\right\|^2}, \quad 0 < \beta_k < 2. \tag{11}$$

2.5   Sellmann's Sub-optimality Observation

Sellmann and Fahle [20] notice that testing (4) for each Lagrange multipliers vector in $(\boldsymbol{\lambda}^{(k)})_{k\geq 0}$ provides stronger filtering than testing (4) at the end of the sequence. Then, Sellmann [19] proves that the optimal Lagrange multipliers for (9) are not optimal for filtering. Essentially, minimizing $V'[x_i = 1 - \bar{x}_i]$ is not the same task as minimizing V'. Sellmann [19] concludes that there are no reasons to discard suboptimal Lagrange multipliers to do cost-based filtering.

Drawback 3 *Sellmann [19] is a paramount observation, but it does not provide any reason as to why the Lagrange multipliers in* $(\boldsymbol{\lambda}^{(k)})_{k\geq 0}$ *are better at filtering.*

2.6   Oscillations of the Objective Function

Inspired by Sellmann [19], Isoart and Régin [11] compare the variation of V' and the amount of filtering detected during the generation of $(\boldsymbol{\lambda}^{(k)})_{k\geq 0}$. They notice that in the early steps the value of V' *oscillates* a lot and more filtering is detected. They suggest that only the early terms in the sequence lead to filtering and that the optimization of (9) should be stopped. Their method is faster by a factor of two than the state-of-the-art; supporting Sellmann [19]'s observation.

2.7   Local Alterations Algorithm

Boudrault and Quimper [5] present a local alterations cost-based Lagrangian filtering algorithm for the WEIGHTEDCIRCUIT constraint. Their algorithm locally explores the Lagrange multipliers space, *while generating the sequence* $(\boldsymbol{\lambda}^{(k)})_{k\geq 0}$, to find optimal filtering Lagrange multipliers, by exploiting the properties of a weighted graph. There is no guarantee that all filtering is applied, but they obtain a good speed up against the state-of-the-art in CP.

Berthiaume and Quimper [3] propose a local alterations filtering algorithm for the ATMOSTNVALUE constraint. In contrast to Boudrault and Quimper [5], the algorithm makes many subgradient descents (potentially one for each binary variable) that start from $\boldsymbol{\lambda}^\star$, the minimizer of V', and minimize $V'[x_i = 1 - \bar{x}_i]$.

To avoid too many time-consuming gradient descents that do not lead to filtering, the descent is performed for a variable x_i only if, for a given threshold τ, this inequality holds: $|V'[x_{i'} = 1 - \bar{x}_{i'}](\boldsymbol{\lambda}^\star) - \min(\mathrm{dom}(P))| < \tau$; i.e., the algorithm minimizes the functions $V'[x_i = 1 - \bar{x}_i]$ that are close to detect filtering.

Berthiaume and Quimper [3] show that the functions $V'[x_i = 1 - \bar{x}_i]$ are generally neither convex nor concave over all $\mathbb{R}^m$, but when restricted to:

$$Q_{i1} = \{\boldsymbol{\lambda} : v_i - \langle \boldsymbol{W}_{*i}, \boldsymbol{\lambda} \rangle \geq 0\} \tag{12}$$

$$Q_{i0} = \{\boldsymbol{\lambda} : v_i - \langle \boldsymbol{W}_{*i}, \boldsymbol{\lambda} \rangle \leq 0\}, \tag{13}$$

the functions $V'[x_i = 1 - \bar{x}_i]$ are convex. Q_{i1} and Q_{i0} are the sets of Lagrange multipliers where the variable $x_i = 1$ and $x_i = 0$ in (7). They also show that the gradient of the functions V' and $V'[x_i = 1 - \bar{x}_i]$ is discontinuous on the set

$$\bigcup_{i=1}^{n} \{\boldsymbol{\lambda} : v_i - \langle \boldsymbol{W}_{*i}, \boldsymbol{\lambda} \rangle = 0\}. \tag{14}$$

Berthiaume and Quimper [3] compare their algorithm with Cambazard and Fages [6]' algorithm. They obtain a significant speed up. They also observe that:

Observation 1 *The algorithm does not cycle due to a decreasing step size.*

Drawback 4 *Their algorithm has three drawbacks.*

1. *They mention the term* filtering zone *without a formal definition;*
2. *They do not check if $\boldsymbol{\lambda}^\star$ is a good point for the threshold test;*
3. *There is no guarantee that every filtering is applied.*

2.8   The RELAXED(λ) Consistency

Sellmann [19] introduces the RELAXED($\boldsymbol{\lambda}$) consistency that filters a value from a domain whenever there exist Lagrange multipliers that can filter it.

Definition 2. *Let $C(\boldsymbol{X}, P)$ be a constraint whose solution space is given by the linear relation $\langle \boldsymbol{c}, \boldsymbol{X} \rangle \geq P \wedge \mathrm{A}\boldsymbol{X} \leq \boldsymbol{b}$. Let $v \in \mathrm{dom}(X_j)$ and $B'[x_j = v]$ be the relaxed upper bound of P:*

$$B'[x_j = v](\boldsymbol{\lambda}) = \max\left\{\langle \boldsymbol{c} - \mathrm{A}^\top\boldsymbol{\lambda}, \boldsymbol{x}\rangle : \forall i,\, x_i \in \mathrm{dom}(X_i),\, x_j = v\right\} + \langle \boldsymbol{\lambda}, \boldsymbol{b}\rangle.$$

The constraint $C(\boldsymbol{X}, P)$ is RELAXED($\boldsymbol{\lambda}$)-consistent if and only if

$$\forall i,\, \forall v \in \mathrm{dom}(X_i),\, \forall \boldsymbol{\lambda} \in \mathbb{R}_+^m \quad B'[x_i = v](\boldsymbol{\lambda}) \geq \min(\mathrm{dom}(P)). \tag{15}$$

3   Contribution

Existing Lagrangian based methods operate without an explicit description of the structure of the Lagrange multiplier space. This paper provides the missing theoretical foundation needed to understand the behavior of most CP-based Lagrangian filtering approaches. We introduce new theoretical concepts. We provide an illustration of these concepts with an instance of a MK constraint.

3.1   New Interpretation of the Lagrange Multipliers Space

Example 1. Consider this instance of a MK constraint:

$$\#\text{items } n = 4 \qquad\qquad \#\text{resources } m = 2$$

$$\boldsymbol{v} = \begin{pmatrix} 10 \ 16 \ 20 \ 8 \end{pmatrix}^{\top} \qquad\qquad \boldsymbol{c} = \begin{pmatrix} 5 \ 5 \end{pmatrix}^{\top}$$

$$W = \begin{pmatrix} 1\ 1\ 3\ 3 \\ 4\ 3\ 2\ 1 \end{pmatrix} \qquad\qquad \mathrm{dom}(P) = [32, 60].$$

There are two resource constraints, so $\boldsymbol{\lambda} \in \mathbb{R}^2$. The valid Lagrange multipliers are in the positive quadrant of Fig. 1.

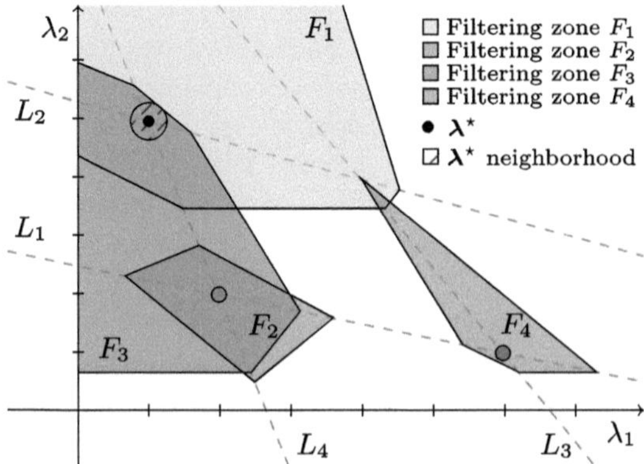

Fig. 1. Lagrange multipliers space.

Definition 3. *The set $L_i = Q_{i1} \cap Q_{i0}$ is called the* decision line *of the variable x_i, because L_i separates the sets Q_{i1} and Q_{i0}.*

Note 1. Observed that $L_i = \{\boldsymbol{\lambda} : v_i - \langle \boldsymbol{W}_{*i}, \boldsymbol{\lambda} \rangle = 0\}$.

In Example 1, there are four items, hence there are four decision lines, one for each variable, and are drawn with dashed lines in Fig. 1. The functions are not differentiable on the decision lines. We see immediately that the decision lines create regions in the Lagrange multipliers space. In these regions, the signs of the coefficients $v_i - \langle \boldsymbol{W}_{*i}, \boldsymbol{\lambda} \rangle$ of each variable remain the same. That is, each Lagrange multipliers vector in the *open interior* of a region enclosed by decision lines produces the same $\bar{\boldsymbol{x}}$.

Definition 4. *Let $F_i = \{\boldsymbol{\lambda} \in \mathbb{R}^m_+ : V'[x_i = 1 - \bar{x}_i](\boldsymbol{\lambda}) < \min(\mathrm{dom}(P))\}$ be the filtering zone of X_i; also known as the level set of $V'[x_i = 1 - \bar{x}_i]$.*

If it exists, any Lagrange multipliers vector $\boldsymbol{\lambda} \in F_i$ satisfies the reduced cost filtering condition (4); which means we can add the value $1 - \bar{x}_i$ to the values to be filtered out from $\mathrm{dom}(X_i)$. Hence, if the filtering zone F_i is non-empty, then filtering can be detected. The filtering zone F_i is a function of the lower bound of P and if $\min(\mathrm{dom}(P)) < \min\{V'[x_i = 1 - \bar{x}_i](\boldsymbol{\lambda}) : \boldsymbol{\lambda} \in \mathbb{R}_+^m\}$ then no filtering is detected for any $\boldsymbol{\lambda} \in \mathbb{R}_+^m$ during the call to the filtering algorithm. So as the solver increases $\min(\mathrm{dom}(P))$ the filtering zones increase in size. Definition 4 is a formal definition, hence resolving Drawback 4.1. The filtering zones of the variables in Example 1 are drawn in Fig. 1.

Hence, the goal of every Lagrangian cost-based filtering algorithm should be to : 1. find a good bound for P and 2. find Lagrange multipliers that are in the filtering zones $F_1, F_2, ..., F_n$. A subgradient descent is a natural choice for the first objective because of the Lagrangian theory. For the second objective, a *unique* subgradient descent might not be the best way to achieve this goal.

Drawback 3 clearly states that there are optimal filtering Lagrange multipliers. In Fig. 1, the blue and green dots are optimal for the variables X_2 and X_4, i.e., they minimize $V'[x_2 = 1 - \bar{x}_2]$ and $V'[x_4 = 1 - \bar{x}_4]$. By using a *unique* subgradient descent, it is very unlikely that any of the optimal filtering Lagrange multipliers will be visited. On the other hand, the sequence of Lagrange multipliers generated by a subgradient descent (9) may contain some Lagrange multipliers in some filtering zones; formalizing the sub-optimality observation.

If we use *many* subgradient descents, at least one in each $Q_{i1} \cap \mathbb{R}_+^m$ and $Q_{i0} \cap \mathbb{R}_+^m$ for all variables, then if a filtering Lagrange multipliers vector exists, we are sure to detect filtering, since $V'[x_i = 1 - \bar{x}_i]$ is a convex function on Q_{ia}.

In the new geometrical interpretation, Isoart and Régin [11]'s observation arises naturally. From Definition 4, we know that filtering zones lie in the Lagrange multipliers space, some are bounded and some not, some cluster together, and some are relatively isolated. The point is that a vast cover of the Lagrange multipliers space is more likely to detect filtering than a small neighborhood around $\boldsymbol{\lambda}^\star$. Furthermore, the objective value V' will vary much more on the vast cover of the Lagrange multipliers space than within a small neighborhood around $\boldsymbol{\lambda}^\star$. Also, the filtering within the small neighborhood will likely be redundant. From this analysis, big variations are a sign of a big cover of $\mathbb{R}_+^m$ which generally leads to better filtering than within a small neighborhood of $\boldsymbol{\lambda}^\star$.

The key insight is that using many subgradient descents appears to be a natural building block for Lagrangian cost-based filtering algorithms. A subgradient descent is used to get a good bound on P. The rest of them are to find filtering zones. The first one naturally appears from the Lagrangian optimization. The filtering subgradient descents are not straightforward without the idea of filtering zones. These subgradient descents do not aim at an optimal filtering $\boldsymbol{\lambda}$, but rather for the filtering zone; hence, they can be stopped rapidly.

3.2 Is $\lambda^\star$ a Good Starting Point to Apply the Threshold Test?

We note some observations from Fig. 1 that we later prove. These observations are at the core of the new filtering algorithm we propose. In Fig. 1, the optimal Lagrange multipliers $\lambda^\star$ lie at the intersection of two decision lines; L_2 and L_4.

Since the function V' is piece-wise linear and the solution to problem (1) is non trivial[1], $\lambda^\star$ can only be in three positions: (1) at the intersection of decision lines, (2) on a segment of a decision line, or (3) in a section enclosed by decision lines. In cases (2) and (3), there would be many optimal Lagrange multipliers.

Definition 5. *Let $\Lambda^\star = \{\lambda : V'(\lambda) = V'(\lambda^\star)\}$ be the set of optimal Lagrange multipliers.*

In Fig. 1, $\Lambda^\star = \{\lambda^\star\}$. In all cases (1), (2) and (3), there are always some decision lines intersecting $\Lambda^\star$.

Definition 6. *Let $\mathcal{J} = \{j : L_j \cap \Lambda^\star \neq \emptyset\}$ be the set of indices of decision lines intersecting the optimal Lagrange multipliers $\Lambda^\star$.*

In Fig. 1, $\mathcal{J} = \{2, 4\}$ and note that the only filtering zones that do not contain $\lambda^\star$ are F_2 and F_4. This suggests that there is a relation between the position of $\lambda^\star$, with respect to the decision lines, and the filtering zones. We could even say that $\lambda^\star$ are bad Lagrange multipliers for filtering or even for applying the threshold test for the variables X_2 and X_4, but not for the filtering of X_1 and X_3. This section formalizes these observations.

Definition 7. *Let $T_i(\tau) = \{\lambda \in \mathbb{R}_+^m : V'[x_i = 1 - \bar{x}_i](\lambda) < \min(\mathrm{dom}(P)) + \tau\}$ be the threshold zone of the variable X_i with threshold τ.*

Threshold zones are bloated filtering zones and are not drawn in Fig. 1.

Lemma 1. *If filtering can be detected ($F_i \neq \emptyset$), then $T_i(\tau) \neq \emptyset$ for any $\tau \geq 0$.*

Proof. It follows from Definitions 4 and 7, that $\forall \tau \geq 0$ we have $F_i \subset T_i(\tau)$. So by the assumption that $F_i \neq \emptyset$, then $T_i(\tau) \neq \emptyset$. □

In Fig. 1, all filtering zones are present (hence non-empty), so any bloated version of these zones, like the threshold zones $T_i(\tau)$, are also non-empty.

Lemma 2. $\exists \tau > 0$ *such that the threshold zone is non-empty; i.e., $T_i(\tau) \neq \emptyset$.*

Proof. $\forall \tau > \min_{\lambda \in \mathbb{R}_+^m} \{V'[x_i = 1 - \bar{x}_i](\lambda)\} - \min(\mathrm{dom}(P))$, $T_i(\tau) \neq \emptyset$. □

Suppose that, in Fig. 1, the filtering zone F_4 was empty, which means that no filtering could be detected. The green dot, the minimizer of $V'[x_4 = 1 - \bar{x}_4]$ in $\mathbb{R}_+^m$, would still be there. Hence, for any threshold $\tau > V'[x_4 = 1 - \bar{x}_4](\lambda) - \min(\mathrm{dom}(P))$, the threshold zone $T_4(\tau)$ would be non-empty. This is a case where the algorithm would trigger an unfruitful descent that leads to no filtering.

---

[1] Trivial solutions are $x = 1$ and $x = 0$ for which $\lambda^\star = 0$ and $\lambda^\star$ is unbounded.

Theorem 1. *If the current problem has a non-trivial solution, then there is a non-instantiated variable X_i, such that some optimal Lagrange multipliers lie on its decision line L_i, in other words, $\exists i$ such that $\Lambda^\star \cap L_i \neq \emptyset$.*

Proof. Since the current problem has a non-trivial solution, we know that $\lambda^\star \in \Lambda^\star$ implies that $\mathcal{J} \neq \emptyset$ and by Definitions 5 and 6, $\Lambda^\star = \bigcap_{j \in \mathcal{J}} Q_{j\bar{x}_j}$. Hence, the theorem holds for all $i \in \mathcal{J}$. $\qquad\square$

As mentioned before, Theorem 1 is a consequence of V' being piece-wise linear. The implication is now clear: as long as there are non-instantiated variables, there are decision lines intersecting $\Lambda^\star$. With Theorem 1, the observation that $\lambda^\star$ is at the intersection of the decision lines L_2 and L_4 in Fig. 1 is not surprising.

Lemma 3. $\forall \lambda \in L_i$, *we have* $V'[x_i = 1 - \bar{x}_i](\lambda) = V'(\lambda)$.

Proof. From Note 1, $\forall \lambda \in L_i$, $V'[x_i = 1 - \bar{x}_i](\lambda) = V'(\lambda) - |v_i - \langle \boldsymbol{W}_{*i}, \lambda \rangle| = V'(\lambda) - 0 = V'(\lambda)$. $\qquad\square$

From the definition of $V'[x_i = 1 - \bar{x}_i](\lambda) = V'(\lambda) - |v_i - \langle \boldsymbol{W}_{*i}, \lambda \rangle|$ for any $\lambda \notin L_i$, $V'[x_i = 1 - \bar{x}_i](\lambda) < V'(\lambda)$. We also know that $V'[x_i = 1 - \bar{x}_i]$ decreases at most linearly with the distance to the decision lines L_i. Hence, in Fig. 1, it is not surprising that the filtering zones are far from their decision lines (e.g., F_1 is far from L_1). For the following, let $\tau_0 = V'(\lambda^\star) - \min(\mathrm{dom}(P))$.

Theorem 2. *Any λ that is optimal and on a decision line L_i ($\lambda \in \Lambda^\star \cap L_i$) is in the threshold zone of the variable X_i ($\lambda \in T_i(\tau)$) if and only if $\tau > \tau_0$.*

Proof. [$\Rightarrow$] If $\lambda \in T_i(\tau)$, then $V'[x_i = 1 - \bar{x}_i](\lambda) < \min(\mathrm{dom}(P)) + \tau$. Lemma 3 and Definition 5 ($\Lambda^\star$) ensure that $V'[x_i = 1 - \bar{x}_i](\lambda) = V'(\lambda) = V'(\lambda^\star)$. So, $V'(\lambda^\star) < \min(\mathrm{dom}(P)) + \tau$, hence $\tau_0 < \tau$.
[$\Leftarrow$] We still have $V'(\lambda^\star) = V'(\lambda) = V'[x_i = 1 - \bar{x}_i](\lambda)$, so $\tau > \tau_0 = V'(\lambda^\star) - \min(\mathrm{dom}(P)) = V'[x_i = 1 - \bar{x}_i](\lambda) - \min(\mathrm{dom}(P))$, meaning $\lambda \in T_i(\tau)$. $\qquad\square$

In Fig. 1, $V'(\lambda^\star) = 37$ and $\min(\mathrm{dom}(P)) = 32$, hence for any threshold $\tau > \tau_0 = 37 - 32 = 5$, $\lambda^\star$ is in the threshold zone of every variable. However, if the threshold $\tau \leq 5$, then $\lambda^\star$ would not be in the threshold zones $T_2(\tau)$ and $T_4(\tau)$. In this case, even if filtering *can* be detected (because the filtering zones are non-empty), the algorithm can *miss* the filtering. This is alarming. The threshold test is supposed to reduce the number of descents that does not lead to filtering, but could prevent the detection of filtering.

Not all is lost, the problematic Lagrange multipliers are only the ones that are optimal and on a decision line. Consider $j \in \mathcal{J}$, since $V'[x_j = 1 - \bar{x}_j]$ deteriorates at most linearly with the distance to the decision line L_j, we know that any λ in a neighborhood of $\lambda^\star$ such that $\lambda \notin L_j$ is more likely to pass the threshold test. One can consider the subgradient descents as projectors on the non-empty filtering zones. By performing the threshold test after a few steps of the gradient descent, rather than at the very beginning, the algorithm makes a more educated guess as to whether there is a filtering zone or not. Furthermore, performing more steps before the threshold test should lessen the impact of the parameter τ on the overall performance since, at each step, we get closer to the filtering zone.

Hence, to answer the question in the title of the section, $\boldsymbol{\lambda}^{\star}$ is bad for the threshold test of the variable j if and only if $\boldsymbol{\lambda}^{\star}$ lies on its decision line L_j.

3.3   Algorithm

Algorithm 1 is our filtering algorithm for the MK constraint. The decision variables subject to the MK constraint are $\boldsymbol{X}$ and P, while the constant parameters are $\boldsymbol{v}, \mathrm{W},$ and $\boldsymbol{c}$, as defined in (1). β_0 and γ_0 are the initial step sizes for the Lagrangian optimization and local alterations. maxItX is the maximal number of steps we perform in our local alteration phase. τ is the threshold. minItX is the number of steps before applying the threshold test. lF is an array of sets of values that were flagged for filtering. lF is short for *Lagrangian forbidden value*. maxIt is the maximal length of the sequence. $\boldsymbol{\lambda}$ is a Lagrange multiplier vector. V' is the relaxed valid upper bound on V in (1) and V_{best} its best known value.

Algorithm 1 filters values using Lagrangian cost-based filtering. First, the function V' is minimized in the repeat loop. Line 5 solves the Lagrangian subproblem. Line 6 computes the upper bound. For all non-instantiated variables, Line 8 performs cost-based filtering by checking the inequality in Definition 4. Line 9 computes the step size. Line 10 updates the Lagrange multiplier. Line 11 stops the minimization of V' if there are maxIt steps or if the greatest difference between consecutive Lagrange multipliers is smaller than a certain factor. The local alteration phase is triggered on Line 13. For each non-instantiated variable, Lines 14 and 15 search for filtering Lagrange multipliers on both sides of L_i. Algorithms 3 and 1 are similar. The difference is that if the Lagrange multiplier is not in the set Q, then Line 8 calls a quadratic solver to project $\boldsymbol{\lambda}$ onto Q.

Theorem 3. *There exist a threshold τ such that Algorithm 1, with* maxItX $\rightarrow$ ∞, *enforces* RELAXED($\boldsymbol{\lambda}$) *on* MK.

Proof. With $\tau > \max_{1 \leq i \leq n}\{V'[x_i = 1 - \bar{x}_i](\boldsymbol{\lambda}^{\star})\} - \min(\mathrm{dom}(P))$, the condition on Line 4 of Algorithm 3 is always false; i.e., $\forall i, \boldsymbol{\lambda}^{\star} \in T_i(\tau)$. Algorithm 3 is guaranteed to converge to a minimizer in $\mathbb{R}_+^m \cap Q_{i0}$ and in $\mathbb{R}_+^m \cap Q_{i1}$ with the chosen step size and maxItX $\rightarrow \infty$. If the minimizers are in F_i, the value is flagged to be filtered on Line 12 in Aglorithm 3 and removed on Line 16 of Algorithm 1. Hence, $\forall i$ and $\forall v \in \mathrm{dom}(X_i)$, $\forall \boldsymbol{\lambda} \in \mathbb{R}_+^m, V'[x_i = 1 - \bar{x}_i](\boldsymbol{\lambda}) \geq \min(\mathrm{dom}(P))$. $\square$

Theorem 3 ensures that Algorithm 1 can enforce RELAXED($\boldsymbol{\lambda}$).

The computation time is improved by calling Algorithm 3 once and avoiding the quadratic program. We replace, in Algorithm 1, lines 14 and 15 by Update&Flag($i, \mathbb{R}_+^m$). Due to Observation 1, the Algorithm 1 does not cycle. We observe empirically that the cases where this modification fails to filter are rare.

Algorithm 1: Propagator for the MK constraint.

1 **Input:** $X, P, v, \mathrm{W}, c, \beta_0, \text{maxIt}, \text{maxItX}, \tau, \gamma_0, \text{minItX}$
2 **Output:** Filtered domains
3 $lF \leftarrow [\emptyset_1, ..., \emptyset_n], \lambda \leftarrow 0, k \leftarrow 0, \ \beta \leftarrow 0, x \leftarrow 0, V_{best} \leftarrow \infty$
4 **repeat**
5   SolveSubProblem$(v - \mathrm{W}^\top \lambda)$ //Algorithm 2
6      $V' \leftarrow \left\langle v - \mathrm{W}^\top \lambda, x \right\rangle + \langle \lambda, c \rangle, \ V_{best} \leftarrow \min(V', V_{best})$
7   **for** $i = 1$ **to** n **where** $|\operatorname{dom}(X_i)| = 2$ **do**
8   **if** $\lambda \in F_i$ **then** $lF[i] \leftarrow lF[i] \cup \{1 - \bar{x}_i\}$.
9      $\beta \leftarrow \dfrac{\beta_0}{2^{\lfloor k/10 \rfloor}} \dfrac{(V' - \min(\operatorname{dom}(P)))}{\|c - \mathrm{W}x\|^2}, \ \lambda' \leftarrow \lambda$
10     $\lambda \leftarrow \max(0, \lambda' - \beta(c - \mathrm{W}x)), \ k \leftarrow k + 1$
11 **until** $(k = \text{maxIt}) \vee (\max_{0 \le j \le m} |\lambda_j - \lambda'_j| \le 10^{-4})$
12 **if** maxItX> 0 **then**
13   **for** $i = 1$ **to** n **where** $|\operatorname{dom}(X_i)| = 2$ **do**
14   Update&Flag(i, Q_{i0}) // Algorithm 3
15   Update&Flag(i, Q_{i1})// Algorithm 3
16 **for** $i = 1$ **to** n **do** $\operatorname{dom}(X_i) \leftarrow \operatorname{dom}(X_i) \setminus lF[i] \ \operatorname{dom}(P) \leftarrow \{p \in \operatorname{dom}(P) : p \le V_{best}\}$

Algorithm 2: SolveSubProblem(q)

1 **for** $i = 1$ **to** n **do**
2   **if** $|\operatorname{dom}(X_i)| = 1$ **then** $x_i \leftarrow \min(\operatorname{dom}(X_i))$ **else if** $q_i > 0$ **then** $x_i \leftarrow 1$ **else** $x_i \leftarrow 0$

Algorithm 3: Update&Flag(i, Q)

1 **Input:** the index of a variable i, Q a set.
2 save$(\lambda, x, V'), \ k \leftarrow 0, \ \gamma \leftarrow 0$
3 **for** $k = 0$ **to** maxItX $- 1$ **do**
4   **if** $k = \text{minItX} \wedge \lambda \notin T_i(\tau)$ **then break**
5      $q_i \leftarrow v_i - \langle \mathrm{W}_{*i}, \lambda \rangle, \ \gamma \leftarrow \dfrac{\gamma_0}{2^{\lfloor k/5 \rfloor}} \dfrac{V' - |q_i| - \min(\operatorname{dom}(P))}{\|c - \mathrm{W}x + \operatorname{sgn}(q_i)\mathrm{W}_{*i}\|^2}, \ \lambda' \leftarrow \lambda$
6      $\lambda \leftarrow \max\left(0, \lambda' - \gamma(c - \mathrm{W}x + \operatorname{sgn}(q_i)\mathrm{W}_{*i})\right)$
7   **if** $Q \ne \mathbb{R}_+^m$ **and** $\lambda \notin \mathbb{R}_+^m \cap Q$ **then**
8          $\lambda \leftarrow \operatorname{argmin}\{\|\lambda - \lambda'\|^2 : \lambda' \in \mathbb{R}_+^m \cap Q\}$
9   SolveSubProblem$(v - \mathrm{W}^\top \lambda)$ //Algorithm 2
10     $V' \leftarrow \left\langle v - \mathrm{W}^\top \lambda, x \right\rangle + \langle \lambda, c \rangle$
11   **if** $\lambda \in F_i$ **then**
12         $lF[i] \leftarrow lF[i] \cup \{1 - \bar{x}_i\}$
13   **break**
14 restore(λ, x, V')

---

4 Experiments and Results

4.1 Multidimensional Knapsack Problem

We implemented our filtering algorithm to test it over instances of the multidimensional knapsack problem. We used three benchmarks for our experiments from Beasley's library[2], which has been widely used to test the performance of

---

[2] https://people.brunel.ac.uk/~mastjjb/jeb/orlib/mknapinfo.html.

different algorithms Lu et al. [14]. Chu and Beasley [7] randomly generated 9 classes of 30 instances. All instances in a given class have the same number of resources and items. They chose 5, 10, and 30 resources, and 100, 250, and 500 items. We call this benchmark Chu and Beasley [7]'s instances.

The second benchmark from Beasley's library contains 48 instances drawn from the literature. We call this benchmark: Beasley's instances.

The third benchmark contains seven instances designed by Petersen [15]. We call this benchmark Petersen [15]'s instances.

We implemented Algorithm 1 in Java in the Choco solver [18]. The code and benchmarks are available on GitHub[3]. We use the quadratic solver ojAlgo [16] to implement Line 8 of Algorithm 3. The experiments were carried out on a MacBook Pro with a M2 chip and 8 Gb of RAM. $\beta_0 = 5.0$ and $\gamma_0 = 15.0$, maxIt $= 1000$ is the maximal number of steps in Algorithm 1 and maxItX $= 60$ the maximum number of additional steps in the local alterations phase. The threshold is $\tau = \frac{1}{n} \sum_{i=1}^{n} v_i$ (the average of the values). minItX is the number of steps prior to the threshold test. A 300-second timeout is set.

We used two models to encode the problem. X is a vector of Boolean variables, P is an integer variable with $\mathrm{dom}(P) = [0, \sum_{i=1}^{n} v_i]$.

$$\text{Maximize } P \qquad \text{(Model 1)} \qquad \text{Maximize } P \qquad \text{(Model 2)}$$
$$\mathrm{MK}(X, P, v, \mathrm{W}, c) \qquad\qquad P = \langle v, X \rangle \,;\, \langle W_{j*}, X \rangle \leq c_j, \forall j$$

Model 1 contains a single MK constraint. Model 2 decomposes the MK constraint into m linear inequalities. Solvers usually enforce bound consistency on the constraints of Model 2. We also implemented Trick [21]'s propagator that enforces domain consistency. We compare ten approaches :

- LIN: Model 2 with bound consistency enforced;
- DC: Model 2 with domain consistency enforced;
- LR: Model 1, the standard Lagrangian relaxation proposed by Cambazard and Fages [6], with maxItX $= 0$, i.e., without local alterations;
- LR+0: Model 1, the Lagrangian relaxation with the local addition proposed by Berthiaume and Quimper [3], with minItX $= 0$, i.e., without additional steps before the threshold test;
- PROJ+1, PROJ+2, PROJ+3: Model 1, our contribution, with minItX $\in \{1, 2, 3\}$ with projections on both sides (the quadratic program);
- LR+1, LR+2, LR+3: Model 1, our contribution, with minItX $\in \{1, 2, 3\}$ without projections on both sides (no quadratic program).

Table 1 presents the times to solve Beasley's instances and Petersen's instances and the node count. The DC approach was discarded from the table as it reaches the timeout too often without finding any feasible solution. PROJ+$\{1, 2, 3\}$ were discarded because in average they were at least 1921% slower than LR+$\{1, 2, 3\}$. Instances solved in 2 s by all the remaining

---

[3] https://github.com/frbert3/knapsackConstraintCPAIOR2026.

Table 1. Multidimensional knapsack problem experiments. Arabic and Roman numerals denote Beasley's and Petersen's instances. Instances solved within 2 s by all methods are omitted. The symbol – indicates that the 300-second timeout is reached.

#	Time [seconds]						Nodes [-]					
	LIN	LR	LR+0	LR+1	LR+2	LR+3	LIN	LR	LR+0	LR+1	LR+2	LR+3
0	67.73	–	–	0.81	**0.70**	0.76	2×10^6	–	–	857	**720**	765
1	–	–	–	**5.51**	5.61	6.45	–	–	–	6946	7106	**6686**
8	41.95	44.84	40.20	47.82	**22.51**	22.93	5×10^6	68928	68928	65602	**39953**	40684
9	–	3.21	3.08	3.33	5.84	**1.54**	–	4754	4754	3897	7062	**2093**
18	0.47	–	–	**0.01**	**0.01**	**0.01**	80255	–	–	**39**	**39**	**39**
19	2.52	5.19	4.93	1.01	0.80	**0.70**	4×10^5	12913	12913	2660	2153	**1909**
20	**0.57**	–	–	9.71	7.81	4.44	84632	–	–	23885	18628	**11441**
21	2.75	75.61	61.07	**0.04**	**0.04**	**0.04**	4×10^5	2×10^5	2×10^5	101	**91**	**91**
22	2.25	–	–	1.52	1.36	**1.29**	3×10^5	–	–	3860	3495	**3214**
23	51.98	–	–	**0.06**	**0.06**	0.07	7×10^6	–	–	**127**	131	131
24	10.13	–	–	1.11	1.21	**0.98**	1×10^6	–	–	2734	2487	**2287**
25	–	–	–	5.51	4.37	**2.36**	–	–	–	2251	**2077**	2513
26	118.54	–	–	**0.04**	**0.04**	0.05	2×10^7	–	–	**92**	**92**	92
27	–	0.15	0.14	**0.13**	**0.13**	**0.13**	–	529	520	412	**410**	**410**
28	–	–	–	0.13	**0.12**	0.13	–	–	–	377	361	**360**
29	–	–	–	3.00	**1.89**	2.40	–	–	–	5021	**3061**	3879

#	Time [seconds]						Nodes [-]					
	LIN	LR	LR+0	LR+1	LR+2	LR+3	LIN	LR	LR+0	LR+1	LR+2	LR+3
30	–	–	–	38.90	37.02	**31.80**	–	–	–	65178	60766	**48460**
31	–	–	–	0.27	0.23	**0.21**	–	–	–	531	430	**367**
32	–	–	–	**0.81**	0.91	1.04	–	–	–	**1622**	1729	1845
33	–	–	–	**0.09**	0.10	0.11	–	–	–	**198**	202	202
34	–	0.58	0.55	**0.10**	0.11	0.12	–	933	933	**141**	**141**	**141**
35	–	0.21	0.20	**0.13**	0.14	0.15	–	491	491	257	244	**243**
36	–	**0.12**	**0.12**	**0.12**	0.13	0.14	–	343	343	**276**	**276**	276
37	–	0.07	**0.06**	0.08	0.09	0.10	–	143	143	**114**	114	114
38	–	–	–	**0.08**	0.09	0.11	–	–	–	**128**	128	128
39	–	0.06	**0.05**	0.07	0.08	0.09	–	168	168	**120**	120	120
44	**0.42**	77.12	62.55	0.95	0.90	0.95	21475	1×10^5	91304	1549	**1422**	1449
45	23.10	–	–	**1.72**	1.83	1.95	2×10^6	–	–	3461	3402	**3287**
iii	**0.01**	2.39	2.30	0.18	0.20	0.08	**69**	22721	22721	746	835	356
vi	0.49	–	–	**0.16**	0.20	0.20	99594	–	–	**543**	677	630
vii	20.34	**0.38**	0.66	0.39	0.41	0.39	4×10^6	2232	3276	1027	1071	**981**

six approaches were omitted. Table 1 shows that our methods LR+$\{1, 2, 3\}$ are significantly faster than LIN, LR, and LR+0. There are 11 (10) instances solved under one second by all of our methods while LIN (LR and LR+0) reached the timeout.

The variation between the columns LR+$\{1, 2, 3\}$ in Table 1 is at most 25 s (for instance 8); otherwise they are in the range of 1 s.

All methods LR+$\{1, 2, 3\}$ close instances with fewer nodes than all other methods except for instance (iii); in some cases *several* orders of magnitude. Among the methods LR+$\{1, 2, 3\}$, the number of nodes is generally in the same order of magnitude. We can see a small correlation with the number of nodes and the time to solve an instance; generally, the fastest method also develops fewer nodes. From the node count, we see that Lagrangian methods explore a completely different search tree than the LIN model.

We tested six different thresholds to study the impact of this parameter on the performance of the methods LR+$\{1, 2, 3\}$; $\tau \in \{10^{-2}, 10^{-1}, 10^0, 10^1, 10^2, 10^3\}$. We compared the average of the average time differences between Table 1 and the thresholds studied. On average, with the threshold chosen in Table 1, LR+1 is 29.31 s faster with a standard deviation of 84.12 s. LR+2 is 0.56 s faster with a standard deviation of 3.80 s. LR+3 is 0.22 s faster with a standard deviation of 0.99 s. LR+1 is much more sensitive to threshold variations than LR+$\{2, 3\}$. We experimentally see that when more additional steps are considered, the more robust the algorithm becomes.

The lines in Table 2 correspond to the nine problem classes from Chu and Beasley [7]. The entries are the number of times a given method finds the best objective value within the 300-second timeout. Except for instances with 5 resources and 100 items, and 10 resources and 100 items, all methods reach the timeout. Table 2 shows that for all instance classes (except the 5 resources and 250 items) LR+$\{1, 2, 3\}$ find the best objective value within the timeout.

Table 2. The number of times that a given method provides the best solution within a time out of five minutes for Chu and Beasley [7]'s instances.

# Res.	# items	LIN	LR	LR+0	LR+1	LR+2	LR+3
5	100	0	26	25	**30**	**30**	**30**
5	250	0	**12**	4	6	9	7
5	500	1	11	3	**14**	11	8
10	100	0	5	3	16	17	**18**
10	250	1	6	5	12	**17**	**17**
10	500	3	6	3	**14**	12	12
30	100	1	5	7	19	**23**	19
30	250	4	10	8	**19**	17	18
30	500	4	10	9	**19**	17	17

4.2 Uncapacitated Facility Location Problem

In the uncapacitated facility location problem, there is a set of facility locations $\mathcal{L}$ and a set of cities $\mathcal{C}$. The goal is to choose which facility to build, and, among these, which to connect to every city to supply them while minimizing the transportation cost and building cost. Let $c_{i,j}$ be the transportation cost between the facility location $i \in \mathcal{L}$ and the city $j \in \mathcal{C}$, and let b_i be the building cost of the facility $i \in \mathcal{L}$. $x_{i,j}$ is a binary variable that encodes if the facility $i \in \mathcal{L}$ is connected to the city $j \in \mathcal{C}$, and y_i is a binary variable that encodes if a facility is built at the location $i \in \mathcal{L}$. The problem is formulated as

$$\min \quad \sum_{i \in \mathcal{L}} \sum_{j \in \mathcal{C}} c_{i,j} x_{i,j} + \sum_{i \in \mathcal{L}} b_i y_i$$
$$\text{s.t.} \quad \sum_{i \in \mathcal{L}} x_{i,j} \geq 1, \quad \forall j \in \mathcal{C}, \quad x_{i,j} \leq y_i, \quad \forall i \in \mathcal{L}, \forall j \in \mathcal{C}.$$

We tested Kuehn and Hamburger [13] benchmark and observed that all Lagrangian based method achieve the same filtering, we do not present these results.

We generated 24 instances with 47 cities from France and 55 cities from the United Kingdom, from which we randomly selected $|\mathcal{L}| \in \{5, 8\}$ facility locations and let the remaining cities form the set $\mathcal{C}$. The cost $c_{i,j}$ is the distance (in km) as a crow flies between the facility location and the city. For France (UK), two instances are created, one where all building costs are set to 1500 (150) and one where they are all set to 3000 (500). We set the timeout to 600 s.

In Table 3, only the 18 instances where at least one method closed the instance within the ten-minute timeout are presented; the other 6 instances are presented in Table 4. The method LR+3 is the fastest method for six instances and LR+2 for five instances. When LR is faster, the number of nodes is the same for every method, see Table 3, which indicates that LR+$\{0, 1, 2, 3\}$ do the same filtering as LR. On two occasions LR+3 is the only method to close the instance.

For instances where all methods reached the timeout, we report in Table 4 the value of the best solution found. We see that on two instances, LR finds the best solution. The difference, when LR is better than our methods, is smaller than when our methods are better. All methods are equivalent for two instances.

5   Analysis

In both experiments, all LR+$\{1, 2, 3\}$ outperform the traditional Lagrangian method and also the other local alterations method. Compared with the traditional Lagrangian method, this performance difference indicates that some filtering zones were missed while developing the sequence of Lagrange multipliers. However, the local alteration method LR+0 performs similarly to the traditional Lagrangian method. Hence, not all local alterations method aid for the overall performance. Even adding one step before the threshold test significantly helps the time performance; more specifically with the multidimensional knapsack problem. However, adding two or three steps makes the algorithm more robust to threshold variations, as we hypothesized at the end of Sect. 3.2. With the uncapacitated facility location problem, the speed up happened after adding two and three steps. The reason for these differences might be the entries of the weight matrix. In the multidimensional knapsack problem, $w_{ji} \in \mathbb{N}$ with no specific structure versus the uncapacitated facility location problem, where $w_{ji} \in \{\pm 1, 0\}$ and with a specific structure. The position and shape of the filtering and threshold zones vary drastically. Still, the performance differences of methods LR+$\{1, 2, 3\}$ with the local alterations method LR+0 are mainly a consequences of Theorems 1 and 2. We tested all the instances, and every underlying

Table 3. Time (in secondes) (left) and number of nodes visited (right) to solves instances of the UFLP for France and UK with a timeout of 600 s.

n	m	Instance Name	LR	LR+0	LR+1	LR+2	LR+3
5	42	france_0_1500	60.61	62.35	184.11	60.51	**24.37**
		france_1_1500	317.70	148.10	283.72	**65.54**	90.32
		france_2_1500	264.70	217.23	**52.80**	53.62	84.98
		france_0_3000	273.51	191.56	145.62	**96.88**	118.50
		france_2_3000	117.09	167.47	149.73	104.44	**66.25**
	50	UK_0_150	582.60	–	86.15	38.37	**35.15**
		UK_1_150	213.68	220.50	56.92	31.09	**22.52**
		UK_2_150	–	–	–	–	**386.12**
		UK_1_500	238.12	197.50	**53.82**	121.31	269.04
8	39	france_0_1500	137.57	110.62	175.30	**70.28**	112.50
		france_1_1500	**0.30**	0.33	0.76	1.19	1.63
		france_2_1500	177.97	208.40	194.94	**68.35**	69.18
		france_0_3000	–	–	–	**343.54**	422.24
	47	UK_0_150	–	–	–	–	**596.18**
		UK_1_150	**0.52**	0.58	1.32	2.08	2.83
		UK_2_150	**0.95**	1.01	1.60	2.21	2.81
		UK_0_500	**0.49**	0.53	1.12	1.71	2.29
		UK_2_500	**0.48**	0.54	1.28	2.03	2.78

n	m	Instance Name	LR	LR+0	LR+1	LR+2	LR+3
5	42	france_0_1500	33572	29453	73039	21478	*7762*
		france_1_1500	145286	64671	114199	*24134*	31442
		france_2_1500	119090	93385	**21810**	*18849*	24940
		france_0_3000	135362	88490	60524	*34874*	37838
		france_2_3000	56282	71914	61150	39446	*23643*
	50	UK_0_150	231106	–	28317	11999	*10412*
		UK_1_150	84805	78840	19126	9875	*7030*
		UK_2_150	–	–	–	–	*124924*
		UK_1_500	104836	77681	*18819*	36184	78545
8	39	france_0_1500	44354	34193	48717	*18052*	27830
		france_1_1500	**322**	*322*	*322*	*322*	*322*
		france_2_1500	54000	58877	52922	**17590**	*17185*
		france_0_3000	–	–	–	*90339*	107511
	47	UK_0_150	–	–	–	–	*118982*
		UK_1_150	*386*	*386*	*386*	*386*	*386*
		UK_2_150	*344*	*344*	*344*	*344*	*344*
		UK_0_500	*342*	*342*	*342*	*342*	*342*
		UK_2_500	*386*	*386*	*386*	*386*	*386*

linear program optimal solutions had some non-integer variables. This indicates that $\lambda^\star$ was always at the intersection of decision lines. Due to Theorem 2, we know that we must move away from $\lambda^\star$ to get a relevant threshold test.

Table 4. Cost of the solution for instances that reached the timeout of 600 s.

n	m	Instance	LR	LR+0	LR+1	LR+2	LR+3
5	42	france_1_3000	**23551**	23612	23632	23612	23612
	50	UK_0_500	6501	6546	6365	**6183**	6726
		UK_2_500	8015	8015	8018	8022	**7319**
8	39	france_1_3000	**22993**	**22993**	**22993**	**22993**	**22993**
		france_2_3000	**23804**	**23804**	**23804**	**23804**	**23804**
	47	UK_1_500	**5804**	5837	5954	5893	5971

We observed a significant decrease in the number of nodes explored in both problems. The method LR has no guarantee to encounter all the filtering zones encountered by methods LR+$\{1, 2, 3\}$. The method LR+0 additional explorations depend completely on the threshold test, which has been shown to be a poor test in general. Because the multidimensional knapsack problem and the uncapacitated facility location problem aver very different problems, we conjecture that LR and LR+$\{1, 2, 3\}$ behave similarly on other problems.

6   Conclusion

We formalized some important observations in cost-based Lagrangian filtering theory. We designed a cost-based filtering algorithm based on a Lagrangian relaxation with enhanced filtering, which comes from the careful addition of a local alterations phase, for the MK constraint. Our methods have proven to be faster, and found better solutions within a timeout than the state-of-the-art filtering algorithms on two well-known problems.

References

1. Bajgiran, O.S., Cire, A.A., Rousseau, L.M.: A first look at picking dual variables for maximizing reduced cost fixing. In: Integration of AI and OR Techniques in Constraint Programming (2017)
2. Benchimol, P., Hoeve, W.J.V., Régin, J.C., Rousseau, L.M., Rueher, M.: Improved filtering for weighted circuit constraints. Constraints (2012)
3. Berthiaume, F., Quimper, C.G.: Local alterations of the lagrange multipliers for enhancing the filtering of the atmostnvalue constraint. In: International Conference on the Integration of Constraint Programming, Artificial Intelligence, and Operations Research (2024)

4. Bessa, S., Dabert, D., Bourgeat, M., Rousseau, L.M., Cappart, Q.: Learning valid dual bounds in constraint programming: Boosted lagrangian decomposition with self-supervised learning. In: Proceedings of the 39th AAAI Conference on Artificial Intelligence and 37th Conference on Innovative Applications of Artificial Intelligence and 15th Symposium on Educational Advances in Artificial Intelligence (2025)

5. Boudreault, R., Quimper, C.G.: Improved cp-based lagrangian relaxation approach with an application to the tsp. In: Proceedings of the 30th International Joint Conference on Artificial Intelligence (2021)

6. Cambazard, H., Fages, J.G.: New filtering for atmostnvalue and its weighted variant: A lagrangian approach. Constraints (2015)

7. Chu, P.C., Beasley, J.E.: A genetic algorithm for the multidimensional knapsack problem. J. Heuristics (1998)

8. Fahle, T., Sellmann, M.: Cost based filtering for the constrained knapsack problem. Ann. Oper. Res. (2002)

9. Focacci, F., Lodi, A., Milano, M.: Cost-based domain filtering. In: Principles and Practice of Constraint Programming (1999)

10. Held, M., Karp, R.M.: The traveling-salesman problem and minimum spanning trees. Oper. Res. (1970)

11. Isoart, N., Régin, J.C.: Adaptive cp-based lagrangian relaxation for tsp solving. In: Integration of Constraint Programming, Artificial Intelligence, and Operations Research (2020)

12. Katriel, I., Sellmann, M., Upfal, E., Van Hentenryck, P.: Propagating knapsack constraints in sublinear time. In: Proceedings of the 22nd National Conference on Artificial Intelligence (2007)

13. Kuehn, A.A., Hamburger, M.J.: A heuristic program for locating warehouses. Manage. Sci. (1963)

14. Lu, Y., Vasko, F.J.: A comprehensive empirical demonstration of the impact of choice constraints on solving generalizations of the 0–1 knapsack problem using the integer programming option of cplex®. Engineering Optimization (2020)

15. Petersen, C.C.: Computational experience with variants of the balas algorithm applied to the selection of r&d projects. Manag. Sci. (1967)

16. Peterson, A.: ojalgo: Open-source java tools for linear and mathematical programming. http://ojalgo.org (2003)

17. Polyak, B.T.: Minimization of unsmooth functionals. USSR Comput. Math. Math. Phys. (1969)

18. Prud'homme, C., Fages, J.G.: Choco-solver: A java library for constraint programming. J. Open Sour. Softw. (2022)

19. Sellmann, M.: Theoretical foundations of cp-based lagrangian relaxation. In: Principles and Practice of Constraint Programming (2004)

20. Sellmann, M., Fahle, T.: Constraint programming based lagrangian relaxation for the automatic recording problem. Ann. Oper. Res. (2003)

21. Trick, M.A.: A dynamic programming approach for consistency and propagation for knapsack constraints. Ann. Oper. Res. (2003)

Learning to Choose Branching Rules for Nonconvex MINLPs

Timo Berthold[1]([✉])([iD]) and Fritz Geis[2]

[1] Fair Isaac Germany GmbH, Takustr. 7, 14195 Berlin, Germany
`timoberthold@fico.com`
[2] Zuse Institute Berlin, Takustr. 7, 14195 Berlin, Germany

Abstract. Outer-approximation-based branch-and-bound is a common algorithmic framework for solving MINLPs (mixed-integer nonlinear programs) to global optimality, with branching variable selection critically influencing overall performance. In modern global MINLP solvers, it is unclear whether branching on fractional integer variables should be prioritized over spatial branching on variables, potentially continuous, that show constraint violations, with different solvers following different defaults. We address this question using a data-driven approach. Based on a test set of hundreds of heterogeneous public and industrial MINLP instances, we train linear and random forest regression models to predict the relative speedup of the FICO® Xpress Global solver when using a branching rule that always prioritizes variables with violated integralities versus a mixed rule, allowing for early spatial branches.

We introduce a practical evaluation methodology that measures the effect of the learned model directly in terms of the shifted geometric mean runtime. Using only four features derived from strong branching and the nonlinear structure, our linear regression model achieves an 8–9% reduction in geometric-mean solving time for the Xpress solver, with over 10% improvement on hard instances. We also analyze a random regression forest model. Experiments across solver versions show that a model trained on Xpress 9.6 still yields significant improvements on Xpress 9.8 without retraining.

Our results demonstrate how regression models can successfully guide the branching-rule selection and improve the performance of a state-of-the-art commercial MINLP solver.

Keywords: Nonlinear Optimization · Machine Learning · Branching

1 Introduction

We consider *MINLPs (mixed-integer nonlinear programs)* of the form

$$min\{c^T x \mid g_k(x) \leq 0, \forall k \in \mathcal{K}, l \leq x \leq u, x_j \in \mathbb{Z}, \forall j \in \mathcal{J}\}, \tag{1}$$

where all constraint functions $g_k : \mathbb{R}^n \to \mathbb{R}$ are factorable and all variable bounds $l, u \in \bar{\mathbb{R}} := \mathbb{R} \cup \{\pm\infty\}$. The set $\mathcal{K} = \{1, \dots, m\}, m \in \mathbb{N}$, indexes the constraints

© The Author(s), under exclusive license to Springer Nature Switzerland AG 2026
T. Guns (Ed.): CPAIOR 2026, LNCS 16595, pp. 71–81, 2026.
https://doi.org/10.1007/978-3-032-27242-3_5

and $\mathcal{J} \subseteq \{1,\ldots,n\}$ the integer variables. A nonlinear objective can be easily modeled by introducing an auxiliary variable and an objective-transfer constraint, see, e.g., [25]. If all g_k are linear, and $\mathcal{J} = \emptyset$ we call (1) a *linear program (LP)*. This work focuses on *nonconvex MINLPs*, i.e., problems of form (1), where at least one g_k is nonconvex.

Note that factorable functions can be represented via a directed acyclic *expression graph*, with nodes representing operators or variables, and arcs representing the data flow of the computation. In this paper, we refer to this representation as the *DAG*, for a good overview on the use of the DAG in MINLP solving, we recommend [25].

For solving problems of the form (1), we use the FICO® Xpress Global [11] MINLP solver, which we will refer to as *Xpress*. Xpress is based on the *branch-and-bound* method (*B&B*), which recursively partitions the problem by splitting the domain of selected variables, which is called *branching*. Selecting good branching variables is crucial for the performance of B&B-based MINLP solvers, see, e.g., [4]. For more details on the implementation in Xpress, see [3].

This paper studies a fundamental question: should we always branch on fractional integer variables first, or consider spatial branches even when there are fractional integers? Unlike most prior work on using ML for branching [2,15,17,20,22], we do not learn individual branching decisions or attempt to mimic existing strategies such as strong branching; instead, we perform algorithm selection between two established branching rules. Different from most prior work, we consider a heterogeneous set of instances. The resulting model can be integrated directly into solver code and does not require any pre-training on the user side. This is akin to prior ML-based algorithm-selection work to choose between scaling procedures [6]. Local cut selection rules [5], linearization techniques [8], or spatial branching strategies in the context of RLT for polynomial optimization [14], respectively, and deliberately different from solver-free learning approaches for MINLP as in [23].

2   A Quick Recap of Branching for MINLPs

The B&B algorithm recursively partitions the problem into smaller subproblems (*branching*) and solves LP relaxations to obtain bounds (*bounding*) until an optimal solution or infeasibility proof is found.

An *LP relaxation* of an MINLP is obtained by dropping integrality constraints and replacing nonlinear constraints with linear underestimators where possible. This relaxation is successively strengthened by *outer-approximation cuts* [10]. A well-designed cutting plane separation procedure often helps to reduce the branch-and-bound tree size while accelerating the overall solving process [24]. Unlike in MIP solving, cutting planes are often additionally separated immediately during branching-node creation in MINLPs.

In this paper, we focus on *variable branching*, in which the domain of a single variable is split into two or more intervals.

Two key types of variable branching are:

1. *Integer branching*, which is applied when an integer variable has a fractional value $\check{x}_j \in \mathbb{R} \setminus \mathbb{Z}$ in the solution $\check{x}$ of the current LP relaxation. Two subproblems are created that enforce $x_j \leq \lfloor \check{x}_j \rfloor$ and $x_j \geq \lceil \check{x}_j \rceil$, respectively.
2. *Spatial branching* [16] is applied when the violation of a nonconvex constraint cannot be resolved by an outer-approximation cut, but requires partitioning variable domains. Spatial branching candidates are often continuous variables, but can also include integer variables whose LP value happens to be integral. Two created subproblems enforce $x_j \leq \lfloor \check{x}_j \rfloor$ and $x_j \geq \lceil \check{x}_j \rceil$, respectively, for a branching point $\check{x}_j \in \mathbb{R}$. Though the LP solution is not explicitly excluded, subsequent outer-approximation cuts typically remove it.

3   Machine Learning Methodological Approach

Learning Task/Feature Space. Our learning task consists of choosing, after root node processing and right before the first branch, one of two rules of how to combine integer branching and spatial branching for the remainder of the branch-and-bound search: Either, always branch on integer candidates and conduct spatial branches only when there is no integer branching candidate, which we will refer to as "PreferInt" (this is the default, e.g., in the SCIP MINLP solver). Or mix both candidate sets and always allow the choice of either type of candidate (which is the default, e.g., in the Xpress solver), which we refer to as "Mixed".[1]

Although this is inherently a binary decision, we frame it as a regression problem. This choice is motivated by two considerations. Firstly, our ultimate goal is to improve the average runtime of the solver, which is a metric that is numerical and not categorical. Secondly, our focus is on getting the prediction right for those instances on which the performance of selecting "Mixed" and "PreferInt" significantly differs, see also [5]. Regression allows us to model the magnitude of this difference directly and thereby focus the learning on the cases where it matters most.

To this end, we train regression models $y_i : \mathbb{R}^d \rightarrow \mathbb{R}$ that map a d-dimensional feature vector $f = (f_1, \ldots, f_d)$ onto the speedup or slowdown factor (the *label*) in runtime by using "PreferInt" instead of "Mixed". We initially used 17 features, see Table 1.

This includes features related to strong branching at the root node, such as the average change in the dual bound resulting from integer and spatial strong branching, `AvgRelBndChngSBLPInt` and `AvgRelBndChngSBLPSpat`, respectively, the number of variables fixed from strong branching on spatial branching candidates `#SpatBranchEntFixed`[2], and the amount of deterministic *work* invested in either strong branching, `AvgWorkSBLPInt` and `AvgWorkSBLPSpat`. Work is a deterministic measure of computational effort implemented in Xpress. These features give us an indication of how effective (and expensive) strong branching on

---

[1] We ruled out always preferring spatial branches in a preliminary experiment, since this option was a factor eight slower on average and rarely won against the others.

[2] There were only a few instances where integer strong branching fixed variables; hence, a corresponding feature would have been almost flat zero.

integer or spatial variables is. Relatedly, `#IntViols` and `#NonlinViols` refer to the number of integer and spatial branching candidates.

As problem structure features, we include the percentage of variables that are integer, `%IntVars`, the percentage of constraints that are equations, `%EqCons`, the ratio of quadratic elements in the problem to variables, `%QuadrElements`, and the percentage of constraints that contain nonlinearities, `%NonlinCons`. Further, to measure the nonlinearity of the problem, we use information about the DAG, in particular, the percentage of variables that are part of any nonlinearity, `%VarsDAG`, and the ratio between nodes in the DAG and nonzeros in the linear part of the problem, `NodesInDAG`. We further include the percentage of integer and unbounded variables among all variables in the DAG, `%VarsDAGInt` and `%VarsDAGUnbnd`, as for integer DAG variables, we "hit two birds with one stone" and branching on unbounded variables can be crucial to get efficient dual bounds. Finally, we consider `%QuadrNodesDAG` to measure whether the nonlinearities in the problem are mostly quadratic.

Table 1. m and n are the number of constraints and variables before presolving, respectively; $\tilde{n}$ and $\tilde{M}$ the number of variables and linear nonzeros after presolving.

Feature	Feature Scaling
Problem Structure	
`%QuadrElements`	number quadratic elements over n
`%IntVars`	#Integer variables after presolve over $\tilde{n}$
`%EqCons`	#equality constraints over m
`%NonlinCons`	#nonlinear constraints over m
Effect of Branching	
`#IntViols`	
`#NonlinViols`	
`#SpatBranchEntFixed`	
`AvgWorkSBLPInt`	
`AvgWorkSBLPSpat`	$\log_{10}(\text{Value}+1)$
`AvgRelBndChngSBLPInt`	
`AvgRelBndChngSBLPSpat`	
`AvgCoeffSpreadConvCuts`	
DAG	
`NodesInDAG`	NodesInDAG over NodesInDAG$+\tilde{M}$
`%VarsDAG`	#vars in DAG over $\tilde{n}$
`%VarsDAGUnbnd`	#unbounded vars over #vars in DAG
`%VarsDAGInt`	#integer vars over #vars in DAG
`%QuadrNodesDAG`	#quadratic operator nodes in DAG over all nonlinear operator nodes in DAG

Data. The data on which the models are trained comes from running Xpress 9.6[3] twice on a heterogeneous benchmark of 683 public and industrial MINLP instances, each with two permutations to mitigate the effect of performance variability [12, 19], yielding 2049 data points. For each instance, we record the runtimes produced by both branching rules and the complete feature set. Instances solved at the root or otherwise unsuitable for comparison are filtered out, resulting in a final dataset of 797 data points, see [13] for details. Solving at the root node was by far the most common reason for filtering.

Training. For training the models, we split the data randomly into 80% training and 20% test set. The models we train on the training set are a linear regressor [9] and a random forest regressor, RF, [18]. We use the python library scikit-learn [21], which provides us with the linear regressor by the function *LinearRegression* and the random forest regressor by the function *RandomForestRegressor.*

Testing. Instead of training one linear regressor and one random forest regressor, we opted for training and testing one hundred models each with different random seeds and average their performance scores to evaluate how promising this ML-based approach is.

To measure the performance of the regression models, we use the *accuracy* and the shifted geometric mean of the runtime (*sgm_runtime*). The accuracy is defined as the percentage of times the model predicted the faster rule. The sgm_runtime is the shifted geometric mean of runtimes when solving each test instance using the predicted branching rule over the shifted geometric mean time using always the default rule.

Hence, accuracy is always between 0 and 100%, with larger values being better. Sgm_runtime can be smaller or larger than 1, with values larger than 1 indicating a deterioration and values smaller than 1 indicating an improvement: the smaller the number, the better. This is the primary performance indicator for solver development in practice.

To compute the shifted geometric mean [1] with a shift of 10, measurements $X = (X_1, \ldots, X_n)$ are aggregated via $sgm(X) = -10 + \prod_{i=1}^{n} (X_i + 10)^{\frac{1}{n}}$. The use of the shifted geometric mean is a commonly used method to aggregate performance measures, in particular running time, inn mathematical optimization [6].

For the linear model, *feature importance* is given by the absolute value of the learned coefficients, whereas for the random forest it is measured as the normalized total reduction in mean squared error induced by splits on that feature (mean decrease impurity, MDI), which are the default importance metrics in scikit-learn. For each random seed and each type of model, we computed the feature importance for all features and sorted them from most important to least important. Then we assigned a score of zero to the most important

---

[3] More precisely: An internal version of the Xpress 9.6 that exposes those features that are otherwise not available as public attributes.

one, a score of one to the second most important one, and so on. Finally, we added, for each model type, the scores across all one hundred runs together. The four most important features per model type (as by this score sum) are listed in Table 2. Although the top-ranked features differ between the linear regression and random forest models, there is overlap in terms of the underlying information captured. In particular, `AvgRelBndChngSBLPSpat` ranks first for the linear model and second for the random forest, and `%NonlinCons`, ranked third for the linear model, is a very close fifth for the random forest. Overall, five of the eight highest-ranked features coincide across the two models. Differences are expected given the different nature of the models: linear regression emphasizes globally predictive, approximately linear effects with low collinearity, whereas random forests prioritize features that enable strong local splits and nonlinear interactions, for instance, capturing cases where a branching rule is beneficial for either extreme but not for intermediate feature values.

Further Approaches. In earlier versions, we tested the algorithm with different features, unscaled or differently scaled features, on an earlier version of Xpress and for the SCIP solver (where it also improved performance, but not as much as in the Xpress case). Details can be found in the thesis [13]. This thesis also contains a detailed description of how we selected the scaler and imputer for the data set and a discussion of restricting the decision tree depth to five in the random forest models.

Table 2. Four most important features for either model type.

Ranking	Linear	Forest
1.	`AvgRelBndChngSBLPSpat`	`AvgCoeffSpreadConvCuts`
2.	`%IntVars`	`AvgRelBndChngSBLPSpat`
3.	`%NonlinCons`	`#NonlinViols`
4.	`%VarsDAGInt`	`%EqCons`

4   Computational Experiments

Our computational experiments consist of three parts: Firstly, we evaluate the regression models trained on the full 17-feature set, and then examine how their performance evolves as we iteratively remove the least important features. This reduction process provides insights into which features drive prediction quality and whether a more compact feature set can yield comparable performance with presumably better robustness. Secondly, we analyze the final reduced models in more detail and provide an analysis of their performance with respect to accuracy and runtime. Finally, we compare how those models continue to perform as the underlying branch-and-bound method is improved (in this case, through a solver version update).

For feature-reduction in the first experiment, we use the ranking by 100 different seeds described in Sect. 3. We remove one feature at a time (the least important) and retrain after each step. The impact on both accuracy and sgm_runtime for the linear models and the random forest models is shown in Figs. 1 and 2, respectively. The x-axis shows the number of features used, the y-axis shows the two performance measures. The dotted red line represents the sgm_runtime, and the dotted salmon-colored constant represents the best possible sgm_runtime score, achievable only if a model made a correct prediction in all cases. The green solid line represents the accuracy score on all test instances, while the blue shows the accuracy score on instances with a label, i.e., a speedup or slowdown factor, larger than four, here called *LargeLabel Accuracy*. These are the most important instances to predict accurately when the goal is to reduce the mean runtime.

In Fig. 1 we can see that the performance of the initial linear model with all features is at about 85% overall accuracy, more than 90% LargeLabel accuracy, and a sgm_runtime factor of a little over 0.91. The accuracy and the sgm_runtime factor of the random forest is a bit better, at 87% and a little over 0.9, respectively. LargeLabel accuracy is about the same. The feature-reduction experiments reveal a remarkably stable behavior: both model types maintain essentially constant accuracy and runtime performance as long as at least four features are retained. The LargeLabel accuracy drops slightly when going down from 17 to 4 features. Once the feature count drops below this threshold, all measures deteriorate noticeably. Based on the aggregated importance scores, we identified the four most influential features for each model type, see Table 2, and use them for the subsequent experiments.

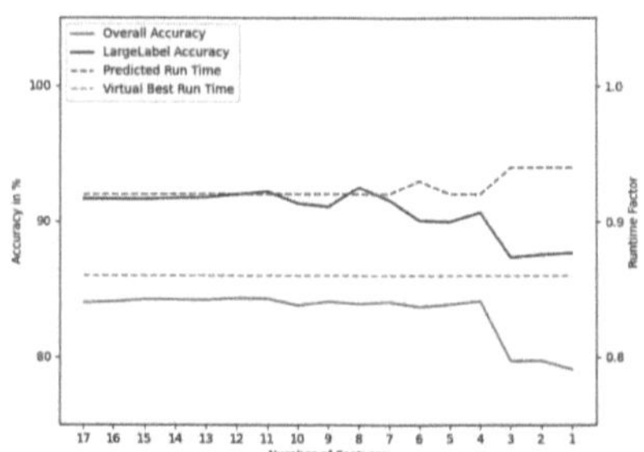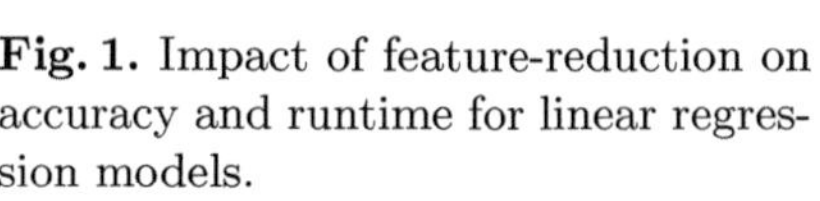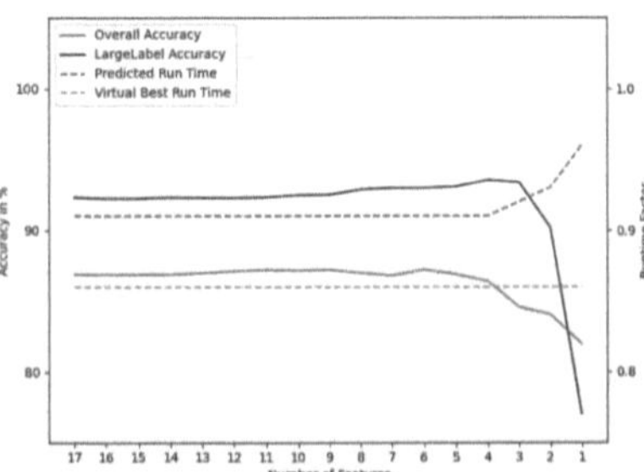

Fig. 1. Impact of feature-reduction on accuracy and runtime for linear regression models.

Fig. 2. Impact of feature-reduction on accuracy and runtime for random forest regression models.

Training and testing both model types using only their respective four most important features yields the performance scores depicted in Table 3. The scores are divided by model type and, for checking possible overfitting, by training and test set. Again, the accuracy score is calculated *Overall* and on LargeLabel instances. The two bottom rows contain the sgm_runtime scores. As before, all scores are averages across one hundred runs for each model type.

The linear model performs almost identical on training and test set; there is no indication of overfitting. On the test set, it predicts the better branching rule with an accuracy of about 84% overall and more than 90% on the subset of instances with the largest performance differences. Most importantly, the sgm_runtime score is 0.919 on the test set, indicating a significant 8% speedup. The standard deviation was 2.8%. On models that take more than 100 s to solve by either of the two variants, we observed a speedup of over 10% with a standard deviation of 6.8%. These results show that a linear model, based on a handful of carefully designed features, provides meaningful and robust performance gains, with an easy-to-implement and interpretable model, which is highly desirable in practice.

For the random forest, however, there is a significant difference between the training and the test set. The almost perfect accuracy on the training set drops to 93.6% on the test set, a clear sign of overfitting, and the sgm_runtime clearly deteriorates as well. With a value of 3.2%, the standard deviation was also higher compared to the linear model. From a practical perspective, the linear model seems preferable, even though its performance is slightly worse even on the test set.

The linear model predicted the "Mixed" rule to be faster for around 60% of the instances and "PreferInt" to be faster for 40% of the instances. For the random forest, the split is close to 50:50. For the linear model, in 51% of the cases, switching to the "PreferInt" rule improves performance by at least 10%, in 11% of the cases performance got at least 10% worse, in the remaining 38%, it stayed roughly the same. For the random forest, this split was 43% wins, 11% losses, 36% within ±10% runtime.

With our final experiment, we approach the question of how sensitive the results are to the changes in the underlying MINLP solving algorithm, more specifically, the robustness across solver versions. Therefore, we apply the models trained on Xpress 9.6 to data generated by Xpress 9.8. The latter version introduces substantial algorithmic improvements that make it roughly 50% faster on average for difficult MINLP instances, including changes to presolving, cut generation, heuristics, and branching logic. Despite these differences, both learned models continue to yield computational benefits: the linear model achieves an average speedup of about 3.3% and the random forest about 4.5%.

This leaves the question of whether the smaller performance gain is inherent to the other changes in the solver or to the trained models not being as accurate for predicting performance of the newer solver version. In an additional experiment, we trained new models from scratch on the data of Xpress 9.8, and to our positive surprise, both the accuracy and the time factors were very similar to the ones shown in Table 4. The linear model trained on Xpress 9.8 data achieved the same sgm_runtime score of 0.967, whereas the random forest slightly improved, presumably again a result of overfitting. Furthermore, the sets of the four most important features were almost identical, with only one feature

Table 3. Comparison of accuracy and runtime factors on training and test sets for final linear and random forest models

	Linear		Forest	
Accuracy	Train	Test	Train	Test
Overall	84.2%	84.1%	89.1%	86.4%
LargeLabel	91.9%	90.7%	99.7%	93.6%
Time Factor				
Predicted	0.914	0.919	0.884	0.910
Virtual Best	0.862	0.863	0.862	0.863

Table 4. Results when using models trained on Xpress 9.6 for predicting 9.8 performance

	Linear	Forest
Accuracy		
Overall	80.8%	85.2%
LargeLabel	77.0%	82.2%
Time Factor		
Predicted	0.967	0.955
Virtual Best	0.866	0.866

differing in the linear model and none in the random forest. This robustness is particularly valuable for practical deployment, as it suggests that models need not be retrained for every release cycle.

5   Conclusion

In this paper, we investigated whether a global MINLP solver should always prioritize branching on fractional integer variables or whether allowing spatial branching earlier can lead to faster overall performance. Using a heterogeneous dataset of public and industrial MINLP instances, we trained linear and random forest regression models to predict the relative performance of two established branching rules. Our experiments demonstrated that linear models can achieve an 8% reduction in mean solving time. We further showed that the learned models remain effective across solver versions, indicating robustness to underlying algorithmic changes.

A natural next step is to implement the regressor directly inside the FICO® Xpress Global solver to validate its impact in a production setting and to extend them to other solvers, like the SCIP open-source MINLP solver. Finally, while our models select a single branching rule for the entire branch-and-bound tree, a more fine-grained approach, such as dynamically choosing rules for different phases of the solve [7] or adaptively at individual nodes of the branch-and-bound tree, represents a promising direction for further research.

Acknowledgements. We thank Tristan Gally for his support with Xpress implementations and Ksenia Bestuzheva and Stefan Vigerske for the valuable discussions on SCIP features. The work for this article was supported through the Research Campus Modal funded by the German Federal Ministry of Education and Research (fund numbers 05M14ZAM,05M20ZBM).

Disclosure of Interests. Timo Berthold is an employee of FICO.

References

1. Achterberg, T.: Constraint integer programming, Ph.D. thesis, Technische Universität Berlin (2007). https://doi.org/10.14279/depositonce-1634
2. Alvarez, A.M., Louveaux, Q., Wehenkel, L.: A machine learning-based approximation of strong branching. INFORMS J. Comput. **29**(1), 185–195 (2017). https://doi.org/10.1287/ijoc.2016.0723
3. Belotti, P., Berthold, T., Gally, T., Gottwald, L., Pólik, I.: Solving MINLPs to global optimality with FICO Xpress Global. Optimization 1–19 (2025). https://doi.org/10.1080/02331934.2025.2595437
4. Belotti, P., Kirches, C., Leyffer, S., Linderoth, J., Luedtke, J., Mahajan, A.: Mixed-integer nonlinear optimization. Acta Numer **22**, 1–131 (2013). https://doi.org/10.1017/S0962492913000032
5. Berthold, T., Francobaldi, M., Hendel, G.: Learning to use local cuts. Math. Program. Comput. **17**(3), 437–450 (2025). https://doi.org/10.1007/s12532-025-00278-y
6. Berthold, T., Hendel, G.: Learning to scale mixed-integer programs. In: Proceedings of the AAAI Conference on Artificial Intelligence, Vol. 35, no. (5), pp. 3661–3668 (2021). https://doi.org/10.1609/aaai.v35i5.16482
7. Berthold, T., Hendel, G., Koch, T.: From feasibility to improvement to proof: three phases of solving mixed-integer programs. Optim. Methods Softw. **33**(3), 499–517 (2018)
8. Bonami, P., Lodi, A., Zarpellon, G.: A classifier to decide on the linearization of mixed-integer quadratic problems in cplex. Oper. Res. **70**(6), 3303–3320 (2022)
9. Chambers, J.M., Hastie, T.: Statistical models in S. Chapman & Hall (1992)
10. Duran, M.A., Grossmann, I.E.: An outer-approximation algorithm for a class of mixed-integer nonlinear programs. Math. Program. **36**(3), 307–339 (1986). https://doi.org/10.1007/BF02592064
11. FICO Xpress Optimizer. https://www.fico.com/en/products/fico-xpress-solver
12. Gamrath, G., Berthold, T., Salvagnin, D.: An exploratory computational analysis of dual degeneracy in mixed-integer programming. EURO J. Comput. Optim. **8**(3), 241–261 (2020)
13. Geis, F.: Selecting branching rules for MINLPs via machine learning. Master's thesis Technische Universität Berlin (2025)
14. Ghaddar, B., Gómez-Casares, I., González-Díaz, J., González-Rodríguez, B., Pateiro-López, B., Rodríguez-Ballesteros, S.: Learning for spatial branching: an algorithm selection approach. INFORMS J. Comput. **35**(5), 1024–1043 (2023)
15. Gupta, P., et al.: Hybrid models for learning to branch. In: Lin, H. (ed.) Advances in Neural Information Processing Systems, vol. 33, pp. 18087–18097. Curran Associates, Inc (2020)
16. Horst, R., Tuy, H.: Global optimization: Deterministic approaches. Springer, Cham (1996). https://doi.org/10.1007/978-3-662-03199-5
17. Khalil, Elias B. and Le Bodic, P., Song, L., Nemhauser, G.L., Dilkina, B.: Learning to branch in mixed integer programming. Proc. AAAI Conf. Artif. Int. **30**(1), 724–731 (2016). https://doi.org/10.1609/aaai.v30i1.10080
18. Liaw, A., Wiener, M.: Classification and regression by randomforest. R News **2**(3), 18–22 (2002)
19. Lodi, A., Tramontani, A.: Performance variability in mixed-integer programming. In: Theory Driven by Influential Applications, pp. 1–12. INFORMS (2013)

20. Nair, V., et al.: Solving mixed integer programs using neural networks. arXiv preprint arXiv:2012.13349 (2020). https://doi.org/10.48550/arXiv.2012.13349
21. Pedregosa, F., et al.: Édouard Duchesnay: Scikit-learn: machine learning in Python. J. Mach. Learn. Res. **12**, 2825–2830 (2011)
22. Scavuzzo, L., Aardal, K., Lodi, A., Yorke-Smith, N.: Machine learning augmented branch and bound for mixed integer linear programming. Math. Program. (2024). https://doi.org/10.1007/s10107-024-02130-y
23. Tang, B., Khalil, E.B., Drgoňa, J.: Learning to optimize for mixed-integer nonlinear programming with feasibility guarantees, arXiv preprint arXiv:2410.11061 (2024)
24. Turner, M., Berthold, T., Besançon, M., Koch, T.: Cutting plane selection with analytic centers and multiregression. In: Cire, A.A. (ed.) Integration of Constraint Programming, Artificial Intelligence, and Operations Research, pp. 52–68. Springer, Cham (2023). https://doi.org/10.1007/978-3-031-33271-5_4
25. Vigerske, S.: Decomposition in multistage stochastic programming and a constraint integer programming approach to mixed-integer nonlinear programming, Ph.D. thesis. Humboldt-Universität zu Berlin, Mathematisch-Naturwissenschaftliche Fakultät II (2013). https://doi.org/10.18452/16704

Imitation-Guided World Models for Multi-agent Train Rescheduling

Max Bourgeat[1,2(✉)] [iD], Antoine Legrain[1,2,3] [iD], and Quentin Cappart[1,2,4] [iD]

[1] Polytechnique Montreal, Montreal, Canada
[2] CIRRELT, Montreal, Canada
[3] GERAD, Montreal, Canada
[4] UCLouvain, Louvain-la-Neuve, Belgium
{max.bourgeat,a.legrain,quentin.cappart}@polymtl.ca

Abstract. Managing railway disruptions is a complex multi-agent routing problem where a single train failure can propagate delays across the network. Traditional approaches rely on heuristic optimization solvers, which are effective but assume access to a global system view and require substantial expert design, limiting their applicability and generalization. Reinforcement learning (RL) offers an alternative by learning adaptive strategies from interactions with the environment. In this paper, we show that none of the available paradigms is sufficient in isolation: (i) heuristic solvers encode valuable global expertise but cannot be deployed directly, (ii) world models improve sample efficiency but struggle to leverage expert knowledge, and (iii) pure RL can adapt policies but often lacks stability without strong guidance. We propose a hybrid framework that integrates the strengths of these approaches. First, imitation learning transfers knowledge from a global expert solver to initialize a neural policy. Then, model-based RL fine-tunes this policy using the DreamerV2 world model to enhance generalization and responsiveness to local perturbations. Our method builds on the Multi-Agent Model-Based Architecture (MAMBA) to model agent interactions and addresses the challenge of transferring expertise from global solvers to decentralized agents operating on local latent observations. Experiments on a train rescheduling problem using the Flatland environment show that our method outperforms MAMBA, improving performance by up to 23% on difficult instances. This highlights the benefit of combining imitation learning with world-model-based multi-agent RL for complex transportation networks. The code is available here https://github.com/corail-research/Imitation-Guided_World_Models.

Keywords: Multi-Agent Reinforcement Learning · Imitation Learning

1 Introduction

Multi-agent routing [41] consists in coordinating several agents moving within a network to reach their destinations while minimizing conflicts and delays. This

© The Author(s), under exclusive license to Springer Nature Switzerland AG 2026
T. Guns (Ed.): CPAIOR 2026, LNCS 16595, pp. 82–100, 2026.
https://doi.org/10.1007/978-3-032-27242-3_6

problem appears in logistics [22, 28], robotics [16], and public transportation [12], and is characterized by strong combinatorial complexity due to agent inter-dependencies. Efficient methods are crucial for reducing operational costs and improving the resilience of large-scale systems. In railway networks, disruptions further increase this complexity. A single train breakdown blocks its track segment, potentially propagating delays across the network. Rescheduling trains to mitigate these cascading effects is, therefore, central to modern railway operations. The *Flatland* environment [30], developed with several European railway companies, provides a realistic platform for studying such scenarios. Traditional approaches rely on heuristic or search-based solvers [3, 37, 42]. While effective, they require handcrafted heuristics tailored to each instance and assume access to a global system view, an unrealistic condition since trains typically observe only local surroundings. Reinforcement learning (RL) [43] offers an alternative by learning from interaction. However, model-free methods often suffer from poor sample efficiency [29], limited generalization [4], and instability in multi-agent settings [11]. Model-based approaches, such as DREAMER [9], improve efficiency by learning latent dynamics but remain limited by model inaccuracies [45] and a performance gap with expert solvers [25]. *Imitation learning* (IL) [15] mitigates exploration challenges by leveraging expert demonstrations. Yet, aligning expert trajectories defined in global real-state space with the local latent space of world models is non-trivial [5].

We propose a hybrid methodology that combines IL and model-based RL for train rescheduling. A neural policy is first initialized from an expert solver [21], then refined using MAMBA system, a multi-agent world model based on Dream-erV2 [10] with Transformer-based communication [47]. Our key contribution is an adaptation strategy enabling the transfer of global solver expertise into a decentralized latent setting. Experiments in Flatland show that our method outperforms state-of-the-art baselines, achieving up to 23% improvement on difficult instances. This demonstrates the value of combining imitation and world-model-based RL for scalable decision-making in complex transportation networks.

2 Related Work

Multi-agent Reinforcement Learning (MARL) has become a central research direction, driven by the need to handle non-stationarity, coordination, and competition among multiple interacting agents. Overviews of deep learning methods for addressing scalability, decentralized decision-making, and efficient communication are provided by few recent surveys [11, 14]. Beyond model-free approaches, several works have investigated the use of predictive models, or world models, in MARL. For instance, MAMBA [6] coupled a Transformer-based communication [47] between agents with DreamerV2 [10]. This world model introduced learned dynamics models to generate imagined rollouts, thereby improving sample efficiency and reducing reliance on expensive environment interactions. Another architecture, referred to as *global-aware world model* [39] adopted centralized training with decentralized execution, leveraging Transformers to aggregate local observations into a global latent representation. Other methods also explored local predictive structures, such as *models as agents* [52] which optimize multi-step predictions of interactive local models, or hierarchical frameworks (e.g., [48], which employs bi-level latent-variable world models). Together, these methods highlight the potential of world models to improve stability, scalability, and coordination. Yet, they often underperform as the number of agents increases.

On the other hand, *imitation learning* offers a complementary paradigm where agents learn from demonstrations rather than sparse or delayed rewards [46]. In the multi-agent setting, imitation learning must also capture inter-agent dependencies. Recent approaches explicitly model joint behaviors, for instance, with copula-based methods [49] or adversarial extensions of *generative adversarial imitation learning* [13] to cooperative and competitive scenarios [40]. Hybrid strategies further combine IL and RL, such as self-imitation frameworks [31] or offline-to-online training strategies like advantage-weighted regression [32], aiming to accelerate convergence while ensuring robustness. However, most methods assume access to large-scale demonstrations or simplified coordination, which may not hold in real-world systems.

Focusing now on the specific railway domain, the Flatland environment was introduced during the NeurIPS 2020 competition [30] as a benchmark for large-scale multi-agent routing and rescheduling. It provides a simplified yet realistic simulation where multiple trains must coordinate under infrastructure and safety constraints. Flatland also includes stochastic disruptions, such as random breakdowns, making it suitable for studying robust decision-making under uncertainty. Beyond the baseline methods proposed in the competition, the winning solution, *scalable rail planning and replanning* [21], combines Multi-Agent PathFinding (MAPF) techniques, large neighborhood search, safe interval path planning, simulated annealing, minimum-communication policies, and parallel computation, demonstrating strong performance with up to 400 trains. Since then, subsequent research has extended Flatland to practical scheduling. For instance, LCPPO [53] introduces a policy-gradient algorithm tailored to large-scale railway networks, while *curriculum-driven continual DQN expansion* [17] addresses stability-plasticity trade-offs through continual learning. But there is also a rich literature outside Flatland: for instance, Cappart and Schaus [3] proposes a CP model using time-interval variables for real-time rescheduling in a Belgian station, outperforming greedy operator strategies; *Reinforcement learning in railway timetable rescheduling* explores RL methods off-line to enable fast online dispatching under delays [55]; Liao et al. [23] combined deep learning and genetic algorithms to optimize metro energy consumption under random disturbances [23]. Finally, recent models also integrate learning-based model predictive control to adapt rolling stock and train compositions [26].

Taken together, these works highlight that flexibility, realistic constraints (e.g. variable speeds, train composition, or disruptions) and the ability to compute solutions rapidly in real-world contexts are crucial requirements for MARL or RL applied to this domain.

3   Background

World models have emerged as a powerful paradigm in model-based reinforcement learning (MBRL) [43], where an agent learns a latent dynamics model of the environment and uses it for planning and policy optimization. Formally, the model consists of a latent state-space model defined by the following conditional distributions:

$$z_t \sim p_\theta(\cdot \mid z_{t-1}, a_{t-1}), \tag{1}$$

$$o_t \sim p_\theta(\cdot \mid z_t), \tag{2}$$

$$r_t \sim p_\theta(\cdot \mid z_t, a_t), \tag{3}$$

where a_t is an action, z_t is a latent state, o_t an observation, and r_t a reward obtained at a timestep t. Each component is implemented as a neural network p_θ, where θ denotes the joint vector of all their parameters. A policy $\pi_\phi(a_t \mid z_t)$ is trained entirely in latent space by maximizing the expected discounted return:

$$J(\pi_\phi) = \mathbb{E}_{z_0 \sim p_\theta, a_t \sim \pi_\phi}\left[\sum_{t=0}^{T} \gamma^t r(z_t, a_t)\right], \tag{4}$$

where ϕ are our policy's parameters, z_0 the initial latent space following the initial distribution p_θ, $\gamma \in (0, 1]$ the discount factor which indicates how much weight we give to short-term reward compared to long-term reward and T the time horizon in the latent space. The *Dreamer* family of algorithms [9,10] demonstrated that training on imagined trajectories in latent space yields state-of-the-art performance while being computationally efficient. In particular, a *recurrent state-space model* [8] with categorical latent variables stabilizing long-horizon predictions and improving scalability is employed by DreamerV2.

3.1   Multi-agent Reinforcement Learning

Multi-agent Reinforcement Learning. (MARL) studies sequential decision-making with multiple agents interacting in a shared environment. This setting can be formalized as a Markov game featuring n agents:

$$\mathcal{M} = \langle \mathcal{S}, \{\mathcal{A}_i\}_{i=1}^{n}, T, \{r_i\}_{i=1}^{n}, \gamma\rangle, \tag{5}$$

where $\mathcal{S}$ is the state space, $\mathcal{A}_i$ is the action space of agent i, and the transition dynamics follow

$$s' \sim T(s' \mid s, a_1, \ldots, a_n), \tag{6}$$

where $s \in \mathcal{S}$ and $a_i \in \mathcal{A}_i, \forall i \in \{1, ..., n\}$. Each agent i receives a reward $r_i(s, a_1, \ldots, a_n)$ and optimizes its policy $\pi_\phi^i(a_i \mid o_i)$, a neural network parameterized by ϕ, based on local observations o_i. As before, γ is the discount factor. Most of the time, if a world model is used then the policy is trained using the actor-critic approach [20]. In actor-critic methods, the policy (actor) is updated using feedback provided by a value function (critic), which estimates the expected return of stateaction pairs. This framework combines the exploration capabilities of policy gradient methods [36,44,51] with the stability of value-based estimation [2,29,50]. In the *centralized training with decentralized execution* (CTDE) paradigm [7], critics are trained with global information while policies rely solely on local observations. A typical policy gradient update is given by:

$$\nabla_\phi J(\pi_\phi) = \mathbb{E}_{s\sim\mathcal{D},a\sim\pi_\theta}\left[\nabla_\phi \log \pi_\phi(a \mid o)\, Q(s,a)\right], \tag{7}$$

where $Q(s,a)$ is the value function output by the critic, it may be learned with centralized knowledge during training and $\mathcal{D}$ is the dataset from which the states are sampled [24,27,44]. To improve sample efficiency, model-based MARL approaches extend latent dynamics models to multi-agent settings. For example, scalable frameworks such as MAMBA [6] introduce communication mechanisms where an agent i shares latent states and actions at step t through a message m_i^t defined as follows:

$$m_i^t = c_\omega\left(z_i^{t-1}, a_i^{t-1}\right), \tag{8}$$

This communication is commonly restricted to neighbors, thereby reducing complexity from $\mathcal{O}(n^2)$ to $\mathcal{O}(n)$. With MAMBA, c_ω is a Transformer neural network.

3.2 Imitation Learning

Imitation Learning (IL) leverages expert demonstrations $\mathcal{D}_E = \{(s_i, a_i^{expert})\}_{i=1}^M$, M being the number of samples available, to accelerate RL. The simplest approach, *behavioral cloning* (BC) [15], minimizes the negative log-likelihood of expert actions:

$$\mathcal{L}_{\mathrm{BC}}(\pi_\phi) = \mathbb{E}_{(s,a^{expert})\sim\mathcal{D}_E}\left[-\log \pi_\phi(a^{expert} \mid s)\right]. \tag{9}$$

However, combining IL with world models is non-trivial. In world models, the actor operates in the latent space z, while demonstrations are expressed in the true state space s. A mapping through the learned encoder e_θ is required:

$$z = e_\theta(s), \tag{10}$$

but this mapping is optimized for compression and prediction, not alignment with expert trajectories. Although the explicit alignment of expert trajectories with latent spaces is uncommon, several recent works tackle this challenge. DITTO [5] introduces an intrinsic latent reward to align agent and expert behaviors, FLARE [54] aligns future latent representations during policy learning, *latent diffusion planning* [1] generates transformed demonstrations to better cover the latent space. These works highlight the open challenge of transferring expert knowledge into latent representations, which motivates our approach to bridge expert solvers with local latent policies.

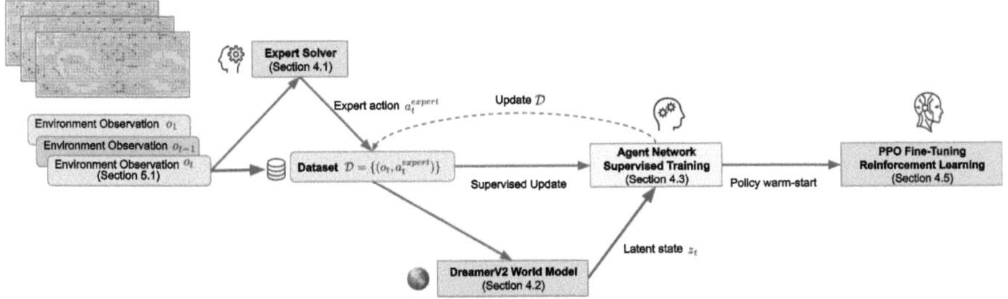

Fig. 1. Overview of our approach: (1) pretraining DreamerV2 world model, (2) collecting expert-labeled trajectories, (3) iterative supervised pretraining, and (4) fine-tuning with RL. Each main step is detailed in a specific subsection.

4 Our Imitation-Guided World Model

In this section, we describe our approach to warm-start both the world model and the agent in a sequential, iterative manner, using expert demonstrations from a Flatland solver to initialize the neural network parameters before RL to fine-tune it. Our approach is illustrated in Fig. 1.

4.1 Expert Demonstrator

A strong reference for large-scale railway rescheduling is the solver of Li et al. [21], the winner of the NeurIPS 2020 Flatland Challenge. It showed that well-engineered planning and optimization techniques can outperform learning-based methods in terms of scalability and robustness under stochastic disruptions.

The solver combines key components from multi-agent pathfinding: an initial feasible solution is built through *prioritized planning* [37], where agents are ordered and planned sequentially; each path is generated with *safe interval path planning* (SIPP) [33] to respect spatiotemporal constraints; and the solution is refined via large neighborhood search (LNS) [38], guided by simulated annealing heuristics [19] to balance exploration and exploitation.

A notable strength is its adaptive replanning: when a breakdown occurs, only the trains affected by the blocked segment are replanned, limiting the propagation of delays. Minimum communication policies [28] further prevent deadlocks in congested areas. The solver scales to 400 trains and achieves top performance in Flatland without using RL, confirming the competitiveness of combinatorial optimization in this domain.

However, this approach assumes global knowledge of the full network and plans paths centrally. This is an unrealistic requirement for real railway operations, where trains rely on local observations and limited communication. In contrast, our method leverages imitation learning and world models to enable decentralized decision-making under partial observability. Using this expert, we collect a dataset $\mathcal{D}_E$ of 800K samples, though our approach remains compatible with any alternative expert, including those reflecting real operational practices.

4.2 DreamerV2 World Model Pretraining

In addition to the expert's policy, we train a latent dynamics model following the DreamerV2 framework [10] as in MAMBA [6] without any modification. DreamerV2 learns a compact latent representation z_t of the environment from observations o_t and actions a_t where z_t is composed of a deterministic recurrent state h_t and a stochastic state s_t. This structured representation enables the model to predict future latent states, rewards, and discount factors. The model consists of 6 components. First, we have a recurrent state-space model (RSSM) composed of three models:

$$\text{Recurrent Model} \quad h_t = f_\theta(h_{t-1}, s_{t-1}, a_{t-1}), \tag{11}$$

$$\text{Representation Model} \quad s_t \sim q_\theta(s_t|h_t, o_t), \tag{12}$$

$$\text{Transition Predictor} \quad \hat{s}_t \sim p_\theta(\hat{s}_t|h_t). \tag{13}$$

Additionally, we have three more predictors:

$$\text{Observation Predictor} \quad \hat{o}_t \sim p_\theta(\hat{o}_t|h_t, s_t), \tag{14}$$

$$\text{Reward Predictor} \quad \hat{r}_t \sim p_\theta(\hat{r}_t|h_t, s_t), \tag{15}$$

$$\text{Discount Predictor} \quad \hat{\gamma}_t \sim p_\theta(\hat{\gamma}_t|h_t, s_t). \tag{16}$$

We denote by q_θ the neural network parameterizing the generative distributions of the real environment, and by p_θ the inference networks that approximate them to enable latent imagination. In our case, we reuse the MAMBA architecture: our observation predictor, reward predictor and discount predictor are 2 fully-connected layers of 400 neurons each; The transition predictor is a single fully-connected layer of 400 neurons, followed by an attention encoder (3 layers, 8 heads), a gated recurrent unit (GRU) (1 layer, 600 neurons), and two fully-connected layers of 400 neurons; The representation model has the same architecture with an additional MLP (2 layers, 400 neurons per layer) at the end. Concerning the recurrent model (Eq. 11), h_t is computed using a GRU referred to as f_θ. The GRU is chosen for its ability to efficiently capture temporal dependencies in sequential data, while being less prone to vanishing gradient issues compared to vanilla RNNs, and is computationally lighter than LSTMs. This is particularly relevant in our multi-agent railway environment, where long-term dependencies matter (e.g., train interactions over multiple time steps) but efficiency is critical due to the large number of agents.

The representation model q_θ (Eq. 12) infers the stochastic latent state from the deterministic hidden state and the current observation. This stochastic component captures uncertainty in the environment and allows the model to represent multiple plausible futures, which is critical in partially observable multi-agent settings such as ours because trains can have information (i.e. their previous actions and states) only from nearby other trains.

The transition predictor p_θ (Eq. 13) predicts the next stochastic latent state purely from the deterministic hidden state, enabling latent imagination without direct access to future observations. This facilitates planning and policy learning entirely in latent space. The observation predictor (Eq. 14) reconstructs observations from the latent state, ensuring that the latent representation retains sufficient information about the environment for downstream tasks, such as training policies or critics. The reward predictor (Eq. 15) predicts rewards in the latent space, providing the signal necessary for RL. By predicting rewards from latent states rather than raw observations, the model enables sample-efficient policy updates. The discount predictor (Eq. 16) estimates the expected discount factor for the next time step, allowing the model to handle stochastic episode terminations or partial observability in a principled manner.

The model is trained by minimizing the negative log-likelihood of observations, rewards, and discounts, along with a KL divergence term to regularize the latent distribution:

$$\mathcal{L}_{\text{world}}(\theta) = -\mathbb{E}_{q_\theta(s_{1:t}|h_{1:t},o_{1:t})}[\log p_\theta(\hat{o}_t, \hat{r}_t, \hat{\gamma}_t|h_t, s_t)]$$
$$+ \text{KL}\big[q_\theta(s_t|h_t, o_t) \,\|\, p_\theta(\hat{s}_t|h_t)\big] \tag{17}$$

where $s_{1:t}, h_{1:t}, o_{1:t}$ are sequences of stochastic states, deterministic states and observations. This training ensures that the latent representation $z_t = (h_t, s_t)$ captures sufficient information for both planning and policy learning, enabling sample-efficient decision-making in downstream imitation learning and RL stages.

4.3 Supervised Agent Pretraining

Using the expert dataset $\mathcal{D}_E$ and the latent states z_t from the world model, we train the agent network $\pi_\phi(a \mid z)$ in a supervised manner by minimizing the cross-entropy loss over discrete actions:

$$\mathcal{L}_{\text{IL}}(\phi) = -\frac{1}{|\mathcal{D}_E|} \sum_{\substack{a_t^{\text{expert}} \in \mathcal{D}_E \\ z_t \sim q_\theta, f_\theta}} \log \pi_\phi\big(a_t = a_t^{\text{expert}} \mid z_t\big) \tag{18}$$

This loss encourages the policy to mimic the expert in the latent space learned by the world model. Importantly, trajectories are collected by rolling out the agent's current policy π_ϕ. At each timestep, we query the expert for the action a_t^{expert} they would have taken, but we do not follow the expert's policy. Instead, the agent executes its own actions, while the expert provides labels used for supervision. This setup ensures that the training distribution matches the states actually encountered by the agent, mitigating covariate shift and making the learned policy more robust during deployment. This pre-training phase is carried out using the Adam algorithm [18] with a learning rate $lr = 5e - 4$, $\beta_1 = 0.9$, $\beta_2 = 0.999$ and a weight decay of 0.00001.

This supervised pretraining aligns the agent's initial policy with expert behaviors, while still grounding learning in the agent's own state-action distribution. As a result, the agent benefits from expert guidance without becoming overly reliant on demonstrations, providing a strong initialization for subsequent

RL fine-tuning. The full pseudocode of the supervised agent pretraining can be found in Algorithm 1. It simply consists of iteratively updating the world model and policy weights by backpropagating the loss functions using a batch $\mathcal{B}$ of k samples drawn from the expert dataset (lines 3–4).

Algorithm 1. `pretraining(.)` - Supervised Agent Pretraining

Require: World model $f_\theta, p_\theta, q_\theta$.
Require: Policy π_ϕ.
Require: Dataset $\mathcal{D}$.
Require: Batch size k, number of epochs E.
  1: **for** step 1 **to** E **do**
  2:      $\mathcal{B} := $ `randomSampling`$(\mathcal{D}, k)$
  3:      $\langle f_\theta, p_\theta, q_\theta \rangle := $ `updateWorldModel`$(\mathcal{B}, \mathcal{L}_{\text{world}}, f_\theta, p_\theta, q_\theta)$
  4:      $\pi_\phi := $ `updateActor`$(\mathcal{B}, \mathcal{L}_{\text{IL}}, f_\theta, p_\theta, q_\theta)$
  5: **end for**
  6: **return** $f_\theta, p_\theta, q_\theta, \pi_\phi$

4.4   Fine-Tuning with Reinforcement Learning

After the pretraining phase, we continue to train the agent using RL to adapt to situations not perfectly captured by the expert. We employ the *proximal policy optimization* (PPO) algorithm [36], an actor-critic method consisting of two main components: (1) the *actor* $\pi_\phi(a|z)$, which outputs a policy over actions given the latent state z produced by the world model, and (2) the *critic* $V_\psi(z)$, which estimates the expected return (value) from the latent state.

Both neural networks are updated by minimizing the actor loss $\mathcal{L}_{\text{actor}}$ and the critic loss $\mathcal{L}_{\text{critic}}$ as defined in the standard PPO algorithm [36]. In our case, the overall update alternates between minimizing $\mathcal{L}_{\text{actor}}$ with respect to ϕ and $\mathcal{L}_{\text{critic}}$ with respect to ψ. This separation ensures that the actor is guided by the advantage signal without being biased by value prediction errors, while the critic learns an accurate baseline for stable advantage estimation. Combined with the world model, this actor-critic scheme enables the agent to make effective decisions under partial observability and stochastic disruptions in railway networks.

4.5   Pseudo-Code of the Full Pipeline

The full training pipeline is described in the Algorithm 2. Starting from an empty training set, we first execute the pretraining phase. At each iteration, new trajectories are collected using the updated policy π_ϕ (line 4). Then, for each observation, the expert is queried about the action they would have taken (line 5). The resulting action is added to the dataset (line 6). The world model and the actor are then retrained to minimize their respective losses on the expanded dataset

---

Algorithm 2. `fullTraining(.)` - Imitation-Guided World Model

Require: World model $f_\theta, p_\theta, q_\theta$.
Require: Actor π_ϕ, critic V_ψ.
Require: IL steps E_{IL}, RL steps E_{RL}, batch size k.
1: Initialize randomly $f_\theta, p_\theta, q_\theta, \pi_\phi, V_\psi$.
2: $\mathcal{D} = \emptyset$
3: **for** step 1 **to** E_{IL} **do**
4:      $\langle o, a, r \rangle := \mathtt{envStep}(\pi_\phi)$ ▷ Following the current policy
5:      $a^{\mathrm{expert}} := \mathtt{getExpertAction}(o)$ ▷ Section 4.1
6:      $\mathcal{D} := \mathcal{D} \cup \{o, a, r, a^{\mathrm{expert}}\}$ ▷ Updating the dataset
7:      $\langle f_\theta, p_\theta, q_\theta, \pi_\phi \rangle := \mathtt{pretraining}(f_\theta, p_\theta, q_\theta, \pi_\phi, \mathcal{D})$ ▷ Alg. 1
8: **end for**
9: **for** step 1 **to** E_{RL} **do**   ▷ Section 4.4
10:      $\mathcal{B} := \mathtt{randomSampling}(\mathcal{D}, k)$
11:      $\langle f_\theta, p_\theta, q_\theta \rangle := \mathtt{updateWorldModel}(\mathcal{B}, \mathcal{L}_{\mathrm{world}}, f_\theta, p_\theta, q_\theta)$
12:      $\pi_\phi := \mathtt{updateActor}(\mathcal{B}, \mathcal{L}_{\mathrm{actor}}, f_\theta, p_\theta, q_\theta, \pi_\phi, V_\psi)$
13:      $V_\psi := \mathtt{updateCritic}(\mathcal{B}, \mathcal{L}_{\mathrm{critic}}, f_\theta, p_\theta, q_\theta, \pi_\phi, V_\psi)$
14: **end for**
15: **return** $f_\theta, p_\theta, q_\theta, \pi_\phi$

(line 7). This iterative procedure ensures that the training distribution continually reflects the states actually encountered by the agent, thereby reducing covariate shift compared to one-shot behavioral cloning. As the policy improves, it explores more diverse regions of the state space, which, in turn, allows the expert to provide guidance in situations that were not present in the initial dataset. The world model also benefits from this process, since it is periodically retrained on the newly collected trajectories, improving its predictive accuracy in regions of the state-action space that become more relevant as the agent evolves.

The process is repeated for E_{IL} steps. We set this number to be high enough to reach convergence, defined as the point where the supervised loss $\mathcal{L}_{\mathrm{IL}}$ no longer decreases significantly or the policy performance on a held-out validation set saturates. In practice, this iterative pretraining acts as a form of data aggregation, similar in spirit to DAgger [34], but adapted to the latent world model setting. It provides a stronger initialization for RL fine-tuning (lines 9-14) by combining expert knowledge with the agent's own evolving state distribution. We finally return the policy (π_ϕ) and the world model $(f_\theta, p_\theta, q_\theta)$.

5   Experiments

This section outlines the experimental protocol used to evaluate the efficiency and reliability of our approach and the obtained results.

5.1   Flatland Environment

Flatland simulates a railway network on a $w \times h$ grid representing n cities. Each unblocked cell contains a rail type that specifies the allowed movement directions. We consider m trains $a_1, \ldots, a_m$, each with a starting cell, initial orientation, and destination (see Fig. 2). Time is discretized from 0 to $T_{\max}$, which depends on the environment size and the number of trains. The objective is to issue commands so that the maximum number of trains reach their destinations before $T_{\max}$, i.e., to maximize the success rate. Rewards are provided directly by the environment. At timestep 0, trains are outside the grid; each is inserted by issuing an entry command, after which they occupy exactly one cell per timestep until reaching their destination. At every step, a train chooses among five actions: *move forward, stop, turn left, turn right*, or *continue* (repeating the previous action). Two actions are in conflict if they cause a collision. Breakdowns interrupt a train at a random moment for a random duration. For each train, breakdown times follow a Poisson process with an unknown rate λ, and durations are drawn uniformly between 20 and 50 timesteps, becoming known only when the breakdown occurs.

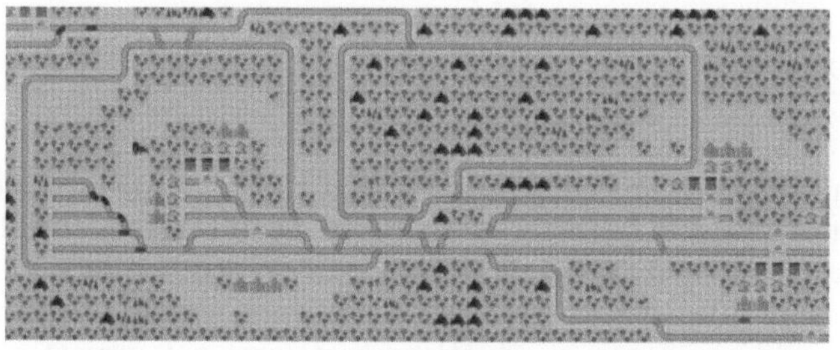

Fig. 2. Illustration of the Flatland Environment.

5.2   Protocol

We evaluate our approach against four baselines:

1. RANDOM, a policy that selects actions uniformly at random for each train.
2. GREEDY, a policy that chooses the action minimizing the immediate distance to the destination, ignoring other trains.
3. EXPERT, a policy that follows the expert solver described in Sect. 4.1.
4. MAMBA [6], the state-of-the-art approach based on a learning algorithm.

All methods are evaluated on three levels of difficulty: 5, 10, and 15 trains, as we restrict ourselves to instance sizes commonly used in the learning-based literature. Learning-based models were trained on a single Tesla V100-PCIE-32GB GPU for a maximum of 220 h.

All models were trained until convergence. For the training of our imitation-guided world model, the supervised learning phase was limited to 800K environment steps, corresponding to approximately 196 h (almost 90% of the total training time) due to the repeated calls to the expert solver, while the RL phase was carried out for 950K steps, taking around 24 h. We note that, unlike the vanilla MAMBA, which trains a separate policy for each environment configuration (5, 10, and 15 trains with breakdown ratios of 1/100, 1/150, and 1/200 respectively), our method requires only a single training session on the intermediate 10-train environment. We refer to MAMBA(10) for version trained only on the 10-train environments and to MAMBA($\star$) for the version trained on the specific scenario. We evaluate the performances using two metrics: $\%r$, the proportion of trains that reach their destination, and $\%d$, the proportion of trains that become deadlocked. We note that these metrics do not necessarily sum to 100%, as some trains may remain on the grid without arriving or becoming deadlocked, a situation that is less critical than deadlock. All results are averaged over 100 environment instances. All methods meet the 3-seconds per-step runtime constraint ($\sim 10^{-5}$s for classical baselines, $\sim 10^{-2}$s for learning-based methods).

5.3 Main Results

Performances with a Fixed Breakdown Ratio. The results are summarized in Table 1. First, we observe that the expert achieves near-perfect performance across all scenarios, which is expected given its access to global state information and centralized optimization. Although available at the planning level and during the training phase, such global knowledge cannot be leveraged for real-time operations at the level of individual trains or restricted areas (e.g., a station). This is why the core of our methodology lies in mimicking the expert's behavior, without prompting it at inference time. Random and greedy policies perform poorly, with low arrival rates and high deadlock percentages. Focusing now on learning-based approaches, our sequential method competes with MAMBA in the 5-train environment, and significantly outperforms it in the 10- and 15-train scenarios, with improvements of 14.7% and 23.7% in arrival rates, respectively. These gains come at the cost of a slight increase in deadlocks (3% for 10 trains and 3.1% for 15 trains), which remains within acceptable bounds. We note that deadlock avoidance is not included in the reward function, and the agent is therefore not explicitly trained to minimize this metric.

Table 1. Results of our method and the baselines (Top: non learning methods, Bottom: learning-based methods).

Policy	5 trains		10 trains		15 trains	
	$\%r$	$\%d$	$\%r$	$\%d$	$\%r$	$\%d$
RANDOM	25.0 ± 0.8	63.0 ± 0.8	11.8 ± 0.9	80.7 ± 0.9	8.2 ± 1.1	84.2 ± 1.1
GREEDY	18.4 ± 0.8	55.0 ± 0.8	10.7 ± 0.9	79.1 ± 0.9	8.5 ± 1.1	81.3 ± 1.1
EXPERT	$\mathbf{100.0 \pm 0.0}$	$\mathbf{0.0 \pm 0.0}$	$\mathbf{98.0 \pm 1.1}$	$\mathbf{1.8 \pm 1.0}$	$\mathbf{98.1 \pm 1.4}$	$\mathbf{1.5 \pm 1.1}$
MAMBA(10)	48.4 ± 0.6	0.4 ± 0.2	66.1 ± 1.9	$\mathbf{0.2 \pm 0.1}$	30.9 ± 1.7	$\mathbf{1.6 \pm 0.5}$
MAMBA($\star$)	97.2 ± 0.6	2.0 ± 0.5	66.1 ± 1.9	$\mathbf{0.2 \pm 0.1}$	47.6 ± 2.4	4.6 ± 1.7
Our approach	$\mathbf{98.2 \pm 0.3}$	$\mathbf{0.0 \pm 0}$	$\mathbf{80.8 \pm 2.2}$	3.2 ± 0.6	$\mathbf{71.3 \pm 3.4}$	7.7 ± 1.8

Performances When Varying the Breakdown Ratio. To assess generalization to stochastic disruptions, we also vary the train breakdown ratio between $1/10$ (i.e., λ from the Poisson law) and 0 for each difficulty level. For this experiment, our model is always the one trained on the 10-train environment with a breakdown ratio of $1/150$, while MAMBA is trained on the original breakdown ratios associated with the instances (5, 10, and 15 trains with breakdown ratios of $1/100$, $1/150$, and $1/200$ respectively). Results are reported in Table 2. Interestingly, our approach shows stable and consistent performance across configurations. While MAMBA performs best on the specific breakdown ratio it was trained on, its performance deteriorates on unseen ratios. In contrast, the performance of our approach tends to increase smoothly as breakdowns decrease, demonstrating better adaptability in practical scenarios where disruptions vary seasonally or across lines.

Main Take-Away from the Results. A key advantage of our method is its ability to transfer knowledge from the global solver to a decentralized local policy. Only a single training on the intermediary 10-train environment is sufficient to achieve good and consistent performance across all difficulty levels, whereas MAMBA requires separate training for each configuration, as it fails to generalize without retraining. Overall, these results show that our imitation-guided world model approach not only improves performance compared to standard model-based MARL, but also generalizes better to new configurations and breakdown ratios, while requiring fewer training resources, as it needs only a single 220 h training session, whereas MAMBA requires three separate runs of 220 h each (one per instance type), totaling 660 h.

5.4 Ablation Study

To better understand the contribution of each component of our method, we perform an ablation study. All results (still averaged over 100 scenarios and with the model trained on the 10-train environment) are reported in Table 3 and are commented in the following paragraphs.

Ablation 1: Degrading the Expert Quality. We investigate the importance of the expert used for imitation. To do so, we replace the solver expert with the random policy and the greedy policy, while keeping the rest of the methodology unchanged. As seen in Table 3, the sequential approach trained with a random or greedy expert still achieves decent performance, though it is lower than when using the solver. Interestingly, the random-expert model can still solve a substantial fraction of the simpler environments, suggesting that the imitation phase, even with weak guidance, can help the policy discover basic navigation and coordination patterns. The greedy-expert model performs better than the random one on smaller environments but deteriorates on larger ones, indicating that partial heuristic knowledge can be beneficial but insufficient to capture the complex interdependencies present in high-density traffic.

Table 2. Results of varying the breakdown ratio for 5, 10, and 15 trains.

| 5 trains (original environment is 1/100) | | | | | | | | | | | | | |
| Policy | 1/10 | | 1/50 | | 1/100 | | 1/500 | | 1/1000 | | 1/10000 | | 0 | |
	%r	%d	%r	%d	%r	%d	%r	%d	%r	%d	%r	%d	%r	%d
Mamba(10)	39.4	**0.0**	47.8	**0.0**	48.4	0.4	52.8	**0.0**	53.6	**0.0**	52.4	**0.0**	50.8	**0.0**
Mamba($\star$)	**94.8**	**0.0**	**99.2**	0.8	97.2	2.0	98.8	1.2	99.2	0.8	97.6	2.4	**99.2**	0.8
Our approach	50.6	0.4	96.6	**0.0**	**98.2**	**0.0**	**99.0**	0.4	**100.0**	**0.0**	**100.0**	**0.0**	**99.2**	0.4

| 10 trains (original environment is 1/150) | | | | | | | | | | | | | |
| Policy | 1/10 | | 1/50 | | 1/150 | | 1/500 | | 1/1000 | | 1/10000 | | 0 | |
	%r	%d	%r	%d	%r	%d	%r	%d	%r	%d	%r	%d	%r	%d
Mamba(10)	30.0	**0.2**	53.3	**0.0**	66.1	**0.2**	56.1	**0.2**	58.6	**0.0**	59.1	**0.0**	56.8	**0.2**
Mamba($\star$)	30.0	**0.2**	53.3	**0.0**	66.1	**0.2**	56.1	**0.2**	58.6	**0.0**	59.1	**0.0**	56.8	**0.2**
Our approach	**42.3**	1.4	**68.2**	2.4	**80.8**	3.2	**83.7**	2.8	**78.9**	3.4	**84.7**	2.6	**82.7**	2.4

| 15 trains (original environment is 1/200) | | | | | | | | | | | | | |
| Policy | 1/10 | | 1/50 | | 1/200 | | 1/500 | | 1/1000 | | 1/10000 | | 0 | |
	%r	%d	%r	%d	%r	%d	%r	%d	%r	%d	%r	%d	%r	%d
Mamba(10)	19.0	2.9	30.1	**0.7**	30.9	**1.6**	35.7	**0.7**	34.5	**1.3**	35.3	**1.3**	36.0	**2.3**
Mamba($\star$)	30.9	**2.7**	39.3	6.2	47.6	4.6	48.8	5.0	49.5	6.3	48.3	6.0	47.3	7.4
Our approach	**31.8**	5.1	**52.7**	5.8	**71.3**	7.7	**72.3**	8.0	**71.0**	12.1	**77.7**	7.8	**70.8**	9.8

Ablation 2: Disabling the World Model. Next, we evaluate the role of the world model by re-training the agent without it. Without a learned dynamics model, the agent is unable to capture temporal dependencies or plan ahead, resulting in poor results (see Table 3) on the training configuration (10 agents) and a complete failure to generalize to smaller (5 agents) or larger (15 agents) scenarios. These results underline the crucial role of the world model in enabling generalization beyond the training distribution. This confirms that the world model is a critical component for learning in complex multi-agent environments such as Flatland. The result highlights that local observation alone, without the latent predictive model, does not provide sufficient information for coordination, especially when trains must anticipate interactions many steps in advance.

Ablation 3: Disabling RL Finetuning. We then assess the necessity of the RL fine-tuning stage by training a policy purely with supervised imitation for the same budget as our full approach. Pure imitation alone yields very poor performance, particularly in larger environments with many agents (see Table 3). This demonstrates that expert-guided initialization is insufficient to handle the stochasticity of the environment, the variability of train breakdowns, and the emergent interactions between agents. It further emphasizes that RL is essential for adapting the policy to situations not captured in the expert trajectories.

Take-Away from the Ablation Study. Overall, the experiments confirm that each component (expert demonstrations, predictive world model, and RL) plays a complementary role. The combination enables efficient policy learning and robust coordination under uncertainty for multi-agent routing problems in Flatland.

Table 3. Results of the ablation study.

Policy	5 trains		10 trains		15 trains	
	%r	Δ	%r	Δ	%r	Δ
Our approach	**98.2**	-	**80.8**	-	**71.3**	-
With RANDOM	75.4	-22.8	76.6	-3.4	58.1	-13.2
With GREEDY	92.0	-6.2	65.2	-14.8	43.7	-27.6
No world model	0.0	-98.2	30.0	-50.8	0.0	-71.3
Pure imitation	2.8	-95.4	0.4	-80.4	0.0	-71.3

6   Conclusion and Future Work

In this work, we proposed an imitation-guided world model to tackle a train rescheduling problem, using the Flatland environment, a simplified simulation of railway networks. First, we leverage imitation learning to mimic the behavior of an expert solver specifically designed for this task. This solver has access to a global view of the environment, which can be used during a training phase but is impractical for real-time operations. The second phase employs model-based RL, allowing each train to make decisions based solely on its local observations, reflecting realistic operational constraints. Our approach demonstrates superior performance compared to baseline methods, requires significantly fewer computational resources for training and exhibits robust generalization to previously unseen breakdown ratios. As future work, we plan to extend beyond the Flatland environment and address more realistic settings and constraints, such as strict arrival time windows, and priority rules reflecting passengers or cargo importance. Constrained reinforcement learning [35], which allows soft penalization of constraint non-adherence, appears to be a promising direction in this regard.

Acknowledgments. This work was supported by Mitacs Grants IT29211 and IT35069. The authors gratefully acknowledge the discussions with SNCF researchers.

References

1. Barcellona, L., Zadaianchuk, A., Allegro, D., Papa, S., Ghidoni, S., Gavves, E.: Dream to manipulate: Compositional world models empowering robot imitation learning with imagination. In: The Thirteenth International Conference on Learning Representations (2025). https://openreview.net/forum?id=3RSLW9YSgk
2. Bellman, R.: Dynamic programming. Science **153**(3731), 34–37 (1966)
3. Cappart, Q., Schaus, P.: Rescheduling Railway Traffic on Real Time Situations Using Time-Interval Variables. In: Salvagnin, D., Lombardi, M. (eds.) CPAIOR 2017. LNCS, vol. 10335, pp. 312–327. Springer, Cham (2017). https://doi.org/10.1007/978-3-319-59776-8_26
4. Cobbe, K., Klimov, O., Hesse, C., Kim, T., Schulman, J.: Quantifying generalization in reinforcement learning. In: International Conference on Machine Learning, pp. 1282–1289. PMLR (2019)
5. DeMoss, B., Duckworth, P., Hawes, N., Posner, I.: Ditto: Offline imitation learning with world models. arXiv preprint arXiv:2302.03086 (2023)
6. Egorov, V., Shpilman, A.: Scalable multi-agent model-based reinforcement learning. In: Proceedings of the 21st International Conference on Autonomous Agents and MultiAgent Systems (AAMAS 2022), pp. 381–389. International Foundation for Autonomous Agents and Multiagent Systems (IFAAMAS) (2022). https://arxiv.org/abs/2205.15023, online 9–13 May 2022
7. Foerster, J., Farquhar, G., Afouras, T., Nardelli, N., Whiteson, S.: Counterfactual multi-agent policy gradients. In: Proceedings of the Thirty-Second AAAI Conference on Artificial Intelligence (AAAI 2018), pp. 2974–2982 (2018), https://arxiv.org/abs/1705.08926
8. Hafner, D., et al.: Learning latent dynamics for planning from pixels. In: Proceedings of the 36th International Conference on Machine Learning (ICML), pp. 2555–2565 (2019). https://arxiv.org/abs/1811.04551
9. Hafner, D., Lillicrap, T.P., Ba, J., Norouzi, M.: Dream to control: Learning behaviors by latent imagination. In: International Conference on Learning Representations (ICLR) (2020). https://arxiv.org/abs/1912.01603, spotlight
10. Hafner, D., Lillicrap, T.P., Norouzi, M., Ba, J.: Mastering atari with discrete world models. In: International Conference on Learning Representations (ICLR) (2021). https://openreview.net/forum?id=5r6uxv1h5V, accepted as spotlight
11. Hernandez-Leal, P., Kartal, B., Taylor, M.E.: A survey and critique of multiagent deep reinforcement learning. Auton. Agent. Multi-Agent Syst. **33**(6), 750–797 (2019)
12. Ho, F., Goncalves, A., Salta, A., Cavazza, M., Geraldes, R., Prendinger, H.: Multiagent path finding for UAV traffic management: Robotics track, pp. 131–139 (2019)
13. Ho, J., Ermon, S.: Generative adversarial imitation learning. Adv. Neural. Inf. Process. Syst. **29** (2016)
14. Huh, D., Mohapatra, P.: Multi-agent reinforcement learning: A comprehensive survey, arXiv preprint arXiv:2312.10256 (2023)
15. Hussein, A., Gaber, M.M., Elyan, E., Jayne, C.: Imitation learning: a survey of learning methods. ACM Comput. Surv. (CSUR) **50**(2), 1–35 (2017)
16. Hönig, W., Preiss, J.A., Kumar, T.K.S., Sukhatme, G.S., Ayanian, N.: Trajectory planning for quadrotor swarms. IEEE Trans. Rob. **34**(4), 856–869 (2018). https://doi.org/10.1109/TRO.2018.2853613
17. Jaziri, A., Künzel, E., Ramesh, V.: Mitigating the stability-plasticity dilemma in adaptive train scheduling with curriculum-driven continual dqn expansion. CoRR abs/2408.09838 (2024). https://doi.org/10.48550/arXiv.2408.09838
18. Kinga, D., Adam, J.B.: A method for stochastic optimization. In: International Conference on Learning Representations (ICLR), vol. 5, California (2015)

19. Kirkpatrick, S., Gelatt, C., Vecchi, M.: Optimization by simulated annealing. Science **220**(4598), 671–680 (1983). https://doi.org/10.1126/science.220.4598.671

20. Konda, V., Tsitsiklis, J.: Actor-critic algorithms. Adv. Neural. Inf. Process. Syst. **12** (1999)

21. Li, J., et al.: Scalable rail planning and replanning: Winning the 2020 flatland challenge. Proc. Int. Conf. Autom. Plan. Sched. **31**(1), 477–485 (2021). https://doi.org/10.1609/icaps.v31i1.15994, https://ojs.aaai.org/index.php/ICAPS/article/view/15994

22. Li, J., Tinka, A., Kiesel, S., Durham, J.W., Kumar, T.S., Koenig, S.: Lifelong multi-agent path finding in large-scale warehouses, pp. 11272–11281 (2021)

23. Liao, J., Zhang, F., Zhang, S., Gong, C.: A real-time train timetable rescheduling method based on deep learning for metro systems energy optimization under random disturbances. J. Adv. Transp. **2020**(1), 8882554 (2020). https://doi.org/10.1155/2020/8882554, https://onlinelibrary.wiley.com/doi/abs/10.1155/2020/8882554

24. Lillicrap, T.P., et al.: Continuous control with deep reinforcement learning, arXiv preprint arXiv:1509.02971 (2015)

25. Liu, S., Zhang, Y., Tang, K., Yao, X.: How good is neural combinatorial optimization? a systematic evaluation on the traveling salesman problem. IEEE Comput. Intell. Mag. **18**(3), 14–28 (2023)

26. Liu, X., da Silva, C.F.O., Dabiri, A., Wang, Y., De Schutter, B.: Learning-based model predictive control for passenger-oriented train rescheduling with flexible train composition. arXiv preprint arXiv:2502.15544 (2025)

27. Lowe, R., Wu, Y.I., Tamar, A., Harb, J., Pieter Abbeel, O., Mordatch, I.: Multi-agent actor-critic for mixed cooperative-competitive environments. Adv. Neural Inf. Process. Syst. 30 (2017)

28. Ma, H., Li, J., Kumar, T.S., Koenig, S.: Lifelong multi-agent path finding for online pickup and delivery tasks, pp. 837–845 (2017)

29. Mnih, V., et al.: Human-level control through deep reinforcement learning. Nature **518**(7540), 529–533 (2015)

30. Mohanty, S., et al.: Flatland-rl: Multi-agent reinforcement learning on trains (2020)

31. Oh, J., Guo, Y., Singh, S., Lee, H.: Self-imitation learning. In: Proceedings of the 35th International Conference on Machine Learning (ICML 2018), pp. 3878–3887. PMLR (2018). https://proceedings.mlr.press/v80/oh18b.html

32. Peng, X.B., Kumar, A., Zhang, G., Levine, S.: Advantage-weighted regression: Simple and scalable off-policy reinforcement learning, arXiv preprint arXiv:1910.00177 (2019)

33. Phillips, M., Likhachev, M.: SIPP: Safe interval path planning for dynamic environments. In: 2011 IEEE International Conference on Robotics and Automation, pp. 5628–5635. IEEE (2011)

34. Ross, S., Gordon, G., Bagnell, D.: A reduction of imitation learning and structured prediction to no-regret online learning. In: Proceedings of the Fourteenth International Conference on Artificial Intelligence and Statistics, pp. 627–635. JMLR Workshop and Conference Proceedings (2011)

35. Roy, J., Girgis, R., Romoff, J., Bacon, P.L., Pal, C.J.: Direct behavior specification via constrained reinforcement learning. In: International Conference on Machine Learning, pp. 18828–18843. PMLR (2022)

36. Schulman, J., Wolski, F., Dhariwal, P., Radford, A., Klimov, O.: Proximal policy optimization algorithms. arXiv preprint arXiv:1707.06347 (2017). https://arxiv.org/abs/1707.06347

37. Sharon, G., Stern, R., Felner, A., Sturtevant, N.: Conflict-based search for optimal multi-agent path finding. In: Proceedings of the Twenty-Sixth AAAI Conference on Artificial Intelligence. pp. 563–569. AAAI'12, AAAI Press (2012)
38. Shaw, P.: Using constraint programming and local search methods to solve vehicle routing problems. In: Proceedings of the 4th International Conference on Principles and Practice of Constraint Programming (CP'98), pp. 417–431. Springer (1998). https://doi.org/10.1007/BFb0054171
39. Shi, Z., Liu, M., Zhang, S., Zheng, R., Dong, S., Wei, P.: Gawm: Global-aware world model for multi-agent reinforcement learning. arXiv preprint arXiv:2501.10116 (2025). https://arxiv.org/abs/2501.10116
40. Song, J., Ren, H., Sadigh, D., Ermon, S.: Multi-agent generative adversarial imitation learning. In: Advances in Neural Information Processing Systems, p. 31 (2018)
41. Stern, R., et al.: Multi-agent pathfinding: Definitions, variants, and benchmarks. In: Proceedings of the International Symposium on Combinatorial Search, vol. 10, pp. 151–158 (2019)
42. Surynek, P.: Problem compilation for multi-agent path finding: a survey. In: Raedt, L.D. (ed.) Proceedings of the Thirty-First International Joint Conference on Artificial Intelligence, IJCAI-22, pp. 5615–5622. International Joint Conferences on Artificial Intelligence Organization (2022). https://doi.org/10.24963/ijcai.2022/783, survey Track
43. Sutton, R.S., Barto, A.G.: Reinforcement Learning: An Introduction, vol. 1, MIT press Cambridge (1998)
44. Sutton, R.S., McAllester, D., Singh, S., Mansour, Y.: Policy gradient methods for reinforcement learning with function approximation. Adv. Neural Inf. Process. Syst. **12** (1999)
45. Talvitie, E.: Self-correcting models for model-based reinforcement learning. In: Proceedings of the AAAI conference on artificial intelligence, vol. 31 (2017)
46. Tang, J., Swamy, G., Fang, F., Wu, S.: Multi-agent imitation learning: Value is easy, regret is hard. In: The Thirty-eighth Annual Conference on Neural Information Processing Systems (2024)
47. Vaswani, A., et al.: Attention is all you need. In: Advances in Neural Information Processing Systems (NeurIPS), pp. 5998–6008 (2017). https://doi.org/10.5555/3295222.3295349, https://arxiv.org/abs/1706.03762
48. Venugopal, A., Milani, S., Fang, F., Ravindran, B.: Mabl: Bi-level latent-variable world model for sample-efficient multi-agent reinforcement learning. In: Proceedings of the 23rd International Conference on Autonomous Agents and MultiAgent Systems (AAMAS 2024), pp. 1865–1873. International Foundation for Autonomous Agents and Multiagent Systems (IFAAMAS) (2024). https://www.ifaamas.org/Proceedings/aamas2024/pdfs/p1865.pdf
49. Wang, H., Yu, L., Cao, Z., Ermon, S.: Multi-agent Imitation Learning with Copulas. In: Oliver, N., P erez-Cruz, F., Kramer, S., Read, J., Lozano, J.A. (eds.) ECML PKDD 2021. LNCS (LNAI), vol. 12975, pp. 139–156. Springer, Cham (2021).
50. Watkins, C.J., Dayan, P.: Q-learning. Mach. Learn. **8**(3), 279–292 (1992)
51. Williams, R.J.: Simple statistical gradient-following algorithms for connectionist reinforcement learning. Mach. Learn. **8**(3), 229–256 (1992)
52. Wu, Z., Yu, C., Chen, C., Hao, J., Zhuo, H.H.: Models as agents: optimizing multi-step predictions of interactive local models in model-based multi-agent reinforcement learning. In: Proceedings of the 37th AAAI Conference on Artificial Intelligence (AAAI 2023). pp. 26241–26249. AAAI Press (2023). https://ojs.aaai.org/index.php/AAAI/article/view/26241

53. Zhang, Y., Deekshith, U., Wang, J., Boedecker, J.: Improving the efficiency and efficacy of multi-agent reinforcement learning on complex railway networks with a local-critic approach. In: Proceedings of the International Conference on Automated Planning and Scheduling, Vol. 34, no. (1), pp. 698–706 (2024). https://doi.org/10.1609/icaps.v34i1.31533, https://ojs.aaai.org/index.php/ICAPS/article/view/31533
54. Zheng, R., et al.: FLARE: Robot learning with implicit world modeling. In: Structured World Models for Robotic Manipulation (2025). https://openreview.net/forum?id=rFiLBM4YCh
55. Zhu, Y., Wang, H., Goverde, R.M.: Reinforcement learning in railway timetable rescheduling. In: 2020 IEEE 23rd International Conference on Intelligent Transportation Systems (ITSC), pp. 1–6. IEEE Press (2020). https://doi.org/10.1109/ITSC45102.2020.9294188

Clustering for Relaxed and Restricted Decision Diagram Bounds: When It Works and Why

Alice Burlats$^{(\boxtimes)}$, Roger Kameugne , Cristel Pelsser , and Pierre Schaus

UCLouvain/ICTEAM, Louvain-la-Neuve, Belgium
`alice.burlats@uclouvain.be`

Abstract. Decision-Diagram-based Branch-and-Bound solves discrete optimization problems by exploiting bounds provided by two types of bounded-width decision diagram. The first is the *restricted decision diagram* obtained by discarding less promising states that provide primal bounds. The second is the *relaxed decision diagram* obtained by state merging that yields a dual bound. Their performance depends heavily on the heuristic used to discard or merge nodes. While traditional methods discard or merge nodes based on the cost, recent research suggests that clustering nodes based on state similarity (e.g., via k-means) can help produce tighter bounds. However, current clustering methods are difficult to apply to complex, non-vector states. We propose to use a more general clustering framework that accepts user-defined distance metrics, allowing it to be applied to any state definition and scales to very large state spaces. We test this approach against standard cost-based strategies on three distinct problems exhibiting different merge-function properties. We additionally introduce a definition framework that characterizes these properties. Our results show that, counter-intuitively, sophisticated clustering does not always pay off, especially when the merge operator produces states that differ greatly from the originals. We provide a detailed analysis explaining when clustering is beneficial versus when simple cost-based strategies suffice, offering some guidelines for solver configuration.

Keywords: Decision Diagrams · Clustering · Branch-and-Bound · Generalized Hyperplan Partitioning · Discrete Optimization

1 Introduction

An elegant optimization framework called DDO, introduced in [4], solves discrete optimization problems formulated as dynamic programs (DPs) through a branch-and-bound (B&B) search enhanced with decision diagrams (DDs). At every explored node, DDO compiles a DD to obtain both primal and dual bounds. To prevent the combinatorial explosion inherent to exact DDs, their width is bounded, producing limited DDs created by dropping or merging nodes according to heuristics. Primal bounds come from restricted DDs [5], where

T. Guns (Ed.): CPAIOR 2026, LNCS 16595, pp. 101–118, 2026.
https://doi.org/10.1007/978-3-032-27242-3_7

nodes in oversized layers are dropped, yielding a diagram that contains only a subset of the feasible solutions of the exact DD. Dual bounds are obtained by merging nodes via a problem-specific operator, which may introduce infeasible paths [1,3,7,9,26]. The effectiveness of DDO strongly depends on the quality of these bounds and the efficiency with which they are computed. The node selection heuristic for merging or discarding is typically based on cost[1] as originally proposed in [4] and reused in [9,12,17,18]. This *Cost* based strategy is both cheap to compute and good at preserving promising paths from being discarded or merged. Recently, Nafar et al. [23] reconsidered this default choice and showed that clustering the nodes with k-means generally helps to produce tighter primal (resp. dual) bounds for restricted (resp. relaxed) DDs than *Cost* when applied at the root state. This paper is a follow-up work building on the *Clustering* strategy and aims to answer two questions:

- The approach in [23] assumes that states admit a Euclidean embedding (as in the multi-knapsack problem where the states are fixed-size vectors of remaining capacities). Some problems have states with a more complex structure (such as for the Maximum Coverage problem, where the state is naturally defined as a set of varying cardinality). The first question is thus whether the clustering approach of [23] can be generalized to richer state representations than Euclidean spaces?
- The analysis in [23] only focuses on root DDs. But DDs compiled deeper in the search space may exhibit different structures. A second question is whether the *Clustering* strategy consistently outperform the *Cost* strategy throughout the entire B&B search?

We answer the first question positively by introducing a new clustering method for DD compilation, based on Generalized Hyperplane Partitioning [28] (GHP). It supports clustering using a problem-specific state distance without requiring a Euclidean vector representation and also greatly reduces the computation time over the k-means clustering. If GHP is not a novel algorithm, to our knowledge, it is the first time that it is applied to DDs.

To address the second question, we evaluate the method on three problems with distinct state structures: the knapsack, the multidimensional knapsack, and the maximum coverage problems. Results show that the benefit of *Clustering* over *Cost* strategy is highly problem-dependent; we provide guidelines indicating when it is likely to help based on the state and merge operator structure.

The paper is structured as follows. Section 2 introduces the optimization problems and their DD formulations. Section 3 reviews the related work. Section 4 presents the generalized hyperplane partitioning approach for DD layer clustering. Section 5 reports experimental results and evaluates the different strategies. We finally conclude in Sect. 6.

---

[1] The total cost of the lightest path from the root to the node.

2   Background

A discrete optimization problem $\mathcal{P}$ is defined by a vector of variables $x = \langle x_0, x_1, \ldots, x_{n-1} \rangle$ where each variable x_j takes its value in a finite domain D_j. Let $\mathcal{D} = D_0 \times D_1 \times \ldots \times D_{n-1}$ denote the space of complete assignments. The problem is subject to a collection of constraints $\mathcal{C} = \{C_1, \ldots, C_m\}$, where each constraint C_i is defined by a pair (S_i, R_i) with $S_i \subseteq \{0, \ldots, n-1\}$ a subset of variable indices and $R_i \subseteq \prod_{j \in S_i} D_j$ a relation specifying the allowed tuples of values for the variables in S_i. An assignment $x \in \mathcal{D}$ satisfies constraint C_i if the projection of x on S_i belongs to R_i. A feasible solution of $\mathcal{P}$ is any assignment $x \in \mathcal{D}$ that satisfies all constraints in $\mathcal{C}$. Let $Sol(\mathcal{P}) \subseteq \mathcal{D}$ denote the set of feasible solutions. The problem also includes an objective function $f : \mathcal{D} \to \mathbb{R}$. The goal of the problem is to find a feasible assignment $x \in Sol(\mathcal{P})$ that optimizes the objective function. Here, we assume minimization without loss of generality.

Based on the divide-and-conquer strategy, dynamic programming (DP) was introduced in [2] and used to solve discrete optimization problems. The problem is decomposed into small and overlapping subproblems solved recursively. To avoid multiple resolutions of the same subproblem, the intermediate results are stored in a cache. A discrete optimization problem $\mathcal{P}$ can be formulated in the DP formalism as a labeled transition system. It consists of:

- The control variables $x_j \in D_j$ used to specify the decision taken in the domain D_j with $j = 0, \ldots, n-1$ and n the number of variables of the problem.
- The state space $\mathcal{S}$ contains partial assignments of the variables x. This space is divided in $n+1$ subset $S_0, S_1, \ldots, S_n$, where S_j contains the states where exactly j variables among x are assigned. It also contains three special states: the *root* state $\hat{r}$, the *terminal* state $\hat{t}$ and the *infeasible* state $\hat{0}$.
- The *transition function* $t : S_j \times D_j \to S_{j+1}$ that maps each state of S_j and a decision in D_j to the corresponding state in S_{j+1}.
- The *transition value function* $h : S_j \times D_j \to \mathbb{R}$ that affects a value to each transition.

The optimization problem $\mathcal{P}$ can then be defined by:

$$\text{minimize } f(x) = \sum_{j=0}^{n-1} h(s^j, x_j) \tag{1}$$

$$\text{subject to } s^{j+1} = t(s^j, x_j), \forall j = 0, \ldots, n-1, x_j \in D_j \tag{2}$$

$$x \in \mathcal{C}, s^j \in S_j, j = 0, \ldots, n \tag{3}$$

2.1   Decision Diagrams-Based Optimization

Solving an NP-hard optimization problem formulated as a labeled transition system using only caching is likely to fail on difficult instances, because the cache may grow exponentially and no bounding mechanism is available to prune states. The B&B framework based on decision diagrams proposed in [4] answers these two issues. In this framework, the search space of the labeled transition system is explored using a classical B&B scheme in which the transition function serves as the node generator. At each node, both primal and dual bounds are computed in a generic way by exploiting the structure of the labeled transition system. These bounds are obtained by compiling it into two memory-bounded Decision Diagrams (DD): the Restricted DD and the Relaxed DD. The first one aims at finding primal bounds, while the second aims at finding a dual bound for the node. A search node is then fathomed whenever the dual bound is worse than the best so far incumbent. For the sake of completeness, we briefly recall the main ingredients on DD compilation from [4], adopting the same notation. A DD is a layered directed acyclic graph $\mathcal{B} = (N, A, \sigma, l, v)$ where N is the set of nodes interconnected by the set of arcs A. Each node u is mapped to a state $\sigma(u)$ by the function σ. The set of nodes N are partitioned as a collection of layers $L = \{L_0, L_1, \ldots, L_n\}$, where each layer usually corresponds to a decision on a variable (n is the number of decision variables in the DP model). The set of arcs A represents the transitions between consecutive states and an arc a is labeled by the function l which maps each arc a to a decision d i.e., $l(a) = d$ and a value function $v(a)$ assigns a cost to the corresponding transition. The layer L_0 contains only the node corresponding to the root state of the DP model, where L_n contains only the terminal nodes. The DD is compiled top-down, layer by layer, starting with L_0.

Algorithm 1: Compilation of a DD rooted at $u_{\hat{r}}$ with a W width limit

---

1   $i \leftarrow 0$
2   $L_i \leftarrow \{u_{\hat{r}}\}$
3   **for** $j = i$ *to* $n - 1$ **do**
4   **if** $|L_j| > W$ **then**
5 | Restriction or relaxation of layer L_j with W
6      $L_{j+1} \leftarrow \emptyset$
7   **forall the** $u \in L_i$ **do**
8   **forall the** $d \in D_j$ **do**
9   create node u' with $\sigma(u') = t_j(\sigma(u), d)$ or retrieve it from L_{j+1}
10   create arc $a = (u \xrightarrow{d} u')$ with $v(a) = h_j(\sigma(u), d)$ and $l(a) = d$
11   add u' to L_{j+1} and add a to A
12   merge nodes in L_n into terminal node $u_{\hat{t}}$

Because the width limit W restricts the layer sizes, the minimum-cost path in the resulting layered acyclic graph does not necessarily correspond to the exact optimal solution to the problem. The restricted strategy is used to obtain a primal bound, while the relaxed strategy is used to obtain a dual bound.

The Restricted DD compilation simply removes nodes from the current layer together with their incoming arcs. If a root-to-terminal path remains in the restricted DD, it can be used to compute a primal bound.

The Relaxed DD compilation limits the width of a layer L_j by merging nodes to create a *relaxed state* using a problem specific binary *merge operator* $\oplus : S \times S \to S$. The application of the operator also redirects all the arcs to the new node. This operator should be such that no solution is removed (relaxation), although some infeasible solution might be introduced. Therefore, the shortest path on the relaxed decision diagram allows one to obtain a dual bound on the optimal cost. The infeasible state is absorbing for the merge operator, i.e., $s \oplus \hat{0} = \hat{0} \; \forall s \in \mathcal{S}$.

Cost-Based Heuristic. The standard way to select the nodes that should be merged/dropped (introduced in [4], then also used in [9,12,17,18]) is based on their costs. The $|L_j| - W$ nodes with the worst costs are dropped or merged. This heuristic has the advantages of being quick to compute and tending to preserve the best paths. But merging so many nodes as a single state may introduce a state that is quite different and much more relaxed from the others of the layer. Also, being a very greedy strategy, it reduces diversity among the states kept, which could be detrimental for the subsequent decisions at deeper layers.

Clustering-Based Heuristic. To address the limitations of the cost-based heuristic, [23] experimented with clustering the nodes based on their state similarity. In a relaxed DD, for each cluster, all its nodes are merged together, while for restricted DD, the best node of each cluster is selected based on its cost and the other ones are dropped. The clustering algorithm used in [23] is *k-means*. The drawback of this approach is that *k-means* requires the state to be of fixed numerical dimensions in a Euclidean space.

In the rest of this paper, we denote by *DDO Model* the pair composed of the DP formulation of a problem and a merge operator.

2.2   DDO Models Characterization

We attempt to classify the DDO model of optimization problems. This characterization depends on the *nontrivial states* and the merge operator of the model.

Definition 1. *A nontrivial state s is a state different from the root state, the terminal state, and the infeasible state, i.e., $s \notin \{\hat{r}, \hat{t}, \hat{0}\}$.*

Definition 2. *Let $\mathcal{M}$ be a DDO model, $\mathcal{S}$ and $\oplus$ be the state space and the merge operator associated to $\mathcal{M}$. A nontrivial state $s \in \mathcal{S}$ is said to be strongly preserved by $\oplus$ if there is no nontrivial state $s' \in \mathcal{S}$ such that $s \oplus s' \in \{\hat{r}, \hat{t}\}$. Conversely, a state s is said to be weakly preserved by $\oplus$ if it is not strong, i.e., there exist a state $s' \in \mathcal{S}$ such as $s \oplus s' \in \{\hat{r}, \hat{t}\}$*

In this definition, the frequency with which the merge operator will bring the merged states back to trivial states can be estimated for each model. These frequencies can be grouped into three broad categories: never, often, and always.

Definition 3. *A DDO model of an optimization problem is said to be*

- state-preserving *if all its nontrivial states are strongly preserved by its merge operator, or*
- slightly state-preserving *if some of its nontrivial states are strongly preserved by its merge operator while others are weakly preserved, or*
- state-altering *if all its nontrivial states are weakly preserved by its merge operator.*

2.3   Problems of Interest

We formalize DP models for three different discrete optimization problems selected for covering the three situations of Definition 3.

The Knapsack Problem (KP) is defined by the given of n items, each item j has a value v_j and a weight w_j. The goal is to select a subset of these to maximize the total value without exceeding a capacity C. In DP formulation of KP, the state is a positive integer that represents the remaining capacity:

- $x_j \in \{0,1\}$ with $j \in \{0,\ldots,n-1\}$ is set to 1 when the jth item is taken $(s \geq w_j)$ and 0 otherwise.
- $S_j = \{s \mid 0 \leq s \leq C\}$ where s is a positive integer representing the remaining capacity. The root state $\hat{r} = C$ is the state of initial capacity and the terminal state is any state $\hat{t} = c$ with $c < \min\{w_i \mid 0 \leq i < n\}$.
- If the remaining capacity of a state s does not allow including the jth item, the transition is redirected to infeasible state $\hat{0}$.

$$t(s, x_j) = \begin{cases} s - w_j \cdot x_j & \text{if } s \geq w_j \cdot x_j \\ \hat{0} & \text{otherwise.} \end{cases}$$

- $h(s, x_j) = -v_j \cdot x_j$ is the value added to the objective for each transition.

The operator used to merge states of this model is the maximum remaining capacity of states, i.e., $s \oplus s' = \max(s, s')$. This model of KP is *state-preserving* because $s \oplus s' = \hat{r} \iff s = \hat{r}$ or $s' = \hat{r}$.

The Multidimensional Knapsack Problem (MKP) is a generalization of the KP to multiple capacity constraints: n items and m dimensions of the knapsack are given, each dimension with capacity bound $(C_1, \ldots, C_m)$. An item j occupies a weight dimension w_j^i in the knapsack for $i \in \{1, \ldots, m\}$ for a value of v_j. The goal is to select a subset of items whose sum of profits is maximized so that all the capacity bound constraints hold simultaneously. The state in the MKP is an m-tuple $(c_1, \ldots, c_m)$ of positive integers that represents the remaining capacity of each dimension of the knapsack.

- $x_j \in \{0,1\}$ with $j \in \{0,\ldots,n-1\}$ is set to 1 when the jth item is taken $(s_i \geq w_j^i$ for all $i \in \{1,\ldots,m\})$ and 0 otherwise.

- $S_j = \{(s_1,\ldots,s_m) \mid 0 \le s_i \le C_i \text{ with } i = 1,\ldots,m\}$ where s_i is a positive integer representing the remaining capacity of the ith dimension of the knapsack where $i \in \{1,\ldots,m\}$. The root state is $\hat{r} = (C_1,\ldots,C_m)$ and the terminal state is any state $\hat{t} = (c_1,\ldots,c_m)$ with $c_i < \min\{w_j^i \mid 0 \le j < n\}$ and $i \in \{1,\ldots,m\}$.
- If the remaining capacity of at least one dimension s_i with $i \in \{1,\ldots,m\}$ does not allow inclusion of the jth item, the transition is redirected to infeasible state $\hat{0}$.

$$t(s, x_j) = \begin{cases} (s_1 - w_j^1 \cdot x_j,\ldots,s_m - w_j^m \cdot x_j) & \text{if } s_i \ge w_j^i \cdot x_j \forall i \in \{1,\ldots,m\} \\ \hat{0} & \text{otherwise.} \end{cases}$$

- $h(s, x_j) = -v_j \cdot x_j$ is the value added to the objective for each transition.

In this model, the operator used to merge states is the maximum remaining capacity for each dimension, i.e., $s \oplus s' = (\max(s_1, s_1'),\ldots,\max(s_m, s_m'))$. Without loss of generality, we consider only two dimensions of capacity C_1 and C_2. A nontrivial state (c_1, c_2) with $c_1 < C_1$ and $c_2 < C_2$ is *strongly preserved* by $\oplus$. However, any nontrivial state of the form (c_1, C_2) or (C_1, c_2) where $c_1 < C_1$ or $c_2 < C_2$, is *weakly preserved* since $(c_1, C_2) \oplus (C_1, c') = (C_1, C_2)$ or $(C_1, c_2) \oplus (c', C_2) = (C_1, C_2)$ for all $c' < C_2$ or $c' < C_1$. Thus, the DDO model of MKP proposed in the paper is *slightly state-preserving*.

The Maximum Coverage Problem (MCP) is defined by giving a finite set U of n elements, a finite collection of m subsets $V = \{V_1,\ldots,V_m\}$ of U, and an integer $1 \le k \le m$. The goal is to find a sub-collection $C \subset V$ such that $|C| = k$ that maximizes the total number of elements covered by this subset. In DP formulation, the states are simply the set of covered elements and the DD has exactly $k+1$ layers:

- $x_j \in \{1,\ldots,m\}$ where x_j is the index of a subset that covers at least one new element not present in the current state s, i.e., $V_{x_j} \setminus s \neq \emptyset$.
- $S_j = 2^U$ for $j = 0,\ldots,k$, $\hat{r} = \emptyset$ and $\hat{t}$ is any state union of k subsets of V.
- $t(s, x_j) = s \cup V_{x_j}$.
- $h(s, x_j) = -|V_{x_j} \setminus s|$, that is the new added elements to the state.

In this model states are merged by computing the intersection of their sets of covered elements, i.e., $s \oplus s' = s \cap s'$. For every nontrivial state s there exists a state s' such as $s \cap s' = \emptyset = \hat{r}$, hence this model is *state-altering*.

MCP can also be modeled as a BDD, where each layer corresponds to the decision of selecting or discarding a particular set, containing exactly $m+1$ layers. However, early results showed that the MDD formulation yields tighter bounds despite its larger branching factor. In practice, $k \ll m$, meaning that the BDD is significantly deeper than the MDD. Hence, paths in the BDD tend to accumulate more relaxation error, making them more susceptible to bound deterioration.

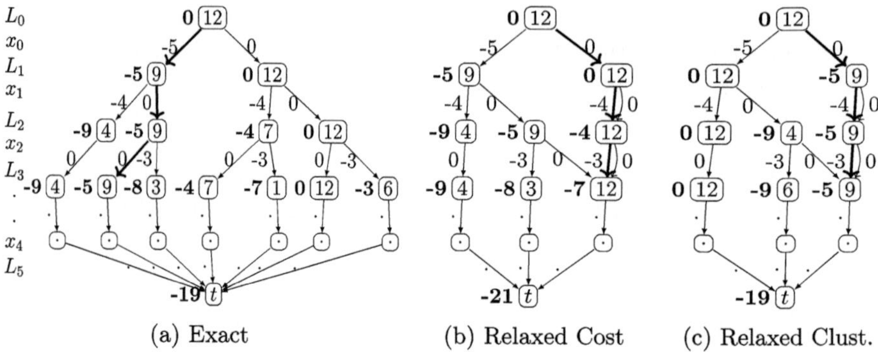

Fig. 1. Decision diagrams of the KP described in the Example 1. For the relaxed DDs $W = 3$.

Example 1. Consider an instance of the KP with $n = 5$, $C = 12$, $v = (5, 4, 3, 6, 8)$ and $w = (3, 5, 6, 4, 5)$. The first four layers of exact DD (Fig. 1a) and relaxed cost DD (Fig. 1b) for a maximum width of 3 are shown in Fig. 1. Each node represents a state, with transition costs assigned to the arcs. The bold number near a node indicates the path length from the root to that node, while the shortest path is highlighted by bold arrows. The value at a node and the transition cost are negative, since the KP is a maximization problem and we model it as a minimization one. Layer 2 of the exact DD ($L_2 = \{4, 9, 7, 12\}$) of Fig. 1a is sorted according to the node cost. For a maximum width of 3, nodes 7 and 12 are merged, and the resulting upper bound is 21. The clusters obtained in layer 2 ($L_2 = \{4, 9, 7, 12\}$) of the exact DD of Fig. 1a based on the *k-means* are $\{\{4\}, \{9, 7\}, \{12\}\}$ resulting in the relaxed clustering DD (Fig. 1c) of upper bound 19 after compilation. In this relaxed DD, the path $\hat{r} \rightsquigarrow \hat{t}$ of the minimal length is a feasible solution, since the merged states are exact, which proves the optimality of the solution.

3 Related Work

Alternative node selection heuristics have been explored: in [16], the authors break ties in cost-based merging using a distance metric. It improves on the basic cost strategy but still relies on it, unlike the clustering-based approach we evaluate. [22] merge nodes following their shared children in the next layer. In [14] the Quadratic Knapsack is solved with a DD using Kruskal-based clustering with a distance metric, but layers may exceed the maximal width, unlike the approaches we evaluate. Their results don't isolate the merging strategy's impact. In [15] the authors select the merging heuristic to use on a layer based on a lookahead of the next layers. Some other works also focused on alternatives to compute dual bounds: in [19,20], the authors proposed the so-called *Domain-Independent Dynamic Programming* paradigm to solve combinatorial optimization problems by avoiding the need for domain-specific state merging but relying on domain-specific lower-bound heuristics to guide the search using A* like algorithms [19,20]. In contrast, traditional Decision Diagram Optimization (DDO) methods [4] require explicit specification of state merging operators to construct relaxations. In [17], the authors also include the possibility of domain-specific lower-bounds. Some other related works have focused on enhancing bounds derived from limited decision diagrams. In [27], the diagram is not compiled from scratch at every node but rather refined incrementally to infer the dual bound and save computation efforts. In [26] a local search framework is proposed to strengthen the dual bound of the relaxed DD. In [6] a MIP model is proposed to discover the provably tightest dual-bound from a relaxed DD. In [11] the authors exploit an aggregate version of the problem to perform additional pruning in the branch-and-bound and a node-selection heuristic to compile restricted DDs.

4   Using Generalized Hyperplane Partitioning for DD

The *k-means* algorithm used by [23] requires mapping states to fixed-size coordinate vectors so that it can compute new centroids with coordinate averaging and also use the Euclidian distance metric to assign points to the centroids. We propose to replace *k-means* to address two primary limitations: (i) Computational Cost: The overhead of *k-means* becomes significant as the number of state increases. For combinatorial problems like the MCP, this cost is heavy, as reported in the experimental section. (ii) Lack of Flexibility: The requirement for vector representations and Euclidean distances is a limitation in generic solvers such as [18,21] since this representation is not natural for problems with more complex states such as sets for the MCP. Adapting set-based states to *k-means* requires embedding them into fixed-size vectors over the entire universe, leading to high-dimensional encodings and high Euclidean distance computation cost. Instead, a general approach should support arbitrary, problem-specific metrics (we refer to [13] for a set of metrics on sets). To meet these requirements, we use the *Generalized Hyperplane Partitioning* (GHP) algorithm introduced in [28] and improved in [8]. GHP is presented in Algorithm 2 in the context of DD. The algorithm assumes $|L_i| > W$. GHP starts from a single cluster containing all the states of L_i, and iteratively selects a cluster to split into two until the desired number of clusters W is obtained. The distance metric used is denoted *dist*. The set of clusters C is stored as a priority queue sorted by decreasing cluster diameter (denoted $\delta(c)$), so the most spread-out clusters are split first. When a cluster c is extracted from C (Line 3), the auxiliary function *computePivots* (Line 5) identifies two states p_1 and p_2 whose distance is ideally close to the diameter of c ($\mathrm{dist}(p_1, p_2) \approx \delta(c)$). The algorithm then partitions the states of c into two subclusters according to their proximity to the pivot states (Line 7).

Algorithm 2: Generalized Hyperplane Partitioning

```
1  C ← {L_i}
2  while |C| < W do
3      c ← extractMax(C)     // Pull the cluster of maximal diameter
4      c_1, c_2 ← ∅
5      p_1, p_2 ← computePivots(c)
6      for v ∈ c do
7          if dist(p_1, v) ≤ dist(p_2, v) then
8              c_1 ← c_1 ∪ {v}
9          else
10             c_2 ← c_2 ∪ {v}
11     C ← C ∪ {c_1, c_2}
12 return C
```

The function *computePivots* in Algorithm 3 returns two states p_1, p_2 such that $dist(p_1, p_2) \approx \delta(c)$. To optimize the computation speed, this is done in three consecutive steps. First a state is randomly selected (Line 1). The first pivot p_1 is the one farthest from it (Line 2). Finally, the second pivot p_2 (Line 3) is one farthest from p_1. Again, to speed up the computation, we use an approximation of the diameter to prioritize the clusters in C. It is the largest distance between the pivot and any state in the cluster. This incurs no additional computational overhead, as it is done while executing the inner loop of Algorithm 2.

Algorithm 3: Compute Pivots

```
1  v ← RandomSelect(c)
2  p_1 ← arg max{dist(u, v) | u ∈ c}
3  p_2 ← arg max{dist(u, p_1) | u ∈ c}
4  return p_1, p_2
```

If we denote by $\mathcal{O}(dist)$ the complexity used to compute the distance between two states, then the complexity of Algorithm 3 is $\mathcal{O}(dist \times |L_i|)$. In Algorithm 2, the inner loop calls the metric of $\mathcal{S}$. The outer loop is repeated W times, thus W extractions and $2W$ insertions are performed on the priority queue, each with a $\mathcal{O}(\log(W))$ complexity. Therefore, the overall complexity of the Algorithm 2 is $\mathcal{O}(W(\log(W) + dist \times |L_i| + dist \times |L_i| + 2\log(W))) = \mathcal{O}(W(\log(W) + dist \times |L_i|))$.

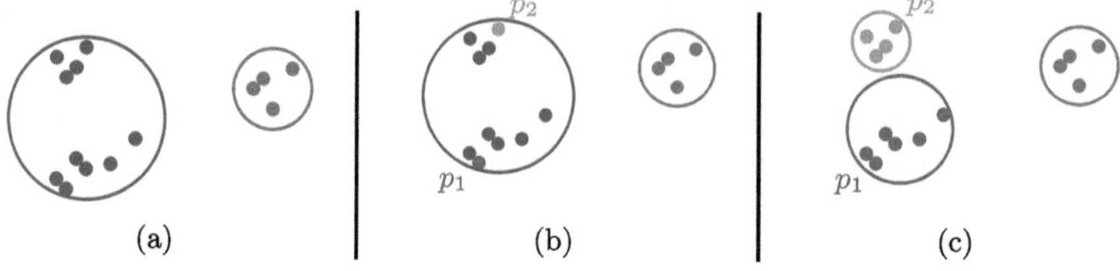

Fig. 2. Example of the GHP procedure.

Figure 2 illustrates the different steps of the GHP function. The set C currently contains two clusters (blue and orange), and the largest is pulled (Fig. 2a). At the second step (Fig. 2b), pivot states are computed and the last step (Fig. 2c) splits the blue cluster into yellow and green clusters.

Example 2. Let us consider an instance of MCP with $U = \{0, 1, 2, 3, 4\}$, $V = \{\{0, 2\}, \{1, 3\}, \{3\}, \{1, 4\}\}$ and $k = 3$. Figure 3a illustrates the exact DD, the relaxed *Cost* DD (Figs. 3b) and the relaxed *clustering* DD (3c) for a maximum width $W = 2$ of the instance. For GHP, the distance between two states is the size of their symmetric difference, normalized by the size of the universe[2]. Merged nodes have a double border. In the relaxed diagrams, several relaxed nodes are *trivial nodes* regardless of the heuristic used for relaxation. This reflects the *state-altering* nature of the DDO model for the MCP.

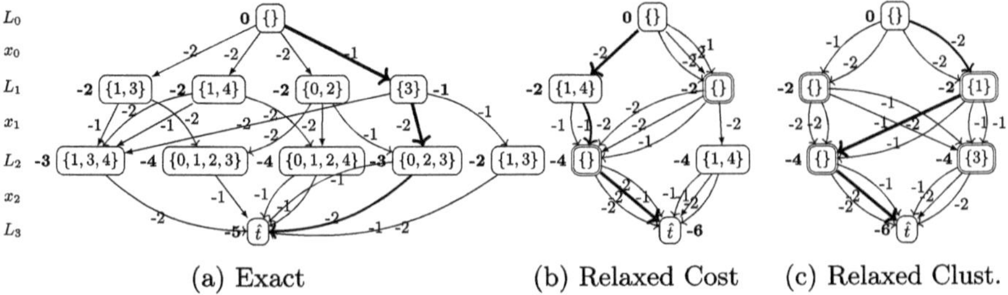

(a) Exact   (b) Relaxed Cost   (c) Relaxed Clust.

Fig. 3. Decision diagrams of the MCP described in Example 2. For the relaxed DDs $W = 2$.

5   Experimental Evaluation

Here, we evaluate the behavior of clustering heuristics on the three problems of interests KP, MKP, and MCP. We denote by GHP and KMEANS the clustering approaches based on GHP and on k-means (as in [23]), respectively. Additionally two heuristics are added. One is a generalization of COST and GHP in which $\alpha \cdot W$ nodes are selected according to their cost and the remaining nodes are grouped into $(1 - \alpha) \cdot W$ clusters using GHP with $\alpha \in [0, 1]$. It is denoted α-HYBRID. For KP and MCP, its bounds lie between COST and GHP: shifting toward COST as α is closer to 1, and toward GHP as α is closer to 0. But for MKP 0.6-HYBRID shows better performances. For readability, we thus display only the performances of 0.6-HYBRID. Finally, a baseline strategy, where W nodes are randomly selected, denoted RANDOM, is also considered to evaluate if an extreme diversification of the states retained in the DD is useful.

---

[2] $d(A, B) = \frac{|(A \cup B) - (A \cap B)|}{|U|}$, where A and B are two subsets of U.

Regarding the metric applied to each problem: for KP and MKP, we use the Euclidean distance, normalized by the maximum distance possible for the instance, i.e., the distance from $\hat{r}$ to a state where all the remaining capacities are null. For the MCP, we experimented with several distance metrics: the Jaccard distance, the Dice distance [13] and the size of the symmetric difference, normalized by the size of the universe. Preliminary results showed that the best bounds were obtained with normalized symmetric difference. Concerning k-$means$ on MCP, states are represented with boolean vectors where the value at coordinate i is 1 if i is covered in the state, 0 otherwise. For MKP, as done in [23], we add the objective value associated with a node to the features considered by k-$means$.

Our experiments aim to answer the following questions, for which we already provide synthetic answers based on the detailed results presented after:

Question 1 Does clustering help obtain tighter lower bounds in relaxed DD?
Answer: Yes for *state-preserving* models (e.g., KP), but not for *state-altering* ones (e.g., MCP).
Question 2 Do GHP and KMEANS obtain the same bound quality and do they execute at the same speed?
Answer: In terms of speed, there is no major difference with small branching factors (KP, MKP), but GHP is superior when the branching factor is large (MCP). The bound quality is roughly the same with GHP and KMEANS.
Question 3 Does clustering help obtain tighter upper bounds in restricted DD?
Answer: Yes, for all three problems.
Question 4 Is clustering worthwhile in a complete B&B search despite its heavier computational cost?
Answer: Yes for *state-preserving* models; but the performances observed for the MCP suggest that for *state-altering* models the tradeoff is worthless, as the performance gain obtained by the stronger pruning depends on the bound quality. The root bound quality seems to be a good estimator of it.

We implemented the approach using an open-source Java version[3] of [18]. Our source code and instances are available on the same repository. All the experiments were executed on an Intel Skylake 16-cores Xeon 6142 processor at 2.6 GHz with up to 192 GB of RAM[4]. For the k-$means$ algorithm, we use the implementation available in the SMILE library[5]. Concerning the maximum number of iterations of k-$means$, we experimented with several values, but observed that a high number of iterations increases runtime without significantly improving bound quality. We thus set this limit to 5 in our experiments. We use the 100 KP instances from [24] with 200 items per instance and the 270 MKP

---

[3] Available at https://github.com/DDOLIB-CETIC-UCL/DDOLib.

[4] Computational resources have been provided by the Consortium des Équipements de Calcul Intensif (CÉCI), funded by the Fonds de la Recherche Scientifique de Belgique (F.R.S.-FNRS) under Grant No. 2.5020.11 and by the Walloon Region.

[5] https://github.com/haifengl/smile.

instances from [10][6] with 100 items and 5 dimensions. For the MCP benchmark, 300 instances of $|U| \in \{100, 150, 200\}$ elements were generated following [25]. For all problems, variables are sorted in lexicographical order.

5.1 Relaxed DD

Figure 4 compares, for the different heuristics, the average normalized bound (wrt to the optimal solution) at the root node obtained with the relaxed DD and the runtime used to compute it for different maximum width W (x-axis).

KP Both KMEANS and GHP allow obtaining significantly better bounds than COST. The quality of the bounds is similar for KMEANS and GHP. KMEANS has a slightly higher execution time.

MKP The three strategies give very similar bounds[7], but the ones offered by the COST are slightly better. Interestingly, the 0.6-HYBRID obtains the best bounds among all strategies. In terms of runtime KMEANS is slightly slower.

MCP The COST strategy offers a significantly better bound than the clustering approaches. KMEANS is much slower than GHP that keeps a runtime close to COST. KMEANS has a slightly better bound quality than GHP though.

To understand heuristic impact on relaxed DD, we measure, for each instance, the proportion of exact nodes and the average degradation of each state defined as its distance with respect to the result of the merge. When the node is exact, its degradation is equal to 0. We also measure the average distance between

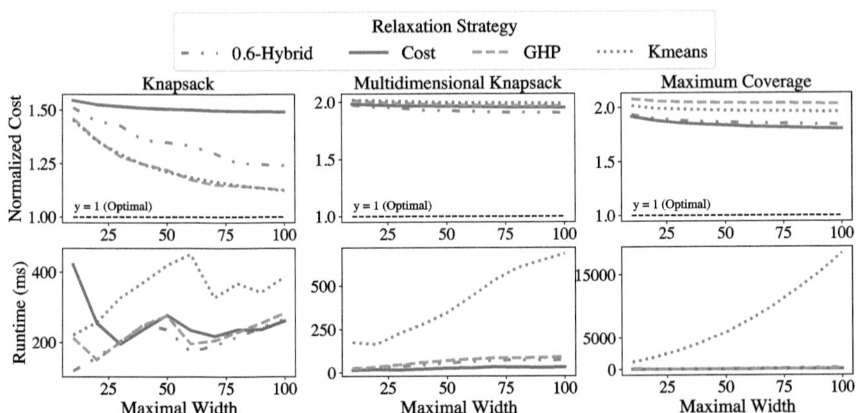

Fig. 4. Average bound quality and runtime versus the maximal width for Relaxed DDs, depending on the heuristic.

---

[6] Available at https://people.brunel.ac.uk/~mastjjb/jeb/orlib/mdmkpinfo.html.

[7] These results differ slightly from those reported in [23], where KMEANS outperforms COST. When run on the same set of instances, our implementation produces comparable outcomes. However, as our instance pool is a superset of theirs, we believe that our results provide a more comprehensive view of the heuristics' performance.

Table 1. Statistics on exact node proportion, average state degradation, and average root distance in relaxed DDs (max width 60), across problem instances.

Problem	Strategy	Proportion of Exact Nodes			Avg State Degradation			Avg Distance with Root		
		Min	Med	Max	Min	Med	Max	Min	Med	Max
KP	Cost	**0.05**	**0.51**	**0.98**	0.12	0.24	0.30	0.19	0.51	0.57
	GHP	0.01	0.01	0.07	**0.00**	**0.00**	**0.00**	**0.45**	**0.63**	**0.80**
	Kmeans	0.01	0.01	0.07	**0.00**	**0.00**	**0.00**	0.43	**0.63**	**0.80**
MKP	Cost	**0.20**	**0.48**	**0.77**	0.19	0.22	0.25	0.44	0.53	0.60
	GHP	0.00	0.01	0.04	**0.00**	**0.01**	**0.03**	0.46	0.77	0.92
	Kmeans	0.00	0.01	0.04	**0.00**	**0.01**	**0.03**	**0.56**	**0.83**	**0.94**
MCP	Cost	**0.13**	**0.31**	**0.98**	0.13	0.29	0.45	**0.11**	**0.24**	**0.39**
	GHP	0.12	0.21	0.37	0.05	0.11	0.18	0.01	0.04	0.09
	Kmeans	0.08	0.13	0.24	**0.03**	**0.07**	**0.12**	0.02	0.06	0.12

each state of the DD and the root state. By design, the merge operators for the three problems produce states closer to the root state. Hence, a relaxed DD with highly deteriorated states will contain states closer to the root state. Table 1 shows statistics on the distribution across the different instances of these measures, for relaxed DDs with a 60 maximal width.

For KP and MKP, GHP and KMEANS greatly reduce state deterioration compared to COST. However, for MCP, the distance from the states to the root is very small when clustering is used, which explains why the lower bounds are weaker in this problem with clustering-based merge strategies.

5.2   Restricted DD

Figure 5 compares the average normalized primal bound obtained with restricted DDs and the runtime, using the different heuristics on the benchmarks.

KP Using CLUSTERING significantly improves (near-optimal solutions found) the primal bounds obtained with restricted DDs. Random selection (RANDOM) performs worse, confirming that the extra cost of CLUSTERING is worthwhile.

MKP The gap between COST and CLUSTERING is smaller than in KP. However, KMEANS and GHP return significantly better bounds than COST. In terms of computational cost, KMEANS is more expensive than GHP and COST.

MCP All heuristics return close to optimal solutions, except for RANDOM which is the worst heuristic. KMEANS is notably more costly, while GHP and COST have similar performance.

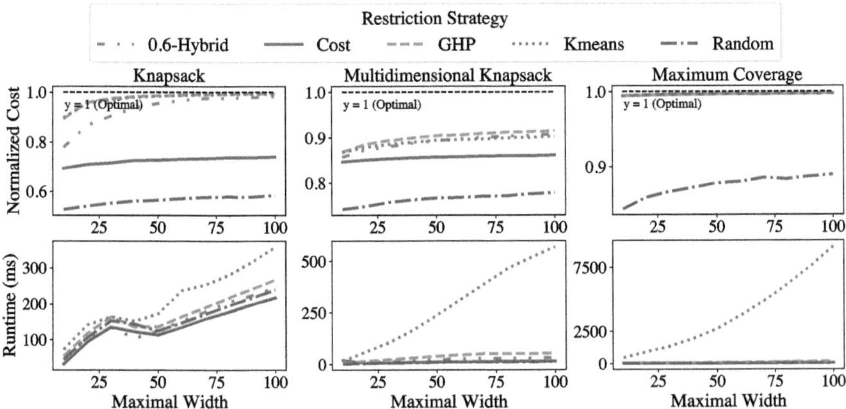

Fig. 5. Average bound quality and runtime versus the maximal width for Restricted DDs, depending on the heuristic.

5.3   Complete Branch-and-Bound

Figure 6 compares the different heuristics on a complete B&B search. A time limit of 5 min is set per instance, and the maximum width is fixed at $W = 60$ for both relaxed and restricted DDs. The cumulative number of instances solved over time (x-axis) is reported for the KP, and for MKP and MCP it shows the number of instances solved with an optimality gap below the threshold (x-axis) at the timeout, as these problems are mostly not solved optimally.

KP The heuristic used for the restricted DD has a strong impact on the runtime. Using CLUSTERING significantly reduces runtime compared to COST. Overall, the best configuration is GHP for both relaxed and restricted DDs, as it improves bounds at the root and reduces solution time.

MKP All configurations show similar performances. Only the heuristic used for the restricted DD affects performance slightly, with COST being slightly disadvantaged and CLUSTERING slightly reducing the optimality gap. The relaxed DD heuristic has minimal impact.

MCP All heuristics produce near-optimal bounds for the restricted DD, so the choice of restricted DD heuristic has little effect on performance. The relaxed DD heuristic affects performance slightly, with COST giving the best bound and slightly improving overall performance compared to GHP or KMEANS.

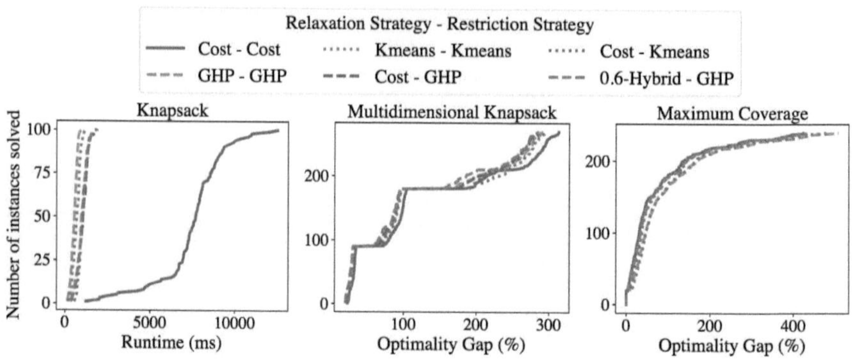

Fig. 6. Number of instances solved by each configuration for the three different problems with respect to the runtime or the optimality gap.

6   Conclusion

We proposed using the Generalized Hyperplane Partitioning (GHP) as a faster, more flexible alternative to k-means proposed in [23] for clustering states in bounded-width decision diagrams (DD). It is particularly efficient for problems with high branching factors (typically MDD rather than BDD). We also studied the impact of using clustering as a heuristic for node selection when computing bounded-width DDs. We observed that for Restricted DDs (primal bounds), it offers tighter bounds than cost-based methods. However, for Relaxed DDs (dual bounds), its success is tied to the merge operator: clustering outperforms cost-based methods on Knapsack, a *state-preserving* model, but underperforms on Maximum Coverage, a *state-altering* model, where merging strongly dilutes the bounds. We thus recommend clustering for primal bounds, and for dual bounds only when merging preserves states well; otherwise, the cost-based heuristic of [4] is preferable. In future work, we intend to evaluate additional models across each problem class to strengthen our conclusions.

References

1. Andersen, H.R., Hadzic, T., Hooker, J.N., Tiedemann, P.: A Constraint Store Based on Multivalued Decision Diagrams. In: Bessière, C. (ed.) CP 2007. LNCS, vol. 4741, pp. 118–132. Springer, Heidelberg (2007). https://doi.org/10.1007/978-3-540-74970-7_11
2. Bellman, R.: The theory of dynamic programming. Bull. Am. Math. Soc. **60**(6), 503–515 (1954)
3. Bergman, D., Cire, A.A., van Hoeve, W.-J., Hooker, J.N.: Optimization Bounds from Binary Decision Diagrams. INFORMS J. Comput. Publisher: INFORMS (2013)
4. Bergman, D., Cire, A.A., van Hoeve, W.-J., Hooker, J.N.: Discrete optimization with decision diagrams. INFORMS J. Comput. **28**(1), 47–66. Publisher: INFORMS (2016)

5. Bergman, D., Cire, A.A., van Hoeve, W.-J., Yunes, T.: BDD-based heuristics for binary optimization. J. Heuristics **20**(2), 211–234 (2014)

6. Bergman, D., Cire, A.A.: On Finding the Optimal BDD Relaxation. In: Salvagnin, D., Lombardi, M. (eds.) CPAIOR 2017. LNCS, vol. 10335, pp. 41–50. Springer, Cham (2017). https://doi.org/10.1007/978-3-319-59776-8_4

7. Bergman, D., van Hoeve, W.-J., Hooker, J.N.: Manipulating MDD relaxations for combinatorial optimization. In: Achterberg, T., Christopher Beck, J., (eds.) Integration of AI and OR Techniques in Constraint Programming for Combinatorial Optimization Problems, pp. 20–35. Springer, Cham (2011)

8. Bugnion, E., Roos, T., Shi, F., Widmayer, P., Widmer, F.: A spatial index for approximate multiple string matching. In: Proceedings of the 1st South American Workshop on String Processing (SP 1993), Belo Horizonte, Brazil, pp. 43–53. First South American Workshop on String Processing (1993)

9. Castro, M.P., Cire, A.A., Christopher Beck, J.: Decision diagrams for discrete optimization: a survey of recent advances. INFORMS J. Comput. **34**(4), 2271–2295. INFORMS (2022)

10. Paul, C., Chu, J.E., Beasley: A genetic algorithm for the multidimensional knapsack problem. J. Heuristics **4**(1), 63–86 (1998)

11. Coppé, V., Gillard, X., Schaus, P.: Boosting decision diagram-based branch-and-bound by pre-solving with aggregate dynamic programming. In: Yap, R.H.C. (ed.) 29th International Conference on Principles and Practice of Constraint Programming (CP 2023), vol. 280 of Leibniz International Proceedings in Informatics (LIPIcs), pp. 13:1–13:17, Dagstuhl, Germany, Schloss Dagstuhl – Leibniz-Zentrum für Informatik (2023). ISSN: 1868–8969

12. Coppé, V., Gillard, X., Schaus, P.: Decision diagram-based branch-and-bound with caching for dominance and suboptimality detection. INFORMS J. Comput. **36**(6), 1522–1542. INFORMS (2024)

13. Deza, M.M., Deza,E.: Encyclopedia of distances. In: Encyclopedia of Distances, pp. 1–583. Springer, Cham (2009)

14. Eliass Fennich, M., Coelho, L.C., Fomeni, F.D.: Tight upper and lower bounds for the quadratic knapsack problem through binary decision diagrams. Comput. Oper. Res. **184**, 107197 (2025)

15. Frohner, N., Raidl, G.R.: Merging Quality Estimation for Binary Decision Diagrams with Binary Classifiers. In: Nicosia, G., Pardalos, P., Umeton, R., Giuffrida, G., Sciacca, V. (eds.) LOD 2019. LNCS, vol. 11943, pp. 445–457. Springer, Cham (2019). https://doi.org/10.1007/978-3-030-37599-7_37

16. Frohner, N., Raidl, G.R.: Towards Improving Merging Heuristics for Binary Decision Diagrams. In: Matsatsinis, N.F., Marinakis, Y., Pardalos, P. (eds.) LION 2019. LNCS, vol. 11968, pp. 30–45. Springer, Cham (2020). https://doi.org/10.1007/978-3-030-38629-0_3

17. Gillard, X., Coppé, V., Schaus, P., Augusto Cire, A.: Improving the filtering of branch-and-bound MDD solver. In: Stuckey, P.J. (ed.) Integration of Constraint Programming, Artificial Intelligence, and Operations Research, pp. 231–247. Springer, Cham (2021)

18. Gillard, X., Schaus, P., Coppé, V.: Ddo, a generic and efficient framework for MDD-based optimization. In: Proceedings of the Twenty-Ninth International Joint Conference on Artificial Intelligence, pp. 5243–5245. Yokohama, Japan (2020). International Joint Conferences on Artificial Intelligence Organization

19. Kuroiwa, R., Christopher Beck, J.: Generic state space search for combinatorial optimization: Domain-independent dynamic programming. In: Proceedings of the International Conference on Automated Planning and Scheduling, Vol. 33, pp. 236–244 (2023)
20. Kuroiwa, R., Christopher Beck, J.: Solving domain-independent dynamic programming problems with anytime heuristic search. In: Proceedings of the International Conference on Automated Planning and Scheduling, vol. 33, pp. 245–253 (2023)
21. Michel, L., van_Hoeve, W.-J.: Codd: A decision diagram-based solver for combinatorial optimization (2024)
22. Nafar, M., Römer, M.: Lookahead, merge and reduce for compiling relaxed decision diagrams for optimization. In: Dilkina, B. (ed.) Integration of Constraint Programming, Artificial Intelligence, and Operations Research, pp. 74–82. Springer, Cham (2024)
23. Nafar, M., Römer, M.: Using clustering to strengthen decision diagram bounds for discrete optimization. In: Proceedings of the AAAI Conference on Artificial Intelligence, Vol. 38, no. (8), pp. 8082–8089 (2024)
24. Pisinger, D.: Where are the hard knapsack problems? Comput. Oper. Res. $32(9)$, 2271–2284 (2005)
25. Mauricio, G.C., Resende: Computing approximate solutions of the maximum covering problem with GRASP. J. Heuristics $4(2)$, 161–177 (1998)
26. Römer, M., Cire, A.A., Rousseau, L.-M.: A Local Search Framework for Compiling Relaxed Decision Diagrams. In: van Hoeve, W.-J. (ed.) CPAIOR 2018. LNCS, vol. 10848, pp. 512–520. Springer, Cham (2018). https://doi.org/10.1007/978-3-319-93031-2_36
27. Rudich, I., Cappart, Q., Rousseau, L.-M.: Improved peel-and-bound: methods for generating dual bounds with multivalued decision diagrams. J. Artif. Intell. Res. 77, 1489–1538 (2023)
28. Uhlmann, J.K.: Satisfying general proximity / similarity queries with metric trees. Inf. Process. Lett. $40(4)$, 175–179 (1991)

A Scalable Learning Approach for Efficient Computation of Independent Set and Cover Variants

Ryan O. Connor[1]($\boxtimes$), Noah Coleman[1], Darren Strash[2], Saurabh Ray[3], and Deepak Ajwani[1]

[1] University College Dublin, Dublin, Ireland
`ryan.o-conner@ucdconnect.ie`
[2] Hamilton College, Clinton, NY, USA
[3] Center for Interdisciplinary Data Science and AI, NYUAD Research Institute, New York University Abu Dhabi, Box 129188, Abu Dhabi, UAE

Abstract. The maximum independent set (MIS) problem is a fundamental NP-hard optimization problem that remains challenging on large graphs. Machine learning (ML) offers the potential to aid algorithm designers in rapidly developing effective heuristics across problem variants and input distributions. However, existing end-to-end ML approaches often struggle with generalization, require extensive training data, and are rarely designed to scale to extremely large problem instances.

We propose a hybrid MLalgorithmic framework that follows the Learning to Prune (LTP) paradigm: a classifier predicts vertices to fix (or prune) and the instance is simplified, before applying a state-of-the-art solver.

A key challenge in this setting is due to the fact that Linear Programming Relaxation-derived features—crucial in many LTP pipelines—are often too slow and too coarse to be practical for MIS at scale. We overcome this by adapting the multiplicative weights method from the theoretical computer science literature, yielding fast, high-quality surrogate features that preserve the key structural signal of the linear programming relaxation. We showcase the flexibility of this generic technique by extending our approach to the 3-path vertex cover problem (VCP_3).

For MIS experiments, we utilize the state-of-the-art ReduMIS solver, which is capable of producing high quality solutions even on massive graphs. Results show that training on only about one hundred graph instances with ReduMIS solutions suffices for our method to achieve solutions within 10% of those obtained by ReduMIS on the test set, while running in roughly half the time, especially on dense graphs.

In experiments on VCP_3, the learned models yield even stronger scalability and practical gains. We adopt the highest ranked heuristic solver from the PACE 2025 challenge for this problem. We show that on large test instances, our classifiers are powerful enough to admit aggressive vertex pruning, yielding solutions that are on average 5% better than the state-of-the-art PACE heuristic baseline in half of the runtime.

Keywords: Algorithm Engineering · Machine Learning

1   Introduction

The maximum independent set (MIS) problem is one of the most fundamental problems in graph theory and combinatorial optimization, and has been extensively studied since its formalization as a core NP-complete problem in 1972 [23]. Given a graph $G = (V, E)$, the objective is to find a maximum cardinality set of vertices $I \subseteq V$ such that no two vertices in I are adjacent. Its combination of theoretical richness and practical relevance has made MIS an enduring focus of algorithmic research, with applications ranging from network analysis to resource allocation. Despite decades of progress and the availability of sophisticated solvers, efficiently computing MIS on large graphs remains a significant practical challenge.

The literature explores many variants of the Maximum Independent Set (MIS) problem. MIS is complementary to the *Minimum Vertex Cover* (MVC) problem: a set $C \subseteq V$ is a vertex cover if every edge of G has at least one endpoint in C, and it holds that I is an independent set if and only if $V \setminus I$ is a vertex cover. Consequently, a maximum independent set in G corresponds exactly to the complement of a minimum vertex cover in G. Other notable variants include the k-independent set problem, which seeks a maximum set of vertices with pairwise distances of at least $k + 1$, and the minimum k-path vertex cover, where the objective is to cover all paths of length k. Like MIS, these problems are computationally demanding on large instances, underscoring the need for scalable algorithmic techniques.

Recent advances in machine learning (ML) offer new opportunities to augment classical solvers by automatically discovering effective heuristics. However, purely end-to-end ML approaches often require large training datasets and struggle to generalize across diverse graph distributions. A promising alternative is the Learning to Prune (LTP) paradigm [29, 30, 44], in which a trained model predicts variables to fix, thereby reducing the problem size before applying a traditional solver. This effectively combines the predictive power of ML with the reliability of modern solvers. Crucially, this approach extends the solvability frontier to large-scale instances that are otherwise intractable for state-of-the-art solvers, trading a slight reduction in solution quality for computational feasibility.

LTP has been effective for several combinatorial optimization problems [17, 44, 48], but adapting it to MIS raises unique challenges. Firstly, features derived from linear programming relaxations (LPR) – commonly used on these other problems – are computationally expensive to compute on large graphs and provide limited predictive signal for MIS as the values are half-integral. Secondly, it is computationally expensive to solve even instances with a few thousand vertices to optimality in order to obtain training data.

In this work, we introduce a hybrid MLalgorithmic framework that addresses these challenges and provides a fast, effective heuristic for both MIS and VCP_3.

Our approach builds on the LTP paradigm while introducing two key innovations:

1. Instead of costly LPR computation, we employ the multiplicative weights update method to generate a fast and reliable predictive signal.
2. We use the state-of-the-art heuristic solvers to generate reasonably accurate training labels instead of exact solutions.

Additionally for MIS, we integrate powerful existing reduction rules directly into our pipeline: models are trained and evaluated on graphs that have already been simplified by these rules, so predictions build on established algorithmic insights rather than attempting to learn them from scratch.

To solve the reduced instances, we use the state-of-the-art solver ReduMIS [25] for the MIS problem and the winner [35] (heuristic track of the hitting set problem) in the recently conducted PACE challenge [37] for the VCP_3 problem. Our results show that by training on only ~ 100 graphs, trained models generalize to much larger instances, and suffice to reach solutions within 10% of ReduMIS for MIS while halving runtime (especially on dense graphs). Results on $k = 3$ path vertex cover are even stronger - we show that our learned models can aggressively prune instances, halving runtime and facilitating solution improvements of $\sim 5\%$ on average over the original PACE heuristic.

2 Preliminaries

Let $G = (V, E)$ be an undirected simple graph. An *independent set* in G is a subset $I \subseteq V$ such that no two vertices in I are adjacent in G. Equivalently, for every edge $\{u, v\} \in E$, at most one of u, v belongs to I. The *Maximum Independent Set* (MIS) problem asks for an independent set of maximum cardinality in G.

One can formulate the MIS problem as an integer linear program (ILP). Introduce a binary variable $x_v \in \{0, 1\}$ for each vertex $v \in V$, where $x_v = 1$ indicates that v is chosen in the independent set. The ILP is

$$\max \quad \sum_{v \in V} x_v$$
$$\text{s.t.} \quad x_u + x_v \leq 1 \quad \text{for each edge } \{u, v\} \in E,$$
$$x_v \in \{0, 1\} \quad \text{for all } v \in V.$$

Note that this standard MIS formulation produces a solution equivalent to the standard vertex cover ILP by simply flipping all vertex assignments.

The *natural LP relaxation* (later denoted LPR) of either ILP is obtained by relaxing the integrality constraints $x_v \in \{0, 1\}$ to $0 \leq x_v \leq 1$ for all $v \in V$, leaving the other constraints unchanged. All feasible solutions of this relaxation are called *fractional independent sets*.

We also consider the generalized problem, *k-independent set*, defined as follows. A set $S \subseteq V$ is a *k-independent set* if the distance between any two distinct vertices in S is strictly greater than k. In particular, a 1-independent set is exactly an ordinary independent set (since distance > 1 forbids adjacency), and a 2-independent set requires that every two vertices in S be at distance at least 3 in G.

The k-independent set problem can be solved via standard independent set methods through the *graph power* transformation of input graphs. The kth power of G, denoted G^k, is the graph on the same vertex set V in which two vertices u, v are adjacent in G^k if and only if their distance in G is at most k. By construction, an independent set in G^k is a set of vertices such that no two vertices in the set are at distance $\leq k$ in G, i.e., with pairwise distance $> k$. Therefore, a maximum independent set in G^k is exactly a maximum k-independent set in the original graph G.

The analogous generalised cover problem is *minimum k-path vertex cover* (VCP_k). Rather than choosing a set of vertices such that all edges in the graph are covered, we instead cover all paths of k vertices. We can describe the ILP for VCP_k as follows:

$$\min \quad \sum_{v \in V} x_v$$

$$\text{s.t.} \quad x_1 + x_2... + x_k \geq 1 \quad \text{for each path of length } k \text{ in } G,$$
$$x_v \in \{0, 1\} \qquad \qquad \text{for all } v \in V.$$

Note that for $k = 2$, this is simply the standard vertex cover formulation. In this paper, we focus on solving variants where $k = 3$.

3 Related Work

The maximum independent set problem has a rich history of theoretical and practical results. Here we focus on the most relevant results to our present work.

Theoretical Results for MIS. Not only is the MIS problem NP-hard, but it is hard to approximate within a factor of $n^{1-\epsilon}$ for all $\epsilon > 0$ [49]. Due to its NP-hardness, the current theoretically fastest worst-case algorithms for solving MIS are exponential-time branching algorithms, the fastest of which is the polynomial-space algorithm due to Xiao and Nagamochi [46] with time $O^*(1.1996^n)$.

Data reduction rules—which transform a problem instance into another instance of the same problem—are powerful for solving MIS and the complementary minimum vertex cover (MVC) problem. A famous result due to Nemhauser and Trotter [36] is that the linear programming relaxation (LPR) for MVC has a half-integral solution, partitioning the vertices V into sets V_0, V_1, and $V_{1/2}$, and furthermore there is an MVC excluding V_0 and including V_1. Thus we can remove $V_0 \cup V_1$ and solve MVC on the graph induced by $V_{1/2}$. This powerful rule leaves $|V_{1/2}| \leq 2k$ vertices, where k is the size of an MVC of G [12]. In practice, the LPR can be solved via maximum matching in a bipartite graph [14, Section 2.5] using the Hopcroft-Karp algorithm which runs in $O(m\sqrt{n})$ time [20].

Theoretical results for VCP_k. The k-path vertex cover problem (VCP_k) generalizes standard vertex cover (Sect. 2) and also arises as a vertex-deletion and hitting-set problem [10]. Consequently, relevant theoretical results appear across these formulations. NP-completeness was shown explicitly by Brešar [8], with an earlier implicit proof for the node-deletion variant given by Lewis [32]. For any $k \geq 3$, the problem admits no c-approximation for $c < 2$ assuming the Unique Games Conjecture [8]. An $O(\log k)$-approximation running in $2^{O(k^3 \log k)} n^{O(1)}$ time is provided by Lee et al. [31]. When parameterized by the solution size t, VCP_k admits an $O(1.2738^t)$-time algorithm; a broader collection of FPT running times for various k is surveyed by Tu et al. [45].

Our experiments focus on $k = 3$, chiefly because enumerating k-paths yields $O(n^k)$ number of constraints, and $k = 3$ keeps enumeration tractable for most of our large graphs. VCP_3 is solvable in linear time on trees [8], admits an exact $O(1.3659^n)$-time algorithm [46], and has been analyzed across further graph classes [22]. A comprehensive list of results for VCP_k appears in the survey of Tu et al. [45].

Experimental Work on MIS. In practice, sparse graphs of up to millions of vertices can be solved exactly in practice using a rich collection of data reductions and branching rules included in the branch-and-reduce framework [2], which can be improved with SAT-based pruning [40]; alternatively, solvers can first apply data reduction rules and then run a fast branch-and-bound solver, or do both [19].

For graphs not solvable exactly, local search [3,11] and evolutionary computing [26] are most effective at finding high quality solutions in practice. Many of these incorporate data reduction rules to ensure search does not get stuck in "easy" parts of the graph. Of these, the evolutionary algorithm ReduMIS has been shown to produce the highest quality solutions [26], however it may require long running times to achieve the best answers on huge instances.

Experimental Work on k-path Vertex Cover. Despite extensive theory, we found no publicly available solvers or ML methods tailored to VCP_k. Treating k-path vertex cover as a special case of hitting set allows us to leverage implementations from the PACE 2025 competition. We therefore use the winning heuristic solver from PACE 2025 [35] as our baseline for this problem.

Learning Approaches for MIS. Two influential learning-based MIS methods are the graph-embedding RL approach of Khalil et al. [24] and the "learn what to defer" (LwD) reinforcement-learning framework of Ahn et al. [1]. Another widely used line of work is the guided tree search of Li et al. [33], which expands partial solutions using vertex-selection predictions from a neural model. Although the original GCN used by Li et al. was later shown to be no stronger than random by Böther et al. [6], the LwD improvement was attributed to genuine learning, and guided-search approaches continue to be competitive with more recent architectures [9,47]. Recent algorithm engineering work focuses on speeding up exact or heuristic solvers by applying graph neural networks (GNNs). For weighted MIS, Langedal et al. [27] use GNNs to select branching vertices speed up branch-

and-reduce, with moderate speed gains. More attention is being paid to learning techniques for the maximum *weight* independent set problem e.g., to select a vertex in reduce-and-peel [28] or to filter ineffective data reductions [18] during preprocessing.

Learning-to-Prune Combinatorial Optimization Problems. This learning technique emerged from early work on fine-grained search space pruning for maximum clique enumeration [29,30]. It proved to be very effective on a range of combinatorial optimization problems such as k-median, facility location, set cover [44], travelling salesperson [17,43], and Steiner tree [48]. We note that this framework has many similarities to Neural Diving [15] and Predict-and-Search [21].

4   Methodology Learning to Prune Problem Instances

Here, we outline our approach which utilizes machine learning for efficient independent set and cover computation. There are two aspects to this approach:
1 - We first design and train machine learning classifiers, which can reliably predict independent set/cover membership for each vertex in an input graph.
2 - We leverage these predictions to improve solver performance, providing a range of tradeoffs between solve time and solution quality.

4.1   Vertex Classifiers

We use supervised classification models to predict independent-set/cover membership for each vertex from efficiently computable graph features, producing both a membership estimate and a confidence score. Since obtaining optimal MIS labels on large graphs is infeasible, we generate training labels using Redu-MIS with a one-hour time limit. For VCP_3, we label instances with the winning solver from the PACE 2025 hitting-set heuristic track. While PACE allocates 300 seconds per instance, we permit 1000 seconds for labeling, chosen empirically as solutions stabilize beyond this point.

Feature selection must balance predictive value and computational cost. We therefore restrict attention to lightweight structural features, excluding more expensive options such as betweenness centrality or graph neural network embeddings, whose computational overhead made them impractical for our experiments. The final feature set is as follows (Table 1):

Instead of raw degrees we use a normalized degree rank: sort the unique degree values, assign each vertex the rank of its degree and scale to $[0, 1]$. We derive min/max/avg neighbor degree rank from these ranks. We also quickly produce multiple maximal independent sets via Luby's algorithm [34], and record each vertex's normalized occurrence frequency across those runs.

Degree-based features: PageRank [38], core number (k-shell) [42], local clustering coefficient, and degree centrality—are computed with the Networkit [4] Python package.

Table 1. Vertex Features

Features
Local clustering coefficient
Degree centrality
Core number
Degree rank
Minimum neighbor degree rank
Maximum neighbor degree rank
Average neighbor degree rank
PageRank score
Maximal Independent Set Membership Frequency
Linear Programming Relaxation Derived Features[a]

[a]Further engineering details on this feature in Sect. 4.2

The uniformity of mesh graphs raised some issues for prediction, as many vertices can share identical feature vectors. The MIS-frequency and LPR-based features reduce this degeneracy by providing a more diverse range of values despite the homogeneity of this graph class.

For prediction we use Random Forests [7] and Support Vector Machines [13], implemented in scikit-learn [39]. Hyperparameters were tuned by grid search over a wide range.

4.2   Linear Programming Relaxation Approximation

As described in Sect. 2, the LP relaxations of MIS and VCP_k are solvable in polynomial time and yield fractional solution values that often serve as powerful structural features in ML-augmented optimization (e.g., [44, 48]). In principle, these values provide a useful signal: vertices with larger LP values are more likely to appear in an optimal cover (or be excluded from an independent set).

In our preliminary investigations on MIS, solving the standard linear programming relaxation (LPR) to optimality proved computationally prohibitive, often exceeding our 1000-second budget even with commercial solvers like Xpress [16]. Furthermore, as established by Nemhauser and Trotter [36], extreme point solutions to the MIS LPR are inherently half-integral, taking values in $\{0, 0.5, 1\}$. In practice, this assigns 0.5 to a vast majority of vertices, providing very little variety or structural signal for ML classification. To address this, we use the Multiplicative Weights Update algorithm (MWUA) [5, 41]. MWUA converges to the optimal solution of the LP relaxation, which, as mentioned before, is not very informative, especially when a large number of vertices have a value of 0.5. However, if we stop the algorithm early, well before convergence, we capture the initial bias of the algorithm towards the vertices it considers more promising. We have empirically found that this useful for ML-based pruning. It would be interesting to find a theoretical explanation for this. Importantly for our use case, using MWUA with early stopping also reduces the running time.

We prepared MWUA to solve general VCP_k instances; for $k = 2$ the complement of our solution approximates the MIS LPR. Algorithm 1 outlines the procedure. All constraints (edges/k-paths) begin with equal weight; each iteration computes a weighted score for each vertex, greedily constructs a fractional solution based on these weights, updates constraint weights based on corresponding violations, and repeats until convergence or a strict time cap. The cap is 90 seconds for MIS, and 60 seconds for VCP_k as the overall solver budget is smaller. The final feature vector consists of the average fractional solution across iterations as well as some statistics over the updated constraint weights incident to each vertex. These combine to yield a range of rich features for our ML classifier.

Algorithm 1. MWUA for Independent Set LPR

Require: Graph $G = (V, E)$, runtime cap $T_{\max}$
 1: Initialize uniform weights $w_e = 1$ for all $e \in E$
 2: Initialize iteration list $\mathcal{S} = \{\}$
 3: **while** runtime $< T_{\max}$ **and not converged do**
 4:     **for** each vertex $v \in V$ **do**
 5:         Compute normalized weighted sum $s_v = \sum_{e \in \text{incident}(v)} w_e$
 6:     **end for**
 7:     **for** each vertex $i \in V$, sorted by s_v **do**
 8:         Assign $x_v = \min 1.0, f(s_v)$
 9:     **end for**
10:     Add current solution $X = \{x_v\}$ to $\mathcal{S}$
11:     **for** each edge $e = (u, v) \in E$ **do**
12:         Update weight w_e based on constraint violation
13:     **end for**
14: **end while**
Ensure: Feature vector: average of $\mathcal{S}$ and final weights $\{w_e\}$

By enforcing a fixed and predictable time budget, MWUA reduces slow LPR computations on MIS by over 90% while remaining competitive on easy instances. By returning both the LPR approximation values and an aggregation of the weights intrinsic to the algorithm, MWUA produces a rich and informative spectrum of feature values. In practice, these features are among the strongest predictors in our models.

We examine the flexibility of this approach in a similar setting, and apply the same MWUA framework directly to VCP_3. The only difference is that the constraints in this problem correspond to paths of three vertices rather than edges (here each constraint contains three variables rather than two), and the number of constraints grows quadratically with graph density. Despite this increased scale, MWUA remains tractable under our time cap, and feature-selection experiments consistently ranked its outputs as among the most informative.

MWUA is a crucial enabler of our overall solver design for both MIS and VCP_3, as it provides problem-specific insight to the classifier through the straightforward application of this generic technique.

4.3 ML-Augmented Independent Set Solver

Our ML models predict independent set membership for each vertex using the features described above. We exploit these predictions to construct an ML-augmented heuristic for solving vertex cover instances.

Each prediction includes both a membership estimate and a confidence score. We sort vertices by confidence and fix the status of the top $1-\theta$ proportion based on their predicted class. This reduces the graph size through two mechanisms:

- If a vertex u is predicted to be in the independent set it implies that any neighbors of this vertex should not be in the independent set. As such, we delete all neighbors of u. Vertex u in the preprocessed graph is now a singleton vertex, which will certainly be included in the resultant independent set
- If a vertex v is predicted to lie outside of the independent set, it is removed from the graph.

We then apply the state-of-the-art ReduMIS solver [26] to the pruned instance. The θ parameter can be changed to control how aggressively we prune the problem instance.

4.4 ML-Augmented k-Path Vertex Cover Solver

As outlined in Sect. 3, we adapt the highest ranked hitting set heuristic from the PACE 2025 challenge [35] in our framework. This highly-engineered solver has been fine-tuned in a competitive setting, and we aim to further improve its performance on large instances with our ML-augmentations. The solver combines three phases: constraint reductions, a fast "iterated PageRank" greedy pass, and finally a customized local search that consumes most of the runtime.

In order to use the PACE hitting set solver on our VCP_3 instances, we must first enumerate all 3-paths in the graph and pass them to the solver as hitting set constraints. Since this operation is common to all approaches, we do not include it in our running time evaluation. On the largest and densest graphs in our experiments this enumeration can exceed 1000 seconds. Although pruning can greatly speed this up, we decide to focus only on how pruning affects the subsequent solving steps.

VCP_k does not admit a natural operation that we can perform directly on the graph to prune the problem instance. Instead, we operate on the constraints as follows:

- If vertex u is predicted to be in the cover, we fix u in the solution (add a singleton constraint) and remove all constraints containing u.
- If vertex v is predicted to be out of the cover, we delete v from all relevant path constraints; if a path ends up with all k vertices predicted out, we force the least-confident vertex of that path into the cover.

We then pass this pruned collection of constraints to the PACE heuristic solver. Empirically, pruning based only on *predictions-in-cover* (the minority class) yielded best results. Majority-class "out" predictions did not meaningfully improve performance enough to justify their use. Similar minority-only pruning strategies have been used elsewhere in LTP for facility location problems [44]. We use the same θ parameter as in Sect. 4.3 to control the proportion of predictions used for pruning.

5   Experimental Results

For MIS our baseline is REDUMIS with a 1000 s time limit. For k-path vertex cover the baseline is the PACE heuristic with a timeout at 300 s. We use the 300 second timeout here because this is the framework for which the PACE solvers were designed, and solver subroutines are often highly tuned to the PACE test conditions. We added simple functionality to report the time the best solution was found. We run our ML-augmented solvers as described in Sects. 4.3 and 4.4 over a range of pruning parameters.

Preprocessing (feature computation, prediction, pruning) is subtracted from the solver time budget to ensure fairness, and for every run we record the time when the best solution was obtained, producing a spectrum of trade-offs between runtime and solution quality. We also provide results from the Learning what to Defer (LwD) solver [1] as a relevant ML baseline for MIS. We only present this in the reduced $k = 1$ setting as it produced memory issues on larger graphs, as we outline in Sect. 5.1.

We also test the $k = 2$ Independent Set via the graph transformation outlined in Sect. 2 to probe solver performance on denser, less-reducible instances. Finally, we present corresponding VCP_3 experiments to determine how powerful and flexible our approach is in an analogous problem setting.

We also evaluated using the MWUA feature directly as an ordering heuristic. However, this approach yielded lower solution quality than ML pruning and increased running time by $50 - 70\%$. This outcome is to be expected: relying on the raw values fails to capture the feature's full predictive power, while also incurring the computational overhead of the pipeline's most expensive component.

All experiments use the ReduMIS dataset [26], covering social, road, mesh, sparse-matrix and Walshaw graphs with 2,000 to $\sim$ 4M vertices. Graphs are randomly split into training and test sets with disjoint classifier and solver hold-out test sets. For solver evaluation we withhold 25% of the 103 graphs, biased toward larger instances while preserving class proportionality. This ensures a meaningful assessment of both ML model generalization and method scalability. From the remaining graphs, 15% are sampled for classifier evaluation.

All experiments were conducted on a server equipped with an AMD EPYC 7281 CPU and 94 GB RAM. The software environment consisted of Ubuntu Linux, Python 3.7.16, and both GCC 10.4.1 and Clang 14.0.6 for compiling solver binaries. Before delving into experimental results, we outline some important preprocessing considerations related to MIS reduction rules.

5.1 Reduction Rules for MIS and Their Experimental Effects

While investigating different solver configurations for MIS, we considered two settings: applying ReduMIS reduction rules before ML pruning, and applying ML pruning directly to unreduced graphs.

Our method performs well on classification metrics and final MIS size in both settings, except on highly reducible instances. On some graphs, reductions remove $> 90\%$ of vertices, and although the classifier effectively mimics these reductions and final MIS size is good, the optimized reduction routines are much faster than computing features for every vertex in the original graph.

Accordingly, we preprocess with ReduMIS reductions for both training and evaluation to focus the classifier on the irreducible core of problem instances, which yields more stable results and better runtimequality trade-offs. This preprocessing also enables us to compare to a state-of-the-art ML baseline: Learning-what-to-Defer (LwD). Without prior reductions, LwD runs into memory errors on many of the unreduced test graphs (the largest of which contain over 4M vertices, 40M edges). Reduced graphs extend LwDs scalability by avoiding these failures and allow us to train and test the method on reduced instances for fair comparison. We also tested the GNN-based weighted MIS solver by Langedal et al. [27]. This solver too, yielded memory errors on unreduced graphs, and could only solve a small number of the reduced test instances.

Implementing these reductions in preprocessing has knock-on effects on the training and test distributions. Reductions are highly effective on sparse matrices, road graphs, and some social networks, while mesh graphs are the least reducible. As classification is on a per-vertex basis, post-reduction data is dominated by mesh graphs. The implications of this shift are discussed in Sects. 5.2 and 5.3.

Reductions are far less effective for k=2 independent-set instances, and comparatively effective reduction techniques are not available for VCP_k in general. Empirically, the PACE hitting-set reductions did not substantially shrink our tested VCP_3 problems, so training and evaluation are performed on original graphs in this setting.

5.2 Classification Results

Here we report results for our random forest classifier. Support vector machines achieved similar accuracy but random forests offered faster inference and better interpretability, making them preferable for solver integration. All results are reported on a hold-out set of test graphs.

Balanced accuracy is our chosen metric due to class imbalance. For MIS with $k = 1$ and $k = 2$, as well as for 3-path vertex cover, balanced accuracies across random train–test splits fall in the 52%–58% range. Although only modestly above random on the low end, this metric has some key limitations: solver performance is our primary target, and label variability arises due to the presence of equivalent solutions.

Any single vertex can have a different label depending on solver initialization conditions. This is particularly pronounced in mesh graphs where many vertices have identical feature vectors, and can receive different labels depending on solver initialization. These results highlight the intricacies in training classifiers for this purpose, and we examine how these classifiers provide powerful inference despite these difficulties in the solver results in Sect. 5.3.

Analyzing feature importance was key in guiding the design and refinement of our ML-augmented solver. Random forests allow direct assessment of feature importance, enabling removal of low-importance features to reduce runtime and solver overhead. Table 2 reports feature importances for all classifiers.

Table 2. Features Importances for All Problem Variants

Feature	k=1	k=2	VCP_3
Approximate LPR Solution Value (MWUA)	0.3831	0.1636	0.2375
PageRank score	0.3491	0.1792	0.0821
Average Incident Constraint Weight (MWUA)	0.1316	0.2134	0.0184
Minimum Incident Constraint Weight (MWUA)	0.0416	0.0214	0.1100
Maximum Incident Constraint Weight (MWUA)	0.0400	0.0891	0.0112
Maximal Independent Set Membership Frequency	0.0208	0.0236	0.0164
Average neighbor degree rank	0.0079	0.0610	0.0253
Local clustering coefficient	0.0068	0.0535	0.0080
Degree centrality	0.0053	0.0634	0.2997
Maximum Neighbor Degree Rank	0.0048	0.0280	0.0135
Degree rank	0.0035	0.0209	0.0210
Minimum Neighbor Degree Rank	0.0027	0.0368	0.0148
Core Number	0.0027	0.0460	0.0143

We see similar results across problem variants, although some reordering is present among the LPR-derived features. One difference of note lies in the preferred generic centrality measure between MIS and VCP_3. For MIS, PageRank proved useful as it reflects global centrality: vertices with lower PageRank tend to be peripheral and are more likely to belong to independent sets. In a similar vein, we see that degree centrality is the most useful feature for inference on VCP_3. Beyond these two, other centrality measurements were of diminishing usefulness. The engineering effort invested in the MWUA is clearly justified here. For MIS, the importance of the approximate LPR solution value is similar to the importance assigned when we used the exact LPR solution in preliminary experiments, but the feature is now consistently and quickly calculated. Further useful inference is drawn from other MWUA-derived features in all settings, with average associated constraint weight being the most useful feature overall in the $k = 2$ MIS setting.

5.3   MIS Solver Results

We present solver results for both independent set variants, focusing on the effect of the pruning parameter θ. Lower values result in more aggressive pruning, yielding a spectrum of trade-offs between runtime and solution quality, as depicted in Figs. 1 and 2. These results are reported on a second hold-out set of test graphs, completely separate from those in Sect. 5.2.

We compare the ML-based approach to a simple degree-based heuristic: vertices are sorted by degree and the highest-degree vertices are pruned (e.g., at $\theta = 0.95$, the top 5% of vertices are removed). As with ML pruning, θ is varied to explore trade-offs. The choice of a desirable operating point along these tradeoff curves ultimately depends on the practitioner's priorities. We also present results from LwD as described in Sect. 5.1.

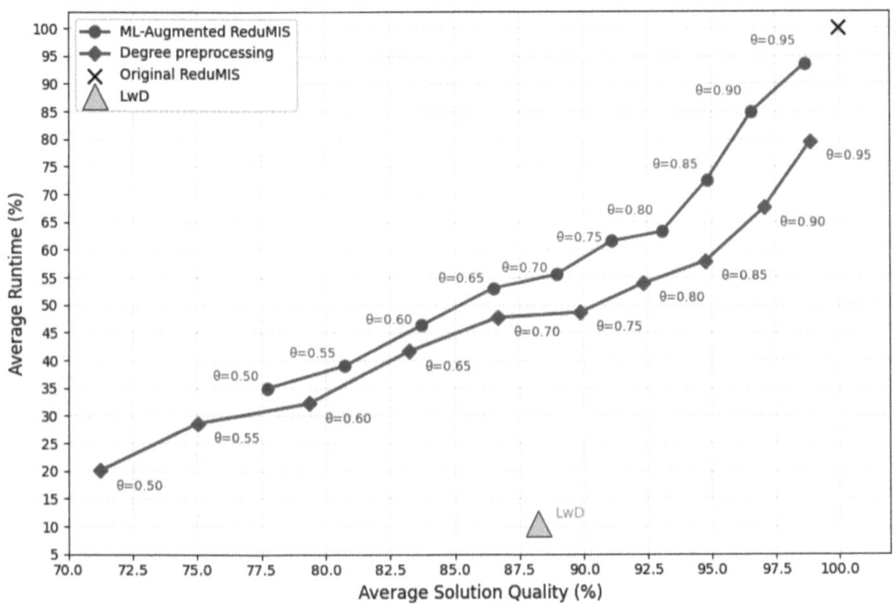

Fig. 1. $k = 1$ Independent Set: Average Runtime vs Solution Quality.

Figure 1 shows that LwD on average provides a solution almost 90% as large as ReduMIS in roughly 10% of the runtime, representing an extremely favorable trade-off when compared to all other methods, except in use cases particularly sensitive to solution quality. We also see that despite the extra cost of feature computation and prediction evaluation, ML-based pruning provides an overall speedup across all θ values while maintaining solution quality above 80% of the original ReduMIS solution, even when pruning up to 45% of vertices. However, simpler degree-based pruning generally offers a better solution-quality versus runtime trade-off in comparison, especially when fewer vertices are pruned. Figure 2 illustrates how this dynamic changes for the $k = 2$ MIS setting.

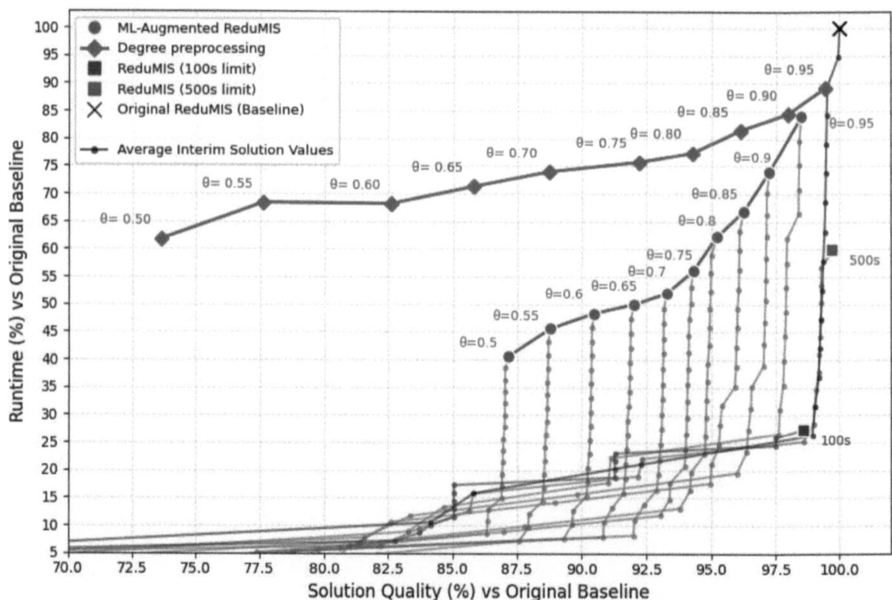

Fig. 2. $k = 2$ Independent Set: Average Runtime vs Solution Quality.

The relative performance of these approaches is completely flipped in this setting. On the larger, denser, and less reducible graphs, LwD is again unusable due to memory errors. To give a view of how solution values evolve over time, we extract interim solution values from the ReduMIS solver (accounting for preprocessing time), as well as plotting results from ReduMIS with shorter timeouts. For visual clarity, we omit degree pruning interim solutions, which analysis confirms consistently underperform alternative methods.

ML-based pruning limits peak quality but converges rapidly, yielding excellent runtime-quality trade-offs for specific configurations. Comparing the time required to hit specific solution quality milestones, a threshold of $\theta = 0.7$ allows ML pruning to reach 85% of the best ReduMIS solution size (found with full $1000s$ runtime budget) on average 3.5 times faster than the original solver reaches this quality. The only behavioral exception occurs on mesh graphs, where pruning restricts peak quality slightly more but accelerates runtimes significantly. Excluding these, ML pruning ($\theta = 0.7$) surpasses 90% of best solution quality on average 3 times faster than the original solver. Importantly, ML reaches these milestones faster than the baseline across all test instances, demonstrating highly consistent performance. This runtime advantage diminishes at peak quality milestones, where the ML approach can even reach them.

5.4 3-Path Vertex Cover Solver Results

Similarly to the MIS experiments, Fig. 3 compares ML-based pruning (Sect. 4.4) against a degree-based baseline for VCP_3. Note that VCP_3 is a minimization problem, contrary to our previous results. We again focus on the pruning parameter θ (where lower values equal more aggressive pruning), and average all results over 5 runs to offset the randomness present in the PACE heuristic local search routine.

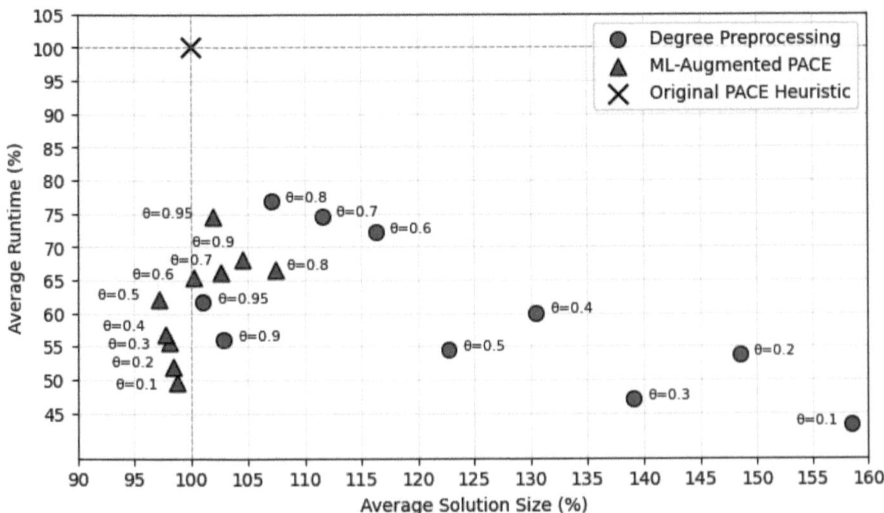

Fig. 3. VCP_3: Average Runtime vs Solution Size.

Aggressive degree-based pruning cuts runtime by almost 60%, but severely degrades solution quality. As pruning is tempered, performance trends toward the baseline, with a notable exception: conservative thresholds ($\theta = 0.9$ and 0.95) maintain solution quality within 5% while keeping runtimes between $55 - 65\%$.

The trend across θ values for ML pruning is more nuanced. Conservative pruning approaches do not prune enough of the problem size away to overcome the cost of feature calculation, so although they may select good vertices to prune, this does not ultimately confer an advantage to the ML approach. On the other end of the spectrum where we prune much more aggressively, we see a slight improvement in solution quality over the original solver in half of the running time, indicating strong predictive performance. As θ increases, so does solution quality, with the best solution quality seen at $\theta = 0.5$. Beyond this, as problem size grows, solutions converge towards the more conservative pruning results.

6 Discussion

We present an ML-augmented solver framework for independent set and vertex-cover variants that enables practitioners to scale to large & difficult problem instances, while also providing a range of different solution quality versus runtime tradeoffs which can be selected according to the relevant use case. In doing so, we address a number of challenges specific to these problems.

Firstly, we outline the difficulties of training a useful classifier in this setting: optimal labels are costly to obtain, and the presence of many equivalent optimal solutions introduces noise that can obfuscate the learning goals. Despite this, our classifiers provide effective pruning in the solver pipeline. While effective reduction rules exist for MIS, our experiments demonstrate that ML can build on this established algorithmic knowledge.

To strengthen predictions in time-sensitive settings, we introduce a lightweight LPR surrogate based on the multiplicative weight update algorithm (MWUA). MWUA yields an approximate fractional solution and informative, fast-to-compute constraint-weight features that admit a fixed-time budget. Consequently, MWUA is a promising, broadly applicable feature for combinatorial problems where LPRs are too costly or provide limited inferential power.

Finally, we demonstrate that our ML-augmented MIS solver can deliver significant runtime gains without sacrificing solution quality, though performance remains somewhat instance-dependent. Results on the G^2 graphs are strongest, highlighting that ML pruning is most effective on dense, less reducible inputs – instances long regarded among the most challenging for MIS solvers. However, ML pruning is limited at the higher end of solution quality, due to some potentially optimal parts of the graph being pruned by the classifier. Importantly, the same pipeline yields even stronger solver outcomes on 3-path vertex cover. ML pruning enables consistent improvements in both runtime and solution quality on very large instances, underscoring the quality and flexibility of this approach across problem domains.

Acknowledgements. We would like to thank Syed Mahmudul Kabir Ratul for providing his implementation of the MWUA, which was a valuable addition to our codebase. This publication has emanated from research supported in part by a grant from Science Foundation Ireland (Grant number 18/CRT/6183). For the purpose of Open Access, the authors have applied a CC BY public copyright licence to any author-accepted manuscript version arising from this submission.

References

1. Ahn, S., Seo, Y., Shin, J.: Learning what to defer for maximum independent sets. In: ICML, Volume 119 of Proceedings of Machine Learning Research, pp. 134–144. PMLR (2020)
2. Akiba, T., Iwata, Y.: Branch-and-reduce exponential/FPT algorithms in practice: a case study of vertex cover. Theor. Comput. Sci. **609**, 211–225 (2016)

3. Andrade, D.V., Resende, M.G.C., Werneck, R.F.: Fast local search for the maximum independent set problem. J. Heuristics, 18(4), 525–547 (2012). https://doi.org/10.1007/S10732-012-9196-4

4. Angriman, E., van der Grinten, A., Hamann, M., Meyerhenke, H., Penschuck, M.: Algorithms for large-scale network analysis and the networkit toolkit. In: Bast, H., Korzen, C., Meyer, U., Penschuck, M. (eds.) Algorithms for Big Data. Lecture Notes in Computer Science, vol. 13201. Springer, Cham (2022). https://doi.org/10.1007/978-3-031-21534-6_1

5. Arora, S., Hazan, E., Kale, S.: The multiplicative weights update method: a meta-algorithm and applications. Theory Comput. 8(1), 121–164 (2012)

6. Böther, M., Kißig, O., Taraz, M., Cohen, S., Seidel, K., Friedrich, T.: What's wrong with deep learning in tree search for combinatorial optimization. In: Proceedings of ICLR 2022 (2022)

7. Breiman, L.: Random forests. Mach. Learn. **45**, 5–32 (2001)

8. Brešar, B., Kardoš, F., Katrenič, J., Semanišin, G.: Minimum k-path vertex cover. Discret. Appl. Math. **159**(12), 1189–1195 (2011)

9. Brusca, L., Quaedvlieg, L.C.P.M., Skoulakis, S., Chrysos, G., Cevher, V.: Self-training through dynamic programming: Maximum independent set. NeurIPS **36**, 40637–40658 (2023)

10. Brüstle, N., Elbaz, T., Hatami, H., Kocer, O., Ma, B.: Approximation Algorithms for Hitting Subgraphs. In: Flocchini, P., Moura, L. (eds.) IWOCA 2021. LNCS, vol. 12757, pp. 370–384. Springer, Cham (2021). https://doi.org/10.1007/978-3-030-79987-8_26

11. Chang, L., Li, W., Zhang, W.: Computing A near-maximum independent set in linear time by reducing-peeling. In Proceedings of SIGMOD 2017, pp. 1181–1196. ACM (2017). https://doi.org/10.1145/3035918.3035939

12. Chen, J., Kanj, I.A., Jia, W.: Vertex cover: Further observations and further improvements. J. Algorithms **41**(2), 280–301 (2001)

13. Cortes, C., Vapnik, V.: Support-vector networks. Mach. Learn. **20**, 273–297 (1995)

14. Cygan, M., Fomin, F.V., Kowalik, Ł, Lokshtanov, D., Marx, D., Pilipczuk, M., Pilipczuk, M., Saurabh, S.: Parameterized Algorithms. Springer, Cham (2015). https://doi.org/10.1007/978-3-319-21275-3

15. Nair, et al.: Solving mixed integer programs using neural networks. CoRR, abs/2012.13349, (2020)

16. FICO. FICO Xpress Optimization, (2025). https://www.fico.com/xpress

17. Fitzpatrick, J., Ajwani, D., Carroll, P.: Learning to sparsify travelling salesman problem instances. In: CPAIOR (2021)

18. Großmann, E., Langedal, K., Schulz, C.: Concurrent iterated local search for the maximum weight independent set problem. In: SEA, volume 338 of LIPIcs, pp. 22:1–22:18 (2025). https://doi.org/10.4230/LIPICS.SEA.2025.22

19. Hespe, D., Lamm, S., Schulz, C., Strash, D.: WeGotYouCovered: the winning solver from the PACE 2019 challenge, vertex cover track. In: Proceedings of CSC 2020, pp. 1–11. SIAM (2020). https://doi.org/10.1137/1.9781611976229.1

20. Hopcroft, J.E., Karp, R.M.: An $O(n^{5/2})$ algorithm for maximum matchings in bipartite graphs. SIAM J. Comput. **2**(4), 225–231 (1973)

21. Huang, T., Ferber, A.M., Zharmagambetov, A., Tian, Y., Dilkina, B.: Contrastive predict-and-search for mixed integer linear programs. In ICML. OpenReview.net (2024)

22. Jena, S.K., Subramani, K.: Analyzing the 3-path vertex cover problem in selected graph classes. J. Comb. Optim. **49**(4), 1–24 (2025)

23. Karp, R.M.: Reducibility Among Combinatorial Problems. In: Jünger, M., Liebling, T.M., Naddef, D., Nemhauser, G.L., Pulleyblank, W.R., Reinelt, G., Rinaldi, G., Wolsey, L.A. (eds.) 50 Years of Integer Programming 1958-2008. LNCS, pp. 219–241. Springer, Heidelberg (2010). https://doi.org/10.1007/978-3-540-68279-0_8
24. Khalil, E.B., Dai, H., Zhang, Y., Dilkina, B., Song, L.: Learning combinatorial optimization algorithms over graphs. In: NeurIPS, pp. 6348–6358 (2017)
25. Lamm, S., Sanders, P., Schulz, C., Strash, D., Werneck, R.F.: Finding near-optimal independent sets at scale. In: 2016 Proceedings of the eighteenth workshop on algorithm engineering and experiments (ALENEX), pp. 138–150. SIAM (2016)
26. Lamm, S., Sanders, P., Schulz, C., Strash, D., Werneck, R.F.: Finding near-optimal independent sets at scale. J. Heuristics **23**(4), 207–229 (2017). https://doi.org/10.1007/S10732-017-9337-X
27. Langedal, K., Hespe, D., Sandersm P.: Targeted branching for the maximum independent set problem using graph neural networks. In SEA, volume 301 of LIPIcs, pp. 20:1–20:21 (2024). https://doi.org/10.4230/LIPICS.SEA.2024.20
28. Langedal, K., Langguth, J., Manne, F., Schroederl, D.T.: Efficient minimum weight vertex cover heuristics using graph neural networks. In: SEA, volume 233 of LIPIcs, pp. 12:1–12:17 (2022)
29. Lauri, J., Dutta, S.: Fine-grained search space classification for hard enumeration variants of subset problems. In: AAAI, pp. 2314–2321. AAAI Press (2019). https://doi.org/10.1609/AAAI.V33I01.33012314
30. Lauri, J., Dutta, S., Grassia, M., Ajwani, D.: Learning fine-grained search space pruning and heuristics for combinatorial optimization. J. Heuristics **29**(2), 313–347 (2023)
31. Lee, E.: Partitioning a graph into small pieces with applications to path transversal. In: Proceedings of the Twenty-Eighth Annual ACM-SIAM Symposium on Discrete Algorithms, pp. 1546–1558. SIAM (2017)
32. John, M., Lewis, M., Yannakakis: The node-deletion problem for hereditary properties is np-complete. J. Comput. Syst. Sci. **20**(2), 219–230 (1980)
33. Li, Z., Chen, Q., Koltun, V.: Combinatorial optimization with graph convolutional networks and guided tree search. NeurIPS **2018**, 537–546 (2018)
34. Luby, M.: A simple parallel algorithm for the maximal independent set problem. SIAM J. Comput. **15**(4), 1036–1053 (1985)
35. Luo, C., Zhang, Q., Su, Z., Lü, Z.: Pace solver description: Weighting-based local search heuristic for the hitting set problem. In: 20th International Symposium on Parameterized and Exact Computation (IPEC 2025), volume 358 of LIPIcs, pp. 40:1–40:4 (2025)
36. Nemhauser, G.L., Trotter, Jr., L.E. : Vertex packings: structural properties and algorithms. Math. Program. **8**(1), 232–248 (1975)
37. PACE-Challenge. https://pacechallenge.org/about/. Accessed 16 Mar 2026
38. Page, L., Brin, S., Motwani, R., Winograd, T.: The Pagerank Citation Ranking: Bringing Order to the Web. Technical report, Stanford (1999)
39. Pedregosa, F., et al.: Scikit-learn: Machine learning in python. JMLR **12** (2011)
40. Plachetta, R., van der Grinten, A.: Sat-and-reduce for vertex cover: Accelerating branch-and-reduce by SAT solving. In: Proceedings of ALENEX 2021, pp. 169–180. SIAM (2021). https://doi.org/10.1137/1.9781611976472.13
41. Plotkin, S.A., David, B., Shmoys, É.: Tardos: fast approximation algorithms for fractional packing and covering problems. Math. Oper. Res. **20**(2), 257–301 (1995)
42. Stephen, B.: Seidman: Network structure and minimum degree. Social Netw. **5**(3), 269–287 (1983)

43. Sun, Y., Ernst, A., Li, X., Weiner, J.: Generalization of machine learning for problem reduction: a case study on travelling salesman problems. OR Spectrum **43**, 607–633 (2021)
44. Tayebi, D., Ray, S., Ajwani, D.: Learning to prune instances of k-median and related problems. In: ALENEX, pp. 184–194 (2022)
45. Jianhua, T.: A survey on the k-path vertex cover problem. Axioms **11**(5), 191 (2022)
46. Xiao, M., Nagamochi, H.: Exact algorithms for maximum independent set. Inf. Comput. **255**, 126–146 (2017). https://doi.org/10.1016/J.IC.2017.06.001
47. Zhang, D., Dai, H., Malkin, N., Courville, A.C., Bengio, Y., Pan, L.: Let the flows tell: solving graph combinatorial problems with GFlowNets. In: Oh, A., Naumann, T., Globerson, A., Saenko, K., Hardt, M., Levine, S. (eds.) NeurIPS 2023, vol. 36, pp. 11952–11969 (2023)
48. Zhang, J., Tayebi, D., Ray, S., Ajwani,D.: Learning to prune instances of Steiner Tree Problem in graphs. In: INOC, pp. 40–45. OpenProceedings.org (2024)
49. Zuckerman, D.: Linear degree extractors and the inapproximability of max clique and chromatic number. In: STOC, pp. 681–690. ACM (2006)

Computing Minimax Regret by Bounding the Weight Space From Within and Without

Guillaume Escamocher[1]([✉]), Paolo Viappiani[2], and Nic Wilson[1]

[1] Insight Centre for Data Analytics, School of Computer Science and IT,
University College Cork, Cork, Ireland
`guillaume.escamocher@insight-centre.org`

[2] LAMSADE, CNRS and Université Paris-Dauphine, PSL, 75016 Paris, France

Abstract. We consider the problem of determining the minimax regret of a set of utility vectors, where the minimax regret is defined with respect to a given space of possible weight vectors. We present two novel approaches for this problem. The first one improves on an existing method based on the set of undominated utility vectors, by adding maximum regret upper bounds in order to reduce the number of linear programming computations. Our second approach uses a branch-and-bound algorithm that can be applied when the utility vectors are given in the form of sub-utility functions over a set of Boolean variables. Both of our approaches make heavy use of two particular sets of weight vectors, one that approximates the convex hull of the weight space from within, and the other from without. We show that our two approaches complement each other well, with each one having its own combinations of parameters for which it outperforms the other.

1 Introduction

Reasoning with preferences and supporting decision making is a key topic in artificial intelligence applications [15]. AI applications need to have a preference mode in order to adapt to the user and propose personalized solutions.

Since acquiring preferences from users is costly, it is often necessary to make choices with partial preference information. The minimax regret criterion [16] has been successfully adopted as a criterion for making robust decisions in presence of preference uncertainty in a variety of settings including multi attribute domains, social choice problems [12], search problems [3], complex aggregators in multi-criteria settings [4], and decisions under risk [11,14].

Preferences such as *"a is as least as good as b"* are modeled using strict (i.e., non-probabilistic) uncertainty, that is, allowing any utility function that gives at least as high utility to a than to b. The minimax regret criterion can be thought of as a hypothetical game between the user, who makes a choice x, and an adversary that will choose the utility function $\bar{u}$ imposing the highest regret, where the regret is defined as the difference between the utility of the best alternative and that of x.

© The Author(s), under exclusive license to Springer Nature Switzerland AG 2026
T. Guns (Ed.): CPAIOR 2026, LNCS 16595, pp. 138–154, 2026.
https://doi.org/10.1007/978-3-032-27242-3_9

In addition to supporting decision making, minimax regret can be used in an interactive process [6] where questions are asked to acquire additional information. The process continues until the minimax regret is zero (a necessary winner is identified) or is less than a given threshold.

The computation of minimax regret depends on the type of decision problem. In configuration problems where the underlying decision problem can be framed as a linear programming problem, the computation of minimax regret can be performed [6] by using mixed integer linear programming using Benders decomposition and constraint generation. For "database" problems where the list of possible choices is explicitly given, minimax regret computation has been tackled as a search problem [7] where the tree search has depth 2.

Computation of minimax regret remains challenging when the number of alternatives or the number of objectives is high. Moreover, the problem of optimizing regret in discrete optimization settings, where the utility is decomposed into several sub-utilities that are defined on a subset of attributes, has been understudied. In this paper, we propose efficient methods to compute minimax regret by considering both an approximation of the space of possible utility weights from within (providing a lower bound of regret) and from without (providing an upper bound). The proposed methods are based on search and aim at pruning the search as much as possible. We show with experiments that our methods allow one to compute minimax regret faster than the state of the art, and therefore the proposed algorithms are suitable to compute minimax regret in large outcome spaces.

While our work focuses on the computation of minimax regret and not on preference elicitation, we stress that our algorithms could be used in interactive elicitation schemes following query strategies that ask queries to reduce regret [6,7,19].

Before continuing, we mention some alternative approaches to decision making under uncertain utility and preference elicitation; several other works addressed the problem by considering Bayesian approaches [5,19] to represent uncertainty in utility functions and selecting informative questions, methods based on picking a representative utility function that maximize the margins in the constraints that represent user preferences [1,17], and a non Bayesian probabilistic approach [9].

The paper is organized as follows. We first review some background in Sect. 2; then, in Sect. 3, we describe our algorithms for computing minimax regret. Experimental results are presented in Sect. 4; finally, Sect. 5 provides some concluding remarks.

2   Technical Background

Throughout the paper, we denote by $v(i)$ the ith component of a vector v. For example, if v is the vector $\langle 3, 5, 8, 13 \rangle$, then $v(3) = 8$.

2.1 Multi-objective Optimization Problem

An instance of the d-Objective Optimization Problem is composed of two parts: a set $\mathcal{A}$ of alternatives and a set $\mathcal{I}$ of preference inequalities.

Each alternative α in $\mathcal{A}$ is a d-dimensional vector, with d being the number of objectives of the instance. The alternatives are *utility* vectors, in that $\alpha(i)$ represents the utility of picking α with respect to the ith objective.

We say that an alternative α is *Pareto-dominated* by a different alternative β if we have $\beta(i) \geq \alpha(i)$ for each i such that $1 \leq i \leq d$. We say that an alternative of $\mathcal{A}$ is *Pareto-undominated* (abbreviated to *undominated*) if it is not Pareto-dominated by any alternative from $\mathcal{A}$.

The *total utility* of an alternative depends on how much weight is given to each objective. If w is a d-dimensional weight vector and α is a d-dimensional alternative then $f_w(\alpha)$, the total utility of α, is equal to $w \cdot \alpha$, the scalar product of w and α. In this paper, as is commonly done, we assume that all weight vectors belong to $\mathcal{N}$, the set of weight vectors that have weights that are real numbers in $[0, 1]$ and that sum to 1. We also assume that there exists a weight vector $w^\star$ that represents the actual preferences of the user. Ideally, we would want to return to the user the alternative α that maximizes $f_{w^\star}(\alpha)$. However, $w^\star$ is not part of the instance input, and in fact, it might even be unknown to the user. The only information available about the user's preferences is provided by the preference inequalities in the set $\mathcal{I}$.

Each preference inequality in $\mathcal{I}$ is a comparison between two d-dimensional utility vectors. The *weight space* $\mathcal{W}$ is the set of weight vectors that satisfy all inequalities in $\mathcal{I}$. Consider for example a preference inequality over three dimensions that states that $\langle 0, 7, 0 \rangle \leq \langle 8, 0, 3 \rangle$. If the dimensions are x, y and z respectively, then the space of vectors that satisfy this inequality is bounded by the plane $8x - 7y + 3z = 0$. This plane intersects $\mathcal{N}$ in $\langle 0, .3, .7 \rangle$ and $\langle \frac{7}{15}, \frac{8}{15}, 0 \rangle$, as well as in all the points between these two vectors. Note that because the weights of every element of $\mathcal{N}$ sum to 1, each vector in $\mathcal{N}$ can be uniquely identified with only its first two coordinates. We visually represent this projection in Fig. 1a. The dashed line in the figure represents the projection of the intersection between $\mathcal{N}$ and the plane $8x - 7y + 3z = 0$. The projection of $\mathcal{N}$ itself is the bottom left half of the figure, the triangle with vertices $(0, 0)$, $(1, 0)$ and $(0, 1)$. The vectors in $\mathcal{N}$ that are projected below the dashed line satisfy the preference inequality $\langle 0, 7, 0 \rangle \leq \langle 8, 0, 3 \rangle$, they form $\mathcal{W}$. Another example with more preference inequalities is given in Fig. 1b.

In general, the only restriction on the utility vectors that are compared in $\mathcal{I}$ is that they both are d-dimensional, with d being the number of objectives of the instance. In practice, preference inequalities are often obtained from preference elicitation, so the utility vectors that they compare directly or indirectly originate from $\mathcal{A}$.

2.2 Regret

The preference inequalities in $\mathcal{I}$ may very well not provide enough information to determine which alternative of $\mathcal{A}$ maximizes $f_{w^\star}(\alpha)$. If this is the case, in

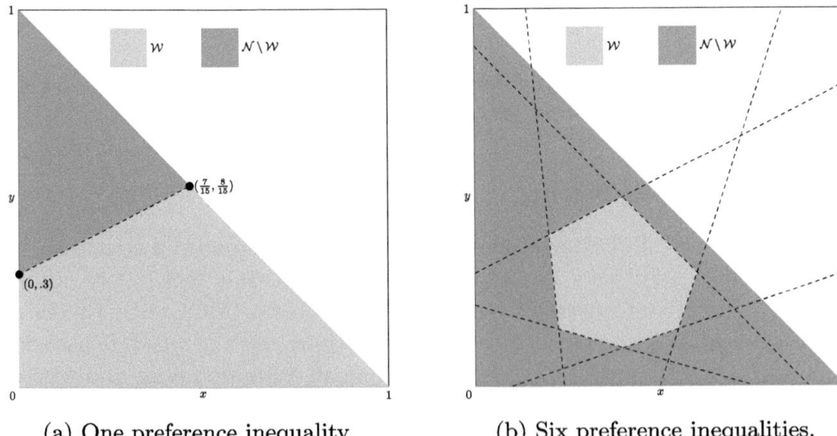

(a) One preference inequality.   (b) Six preference inequalities.

Fig. 1. How preference inequalities (dashed lines) partition weight vectors.

order to make a recommendation to the user we choose the alternative that achieves the minimax regret [16], which is the lowest maximum regret, where the maximum regret of an alternative α is the highest pairwise maximum regret between α and another alternative of $\mathcal{A}$.

Let us properly define all of these concepts, starting with the pairwise maximum regret (PMR):

$$PMR(\alpha, \beta; \mathcal{W}) = \max_{w \in \mathcal{W}} (f_w(\beta) - f_w(\alpha)).$$

Notice that for any alternative α, we have $PMR(\alpha, \alpha; \mathcal{W}) = 0$.

Now that we have defined the PMR, we can define the maximum regret (MR) of an alternative:

$$MR(\alpha, \mathcal{A}; \mathcal{W}) = \max_{\beta \in \mathcal{A}} PMR(\alpha, \beta; \mathcal{W}).$$

Finally, the minimax regret (MMR) of an instance is the minimum MR:

$$MMR(\mathcal{A}; \mathcal{W}) = \min_{\alpha \in \mathcal{A}} MR(\alpha, \mathcal{A}; \mathcal{W})$$

If $\mathcal{I}$ is a set of preference inequalities, and $\mathcal{W}$ is the set of weight vectors that satisfy $\mathcal{I}$, then we will accept either $\mathcal{I}$ or $\mathcal{W}$ as an argument of PMR, MR and MMR. So for example, we will have $MMR(\mathcal{A}; \mathcal{I}) = MMR(\mathcal{A}; \mathcal{W})$.

If $w_1, w_2 \in \mathcal{W}$ and w is on the line segment between w_1 and w_2 then $w \cdot (\beta - \alpha)$ is between $w_1 \cdot (\beta - \alpha)$ and $w_2 \cdot (\beta - \alpha)$, and so adding w to $\mathcal{W}$ does not change $PMR(\alpha, \beta; \mathcal{W})$. Hence, if $\mathcal{W}$ is equal to the convex hull of $\mathcal{T}$ then

$$PMR(\alpha, \beta; \mathcal{W}) = PMR(\alpha, \beta; \mathcal{T}), \tag{1}$$

and thus, $MR(\alpha, \mathcal{A}; \mathcal{W}) = MR(\alpha, \mathcal{A}; \mathcal{T})$, and $MMR(\mathcal{A}; \mathcal{W}) = MMR(\mathcal{A}; \mathcal{T})$. This holds, in particular, if $\mathcal{T}$ equals the set of extreme points of $\mathcal{W}$. Also, it can be easily seen that $MMR(\mathcal{A}; \mathcal{W}) = MMR(\mathcal{A}'; \mathcal{W})$, where $\mathcal{A}'$ is the set of undominated elements of $\mathcal{A}$.

For a given $w \in \mathcal{W}$, we define the highest total utility with respect to w among all alternatives of $\mathcal{A}$ as $MU_w(\mathcal{A}) = \max_{\alpha \in \mathcal{A}} f_w(\alpha)$.

2.3 Combinatorial Setting

Sometimes the set $\mathcal{A}$ of alternatives arises from a combinatorial structure, such as a generalized additively decomposable utility function [2,8,10]; or similarly to a Multi-Objective Constraint Optimization Problem (MOCOP) [13]. In such situations we have a number of sub-utility functions, each of which is associated with a set (called its *scope*) of variables with small finite domains, and where an assignment to the scope of the function determines a multi-objective vector. A complete assignment $\mathbf{x}$ to all the variables determines a multi-objective vector for each of the functions, with the overall utility vector $u(\mathbf{x})$ associated with a complete assignment $\mathbf{x}$ being the sum of these vectors. The set $\mathcal{A}$ in such a situation is then equal to the set of all vectors $u(\mathbf{x})$ over all complete assignments $\mathbf{x}$ to the variables.

We show in Fig. 2 an example of a combinatorial instance with three objectives, three Boolean variables x_1, x_2 and x_3, and two sub-utility functions u_1 and u_2, both of arity 2. Because there are three variables and all domains are Boolean, the total number of alternatives in $\mathcal{A}$ will be $2^3 = 8$. The alternative corresponding to assigning 1 to x_1, 0 to x_2 and 1 to x_3 will be the sum of the utility vector returned by u_1 when assigning 1 to x_1 and 0 to x_2 (the two variables in the scope of u_1) and of the utility vector returned by u_2 when assigning 0 to x_2 and 1 to x_3 (the two variables in the scope of u_2), in this case $\langle 8, 6, 10 \rangle + \langle 4, 9, 8 \rangle = \langle 12, 15, 18 \rangle$.

x_1 x_2	u_1
0 0	$\langle 4, 7, 7 \rangle$
0 1	$\langle 7, 10, 4 \rangle$
1 0	$\langle 8, 6, 10 \rangle$
1 1	$\langle 6, 1, 10 \rangle$

x_2 x_3	u_2
0 0	$\langle 7, 7, 6 \rangle$
0 1	$\langle 4, 9, 8 \rangle$
1 0	$\langle 2, 3, 8 \rangle$
1 1	$\langle 8, 9, 1 \rangle$

Fig. 2. Two sub-utility functions u_1 (with scope $\{x_1, x_2\}$) and u_2 (with scope $\{x_2, x_3\}$).

In general, a combinatorial structure allows for a more compact representation of a multi-objective optimization instance than listing all alternatives. Consider, for example, a combinatorial instance with 4 objectives, 20 Boolean variables, and 30 sub-utility functions of arity 3. The number of utility vectors to represent for a given sub-utility function is the domain size to the power of the arity of the sub-utility function, so in this case each sub-utility function can be stored using only 35 integers (8 vectors times 4 dimensions plus 3 additional integers for the scope), for a total of 1050 integers for the whole instance. Compare with the more than 4 million integers that it would have required to enumerate all 2^{20} four-dimensional utility vectors in $\mathcal{A}$.

Another benefit of a combinatorial structure is to allow search tree operations, like in a Constraint Satisfaction Problem (CSP), over the complete assignments $\mathbf{x}$ and the associated utility vectors, where a leaf of the search tree is associated with a single multi-attribute vector.

3   Finding the Minimax Regret

3.1   Bounding the Weight Space $\mathcal{W}$

Let $\langle \mathcal{A}, \mathcal{I} \rangle$ be a d-objective optimization instance. Let $\mathcal{W}$ be the set of weight vectors that satisfy $\mathcal{I}$. In general $\mathcal{W}$ is infinite, so computing the exact value of $PMR(\alpha, \beta; \mathcal{W})$ for two alternatives α and β from $\mathcal{A}$ requires a call to a Linear Programming solver (LP call). To avoid making the $(|\mathcal{A}| - 1)^2$ LP calls that the brute-force approach requires, we instead approximate $\mathcal{W}$ by two finite sets $\mathcal{W}_{in}$ and $\mathcal{W}_{out}$. The *inner set* $\mathcal{W}_{in}$ will contain vectors that are within $\mathcal{W}$ and will be used to obtain lower bounds of PMR and MR values, while the *outer set* $\mathcal{W}_{out}$ will contain vectors outside, or on the edge of, $\mathcal{W}$ and will be used to obtain upper bounds of PMR and MR values.

We present in Algorithm 1 how we compute $\mathcal{W}_{in}$, using the fact that $\mathcal{W}$ is a compact set and so $w(i)$ attains its bounds on $\mathcal{W}$. Each weight vector added to $\mathcal{W}_{in}$ can be computed with one LP call, so the total number of LP calls made is $2d$. Since $\mathcal{W}_{in}$ is a subset of $\mathcal{W}$, we have for every α and β in $\mathcal{A}$: $PMR(\alpha, \beta; \mathcal{W}_{in}) \leq PMR(\alpha, \beta; \mathcal{W})$.

Data: A set $\mathcal{I}$ of preference inequalities.
Result: The inner set $\mathcal{W}_{in}$ for $\mathcal{I}$.

1 $\mathcal{W}_{in} \leftarrow \emptyset$;
2 **for** $i \leftarrow 1$ **to** d **do**
3   Add to $\mathcal{W}_{in}$ a weight vector w that satisfies $\mathcal{I}$ while minimizing $w(i)$;
4   Add to $\mathcal{W}_{in}$ a weight vector w that satisfies $\mathcal{I}$ while maximizing $w(i)$;
5 **end**
6 Remove duplicate vectors from $\mathcal{W}_{in}$;
7 **return** $\mathcal{W}_{in}$;

Algorithm 1: ComputeInnerSet($\mathcal{I}$)

We visually illustrate in Fig. 3 how to construct $\mathcal{W}_{in}$. This figure is a continuation of the visual example from Fig. 1b, which showed the two-dimensional projection of a three-dimensional set $\mathcal{W}$ defined from six preference inequalities. In this example, the minimum value that $w(1)$ can reach while satisfying all inequalities of $\mathcal{I}$ is .2, attained by weight vector $\langle .2, .4, .4 \rangle$ which projects on $(.2, .4)$, while the corresponding maximum value is .6, attained by $\langle .6, .3, .1 \rangle$ which projects on $(.6, .3)$. Similarly, the extreme values for $w(2)$ are .1 and .5, attained by vectors that project on $(.4, .1)$ and $(.4, .5)$ respectively.

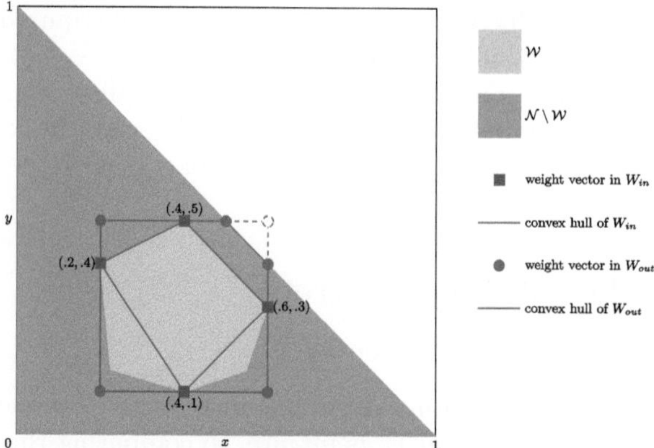

Fig. 3. Constructing $\mathcal{W}_{in}$ and $\mathcal{W}_{out}$.

We present in Algorithm 2 how we compute the outer set $\mathcal{W}_{out}$. Let hyperrectangle $\mathcal{R}$ be the set of elements w of $\mathbb{R}^d$ such that $L_i \leq w(i) \leq U_i$ for all $i = 1, \ldots, d$, where L_i and U_i are, respectively, the minimum and maximum values of $w(i)$ over $w \in \mathcal{W}$. On Line 6, two vectors $w_{i,1}$ and $w_{i,2}$ form an *edge* if $w_{i,1}(i) = L_i$, $w_{i,2}(i) = U_i$, and for each $j \neq i$, either $w_{i,1}(j) = w_{i,2}(j) = L_j$ or $w_{i,1}(j) = w_{i,2}(j) = U_j$. This thus corresponds with an edge of the hyperrectangle $\mathcal{R}$. On Line 7, we define the *norm* as the vector w such that $w(j) = w_{i,1}(j) = w_{i,2}(j)$ for $j \neq i$, and with $w(i)$ such that the components of w sum to 1. Since the number of edges is $d \times 2^{d-1}$, and at most one weight vector per edge is added to $\mathcal{W}_{out}$, we have $|\mathcal{W}_{out}| \leq d \times 2^{d-1}$. Since the values for L_i and U_i can be obtained from $\mathcal{W}_{in}$, the construction of $\mathcal{W}_{out}$ does not require any additional LP call.

Data: A set $\mathcal{I}$ of preference inequalities and the inner set $\mathcal{W}_{in}$ for $\mathcal{I}$.
Result: The outer set $\mathcal{W}_{out}$ for $\mathcal{I}$.

```
1  for i ← 1 to d do
2  │   L_i ← min{w(i) : w ∈ W};
3  │   U_i ← max{w(i) : w ∈ W};
4  end
5  W_out ← ∅;
6  foreach edge (w_{i,1}, w_{i,2}) do
7  │   w ← norm(w_{i,1}, w_{i,2});
8  │   if w_{i,1}(i) ≤ w(i) ≤ w_{i,2}(i) then
9  │   │   Add w to W_out;
10 │   end
11 end
12 Remove duplicate vectors from W_out;
13 return W_out;
```

Algorithm 2: ComputeOuterSet($\mathcal{I}, \mathcal{W}_{in}$)

Figure 3 shows how $\mathcal{W}_{out}$ is obtained from $\mathcal{W}_{in}$. The vertices of the hyper-rectangle $\mathcal{R}$ project to coordinates $(.2, .1)$, $(.6, .1)$, $(.2, .5)$ and $(.6, .5)$. However, this last point is outside of the projection of $\mathcal{N}$, so the two (one for each extreme value of the third objective $w(3)$) vertices that project here will be replaced by the intersections between their adjacent edges and $\mathcal{N}$.

Theorem 1 below implies that $\mathcal{W}_{out}$ gives an upper bound for max regret. To show this, we prove a technical lemma. Let $\mathcal{N}$ be the set of elements w of $\mathbb{R}^d$ such that $\sum_{i=1}^{d} w(i) = 1$ and $w(i) \geq 0$ for all $i = 1, \ldots, d$. Part (i) follows easily from the definitions. Part (ii) follows because $\mathcal{W}_{out}$ consists of all the norms of edges that are in the hyper-rectangle $\mathcal{R}$, and thus in $\mathcal{W}'$. Since each such norm is on an edge of $\mathcal{R}$, and is the only point on that edge in $\mathcal{W}'$, each element of $\mathcal{W}_{out}$ is an extreme point of $\mathcal{W}'$; furthermore, they can be shown to be the only extreme points of $\mathcal{W}'$, giving (iii).

Lemma 1. *Let $\mathcal{W}' = \mathcal{R} \cap \mathcal{N}$. Then (i) $\mathcal{W} \subseteq \mathcal{W}'$; (ii) $\mathcal{W}_{out}$ is the set of all elements w of $\mathcal{W}'$ such that there exists at most one $i \in \{1, \ldots, d\}$ such that $L_i < w(i) < U_i$; and (iii) $\mathcal{W}_{out}$ equals $Ext(\mathcal{W}')$, the set of extreme points of $\mathcal{W}'$.*

Theorem 1. *For every α and β in $\mathcal{A}$, $PMR(\alpha, \beta; \mathcal{W}_{in}) \leq PMR(\alpha, \beta; \mathcal{W}) \leq PMR(\alpha, \beta; \mathcal{W}_{out})$; and $MR(\alpha, \mathcal{A}; \mathcal{W}_{in}) \leq MR(\alpha, \mathcal{A}; \mathcal{W}) \leq MR(\alpha, \mathcal{A}; \mathcal{W}_{out})$.*

Proof. The first inequality for PMR follows from the fact that $\mathcal{W}_{in} \subseteq \mathcal{W}$. For the second inequality, by Lemma 1(i), we have $PMR(\alpha, \beta; \mathcal{W}) \leq PMR(\alpha, \beta; \mathcal{W}')$, and by Lemma 1(iii) and Eq. 1, $PMR(\alpha, \beta; \mathcal{W}') = PMR(\alpha, \beta; \mathcal{W}_{out})$. The MR inequalities follow immediately from the PMR inequalities, by maximising over all $\beta \in \mathcal{A}$. $\square$

Lower and upper bounds r_α and r^α for $MR(\alpha, \mathcal{A}; \mathcal{W})$: We can compute the maximum utility $MU_w(\mathcal{A})$ for each $w \in \mathcal{W}_{in} \cup \mathcal{W}_{out}$, and then compute

$$r_\alpha = MR(\alpha, \mathcal{A}; \mathcal{W}_{in}) = \max_{w \in \mathcal{W}_{in}} MU_w(\mathcal{A}) - f_w(\alpha), \text{and} \qquad (2)$$

$$r^\alpha = MR(\alpha, \mathcal{A}; \mathcal{W}_{out}) = \max_{w \in \mathcal{W}_{out}} MU_w(\mathcal{A}) - f_w(\alpha). \qquad (3)$$

Theorem 1 states that, for every $\alpha \in \mathcal{A}$, $r_\alpha \leq MR(\alpha, \mathcal{A}; \mathcal{W}) \leq r^\alpha$.

3.2   Undominated Approach

We describe in Algorithm 3 (in black, excluding parts in magenta) an approach that has been previously used to compute minimax regret [18]. It relies on removing Pareto-dominated alternatives (Line 3), before computing a lower bound r_α on the maximum regret of each remaining alternative α (Lines 4–6). There are several possible ways to obtain r_α. In Sect. 3.1 we discussed how this can be accomplished using $\mathcal{W}_{in}$ (Line 1). For each undominated alternative α (Line 8), if r_α is low enough (Line 9), then the maximum regret of α is determined by computing the pairwise maximum regret between α and every other alternative β (Lines 11–22), otherwise α is pruned. Each PMR computation (Line

14) is done with a single call to a linear programming solver that computes $\max_{w \in \mathcal{W}} f_w(\beta) - f_w(\alpha) = \max_{w \in \mathcal{W}} w \cdot (\beta - \alpha)$.

Thanks to r_α, the maximum regret does not need to be computed for all alternatives. But even in the best case, when the maximum regret of only one alternative α is computed, the approach still determines $PMR(\alpha, \beta; \mathcal{I})$ for every other undominated alternative β, so the number of LP calls for this approach will always be at least $|\mathcal{A}'| - 1$, where $\mathcal{A}'$ is the set of undominated alternatives in $\mathcal{A}$.

We now introduce UnDominated+$\mathcal{W}_{out}$ (UD+WO), the first of our two novel approaches for computing minimax regret. It extends the UnDominated approach by using the set $\mathcal{W}_{out}$ to avoid many of the PMR computations. The main differences between the UD+WO and the UnDominated approach, highlighted in magenta in Algorithm 3, are:

(a) UD+WO computes the outer set $\mathcal{W}_{out}$ (Line 2 of Algorithm 3).
(b) Before explicitly computing the exact PMR of alternatives α and β (Line 14), UD+WO computes $PMR(\alpha, \beta; \mathcal{W}_{out})$ (Line 12). We know from Theorem 1 that $PMR(\alpha, \beta; \mathcal{I}) \leq PMR(\alpha, \beta; \mathcal{W}_{out})$, so if $PMR(\alpha, \beta; \mathcal{W}_{out})$ is lower than the highest pairwise maximum regret seen so far for α (Line 13), then we know that the maximum regret for α will not be achieved with β, so we do not need to run a linear programming solver to compute $PMR(\alpha, \beta; \mathcal{I})$.

The UnDominated approach uses $\mathcal{W}_{in}$ to find lower bounds for maximum regret, leading to pruning alternatives in the outer loop (Line 8 of Algorithm 3). The UnDominated+$\mathcal{W}_{out}$ approach complements this by also using $\mathcal{W}_{out}$ to find upper bounds for pairwise maximum regret, leading to pruning alternatives in the inner loop (Line 11 of Algorithm 3).

3.3   Branch-and-Bound Approach

In this section we explain how our *Branch-and-Bound (BrBo)* approach works. This approach can be used when the alternatives in $\mathcal{A}$ are not explicitly listed, but are instead implicitly given through sub-utility functions (see Sect. 2.3). The set $\mathcal{W}$ of possible weight vectors is still given through a set $\mathcal{I}$ of preference inequalities.

Use of Search Tree Representation. While the UD and UD+WO approaches loop through all undominated alternatives, the search in our BrBo approach instead follows a search tree, where each node σ of $\mathcal{A}$ can be seen as a set of assignments of the variables of the instance. If σ is a leaf node, then all variables have been assigned. If σ is a non-leaf node, then only some of the variables have been assigned.

Expressing the set $\mathcal{A}$ of alternatives in terms of a search tree offers several advantages compared to individually listing all alternatives. Because non-leaf nodes represent multiple alternatives, more than one alternative can be pruned at once by the BrBo approach. In contrast, only one alternative can be pruned

Data: A multi-objective optimization instance $\langle \mathcal{A}, \mathcal{I} \rangle$.
Result: The minimax regret $MMR(\mathcal{A}; \mathcal{I})$ of the instance.

1 $\mathcal{W}_{in} \leftarrow$ ComputeInnerSet$(\mathcal{I})$;
2 $\mathcal{W}_{out} \leftarrow$ ComputeOuterSet$(\mathcal{I}, \mathcal{W}_{in})$;
3 $\mathcal{A}' \leftarrow$ undominated elements of $\mathcal{A}$;
4 **foreach** $\alpha \in \mathcal{A}'$ **do**
5 | $r_\alpha \leftarrow$ lower bound for $MR(\alpha, \mathcal{A}'; \mathcal{I})$ obtained with $\mathcal{W}_{in}$;
6 **end**
7 $ubound \leftarrow \infty$;
8 **foreach** $\alpha \in \mathcal{A}'$ *in increasing order of* r_α **do**
9   **if** $r_\alpha < ubound$ **then**
10      $lbound \leftarrow 0$;
11   **foreach** $\beta \in \mathcal{A}'$ **do**
12        $pw \leftarrow PMR(\alpha, \beta; \mathcal{W}_{out})$;
13   **if** $pw > lbound$ **then**
14          $r \leftarrow PMR(\alpha, \beta; \mathcal{I})$;
15   **if** $r > lbound$ **then**
16 | $lbound \leftarrow r$;
17   **end**
18   **if** $r \geq ubound$ **then**
19   **break** (out of Line 11 **foreach** loop)
20   **end**
21   **end**
22   **end**
23   **if** $lbound < ubound$ **then**
24 | $ubound \leftarrow lbound$;
25   **end**
26 **end**
27 **end**
28 **return** $ubound$;

Algorithm 3: UnDominated (UD, in black) and UnDominated+$\mathcal{W}_{out}$ (UD+WO) approaches

at a time by the UD and UD+WO approaches. Furthermore, representing the alternatives by sub-utility functions is often far more efficient in terms of space than keeping all the undominated alternatives in memory, as UD and UD+WO need to do.

Let $\mathcal{L}$ be the set of leaf nodes. Each leaf node $\sigma \in \mathcal{L}$ is associated with a multi-attribute utility vector α_σ. Then, $\mathcal{A} = \{\alpha_\sigma : \sigma \in \mathcal{L}\}$. Let $\mathcal{B}$ be the set of non-leaf nodes in the tree. Each non-leaf node $\sigma \in \mathcal{B}$ is (implicitly) associated with a set $\mathcal{A}_\sigma$ of multi-attribute utility vectors, i.e., the set of all $\alpha_{\sigma'}$ for all leaf nodes σ' below σ. Thus, when σ is the root node of the tree we have $\mathcal{A}_\sigma = \mathcal{A}$.

Our BrBo approach explores the tree of alternatives with two kinds of tree searches: the *outer* tree search and the *inner* tree search. The outer tree search looks for an alternative that achieves the lowest maximum regret. From a high-level perspective, it serves the same function as the outer loop of the UD and

UD+WO approaches (Line 8 in Algorithm 3). Every time that the outer tree search reaches a leaf node σ that has not been pruned by the precomputed bounds, it then enters the inner tree search to find the alternative β that maximizes $PMR(\alpha_\sigma, \beta; \mathcal{I})$. This corresponds to the inner loop of the UD and UD+WO approaches (Line 11 in Algorithm 3). Once the outer tree search finishes, the inner tree search has already been done, potentially more than once.

Information Accessible at Non-leaf Nodes: At a non-leaf node σ of the tree we want to be able to generate for any weight vector w an upper bound $\bar{s}_\sigma(w)$ for $\{f_w(\alpha) : \alpha \in \mathcal{A}_\sigma\}$. This means that if α is associated with some leaf node below σ then for any chosen weight vector w, we have $f_w(\alpha) \leq \bar{s}_\sigma(w)$.

Given a node σ of the tree and a set V of variables, we say that an assignment $\mathbf{x}$ to the variables of V is *compatible* with σ if for each variable $v \in V$, either v has not been assigned in σ, or v has been assigned in σ the same value that it has been assigned in $\mathbf{x}$.

Given a set U of utility functions, $\bar{s}_\sigma(w)$ is defined as follows:

$$\bar{s}_\sigma(w) = \sum_{u \in U} \max_{\mathbf{x} \in X_\sigma^u} f_w(u(\mathbf{x})),$$

where X_σ^u is the set of assignments to the variables in the scope of u that are compatible with σ.

Single Optimisation to Compute $MU_w(\mathcal{A})$ for Given w: Recall that $MU_w(\mathcal{A}) = \max_{\alpha \in \mathcal{A}} f_w(\alpha)$, i.e., the maximal value of f_w over $\mathcal{A}$. We can relatively efficiently compute $MU_w(\mathcal{A})$ for any given weight vector w by using a branch-and-bound algorithm, making use of the upper bounds $\bar{s}_\sigma(w)$ accessible at any non-leaf node $\sigma \in \mathcal{B}$. We can write $MR(\alpha, \mathcal{A}; \mathcal{I})$ in terms of $MU_w(\mathcal{A})$:

$$MR(\alpha, \mathcal{A}; \mathcal{I}) = \max_{w \in \mathcal{W}} MU_w(\mathcal{A}) - f_w(\alpha).$$

Outer Tree Search. At each point of each search we have a global upper bound $\bar{s}$ which only ever decreases. At the beginning of the first pass we initialise $\bar{s}$ to be r^α (see Sect. 3.1) for some $\alpha \in \mathcal{A}$. At the end of the tree search we will have $\bar{s} = MMR(\mathcal{A}; \mathcal{I})$.

At a Non-leaf Node. $\sigma \in \mathcal{B}$: we generate a lower bound $G(\sigma)$ for $MR(\alpha \in \mathcal{A}_\sigma, \mathcal{A}; \mathcal{I})$. We adapt the expression $MR(\alpha, \mathcal{A}; \mathcal{W}_{in}) = \max_{w \in \mathcal{W}_{in}} MU_w(\mathcal{A}) - f_w(\alpha)$. We define

$$G(\sigma) = \max(0, \max_{w \in \mathcal{W}_{in}} MU_w(\mathcal{A}) - \bar{s}_\sigma(w))$$

(see Sect. 3.3 for explanation of $\bar{s}_\sigma(w)$). We have, for $\alpha \in \mathcal{A}_\sigma$, $f_w(\alpha) \leq \bar{s}_\sigma(w)$, which implies that, for all $\alpha \in \mathcal{A}_\sigma$, $MR(\alpha, \mathcal{A}; \mathcal{W}_{in}) \geq G(\sigma)$, and thus, $MR(\alpha, \mathcal{A}; \mathcal{I}) \geq G(\sigma)$ for all $\alpha \in \mathcal{A}_\sigma$ as required. If $G(\sigma)$ is strictly greater than the global upper bound $\bar{s}$ for $MMR(\mathcal{A}; \mathcal{I})$, then there is no need to explore below node σ, so we backtrack.

At a Leaf Node. $\sigma \in \mathcal{L}$: Let $\alpha = \alpha_\sigma$ be the associated utility vector.

(1) If $r_\alpha > \bar{s}$ then we backtrack (see Eq. 2 for the definition of r_α).
(2) Otherwise, if $r_\alpha \leq \bar{s}$, we compute $c = MR_{\bar{s}}(\alpha, \mathcal{A}; \mathcal{I})$ and then update $\bar{s}$ to be c. The value $MR_{\bar{s}}(\alpha, \mathcal{A}; \mathcal{I})$ is defined to be $\min(\bar{s}, MR(\alpha, \mathcal{A}; \mathcal{I}))$; it is computed using the method described below in Sect. 3.3. It can sometimes be much faster to compute this than $MR(\alpha, \mathcal{A}; \mathcal{I})$: in cases where we can deduce that $MR(\alpha, \mathcal{A}; \mathcal{I}) \geq \bar{s}$ we can then immediately return $MR_{\bar{s}}(\alpha, \mathcal{A}; \mathcal{I}) = \bar{s}$ without having to do any more computation.

Pre-processing Search: We also have a pre-processing tree search that has the same structure except that at Step (2) the algorithm does not evaluate $MR_{\bar{s}}(\alpha, \mathcal{A}; \mathcal{I})$; in that case we compute r^α (see Eq. 3), and if $r^\alpha < \bar{s}$ then we can update the upper bound $\bar{s}$ to be r^α, and backtrack. The advantage of such an extra search can be that the relatively expensive computation of a value $MR(\alpha, \mathcal{A}; \mathcal{I})$ may turn out later to have been unnecessary, because other alternatives may cause the reduction of the upper bound $\bar{s}$.

Inner Tree Search. We describe here how to compute the value of $MR_{\bar{s}}(\alpha, \mathcal{A}; \mathcal{I})$ by using a branch-and-bound search, going through (implicitly or explicitly) elements β of $\mathcal{A}$ sequentially.

At each point of the search we have a lower bound $\underline{s}_\alpha$, which can be initialised as $r_\alpha = MR(\alpha, \mathcal{A}; \mathcal{W}_{in}) = \max_{w \in \mathcal{W}_{in}} MU_w(\mathcal{A}) - w \cdot \alpha$. If the search is not terminated before reaching the end of the tree, then at that point $\underline{s}_\alpha$ will be equal to $MR(\alpha, \mathcal{A}; \mathcal{I})$.

At a Non-leaf Node. $\sigma \in \mathcal{B}$: we generate an upper bound $H_\alpha(\sigma)$ for $MR(\alpha, \mathcal{A}_\sigma; \mathcal{I})$.

Let $\gamma_\sigma(w)$ be such that $f_w(\beta) \leq \gamma_\sigma(w)$ for all $\beta \in \mathcal{A}_\sigma$. There are multiple ways to obtain a valid $\gamma_\sigma(w)$ for every w, for example by setting $\gamma_\sigma(w) = \bar{s}_\sigma(w)$ (see Sect. 3.3). We define $H_\alpha(\sigma) = \max_{w \in \mathcal{W}_{out}}(\gamma_\sigma(w) - f_w(\alpha))$. Thus, $PMR(\alpha, \beta; \mathcal{W}_{out}) = \max_{w \in \mathcal{W}_{out}} f_w(\beta) - f_w(\alpha) \leq \max_{w \in \mathcal{W}_{out}}(\gamma_\sigma(w) - f_w(\alpha)) = H_\alpha(\sigma)$. We therefore have: $MR(\alpha, \mathcal{A}_\sigma; \mathcal{I}) \leq MR(\alpha, \mathcal{A}_\sigma; \mathcal{W}_{out}) \leq H_\alpha(\sigma)$.

If $H_\alpha(\sigma) < \underline{s}_\alpha$ then $MR(\alpha, \mathcal{A}_\sigma; \mathcal{I}) < \underline{s}_\alpha$, and so we can backtrack at node σ, because no β in $\mathcal{A}_\sigma$ can lead to a value $PMR(\alpha, \beta; \mathcal{I})$ which is as high as $\underline{s}_\alpha$.

At a Leaf Node. $\sigma \in \mathcal{L}$: Let $\beta = \alpha_\sigma$ be the associated utility vector.

- We first compute $PMR(\alpha, \beta; \mathcal{W}_{in})$. If $PMR(\alpha, \beta; \mathcal{W}_{in}) \geq \bar{s}$ then we exit the $MR_{\bar{s}}$ procedure, returning $MR_{\bar{s}}(\alpha, \mathcal{A}_\sigma; \mathcal{I}) = \bar{s}$.
- We next compute $PMR(\alpha, \beta; \mathcal{W}_{out}) = \max_{w \in \mathcal{W}_{out}} f_w(\beta) - f_w(\alpha)$. $PMR(\alpha, \beta; \mathcal{W}_{out}) < \underline{s}_\alpha$ then we backtrack, since we then have $PMR(\alpha, \beta; \mathcal{I}) \leq PMR(\alpha, \beta; \mathcal{W}_{out}) < \underline{s}_\alpha$, showing that β is not relevant.
- If $PMR(\alpha, \beta; \mathcal{W}_{in}) = PMR(\alpha, \beta; \mathcal{W}_{out})$, then we have the value of $PMR(\alpha, \beta; \mathcal{I})$ and we can exit the PMR procedure after updating the lower bound $\underline{s}_\alpha$ to $PMR(\alpha, \beta; \mathcal{I})$.

- Otherwise, we obtain $PMR(\alpha, \beta; \mathcal{I})$ by using a linear programming solver to compute $\max_{w \in \mathcal{W}} f_w(\beta) - f_w(\alpha) = \max_{w \in \mathcal{W}} w \cdot (\beta - \alpha)$.
  - If $PMR(\alpha, \beta; \mathcal{I}) \geq \overline{s}$ then we exit the $MR_{\overline{s}}$ procedure, returning $MR_{\overline{s}}(\alpha, \mathcal{A}_\sigma; \mathcal{I}) = \infty$.
  - Otherwise, if $PMR(\alpha, \beta; \mathcal{I}) > \underline{s}_\alpha$ then we update the lower bound $\underline{s}_\alpha$ to $PMR(\alpha, \beta; \mathcal{I})$.

4   Experimental Results

All experiments for this paper were done on a 2.70GHz Intel Xeon E5-2680 CPU. The operating system was Ubuntu 24.04. The linear programming solver used was the GNU Linear Programming Kit 5.0.[1] The source code of our implementation, along with some statistics (how many and how long) about the LP calls made during the experiments, is available at https://zenodo.org/records/18839028.

We generated random combinatorial d-objective optimization instances, structured as described in Sect. 2.3. Each instance is composed of a set of ternary sub-utility functions over Boolean variables, and of a set of preference inequalities. Each sub-utility function is composed of eight random d-dimensional utility vectors, one for each possible assignment to the variables in its scope. The main parameters of the instances are:

- The number d of objectives.
- The number n of variables.
- The number m of sub-utility functions. We set m to be equal to $\lfloor \frac{3n}{2} \rfloor$.
- The number nb_{ineq} of preference inequalities.

The experiments in Sect. 4.1 study the behavior of the approaches when the number n of variables increases. These experiments were run without preference inequalities. The experiments in Sect. 4.2 study what happens when the number of preference inequalities increases. For these experiments, the number of variables was set to 20. For both kinds of experiments, we have one set of instances with 6 objectives, one with 8 objectives, and one with 10 objectives.

For Sect. 4.2, we iteratively added an inequality between α and β, where α was the alternative that achieved the minimax regret for the previous instance, and β was the alternative that maximized the PMR with α. The preference inequalities generated are therefore the same as the ones that would be obtained with preference elicitation.

4.1   Increasing the Number of Alternatives

We compare in Fig. 4 the behavior of the approaches when the number of variables increases and there are no preference inequalities. Each data point in all three figures is the median of five instances. Since the variables are Boolean,

---

[1] https://www.gnu.org/software/glpk/.

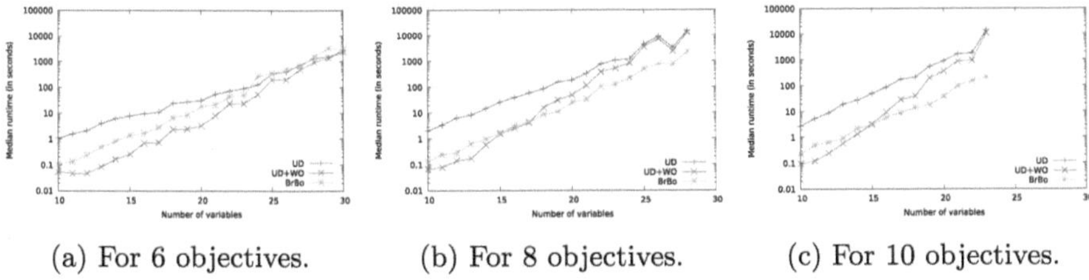

(a) For 6 objectives.   (b) For 8 objectives.   (c) For 10 objectives.

Fig. 4. Increasing the number of variables.

each new variable doubles the number of total alternatives. While there was some amount of variance between the five instances of a given configuration of parameters, the relation between methods was usually conserved: the fastest method for an individual instance was also the fastest method for the median instance of its configuration in 243 out of 270 cases (90%).

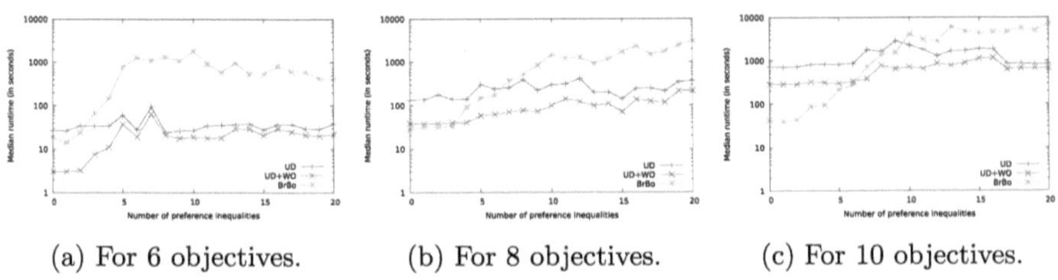

(a) For 6 objectives.   (b) For 8 objectives.   (c) For 10 objectives.

Fig. 5. Increasing the number of preference inequalities for 20 variables.

We make three major observations from these results with no preference inequalities:

1. When there are few alternatives, UD+WO is much faster than UD. When the number of alternatives increases, the difference between the two lessens, until a point is reached where both approaches take the same time to find the minimax regret. This would suggest that UD+WO is either better or equivalent to UD in terms of performance, never worse.
2. When there are few alternatives, UD+WO is faster than BrBo. When the number of alternatives increases, the difference between the two lessens, until a point is reached where BrBo becomes faster, although that point has not been reached for the dataset with 6 objectives. This would suggest that BrBo outperforms UD+WO (and therefore UD too) when the number of alternatives is high.
3. The more objectives there are, the earlier (in terms of number of alternatives) BrBo becomes faster than UD+WO. This would suggest that BrBo outperforms UD+WO when the number of objectives is high, and that UD+WO outperforms BrBo when the number of objectives is low.

4.2   Increasing the Number of Preference Inequalities

We now show in Fig. 5 how the approaches behave when the number of preference inequalities increases. Once again, each data point represents the median of five instances. The fastest method for an individual instance was also the fastest method for the median instance of its configuration in 301 out of 315 cases (95.6%). We make three major observations from these results, some of them confirming the trends seen in the previous experiments:

1. The median for the UD+WO approach is never slower than the median for the UD approach.
2. When there are few preference inequalities, BrBo is faster than (or almost as fast as, in the case of 6 objectives) UD+WO. As inequalities are added, the performance of UD+WO with regard to BrBo improves. This would suggest that UD+WO outperforms BrBo when there are many preference inequalities, and that BrBo performs better when there are few preference inequalities.
3. The number of preference inequalities at which UD+WO becomes faster than BrBo increases as the number of objectives increases: 0 for 6 objectives, 4 for 8 objectives, and 7 for 10 objectives. This would suggest, as already indicated by the previous experiments, that BrBo outperforms UD+WO when the number of objectives is high, and that UD+WO outperforms BrBo when the number of objectives is low.

5   Conclusions

We introduced two approaches to compute minimax regret: UD+WO and BrBo. We empirically showed that UD+WO strictly improves on the state-of-the-art approach, and that whether UD+WO or BrBo is the fastest method depends on the instance parameters. It appears from our experiments that UD+WO is a better choice when there are few variables, few objectives, and many preference inequalities, while it seems that it would be better to pick BrBo when there are many variables, many objectives, and few preference inequalities.

Both of our approaches rely on bounding the space of possible weight vectors to return the exact minimax regret value. Future work could study whether those bounds could instead be used to obtain an approximate result, with a reduced runtime. It could either return an estimation of the minimax regret that is close to its actual value, or return an alternative that has a high likelihood of being the one with the lowest maximum regret.

Acknowledgments. This publication has emanated from research supported in part by a grant from Research Ireland under Grant number [12-RC-2289-P2] which is co-funded under the European Regional Development Fund. For the purpose of Open Access, the authors have applied a CC BY public copyright licence to any Author Accepted Manuscript version arising from this submission.

We're grateful for the reviewers' helpful comments, and also to Patrice Perny for valuable discussions.

References

1. Ah-Pine, J., Mayag, B., Rolland, A.: Identification of a 2-additive bi-capacity by using mathematical programming. In: Perny, P., Pirlot, M., Tsoukiàs, A. (eds.) Algorithmic Decision Theory - Third International Conference, ADT 2013, Bruxelles, Belgium, November 12–14, 2013, Proceedings. Lecture Notes in Computer Science, vol. 8176, pp. 15–29. Springer (2013). https://doi.org/10.1007/978-3-642-41575-3_2
2. Bacchus, F., Grove, A.J.: Graphical models for preference and utility. In: Besnard, P., Hanks, S. (eds.) UAI '95: Proceedings of the Eleventh Annual Conference on Uncertainty in Artificial Intelligence, Montreal, Quebec, Canada, August 18–20, 1995, pp. 3–10. Morgan Kaufmann (1995)
3. Benabbou, N., Perny, P.: Incremental weight elicitation for multiobjective state space search. In: Bonet, B., Koenig, S. (eds.) Proceedings of the Twenty-Ninth AAAI Conference on Artificial Intelligence, January 25–30, 2015, Austin, Texas, USA, pp. 1093–1099. AAAI Press (2015). https://doi.org/10.1609/AAAI.V29I1.9362
4. Benabbou, N., Perny, P., Viappiani, P.: Incremental elicitation of choquet capacities for multicriteria choice, ranking and sorting problems. Artif. Intell. **246**, 152–180 (2017). https://doi.org/10.1016/J.ARTINT.2017.02.001
5. Bourdache, N., Perny, P., Spanjaard, O.: Incremental elicitation of rank-dependent aggregation functions based on bayesian linear regression. In: Kraus, S. (ed.) Proceedings of the Twenty-Eighth International Joint Conference on Artificial Intelligence, IJCAI 2019, Macao, China, August 10–16, 2019, pp. 2023–2029. ijcai.org (2019). https://doi.org/10.24963/IJCAI.2019/280
6. Boutilier, C., Patrascu, R., Poupart, P., Schuurmans, D.: Constraint-based optimization and utility elicitation using the minimax decision criterion. Artif. Intell. **170**(8–9), 686–713 (2006). https://doi.org/10.1016/J.ARTINT.2006.02.003
7. Braziunas, D.: Decision-theoretic elicitation of generalized additive utilities. Ph. D. thesis, University of Toronto, Canada (2012). http://hdl.handle.net/1807/32669
8. Braziunas, D., Boutilier, C.: Minimax regret based elicitation of generalized additive utilities. In: Parr, R., van der Gaag, L.C. (eds.) UAI 2007, Proceedings of the Twenty-Third Conference on Uncertainty in Artificial Intelligence, Vancouver, BC, Canada, July 19–22, 2007, pp. 25–32. AUAI Press (2007). https://doi.org/10.5555/3020488.3020492
9. Escamocher, G., Pourkhajouei, S., Toffano, F., Viappiani, P., Wilson, N.: Interactive preference elicitation under noisy preference models: an efficient non-bayesian approach. Int. J. Approx. Reason. **178**, 109333 (2025). https://doi.org/10.1016/J.IJAR.2024.109333
10. Gonzales, C., Perny, P.: GAI networks for utility elicitation. In: Dubois, D., Welty, C.A., Williams, M. (eds.) Principles of Knowledge Representation and Reasoning: Proceedings of the Ninth International Conference (KR2004), Whistler, Canada, June 2–5, 2004, pp. 224–234. AAAI Press (2004). http://www.aaai.org/Library/KR/2004/kr04-025.php
11. Hines, G., Larson, K.: Preference elicitation for risky prospects. In: van der Hoek, W., Kaminka, G.A., Lespérance, Y., Luck, M., Sen, S. (eds.) 9th International Conference on Autonomous Agents and Multiagent Systems (AAMAS 2010), Toronto, Canada, May 10–14, 2010, vol. 1–3, pp. 889–896. IFAAMAS (2010). https://dl.acm.org/citation.cfm?id=1838325

12. Lu, T., Boutilier, C.: Robust approximation and incremental elicitation in voting protocols. In: Walsh, T. (ed.) IJCAI 2011, Proceedings of the 22nd International Joint Conference on Artificial Intelligence, Barcelona, Catalonia, Spain, July 16–22, 2011, pp. 287–293. IJCAI/AAAI (2011). https://doi.org/10.5591/978-1-57735-516-8/IJCAI11-058

13. Marinescu, R., Razak, A., Wilson, N.: Multi-objective constraint optimization with tradeoffs. In: Schulte, C. (ed.) Principles and Practice of Constraint Programming - 19th International Conference, CP 2013, Uppsala, Sweden, September 16–20, 2013. Proceedings. Lecture Notes in Computer Science, vol. 8124, pp. 497–512. Springer (2013). https://doi.org/10.1007/978-3-642-40627-0_38

14. Perny, P., Viappiani, P., Boukhatem, A.: Incremental preference elicitation for decision making under risk with the rank-dependent utility model. In: Ihler, A., Janzing, D. (eds.) Proceedings of the Thirty-Second Conference on Uncertainty in Artificial Intelligence, UAI 2016, June 25–29, 2016, New York City, NY, USA. AUAI Press (2016). http://auai.org/uai2016/proceedings/papers/204.pdf

15. Pigozzi, G., Tsoukiàs, A., Viappiani, P.: Preferences in artificial intelligence. Ann. Math. Artif. Intell. **77**(3–4), 361–401 (2016). https://doi.org/10.1007/S10472-015-9475-5

16. Savage, L.J.: The Foundations of Statistics. Wiley, New York (1954)

17. Teso, S., Passerini, A., Viappiani, P.: Constructive preference elicitation by setwise max-margin learning. In: Kambhampati, S. (ed.) Proceedings of the Twenty-Fifth International Joint Conference on Artificial Intelligence, IJCAI 2016, New York, NY, USA, 9–15 July 2016, pp. 2067–2073. IJCAI/AAAI Press (2016). http://www.ijcai.org/Abstract/16/295

18. Viappiani, P., Boutilier, C.: Optimal set recommendations based on regret. In: Anand, S.S., Mobasher, B., Kobsa, A., Jannach, D. (eds.) Proceedings of the 7th Workshop on Intelligent Techniques for Web Personalization & Recommender Systems (ITWP'09), Pasadena, California, USA, July 11–17, 2009 in Conjunction with the 21st International Joint Conference on Artificial Intelligence - IJCAI 2009. CEUR Workshop Proceedings, vol. 528. CEUR-WS.org (2009). https://ceur-ws.org/Vol-528/paper3.pdf

19. Viappiani, P., Boutilier, C.: On the equivalence of optimal recommendation sets and myopically optimal query sets. Artif. Intell. **286**, 103328 (2020). https://doi.org/10.1016/J.ARTINT.2020.103328

Resolution Meets Cutting Planes:
Introducing Hypercube Linear Resolution

Maarten Flippo[1]([⊠])([iD]), Peter J. Stuckey[2]([iD]), and Emir Demirović[1]([iD])

[1] Delft University of Technology, Delft, The Netherlands
m.l.flippo@tudelft.nl
[2] Monash University, Melbourne, Australia

Abstract. Modern combinatorial solvers can be understood as searching for proofs of unsatisfiability or optimality. The proof system implemented by a solver, therefore, fundamentally shapes solver performance. While propositional resolution is simple and complete for propositional formulas, no existing practical resolution-based system offers unrestricted complete reasoning over integer linear inequalities.

We introduce hypercube linear resolution, a new proof system that is both sound and complete for integer linear reasoning. Hypercube linear resolution integrates propositional resolution with Fourier resolution through a new constraint type, the hypercube linear constraint, which captures linear relations within a discrete hypercube.

Our main contribution is the theory of the proof system, establishing its soundness and completeness. Our proof system generalises propositional and Fourier resolution, and can be seen as an extended cutting planes proof system. We also provide a conflict-driven search algorithm that exploits the system in practice. Our preliminary experiments demonstrate that the new system reduces conflicts compared to propositional resolution. This shows that the structure captured by hypercube linear constraints can be exploited to improve constraint solving.

Keywords: Proof systems · Resolution · Constraint Solving

1 Introduction

Modern combinatorial solvers have been effective at solving real-world satisfaction and optimisation problems. These solvers operate based on a wide variety of formalisms, including Boolean Satisfiability (SAT) [5], pseudo-Boolean optimisation (PB), and Integer Linear Programming (ILP) [29]. Solving a problem in each of these formalisms is equivalent to finding a proof of unsatisfiability or optimality. The inference rules that make up the proof system implemented by the solver determine how effective a solver is for any given problem.

Propositional resolution [11] on clauses has been effectively used to solve SAT problems. It is a single, simple, logical inference rule that can derive any consequence of a Boolean formula. This simplicity enables clause-learning SAT to

T. Guns (Ed.): CPAIOR 2026, LNCS 16595, pp. 155–172, 2026.
https://doi.org/10.1007/978-3-032-27242-3_10

solve huge formulas [10]. The success of resolution-based reasoning has inspired efforts to extend these ideas to other constraints than clauses as well.

Generalised resolution [17] takes the idea of resolution on clauses and applies it to 0–1 integer linear inequalities, or PB constraints. Since a single PB constraint requires potentially many clauses to express the same solutions, generalised resolution can derive much stronger constraints. However, where resolution over clauses is a single rule that is complete for all propositional formulas, a complete proof system over PB constraints requires additional inference rules to prove every true fact [8].

A further generalisation to a resolution-based proof system over unrestricted integer linear inequalities is non-trivial. Two main efforts have been made in this direction: CutSat [7,19] and IntSat [23]. Both CutSat and IntSat use Fourier resolution as the core resolution operation, which is complete for linear inequalities over *real* variables, but incomplete over integer variables. Given that theoretical shortcoming, CutSat still guarantees completeness by heavily restricting the search procedure. IntSat ignores the completeness and accepts that some facts cannot be derived through Fourier resolution, falling back to propositional resolution when required.

To address these issues, we formulate a new constraint type, called the *hypercube linear* constraint, that generalises both a clause and a linear inequality. We then devise a sound and complete proof system called *hypercube linear resolution* that combines propositional resolution and Fourier resolution. A hypercube linear constraint enforces a linear relation within a hypercube of the solution space, and our proof system can be seen as an extended cutting planes proof system where, the extended variables correspond to variable bounds.

We demonstrate that the structure captured by a hypercube linear constraint is observed in practice through an experimental evaluation. With a proof-of-concept implementation, we show that hypercube linear constraints can be derived in existing benchmarks. Hypercube linear resolution derives constraints that are useful for pruning the search space. Overall, it suggests that the hypercube linear constraint can be a valuable addition to a constraint solver.

Our work is an important step towards elevating constraint learning in combinatorial solvers to a more expressive system, thereby enhancing our ability to exploit the high-level structure present in models of real-world problems.

To summarise, our contributions are three-fold: (1) a new sound and complete proof system that natively reasons on the propositional and algebraic side, (2) a conflict-driven algorithm that exploits our proof system in practice, and (3) a proof-of-concept implementation that shows the potential of our approach.

2   Preliminaries

A *constraint satisfaction problem* (CSP) $\mathcal{P} = (\mathcal{X}, \mathcal{D}, \mathcal{C})$ is a tuple consisting of a set of variables $\mathcal{X}$, a domain mapping each variable $x \in \mathcal{X}$ to a range of integers $\mathcal{D}(x) = \{i \in \mathbb{Z} \mid l_x \leq i \leq u_x\}$ for some $l_x, u_x \in \mathbb{Z}, l_x \leq u_x$, and a set of constraints $\mathcal{C}$. Note that Boolean variables are simply integers x with $l_x = 0$

and $u_x = 1$. For a domain $\mathcal{D}$, we define $lb(\mathcal{D}, x)$ and $ub(\mathcal{D}, x)$ to be the smallest and largest value in $\mathcal{D}(x)$, respectively. A *constraint* $c \in \mathcal{C}$ is a relation over $\mathcal{D}(x_1) \times \cdots \times \mathcal{D}(x_n)$ where $[x_1, \ldots, x_n]$ is the (ordered) scope of the constraint $sc(c) \subseteq \mathcal{X}$. An *assignment* $\theta : \mathcal{X} \mapsto \mathbb{Z}$ maps each variable to a value in its domain: $\theta(x) \in \mathcal{D}(x)$ for all $x \in \mathcal{X}$. Assignment θ *satisfies* a constraint $c \in \mathcal{C}$ if $[\theta(x_1), \ldots, \theta(x_n)] \in c$ where $sc(c) = [x_1, \ldots, x_n]$. If θ satisfies all $c \in \mathcal{C}$, then θ is a *solution* of $\mathcal{P}$. Given two conjunctions of constraints $\mathcal{C}_1$ and $\mathcal{C}_2$ in the context of a CSP, we write $\mathcal{C}_1 \models \mathcal{C}_2$ if all solutions of $(\mathcal{X}, \mathcal{D}, \mathcal{C}_1)$ are solutions of $(\mathcal{X}, \mathcal{D}, \mathcal{C}_2)$, i.e., $\mathcal{C}_1$ implies $\mathcal{C}_2$.

For a variable $x \in \mathcal{X}$, a *bounds constraint* takes the form $\langle x \geq d \rangle$ or $\langle x \leq d \rangle$, where $d \in \mathcal{D}(x)$. For Boolean variables x, $\langle x \geq 1 \rangle$ equates to setting the Boolean to true. Note that $\langle x \geq l_x \rangle$ and $\langle x \leq u_x \rangle$ are trivially true constraints. Given a domain $\mathcal{D}$, we write $\mathcal{D} \models \langle x \geq d \rangle$ (respectively $\mathcal{D} \models \langle x \leq d \rangle$) if and only if $\min \mathcal{D}(x) \geq d$ (respectively $\max \mathcal{D}(x) \leq d$).

Let B be a conjunction of bounds constraints. We extend the use of $\models$ such that $\mathcal{D} \models B \iff \forall c \in B : \mathcal{D} \models c$. Additionally, B is *consistent* if and only if there is no pair $\{\langle x \geq l \rangle, \langle x \leq u \rangle\} \subseteq B$ where $l > u$.

A *hypercube* H is a consistent conjunction of bounds constraints. We can always assume that H includes $x \geq l_x$ and $x \leq u_x$, for all $x \in \mathcal{X}$, without changing its meaning. Then, $\mathcal{D}_H$ denotes the domain that is induced by the bounds in H.

Clauses. A *clause* is a constraint of the form $H \rightarrow \perp$, where H is a conjunction. We use this (negated) form to more closely follow the notation used later. Given current domain $\mathcal{D}$, a clause *propagates* in three ways: if $\mathcal{D} \models H$ then the clause fails (propagates $\perp$); if $H \equiv H' \wedge \langle x \geq d \rangle$ and $\mathcal{D} \models H'$ then the clause propagates $x \leq d - 1$; and if $H \equiv H' \wedge \langle x \leq d \rangle$ and $\mathcal{D} \models H'$ then the clause propagates $x \geq d + 1$.

We can *(propositionally) resolve* two clauses $L_1 \equiv H_1 \wedge \langle x \geq d_1 \rangle \rightarrow \perp$ and $L_2 \equiv H_2 \wedge \langle x \leq d_2 \rangle \rightarrow \perp$, where $d_1 \leq d_2 + 1$, to obtain the consequence clause $H_1 \wedge H_2 \rightarrow \perp$. Note that this generates a trivial clause if $H_1 \wedge H_2$ is inconsistent.

Linear Constraints. An *(integer) linear constraint* is of the form $\sum_{i=1}^{n} w_i x_i \leq r$ where each of $w_i, 1 \leq i \leq n$ and r are integer constants and x_i are variables in $\mathcal{X}$. The *slack* of a linear constraint $R \equiv \sum_{i=1}^{n} w_i x_i \leq r$ captures how far away the constraint is from being violated: $slack(\mathcal{D}, R) = r - \sum_{i=1}^{n} lb(\mathcal{D}, w_i x_i)$ where

$$lb(\mathcal{D}, w_i x_i) = \begin{cases} w_i \times lb(\mathcal{D}, x_i) & \text{if } w_i \geq 0 \\ w_i \times ub(\mathcal{D}, x_i) & \text{if } w_i < 0 \end{cases}.$$

Given domain $\mathcal{D}$, a linear constraint propagates as follows: if $w_j > 0$ then

$$x_j \leq \left\lfloor \frac{slack(\mathcal{D}, R) + lb(\mathcal{D}, w_j x_j)}{w_j} \right\rfloor ;$$

and if $w_j < 0$ then it can propagate

$$x_j \geq \left\lceil \frac{slack(\mathcal{D}, R) + lb(\mathcal{D}, w_j x_j)}{w_j} \right\rceil .$$

Note that it only propagates if the new bound is not a consequence of $\mathcal{D}$ already.

We can *Fourier resolve* two linear constraints $R_1 \equiv \sum_{i=1}^{n} w_i x_i \leq r$ and $R_2 \equiv \sum_{i=1}^{n} w_i' x_i \leq r'$ where $w_j > 0$ and $w_j' < 0$ to obtain the linear consequence constraint

$$FR(R_1, R_2, x_j) \equiv \sum_{i=1, i \neq j}^{n} (-w_j' \times w_i + w_j \times w_i') x_i \leq (-w_j' \times r + w_j \times r')$$

eliminating the variable x_j from the constraints. Over integer variables, the *FR* inference rule by itself does not yield a complete proof system.

3   Related Work

The first algorithm for solving SAT problems (Davis-Putnam) was based on the resolution proof system [11]. Since then, SAT solvers have evolved significantly, especially since the GRASP [21] SAT solver introduced Conflict-Driven Clause Learning (CDCL). However, resolution is still the main proof system that is used. Stronger proof systems, such as extended resolution [27], have been explored for SAT [1,18], but always with the clause as the elementary constraint.

Generalised resolution generalises propositional resolution to work with 0–1 linear inequalities (pseudo-Boolean constraints) [17]. It is used to implement pseudo-Boolean conflict analysis in CDCL-style solvers [5,9,13]. However, even though generalised resolution is sound, it is not complete for discrete variables. As a result, pseudo-Boolean solvers use additional inference steps to maintain the completeness property of the CDCL algorithm.

The generalised resolution rule is also sound for integer variables. This inspired the propagation-based ILP solvers CutSat [19], IntSat [23], and Cut-Sat++ [7]. These solvers also handle the incompleteness of generalised resolution over integer variables. CutSat and CutSat++ restrict how the search space is explored, thereby maintaining completeness. IntSat takes a different approach and accepts that it will not always derive a new linear inequality. A complete conflict analysis procedure that derives linear inequalities and runs without assumptions on the solver search has not been presented yet.

Constraint derivation has also been effective in CP solvers. Early works derive conjunctions or disjunctions of $\langle x \neq v \rangle$ constraints called g-nogoods [20]. Lazy Clause Generation (LCG) [14,24] compresses g-nogoods and directly builds on the CDCL algorithm. Non-clausal constraints can be learned by the HaifaCSP solver [28] for predefined pairs of constraint types. Finally, LCG has recently been extended to enable learning of linear constraints [2], building on the IntSat conflict analysis routine. It shows promising experimental results but suffers from the same drawback as IntSat, in that conflict analysis fails to learn a linear inequality in most cases.

4   Hypercube Linears

Before we describe hypercube linear resolution in Sect. 5, we first introduce our hypercube linear constraint. It can natively express the piecewise nature of constraints found in real-world problems, as well as linear relations between integer variables. Given a hypercube linear constraint, we also define the propagation rules for a hypercube linear propagator. Finally, we define the concept of weakening (Definition 3).

A hypercube linear constraint imposes a linear constraint within a region of the solution space. The region in which it is active is called the hypercube, and that hypercube implies a linear inequality. Example 1 shows an example of the constraint over three integer variables.

Example 1 (Hypercube Linear Constraint). The following is an example of a hypercube linear constraint: $\langle x \geq 4 \rangle \wedge \langle y \leq 7 \rangle \rightarrow 2x - 7z \leq 8$.

There are no restrictions on which variables appear in the hypercube and which variables have non-zero weight in the linear inequality. This can be seen in Example 1, where a bound on x is part of the hypercube as well as x having non-zero weight in the linear component. Definition 1 formalises the definition of a hypercube linear.

Definition 1 (Hypercube Linear). *A hypercube linear constraint has the following form, where H is a hypercube:*

$$H \rightarrow \sum_{i=1}^{n} w_i x_i \leq r.$$

It generalises a clause if $w_i = 0$ for $1 \leq i \leq n$ and $r \leq -1$, simplifying the linear term to $0 \leq -1 \equiv \bot$. Conversely, if $\mathcal{D} \models H$, it is equivalent to $\sum_{i=1}^{n} w_i x_i \leq r$ where $x_i \in \mathcal{X}$ are integer variables.

The hypercube linear constraint is also a generalisation of the indicator constraint [3] commonly supported by MIP solvers. To be an indicator constraint, the hypercube would consist over a single bound constraint over a 0–1 variable.

Corollary 1. *A hypercube linear constraint $L \equiv H \rightarrow R$ is conflicting w.r.t. domain $\mathcal{D}$ if and only if $\mathcal{D} \models H$ and $slack(\mathcal{D}, R) < 0$.*

4.1   Hypercube Linear Slack

We lift the common notion of slack of linear constraints to hypercube linear constraints in Definition 2. When the hypercube linear slack is negative, the linear part of the constraint is conflicting once the hypercube is satisfied.

Definition 2 (Hypercube Linear Slack). *Let $L \equiv H \rightarrow \sum_{i=1}^{n} w_i x_i \leq r$ be a hypercube linear constraint. Then the* slack *of L under domain $\mathcal{D}$, written as $hlslack(\mathcal{D}, L)$ is defined as*

$$hlslack(\mathcal{D}, L) = r - \sum_{i=1}^{n} \max(lb(\mathcal{D}, w_i x_i), lb(\mathcal{D}_H, w_i x_i))$$

Corollary 2. *Let $H \to R$ be a hypercube linear constraint and $\mathcal{D}$ be a domain. If $\mathcal{D} \models H$, then $hlslack(\mathcal{D}, H \to R) = slack(\mathcal{D}, R)$.* $\qquad\square$

Instead of only considering the bounds of the variables in domain $\mathcal{D}$, the hypercube linear slack also looks at the contribution of the bounds in the hypercube. This is illustrated by Example 2. If the hypercube contains a bound $\langle x \geq 5 \rangle$ that is not yet implied by $\mathcal{D}$, then we know that once the linear part needs to be enforced, $x \geq 5$ will be true. This is used in Sect. 4.2 to define the propagation rules for the hypercube linear propagator.

Example 2 (Hypercube Linear Slack). Let $x, y, z \in \mathcal{X}$ be variables with domains $\mathcal{D}(x) = [2, 10]$, $\mathcal{D}(y) = [0, 10]$, and $\mathcal{D}(z) = [2, 5]$. Additionally, consider the hypercube linear constraint $L = \langle x \geq 2 \rangle \wedge \langle y \geq 2 \rangle \to x + y + z \leq 5$. We have $hlslack(\mathcal{D}, L) = 5 - \max(2, 2) - \max(0, 2) - \max(2, -\infty) = -1$, which shows that this domain cannot satisfy the hypercube of L *and* the linear component of L.

4.2 Hypercube Linear Propagator

To incorporate the hypercube linear constraint into a propagation-based solver, we define a propagator for the constraint. Given a current domain $\mathcal{D}$, there are three propagation rules for the hypercube linear $L \equiv H \to R$:

- If $\mathcal{D} \models H$, then the hypercube linear propagates the linear component as a standard linear constraint as described in Sect. 2.
- If $hlslack(\mathcal{D}, L) < 0$ and all but one bounds constraint $c \in H$ are satisfied, then c is propagated to false. That means one of two propagations happens:
  - If $c = \langle x \leq d \rangle$, then this propagates $x \geq d + 1$,
  - Otherwise $c = \langle x \geq d \rangle$, in which case this propagates $x \leq d - 1$.
  Example 3 shows an application of this propagation rule.
- If $hlslack(\mathcal{D}, L) \geq 0$ and all but one bound constraint $c \in H$ are satisfied, then the linear constraint propagates a bound on x with coefficient w in one of two cases:
  - If $c \equiv \langle x \geq d \rangle$, $w > 0$ then we can propagate $x \leq \left\lfloor \dfrac{hlslack(\mathcal{D},L) + wd}{w} \right\rfloor$.
  - If $c \equiv \langle x \leq d \rangle$, $w < 0$ then we can propagate $x \geq \left\lceil \dfrac{hlslack(\mathcal{D},L) + wd}{w} \right\rceil$.
  See Example 4 for an illustration of this propagation rule.

Example 3 (Propagating hypercube). We extend Example 2. $\langle x \geq 2 \rangle$ in the hypercube of L is true because $lb(\mathcal{D}, x) = 2$. Contrarily, $\langle y \geq 2 \rangle$ is unassigned because $\mathcal{D}(y) = [0, 10]$. Therefore, since L has negative slack, the propagator propagates $\neg \langle y \geq 2 \rangle \equiv \langle y \leq 1 \rangle$.

Example 4 (Propagating a weaker hypercube bound). Consider the hypercube linear constraint $L = \langle x \geq 2 \rangle \wedge \langle y \geq 2 \rangle \to x + y + z \leq 7$ with domains $\mathcal{D}(x) = [0, 10]$, $\mathcal{D}(y) = [2, 5]$, $\mathcal{D}(z) = [0, 10]$. In this situation, $hlslack(\mathcal{D}, L) = 3$. Now $\langle y \geq 2 \rangle$ is true, but $\langle x \geq 2 \rangle$ is unassigned. In any assignment, we will either have $\langle x \geq 2 \rangle$ or $\langle x \leq 1 \rangle$. For the first case, $\langle x \geq 2 \rangle$, the hypercube of L is satisfied, so $x + y + z \leq 7$ must be true, enforcing $\langle x \leq 5 \rangle$. For the second case, $\langle x \leq 1 \rangle$ is true, which implies that $\langle x \leq 5 \rangle$ is also true. Since $\langle x \leq 5 \rangle$ is true for both cases, the propagator can propagate $\langle x \leq 5 \rangle$.

Every propagated bound c is associated with a conjunction of bounds B that caused the propagation. We refer to B as the *explanation* of c. For example, in Example 3, the explanation for $\langle y \leq 1 \rangle$ is $\langle x \geq 2 \rangle \wedge \langle z \geq 2 \rangle$. This concept is identical to explanations as used in LCG solvers.

4.3   Weakening the Hypercube Linear Constraint

Weakening allows rewriting a hypercube linear by moving part of a variable coefficient from the linear part of the constraint into the hypercube. We can always weaken a hypercube linear constraint L to create another hypercube linear constraint L' that is implied by L as follows:

Definition 3 (Weakening). *Equation* (1) *implies Eq.* (2) *and Eq.* (3).

$$H \to \sum_{i=1}^{n} w_i x_i \leq r \tag{1}$$

$$H \wedge \langle x_j \geq d \rangle \to \sum_{i=1, i\neq j}^{n} w_i x_i + (w_j - 1)x_j \leq r - d \tag{2}$$

$$H \wedge \langle x_j \leq d \rangle \to \sum_{i=1, i\neq j}^{n} w_i x_i + (w_j + 1)x_j \leq r + d \tag{3}$$

Proposition 1. *Equation* (1) *implies Eq.* (2) *and Eq.* (3).

Proof. We prove the result for Eq. (2); the other result holds similarly. If Eq. (1) is true then $H \wedge \langle x_j \geq d \rangle \to \sum_{i=1}^{n} w_i x_i \leq r$ is also true. Further, $x_j \geq d$ implies $-x_j \leq -d$, which is subtracted from the linear component of Eq. (1) to give the linear component of Eq. (2). Since both inequalities hold in $H \wedge \langle x_j \geq d \rangle$, the result is a correct consequence of Eq. (1). $\square$

Notably, weakening is sound for both continuous and integer variables. Regardless, our focus remains exclusively on integer variables.

5   The Hypercube Linear Proof System

We define the inference rules for the hypercube linear resolution proof system. It consists of four rules: WEAKEN moves a variable from the linear part to the hypercube, DIVIDE divides the linear inequality by a positive integer, RESF combines two hypercube linear constraints to eliminate a variable from the linear inequality, and RESH combines two hypercube linear constraints to eliminate bounds from the hypercube. These operations naturally generalise propositional and Fourier resolution, resulting in a sound and complete proof system. All these rules are used in the next section to implement the conflict analysis for hypercube linear constraints.

Weakening. The weakening rule fully eliminates one linear term of variable x_j by applying the weakening from Definition 3 $|w_j|$ times.

$$\text{WEAKEN}(L, \langle x_j \geq d \rangle) \frac{L \equiv H \to \sum_{i=1}^{n} w_i x_i \leq r \qquad w_j > 0}{H \wedge \langle x_j \geq d \rangle \to \sum_{i=1, i \neq j}^{n} w_i x_i \leq r - w_j d}$$

$$\text{WEAKEN}(L, \langle x_j \leq d \rangle) \frac{L \equiv H \to \sum_{i=1}^{n} w_i x_i \leq r \qquad w_j < 0}{H \wedge \langle x_j \leq d \rangle \to \sum_{i=1, i \neq j}^{n} w_i x_i \leq r + w_j d}$$

Proposition 2. *The* WEAKEN *rules are sound.*

Proof. The results hold by $|w_j|$ applications of Proposition 1. $\qquad\qquad\square$

A very useful observation for Sect. 6 is that weakening on bounds in a domain $\mathcal{D}$ does not impact the slack w.r.t. $\mathcal{D}$. Proposition 3 shows that if we weaken hypercube linear constraint L based on a domain $\mathcal{D}$, then the slack of L under $\mathcal{D}$ does not change.

Proposition 3. *Let* $L \equiv H \to ax + R \leq c$ *be a hypercube linear, with* $a > 0$ *(*$a < 0$*),* $x \in \mathcal{X}$*, and* R *a sum of linear terms. Additionally, let* $\mathcal{D}$ *be a domain such that* $\mathcal{D} \models H$*. Finally, let* $c \equiv \langle x \geq lb(\mathcal{D}, x) \rangle$ *(*$c \equiv \langle x \leq ub(\mathcal{D}, x) \rangle$*) and* $L' \equiv \text{WEAKEN}(L, c)$*. Then,* $hlslack(\mathcal{D}, L) = hlslack(\mathcal{D}, L')$*.*

Proof. We will prove this for weakening on the lower bound of x. The result for upper bounds is symmetric. Therefore, assume that $a > 0$. Since $\mathcal{D} \models H$, we have that $lb(\mathcal{D}, x) \geq lb(\mathcal{D}_H, x)$. By removing x through weakening, we reduce the lower-bound of summation by $a \cdot lb(\mathcal{D}, x_i)$, and we reduce the right-hand side of the linear part by the same amount. Hence the slack is unchanged. $\qquad\square$

Division. Division is a common operation in cutting-plane systems, e.g., in Chvatal and Gomory cuts [16]. Every constant is divided by a scalar $d > 0$, and rounded down.

$$\text{DIVIDE}(L, d) \frac{L \equiv H \to \sum_{i=1}^{n} w_i x_i \leq r \qquad d > 0}{H \to \sum_{i=1}^{n} \lfloor \frac{w_i}{d} \rfloor x_i \leq \lfloor \frac{r}{d} \rfloor}$$

Proposition 4. *The* DIVIDE *rule is sound.*

Proof. Follows from the soundness of division on traditional linear inequalities. $\square$

Fourier Resolution. Fourier resolution captures algebraic reasoning over the linear parts of hypercube linear constraints $L_1 \equiv H_1 \to R_1$ and $L_2 \equiv H_2 \to R_2$. Assume that $H_1 \wedge H_2$ is consistent and the coefficient of some variable $x_j \in \mathcal{X}$ is positive in R_1 and negative in R_2. We can use Fourier resolution to combine R_1 and R_2 into a new constraint $R_3 = FR(R_1, R_2, x_j)$ with the guarantee that $R_1 \wedge R_2 \models R_3$. Fourier resolution also ensures that the weight for x_j in R_3 is 0, removing its contribution to the satisfiability of L_3.

$$\text{ResF}(L_1, L_2, x_j) \quad \frac{H_1 \to R_1 \qquad H_2 \to R_2}{H_1 \wedge H_2 \to FR(R_1, R_2, x_j)}$$

Proposition 5. *Let a be the weight of x_j in R_1, and b be the weight of x_j in R_2. If $a \cdot b < 0$, then $L_1 \wedge L_2 \models \text{ResF}(L_1, L_2, x_j)$.*

Proof. If both $L_1 \wedge L_2$ are true, it follows that $H_1 \wedge H_2 \to R_1 \wedge R_2$. By soundness of Fourier elimination, $R_1 \wedge R_2 \models FR(R_1, R_2, x_j) = R_3$. So $H_1 \wedge H_2 \to R_3$. $\square$

Propositional Resolution. To complement the reasoning over linear inequalities, we also need a rule to perform disjunctive reasoning over $L_1 \equiv H_1 \to R_1$ and $L_2 \equiv H_2 \to R_2$. Let $c_1 \equiv \langle x \geq d_1 \rangle$ and $c_2 \equiv \langle x \leq d_2 \rangle$ be two bound constraints such that $d_1 \leq d_2 + 1$. This means that, for every value that can be assigned to x, one of c_1 or c_2 is true. Both c_1 and c_2 are part of H_1 and H_2 respectively: $H_1 \equiv H_1' \wedge c_1$ and $H_2 \equiv H_2' \wedge c_2$. Additionally, assume that $H_1' \wedge H_2'$ is consistent. Traditional propositional resolution concludes the following:

$$H_1' \wedge H_2' \to (R_1 \vee R_2)$$

Regretfully, the resulting constraint is not a hypercube linear constraint. Since $R_1 \vee R_2$ is not linear, we need to formulate a new constraint $L_3 \equiv H_3 \to R_3$ such that $H_1' \wedge H_2' \to (R_1 \vee R_2) \models L_3$. Given that, we want a resolution rule ResH_F, parameterised by a function F that finds an appropriate L_3. The function F returns additional bounds constraints H_{new} and a linear inequality R_3 such that $L_3 \equiv H_1' \wedge H_2' \wedge H_{\text{new}} \to R_3$.

For this work, we provide an implementation of F that converts L_2 to a clause and then combines that with L_1. Through weakening, we can rewrite any hypercube linear constraint to adhere to this restriction. If the linear component R_2 simplifies to $\bot$, then the linear component of L_3 simplifies to R_1. Substituting that logic into the parameterised ResH_F we obtain ResH as follows:

$$\text{ResH}(L_1, L_2, c_1, c_2) \quad \frac{H_1' \wedge c_1 \to R_1 \qquad H_2' \wedge c_2 \to \bot}{H_1' \wedge H_2' \to R_1}$$

Proposition 6. ResH *is sound.*

Proof. Since $d_1 \leq d_2 + 1$, at least one of $\langle x \geq d_1 \rangle$ or $\langle x \leq d_2 \rangle$ is true for every value of x. In addition, weakening is sound by Proposition 4. Then $R_1 \vee \bot$ simplifies to R_1. So we conclude that ResH is sound. $\square$

6   Conflict-Driven Hypercube Linear Learning

When we view the hypercube linear constraint as an elementary constraint, we can design a solver that derives implied hypercube linear constraints. We substitute the conflict analysis procedure from the CDCL [21] algorithm used in most state-of-the-art SAT solvers to learn hypercube linear constraints rather than clauses. This gives us a sound and complete procedure for determining whether solutions exist for a conjunction of hypercube linear constraints.

6.1   The Conflict Analysis Procedure

Algorithm 1 shows an implementation of the conflict analysis procedure using hypercube linear resolution called HLCA. It expects a trail ρ of bounds constraints that are true. Additionally, it expects a hypercube linear constraint L_{confl} that is conflicting w.r.t. the partial assignment described by ρ, which we will refer to as $\mathcal{D}_\rho$. The algorithm iteratively combines hypercube linear constraints until L_{confl} propagates at a decision level lower than the current decision level.

For each iteration, the last asserted bounds constraint $c \equiv \langle x \diamond v \rangle$ is popped from the trail (line 3). The function EXPLAINCONFLICT weakens L_{confl} to a clause with respect to the bounds in $\mathcal{D}_{\rho \wedge c}$, collecting all bounds in ρ that contribute to the conflict. In case c does not contribute to the current conflict, the algorithm proceeds to the next bound in ρ (line 5). If c does contribute to the conflict, line 6 gets the hypercube linear constraint $L = H \rightarrow R$ that propagated c, as well as the explanation B for the propagation. The algorithm then uses the inference rules from Sect. 5 to combine L_{confl} and L such that c is no longer necessary for L_{confl} to be conflicting.

If a Fourier elimination is possible, then c contributes to the conflict through the linear component of L_{confl}. WEAKENANDDIVIDE (Algorithm 2) removes all the linear terms that are not divisible by the weight w of x, and then divides the linear by $|w|$. This means x will have unit weight and RESF is guaranteed to produce a linear component that is conflicting w.r.t. $\mathcal{D}_\rho$.

If c is implied by the hypercube of L_{confl}, propositional resolution is used. As explained in Sect. 5, propositional resolution is implemented by combining a hypercube linear with a clause. The clausal reason for c is constructed using the explanation B for c, and RESH is applied.

Before continuing to the next iteration, any variables that are fixed by the hypercube of L_{confl} are removed from the linear part. I.e. if H is the hypercube of L_{confl} and we have that $\mathcal{D}_H(x_j) = \{d\}$, then we subtract $w_j d$ from both sides of the linear inequality to get an equivalent constraint.

6.2   Discussion on Extended Cutting Planes, and PB and IntSAT Conflict Analysis

Our proof system may be viewed as an extended cutting-planes proof system. The extended variables correspond to the bound constraints generated by our

---

Algorithm 1. $\mathrm{HLCA}(\rho, L_{\mathrm{confl}})$: Hypercube linear conflict analysis

---

Require: A trail ρ

Require: A hypercube linear L_{confl} that is conflicting w.r.t. ρ.
Ensure: L_{confl} is propagating w.r.t. ρ at the decision level dl.
1: $dl_{\mathrm{confl}} \leftarrow$ the current decision level
2: **while** L_{confl} does not propagate at a decision level $dl' < dl$ **do**
3:     $c \equiv \langle x \diamond v \rangle \leftarrow$ pop last propagated bounds constraint from ρ
4:     **if** $c \notin \mathrm{ExplainConflict}(\mathcal{D}_{\rho \wedge c}, L_{\mathrm{confl}})$ **then**
5:        **continue**
6:     $L, B \leftarrow \mathrm{Reason}(c)$
7:     **if** ResF can be applied to L_{confl} and L eliminating x **then**
8:        $L_{\mathrm{expl}} \leftarrow \mathrm{WeakenAndDivide}(L, x)$
9:        $L_{\mathrm{confl}} \leftarrow \mathrm{ResF}(L_{\mathrm{confl}}, L_{\mathrm{expl}}, x)$
10:     **if** $\exists c'$ in the hypercube of L_{confl} such that $c \models c'$ **then**
11:        $L_{\mathrm{expl}} \leftarrow (B \wedge \neg c \to \bot)$
12:        $L_{\mathrm{confl}} \leftarrow \mathrm{ResH}_F(L_{\mathrm{confl}}, L_{\mathrm{expl}}, c, \neg c)$
13:     $L_{\mathrm{confl}} \leftarrow \mathrm{Simplify}(L_{\mathrm{confl}})$
14: $dl \leftarrow \max \{ \mathrm{DecisionLevel}(c) \mid c \in RS_{\mathrm{confl}} \wedge \mathrm{DecisionLevel}(c) < dl_{\mathrm{confl}} \}$
15: **return** L_{confl}, dl

---

weakening operations, enabling us to capture nonlinear propositional reasoning in a natural way and combine it with algebraic operations.

Algorithm 1 allows us to use this proof system in practice. The reformulation of the explanation hypercube linear generalises PB conflict analysis using a similar Divide inference rule [13]. A key advantage of our hypercube linear weakening is that we retain information on which variables have been weakened in the hypercube, and it enables us to apply generalised resolution to integer linear constraints. In this sense, comparing our resolution approach with PB (where hypercube linear constraints are encoded as big-M constraints [26]), our approach derives constraints at least as strong or stronger.

The situation is slightly different when compared to the Fourier resolution used in IntSAT [23]. Our approach applies weakening and division to ensure the Fourier operation succeeds. In contrast, the IntSAT approach attempts Fourier resolution and uses a fallback mechanism if it fails. In principle, we could also adopt that approach and use our weaken-and-divide approach as the fallback mechanism. However, our current research is focused on developing a unified approach guaranteed to succeed. Further work is needed to understand the relative effectiveness of different operations and how they can be combined effectively.

Note that we do not need the Divide inference rule for the system to be complete. A proof can always fall back on ResH with additional weakening. However, Divide prevents coefficients from overflowing a machine integer.

6.3 Soundness and Completeness of HLCA

Using HLCA in a CDCL search algorithm is both sound and complete, as the conflict analysis from Algorithm 1 satisfies the necessary invariants provided by Theorem 1 and Theorem 2.

Theorem 1 (Soundness). *Let L be derived through HLCA (Algorithm 1), and KB be the conjunction of hypercube linear constraints in the problem. Then $KB \models L$.*

Proof. By Proposition 4 we have that weakening is sound, so we can use it to rewrite L_{confl}. Additionally, Proposition 5 and Proposition 6 show that RESH and RESF are sound. Since the initial L_{confl} and all explanations L are part of KB, and the way we combine them is sound, HLCA is also sound.   □

Algorithm 2. WEAKENANDDIVIDE(L, c): Prepare the linear part of L to explain its propagation of bounds constraint c.

Require: A hypercube linear $L \equiv H \rightarrow R$

Require: A bounds constraint c over variable x that was propagated by R.

Require: Domains $\mathcal{D}$ such that $\mathcal{D} \wedge L \models c$.
Ensure: $hlslack(\mathcal{D}, L_{\mathrm{expl}}) = 0$
Ensure: $L_{\mathrm{expl}} \models L$
Ensure: $\mathcal{D} \wedge L_{\mathrm{expl}} \models c$.
 1: $L_{\mathrm{expl}} \leftarrow L$
 2: $w_j \leftarrow$ the weight of x in R
 3: **for** $(w_i, x_i) \in R$ **do**
 4:     **if** $w_i \neq 0$ and w_i is *not* divisible by $|w_j|$ **then**
 5:         **if** $w_i > 0$ **then**
 6:             $c' \leftarrow \langle x \geq lb(\mathcal{D}, x) \rangle$
 7:         **else if** $w_i < 0$ **then**
 8:             $c' \leftarrow \langle x \leq ub(\mathcal{D}, x) \rangle$
 9:         $L_{\mathrm{expl}} \leftarrow$ WEAKEN(L_{expl}, c')
10: $L_{\mathrm{expl}} \leftarrow$ DIVIDE$(L_{\mathrm{expl}}, |w_j|)$
11: **return** L_{expl}

The proof for completeness consists of showing that L_{confl} remains conflicting w.r.t. $\mathcal{D}_\rho$, and that it terminates with at most one bound at the current decision level contributing to the conflict.

Lemma 1. *Algorithm 1 maintains that L_{confl} is conflicting w.r.t. $\mathcal{D}_\rho$: Let L_{confl} be conflicting w.r.t. $\mathcal{D}_{\rho \wedge c}$ at the beginning of a loop iteration, and let the last element on the trail be the bounds constraint c. The loop body in Algorithm 1 transforms it into the constraint L'_{confl}. Then, L'_{confl} is conflicting w.r.t. $\mathcal{D}_\rho$.*

Proof. In the loop body, RESF is used to eliminate the contribution of c to the conflict in the linear inequality. It will always end up with a conflicting constraint due to the reformulation performed by WEAKENANDDIVIDE (see Proposition 3 and [19]). Then RESH removes the contribution of c from the hypercube, yielding a conflicting constraint. So the invariant is maintained.   □

Theorem 2 (Completeness).   *Let KB be an inconsistent conjunction of hypercube linear constraints. HLCA will derive the constraint $\top \rightarrow \bot$ in a finite number of steps.*

Proof. HLCA is complete for the same reason that CDCL is complete. By Lemma 1, Algorithm 1 derives a constraint that is conflicting w.r.t. the current domain. Every iteration, the contribution of the last bound to the conflict is removed. At some point, there is only one bound on the current decision level, so the constraint will propagate to a previous decision level.   □

7   Experiments

With the theory of hypercube linear constraints established, we aim to investigate whether a solver can leverage their structure. Therefore, we run preliminary experiments to answer one question: How does the use of hypercube linear resolution impact the number of conflicts encountered during search compared to propositional resolution (used by SAT and CP solvers)?

Aside from the main question, we also compare the conflict count to the IntSat method of Fourier resolution (see Sect. 6.2 for a discussion).

7.1   Setup

We added a proof-of-concept version of the hypercube linear propagator and a naive implementation of HLCA to the CP solver Pumpkin [15]. We are only interested in how the conflict count is impacted to assess the potential of hypercube linear resolution. Future work is needed to identify efficient algorithms for the propagator and conflict analysis.

We evaluate our implementation using models from MiniZinc challenges 2020–2024. From the challenge instances, we selected the 64 that Pumpkin solves in five minutes using its default MiniZinc library. All instances were encoded as integer linear problems using the MiniZinc toolchain, so that all models can be expressed using hypercube linear constraints. Any search procedures that assign values or remove individual values (i.e. decide on $\langle x = v \rangle$ or $\langle x \neq v \rangle$) were altered to decide on bounds of variables instead. This is because HLCA expects the trail to consist solely of bound constraints, and it does not support equality or disequality constraints.

Every instance was solved by the default version of Pumpkin, which implements propositional resolution, a version of Pumpkin that learns linear inequalities using IntSat conflict analysis [2], and our version that implements hypercube linear resolution. The baselines ran for five minutes, and the hypercube linear solver ran for 30 min. Since the hypercube linear resolution implementation is roughly 10x slower in terms of conflicts per second, it received more time.

All experiments were conducted on the DelftBlue [12] cluster. Each instance used a single core on an Intel Xeon Gold 6448Y 32C 2.1GHz processor with 4GB RAM. The entire setup with data can be found on Zenodo.[1]

7.2   Results

We solved 33 instances with the hypercube linear resolution solver, and timed out on the remaining 31. Of the 33 instances, 25 are also solved by the propositional configuration, and 26 are solved by the Fourier configuration. 22 instances are solved by all configurations.

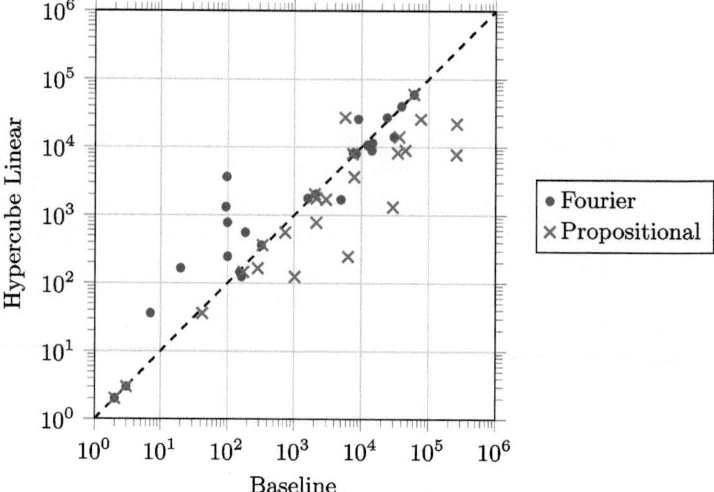

Fig. 1. A scatter plot comparing the conflict count of hypercube linear resolution with one of the two baseline configurations. Every dot is an instance. The dashed line is the diagonal. Below the dashed line means fewer conflicts for hypercube linear resolution, and above the diagonal means fewer conflicts for the respective baseline.

Compared to Propositional. In Fig. 1, we see that most of the instances see fewer conflicts when using hypercube linear resolution compared to propositional resolution. For some instances, the difference is two orders of magnitude. This is unsurprising, as hypercube linear resolution simulates regular resolution when all the linear parts simplify to $\perp$, but its linear component requires potentially many clauses to express. We can take the ratio of the number of learned clauses over the number of learned hypercube linear constraints to see what fraction of

---

[1] See https://zenodo.org/records/17807799.

derived constraints is a clause. The median value of this ratio is 4.3%, meaning that, in most cases, HLCA derives a constraint with a linear inequality over at least two variables. For other instances, the difference is small, and those instances tend to learn constraints that are almost clauses.

However, we observe an interesting case in which the conflict count increases when using hypercube linear resolution. This happens because the order of propagation varies across configurations, so the first learned constraint is based on different model constraints. That impacts how far the solver can backtrack and where the search proceeds, leading to differing conflict counts even with a fixed search strategy. For the other instances, this also happens, but the stronger constraints we learn make up for the difference.

Compared to Fourier. Compared to pure Fourier resolution, we see that hypercube linear resolution needs significantly more conflicts in 8 instances. A difference in propagation order accounts for a small part of the difference, but the benchmarks that we use also contribute.

Ideally, instances would be encoded with hypercube linear constraints. However, available instances are encoded as traditional linear inequalities [4] since hypercube linear constraints are not standard. This encoding hinders our approach, as many constraints have a few large coefficients that are unlikely to be included in the output of the WEAKENANDDIVIDE algorithm. A possible solution is to modify the MiniZinc library used to flatten the instances to use hypercube linear constraints, or do post-processing to detect when a hypercube linear constraint is encoded as a regular inequality. However, that is beyond the scope of our main contributions.

Where hypercube-linear resolution comes out favourably is in the 7 instances that are solved by hypercube linear resolution but not by Fourier resolution. Those seven instances are also not solved by propositional resolution, and both the Fourier configuration and the propositional configuration time out at more conflicts than we necessary for hypercube linear resolution.

This gives the impression that there are instances where Fourier resolution just works very well, but when it does not, it relies on propositional resolution being effective. If that is not the case, the improvement hypercube linear resolution brings over propositional resolution is extremely useful.

8 Conclusion and Future Work

We present a novel resolution-style proof system combining propositional resolution with Fourier resolution. This combination allows us to systematically derive implied constraints in a CDCL-style algorithm that capture both linear relations between integer variables and disjunctive relations between variable bounds.

The key to the proof system is to combine the linear inequality constraint with a clause into a single constraint. We call this constraint the hypercube linear constraint, and it generalises both a linear inequality and a clause. It can therefore be used in place of either and allows for unified reasoning over both.

A proof-of-concept implementation demonstrates that the resolution system can be utilised to solve constraint problems. Compared to propositional resolution, the number of conflicts required to solve an instance can be significantly reduced. Yet, the experiments also show that, compared to an incomplete Fourier-resolution approach, learning is less effective. This result is explained by the benchmarks used in the evaluation.

Looking forward, there are many areas to explore for further evolution of hypercube linear constraints and resolution. On the propagation side, we can implement a global propagator using a watched literal [22] approach. Additionally, the linear inequalities in the hypercube linear that are active could be added to a linear propagator as is done in OR-Tools CP-SAT [25]. On the side of resolution, different instantiations of the $\textsc{ResH}_F$ rule can be explored, to avoid having to always weaken one of the two constraints to a clause. Additionally, the proof system can be expanded to generalise more rules from the VeriPB [6] proof system, potentially allowing effective proof logging and checking with first-class integer variables.

References

1. Audemard, G., Katsirelos, G., Simon, L.: A Restriction of extended resolution for clause learning SAT solvers. In: Proceedings of the AAAI Conference on Artificial Intelligence, vol. 24, no. 1, pp. 15–20 (2010). https://doi.org/10.1609/aaai.v24i1.7553
2. Baauw, R., Flippo, M., Demirović, E.: Conflict analysis based on cutting-planes for constraint programming. In: de la Banda, M.G. (ed.) 31st International Conference on Principles and Practice of Constraint Programming (CP 2025). Leibniz International Proceedings in Informatics (LIPIcs), vol. 340, pp. 4:1–4:19. Schloss Dagstuhl - Leibniz-Zentrum für Informatik, Dagstuhl (2025). https://doi.org/10.4230/LIPIcs.CP.2025.4
3. Belotti, P., et al.: On handling indicator constraints in mixed integer programming. Comput. Optim. Appl. **65**(3), 545–566 (2016). https://doi.org/10.1007/s10589-016-9847-8
4. Belov, G., Stuckey, P.J., Tack, G., Wallace, M.: Improved linearization of constraint programming models. In: Rueher, M. (ed.) CP 2016. LNCS, vol. 9892, pp. 49–65. Springer, Cham (2016). https://doi.org/10.1007/978-3-319-44953-1_4
5. Biere, A., Heule, M., van Maaren, H.: Handbook of Satisfiability. IOS Press (2009)
6. Bogaerts, B., Gocht, S., McCreesh, C., Nordström, J.: Certified dominance and symmetry breaking for combinatorial optimisation. J. Artif. Intell. Res. **77**, 1539–1589 (2023)
7. Bromberger, M., Sturm, T., Weidenbach, C.: A complete and terminating approach to linear integer solving. J. Symb. Comput. **100**, 102–136 (2020). https://doi.org/10.1016/j.jsc.2019.07.021
8. Buss, S., Nordström, J.: Chapter 7. Proof complexity and SAT solving. In: Handbook of Satisfiability, pp. 233–350. IOS Press (2021). https://doi.org/10.3233/FAIA200990
9. Chai, D., Kuehlmann, A.: A fast pseudo-boolean constraint solver. In: Proceedings of the 40th Annual Design Automation Conference, DAC '03, pp. 830–835. Association for Computing Machinery, New York (2003). https://doi.org/10.1145/775832.776041

10. Codel, C., Fazekas, K., Heule, M., Iser, M.: Proceedings of SAT Competition 2025: Solver and Benchmark Descriptions, Report, TU Wien (2025). https://doi.org/10.34726/10379
11. Davis, M., Putnam, H.: A computing procedure for quantification theory. J. ACM **7**(3), 201–215 (1960). https://doi.org/10.1145/321033.321034
12. Delft High Performance Computing Centre (DHPC): DelftBlue Supercomputer (Phase 2) (2024). https://www.tudelft.nl/dhpc/ark:/44463/DelftBluePhase2
13. Elffers, J., Nordström, J.: Divide and conquer: towards faster pseudo-Boolean solving. In: Proceedings of the 27th International Joint Conference on Artificial Intelligence, IJCAI'18, pp. 1291–1299. AAAI Press, Stockholm (2018)
14. Feydy, T., Stuckey, P.J.: Lazy clause generation reengineered. In: Gent, I.P. (ed.) CP 2009. LNCS, vol. 5732, pp. 352–366. Springer, Heidelberg (2009). https://doi.org/10.1007/978-3-642-04244-7_29
15. Flippo, M., Sidorov, K., Marijnissen, I., Smits, J., Demirović, E.: A multi-stage proof logging framework to certify the correctness of CP solvers. In: Shaw, P. (ed.) 30th International Conference on Principles and Practice of Constraint Programming (CP 2024). Leibniz International Proceedings in Informatics (LIPIcs), vol. 307, pp. 11:1–11:20. Schloss Dagstuhl - Leibniz-Zentrum für Informatik, Dagstuhl (2024). https://doi.org/10.4230/LIPIcs.CP.2024.11. https://drops.dagstuhl.de/entities/document/10.4230/LIPIcs.CP.2024.11
16. Gomory, R.E.: Outline of an algorithm for integer solutions to linear programs. Bull. Am. Math. Soc. **64**(5), 275–278 (1958). https://doi.org/10.1090/S0002-9904-1958-10224-4
17. Hooker, J.N.: Generalized resolution and cutting planes. Ann. Oper. Res. **12**(1), 217–239 (1988). https://doi.org/10.1007/BF02186368
18. Huang, J.: Extended clause learning. Artif. Intell. **174**(15), 1277–1284 (2010). https://doi.org/10.1016/j.artint.2010.07.008
19. Jovanović, D., de Moura, L.: Cutting to the Chase. J. Autom. Reason. **51**(1), 79–108 (2013). https://doi.org/10.1007/s10817-013-9281-x
20. Katsirelos, G.: Generalized NoGoods in CSPs. In: Proceedings of the AAAI Conference on Artificial Intelligence, vol. 5, pp. 390–396 (2005)
21. Marques Silva, J., Sakallah, K.: GRASP-a new search algorithm for satisfiability. In: Proceedings of International Conference on Computer Aided Design, pp. 220–227 (1996). https://doi.org/10.1109/ICCAD.1996.569607
22. Moskewicz, M.W., Madigan, C.F., Zhao, Y., Zhang, L., Malik, S.: Chaff: engineering an efficient SAT solver. In: Proceedings of the 38th Annual Design Automation Conference, DAC '01, pp. 530–535. Association for Computing Machinery, New York (2001). https://doi.org/10.1145/378239.379017
23. Nieuwenhuis, R., Oliveras, A., Rodríguez-Carbonell, E.: IntSat: integer linear programming by conflict-driven constraint learning. Optim. Methods Softw. 1–28 (2023). https://doi.org/10.1080/10556788.2023.2246167
24. Ohrimenko, O., Stuckey, P.J., Codish, M.: Propagation via lazy clause generation. Constraints **14**(3), 357–391 (2009). https://doi.org/10.1007/s10601-008-9064-x
25. Perron, L., Didier, F., Gay, S.: The CP-SAT-LP solver. In: Yap, R.H.C. (ed.) 29th International Conference on Principles and Practice of Constraint Programming (CP 2023). Leibniz International Proceedings in Informatics (LIPIcs), vol. 280, pp. 3:1–3:2. Schloss Dagstuhl - Leibniz-Zentrum für Informatik, Dagstuhl (2023). https://doi.org/10.4230/LIPIcs.CP.2023.3
26. Trespalacios, F., Grossmann, I.E.: Improved Big-M reformulation for generalized disjunctive programs. Comput. Chem. Eng. **76**, 98–103 (2015). https://doi.org/10.1016/j.compchemeng.2015.02.013

27. Tseitin, G.S.: On the complexity of derivation in propositional calculus. In: Siekmann, J.H., Wrightson, G. (eds.) Automation of Reasoning: 2: Classical Papers on Computational Logic 1967–1970, pp. 466–483. Springer, Heidelberg (1983). https://doi.org/10.1007/978-3-642-81955-1_28
28. Veksler, M., Strichman, O.: Learning general constraints in CSP. Artif. Intell. **238**, 135–153 (2016). https://doi.org/10.1016/j.artint.2016.06.002
29. Wolsey, L.A.: Integer Programming. John Wiley & Sons, Hoboken (2020)

Transit Network Design with Two-Level Demand Uncertainties: A Machine Learning and Contextual Stochastic Optimization Framework

Hongzhao Guan[1]([envelope])[iD], Beste Basciftci[2][iD], and Pascal Van Hentenryck[1][iD]

[1] H. Milton Stewart School of Industrial and Systems Engineering,
Georgia Institute of Technology, Atlanta, GA 30345, USA
{hguan7,pvh}@gatech.edu
[2] Tippie College of Business, University of Iowa, Iowa City, IA 55242, USA
beste-basciftci@uiowa.edu

Abstract. Transit Network Design is a well-studied problem in the field of transportation, typically addressed by solving optimization models under fixed demand assumptions. Considering the limitations of these assumptions, this paper proposes a new framework, namely the Two-Level Rider Choice Transit Network Design (2LRC-TND), that leverages machine learning and contextual stochastic optimization (CSO) through constraint programming (CP) to incorporate two layers of demand uncertainties into the network design process. The first level identifies travelers who rely on public transit (core demand), while the second level captures the conditional adoption behavior of those who do not (latent demand), based on the availability and quality of transit services. To capture these two types of uncertainties, 2LRC-TND relies on two travel mode choice models, that use multiple machine learning models. To design a network, 2LRC-TND integrates the resulting choice models into a CSO that is solved using a CP-SAT solver. 2LRC-TND is evaluated through a case study involving over 6,600 travel arcs and more than 38,000 trips in the Atlanta metropolitan area. The computational results demonstrate the effectiveness of the 2LRC-TND in designing transit networks that account for demand uncertainties and contextual information, offering a more realistic alternative to fixed-demand models.

Keywords: Constraint Programming · Contextual Stochastic Optimization · Transit Network Design · Travel Behavior

1 Introduction

Transit Network Design is an important problem in transportation and urban planning, that focuses on developing effective public transportation services.

____

Supplementary Information The online version contains supplementary material available at https://doi.org/10.1007/978-3-032-27242-3_11.

Transit agencies, which operate transit networks, may need to redesign their systems when new technologies emerge or when the urban environment undergoes changes. A common approach to re-design a transit network consists in solving an optimization problem that leverages existing transit demand, typically represented by Origin-Destination (O-D) pairs. However, this method often has difficulties to account for the changes in demand, i.e., the number of riders who will actually use the system after the network re-design. For a transit agency, relying solely on transit demand of the existing system is insufficient because the new system may draw additional riders, potentially leading to overcrowding and poor service quality. On the other hand, overestimating transit demand can result in an inefficient and costly design. The transit design can then be thought of as a game between the transit agency and riders, where the transit design influences ridership, which in turn determines the quality and cost of the system. At this point, the critical issues become how to accurately capture travel behaviors from available data and contextual information then integrate them into optimization models.

To address these challenges, this paper provides a novel framework to model the transit network design problem, namely the Two-Level Rider Choice Transit Network Design (2LRC-TND). The 2LRC-TND framework aims at finding a network design as the equilibrium point between the transit agency and its potential riders. To find this equilibrium point, 2LRC-TND introduces a contextual stochastic optimization (CSO) approach that learns contextual probability distributions to capture rider behavior effectively.

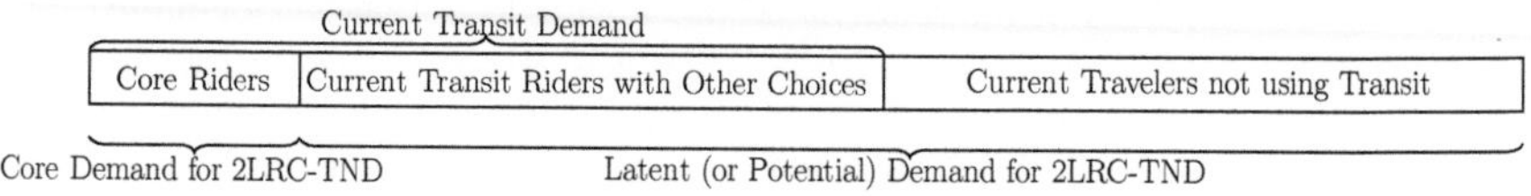

Fig. 1. A Diagram depicting the demand structure in 2LRC-TND.

More precisely, for representing travel demand, 2LRC-TND considers two levels of uncertainty in capturing core and latent demand, as shown in Fig. 1. The first level focuses on identifying "core demand", i.e., travelers who rely on public transit as their primary travel mode and must be served by the transit system; these riders need to be identified from the existing transit demand. The second level captures the "latent demand", i.e., potential riders who may choose to use transit if the system provides good service quality; these riders have alternative travel options and will only adopt transit if the quality meets their needs. Thus, some of these potential riders are already part of the current transit demand, while others are currently rejecting the existing system and using other travel modes. By combining these two levels of uncertainty, the proposed framework introduces a novel approach to represent variable demand.

Compared to prior work in transit network design, the 2LRC-TND framework introduces novel contributions along three axes. First, it incorporates two levels of demand uncertainty; this contrasts with studies, such as [13], which divide trip demand into two fixed subsets: the core transit users and the latent demand. Importantly, 2LRC-TND treats the identification of these two sets as a classification problem. Secondly, to account for uncertainties in the network design optimization, 2LRC-TND models individual decisions as probability distributions rather than as fixed values. Lastly, 2LRC-TND is a flexible and generic framework that integrates learning and optimization in a way that is applicable to various types of network design problems beyond transit systems.

From a technical standpoint, 2LRC-TND framework leverages CSO and the Sample Average Approximation (SAA) methods to explicitly incorporate behavioral uncertainty into the transit network design optimization model. Under this setting, multiple types of machine learning techniques are used to model riders' choices. The optimization component of 2LRC-TND uses CP-SAT to maximize total system coverage while respecting budget and operational constraints.

To demonstrate the practicality of 2LRC-TND, this paper reports a comprehensive real-world case study using over 38,000 observed trips from the Atlanta metropolitan region. The results show that, within a given budget, 2LRC-TND can fully serve core trip demand while attracting additional riders, leading to a ridership increment with up to 30% higher adoption. This showcases the practicality and potential social impact of 2LRC-TND for transit agencies that aim at implementing data-driven planning solutions.

In summary, the contribution of this paper can be characterized as follows:

1. This paper proposes 2LRC-TND, a novel transit network design framework that captures multiple sources of uncertainty in a hierarchical way.
2. To solve 2LRC-TND applications, the paper introduces a (widely applicable) CSO approach that integrates machine learning and stochastic optimization.
3. Computational results from a large-scale case study with real data show that 2LRC-TND is both practical and beneficial for real-world transit network planning.

2   Related Work

The Transit Network Design problem has been extensively studied and is widely recognized as an essential topic for societal development [11,12,16]. A common approach to design a transit network is by solving an optimization problem, which is also the approach used in this paper. To help readers better understand the context of this work, it is useful to categorize existing studies based on how researchers model transit demand. Majority of studies in this field assume demand is fixed [8,9,11,23,25], while others consider uncertain or variable demand that responds to the proposed network design [4,5,10,13,19,20]. This paper belongs to the latter category and adopts more complex yet more realistic assumptions regarding uncertain demand by filling this critical research gap in this area.

A general approach to modeling transit demand is travel mode choice modeling, which predicts who will use transit and when, and is a key research focus

Table 1. Notations

Notation	Definition
Key Terms:	
Rider	A person who takes transit services.
Trip r	Defined by an O-D pair and a group of e_r riders who travel together and share the same travel decisions.
Sets:	
$\mathcal{T}$	The set of all trips (O-D Pairs).
$\mathcal{T}^i_{core}$, $\mathcal{T}^i_{latent}$	The partition of set $\mathcal{T}$ into $\mathcal{T}_{core}$ and $\mathcal{T}_{latent}$ under each scenario i.
$\mathcal{N}$	The set of locations for potential transit stops.
$\mathcal{M}$	The set of transit modes. Default: $\{rail, bus\}$.
$\mathcal{P}_r$	The set of considered paths for trip r.
$\mathcal{A}$, $\mathcal{A}_p$	The set of all arcs and arcs belong to path p.
$\mathcal{A}_{fixed}$	The set of all fixed arcs.
$\mathcal{A}^+_{n,m}$, $\mathcal{A}^-_{n,m}$	The set of out-arcs and in-arcs for node n with mode m.
$\mathcal{A}_{n_1,n_2,m}$	The set of arcs from node n_1 to n_2 with mode m.
$\mathcal{Z}$	The set containing feasible transit network designs.
Decision Vars.:	
z_a	Binary variable indicating if transit arc a is opened in the network design.
u^i_r	Binary variable indicating if trip r uses the transit system, under scenario i.
d^i_r	Binary variable indicating if trip r adopts the transit system, under scenario i.
f_p	Binary variable indicating if path p is feasible, dynamically based on $\mathbf{z}$.
Parameters:	
I, I'	Number of SAA scenarios for solving and evaluation, respectively.
B	Budget of transit agency for a planning horizon (e.g., a day).
e_r	Number of riders in trip r who share the same travel behaviors.
s_a	Cost of operating arc a.
h_a	Number of buses in an hour of arc a.
c^i_r	Indicating if trip r is a core trip under scenario i.
$d^i_{r,p}$	Indicating the adoption decision of trip r on path p under scenario i.
l_r	Penalty term when evaluating a network design for not serving a trip r.
w_r	Amount of paths that are considered for trip r.
Vectors:	
$\mathbf{z}$	Vector of z_a over arcs $a \in \mathcal{A}$, representing a transit network design.
$\mathbf{x}_r$	Contextual information vector for trip r.

in transportation. These studies primarily rely on survey data collected from individuals. Common modeling techniques range from logit models [21] to traditional machine learning model [28,30], as well as more advanced Deep Neural Networks (DNN) [22]. This paper develops its choice models by extending the insights and findings from the studies discussed above, where the existing studies in this area do not consider the subsequent network design problem while solely focusing on learning rider behavior.

The 2LRC-TND problem explored in this paper combines the two previously discussed components—a machine learning problem and an optimization problem. Generally speaking, it is a decision-making problem that incorporates predictive models [6], and a recently emerging framework for addressing such problems is CSO which considers contextual information and learning approaches in representing uncertainty within the optimization problems [24]. The core ideas of these studies have been successfully applied across various domains, such as route planning in transportation [26], order fulfillment [29] and inventory control [7,27] in supply chain management, shelter location selection in disaster management [17], and public school redistricting in computational social science [15]. The next section will formally introduce the 2LRC-TND problem and establish its connection to the CSO framework.

3 Problem Description and Methodology

This section presents the 2LRC-TND problem that integrates two levels of demand uncertainties within a stochastic optimization framework. From a technical perspective, 2LRC-TND adopts CSO as a methodological framework and employs Constraint Programming (CP) to model and solve the resulting optimization problems. The CP model aims at maximizing the expected transit coverage by making decisions on which bus arcs to activate or deactivate, while satisfying several constraints. It is important to note that the primary goal of this study is to demonstrate the practicality of the proposed CSO-based framework. The transit network design problem can be adapted to optimize other objectives, such as cost and quality of service, as long as the formulation remains compatible with the structure of the framework. Table 1 summarizes the notations.

3.1 Demand Uncertainty

To accurately capture demand uncertainties, two key questions must be answered. The first question identifies those riders who needs to use public transit on a daily-basis, i.e., they have no alternative travel options. Following this question, there are two demand sets that can be defined: $\mathcal{T}_{core}$ and $\mathcal{T}_{latent}$. The core set $\mathcal{T}_{core}$ captures those riders who must use transit, while the latent demand set $\mathcal{T}_{latent}$ contains those who have options. Trips in $\mathcal{T}_{latent}$ will adopt the transit system only when certain criteria are met. Together, they form the trip set $\mathcal{T}$. Therefore, the first level of demand uncertainty lies in how $\mathcal{T}$ is divided into $\mathcal{T}_{core}$ and $\mathcal{T}_{latent}$. Once the trips in $\mathcal{T}_{core}$ are identified, the transit agency must provide them with appropriate service.

The second question considers the latent set $\mathcal{T}_{latent}$ and asks when these riders choose to use transit instead of other modes? These riders will likely make a decision on adopting or rejecting transit based on the travel path presented to them, and this is the second level of demand uncertainties. In contrast to $\mathcal{T}_{core}$, the transit agency is not obligated to serve all riders in $\mathcal{T}_{latent}$; however it aims at attracting as many of them as possible by offering them high-quality service.

With the two aforementioned questions in mind, all adoption decisions are defined as random variables coming from (unknown) distributions, which will be learned from contextual information. The context $\mathbf{x}_r$ for rider r typically includes demographic, geographic, and trip-specific information. Consider two binary random variables: (1) C_r which is 1 if trip r is a core trip and 0 otherwise, and (2) D_r is 1 if latent trip r adopts the system, and 0 otherwise. The distributions of these random variables are learned by machine learning models given the contextual information. Let U_r be random variable indicating whether trip r utilizes the transit system, i.e.,

$$U_{\mathbf{r}} = \begin{cases} 1, & if \;\; C_r = 1, \\ 1, & if \;\; C_r = 0 \;\; and \;\; D_r = 1, \\ 0, & if \;\; C_r = 0 \;\; and \;\; D_r = 0, \end{cases} \tag{1}$$

3.2 Network Design with Contextual Stochastic Optimization

This paper defines the 2LRC-TND problem over a directed multi-graph $\mathcal{G} = (\mathcal{N}, \mathcal{A})$. The set of nodes $\mathcal{N}$ includes all locations, where each $n \in \mathcal{N}$ represents a potential stop for transit service. The set of arcs $\mathcal{A}$ contains all allowed connections between pairs of locations. Each arc $a \in \mathcal{A}$ is associated with an origin, a destination, a transit mode a_m such as bus and rail, and a frequency h_a (for example, if the interval between two consecutive buses on this arc is 10 minutes, then h_a is 6 buses per hour). The decision variable z_a indicates whether arc $a \in \mathcal{A}$ is opened. The vector $\mathbf{z}$ collectively represents the network design, and the set $\mathcal{Z}$ contains all feasible network designs. Further modeling assumptions and details are discussed in subsequent sections.

2LRC-TND aims at maximizing expected transit coverage (or ridership) which can be represented in the following compact form:

$$\max_{\mathbf{z} \in \mathcal{Z}} \mathbb{E}_{\mathbf{U} \sim \mathbb{P}(\mathbf{U}|\mathbf{z},\mathbf{x})} \left[g(\mathbf{U}) \right], \tag{2}$$

where the function $g(\mathbf{U})$ represents the transit coverage amount under network design $\mathbf{z}$ and ridership $\mathbf{U}$ as defined for every trip r in (1), and the conditional distribution $\mathbb{P}(\mathbf{U}|\mathbf{z}, \mathbf{x})$ corresponds to the adoption behavior of riders based on the all trips' context information $\mathbf{x}$ and network design $\mathbf{z}$. The transit coverage is defined as

$$g(\mathbf{U}) = \sum_{r \in \mathcal{T}} e_r \cdot U_r \tag{3}$$

where e_r is the set of riders associated with trip r. Assuming that each trip r makes its adoption decision independently from other trips, the resulting CSO can be expressed as

$$\max_{\mathbf{z} \in \mathcal{Z}} \sum_{r \in \mathcal{T}} e_r \cdot \mathbb{E}_{U_r \sim \mathbb{P}(U_r|\mathbf{z},\mathbf{x}_r)}[U_r]. \tag{4}$$

2LRC-TND uses the Sample Average Approximation (SAA) method [18] to approximate the CSO problem (4). This approach generates I independent and identically distributed scenarios by sampling from the distribution $\mathbb{P}(U_r|\mathbf{z}, \mathbf{x}_r)$.

Let $\mathrm{SAA}_r^i(\mathbf{z})$ represents the adoption decision made by trip $r \in \mathcal{T}$ under the network design $\mathbf{z}$ in scenario $i \in I$. Then, the 2LRC-TND problem in (4) can be reformulated as follows:

$$\max_{\mathbf{z} \in \mathcal{Z}} \frac{1}{I} \sum_{i=1}^{I} \sum_{r \in \mathcal{T}} e_r \cdot \mathrm{SAA}_r^i(\mathbf{z}). \tag{5}$$

3.3   Decisions and Paths

To estimate $\mathbb{P}(U_r|\mathbf{z}, \mathbf{x}_r)$, 2LRC-TND uses probabilistic choice models $\mathcal{C}_{\text{core}}(\mathbf{x}_r)$ and $\mathcal{C}_{\text{adopt}}(\mathbf{x}_r, p)$ that capture the distributions of C_r and D_r, respectively. In general, both $\mathcal{C}_{\text{core}}$ and $\mathcal{C}_{\text{adopt}}$ are trained machine learning models based on the contextual information and transit paths suggested to potential riders. Both models capture the probabilities associated with the positive labels. Then based on the probabilities, model $\mathcal{C}_{\text{core}}$ and $\mathcal{C}_{\text{adopt}}$ can be further used for classification task, where $\mathcal{C}_{\text{core}}$ decide whether a trip r is a core trip (label 1) or a latent trip (label 0) and $\mathcal{C}_{\text{adopt}}$ classifies whether a latent trip r would adopt (1) or reject (0) a given transit path p. Following this, C_r can be modeled as:

$$C_r = \begin{cases} 1, & \text{if } \mathcal{C}_{\text{core}}(\mathbf{x}_r) = 1, \\ 0, & \text{otherwise.} \end{cases} \tag{6}$$

To model $\mathcal{C}_{\text{adopt}}$, define a transit path p as a route that links the origin and destination of a trip through transit services. Each path p typically includes two walking trip-legs and one or more transit trip-legs. Consider the network $\mathbf{z}$ where all the arcs (e.g., bus routes) are opened, and define, for each trip $r \in \mathcal{T}$, the set $\mathcal{P}_r$ representing the w_r paths with the shortest travel times, where w_r is a small number. For each trip $r \in \mathcal{T}_{latent}$, the choice model $\mathcal{C}_{\text{adopt}}(\mathbf{x}_r, p)$ takes two inputs, the contextual information $\mathbf{x}_r$ and information associated with a transit path p (e.g., the number of transfers and the transit time); it outputs the adoption decision for trip r. The intuition behind $\mathcal{C}_{\text{adopt}}$ is that individuals decide whether to use transit based on the availability of a favorable path. With these definitions, D_r is modeled as:

$$D_r = \begin{cases} 1, & \text{if } \bigvee_{p \in \mathcal{P}_r} (feasible(\mathbf{z}, p) \ \wedge \ \mathcal{C}_{\text{adopt}}(\mathbf{x}_r, p) = 1), \\ 0, & \text{otherwise} \end{cases} \tag{7}$$

Here, $feasible(\mathbf{z}, p)$ holds if path p is feasible under the network design $\mathbf{z}$, which means if all transit trip-legs in path p are open. Therefore, following Equation (7), trip r selects transit under design $\mathbf{z}$ if there exists a feasible path p that r adopts. Note that, for latent trips, riders are assumed to consider only the paths in $\mathcal{P}_r$, as these represent the fastest options. Any path with longer travel times are treated as rejections.

Both $\mathcal{C}_{\text{core}}$ and $\mathcal{C}_{\text{adopt}}$ are treated as black-boxes within the 2LRC-TND framework and can range from complex models, such as deep neural networks, to simple rule-based approaches. Importantly, under the black-box setting, even without high-quality or sufficient data to construct detailed contextual features $\mathbf{x}_r$, one can still develop reasonable rule-based models based on domain expertise and experiences to make the 2LRC-TND framework function effectively.

The overall 2LRC-TND framework can thus be summarized as follows. Given the set of trips $\mathcal{T}$, for each scenario, the first step identifies the core riders. Each core trip r must be served, which means at least one path in $\mathcal{P}_r$ is available under the network design. The remaining trips are classified as latent in that scenario. For each latent trip, 2LRC-TND checks whether any path in $\mathcal{P}_r$ is available and adopted. If not, trip r continues using other travel modes.

3.4 The Constraint Programming Model

This study adopts CP as the modeling approach due to its strengths in handling complicated logical relationships. The proposed model includes many AND and OR operations, which would require extensive linearization if implemented using Mixed-Integer Programming (MIP), significantly increasing the complexity of the model. Building on the concepts introduced earlier, Fig. 2 presents the CP model. The objective function (8a) sums over the I SAA scenarios to compute the expected objective, where u_r^i denotes the final adoption decision of trip r under scenario i.

The first group of constraints link z_a with the adoption decisions. Constraints (8b) and (8c) implement Equations (1) and (7), where c_r^i and $d_{r,p}^i$ are sampled, using the choice functions $\mathcal{C}_{\mathrm{core}}(\mathbf{x}_r)$ and $\mathcal{C}_{\mathrm{adopt}}(\mathbf{x}_r, p)$, respectively. The definition of $\mathcal{T}_{latent}^i$ and $\mathcal{T}_{core}^i$ are based on c_r^i: $\mathcal{T}_{latent}^i = \{r \in \mathcal{T} : c_r^i = 0\}$ and $\mathcal{T}_{core}^i = \{r \in \mathcal{T} : c_r^i = 1\}$. Constraints (8d) and (8e) together define f_p dynamically based on $\mathbf{z}$,

$$\max \quad \frac{1}{I}\sum_{i=1}^{I}\sum_{r \in \mathcal{T}} e_r \cdot u_r^i \tag{8a}$$

$$\text{s.t.} \quad u_r^i = \begin{cases} 1 & \forall r \in \mathcal{T}_{core}^i, i \in [I] \\ d_r^i & \forall r \in \mathcal{T}_{latent}^i, i \in [I] \end{cases} \tag{8b}$$

$$d_r^i = \bigvee_{p \in \mathcal{P}_r} (f_p \wedge d_{r,p}^i) \ \forall r \in \mathcal{T}_{latent}^i, i \in [I] \tag{8c}$$

$$z_a \ge f_p \ \forall a \in \mathcal{A}_p, p \in \mathcal{P}_r, r \in \mathcal{T} \tag{8d}$$

$$f_p \ge \sum_{a \in \mathcal{A}_p} z_a - |\mathcal{A}_p| + 1 \ \forall p \in \mathcal{P}_r, r \in \mathcal{T} \tag{8e}$$

$$c_r^i \le \sum_{p \in \mathcal{P}_r} f_p \ \forall r \in \mathcal{T}_{core}^i, i \in [I] \tag{8f}$$

$$B \ge \sum_{a \in \mathcal{A}} s_a \cdot z_a \tag{8g}$$

$$z_a = 1 \ \forall a \in \mathcal{A}_{fixed} \tag{8h}$$

$$0 = \sum_{a \in \mathcal{A}_{n,m}^+} h_a \cdot z_a - \sum_{a \in \mathcal{A}_{n,m}^-} h_a \cdot z_a \tag{8i}$$
$$\forall n \in \mathcal{N}, m \in \mathcal{M}$$

$$1 \ge \sum_{a \in \mathcal{A}_{n_1,n_2,m}} z_a \ \forall n_1, n_2 \in \mathcal{N}, m \in \mathcal{M} \tag{8j}$$

$$z_a \in \{0,1\} \ \forall a \in \mathcal{A} \tag{8k}$$

$$u_r^i \in \{0,1\} \ \forall r \in \mathcal{T}, i \in [I] \tag{8l}$$

$$d_r^i \in \{0,1\} \ \forall r \in \mathcal{T}_{latent}^i, i \in [I] \tag{8m}$$

$$f_p \in \{0,1\} \ \forall p \in \mathcal{P}_r, r \in \mathcal{T} \tag{8n}$$

Fig. 2. The Constraint Programming Model.

which is a linearized version of $feasible(\mathbf{z}, p)$. Constraint (8d) ensures that path p is infeasible if any of its transit arcs is closed. Constraint (8e) ensures that path p is set as feasible when all of its transit arcs are open. Using f_p, Constraint (8f) ensures that at least one path is feasible for each core trip, thereby guaranteeing that the core riders have access to the service. At this stage, solving the model does not explicitly determine which path each trip will use; however, the solution reveals the set of feasible path options available to each trip and identifies which latent trips adopt the new system.

Constraints (8g)–(8j) focus on the modeling of the transit network itself and incorporate several realistic assumptions commonly found in transit network design. It is important to note that these constraints are independent of the SAA scenarios. First, Constraint (8g) ensures that the network design remains within the operational budget B, where s_a denotes the cost of operating arc a over the planning horizon (e.g., a typical weekday). Secondly, Constraint (8h) enforces that all arcs in the set $\mathcal{A}_{fixed}$ remain open. The set $\mathcal{A}_{fixed}$ typically represents existing infrastructure that the transit agency intends to preserve in the new system, such as underground subway lines. Next, Constraint (8i) ensures the flow balance for each node in $\mathcal{N}$ and each mode in $\mathcal{M}$, where $\mathcal{A}_{n,m}^{+}$ and $\mathcal{A}_{n,m}^{-}$ include arcs that go in and out of node n with mode m, respectively. In this paper, $\mathcal{M}$ by default is $\{bus, rail\}$; however, one can easily customize it. Lastly, Constraint (8j) assures that any two locations n_1 and n_2 are only connected by one arc per mode, where $\mathcal{A}_{n_1,n_2,m}$ includes all arcs that start from n_1 to n_2 with mode m. Note that the output consists of a set of connected arcs between nodes, whereas in real-world settings, travelers use complete bus lines rather than navigating arc by arc. Post-processing methods can be employed to convert the selected arcs into operational bus lines.

4   The Atlanta Case Study

This section presents a case study based in Atlanta, Georgia, USA. In summary, by solving a large-scale 2LRC-TND problem, this case study redesigns a transit system with 38,179 trips ($\mathcal{T}$) in the city of Atlanta, where services are currently operated by the Metropolitan Atlanta Rapid Transit Authority (MARTA). To ease the processes of constructing meaningful datasets, the entire case study is conducted within a rectangular area (see Fig. 4a that closely aligns with the I-285 interstate and MARTA's service area. Figure 3 illustrates how the case study fits within the 2LRC-TND framework. This case study utilizes three datasets: SURVEY-ARC for training $\mathcal{C}_{\mathrm{core}}$, L-ARC-9 for training $\mathcal{C}_{\mathrm{adopt}}$, and L-ARC-8 for evaluating the complete 2LRC-TND framework.

4.1   Construct $\mathcal{C}_{\mathrm{core}}$ and $\mathcal{C}_{\mathrm{adopt}}$

The model $\mathcal{C}_{\mathrm{core}}$ predicts the probability of a trip r being a core trip. It is trained on the SURVEY-ARC dataset, which is constructed from a 2019 transit rider survey conducted by the Atlanta Regional Commission (ARC) [1]. The

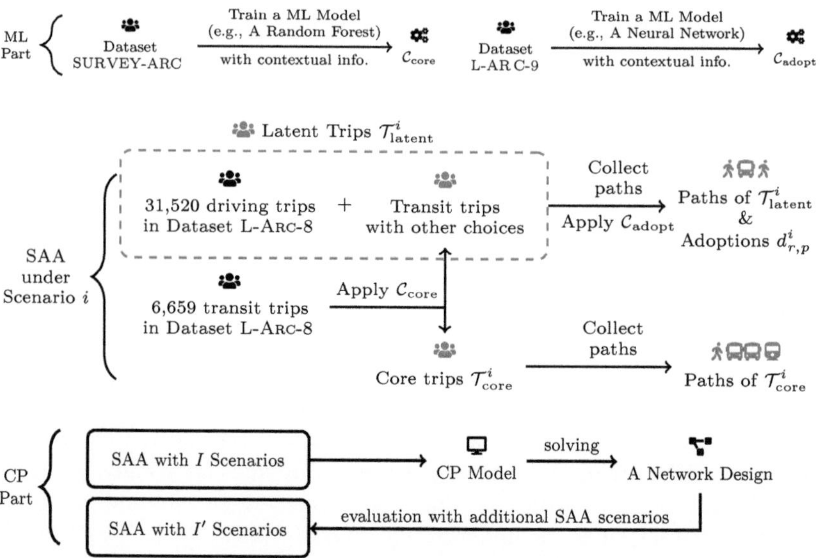

Fig. 3. A Flowchart showing the Atlanta case study within the 2LRC-TND framework. Blue indicates that these components change in each SAA scenario.

ground-truth labels contain 13,607 core trips (class 1) and 5,233 transit trips with other choices (class 0), where each trip represents a single rider. Since no prior study has directly addressed core trip classification, this paper adapts this survey by defining core trips as those made by commuters (trips between home and works) who cannot travel without transit, use transit at least twice weekly, and walk to/from their stops. All other commuters in the dataset are classified to class 0. The contextual features include demographic characteristics (e.g., age, income level), geographic data (e.g., number of bus stops within 5 minutes walking distance from the trip origin), and travel time information (e.g., travel time from origin to destination if driving a car).

A separate dataset, referred as L-Arc-9, is used to train the $\mathcal{C}_{\text{adopt}}$ model. The L-Arc-9 dataset is derived from ARC's 2020 activity-based simulation for the Atlanta metropolitan area [2], detailed data processing procedures are documented in the Appendix of [14]. The original simulation study covers 5 million people and 13 million tours in a regular workday, from which L-Arc-9 selects 15,109 home-to-work trips made between 9-10 AM that overlaps the rectangular area in Fig. 4a. This highly imbalanced dataset contains 1,019 adopting trips (class 1) and 14,090 rejecting trips (class 0). Adopting trips are made by commuters who use transit despite having spare vehicles at home, while rejecting trips are made by those who drive alone despite transit availability in their area. Each trip represents a single rider and includes demographic, geographic, and trip-related contextual features. Since no existing datasets or surveys directly capture adoption behavior, this dateset provides a practical alternative to resource-intensive data collection.

In summary, the SURVEY-ARC and L-ARC-9 datasets are used solely for training the choice models. Although the datasets originate from different sources, both were collected by the same agency within the same geographic region. Additionally, to ensure the trips under study follow relatively fixed patterns, the analysis focuses on commuting trips, as they best represent consistent travel behaviors on a regular day.

4.2   Demand Set $\mathcal{T}$, Network Arcs, and Paths

A third dataset, L-ARC-8, is introduced here to represent the demand used for evaluating the 2LRC-TND framework. A separate evaluation dataset is necessary to assess the framework's performance on unseen data and avoid overfitting from the training sets used for the choice models. This dataset is derived from the same ARC simulation source as L-ARC-9 but focuses on all commuters traveling in the rectangular area during 8-9 AM, the peak morning rush hour.

There are 38,179 trips ($\mathcal{T}$) in L-ARC-8, with each trip having a single rider ($e_r = 1$). Among these trips, 31,520 commute by driving alone, while 6,659 are current transit trips. All trips have contextual information compatible with the trained $\mathcal{C}_{\text{core}}$ and $\mathcal{C}_{\text{adopt}}$ models. Following the steps illustrated in Fig. 3, for each SAA scenario, the 6,659 existing transit trips are first split into core trips and current adopting trips using $\mathcal{C}_{\text{core}}$. The current adopting trips and the 31,520 driving trips then form the latent demand for 2LRC-TND.

This case study assumes removal of all existing bus lines while preserving the rail lines (see Fig. 4a), as rail infrastructure is not flexible and difficult to modify. Within the defined rectangular area, there are 933 stops ($\mathcal{N}$), including 38 rail stations. To construct the set $\mathcal{A}$ and keep the CP model computationally manageable, when considering potential bus connection for each stop, only the five nearest stops based on driving distance are considered. For each potential connection between two stops, a fixed frequency of six buses per hour is assumed ($h_a = 6$ for all arcs). In addition to the standard on-road travel time, an extra five minutes is added to account for peak-hour congestion, waiting time, and other potential delays. In total, the network includes 6,004 arcs ($\mathcal{A}$), of which 692 are fixed rail arcs ($\mathcal{A}_{fixed}$). To model the operating cost between 8 a.m. and 9 a.m., the case study assumes a rate of \$72.15 per hour for buses while in motion. Rail operating costs are dropped from the budget B, as all rail arcs are fixed and not subject to redesign. The cost estimation and stops information can be found in an Atlanta case study [3].

To enumerate the paths, a graph is first constructed considering all arcs in $\mathcal{A}$ are open. For each trip r, the origin and destination are connected to their five nearest stops based on walking distance, and the four (w_r) fastest paths are then selected using shortest-path algorithms to construct the set $\mathcal{P}_r$. w_r is set to 4 to maintain reasonable model size. Additionally, real-world navigation apps commonly considers a similar number of transit paths for users. When constructing SAA scenarios i, an adoption decision ($d_{r,p}^i$) for each path in $\mathcal{P}_r$ is sampled based on the predicted adoption probability from the pre-determined model C_{adopt}. Similarly, c_r^i are sampled using the predicted probability from C_{core}. After sampling, for latent trips where $c_r^i = 0$, if all paths in $\mathcal{P}_r$ are rejecting paths, these trips can be excluded from the optimization model since these trips will reject the network regardless.

4.3   Experiments: Choice Modeling and Evaluating 2LRC-TND

The first group of experiments in this case study focuses on evaluating the performance of C_{core} and C_{adopt} when treated as classifiers. Specifically, C_{core} is evaluated on the Survey-Arc dataset, while C_{adopt} is evaluated on L-Arc-9. Three types of machine learning models are used: Multinomial Logit (L), Random Forest (RF), and Deep Neural Network (DNN). The dataset is randomly split using an 80-20 training-test split. This process repeats 100 times.

For each training-test split, hyper-parameters are tuned via a grid-search 5-fold cross-validation of the training set, and the best configuration is evaluated on the test set. Specifically, for the Logit model, the regularization parameter C is tested with values $\{0.1, 0.8, 1.0, 10\}$, maximum iterations with $\{100, 500, 2000\}$, and class weight ratios of $\{1{:}2, 1{:}3, 1{:}5, 1{:}10\}$ are evaluated for C_{adopt} to address class imbalance. The Random Forest models test the number of estimators with values $\{500, 1000, 2000\}$ and maximum tree depth with $\{10, 15, 20, 25\}$, with the class weight ratios of $\{1{:}2, 1{:}3, 1{:}5, 1{:}10\}$ also explored for C_{adopt}. The DNN architecture employs embeddings for categorical and ordinal variables, followed by two fully connected layers with 32 and 16 neurons respectively, each using a dropout rate of 0.2 and the Adam optimizer. Learning rates are tuned with candidate values $\{0.001, 0.01, 0.1, 0.15\}$, and early stopping is applied with a patience of 20 epochs and tolerance of 0.001.

The second group of experiments evaluates the 2LRC-TND framework by comparing it against three benchmarks. The first benchmark, Fixed-Demand (FD), designs a network using only the 6,659 existing transit riders in a deterministic manner, without considering demand uncertainties. The second benchmark, Naive Rule-Based (RB), with no involvement of Machine Learning, uses hand-crafted rules to perform the same functions as C_{core} and C_{adopt}. A simple rule-based model example is:"If both the origin and destination of a trip are within a 5-minute walk distance to a transit station, assign a core trip probability of 0.8; otherwise, assign 0.2." In general, a rule-based approach requires extensive domain expertise, but the design of such rules is beyond the scope of this paper. The third benchmark, Deterministic, employs the same machine learning models as 2LRC-TND but considers them under a fixed threshold of 0.5 to obtain deterministic decisions and designs the network under this setting.

For each of these runs (2LRC-TND and Benchmarks), two budget levels are tested. These values are chosen based on the total cost of activating all available arcs from 8AM to 9AM, which is approximately \$58K. A budget of \$30K represents roughly half of this total, while \$35K corresponds to 60%, allowing for a meaningful comparison under different budget constraints.

All implementations for this paper are programmed in Python 3.11, with the support of machine learning packages PyTorch, Scikit-learn, and imblearn. The CP-SAT solver from OR-Tools is used for solving the optimization model, configured to run with 8 threads with 12 h solving time. All experiments were conducted using the CPUs and RTX-6000 GPUs available on a Linux High Performance Computing cluster.

4.4 Evaluation Metrics for Network Designs

Network designs can be evaluated by standard transportation metrics, such as the number of people covered and the best travel time offered to riders. This study presents an additional evaluation approach based on SAA, where a network design is assessed using out-of-sample scenarios (denoted as I' and $I' >> I$), different than I scenarios used during network design. Since a network is optimized only with respect to a specific set of scenarios, it may not be feasible for scenarios outside that set, i.e., Constraint (8f) might not be satisfied. Function $Eval$ evaluates a network design $\mathbf{z}$ by quantifying the number of riders it successfully serves and by penalizing the core trips r with no feasible paths with cost l_r. The resulting $Eval$ function can be represented as follows by considering the optimization model in Fig. 2 under the given network design $\mathbf{z}$ over I' out-of-sample scenarios:

$$Eval(\mathbf{z}) = \frac{1}{I'} \sum_{i=1}^{I'} \sum_{r \in T} e_r \cdot (u_r^i - l_r \cdot \max(c_r^i - \sum_{p \in \mathcal{P}_r} f_p, 0)) \tag{9}$$

Table 2. Choice modeling results averaged over 100 runs (100 different training-test split), with standard deviations in parentheses.

	$\mathcal{C}_{\text{core}}$ (on Survey-Arc Dataset)			$\mathcal{C}_{adopt}$ (on L-Arc-9 Dataset)		
Models	F1-score	Accuracy	AUC-ROC	Weighted F1	Accuracy	AUC-ROC
L	0.844 (0.005)	0.780 (0.006)	0.805 (0.008)	0.912 (0.003)	0.924 (0.004)	0.820 (0.015)
DNN	0.866 (0.015)	0.803 (0.016)	0.799 (0.011)	0.907 (0.022)	0.897 (0.036)	0.807 (0.029)
RF	0.885 (0.003)	0.821 (0.005)	0.819 (0.008)	0.923 (0.003)	0.933 (0.003)	0.855 (0.015)

5 Computational Results

This section begins by presenting the machine learning results from the two choice models. It then reports the optimization results of the 2LRC-TND framework applied to the aforementioned case study.

5.1   Results on Choice Modeling

Table 2 summarizes the performance of choice models on classification tasks. While the logit model serves as a baseline due to its simplicity, it still produces reasonably consistent results. Tree-based models, such as RF, outperform the other two models. In contrast, DNNs demonstrate comparatively weaker performance across both tasks. This under-performance is likely attributable to the tabular nature of the data, a context in which DNNs usually struggle. For RF's hyper-parameters, the best configuration for C_{core} is 2000 estimators with a maximum depth of 25. For C_{adopt}, RF uses 1000 estimators, a maximum depth of 10, and a class weighting ratio of 1:5 between class 0 and class 1 during training.

Overall, the results are consistent and exhibit low standard deviation across 100 runs. For C_{adopt}, the weighted F1-score is employed to address the highly imbalanced nature of the dataset. Given this imbalance, models achieving a AUC-ROC score above 0.8 can be considered strong. Despite the classification performance, the models also yield valuable insights that usually surpass the capabilities of basic rule-based methods, and they are particularly useful when integrated into larger optimization frameworks. These results further emphasize the need for more comprehensive and balanced data collection to better capture the complexities inherent in both tasks.

5.2   Results on 2LRC-TND

Table 3 summarizes the computational results on optimization. Random Forest is employed as C_{core} and C_{adopt} in 2LRC-TND because its simplicity for implementation (compared to DNNs) and superior performance in prediction results demonstrated in prior experiments. Selected visualizations (due to page limit) of designs are shown in Figs. 4. Examining the network designs, it appears that all runs open a similar number of arcs within the same budget. This outcome occurs because the model tries to utilize the entire available budget. The resulting networks largely preserve existing MARTA corridors while reallocating capacity toward underserved areas with high latent demand.

The first benchmark, FD, assumes no demand uncertainties and solves the CP model using only the existing 6,659 transit trips. However, after the network is designed, C_{core} and C_{adopt} (both using the random forest models in 2LRC-TND) are applied to evaluate the actual number of users who would adopt the service, with results reported in the "coverage" column in the table. Without considering demand uncertainties (run FD), the network design will still attract riders to join the system, and some riders among the 6,659 will reject it. The new design by FD can still attract some riders because the objective does not aim at budget savings, resulting in a quite extensive network. In contrast, 2LRC-TND accounts for demand uncertainties and achieves a much higher coverage by approximately 20%. These results demonstrate that through the redesign of 2LRC-TND, the agency can attract additional riders by offering services to those who previously lacked transit access. The column coverage bound indicates the

Table 3. 2LRC-TND results. For evaluations, the reported metrics are averaged across senarios: 50 (I) of them are used in solving and 1,000 (I') only for additional evaluation.For the Adopted% metric, the denominator is $|\mathcal{T}_{latent}|$.

| Budget ($) | Run | Run Time (hours) | # Opened Bus Arcs | Averaged Results over 50 (I) SAA Senarios | | | | | | Eval. with I' |
				# Core $|\mathcal{T}_{core}|$	# Latent $|\mathcal{T}_{latent}|$	# (%) Adopted	Coverage	Coverage Bound	Travel Time (min)	Eval(z) (# Violated)
30K	FD	0.04	2824	-	-	-	13086	-	57.13	-
	Naive RB	2.23	2844	2139	36040	5676 (15.75%)	7814	8819	56.80	7809.15 (0)
	Deterministic	0.03	2701	4776	33403	966 (2.89%)	5742	5742	53.15	13338.58 (15)
	2LRC-TND	12.00	2871	4259	33920	11132 (32.82%)	15391	16655	56.44	15390.11 (0)
35K	FD	0.05	3148	-	-	-	13718	-	57.10	-
	Naive RB	1.98	3319	2139	36040	6182 (17.15%)	8321	8819	57.03	8313.00 (0)
	Deterministic	0.03	2838	4776	33403	966 (2.89%)	5742	5742	53.12	13474.61 (15)
	2LRC-TND	3.88	3313	4259	33920	11745 (34.63%)	16005	16655	56.56	15999.53 (0)

maximum ridership that an agency can cover (assuming no budget limit). This value is determined prior to solving the CP model, during pre-processing.

The second benchmark, Naive Rule-based (RB), uses nested if-else conditions based on travel time and origin-destination locations to generate probabilities, serving the same purpose as $\mathcal{C}_{\mathrm{core}}$ and $\mathcal{C}_{\mathrm{adopt}}$ in 2LRC-TND but without machine learning. The first observation is, using the rule-based models, the CP can be solved optimally within a short time, whereas 2LRC-TND ($B = 30K$) only reports the best feasible solution after the run time limit of 12 h is reached. This is likely due to the machine learning model predicting a greater number of core riders, which requires the CP model to provide a path for each core rider, thereby increasing computational difficulties. Secondly, although the core riders differ across runs, all experiments demonstrate that 2LRC-TND achieves a notable number of adopted riders. Specifically, RF yields a higher adoption rate by learning directly from data, while Naive RB depends on manually defined rules that lack consideration of geographical and demographic information.

To demonstrate the importance of stochastic modeling, a third benchmark employing deterministic choice models is evaluated against 2LRC-TND. This configuration utilizes the same machine learning models as 2LRC-TND; however, they generate deterministic binary decisions using a typical fixed threshold of 0.5, thus eliminating the need for SAA scenarios in I. In this case study, since the predicted probability of adopting a path for latent trips is typically below 0.5 in most cases, this deterministic approach underestimates potential ridership adoption. As shown in Table 3, the deterministic approach achieves a significantly weaker adoption rate across both budget levels.

For core riders and adopted riders using the final network design, Table 3 also reports their travel times. It should be noted again that the model itself does not assign specific paths to riders. It ensures that core riders have at least one feasible path in P_r and also attempts to increase the number of feasible paths

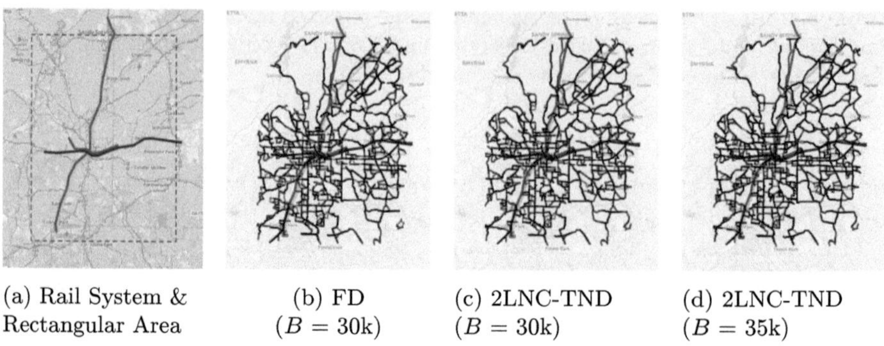

(a) Rail System &   (b) FD   (c) 2LNC-TND   (d) 2LNC-TND
Rectangular Area $(B = 30\text{k})$ $(B = 30\text{k})$ $(B = 35\text{k})$

Fig. 4. Selected Maps Related to the Case Study. (Designs are Hypothetical)

for latent riders. Once the network design is established, the fastest path can be selected for riders, since this approach is intuitive.

Lastly, the designs are evaluated using 1000 (I') additional SAA scenarios. Results show that all core riders are served under the respective designs with no violation in most of the cases, which is likely because each $\mathcal{P}_r$ contains four paths and the fixed rail arcs are effectively utilized by core riders. However, Deterministic case exhibits 15 constraint violations when evaluated on these scenarios indicating the need for the stochastic decision-making models. Additionally, the **Eval(z)** results closely match the coverage values from the evaluation with I for the 2LRC-TND approach, highlighting that the results are highly stable after solving from only 50 SAA scenarios.

6   Conclusion

This paper introduces the 2LRC-TND framework to address the limitations of the traditional transit network design approaches that assume fixed demand. Leveraging CSO, the framework integrates machine learning-based choice models and constraint programming-based optimization models to produce transit network designs. A large-scale case study over 38,000 trips from Atlanta demonstrates that 2LRC-TND can improve ridership within reasonable budget limits, highlighting its practicality and potential social impact. Several promising directions for future research include design of more targeted surveys for enhancing data collection and extension of 2LRC-TND to emerging transportation systems.

Acknowledgments. The work was partially funded based upon work supported by the National Science Foundation under Grant No. 2112533 and Grant No. 2434302.

References

1. Atlanta regional commision: regional on-board transit survey 2019 final report. https://cdn.atlantaregional.org/wp-content/uploads/final-report-arc-2019-regional-transit-on-board-survey.pdf (2020). Accessed 15 Jun 2025

2. Atlanta regional commision: activity based model. https://abmfiles.atlantaregional.com/. Accessed 25 Jul 2025

3. Auad, R., et al.: Resiliency of on-demand multimodal transit systems during a pandemic. Transp. Res. Part C Emerg. Technol. **133**, 103418 (2021)

4. Basciftci, B., Van Hentenryck, P.: Integration of constraint programming, artificial intelligence, and operations research. Presented at the (2020) Bilevel Optimization for On-demand Multimodal Transit Systems

5. Basciftci, B., Van Hentenryck, P.: Capturing travel mode adoption in designing on-demand multimodal transit systems. Transp. Sci. **57**(2), 351–375 (2023)

6. Bertsimas, D., Kallus, N.: From predictive to prescriptive analytics. Manage. Sci. **66**(3), 1025–1044 (2020)

7. Bertsimas, D., Kallus, N., Hussain, A.: Inventory management in the era of big data. Prod. Oper. Manag. **25**(12), 2006–2009 (2016)

8. Bertsimas, D., Ng, Y.S., Yan, J.: Data-driven transit network design at scale. Oper. Res. **69**(4), 1118–1133 (2021)

9. Borndörfer, R., Grötschel, M., Pfetsch, M.E.: A column-generation approach to line planning in public transport. Transp. Sci. **41**(1), 123–132 (2007)

10. Canca, D., De-Los-Santos, A., Laporte, G., Mesa, J.A.: A general rapid network design, line planning and fleet investment integrated model. Ann. Oper. Res. **246**(1), 127–144 (2016)

11. Cipriani, E., Gori, S., Petrelli, M.: Transit network design: a procedure and an application to a large urban area. Transp. Res> Part C Emerg. Technol. **20**(1), 3–14 (2012)

12. Farahani, R.Z., Miandoabchi, E., Szeto, W., Rashidi, H.: A review of urban transportation network design problems. Eur. J. Oper. Res. **229**(2), 281–302 (2013)

13. Guan, H., Basciftci, B., Van Hentenryck, P.: Path-based formulations for the design of on-demand multimodal transit systems with adoption awareness. INFORMS J. Comput. **36**(6), 1359–1756 (2024)

14. Guan, H., Basciftci, B., Van Hentenryck, P.: Bilevel optimization and heuristic algorithms for integrating latent demand into the design of large-scale transit systems. Transp. Sci. (2026)

15. Guan, H., Gillani, N., Simko, T., Mangat, J., Van Hentenryck, P.: Contextual stochastic optimization for school desegregation policymaking. In: Proceedings of the AAAI Conference on Artificial Intelligence, vol. 39(27), pp. 28024–28032 (2025)

16. Guihaire, V., Hao, J.K.: Transit network design and scheduling: a global review. Transp. Res. Part A Policy Pract. **42**(10), 1251–1273 (2008)

17. Jiang, Z., Ji, R.: Optimising hurricane shelter locations with smart predict-then-optimise framework. Int. J. Prod. Res. **63**(8), 2905–2925 (2025)

18. Kleywegt, A.J., Shapiro, A., Homem-de Mello, T.: The sample average approximation method for stochastic discrete optimization. SIAM J. Optim. **12**(2), 479–502 (2002)

19. Klier, M.J., Haase, K.: Line optimization in public transport systems. In: Operations Research Proceedings 2007: Selected Papers of the Annual International Conference of the German Operations Research Society (GOR) Saarbrücken, September 5–7, 2007, pp. 473–478. Springer (2008)

20. Klier, M.J., Haase, K.: Urban public transit network optimization with flexible demand. OR Spectrum **37**, 195–215 (2015)
21. Lee, D., Derrible, S., Pereira, F.C.: Comparison of four types of artificial neural network and a multinomial logit model for travel mode choice modeling. Transp. Res. Rec. **2672**(49), 101–112 (2018)
22. Ma, Y., Zhang, Z.: Travel mode choice prediction using deep neural networks with entity embeddings. IEEE Access **8**, 64959–64970 (2020)
23. Maheo, A., Kilby, P., Van Hentenryck, P.: Benders decomposition for the design of a hub and shuttle public transit system. Transp. Sci. **53**(1), 77–88 (2019)
24. Sadana, U., et al.: A survey of contextual optimization methods for decision-making under uncertainty. Eur. J. Oper. Res. **320**(2), 271–289 (2025)
25. Schöbel, A.: Line planning in public transportation: models and methods. OR Spectrum **34**(3), 491–510 (2012)
26. Tang, L., Luo, R., Zhou, Z., Colombo, N.: Enhanced route planning with calibrated uncertainty set. Mach. Learn. **114**(5), 1–16 (2025)
27. Wang, W., Deng, S., Zhang, Y.: Data-driven ordering policies for target oriented newsvendor with censored demand. Eur. J. Oper. Res. **323**(1), 86–96 (2025)
28. Xie, C., Lu, J., Parkany, E.: Work travel mode choice modeling with data mining: decision trees and neural networks. Transp. Res. Rec. **1854**(1), 50–61 (2003)
29. Ye, T., Cheng, S., Hijazi, A., Van Hentenryck, P.: Contextual stochastic optimization for omnichannel multi-courier order fulfillment under delivery time uncertainty. Manufacturing & Service Operations Management (2025)
30. Zhao, X., Yan, X., Yu, A., Van Hentenryck, P.: Prediction and behavioral analysis of travel mode choice: a comparison of machine learning and logit models. Travel Behaviour Soc. **20**, 22–35 (2020)

Improving the Planning of Stochastic Tasks with Availability Windows Using Prediction

Alexis Guigal[1,3]([✉]) [iD], Emmanuel Hyon[1,2] [iD], and Claire Hanen[1,2] [iD]

[1] Sorbonne Université, CNRS, LIP6, Paris, France
{alexis.guigal,Emmanuel.hyon,claire.hanen}@lip6.fr
[2] Paris Nanterre University, Nanterre, France
[3] Thales DMS, Elancourt, France

Abstract. We consider a dynamic stochastic capacity problem in which tasks arrive according to a Poisson Process. Release times, stochastic deadlines and masses characterize tasks. A task might be scheduled within its time windows at a given time slot of bounded mass capacity. Each non-empty slot induces a deterministic setup cost. We aim to define a dynamic scheduling policy that maximizes the expected mass of scheduled tasks minus the setup costs over a long time horizon. This type of problem has many applications in planning, scheduling, and yield management. We use a rolling horizon approach: at each step, a particular case of the Multiple Knapsack with Restriction Assignment must be solved. We present and benchmark myopic heuristics, lookahead heuristics, and the exact solution provided by integer linear programming. We provide an extensive comparison in our dynamic and stochastic context. Finally, rather than searching for the most robust heuristic, we devise an algorithm selection process, based on random forests, to select the most effective heuristic for a given instance, considering its parameters. This achieves significant improvements.

Keywords: Stochastic Capacity Planning · Rolling Horizon ·
Algorithm Selection · Multiple Knapsack with Assignment Restrictions

1 Introduction

In this work, we address a challenging problem involving a dynamic stochastic capacity process with a finite horizon. The problem entails the arrival of tasks according to a Poisson process, and release times with a constant deterministic distance from their arrival time. The tasks have stochastic deadlines and stochastic masses, all distributed according to known probabilities. A task can be scheduled in one time slot within its time window. Each nonempty slot has a deterministic mass capacity and a setup cost. Upon arrival, each task's stochastic elements are known, and performing a task yields a reward equal to its mass. Decisions are made sequentially at the beginning of each slot to determine the

© The Author(s), under exclusive license to Springer Nature Switzerland AG 2026
T. Guns (Ed.): CPAIOR 2026, LNCS 16595, pp. 191–206, 2026.
https://doi.org/10.1007/978-3-032-27242-3_12

optimal policy that maximizes the expected value (rewards minus costs) over the entire horizon.

The study of a real satellite launch complex is the basis of this work [2,16]. Launching a rocket yields a setup cost, and there is a limited weight of embedded satellites. Demands are made online. But this model finds applications in various fields such as healthcare management [1], resource reservation [12], production [10,13] and transportation [4]. However, the dynamic (online data and decisions with respect to time) and stochastic (random data) settings of our problem with time windows was not considered previously.

Exact methods are acknowledged to lack scalability for large-scale dynamic stochastic problems. Hence, stochastic dynamic programming is used to solve a general stochastic capacity reservation problem, where arriving orders can be placed in one knapsack [13] or in several knapsacks [11]. However, the time horizon is limited to at most five periods. Crainic [4] highlights the challenges faced by state-of-the-art solvers, which struggle to efficiently solve even simple models with a horizon of only two periods.

Given the complexity of dynamic programming or stochastic programming approaches for such problems (see [10]), Rolling Horizon methods have been proposed as an alternative (see [18] for a survey). In these methods, decisions for the dynamic problem are determined at each time step by solving a static version of the problem over a short or medium-term horizon. Nevertheless, Rolling Horizon methods present several limitations. First, the static problem must be solved quickly. Second, solving a static, deterministic version of the problem neglects future uncertainties. Finally, the optimal static policy does not necessarily correspond to the optimal dynamic policy (see [18]). Hence, there is currently no efficient method for solving stochastic and dynamic problems where the decision at each stage is NP-complete.

This motivates a careful analysis of the static case, which, in our context, turns out to be a *Multiple Knapsack with Assignment Restrictions* (MKAR) [5, 6]. It is proven to be NP-complete in [6], which also provides a heuristic with error guarantees (see [3] for Multi Knapsack without restrictions). Another possible approach to the static problem is the *Restricted Bin packing Problem* [9] and its refinement with time windows [8,15]. However, this approach is incompatible with our requirements, since it allows multiple bins to have identical execution dates. At last, based on classical scheduling methods (e.g. Earliest Deadline First, Best Fit), [2] introduces some heuristics to solve MKAR, and [16] presents a dynamic programming algorithm for unit masses. Given these developments, we argue that it is critical to evaluate the practical efficiency of ILP approaches, the performance of existing and novel heuristics, as well as the relevance of adapting literature-based heuristics [3,6,8,9], especially in realistic scenarios with large horizons.

To address the neglect of future uncertainties, *lookahead methods* include future forecast in the rolling horizon [7,19]. However, stochastic programming remains inefficient in this context, whereas the scenario-sampling framework proposed by [12] appears to be the most appropriate way.

From an operational perspective, it is important to know which method is most effective to compute the global optimal instead of the static one for a given instance. As we have at our disposal a large set of heuristics, then, we use *Algorithm selection* (see [14]). It is a framework to select an algorithm from a portfolio on a case-by-case basis. The association between the instance and the selection of the algorithm is based on supervised machine learning.

We propose, in this work, new resolution heuristics, and we conduct a computational evaluation using a large set of diverse instances. Inspired by the work of [12], we devise a lookahead resolution algorithm for this stochastic multiple knapsack with assignment restriction, which we show to perform very well. At last, we propose a new approach of algorithm selection that, up to our knowledge, has never done before for rolling horizon. It allows us to determine the best dynamic policy based on the instance characteristics that improves the performance by 50%.

The rest of the paper is organized as follows: Sect. 2 introduces our problem. Section 3 is focused on static analysis, while Sect. 4 focuses on lookahead heuristics. Then Sect. 5 is devoted to the numerical assessment of the heuristics, and Sect. 6 details the algorithm selection and it benefits.

2   Problem Definition

We consider a stochastic scheduling problem where tasks randomly arrive to be performed on a cumulative resource during a random and discrete time window. A task that misses its deadline is considered to be lost. A rolling horizon decision process is used to decide which tasks are to be performed at each time slot. The objective function adds up a setup cost for each used slot and a gain for each on-time task. In this section, we outline the problem features.

2.1   Time, Tasks, Constraints, Objective

We assume discrete time units, called slots. Slot t corresponds to the continuous time interval $[t, t + 1)$. We call the horizon the N consecutive available slots corresponding to the interval $[0, N]$. For each slot t, the scheduler can decide to open a cumulative resource of capacity C_t, with a resulting setup cost σ_t. We then say that slot t is open. Only one resource can be opened at slot t.

Tasks arrive in the system following a continuous Process of intensity λ. This means that the probability that k tasks arrive during slot t equals

$$Pr(N_t = k) = e^{-\lambda}\frac{\lambda^k}{k!}. \tag{1}$$

An arriving task i is characterized by the following parameters:

arrival time a_i the slot during which it arrives;
release time $r_i = a_i + A$ is the minimal slot in which it can be scheduled. The
   constant delay $A > 0$, does not depend on the task;

deadline $d_i = r_i + D_i$, where D_i is a uniform random variable in $[\underline{D}, \overline{D}]$ with $\underline{D}$ (respectively $\overline{D}$) at the lower bound (respectively upper) of the availability duration. The task i must be scheduled before slot d_i;

mass an integer m_i measures the usage of the cumulative resource. The mass m_i is a random variable according to a discrete uniform distribution μ with support in $[1, M]$.

All random variables are independent. The task i and its parameters are known by the scheduler from slot $a_i + 1$. It can be scheduled in an open slot of its time window $[r_i, d_i)$, or rejected. If i is scheduled, it induces a gain of m_i. In an open slot t, the set $\mathcal{P}[t]$ of scheduled jobs must fit within the resource's capacity. So we can state the following constraints:

- Open slot: $\mathcal{P}[t] \neq \emptyset$ only if t is open.
- Capacity constraint:

$$\sum_{i \in \mathcal{P}[t]} m_i \leq C_t \tag{2}$$

- Time window constraints:

$$\forall i \in \mathcal{P}[t], \quad r_i \leq t < d_i \tag{3}$$

Hence, the decision of the system consists of opening slots and scheduling tasks in their time windows in open slots. Considering a realization I of the random variables on a horizon N, a schedule, or plan, is denoted by $\mathcal{P}^I$, and we model the decision to open a slot t by a binary variable $y_t^{\mathcal{P}^I}$. We denote by $\mathcal{P}^I[t]$ the set of tasks scheduled in slot t according to $\mathcal{P}^I$. A schedule fulfills the constraints (2)(3). The gain of the schedule is the sum of the masses of scheduled tasks minus the setup costs of open slots:

$$\mathcal{G}(\mathcal{P}^I) = \sum_{t=1}^{N} \left(-\sigma_t \cdot y_t^{\mathcal{P}^I} + \sum_{i \in \mathcal{P}^I[t]} m_i \right). \tag{4}$$

By extension if $\mathcal{P}^I$ is a plan, and t a time slot we also denote by $\mathcal{G}(\mathcal{P}^I[t])$ the gain of slot t: 0 if no task is scheduled, $\sum_{i \in \mathcal{P}^I[t]} m_i - \sigma_t$ otherwise. So, our objective would be to design rolling horizon scheduling policies that would attempt to maximize the expected value of the gain: $E\left(\mathcal{G}(\mathcal{P}^I)\right)$.

Notice that in the rest of the paper, when the realization is fixed, the superscript I is omitted. We introduce some notations that will be used in the next sections for a schedule:

- $\mathcal{S}_t = \{i \text{ such that } a_i = t\}$ is the set of tasks that arrive at time t.
- $\mathcal{R}_t = \{i \text{ unscheduled before } t, a_i < t < d_i\}$ is the set of already known tasks to consider from t.
- $\mathcal{F}_t = \{i \in \mathcal{R}_t, r_i \leq t < d_i\}$ the set of unscheduled tasks that can be scheduled in slot t
- $\mathcal{Z}_t = \{i \in \mathcal{F}_t, d_i = t + 1\}$ the set of schedulable tasks whose deadline is $t + 1$.

Notice that we have: $\mathcal{R}_t = \mathcal{R}_{t-1} \cup \mathcal{S}_{t-1} \setminus (\mathcal{P}[t-1] \cup \mathcal{Z}_{t-1})$.

2.2 Rolling Horizon and Uncertainty Consideration

At each time slot t, one must decide whether t is open and which tasks are to be executed. The method to determine this decision can be an online algorithm or an offline algorithm based on a certain amount of information. The efficiency of the resolution method directly depends on the way uncertainty is considered and also on the time horizon used for the calculation.

If we consider the whole set of possible realizations of random variables, and decision-making according to the full knowledge of these realizations to maximize our expected objective, the combinatorial explosion of scenarios makes it impossible to take into account the uncertainty up to a large horizon N in practice.

That is why we would rather consider rolling horizon policies.

Principle of the Rolling Horizon in a Random Setting. A policy with a rolling horizon involves making the decision at the current stage based on information and forecasts that consider the future up to a given finite horizon of length $H < N$, called the *rolling horizon*. For our problem, the rolling horizon H will have maximum size $\overline{H} = \overline{D} + A$. At each time step t, we will calculate a decision in interval $[t, t + H)$ to be taken at the current slot based on the available information and forecast over the rolling horizon. This part is called the static optimization problem, which is solved either exactly or by an approximation algorithm. Then, the decisions issued from the static solutions apply for slot t. The information is then updated for the next time slot, and a new static problem is considered.

In the static problem, uncertainty is either unconsidered (myopic policies) or a part of forecasting is considered (lookahead policies).

No Consideration of Uncertainty: Myopic Policy. For each slot t, in the static problem, we consider only the set of known tasks (here $\mathcal{R}_t = \{i$ unscheduled before $t, a_i < t < d_i\}$) to make the decision.

The generic algorithm for myopic methods is given by Algorithm 1. At each t slot, we update the set $\mathcal{R}_t$ of known tasks (line 4). And we compute the decision $\mathcal{P}[t]$ using the function $SCHEDULE$ (line 5): in practice, this is done with one of the different algorithms solving the deterministic problem for the slot t that we will see in Sect. 3. Then we update planning and gain.

Partial Consideration: Lookahead Policy. In addition to $\mathcal{R}_t$, we consider random evolutions from the current slot t over a small horizon (the *lookahead period*) of several steps after t. The lookahead period is significantly shorter than the rolling horizon. Nevertheless, even with a partial consideration of the future, we do not necessarily avoid a combinatorial explosion. That is why some methods, called *approximation by sampling*, only consider a subset of all possible scenarios in order to evaluate the best action. The challenge is to consider a sufficiently large set of samples for the approximation to be realistic. For our problem, we consider at slot t samples of tasks arrived during slot t (and thus released at $t + A$).

Algorithm 1. Myopic Algorithm

Ensure: $\mathcal{P}$: The schedule, w : Total gain
1: $w \leftarrow 0$
2: $\mathcal{P} \leftarrow \emptyset$
3: **for** $t \in 1, .., N$ **do**
4:     $\mathcal{R}_t \leftarrow \mathcal{R}_{t-1} \cup \mathcal{S}_{t-1} \backslash (\mathcal{P}[t-1] \cup \mathcal{Z}_{t-1})$
5:     $\mathcal{P}' \leftarrow SCHEDULE(t, \mathcal{R}_t, H)$ $\triangleright$ compute a planning of $[t, t+H]$
6:     $\mathcal{P}[t] \leftarrow \mathcal{P}'[t]$
7:     $w \leftarrow w + \mathcal{G}(\mathcal{P}[t])$
8: **end for**

---

3 Algorithms for the Static Problem

In this section, we present the algorithms we tested and used to solve the underlying static deterministic scheduling problem on a fixed horizon starting at time t_0. They must be solved at each time step of the rolling horizon approach. We start by modeling the problem with an integer linear program, and then present the principles of our different heuristics.

3.1 Integer Linear Program

We define the following decision variables:

- X_{it}: is 1 if task i is in slot t, and 0 otherwise
- Y_t: is 1 if slot t is open, and 0 otherwise

At each slot t_0:

$$\max \quad \sum_{i \in \mathcal{R}_{t_0}} \sum_{t \in [r_i, d_i[} m_i \times X_{i,t} - \sum_{t \in [t_0..t_0+H]} \sigma_t \times Y_t$$

$$
\begin{aligned}
1 : \text{s. c.} \quad & \sum_{t \in [r_i, d_i[} X_{i,t} \leq 1, & i \in \mathcal{R}_{t_0} \\
2 : \quad & X_{i,t} \leq Y_t, & i \in \mathcal{R}_{t_0}, t \in [r_i, d_i) \\
3 : \quad & \sum_{i \in F_t} X_{i,t} \geq Y_t, & t \in [t_0..t_0 + H] \\
4 : \quad & \sum_{i \in \mathcal{R}_{t_0}} m_i \times X_{i,t} \leq C_t, & t \in [t_0..t_0 + H]. \\
5 : \quad & X_{it}, Y_t \in \{0, 1\} & i \in \mathcal{R}_{t_0}, t \in [t_0..t_0 + H].
\end{aligned}
$$

Our objective function is to maximize the mass of scheduled tasks while minimizing the cost of open slots. Constraint 1 indicates that each task is scheduled in at most one slot. Constraint 2 expresses that if we have a task in a slot, this slot is open. Constraint 3 indicates that if our slot is open, there is at least one task in this slot. Constraint 4 expresses that the sum of the masses of the tasks in the slot is smaller than the capacity of the slot. At last, we have the integrity constraints.

3.2   Greedy Heuristics

We notice that most of the myopic heuristics (even in [8,9]) are organized into two parts (see Algorithm 2): A first part (1) that sorts out tasks in a chosen order, and a second one (2) which decides where to place tasks, taken in the predefined order, in the different slots.

Finally, we check that the next slot to launch (the slot t) has a positive gain: the gain of all tasks compensates for the loss of the slot opening (3). The goal is to have the best combination of sorting and placement strategies.

Algorithm 2. SCHEDULE$(t, \mathcal{R}_t, H)$ for greedy heuristics

---

Require: t : starting slot, $\mathcal{R}_t$: the set of known unscheduled tasks, H : Rolling horizon
Ensure: $\mathcal{P}[t] : =$ The set of tasks for slot t
 1: $\mathcal{R}_t \leftarrow SortingTasks(\mathcal{R}_t)$
 2: $\mathcal{P} \leftarrow PlacementTasks(t, \mathcal{R}_t, H)$
 3: $\mathcal{P}[t] \leftarrow VerifLaunch(\mathcal{P}[t])$

Tasks Sorting Strategies. We propose six sorting heuristics. When ties appear during sort, they are broken using the release date for all of them except in *FIFO*, which uses the deadline. The heuristics are:
- **FIFO** ascending order of releases dates
- **EDF** ascending order of deadlines
- **Mass** descending order of task masses
- **LCF** [8] sorting according affinity with the other tasks. A task has an affinity with another task if both tasks can fit into the same bin. The sort is done each time a task is placed in a slot.
- **Poly** Ascending order of a score assigned to tasks. The score of a task is the sum of its ranks in the two sorting methods **EDF** and **Mass**.

- **ZigZag** [9] Instead of being an ascending or descending sort, it first takes the task closest to the median mass and then alternates between the larger and smaller masses around the median task.

Placement Strategies. The placement phases should decide in which slot the next task should be placed. Notice that if a task cannot be placed within its time window, it is discarded. For a time slot t, we say that a task i is available at t if it is unscheduled yet (if $r_i \leq t < d_i$ and if the remaining capacity of slot t is less than m_i). We also say in this case that slot t can accommodate task i.

Assume that task i is the current task. The placement strategies are:
- **First-Fit:** We schedule i in the first open slot that can accommodate it.
- **Best-Fit:** we place i in the already open slot with the smallest remaining capacity that can accommodate the task. If no open slot can accommodate it, a

new slot is opened if possible within the task's time window if possible.

– **Store**: The list of tasks is scanned to constitute a subset of tasks that can be performed simultaneously with i (their time windows overlap and their masses do not exceed the maximal capacity). And once we can no longer add a new task, then we schedule all the tasks of the subset in the first slot that can accommodate them together.

– **Fill**: The fill strategy opens the next slot to accommodate the next unscheduled task of the given sorting, and then uses another sorting to fill this slot (according to non-increasing masses).

The next two heuristics are based on the cardinality of the subsets of available tasks at each time slot:

– **Max-Fit** [9]: We schedule i in the slot with the most available tasks.

– **Min-Fit** : We schedule i in the slot with the fewest available tasks.

Below, we list the different heuristics implemented, and define the acronyms that will be used to designate these heuristics later. Acronyms are of the form xx_yy where xx is the placement algorithm and yy the task sorting algorithm, for example: bf_edf is the heuristic that uses *EDF* to sort tasks and *Best Fit* to schedule tasks.

Placing/sort	FIFO	EDF	Mass	LCF	Zigzag	Poly
First-Fit	ff_fifo	ff_edf	ff_m	ff_lcf	ff_zigzag	ff_poly
First-Fit Fill	fff_fifo	fff_edf	fff_m	fff_lcf	fff_zigzag	fff_poly
Best-Fit	bf_fifo	bf_edf	bf_m	bf_lcf	bf_zigzag	bf_poly
Store	s_fifo	s_edf	s_m	s_lcf	s_zigzag	s_poly
Store Fill	sf_fifo	sf_edf	sf_m	sf_lcf	sf_zigzag	sf_poly
Max-Fit	maxf_fifo	maxf_edf	maxf_m	maxf_lcf	maxf_zigzag	maxf_poly
Min-Fit	minf_fifo	minf_edf	minf_m	minf_lcf	minf_zigzag	minf_poly

3.3   Other Myopic Heuristics

Heuristic Based on the Continuous Relaxation of the ILP. We introduced several heuristics based on the continuous relaxation of the ILP. Let us note $\hat{X}$ and $\hat{Y}$ the values of the continuous variables at the optimum. We used three ways to build a feasible integer solution:

– **ContLP1**: We retrieve the most probable open slots (in descending order of $\hat{Y}_t$) in the system and, slot by slot, we put the most likely tasks (in descending order of $\hat{X}_{i,t}$ with t the chosen slot) that could appear in the chosen slot.

– **ContLP2**: We retrieve the set of most likely tasks in the system (in descending order of $\hat{X}_{i,t}$) and put them in the most probable slot in which the chosen task can appear (the slot t).

– **Dawande (DLP):** For an MKAR, Dawande et al. [6] proposed an algorithm to get a feasible integer solution from the continuous relaxation based on an iterated network flow problem. We adapted this heuristic to our problem by considering setup costs and coupled it with a myopic heuristic.

Adaptation of Multiple Knapsack Heuristics of the Literature. We also adapted several heuristics from the literature on multi-knapsack problems.

- **H34**: Adapted from [3] in which the authors studied a multiple knapsack problem. They proposed a polynomial-time heuristic 3/4-approximation algorithm by dividing the tasks into groups according to their mass and placing them so as to fill at least 3/4 of the Knapsacks. We added time windows to their model and slightly modified the gain considered.
- **MKH**: Adapted from [6] in which an MKAR is solved by an algorithm that consists of approximately solving knapsack problems independently of each other for each slot in the horizon. Then to select the slot with the best gain and keep the schedule of this slot. The process is repeated on the remaining slots and tasks until there are no more available tasks or slots. We added setup costs considerations to this framework.

4   Lookahead Algorithms

In this section, we present heuristics that partially consider future tasks by sampling, inspired by the article by Van Hentenryck [12]. In this paper, the authors address a stochastic online reservation problem. At each time slot, a single task may randomly arrive with random parameters and must be allocated to one of the B resources, which have limited capacity. Upon each arrival, an irrevocable decision must be made whether to allocate the request to a resource or not. The objective is to maximize the expected overall payoff by summing the gains from the tasks assigned to the different resources at the horizon time. They use a lookahead heuristic: at each time slot t, if a task arrived at t, the algorithm samples different possible scenarios for slot $t + 1$ and retrieves a final decision for slot t. Several different strategies are proposed, among which we choose to adapt the *consensus* strategy.

The features of our model that differ from those of [12] are that a variable number of tasks may arrive at any given time. Moreover, several tasks may be scheduled at each time step, and our tasks have time windows. Algorithm 3 summarizes our adaptation of a lookahead approach. Starting from a partial schedule $\mathcal{P}$, it outputs the final decision $\mathcal{P}[t]$ at time t. To this purpose, the algorithm will give a score to each possible subset of tasks that can be executed in slot t, recorded in *listCombi*, and then choose the one with the maximum score (line 11). The scores have an initial value 0 (lines 1–5). The scoring of the combinations (lines 5 to 10) is done by repeating a stochastic forecast for the near future a large number (nbScenario) of times. In each Scenario, the algorithm $GETSAMPLE$ simulates the arrival of new tasks during slot t and adds this subset of new tasks to the currently known tasks $\mathcal{R}_t$ to get set $\mathcal{R}'$ (line 6). A schedule $\mathcal{P}'$ is then computed with a myopic algorithm $SCHEDULE$ on this set of tasks in the slots t to $t + H$ (line 7). Notice that at time t, $\mathcal{P}'[t]$ only contains tasks i such that $a_i < t$ since tasks of the sample are not schedulable at t. So it is in the list *listCombi* computed at line 1. The score of this combination is incremented by the overall gain of $\mathcal{P}'$ on the rolling horizon (line 9).

Algorithm 3. Lookahead-Consensus $(\mathcal{P}, \mathcal{R}_t, t)$

Require: $\mathcal{P}, \mathcal{R}_t, t$
Ensure: $\mathcal{P}[t]$
1: Compute the list $listCombi$ of subsets of $\mathcal{F}_t$,
2: **for** $combi \in listCombi$ **do**   ▷ Score initalization
3:     $score(combi) \leftarrow 0$
4: **end for**
5: **for** $i = 1$ to $nbScenario$ **do**   ▷ Run stochastic lookahead scenarii
6:     $\mathcal{R}' \leftarrow \mathcal{R}_t \cup GETSAMPLE(t)$ ▷ Add sample subset of tasks arriving in slot t
7:     $\mathcal{P}' \leftarrow SCHEDULE(t, \mathcal{R}', H)$ ▷ Compute a static schedule on $[t, t + H]$
8:     $combi \leftarrow \mathcal{P}'[t]$ ▷ Retrieve the slot t
9:     $score(combi) \leftarrow score(combi) + \mathcal{G}(\mathcal{P}')$
10: **end for**
11: $\mathcal{P}[t] \leftarrow \underset{c \in listCombi}{\arg\max}\ score(c)$ ▷ Schedule the subset with highest score in slot t

---

5   Measuring the Efficiency

5.1   Test Plan and Simulation

We detail now how we evaluate the expected objective of each of the scheduling algorithms. This is done by Monte Carlo simulations on different relevant scenarios of the problem. Table 1 is our test plan and records all the parameter values we use in our tests. To build an instance (also called a set of parameters) for each field, we pick one value from among those available. By combining them all, we get our 10800 instances. Let us specify that the parameter *Rolling Horizon* describes the length of the horizon with respect to the maximal length of the rolling window.

We compute the expected overall objective for a given algorithm using the Monte Carlo principle. A *Replication* is the realization of a random path over a horizon N of the system governed by the assessed algorithm. We note by $\mathcal{G}(\mathcal{P}^{I_n})$ the objective collected during the running of n-th replication I_n. For a fixed set of parameters j and a given algorithm, we replicate the simulations a large number of times (L) to obtain the expected total objective, that is:

$$f_j = \frac{1}{L} \sum_{\ell=1}^{\ell=L} \mathcal{G}(\mathcal{P}^{I_\ell}).$$

In order to get precise comparisons between heuristics, we also compute the 95% confidence interval, the size of which decreases with the number of replications. We kept the value of L as large as possible ($L = 250$ for Greedy heuristics, $L = 500$ for H34 and MKH heuristics), except when the running times of the simulations were too long ($L = 20$ for continuous relaxation of the ILP and Lookahead heuristics and $L = 10$ for the ILP).

Table 1. Test Plan parameters.

Parameter	Values	Caption
Rolling Horizon	$[2, 15\%\overline{H}, 30\%\overline{H}, 50\%\overline{H}, \overline{H}]$	$\overline{H}$ is the maximal horizon
CSlot	$[8, 16, 24]$	maximum capacity of a slot
Ratio	$[\frac{1}{4}, \frac{1}{2}, \frac{3}{4}, 1]$	parameter used to tune masses
Mass	$[1, \text{Ratio} \cdot \text{CSlot}]$	variation interval of masses
λ	$[\frac{1}{4}, \frac{1}{2}, 1, 3, 5]$	arrival rate of the Poisson process
Delay A	$[1, 6, 12]$	delay between arrival and release
SC	$[0, 10, 50, 90]$	setup cost as % of the slot capacity
$\overline{D}$	$1, 6, 12$	Maximum availability duration $(\underline{D} = 1)$
N	960	general simulation horizon

5.2   Results

For the results, we selected the heuristics that appear to be the best once for a subset of parameters, as well as the best heuristics of the different articles adapted to our model as baselines and the ILP. The main metric that we use to compare the results is the *average deviation rate* from the best. For a resolution method $\pi \in \Pi$, we define this value $e_\pi(J)$ as the Cesaro mean of the average percentage difference, per set of parameters $j \in J$, between (f_j^π) the gain of π and the gain of the best resolution method (f_j^*). When confidence intervals cross, we fix $f_j^\pi - f_j^* = 0$. It is equal to:

$$e_\pi(J) = \frac{1}{|J|} \cdot \sum_{j \in J} \frac{f_j^* - f_j^\pi}{f_j^*} . \tag{5}$$

The lower this value is, the closer, on average, our resolution method is to the best resolution method for sets of parameters.

Results for Static Problems. ILP builds an optimal solution. Myopic heuristics obtain average deviation rates around 7% for *MKH*, around 10% for both *ContLP1*, and greedy with *bf*, *ff*, *minf* placement strategies, around 20% for greedy with *maxf* and *stock* placement strategies, and up to 30% for *H34* and *DLP*.

Results for Dynamic Problems. Figure 1 shows the average deviation rates of the whole set of policies. It can be seen that the *ILP* method has an average deviation rate strictly greater than zero, which means that for several sets of parameters, it is no longer the best method. However, on average, the ILP remains the most efficient resolution method.

Moreover, baselines from the literature, such as *ff_lcf* in [8], or even *H34* in [3] show on average performances lower than those of the *sf_edf* heuristic, recognized as the most robust among myopic heuristics. Finally, *c_sf_edf*, which

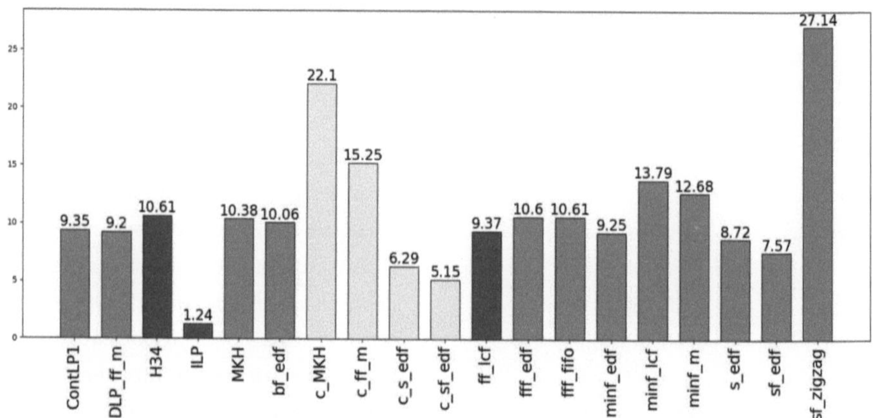

Fig. 1. Comparison of the average deviation rate from the best. In red, baseline algorithms of the literature, in yellow, lookahead algorithms, and in blue, myopic heuristics. (Color figure online)

is *consensus* lookahead with scheduling by *sf_edf*, stands out as the most robust among lookahead heuristics, and more generally among all evaluated heuristics.

Table 2 gives the mean computation time of one replication for different categories of resolution. First of all, ILP often requires more than 2 minutes to solve an instance. This underlines its major drawback: its computation time. The other categories of solving methods include: fast heuristics (such as *Greedy*, both *H34* and *MKH*), which are up to 2000 times faster than *ILP*, and slow heuristics (lookahead and continuous relaxation heuristics), which, in general, offer comparable computation times, except when processing numerous small tasks, which makes lookahead very slow. Thus, apart from this particular case, slow heuristics require on average 2 s per instance, which is 12 times less time than ILP, but remain 160 times slower than fast heuristics.

Limiting the rolling horizon is a way to reduce computation time. Thus, in ILP by limiting the horizon to $30\%\overline{H}$, one can reduce the computation time to 24.62 s per set of parameters, while maintaining the same level of performance (which is not guaranteed for shorter rolling horizons). For the other solving methods, a similar phenomenon is observed: increasing the rolling horizon does not improve the results compared to a horizon of only 2 slots (i.e. considering only one additional slot beyond the current one). This phenomenon seems to be related to the greedy characteristics of the algorithms.

We introduce now, for a given algorithm π and a subset of parameters J', the value $ec_\pi(J')$ which measures the gap between $e_\pi(J')$ and the mean of the average deviation rate of the other methods. We have:

$$ec_\pi(J') = \frac{e_\pi(J') \times |\Pi|}{\sum_{\pi' \in \Pi} e_{\pi'}(J')} \, .$$

Table 2. Mean running time of a replication for each horizon value.

Methods	$H = 2$	$H = 15\%\overline{H}$	$H = 30\%\overline{H}$	$H = 50\%\overline{H}$	$H = \overline{H}$
Greedy	0.011 s	0.015 s	0.015 s	0.020 s	0.028 s
ILP	3.36 s	11.77 s	24.62 s	324.9 s	348.56 s
Lookahead	6.80 s	9.60 s	9.64 s	9.62 s	9.30 s
ContLP	1.29 s	1.48 s	1.65 s	2.00 s	3.83 s
MKH & H34	0.014 s	0.018 s	0.024 s	0.023 s	0.035 s

The heatmap in Fig. 2 exhibits the quality of a method for a subset of parameters. We observe that ILP provides lower results than some heuristics when the arrival rate is low. Also, one sees that ILP clearly surpasses heuristics when the cost of opening a slot is almost equal to the capacity of the slot, that is to say, in situations where it is optimal either to fill the slot completely or not to schedule any task.

6   Heuristic Selection with Random Forest Prediction

We observed that, even if some myopic and lookahead heuristics give good results, their performances still remain lower than those obtained with the ILP. Furthermore, if lookahead policies have better results than myopic ones, they turn out to be much slower. We have therefore not identified a robust heuristic able to reach the results obtained by the ILP on all instances. This is why we present our method for selecting the most appropriate heuristic for an instance

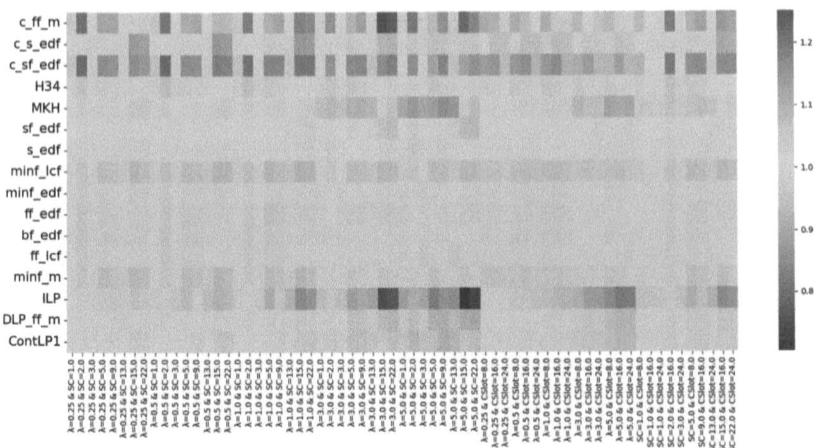

Fig. 2. Comparison of resolution methods with $ec_\pi(J')$. Each column corresponds to a subset of parameters J' with fixed $(\lambda, SC, CSlot)$. Blue color indicates the good quality of the resolution method. (Color figure online)

in order to improve the quality of the solution while maintaining a low execution speed. For this aim, we follow the algorithm selection framework presented in [14]. Let us detail the different steps.

Portfolio Selection. First, we determine the set of algorithms that we will consider as the smallest group of heuristics sufficient to obtain the best payout. We only keep algorithms that have been the best at least once on the whole set of instances. We then retain myopic greedy ones (*bf_edf, ff_lcf, fff_edf, fff_fifo, minf_edf, minf_lcf, s_edf, sf_edf, sf_zigzag*), multiple Knapsack (*MKH, H34*), continued relaxation (*ContLP1, DLP_ff_m*), and Lookahead heuristics (*c_MKH, c_s_edf, c_sf_edf, c_ff_m*). A second portfolio is selected, keeping only faster heuristics (Greedy and multiple Knapsack).

Learning Model. The machine learning model used is a multi-class classification model in which each algorithm is a class, and the label is 1 if, for an instance the algorithm belongs to the best ones and 0 otherwise. The prediction is performed with a random forest algorithm implemented in [17]. We then use a Grid Search algorithm to optimize the hyperparameters of the random forest.

Learning Sets. We proceed with the learning by dividing our data (retrieved from the previous test plan and experimentations) as follows: we randomly pick up 60% for the training set, 20% for the test set. The remaining 20% is the evaluation set and will be used later. We validate our learning since our metrics do not deteriorate on the test set, and the values obtained are very good (macro-avg equals 0.87 for precision, 0.89 for recall, and 0.88 for F1-Score). Note that a classification is considered correct for an instance if it anticipates one of the heuristics with the best gain.

Including testing and hyper-parameterization, the learning process takes 61.6 minutes. Once it is achieved, we get a mapping between a set of parameters and the best of the associated portfolio algorithms. Thus, we obtain an algorithm (called **prediction**) that takes the parameter set as input, applies matching to the best predicted dynamic heuristic, and performs scheduling with it.

We represent the *average deviation rate* computed on instances that belong to the evaluation set. We compare our **prediction** algorithm with the first portfolio in Fig. 3a and with the second portfolio, containing only fast heuristics, in Fig. 3b. We keep the best heuristic (adding *consensus* in Fig. 3a). We also add an additional score *oracle*, which represents the optimal prediction score, obtained in the case where the prediction does not commit any error. We notice that **prediction** reaches a score very close to that of the *oracle* and surpasses all the individual heuristics. For the first portfolio, the **prediction** score is 3.6% and we achieve 29% reduction in error compared to **c_sf_edf**. We are also faster, with a mean running time of a replication equal to 3.41 s. This is because we are replacing a *consensus* heuristic with a fast heuristic in 41, 7% of cases. Furthermore, we have a 13% difference between the error of **prediction** and that of the *oracle*. Considering only fast heuristics, the **prediction** score reaches 4.66%, so we have a 34% reduction in error compared to **sf_edf**. Finally, we are only 5% off between the error of our **prediction** and that of the *oracle* (4.41%).

Thus, this algorithm provides a very robust solution method while retaining the speed of a myopic heuristic. And if we prioritize robustness, we still have a method that is both faster and more efficient than the most robust heuristic available.

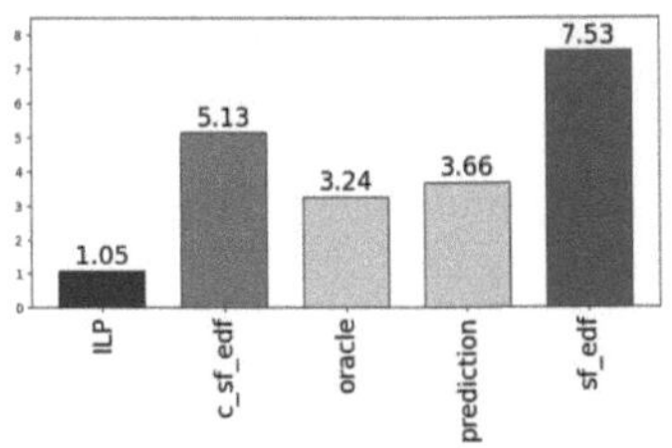

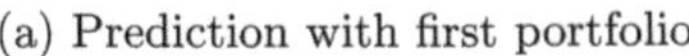

(a) Prediction with first portfolio

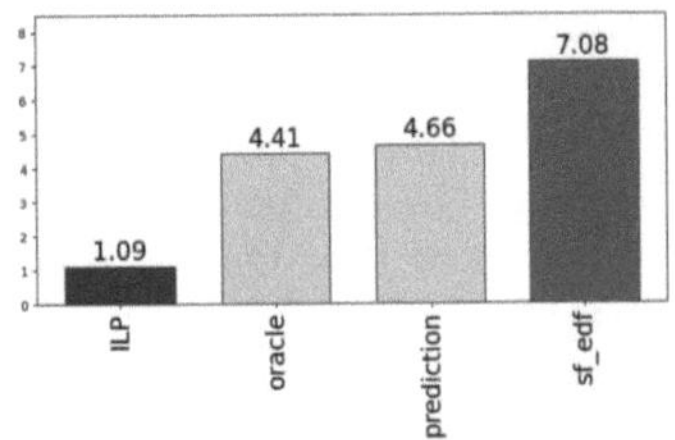

(b) Prediction with faster portfolio

Fig. 3. Comparison of the average deviation rate to the best with prediction and oracle scores (only on evaluated data).

7 Conclusion

In this work, we studied a stochastic capacity problem. We solved it with a Rolling Horizon approach, for which we propose an extensive benchmark of heuristics. We highlighted that some basic heuristics perform very well, while heuristics from the literature are not effective. We also observed that the best results were obtained by integer linear programming, at the cost of significantly larger computational time. We also proposed a lookahead approach that outperforms all the heuristics studied and provides an alternative faster than the usual recourse methods. Finally, we devised an algorithm selection approach based on AI and data visualization that allowed to obtain results similar to linear programming in a far shorter time. Expanding the portfolio of heuristics to cover all cases where ILP is significantly better than other algorithms is an area for improvement of this work.

Acknowledgments. The authors would like to thank Thales DMS for its support and for allowing this work to be carried out during the hours dedicated to the PhD.

References

1. Addis, B., Carello, G., Grosso, A., Tànfani, E.: Operating room scheduling and rescheduling: a rolling horizon approach. Flex. Serv. Manuf. J. (2016)
2. Benammar, N., Chrétienne, P., Hyon, E., Jean-Marie, A.: Approches par horizon roulant pour un problème de planification stochastique. In: ROADEF (2020)

3. Caprara, A.: A 3/4-approximation algorithm for multiple subset sum. J. Heuristics **9** (2003)
4. Crainic, T.: Multi-period bin packing model and effective constructive heuristics for corridor-based logistics capacity planning. Comput. Oper. Res. **132** (2021)
5. Dahl, G., Foldnes, N.: LP based heuristics for the multiple knapsack problem with assignment restrictions. Ann. Oper. Res., 91–104 (2006)
6. Dawande, M., Kalagnanam, M., Keskinocak, M., Salman, F.S., Ravi, R.: Approximation algorithms for the multiple knapsack problem with assignment restrictions. J. Comb. Optim. **4**(2), 171–186 (2000)
7. DeYong, G.D., Cattani, K.D.: Fenced in? Stochastic and deterministic planning models in a time-fenced, rolling horizon scheduling system. Eur. J. Oper. Res. **251**(1), 85–95 (2016)
8. Fleszar, K.: A new MILP model and fast heuristics for the variable-sized bin packing problem with time windows. Comput. Ind. Eng. **175** (2023)
9. Fu, Y.: Heuristic/meta-heuristic methods for restricted bin packing problem. J. Heurist. **26** (2020)
10. Gioia, D.G., Fadda, E., Brandimarte, P.: Rolling horizon policies for multi-stage stochastic assemble-to-order problems. Int. J. Prod. Res. **62**(14), 5108–5126 (2024)
11. Hartman, J., Perry, T.: Approximating the solution of a dynamic, stochastic multiple knapsack problem. Control. Cybern. **35**(3), 535–550 (2006)
12. Hentenryck, P.V., Bent, R., Mercier, L., Vergados, Y.: Online stochastic reservation systems. Ann. Oper. Res. **171**, 101–127 (2009)
13. Kleywegt, A., Papastavrou, J.: The dynamic and stochastic knapsack problem. Oper. Res. **46**(1), 17–35 (1998)
14. Kotthoff, L.: Algorithm selection for combinatorial search problems: a survey. AI Mag. **35**(3), 48–60 (2014)
15. Liu, Q., et al.: Algorithms for the variable-sized bin packing problem with time windows. Comput. Ind. Eng. (2021)
16. Meng-Gérard, J., Chrétienne, P., Baptiste, P., Sourd, F.: On maximizing the profit of a satellite launcher: selecting and scheduling tasks with time windows and setups. Discret. Appl. Math. **156** (2009)
17. Pedregosa, F., et al.: Scikit-learn: machine learning in Python. J. Mach. Learn. Res. **12**, 2825–2830 (2011)
18. Sahin, F., Narayanan, A., Powell Robinson, E.: Rolling horizon planning in supply chains: review, implications and directions for future research. Int. J. Prod. Res. (2013)
19. Samà, M., D'Ariano, A., Pacciarelli, D.: Optimal aircraft traffic flow management at a terminal control area during disturbances. In: 15th EURO Working Group on Transportation (EWGT), pp. 460–469 (2012)

Resource-Constrained Project Scheduling Problem with Transfer Times Using Secondary Resources with Instant Self-transfers

Vilém Heinz[1,2]($\boxtimes$) (iD), Zdeněk Hanzálek[2] (iD), Christian Artigues[3] (iD), and Emmanuel Hebrard[3] (iD)

[1] Faculty of Electrical Engineering, Czech Technical University in Prague, Prague, Czech Republic
`heinzvil@fel.cvut.cz`
[2] Czech Institute of Informatics, Robotics and Cybernetics, Czech Technical University in Prague, Prague, Czech Republic
`Zdenek.Hanzalek@cvut.cz`
[3] Laboratory for Analysis and Architecture of Systems, French National Centre for Scientific Research, Toulouse, France
`{christian.artigues,hebrard}@laas.fr`
`https://fel.cvut.cz/` , `https://www.ciirc.cvut.cz/` , `https://www.laas.fr/`

Abstract. In this paper, we introduce a novel variant of the Resource-Constrained Project Scheduling Problem (RCPSP), called RCPSP with Transfer Times using Secondary Resources with Instant Self-transfers (RCPSPTT-2I). This variant extends the classical RCPSP by modeling resource transfers between activities and hierarchical resource interactions. Specifically, we consider two types of resources: (i) the primary resources used to execute the activities and (ii) secondary resources that help facilitate transfers of primary resources between activities they execute. Transfer lengths depend on both the pair of activities and the primary resource being transferred. Secondary resources can facilitate part or all of the transfer process, and their own transfers without primary resource are assumed instant. This problem definition captures practical scenarios, such as digital or decision-making secondary resources, that (i) do not require physical movement or (ii) have negligible/constant transfers, combining elements of RCPSP with Transfer Times and RCPSP with First- and Second-Tier Transfers variants from the literature. We formalize RCPSPTT-2I using both Integer Linear Programming and Constraint Programming (CP). These models obtain feasible solutions for most 30-activity instances within a 10-minute runtime but struggle to scale to larger instances. To address this limitation, we introduce a heuristic warm-start and a multi-phase procedure that incrementally refines partial solutions. Computational results show that the multi-phase procedure, combined with the Constraint Programming model, significantly improves scalability and solution quality, while preserving optimality guarantees, making it well-suited for practical instance sizes.

T. Guns (Ed.): CPAIOR 2026, LNCS 16595, pp. 207–224, 2026.
https://doi.org/10.1007/978-3-032-27242-3_13

Keywords: Constraint Programming · Scheduling · RCPSP · Transfer times · Multi-phase procedure · Integer Linear Programming

1   Introduction

The Resource-Constrained Project Scheduling Problem with Transfer Times using Secondary Resources with Instant Self-transfers (RCPSPTT-2I) extends the RCPSP with Transfer Times (RCPSPTT) introduced in [23], which itself builds on the classical RCPSP problem [3]. The classical RCPSP is defined by a set of activities and resources with limited capacities. Each activity requires specific resource amounts for processing and has to respect precedence relations. The objective is to minimize the makespan.

The classical RCPSP considers that once an activity is finished, the resources are immediately available—however, reallocation or resource setup may be required before they can be used by the next activity. Furthermore, an additional (secondary) resource might be required to help with the transfer process of the primary resource. Considering these properties is crucial in high-throughput productions or complex project environments, where transfers are of non-negligible lengths and their incorrect handling can cause significant delays.

Generally, RCPSPTT-2I represents real-world settings where secondary resources either: (i) do not require physical transfer or (ii) their transfer without a primary resource is negligible or constant. As an illustration, consider a software development project where developers (primary resources) move between tasks and require preparation time before starting new work. These preparations (transfers) may involve software architects or product owners (secondary resources) who support part or all of the transfer process but whose own transfer between tasks is negligible (they already know the project and require no/negligible physical movement). Similar situations arise whenever different worker capacities are required by activities, and secondary resources represent supervisors with no/negligible movement time of their own, such as in some construction or manufacturing settings. A conceptually different application arises in cloud computing with containerized workloads, where primary resources are servers, and secondary resources are additional orchestrating/automation agents.

1.1   Related Work

Since transfers (also often referred to as setups in the literature) are common not only in industrial applications, they are widely studied in the scheduling literature; see [19] for an overview. However, in many closely related problems, such as parallel machines with sequence-dependent setups [6], the primary resources are usually of unit capacity. The notion of transfers in an RCPSP setting was first considered in [30] where resource transfers across multiple subprojects via a central resource pool were solved by different transfer rules. Subsequent work by [12,20] extended this concept, proposing different solutions using mathematical programming and heuristics, while [4] considered a setting where activities are scattered across multiple physical sites.

While modeling transfers between different subprojects or worksites often suffices, some settings require resource transfers between individual activities—for example, if each activity has a distinct location. The authors of [24] address this by proposing RCPSP with Transfer Delays, where an activity-pair-dependent transfer can be required. A heuristic solution demonstrated on instances of up to 120 activities is proposed. However, this formulation lacks a key aspect—the time dependence of the transfer based on a particular resource. This is addressed by the RCPSPTT problem introduced in [23], where transfer times depend on both pair of activities and the resource being transferred. The authors propose a branch-and-bound method and tabu search, demonstrating its effectivity on extended PSPLib [9] instances of up to 120 activities.

Later, RCPSPTT was studied by [8,17,25]. In [8], an ILP model and an efficient Genetic Algorithm (GA) using two-point crossover were proposed. A more efficient ILP formulation and a tree search heuristic were proposed in [17]. In [25], RCPSPTT is considered in the aircraft assembly setting and an ILP model and branch and bound methods are proposed.

RCPSPTT was also subject to many extensions, capturing additional real-world requirements. The authors of [11] and of [28] study the multi-project variant RCMPSPTT; the former uses rule-based heuristics and an ILP model, while the latter adds priority rules and a GA. Decentralized versions of RCMPSPTT are investigated in [1] using a multi-agent framework and in [31] using a GA. Various extensions also address activity, resource, and objective variants. In [2,29] the authors consider activity preemption/splitting, proposing a GA and a SAT-based solver, respectively. The authors of [15,16] consider efficient approaches for unit-capacity resources using CP models, branch-and-bound, and heuristics. Movable and non-movable resources are explored in [14,27], and partially removable resources in [21]. A robustness objective is examined in [18].

Only a handful of papers consider the use of secondary resources: RCPSP with First- and Second-Tier Resource Transfer in [22], RCMPSPTT-2 (multi-project setting) in [11], and a multi-mode variant considered in [23]. However, the exact methods proposed in said literature are limited to only ILP/CP models, and their effective use was limited to instances of only 10 activities.

1.2   Contribution and Outline

The literature review indicates that RCPSP variants with secondary resources are scarcely studied and existing exact methods scale poorly. Thus, we introduce a new variant of the problem called RCPSPTT-2I along with more efficient exact solution methods. The key difference from [11,22,23] is that in our setting, secondary resources can handle either part or all of the transfer of primary resources, and their own transfers—when they are not moving a primary resource—are treated as instant. As demonstrated in Sect. 1, RCPSPTT-2I captures a number of real-world scenarios.

To address the problem, we propose two exact methods—a CP model and an ILP model—together with a warm-start heuristic. The results show that the problem greatly benefits from the expressivity of the CP, with the CP model clearly outperforming the ILP. We also propose an effective multi-phase procedure, significantly increasing the solvable instance size compared to the related problems in the literature. Computational results show that the multi-phase procedure combined with the CP model scales to instances of 120 activities, lowers the gaps to lower bounds by roughly 20% relative to the warm-started CP model, and still preserves optimality guarantees—yielding provably optimal solutions for more than a third of the 30-activity instances in a 10-minute time limit. Finally, we provide a new benchmark set that extends the set by [22].

The remainder of the paper is organized as follows. In Sect. 2, we formally define the RCPSP with Transfer Times using Secondary Resources with Instant Self-transfers (RCPSPTT-2I). In Sect. 3, we discuss instance preprocessing and present the ILP model (Subsect. 3.1), the CP model (Subsect. 3.2), the warm-start approach (Subsect. 3.3) and the multi-phase procedure (Subsect. 3.4). Section 4 reports the computational results and analyzes scalability. Section 5 concludes the paper and outlines directions for future research.

2   Problem Definition

Formally, RCPSPTT-2I is defined as follows.

- Let $\mathcal{A} = \{0, 1, \ldots, n\}$, $n \in \mathbb{N}$, denote the activity set indexed by $i, j \in \mathcal{A}$, where 0 and $|\mathcal{A}|$ represent dummy start and end activities.
- Let $P \in \mathbb{N}^{|\mathcal{A}|}$ denote the vector of activity processing times.
- Let $\mathcal{E} = \{(i, j) \mid$ activity i immediately precedes activity $j\}$ be the set of precedences.
- Let $\mathcal{R} = \{0, 1, \ldots, m\}$, $m \in \mathbb{N}$, be the set of primary resources indexed by $r \in \mathcal{R}$.
- Let $C \in \mathbb{N}^{|\mathcal{R}|}$ be the vector of primary resource capacities.
- Let $\mathbf{Q} \in \mathbb{Z}_{\geq 0}^{|\mathcal{A}| \times |\mathcal{R}|}$ be the activity-resource consumption matrix, where $\mathbf{Q}_{i,r}$ denotes activity i consumption of the resource r. We consider $i = 0 \vee i = |\mathcal{A}| - 1 \Rightarrow \mathbf{Q}_{i,r} = C_r$, $\forall r \in \mathcal{R}$.
- Let $\Delta \in \mathbb{Z}_{\geq 0}^{|\mathcal{A}| \times |\mathcal{A}| \times |\mathcal{R}|}$ be the transfer time matrix, where $\Delta_{i,j,r}$ denotes the time required for resource r to be transferred between activities i and j.
- Let $\mathcal{R}' = \{0, 1, \ldots, m'\}$, $m' \in \mathbb{N}$, be the set of secondary resources indexed by $r' \in \mathcal{R}'$.
- Let $C' \in \mathbb{N}^{|\mathcal{R}'|}$ be the vector of secondary resource capacities.
- $\mathbf{M} \in \mathbb{Z}_{>0}^{|\mathcal{R}| \times |\mathcal{R}'|}$ be the matrix of primary-secondary resource requirement, where entry $\mathbf{M}_{r,r'}$ denotes the units of secondary resource r' required to transfer one unit of primary resource r. If a unit of secondary resource r' is enough to transfer multiple units of a primary resource r, the capacities and requirements can be rescaled by an appropriate factor and then scaled back after obtaining the solution to keep the solution process integral.
- Let $\Delta' \in \mathbb{Z}_{\geq 0}^{|\mathcal{A}| \times |\mathcal{A}| \times |\mathcal{R}'|}$ be the secondary transfer time matrix, where $\Delta'_{i,j,r'}$ denotes the time required for the resource r' to support the transfer of some primary resource between activities i and j.

All secondary resources are *pure*: they only facilitate primary-resource transfers and never execute activities. Their transfers are instantaneous when no primary resource is involved. Both primary and secondary resources are *renewable*: once an activity or transfer finishes, their capacity becomes available again.

Illustrative Example. To illustrate the problem, an instance is proposed with $|\mathcal{A}| = 7$, $|\mathcal{R}| = 2$, $|\mathcal{R}'| = 1$, $C_0 = 4$, $C_1 = 3$, $C_0' = 3$, $\mathbf{M} = J_{|\mathcal{A}| \times |\mathcal{A}|}$ and the rest of the parameters provided in Fig. 1. Note that no transfers are required from dummy start or to dummy end in this case.

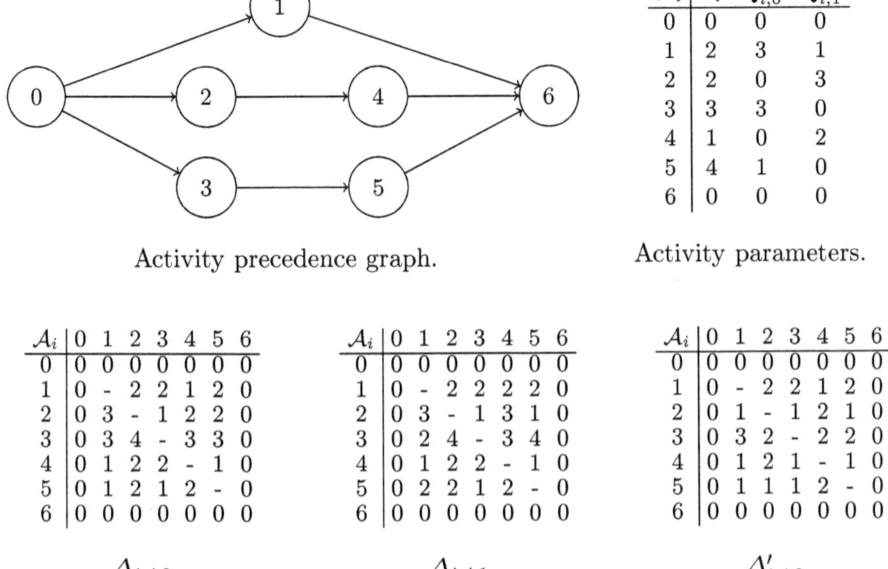

Activity precedence graph.

$\mathcal{A}_i$	P_i	$\mathbf{Q}_{i,0}$	$\mathbf{Q}_{i,1}$
0	0	0	0
1	2	3	1
2	2	0	3
3	3	3	0
4	1	0	2
5	4	1	0
6	0	0	0

Activity parameters.

$\mathcal{A}_i$	0	1	2	3	4	5	6
0	0	0	0	0	0	0	0
1	0	-	2	2	1	2	0
2	0	3	-	1	2	2	0
3	0	3	4	-	3	3	0
4	0	1	2	2	-	1	0
5	0	1	2	1	2	-	0
6	0	0	0	0	0	0	0

$$\Delta_{i,j,0}$$

$\mathcal{A}_i$	0	1	2	3	4	5	6
0	0	0	0	0	0	0	0
1	0	-	2	2	2	2	0
2	0	3	-	1	3	1	0
3	0	2	4	-	3	4	0
4	0	1	2	2	-	1	0
5	0	2	2	1	2	-	0
6	0	0	0	0	0	0	0

$$\Delta_{i,j,1}$$

$\mathcal{A}_i$	0	1	2	3	4	5	6
0	0	0	0	0	0	0	0
1	0	-	2	2	1	2	0
2	0	1	-	1	2	1	0
3	0	3	2	-	2	2	0
4	0	1	2	1	-	1	0
5	0	1	1	1	2	-	0
6	0	0	0	0	0	0	0

$$\Delta'_{i,j,0}$$

Fig. 1. Instance parameter definition.

The Gantt chart in Fig. 2 illustrates the optimal solution, with each graph representing one resource. For primary resources $\mathcal{R}_0$ and $\mathcal{R}_1$, the bars represent the activities executed, while for the secondary resource $\mathcal{R}_0'$, the bars represent supported transfers of the primary resources. The solid arrows represent the primary resource transfers, while the dashed arrows represent the parts of the transfer supported by the secondary resource $\mathcal{R}_0'$. Notice that due to $\mathcal{R}_0'$ capacity, transfer (2,4,1) is delayed, making the schedule length equal to 9 instead of 8. This shows how secondary capacity can influence the schedule and why it has to be considered in such scheduling scenarios. Furthermore, activity 1 has to wait for all incoming resource transfers to finish to start execution.

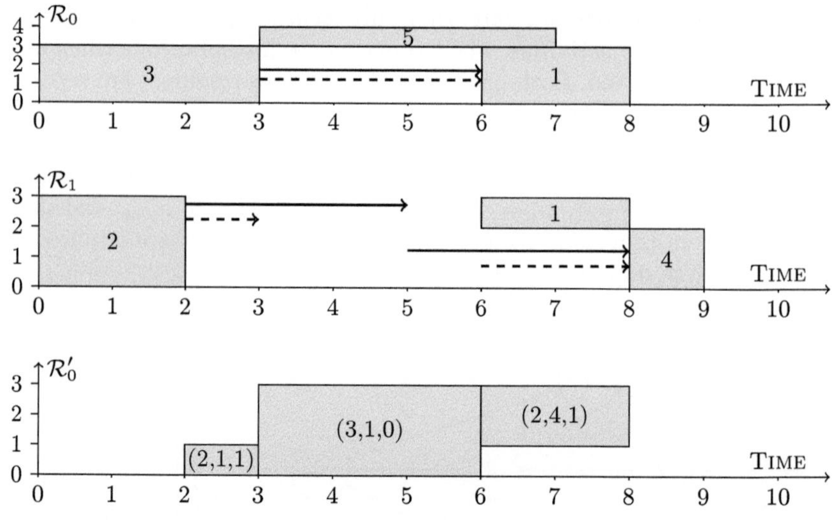

Fig. 2. Solution of the illustrative example.

3   Solution Methodology

To tackle the problem, we introduce two exact models, Integer-Linear Programming (ILP) in Subsect. 3.1 and Constraint Programming (CP) in Subsect. 3.2. However, initial tests showed that the scalability of the models within a practical runtime is limited to instances of 30 or fewer activities. Thus, we propose a simple warm-start heuristic in Subsect. 3.3 and a multi-phase procedure in Subsect. 3.4 to get the best possible result within the time limit while preserving optimality guarantees. Before any of the aforementioned methods is used to solve a problem instance, the following steps are also performed.

1. The transitive closure of instance-defined precedences $\mathcal{E}$ containing all implied precedences is calculated and used to replace the original $\mathcal{E}$.
2. The vectors containing earliest starts (ES), latest starts (LS), earliest finishes (EF), and latest finishes (LF), $\forall i \in \mathcal{A}$ are calculated using the critical path method lower bound, the serial execution upper bound and $\mathcal{E}$ from step 1. A value H is set equal to $\text{LS}_{|\mathcal{A}|-1}$.
3. A feasible resource-transfer set $\mathcal{T}$, where each transfer (i, j, r) satisfies $i, j \in \mathcal{A}$ and $r \in \mathcal{R}$, with upperbounds $U = \{U_{i,j,r}\} \in \mathbb{Z}_{>0}$, is calculated from the instance parameters, updated $\mathcal{E}$ from step 1, and ES/LS/EF/LF vectors from step 2. The pseudocode is provided in Algorithm 1.

3.1   Integer Linear Programming Formulation

There is a variety of ILP formulations for the RCPSP, with time-indexed, flow-based, and event-based models being the most common; see [10] for an overview. In a flow-based model, resource units are treated as a flow moving along arcs between activities, ensuring that each activity receives the exact amount it needs. Since flows already represent the primary-resource transfers between activities, the only additional requirement is to reserve sufficient time for transfer whenever a flow occurs. However, the flow structure alone does not account for secondary resources. Specifically, enforcing their capacities requires knowing exactly when the secondary resource supports the transfer, which necessitates adding a time-tracking variable for secondary resources.

Algorithm 1. Possible Resource Transfers Between Activity Pairs

---

1: **function** POSSIBLETRANSFERS($\mathcal{A}, \mathcal{R}, C, \mathbf{Q}, \mathcal{E}, \Delta, \mathrm{EF}, \mathrm{LS}$)
2:    $\mathcal{T} \leftarrow \{\}$
3:    **for** $i \in \mathcal{A} \setminus \{|\mathcal{A}|\} - 1$ **do**
4:       **for** $j \in \mathcal{A} \setminus \{0\}$ **do**
5:          **for** $r \in \mathcal{R}$ **do**
6:             $\texttt{transfer\_capacity} \leftarrow 0$
7:             **if** $(j, i) \in \mathcal{E} \vee (j = i)$ **then**
8:                **continue** $\triangleright$ skip: transfers from ancestor/same activity
9:             **else if** $(i > 0 \wedge j = 0) \vee (i = |\mathcal{A}| - 1 \wedge j < |\mathcal{A}| - 1)$ **then**
10:               **continue** $\triangleright$ skip: transfers to 0 and from $|\mathcal{A}| - 1$
11:            **else if** $\mathrm{LS}[j] < \mathrm{EF}[i] + \Delta_{i,j,r}$ **then**
12:               **continue** $\triangleright$ skip: not enough time for transition
13:            **else if** $i = 0 \wedge j = |\mathcal{A}| - 1$ **then**
14:               $\texttt{transfer\_capacity} \leftarrow C_r$ $\triangleright$ unused resources directly to $|\mathcal{A}| - 1$
15:            **else if** $i = 0 \vee j = |\mathcal{A}| - 1$ **then**
16:               $\texttt{transfer\_capacity} \leftarrow \max\{\mathbf{Q}_{i,r}, \mathbf{Q}_{j,r}\}$
17:            **else**
18:               $\texttt{transfer\_capacity} \leftarrow \min\{\mathbf{Q}_{i,r}, \mathbf{Q}_{j,r}\}$
19:            **if** $\texttt{transfer\_capacity} > 0$ **then**
20:               $\mathcal{T}[(i, j, r)] \leftarrow \texttt{transfer\_capacity}$
21:    **return** $\mathcal{T}$

Our RCPSPTT-2I ILP model extends the flow-based RCPSPTT model proposed in [17]; we also tested the time-indexed variant proposed in [8], but it performed much worse. The complete set of the model's variables is as follows.

- a_i, $\forall i \in \mathcal{A}$: Integer variable representing the start of the activity i, with a lower bound equal to ES_i and an upper bound equal to LS_i.
- $f_{i,j,r}$, $\forall(i, j, r) \in \mathcal{T}$: Integer variable representing the flow volume of resource r between activities i and j, with a lower bound equal to 0 and an upper bound equal to $U_{i,j,r}$.
- $y_{i,j,r}$, $\forall(i, j, r) \in \mathcal{T}$: Binary variable that represents if there is a non-zero flow of resource r between activities i and j.
- $s_{t,i,j,r,r'}$, $\forall t \in \{0, 1, \ldots, H\}$, $(i, j, r) \in \mathcal{T}$, $r' \in \mathcal{R}'$, $\mathbf{M}_{r,r'} > 0$: Binary variable that is true only if the transfer of primary resource r supported by r' between activities i and j begins at time t.

The model is defined in the following way.

$$\min a_{|\mathcal{A}|-1} \tag{1}$$

$$a_i + p_i \le a_j \quad \forall (i,j) \in \mathcal{E} \tag{2}$$

$$\sum_{j:(j,i,k)\in\mathcal{T},\ \Delta_{j,i,k}>0,\ k=r} f_{j,i,k} = \mathbf{Q}_{i,r} \quad \forall i \in \mathcal{A} \setminus \{0\}, \forall r \in \mathcal{R} \tag{3}$$

$$\sum_{j:(i,j,k)\in\mathcal{T},\ \Delta_{i,j,k}>0,\ k=r} f_{i,j,k} = \mathbf{Q}_{i,r} \quad \forall i \in \mathcal{A} \setminus \{|\mathcal{A}|-1\}, \forall r \in \mathcal{R} \tag{4}$$

$$y_{i,j,r} \cdot U_{i,j,r} \ge f_{i,j,r} \quad \forall (i,j,r) \in \mathcal{T} \tag{5}$$

$$a_j - a_i \ge (p_i + \Delta_{i,j,r}) \cdot y_{i,j,r} - H \cdot (1 - y_{i,j,r})$$
$$\forall (i,j,r) \in \mathcal{T} \tag{6}$$

$$\sum_{t\in\{0,1,...,H\}} s_{t,i,j,r,r'} = y_{i,j,r}$$
$$\forall (i,j,r) \in \mathcal{T},\ \forall r' \in \mathcal{R}',\ \mathbf{M}_{r,r'} > 0,\ \Delta'_{i,j,r'} > 0 \tag{7}$$

$$\sum_{t\in\{0,1,...,H\}} s_{t,i,j,r,r'} \cdot t \le a_j - \Delta'_{i,j,r'} \cdot y_{i,j,r}$$
$$\forall (i,j,r) \in \mathcal{T},\ \forall r' \in \mathcal{R}' \tag{8}$$

$$\sum_{t\in\{0,1,...,H\}} s_{t,i,j,r,r'} \cdot t \ge (a_i + p_i) \cdot y_{i,j,r}$$
$$\forall (i,j,r) \in \mathcal{T},\ \forall r' \in \mathcal{R}',\ \mathbf{M}_{r,r'} > 0,\ \Delta'_{i,j,r'} > 0 \tag{9}$$

$$\sum_{(i,j,r)\in\mathcal{T}} \left(f_{i,j,r} \cdot \mathbf{M}_{r,r'} \cdot [\![\Delta_{i,j,r} > 0]\!] \cdot \sum_{t'=max(0,t-\Delta'_{i,j,r'}+1)}^{t} s_{t',i,j,r,r'} \right) \le C'_{r'}$$
$$\forall t \in \{0, 1, \ldots, H\},\ \forall r' \in \mathcal{R}' \tag{10}$$

Equation (1) minimizes the start time of the final dummy activity, which follows all other activities. Equation (2) sets activity start times according to precedence constraints. Equations (3) and (4) ensure that each activity receives the required amounts of primary resources by matching their incoming and outgoing flows. Equation (5) force the flow-existence variable $y_{i,j,r}$ to be 1 whenever a transfer has nonzero flow. Equation (6) enforces the temporal consistency between activities and their associated transfers. Equation (7 ensures that each secondary transfer starts exactly once. Equations (8) and (9) require that the secondary transfer begins only after the previous activity ends and before the next begins. Lastly, Equation (10) ensures that secondary resource capacities are maintained. Please note that Eq. (10) is intentionally nonlinear, as a big-M linearization introducing four-dimensional auxiliary variables performed multiple times worse on the dataset used in Sect. 4 in Gurobi and CPLEX.

We also extend the model with variable $x_{i,j}$ representing the relative order of activities i,j and with the following set of redundant constraints that improve model's performance (referenced by related constraints in the main model part).

$$x_{i,j} = 1 \quad \forall(i,j) \in \mathcal{E} \tag{2b}$$

$$x_{j,i} = 0 \quad \forall(i,j) \in \mathcal{E} \tag{2c}$$

$$p_i \cdot x_{i,j} - H \cdot (1 - x_{i,j}) + a_i \le a_j \quad \forall(i,j) \in \mathcal{A} \times \mathcal{A}, \ i < j \tag{2d}$$

$$y_{i,j,r} \le f_{i,j,r} \quad \forall(i,j,r) \in \mathcal{T} \tag{5b}$$

After solving the model, we can extract the activity start times (a_i), flow amounts $(f_{i,j,r})$, secondary flow amounts $(f_{i,j,r} \cdot \mathbf{M}_{r,r'})$, and secondary transfer start times $(s_{t,i,j,r,r'})$, but primary transfer start times require postprocessing. Specifically, the start time for each primary transfer is computed as $\max\{a_i + p_i, \sum_{t \in \{0,1,\dots,H\}} s_{t,i,j,r,r'} \cdot t + \Delta'_{i,j,r'} - \Delta_{i,j,r}\}$ to ensure that there is enough time to execute it and that the secondary transfer is contained within its limits.

3.2   Constraint Programming Model Formulation

The Constraint Programming (CP) model for a classical RCPSP is very concise as CP offer more complex modeling constructs than ILP; see [13] for its formulation. The model uses three main notions: (i) interval variables that represent activities in the schedule, (ii) constraint ENDBEFORESTART that ensures proper activity ordering based on their precedences, and (iii) function PULSE that represents resource use over activity intervals. This model also performs very fast; CP generally works well on RCPSP as shown in [7], where hundreds of new state-of-the-art lower bounds and few newly closed instances are obtained.

However, since the RCPSP CP model does not implicitly represent flows, a larger extension is required for RCPSPTT-2I. Specifically, sets of optional interval variables for primary and secondary resource transfers are introduced, allowing the solver to decide which transfers occur without explicit selection through additional constraints. Similarly to the activities in the RCPSP model, PULSE can be used over the transfer intervals. The difference is that transfer intervals use a variable amount of resource. This is solved using HEIGHTATSTART to set the height of PULSE to a feasible transfer amount; note that some solvers [26] allow PULSE height to be a variable. The model's variables are as follows.

- $a_i, \ \forall i \in \mathcal{A}$: Interval variable representing activity i, with length p_i, start from interval $[\text{ES}_i, \text{LS}_i]$ and end from interval $[\text{EF}_i, \text{LF}_i]$.
- $z_{i,j,r}, \ \forall(i,j,r) \in \mathcal{T}$: Optional interval variable representing the potential transfer of resources r between activities i and j, with length $\Delta_{i,j,r}$, start from the interval $[\text{EF}_i, \text{LS}_j - \Delta_{i,j,r}]$ and end from $[\text{EF}_i + \Delta_{i,j,r}, \text{LS}_j]$.

- $f_{i,j,r}$, $\forall (i,j,r) \in \mathcal{T}$: Integer variable representing the flow volume of resource r between activities i and j, with a lower bound 0 and an upper bound $U_{i,j,r}$.
- $z'_{i,j,r,r'}$, $\forall (i,j,r) \in \mathcal{T}$, $\Delta_{i,j,r} > 0$, $r' \in \mathcal{R}'$, $\mathbf{M}_{r,r'} > 0$: Optional interval variable representing the secondary transfer of r' between activities i and j supporting the transfer of the resource r, with length $\Delta'_{i,j,r'}$, start from interval $[\mathrm{EF}_i, \mathrm{LS}_j - \Delta'_{i,j,r'}]$ and end from interval $[\mathrm{EF}_i + \Delta'_{i,j,r'}, \mathrm{LS}_j]$.
- $f'_{i,j,r,r'}$, $\forall (i,j,r) \in \mathcal{T}$, $\Delta_{i,j,r} > 0$, $\forall r' \in \mathcal{R}'$, $\mathbf{M}_{r,r'} > 0$: Integer variable representing the flow volume of resource r' between activities i and j supporting the transfer of the resource r, with a lower bound equal to 0 and an upper bound equal to $U_{i,j,r} \cdot \mathbf{M}_{r,r'}$.

Note that the variables $f_{i,j,r}$ and $f'_{i,j,r,r'}$, along with some associated constraints, are required because (i) zero-length transfers cannot be modeled by interval variables alone and (ii) the CPOptimizer Python API does not allow retrieving PULSE height after solving. The complete set of model's constraints is as follows.

$$\min\ \mathrm{end}(a_{|\mathcal{A}|-1}) \tag{11}$$

$$\mathrm{EndBeforeStart}(a_i, a_j)\quad \forall (i,j) \in \mathcal{E} \tag{12}$$

$$\mathrm{IfThen}(f_{i,j,r} \geq 1,\ \mathrm{PresenceOf}(z_{i,j,r}))\quad \forall (i,j,r) \in \mathcal{T}, \Delta_{i,j,r} = 0 \tag{13}$$

$$\mathrm{EndBeforeStart}(a_i, z_{i,j,r})\quad \forall (i,j,r) \in \mathcal{T} \tag{14}$$

$$\mathrm{EndBeforeStart}(z_{i,j,r}, a_j)\quad \forall (i,j,r) \in \mathcal{T} \tag{14b}$$

$$\sum_{\substack{j,k:(j,i,k)\in\mathcal{T},\\ \Delta_{j,i,k}>0,\ k=r}} \mathrm{HeightAtStart}(z_{j,i,k}, \mathrm{Pulse}(z_{j,i,k}, (0, U_{j,i,k}))) = f_{j,i,k}$$
$$\forall i \in \mathcal{A} \setminus \{0\},\ \forall r \in \mathcal{R},\ \Delta_{i,j,r} > 0 \tag{15}$$

$$\sum_{j,k:(j,i,k)\in\mathcal{T},\ k=r} f_{j,i,k} = \mathbf{Q}_{i,r}\quad \forall i \in \mathcal{A} \setminus \{0\},\ \forall r \in \mathcal{R} \tag{16}$$

$$\sum_{\substack{j,k:(i,j,k)\in\mathcal{T},\\ \Delta_{i,j,k}>0,\ k=r}} \mathrm{HeightAtStart}(z_{i,j,k}, \mathrm{Pulse}(z_{i,j,k}, (0, U_{i,j,k}))) = f_{i,j,k}$$
$$\forall i \in \mathcal{A} \setminus \{|\mathcal{A}| - 1\},\ \forall r \in \mathcal{R},\ \Delta_{i,j,r} > 0 \tag{17}$$

$$\sum_{j,k:(i,j,k)\in\mathcal{T},\ k=r} f_{i,j,k} = \mathbf{Q}_{i,r}\quad \forall i \in \mathcal{A} \setminus \{|\mathcal{A}| - 1\},\ \forall r \in \mathcal{R} \tag{18}$$

$$\sum_{(i,j,k)\in\mathcal{T},\ \Delta_{i,j,k}>0,\ k=r} \text{PULSE}(z_{i,j,k},(0,U_{i,j,k}))\ +$$

$$\sum_{i,k:\mathbf{Q},\ \mathbf{Q}_{i,k}>0,\ k=r} \text{PULSE}(a_i,\mathbf{Q}_{i,k}) \le C_r \qquad \forall r \in \mathcal{R} \qquad (19)$$

$$\text{STARTBEFORESTART}(z_{i,j,r}, z'_{i,j,r,r'})$$

$$\forall (i,j,r) \in \mathcal{T},\ \Delta_{i,j,r} > 0,\ r' \in \mathcal{R}',\ \mathbf{M}_{r,r'} > 0,\ \Delta'_{i,j,r'} > 0 \qquad (20)$$

$$\text{ENDBEFOREEND}(z'_{i,j,r,r'}, z_{i,j,r}) \qquad (20b)$$

$$\forall (i,j,r) \in \mathcal{T},\ \Delta_{i,j,r} > 0,\ r' \in \mathcal{R}',\ \mathbf{M}_{r,r'} > 0,\ \Delta'_{i,j,r'} > 0$$

$$f_{i,j,r} \cdot \mathbf{M}_{r,r'} = f'_{i,j,r,r'} \qquad \forall (i,j,r) \in \mathcal{T},\ \Delta_{i,j,r} > 0,\ \forall r' \in \mathcal{R}',\ \Delta'_{i,j,r'} > 0 \qquad (21)$$

$$\text{HEIGHTATSTART}(z'_{i,j,r,r'}, \text{PULSE}(z'_{i,j,r,r'},(0,U_{i,j,r} \cdot \mathbf{M}_{r,r'}))) = f'_{i,j,r,r'}$$

$$\forall (i,j,r) \in \mathcal{T},\ \Delta_{i,j,r} > 0,\ \forall r' \in \mathcal{R}',\ \Delta'_{i,j,r'} > 0 \qquad (22)$$

$$\sum_{(i,j,r)\in\mathcal{T}} \text{PULSE}(z'_{i,j,r,r'},(0,U_{i,j,r} \cdot \mathbf{M}_{r,r'})) \le C'_{r'} \qquad \forall r' \in \mathcal{R}' \qquad (23)$$

Constraint (11) defines the objective. Constraint (12) enforces the activity precedences. For transfers of zero length, Constraint (13) ensures that the optional interval is still present. Constraints (14) and (14b) place each optional transfer interval between its corresponding activities in the schedule. Constraints (16) and (18) ensure the correct flow of the primary resource into and out of the activity, while constraints (15) and (17) connect $f_{i,j,r}$ variables to their respective interval variable PULSE heights. Constraint (19) enforces the capacities of primary resources. Constraints (20) and (20b) ensure that each optional interval that represents secondary transfer is correctly placed within the primary one. Constraint (21) ensures enough secondary resources for each primary transfer. Constraint (22) assigns the pulse height to the variable $f'_{i,j,r,r'}$. Constraint (23) enforces the capacity of secondary resources. Because transfer lengths (interval variables) and transferred amounts (integer variables) are represented explicitly, no postprocessing is needed to extract the solution, unlike in the ILP model.

Note that CP provides a large set of modeling constructs, so several alternative formulations are possible. We tested two additional formulations: (i) using a SPAN constraint over all optional transfers preceding the activity while enforcing ENDBEFORESTART on the resulting spanning interval variable instead of the individual optional interval, and (ii) using STEPATEND on optional transfer intervals together with STEPATSTART on the successor activity to enforce sufficient incoming flow. However, both formulations performed slightly worse, so we do not include them here.

3.3 Warm-Start

Our experiments showed that the models proposed in Subsects. 3.1 and 3.2 require considerable time to obtain a reasonably good initial solution. To mitigate this, we introduce a simple warm-start heuristic: a randomized greedy procedure executed repeatedly for a fixed number of iterations, returning the best solution obtained. The pseudocode is provided in Algorithm 2.

The main algorithm loop (lines 3 to 18) generates one solution per random seed. Within each iteration, the schedule loop (lines 6 to 17) incrementally constructs a feasible schedule by assigning activities to the existing schedule. This loop is iterated over discrete time steps for simplicity. At each time step, the available secondary resource capacities (line 7) and the set of activities ready for scheduling (line 8) are updated based on precedences and completed activities. An activity is selected from a randomly permuted list of ready activities (line 9), ensuring variation between runs. The chosen activity is attempted for scheduling (line 11); if the required resources are available, it is scheduled and removed from the remaining set, otherwise it is reconsidered in the next time step.

Although the heuristic is simple, it is fast and provides sufficiently good initial solutions for both the ILP and the CP models. For both models, all variables used are assigned values based on the heuristic solution. Finally, note that in the absence of resource capacities and transfers, the remaining problem would be similar to the topological ordering problem. In such a case, the heuristic yields an optimal solution, and its runtime could be optimized to be linear.

Algorithm 2. Randomized Greedy Heuristic

```
 1: function RANDGREEDYHEURISTIC(A, P, E, R, C, Q, Δ, M, C', Δ', random_tries)
 2:     best_solution ← ∅
 3:     for run_idx ← 0 to random_tries − 1 do          ▷ main loop (each run new seed)
 4:         state ← INITIALIZERUN(A, P, E, R, C, Q)      ▷ holds all state information
 5:         remaining_activities ← A \ {0}
 6:         while remaining_activities ≠ ∅ do                          ▷ schedule loop
 7:             UPDATEAVAILABLESECONDARYRESOURCECAPACITIES(state)
 8:             UPDATEREADYACTIVITYSET(state, remaining_activities)
 9:             next_ready ← ∅
10:             for all i ∈ SHUFFLE(state.ready_activities) do
11:                 success ← TRYSCHEDULEACTIVITY(state, i, Δ, M, C', Δ')
12:                 if success then
13:                     remaining_activities ← remaining_activities \ {i}
14:                 else
15:                     next_ready ← next_ready ∪ {i}
16:             state.ready_activities ← next_ready
17:             state.time ← state.time + 1
18:         UPDATEBESTSOLUTION(state)
19:     return best_solution
```

3.4   Multi-phase Procedure

To further improve performance beyond warm-started models, we propose a multi-phase procedure that combines the greedy heuristic from Subsect. 3.3 with two mathematical models suitable for different subproblems: the RCPSP CP model for initial activity placement and the simple ILP model for initial feasible flow assignment. The steps of the multi-phase procedure are the following.

1. The classical RCPSP CP model proposed in [13] is used to solve the instance without considering $\Delta, \Delta', \mathcal{R}', C'$ or $\mathbf{M}$.
2. Based on the solution from step 1, the partial order of all activities $\mathcal{P}$ is calculated. Using $\mathcal{P}$, additional precedences are created and merged with $\mathcal{E}$ to obtain an extended set of heuristic precedences $\mathcal{E}'$. A set of inconsistent resource transfers with $\mathcal{P}$ called $\mathcal{I}$ is created. A set of time delays τ between each pair of activities in the solution from step 1 is calculated.
3. A complete RCPSPTT-2I instance is solved using the randomized greedy heuristic proposed in Subsect. 3.3 to obtain an initial upper bound.
4. The vectors Earliest Start (ES), Latest Start (LS), Earliest Finish (EF) and Latest Finish (LF) are calculated using the upper bound obtained in step 3.
5. A new set of allowed transfers called $\mathcal{T}'$ is constructed based on $\mathcal{E}'$, ES, LS, EF and LF while excluding transfers in $\mathcal{I}$. This greatly reduces the number of possible flow transfer assignments, making the problem much easier.
6. The flow assignment $\mathcal{F}$ based on $\mathcal{T}'$ is then calculated using the model proposed below. Note that this subproblem is considerably easier, as acyclicity no longer needs to be enforced explicitly—$\mathcal{E}'$ and $\mathcal{I}$ already eliminate all cyclic possibilities.
7. Based on $\mathcal{F}$ and the activity start times computed in step 1, warm-start is produced by right-shifting activities. Activities are processed in order of their start times, and any activity lacking sufficient resources at its scheduled start is shifted to the right along with its successors and associated transfers. (Pseudocode omitted due to space limitations.)
8. The vectors ES, LS, EF, and LF are updated based on warm-start in step 7.
9. The RCPSPTT-2I model (ILP or CP) is then solved using the warm-start from step 7 together with the vectors ES, LS, EF, and LF from step 8. A lower bound is obtained during the solve and compared with the schedule length from step 1; the larger of the two is taken as the final lower bound.

The variables of the flow model proposed in step 6 are as follows.

- $f_{i,j,r}$, $\forall (i,j,r) \in \mathcal{T}'$: Integer variable representing the amount of resource flow r between activities i and j, with a lower bound equal to 0 and an upper bound equal to $U_{i,j,r}$.
- $y_{i,j,r}$, $\forall (i,j,r) \in \mathcal{T}'$: Binary variable that represents if there exists a non-zero resource flow r between activities i and j.

The model constraints are as follows.

$$\min \sum_{(i,j,r)\in \mathcal{T}'} y_{i,j,r} \cdot \max\{0, \Delta_{i,j,r} - \tau_{i,j}\} \qquad (24)$$

$$\sum_{\substack{j:(j,i,k)\in\mathcal{T}',\\ \Delta_{j,i,k}>0,\ k=r}} f_{j,i,k} = \mathbf{Q}_{i,r} \quad \forall i \in \{0,\dots,|\mathcal{A}|-1\}, \forall r \in \{0,\dots,|\mathcal{R}|-1\} \qquad (25)$$

$$\sum_{\substack{j:(i,j,k)\in\mathcal{T}',\\ \Delta_{i,j,k}>0,\ k=r}} f_{i,j,k} = \mathbf{Q}_{i,r} \quad \forall i \in \{0,\dots,|\mathcal{A}|-1\}, \forall r \in \{0,\dots,|\mathcal{R}|-1\} \qquad (26)$$

$$f_{i,j,r} \leq y_{i,j,r} \cdot U_{i,j,r} \quad \forall(i,j,r) \in \mathcal{T}' \qquad (27)$$

The objective in Eq. (24) minimizes the positive excess of the transfer length over the delay τ between each pair of activities in solution from step 1. Thus, a transfer of resource r from activity i to j is penalized only when $\Delta_{i,j,r} > \tau_{i,j}$. If $\Delta_{i,j,r} \leq \tau_{i,j}$, the transfer is effectively "free", as the existing delay guarantees that the activity j will not have to be right-shifted. Several other alternative objective functions were tested, but this formulation performed the best. Equations (25) and (26) ensure correct flow into and out of the activity. Equation (27) links flow variables with their existence indicators, forcing the latter to be true only when a positive flow is assigned. Finally, note that the correct precedences are already enforced by the restricted set of available transfers $\mathcal{T}'$.

4   Experiments

For evaluation purposes, we created four datasets with different instance sizes (30, 60, 90 and 120 activities), available at [5]. These datasets extend the datasets from [23], which extend the well-known PSPLib [9] for RCPSP. Each instance of [23] was modified in the following way.

- The set $\mathcal{R}'$ is generated by drawing a random number of secondary resources between 1 and $|\mathcal{R}|$.
- The matrix $\mathbf{M}$ is created by randomly assigning one supporting secondary resource to each primary resource, while the required amount of secondary resource per unit of supported primary resource is 1, 2, or 4. Please note that integers are used due to modeling limitations of CP but can be arbitrarily rescaled after solving.
- To ensure that every activity can be executed, each secondary resource capacity $C'_{r'}$ was calculated as
  $\max\{rand(\min\{C\}, \max\{C\}), \min_{i\in\mathcal{A}}\{\sum_{r\in\mathcal{R},\ r'\in\mathcal{R}'} \mathbf{Q}_{i,r} \cdot \mathbf{M}_{r,r'}\}\}$.
- The matrix Δ' is generated randomly, with the constraint that $\forall(i,j,r) \in \mathcal{T}, \Delta'_{i,j,r'} \leq \Delta_{i,j,r}$. Note that $\Delta_{0,i,r} = 0 \wedge \Delta_{i,|\mathcal{A}|,r} = 0$ and therefore $\Delta'_{0,i,r'} = 0 \wedge \Delta'_{i,|\mathcal{A}|,r'} = 0$, $\forall i \in \mathcal{A}, \forall r \in \mathcal{R}, \forall r' \in \mathcal{R}'$.

For each dataset of different activity sizes, we evaluated several methods and configurations. We tested 2 ILP models: a more effective flow-based model

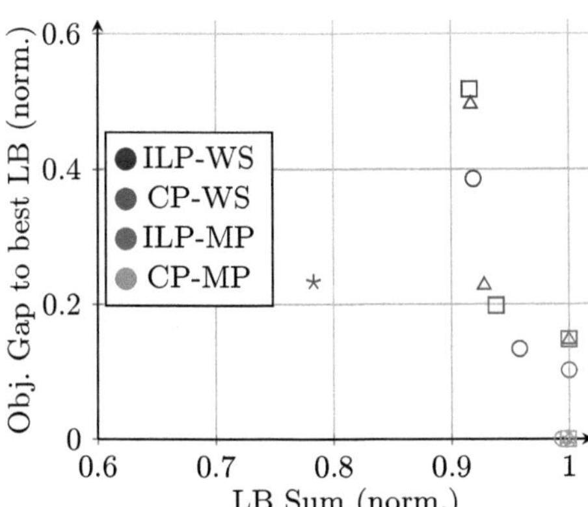

Fig. 3. Graph comparing lower bound and objective qualities. Colors denote methods, shapes different datasets.

Table 1. Objective gap to lower bound, relative to the best method

	○	□	△	⋆
Approach	j30	j60	j90	j120
ILP-WS	0.39	0.52	0.50	–
CP-WS	0.13	0.20	0.23	0.23
ILP-MP	0.10	0.15	0.15	–
CP-MP	0.00	0.00	0.00	0.00

Table 2. Lower bound sum relative to the best method.

Approach	j30	j60	j90	j120
ILP-WS	0.92	0.92	0.92	–
CP-WS	0.96	0.94	0.93	0.78
ILP-MP	1.00	1.00	1.00	–
CP-MP	0.99	1.00	1.00	1.00

detailed in Subsect. 3.1 and a time-indexed model, extending the RCPSPTT model of [8]. We also evaluated 3 variants of the CP model, but only consider the best version detailed in Subsect. 3.2) in the evaluation. As discussed in Subsect. 3.3, warm-start significantly improves all models. We report results only for the best warm-started models, denoting the warm-started ILP model as ILP-WS and the CP model as CP-WS. Finally, we evaluated the multi-phase procedure using both the ILP and CP models as the final step, denoted ILP-MP and CP-MP, respectively.

All tests were executed on a single thread of an AMD Ryzen 7700X with 64 GB RAM with a 10-minute time limit per instance. We tested Gurobi 12.0.3 and CPLEX 22.1.1 for the ILP models; Gurobi performed slightly better, so all reported ILP results use it. For CP models, we used IBM CPOptimizer 22.1.1.

In Fig. 3, we compare all methods across all datasets of different instance sizes; the symbols used for each dataset are listed in Table 1. Table 1 reports each method's objective-sum gap to the best lower bound sum normalized by the best method, while Table 2 reports each method's lower bound sum normalized by the best lower bound sum overall. Note that the ILP-based methods do not scale well to 120-activity instances with several models failing to build within the 10-minute time limit, so these results are omitted. Results on remaining datasets indicate that CP-MP is clearly the best method, with gaps roughly 15% smaller than the next best method ILP-MP, demonstrating that CP is a much more effective paradigm for this problem. Furthermore, CP-MP finds optimal solutions in more than a third of the 30-activity instances, while a separate CP model fails to find even a feasible solution for some of them.

5   Conclusion

This work introduced the Resource-Constrained Project Scheduling Problem with Transfer Times using Secondary Resources with Instant Self-transfers (RCPSPTT-2I), a generalization of RCPSPTT [23] that incorporates secondary resources facilitating primary-resource transfers between activities. Practical applications of RCPSPTT-2I were illustrated, and it was shown that the exact methods in the literature cannot handle similar scenarios effectively. Two exact formulations addressing the problem were developed: a flow-based ILP and a CP model with optional intervals for primary and secondary transfers; with the CP model performing noticeably better. To improve overall performance, we introduced warm-started models using a randomized greedy heuristic (ILP-WS and CP-WS) and a multi-phase procedure (ILP-MP and CP-MP) that separates activity ordering, flow construction, and final refinement via a monolithic ILP or CP model, respectively. While models without warmstart struggle to find feasible solutions for some of the 30-activity instances in 10-minute time limit, the multi-phase procedure CP-MP solves all of them, finding optimal solutions for more than one third of the instances. Both CP-WS and CP-MP scale to instances of 120 activities (with CP-MP showing about a 20% smaller average gap to the lower bound), demonstrating that RCPSPTT-2I can be handled effectively despite its added complexity. Future work may consider multi-project or multi-mode settings or more powerful heuristics for the warm-start.

Acknowledgments. This work was co-funded by the European Union under the ROBOPROX project (reg. no. CZ.02.01.01/00/22_008/0004590), by the Grant Agency of the Czech Republic under Project GACR 25-17904S and by the Czech Technical University in Prague, grant No. SGS25/144/OHK3/3T/13. This work benefited from ANITI, funded by the France 2030 program under Grant Agreement No. ANR-23-IACL-0002.

Disclosure of Interests. The authors have no competing interests to declare that they are relevant to the content of this article.

References

1. Adhau, S., Mittal, M., Mittal, A.: A multi-agent system for decentralized multi-project scheduling with resource transfers. Int. J. Prod. Econ. **146**(2), 646–661 (2013). https://doi.org/10.1016/j.ijpe.2013.08.013
2. Afshar-Nadjafi, B., Majlesi, M.: Resource constrained project scheduling problem with setup times after preemptive processes. Comput. Chem. Eng. **69**, 16–25 (2014). https://doi.org/10.1016/j.compchemeng.2014.06.012
3. Artigues, C., Hartmann, S., Vanhoucke, M.: Fifty years of research on resource-constrained project scheduling explored from different perspectives. Eur. J. Oper. Res. **328**(2), 367–389 (2026). https://doi.org/10.1016/j.ejor.2025.03.024

4. Gnägi, M., Trautmann, N.: A continuous-time mixed-binary linear programming formulation for the multi-site resource-constrained project scheduling problem. In: 2019 IEEE International Conference on Industrial Engineering and Engineering Management (IEEM), pp. 611–614 (2019). https://doi.org/10.1109/IEEM44572.2019.8978811

5. Heinz, V.: RCPSPTT-2I instance sets. https://gitlab.com/vilem_heinz/rcpsptt-2i-instances

6. Heinz, V., Novák, A., Vlk, M., Hanzálek, Z.: Constraint programming and constructive heuristics for parallel machine scheduling with sequence-dependent setups and common servers. Comput. Ind. Eng. **172**, 108586 (2022). https://doi.org/10.1016/j.cie.2022.108586

7. Heinz, V., Vilím, P., Hanzálek, Z.: Reinforcement learning for search tree size minimization in constraint programming: new results on scheduling benchmarks. Comput. Ind. Eng. **209**, 111413 (2025). https://doi.org/10.1016/j.cie.2025.111413

8. Kadri, R.L., Boctor, F.F.: An efficient genetic algorithm to solve the resource-constrained project scheduling problem with transfer times: the single mode case. Eur. J. Oper. Res. **265**(2), 454–462 (2018). https://doi.org/10.1016/j.ejor.2017.07.027

9. Kolisch, R., Sprecher, A.: PSPLIB - a project scheduling problem library: OR software - ORSEP operations research software exchange program. Eur. J. Oper. Res. **96**(1), 205–216 (1997). https://doi.org/10.1016/S0377-2217(96)00170-1

10. Koné, O., Artigues, C., Lopez, P., Mongeau, M.: Event-based MILP models for resource-constrained project scheduling problems. Comput. Oper. Res. **38**(1), 3–13 (2011). https://doi.org/10.1016/j.cor.2009.12.011, https://www.sciencedirect.com/science/article/pii/S0305054809003360, project Management and Scheduling

11. Krüger, D., Scholl, A.: Managing and modelling general resource transfers in (multi-)project scheduling. OR Spectrum **32**(2), 369–394 (2010). https://doi.org/10.1007/s00291-008-0144-5

12. Krüger, D., Scholl, A.: A heuristic solution framework for the resource constrained (multi-)project scheduling problem with sequence-dependent transfer times. Eur. J. Oper. Res. **197**(2), 492–508 (2009). https://doi.org/10.1016/j.ejor.2008.07.036

13. Laborie, P.: Industrial project and machine scheduling with constraint programming. In: Conference Presentation at PMS 2021 (2021). https://doi.org/10.13140/RG.2.2.30470.70726

14. Laurent, A., Deroussi, L., Grangeon, N., Norre, S.: A new extension of the RCPSP in a multi-site context: mathematical model and metaheuristics. Comput. Ind. Eng. **112**, 634–644 (2017). https://doi.org/10.1016/j.cie.2017.07.028

15. Liu, Y., Jin, S., Zhou, J., Hu, Q.: A branch-and-bound algorithm for the unit-capacity resource constrained project scheduling problem with transfer times. Comput. Oper. Res. **151**, 106097 (2023). https://doi.org/10.1016/j.cor.2022.106097

16. Liu, Y., Zhou, J., Lim, A., Hu, Q.: Lower bounds and heuristics for the unit-capacity resource constrained project scheduling problem with transfer times. Comput. Ind. Eng. **161**, 107605 (2021). https://doi.org/10.1016/j.cie.2021.107605

17. Liu, Y., Zhou, J., Lim, A., Hu, Q.: A tree search heuristic for the resource constrained project scheduling problem with transfer times. Eur. J. Oper. Res. **304**(3), 939–951 (2023). https://doi.org/10.1016/j.ejor.2022.05.014

18. Ma, Z., He, Z., Wang, N.: The proactive resource-constrained project scheduling problem with resource transfer times. In: 2019 International Conference on Industrial Engineering and Systems Management (IESM), pp. 1–6 (2019). https://doi.org/10.1109/IESM45758.2019.8948089

19. Mika, M., Waligóra, G., Weglarz, J.: Modelling Setup Times in Project Scheduling. In: Józefowska, J., Weglarz, J. (eds.) Perspectives in Modern Project Scheduling. International Series in Operations Research & Management Science, vol. 92, pp. 131–163. Springer, Boston, MA (2006). https://doi.org/10.1007/978-0-387-33768-5_6

20. Mittal, M.L., Kanda, A.: Scheduling of multiple projects with resource transfers. Int. J. Math. Model. Oper. Res. **1**(3), 303–325 (2009). https://doi.org/10.1504/IJMOR.2009.024288

21. Okubo, H., Miyamoto, T., Yoshida, S., Mori, K., Kitamura, S., Izui, Y.: Project scheduling under partially renewable resources and resource consumption during setup operations. Comput. Ind. Eng. **83**, 91–99 (2015). https://doi.org/10.1016/j.cie.2015.02.006

22. Poppenborg, J.: Modeling and optimizing the evacuation of hospitals based on the rcpsp with resource transfers. Doctor rerum naturalium (dr. rer. nat.), Clausthal University of Technology, Clausthal-Zellerfeld, Germany (2014), date of oral examination: 11 Mar 2014

23. Poppenborg, J., Knust, S.: A flow-based tabu search algorithm for the RCPSP with transfer times. OR Spectrum **38**(2), 305–334 (2016). https://doi.org/10.1007/s00291-015-0402-2

24. Quilliot, A., Toussaint, H.: Resource constrained project scheduling with transportation delays. IFAC Proc. Volumes **45**(6), 1481–1486 (2012). https://doi.org/10.3182/20120523-3-RO-2023.00061

25. Ren, Y., Lu, Z., Liu, X.: A branch-and-bound embedded genetic algorithm for resource-constrained project scheduling problem with resource transfer time of aircraft moving assembly line. Optim. Lett. **14**(8), 2161–2195 (2020)

26. ScheduleOpt: Optalcp (2023). https://scheduleopt.com/, optalCP's solver landing page

27. Stiti, C., Driss, O.B.: A new approach for the multi-site resource-constrained project scheduling problem. Procedia Comput. Sci. **164**, 478–484 (2019). https://doi.org/10.1016/j.procs.2019.12.209, https://www.sciencedirect.com/science/article/pii/S1877050919322562, cENTERIS 2019 - International Conference on ENTERprise Information Systems / ProjMAN 2019 - International Conference on Project MANagement / HCist 2019 - International Conference on Health and Social Care Information Systems and Technologies, CENTERIS/ProjMAN/HCist 2019

28. Suresh, M., Dutta, P., Jain, K.: Resource constrained multi-project scheduling problem with resource transfer times. Asia Pac. J. Oper. Res. **32**(6) (2015). https://doi.org/10.1142/S0217595915500487

29. Vanhoucke, M., Coelho, J.: Resource-constrained project scheduling with activity splitting and setup times. Comput. Oper. Res. **109**, 230–249 (2019). https://doi.org/10.1016/j.cor.2019.05.004

30. Yang, K.K., Sum, C.C.: A comparison of resource allocation and activity scheduling rules in a dynamic multi-project environment. J. Oper. Manag. **11**(2), 207–218 (1993). https://doi.org/10.1016/0272-6963(93)90023-I

31. Zhang, J., Liu, W., Liu, W.: An efficient genetic algorithm for decentralized multi-project scheduling with resource transfers. J. Ind. Manage. Optim. **18**(1), 1–24 (2022). https://doi.org/10.3934/jimo.2020140

No-Opponent-Cycle Propagators for Solving Parity Games

Gonzalo Hernandez[1,3]([✉]), Julian Garcia[1], Julian Gutierrez[2], and Guido Tack[1]

[1] Monash University, Melbourne, Australia
{gonzalo.hernandez,julian.garcia,guido.tack}@monash.edu
[2] University of Sussex, Brighton and Hove, UK
J.Gutierrez@sussex.ac.uk
[3] Universidad de Nariño, Pasto, Colombia
gonzalo.hernandez@udenar.edu.co

Abstract. Solving parity games is a core problem in formal verification, specifically when using model checking and synthesis methods, which have several industrial-scale applications. In this paper, we propose a constraint-based approach for parity games, built upon a new propagation algorithm that eliminates opponent cycles from the game graph. This approach results in an efficient method that exploits the duality between the players in a parity game. Our implementation within a Lazy Clause Generation framework outperforms all existing constraint-based methods for parity games and performs competitively against the most specialized non-CP-based algorithms—often matching and sometimes exceeding their performance—while retaining the flexibility inherent to general constraint-based approaches. Since new constraints can be easily added, our method provides a flexible framework for refining existing problem formulations to search for preferred solutions by satisfying additional constraints. This feature can also be used to solve parity games where additional quantitative constraints (*e.g.*, cost or resource-bounded requirements) need to be satisfied. We present the theoretical foundations of the approach, an experimental evaluation, and a discussion of the results, including an analysis of different versions of the new propagator.

Keywords: Constraint programming · Parity games · Verification

1 Introduction

Parity games are two-player zero-sum infinite-duration games played on a labeled directed graph, where vertices represent states of the players in the game and edges represent their possible available actions at every state. Solving these games is central to several techniques in automated formal verification, in particular for solving model checking and automated synthesis problems.

Since parity games are zero-sum, perfect-information games, solving them amounts to finding which player (usually called Player Even, or $\bigcirc$, and Player

T. Guns (Ed.): CPAIOR 2026, LNCS 16595, pp. 225–242, 2026.
https://doi.org/10.1007/978-3-032-27242-3_14

Odd, or □) has a winning strategy from a given starting vertex in the graph. It is known that deciding the winner in parity games is an NP ∩ co-NP problem [22], but it is not known whether or not they can be solved in polynomial time; a problem that has remained open for nearly 50 years. Due to the importance of parity games in formal verification, several algorithms have been developed [1,3,21,23,32], all of which suffer from exponential running time limitations for specific classes of games. One such approach involves reducing the solution of a parity game to the solution of a SAT instance representing the game. The generated SAT instance is satisfiable if and only if Player Even has a winning strategy in the original game. This idea has a lot of potential given the advances in SAT-solving technology, but, as we will show later, it suffers from poor scalability due to the large size of the generated CNF formula for a given game.

In this paper, we address these scalability issues by developing a specialised propagator that takes into account the structure of parity games, resulting in significant improvements in performance and modelling power. In particular, we show that our method outperforms all other state-of-the-art constraint-based solvers. Specifically, we propose replacing the SAT-based encoding used to forbid cycles corresponding to Player Odd wins in the game graph [18] with a set of efficient global constraint propagators; this significantly reduces the encoding size of the problem. Two of the proposed propagators are based on depth-first search, and a third is a checker that analyses the strongly connected components (SCCs) of the game graph. These propagators filter edges that would close winning cycles for a selected opponent player and generate explanations for each filtering step, enabling their integration into a Lazy Clause Generation solver [29].

We show that our implementation not only outperforms existing SAT and CP-based approaches but also achieves performance comparable to that of state-of-the-art specialized non-CP-based algorithms, often matching or exceeding their results. This work contributes to establishing the CP framework as a practical tool for solving both general and specialized classes of parity games.

The remainder of this paper is organized as follows. We first present the theoretical foundations and necessary definitions used in further sections, and then introduce our proposed approach – including the design and implementation of the No-Opponent-Cycle propagators. After that, we provide different experimental results, a comparative analysis against state-of-the-art solvers, and present the extension of our method with additional constraints. At the end of the paper, we focus on the presentation of a number of concluding remarks, a discussion of our main findings, and some proposals for potential future research directions.

2   Preliminaries and Related Work

This section introduces parity games and overviews the existing translations into Boolean Satisfiability (SAT) and Satisfiability Modulo Theories (SMT).

2.1   Parity Games

Parity games are two-player zero-sum games played on a labeled directed graph, called the *arena*, where two players, *Even* and *Odd* (denoted $\bigcirc$ or 0, and $\square$ or 1, respectively), move a token along adjacent vertices depending on which player controls the current vertex. This process generates a path of states visited by the token. In a parity game, each path is winning for either Player $\bigcirc$ or Player $\square$. The goal of each player is to enforce a winning path in their favour. We now formalize parity games, including the definition of when a given path in the graph is winning for either player.

Definition 1 ([7]). *An arena G is a finite directed graph $G = (V = V_{\bigcirc} \cup V_{\square}, E, \Omega)$, where: V is a finite set of vertices, partitioned into $V_{\bigcirc}$ (controlled by Player $\bigcirc$) and $V_{\square}$ (controlled by Player $\square$); $E \subseteq V \times V$ is a set of directed edges such that every vertex has at least one outgoing edge; and $\Omega : V \to \mathbb{N}$ is a labeling function that assigns to each vertex a non-negative integer, called its priority. For any vertex $v \in V$, we write $vE = \{w \in V \mid (v,w) \in E\}$ for its set of successors (adjacent to outgoing edges from v), and $Ev = \{u \in V \mid (u,v) \in E\}$ for its set of predecessors (adjacent to incoming edges to v).*

The game starts with a token at an initial vertex $v_0 \in V$. Player Even is responsible for moving the token when it is on a vertex in $V_{\bigcirc}$, and Player Odd is responsible for moving the token when it is on a vertex in $V_{\square}$. Because every node is required to have at least one outgoing edge, the token travels along the graph generating a path of infinite length. The winner is determined by the highest priority that is visited by the token infinitely often. Figure 1 illustrates an instance of a parity game arena.

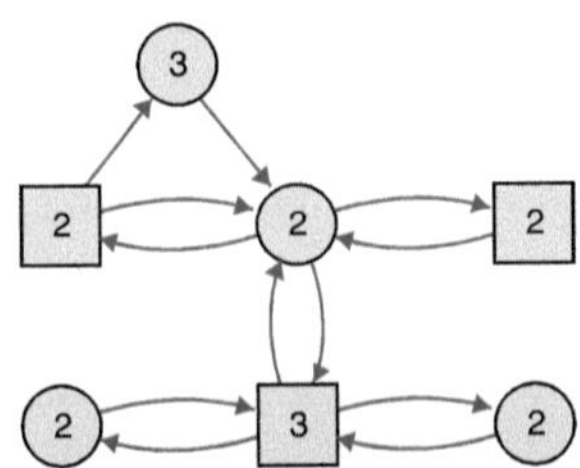

Fig. 1. A parity game with 7 vertices and 12 edges. Each vertex carries a priority from the set $\{2, 3\}$. Vertices controlled by Player $\bigcirc$ are shown as circles, while vertices controlled by Player $\square$ are shown as squares.

Strategies, Plays, and Winning Conditions

A memoryless or *positional strategy* for player Even (player Odd) is a function $f_{\bigcirc} : V_{\bigcirc} \to V$ ($f_{\square} : V_{\square} \to V$) that dictates how the player should move the token when it is in one of the vertices the player controls. Every pair of memoryless strategies, one for each player, uniquely determines an infinite path π, called a *play*, in the game. Accordingly, $\Omega\pi$ is an infinite sequence of priorities. Let π_k be the $(k+1)$-th element in such a sequence of vertices, that is, in such a path. Because arenas are finite, such an infinite sequence of priorities has a (possibly empty) finite prefix $\Omega(\pi_0)\Omega(\pi_1)\ldots\Omega(\pi_h)$ of length h and an infinite tail $\Omega(\pi_{h+1})\Omega(\pi_{h+2})\ldots$ of priorities that are seen infinitely often (given that the graph is finite and the strategies are memoryless/positional). We say that a play π is *winning* for player $\bigcirc$ (player $\square$) if the highest priority that appears infinitely often in its associated sequence of priorities $\Omega\pi$ is even (odd).

Given an initial vertex v_0, a strategy is said to be winning from v_0 for player $\bigcirc$ (player $\square$) if it only generates winning plays for player $\bigcirc$ (player $\square$) starting at v_0 for all strategies of the other player. It is known that for every vertex $v \in V$ in a parity game, one of the two players always has a memoryless winning strategy from that vertex v [32]. Solving a parity game means deciding, for every vertex $v \in V$, which player has a memoryless winning strategy from that vertex.

2.2   Solving Parity Games Using SAT-Modulo-Theories

Besides dedicated methods for parity games, algorithms have been developed to *encode* a given game into constraint-based formalisms, in order to apply off-the-shelf solvers. One such approach [18] uses Satisfiability Modulo Theories (SMT) solvers to do so. The encoding is based on the following observation: A game is won by player $\bigcirc$ if, starting at v_0, every $v \in V_\bigcirc$ has at least one outgoing edge leading to an infinite path where the highest priority is even. Furthermore, every $v \in V_\square$ must have all of its outgoing edges leading to infinite paths with the same condition. The idea, then, is to construct a CNF formula that is satisfiable if and only if the above condition for winning holds. Determining satisfiability is therefore equivalent to proving that Player $\bigcirc$ has a winning strategy, while unsatisfiability proves the same but for Player $\square$. This approach introduces Boolean variables for the vertices and edges of the arena: S_v for every $v \in V$ and $T_{v,w}$ for every $(v,w) \in E$. Variables S_v and $T_{v,w}$ represent whether the corresponding vertices and edges lie on a winning path for $\bigcirc$. A CNF formula can then capture the above simple path conditions for winning a game in the following way:

- $\Phi_\exists = \bigwedge_{v \in V_\bigcirc} (S_v \rightarrow \bigvee_{w \in vE} T_{(v,w)})$
- $\Phi_\forall = \bigwedge_{v \in V_\square} (S_v \rightarrow \bigwedge_{w \in vE} T_{(v,w)})$
- $\Phi_V = \bigwedge_{w \in V, w \neq v_0} ((\bigvee_{v \in Ev} T_{(v,w)}) \rightarrow S_w)$

These constraints state that for all vertices controlled by player $\bigcirc$, if a vertex is on the winning path, then at least one outgoing edge must be on the winning path ($\Phi_\exists$); for all vertices controlled by player $\square$, if the vertex is on the winning path (for $\bigcirc$), then all outgoing edges must be on the winning path ($\Phi_\forall$); and for all vertices, if an edge leading to the vertex is on the winning path for player $\bigcirc$, then the vertex itself is on the winning path (Φ_V).

The final condition is slightly more difficult to implement: every *infinite* path must satisfy the parity condition, that is, the highest priority that occurs infinitely often along the path must be even. This can be locally encoded using a *progress measure* [18,21], expressed with additional integer variables. A variable x_p^v is introduced for every $v \in V$ and every $p \in Odd(G)$, corresponding to the set of all odd priorities in the game, defined as $Odd(G) = \{p \in \mathbb{N} \mid p$ *is odd and* $\Omega(v) = p$, *for some* $v \in V\}$. These variables track *progress* along every $t \in Inf(\pi)$, the set of priorities that appear infinitely often in π, ensuring that $x_p^v > x_p^w$ if the priority is odd, or $x_p^v \geq x_p^w$ if the priority is even. Since odd priorities require a strictly increasing sequence of integers, cycles with odd priorities are ruled out by this constraint. On the other hand, for even priorities, only a non-strict inequality is required. The definition of the progress measure constraints is as follows [12,18]:

- $\Phi_A = \bigwedge_{(v,w) \in E} \left(T_{(v,w)} \rightarrow \Psi_{(v,w)} \right)$
- $\Psi_{(v,w)} = \bigwedge_{p \in Odd^{> \Omega(w)}(G)} (x_p^v \geq x_p^w)$, *if* $\Omega(w)$ *is even*
- $\Psi_{(v,w)} = \bigwedge_{p \in Odd^{> \Omega(w)}(G)} (x_p^v \geq x_p^w) \wedge (x_{\Omega(w)}^v > x_{\Omega(w)}^w)$, *if* $\Omega(w)$ *is odd*

These constraints are instances of *difference logic*, which is supported by some SMT solvers. Thus, an SMT solver can determine the satisfiability of $(S_{v_0} \wedge \Phi_\exists \wedge \Phi_\forall \wedge \Phi_V \wedge \Phi_A)$, and, as such, decide whether player $\bigcirc$ has a winning strategy.

The difference logic constraints of the progress measure can also be encoded into pure CNF, which enables the use of modern SAT solvers to decide the winner of a parity game. This was the approach chosen in the paper that introduced the SMT encoding [18].

3 A Constraint-Based Approach Using an Efficient No-Opponent-Cycle Propagator

In this section, we present our approach to solving parity games by encoding them into a Constraint Satisfaction Problem (CSP) with three specialised propagators for the winning condition. Our method closely follows the framework presented in the previous section, determining satisfiability by evaluating whether a game starting at v_0 is won by player $\bigcirc$.

However, while we retain the constraints $\Phi_\exists$, $\Phi_\forall$, and Φ_V to ensure connectivity within the arena, we observe that the Φ_A constraints—based on progress measures—significantly hinder the scalability of SAT-based approaches. In the worst case, the size of the encoding grows cubically with the size of the arena. In practice, although the encoding is not always cubic, it is often still prohibitively large. For example, a Steady Game with 5,000 nodes may generate more than 215 million clauses. Such prohibitive instances are common in practice, even when the set $Odd(G)$ is small.

Our approach therefore replaces the progress measure with a global constraint that avoids the use of the integer variables x_p^v. We also define our model to enable search from both players' perspectives, aiming to determine, efficiently, if the instance is unsatisfiable for either Player $\bigcirc$ or Player $\square$.

3.1 Characterising Winning Strategies Globally

We first translate the game into a CSP. Thus, let $\mathsf{p} \in \{\bigcirc, \square\}$ be a player and $\bar{\mathsf{p}} \in \{\bigcirc, \square\}$, with $\bar{\mathsf{p}} \neq \mathsf{p}$, be its opponent.

Definition 2. *The CSP encoding of the parity game with arena $G = (V, E, \Omega)$, starting vertex v_0 and searching for the perspective of player p is defined as $\mathcal{CSP} = (\mathcal{X}, \mathcal{D}, \mathcal{C})$, where the set of variables is given by $\mathcal{X} = (\mathcal{V}_v)_{v \in V} \cup (\mathcal{E}_e)_{e \in E}$. All of these variables are Boolean, thus, $\mathcal{D} = \{\bot, \top\}$. The set of constraints, denoted by C, is defined as:*

- $\mathcal{C}_{v_0} : \mathcal{V}_{v_0} = \top$
- $\mathcal{C}_\mathsf{p} : \bigwedge_{v \in V_\mathsf{p}} (\mathcal{V}_v \rightarrow (\sum_{w \in vE} \mathcal{E}_{(v,w)} = 1))$
- $\mathcal{C}_{\bar{\mathsf{p}}} : \bigwedge_{v \in V_{\bar{\mathsf{p}}}} (\mathcal{V}_v \rightarrow \bigwedge_{w \in vE} \mathcal{E}_{(v,w)})$
- $\mathcal{C}_E : \bigwedge_{w \in V, w \neq v_0} ((\bigvee_{v \in Ew} \mathcal{E}_{(v,w)}) \rightarrow \mathcal{V}_w)$
- $\mathcal{C}_{NOC} : \bigwedge_{v_0, \ldots, v_k} ((\bigwedge_{i \in 0 \ldots k-1} \mathcal{E}_{(v_i, v_{i+1})}) \rightarrow \bigwedge_{j \in 0 \ldots k-1} (v_j = v_k \rightarrow \max\{\Omega(v_i) \mid i \in j \ldots k\} \bmod 2 \neq \bar{\mathsf{p}}))$

Constraints $\mathcal{C}_{v_0}$, $\mathcal{C}_{\bar{\mathsf{p}}}$ and $\mathcal{C}_E$ directly correspond to the S_{v_0}, $\Phi_\forall$ and Φ_V constraints in the SMT encoding. We have adapted the $\Phi_\exists$ constraint into $\mathcal{C}_\mathsf{p}$ to enforce that *exactly one* outgoing edge must be activated for each vertex of player p. Additionally, we replace the progress measure constraint Φ_A with a global constraint $\mathcal{C}_{NOC}$, where *NOC* stands for No-Opponent-Cycle. This constraint holds iff for every path $v_0, \ldots, v_k$ where $1 \leq k \leq |V|$, if a cycle is formed by $v_j = v_k$ (for any $j < k$), then the parity of the maximum priority value on that cycle must differ from the $\bar{\mathsf{p}}$ value. This constraint forbids any cycle in which the highest priority that occurs infinitely often has the same parity as the opponent. Unlike previous formulations that assumed player $\square$ as the default opponent, our construction generalizes to any arbitrary pair of players p and $\bar{\mathsf{p}}$. This ensures that player p is guaranteed to win if, after ruling out all possible winning strategies for $\bar{\mathsf{p}}$, the CSP remains satisfiable with p as the winner. Let us illustrate the filtering we want to achieve using this global constraint with an example.

Example 1. Figure 2 shows the same arena as in Fig. 1, but with additional labels for the edges and vertices. Searching from the perspective of Player $\mathsf{p} = \bigcirc$ and starting at vertex V_0, after enforcing constraints $\mathcal{C}_{v_0}, \mathcal{C}_\mathsf{p}, \mathcal{C}_{\bar{\mathsf{p}}}, \mathcal{C}_E$ from Definition 2, assume that the solver makes an initial assignment of $\mathcal{V}_0 = \mathcal{V}_1 = \mathcal{V}_2 = \mathcal{E}_0 = \mathcal{E}_1 = \mathcal{E}_2 = \mathcal{E}_3 = \top$ (see Fig. 2a), while the remaining variables are unassigned. At this point, Player $\square$ has a winning strategy because edges $\mathcal{E}_0$, $\mathcal{E}_2$ and $\mathcal{E}_3$ form a cycle where the highest priority occurring infinitely often is 3.

A checker of $\mathcal{C}_{NOC}$ would, in this situation, report unsatisfiability of the problem, triggering a backtrack and a search for an alternative assignment. The solver would continue the search, possibly exploring another assignment, $\mathcal{V}_0 = \mathcal{V}_1 = \mathcal{V}_2 = \mathcal{V}_3 = \mathcal{E}_0 = \mathcal{E}_1 = \mathcal{E}_2 = \mathcal{E}_4 = \mathcal{E}_6 = \top$ (see Fig. 2b). Here, another cycle is formed, but the highest priority occurring infinitely often is 2, which satisfies all constraints, making the problem satisfiable. As a result, starting from V_0, this game is won by Player $\bigcirc$.

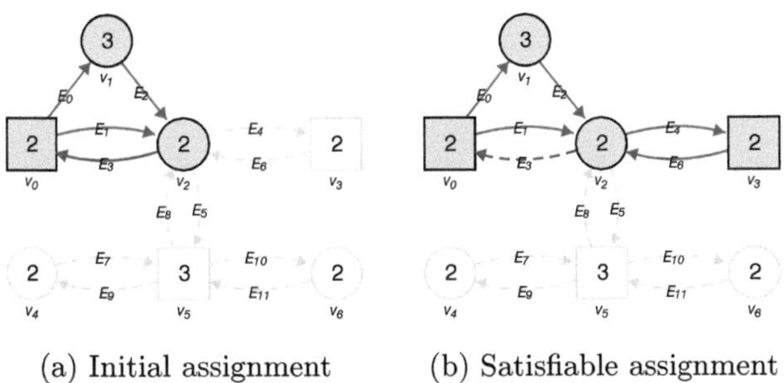

(a) Initial assignment   (b) Satisfiable assignment

Fig. 2. A parity game with vertices $\{V_0 \ldots V_6\}$, edges $\{E_0 \ldots E_{11}\}$, and priorities $\{2, 3\}$. Dashed edges represent unassigned edges; red edges represent edges falsified by the propagator.

Stronger versions of the C_{NOC} constraint can filter out assignments that necessarily lead to cycles with an odd maximum priority. For instance, if the former assignment had been considered first, the constraint could have determined in advance that $\mathcal{E}_3$ *must* be $\bot$, since otherwise an odd cycle would be present.

3.2   A No-Opponent-Cycle Checker

To enforce C_{NOC}, we propose a propagator based on recursively decomposing the graph into Strongly Connected Components (SCCs). By definition, all vertices inside an SCC of size greater than 1 are part of at least one cycle, while single-vertex SCCs may still contain self-loops. Each resulting subgraph is then evaluated: if it contains a cycle in which the highest priority occurring infinitely often favors $\bar{\mathsf{p}}$, the current assignment is rejected. Otherwise, the maximum priorities favouring p are removed from the SCC, and new SCCs are computed on the reduced graph. If, in the end, all remaining single vertices lack both self-loops and $\bar{\mathsf{p}}$ priorities, the assignment is accepted.

Algorithm 1. *NOCChecker*

Input: G
Output: $\top$ is valid; $\bot$ otherwise

 1: **for** $scc \in Tarjan(G)$ **do**
 2:    **if** $|scc| = 1$ **then**
 3:       $v \leftarrow$ the unique vertex in scc
 4:       **if** $priority(v) \bmod 2 = \bar{\mathsf{p}} \wedge vEv \wedge \mathcal{E}[vEv]$ **then return** $\bot$
 5:    **else**
 6:       $bestSet \leftarrow bestVertices(scc)$
 7:       $v \leftarrow$ an arbitrary vertex in $bestSet$
 8:       **if** $priority(v) \bmod 2 = \bar{\mathsf{p}}$ **then return** $\bot$
 9:       $NOCChecker(G \setminus bestSet)$
10: **return** $\top$

Algorithm 1 is implemented recursively, with an argument corresponding to a subgraph induced by the vertices assigned $\top$. Before invoking the function, all variables are checked to ensure they are fixed. The function is invoked as $NOCChecker([\ v \mid \mathcal{V}[v] = \top\ ])$. The current subgraph G is decomposed into SCCs using Tarjan's algorithm [30], so that each SCC is analyzed independently (line 1). In each iteration, the algorithm checks whether the current SCC consists of a single vertex (line 2). If so, it verifies whether the vertex has a self-loop and whether its priority matches that of $\bar{p}$; in such a case, the assignment is rejected (line 4). If the SCC contains more than one vertex, the algorithm identifies the set of vertices with the highest priority (line 6) and selects one arbitrarily to check its parity. If the selected vertex's priority corresponds to $\bar{p}$, the assignment is also rejected (line 8). Otherwise, the vertices in this highest-priority set are removed from the SCC, and the function is recursively called on the resulting subgraph (line 9).

The complexity of the NOC-Checker propagator is polynomial in the size of the input graph. Each call to Tarjan's algorithm computes the SCCs in linear time, i.e., $\mathcal{O}(|V| + |E|)$. In the worst case, the algorithm may recursively remove and reprocess subsets of vertices with maximal priority multiple times. However, each vertex can only be part of an attracting set and removed once per recursive level, ensuring that the total number of Tarjan calls remains linear in the number of vertices. As a result, the overall time complexity of the propagator remains $\mathcal{O}(|V| \cdot (|V| + |E|))$ in the worst case.

3.3   A No-Opponent-Cycle Filter

In addition to NOC-Checker, we also propose a complete propagator based on Depth-first Search (DFS) to explore the arena. The resulting DFS tree captures all possible paths from v_0, and can be used to detect all existing cycles. We define the propagator in the context of a *Lazy Clause Generation* solver [29], to take advantage of the strong performance of this type of solver, in particular for problems that contain a large number of Boolean clauses. Algorithm 2 presents this version of the propagator, which checks for cycles and reports failure if a cycle with highest priority in favour of $\bar{p}$ is detected.

Algorithm 2 is also written in a recursive style. The arguments $path_V$ and $path_E$ represent the currently explored path in the DFS procedure, expressed in terms of both vertices and edges. The parameter v denotes the current vertex, e is the currently explored edge, and $\mathcal{E}$ is the array of Boolean variables associated with the edges of the graph. The No-Opponent-Cycle propagator invokes this function as $NOCEager(\emptyset, \emptyset, start, -1, \mathcal{E}, \top)$, where -1 indicates that no edge has been selected yet from $\mathcal{E}$, as valid edges are indexed with $e \geq 0$.

First, consider the case where the current vertex v is not already in the current path (i.e., the else case in line 7). This indicates that no cycle has been detected so far, and the algorithm recursively explores each outgoing edge from v whose value is not $\bot$, extending the path accordingly. A basic implementation of NOC-Eager would only explore edges for which $\mathcal{E}[e'] = \top$, thus limiting the search to assigned edges. However, the goal of this propagator is to allow the

Algorithm 2. *NOCEager*

Input: $path_V$, $path_E$, v, e, $\mathcal{E}$, *defined*
Output: $\top$ is valid; $\bot$ otherwise

```
 1: if v ∈ pathV then
 2:     iv ← index(v, pathV)
 3:     m ← max{f(v') | v' ∈ pathV[iv :]}
 4:     if m mod 2 = p̄ then
 5:         reason ← clausify(¬pathE[iv :])
 6:         if not (E[e] ← (⊥, reason)) then return ⊥
 7: else if defined then
 8:     for e' ∈ edges(v) where E[e'] ≠ ⊥ do
 9:         pV ← pathV ∪ {v}
10:         pE ← pathE ∪ {e'}
11:         w ← target(e')
12:         def ← E[e'] = ⊤
13:         NOCEager( pV, pE, w, e', E, def )
14: return ⊤
```

exploration of at most one unassigned edge, in order to proactively eliminate potential cycles before they become committed.

In contrast, if the current vertex v has appeared in the path (line 1), this implies that the current edge closes a cycle. The algorithm then determines the highest priority among the vertices in the cycle, and if this priority favors $\bar{p}$ (line 4), the current edge is set to $\bot$. If the edge had already been assigned $\top$, this causes a conflict and triggers backtracking.

Since Lazy Clause Generation (LCG) solvers require a justification for each propagation step (to enable clause learning), all edges in the detected cycle are collected to form a clause stating that at least one must be assigned $\bot$ (line 5). For instance, in the scenario shown in Fig. 2b, the clause ($\neg\mathcal{E}_0 \vee \neg\mathcal{E}_2 \vee \neg\mathcal{E}_4 \vee \neg\mathcal{E}_6 \vee \neg\mathcal{E}_3$) would be added to prevent the opponent cycle.

The complexity NOC-Eager is primarily driven by the DFS traversal over the graph, with additional work performed when a cycle is detected. In the worst case, the algorithm explores every simple path in the graph starting from the initial vertex. This may result in an exponential number of recursive calls in the number of vertices since each branching decision adds a new path to explore. However, in practice, the number of recursive calls is bounded by the number of edges assigned to $\top$ and at most one unassigned edge, significantly reducing the branching factor. Each recursive call performs a constant number of operations, except when a cycle is found (line 1). Nevertheless, because the algorithm may visit multiple paths and cycles, and because each edge can potentially trigger multiple clause generation steps before being fixed, the overall worst-case complexity remains exponential in the size of the graph. This trade-off is justified by the early pruning capabilities enabled by detecting potentially winning opponent cycles before they are fully formed, especially in graphs with many unassigned edges. In typical solver applications, where assignments propagate quickly and many edges are fixed early, the practical runtime of the propagator is significantly better than the theoretical worst case—as demonstrated by the experimental results presented in Sect. 4, where this approach consistently outperforms alternatives in both speed and effectiveness.

3.4 Improved Filtering Strategy for the No-Opponent-Cycle Propagator

In order to mitigate the exponential behavior of NOC-Eager, we have developed a variant that memorizes cycles and can quickly remove some new cycles. To do so, one can use Algorithm 3, which incorporates a parameter *memo* that stores, for each edge, the starting node of an infinite path and the associated highest priority. This information is updated whenever a cycle is found (line 4). If *memo* does not contain information about the next edge e', or a new highest priority was found, then the propagator proceeds recursively as before (lines 16 and 21). Otherwise, it uses the stored information for the new edge (line 4). This results in polynomial worst-case runtime, since the recursion is now bounded by the number of edges and the number of different priorities. However, the memoization only captures cycles with the same starting node. The propagator may therefore miss some cycles, rendering it incomplete. To preserve completeness, this variant is therefore used in conjunction with the NOC-Checker introduced earlier.

For the example of Fig. 2, the updated version of No-Opponent-Cycle stores a tuple $(v = 5, m = 3)$ on edge E_5, indicating that the best priority reachable from this edge is 3 and that the infinite path starts at V_5. This prevents the exploration of the subgraph $\{V_4, V_5, V_6\}$ in future stages. Remembering cycles is particularly effective in games where the number of cycles favoring p exceeds those favoring $\bar{p}$, as demonstrated below in the section on experiments.

The three versions of the No-Opponent-Cycle propagator can be used to solve parity games, with NOC-Memo requiring the NOC-Checker.

3.5 Solving for p and $\bar{p}$ in Parallel

As our experiments will demonstrate, solver performance is heavily dependent on the game's underlying structure. A key observation is the asymmetry in computational cost inherent in game structure: properties that challenge p's solution often simplify $\bar{p}$'s, and vice versa. This problem duality allows us to transform and simplify the solution. By exploiting this, we can avoid a worst-case scenario for one player by solving the game from the opposite perspective. Consequently, solving the game from both perspectives in parallel significantly improves overall efficiency. This acceleration is largely attributed to the No-Opponent-Cycle propagator's capacity to rapidly prune the search space by eliminating $\bar{p}$'s winning options, effectively proving p has no winning strategy. We execute both perspectives concurrently and use the result obtained by the faster computation.

As confirmed by additional experiments [20], we confirm that state-of-the-art methods cannot exploit this duality, demonstrating consistent performance regardless of which player wins.

4   Experiments and Evaluation

To evaluate our approach, we implemented the No-Opponent-Cycle propagators using the Constraint Programming (CP) solver Chuffed (version 0.12.1) [11]. We used MiniZinc [28] to model additional constraints in selected experiments.
For comparison, we employed several external state-of-the-art solvers:

- SAT Solvers (Reduction SAT): We used **CaDiCaL** (v2.1.3) [4], **MiniSat** (v2.2.0) [14], **Kissat** (v4.0.2) [5], and zChaff (v2007.3.12) [27].
- State-of-the-Art (SOTA) Parity Game Solvers: We incorporated Oink (fresh GitHub version) [13], which includes several high-performance algorithms like Priority Promotion (**PP** and **PP+**) and Parys' improvements (**PAR**). We also included the Zielonka Recursive Algorithm (**ZRA**), with an improved attractor strategy as detailed in [1], which offers better performance.

Algorithm 3. *NOCMemo*

Input: $path_V, path_E, v, e, \mathcal{E}, defined, memo$
Output: $\top$ is valid; $\bot$ otherwise

1: **if** $v \in path_V$ **then**
2:     $i_v \leftarrow index(v, path_V)$
3:     $m \leftarrow \max\{f(v') \mid v' \in path_V[i_v :]\}$
4:     $memo[e] \leftarrow (v, m)$
5:     **if** $m \bmod 2 = \bar{\mathsf{p}}$ **then**
6:        $reason \leftarrow clausify(\neg path_E[i_v :])$
7:        **if** not $(\mathcal{E}[e] \leftarrow (\bot, reason))$ **then return** $\bot$
8: **else if** $defined$ **then**
9:     **for** $e' \in edges(v)$ **where** $\mathcal{E}[e'] \neq \bot$ **do**
10:      $(v_{memo}, m_{memo}) \leftarrow memo[e']$
11:      $p_V \leftarrow path_V \cup \{v\}$
12:      $p_E \leftarrow path_E \cup \{e'\}$
13:      $w \leftarrow target(e')$
14:      $def \leftarrow \mathcal{E}[e'] = \top$
15:      **if** $(v_v, m_{memo}) = (-1, -1)$ **then**
16:        $NOCMemo(p_V, p_E, w, e', \mathcal{E}, def, memo)$
17:      **else**
18:        $i_v \leftarrow index(v_{memo}, path_V)$
19:        $m = \max\{f(v') \mid v' \in path_V[i_v :]\}$
20:        **if** $m > m_{memo}$ **then**
21:          $NOCMemo(p_V, p_E, w, e', \mathcal{E}, def, memo)$
22:        **else**
23:          $memo[e] \leftarrow memo[e']$
24: **return** $\top$

All experiments were conducted on an Intel Xeon 8260 CPU (24 cores, non-hyperthreaded) with 268.55 GB of RAM, running Linux. The complete source code, including implementations and experimental setups, is available at [19]. We evaluated all algorithms using a standard benchmark set commonly employed in parity game research, drawing from the PGSolver Library [17] and the Benchmarks for Parity Games [24]. The games were classified into four categories:

- Equivalence Checking (**EqC**): This group contains 77 games derived from formal equivalence checking problems, including strong bisimulation, weak bisimulation, and stuttering equivalence. These games are characterized by complex, structured connectivity patterns reflecting system design.
- Model Checking (**MoC**): This set consists of 106 games originating from temporal logic model checking (μ-calculus) against system models. These instances are considered organic and reflect real-world verification challenges.
- PGSolver Collection (**PGS**): This collection comprises 161 games sourced from the hardest known families of parity games, including ladder games, recursive ladder games, elevator verification, clique games, steady games, and Jurdziński games [2], likely to expose some of the worst performance.
- Random Games (**RaG**): This category includes 125 randomly generated games with varying sizes and priority distributions, designed primarily to test the general robustness and average-case performance of solving algorithms.

The benchmark sets were partitioned into two distinct size categories for evaluation, optimizing for the feasibility of the Reduction to SAT approach on smaller instances and focusing on scalability for larger instances:

Table 1. Time Performance of Algorithms on Small-Scale Instances, Average measured in seconds. (Max vertices: EqC/MoC $\approx 10^6$, PGS/RaG $\approx 10^3$).

Bench	Oink			ZRA	NOC		Reduction SAT		
	PP	PP+	PAR	ImpAttr	Eager	Memo	CaDiCal	Kissat	MiniSat
EqC	0.663	0.649	0.544	**0.190**	0.399	0.400	16.114	76.424	32.594
MoC	0.327	0.324	0.294	**0.172**	0.289	0.290	15.524	28.924	22.443
PGS	**0.126**	0.126	2.580	7.237	5.043	25.220	292.021	294.794	306.186
RaG	0.001	0.001	**0.001**	0.002	0.003	0.005	1.100	1.113	1.384

In Table 1, we compare the performance of the No-Opponent-Cycle propagators against two groups of competing approaches on small benchmark games: (1) other constraint-based methods that use Reduction to SAT, and (2) non-constraint-based methods, such as the ZRA algorithm (Improved Attractor) and the Oink algorithms (PP, PP+, PAR). The goal of this evaluation is to assess the effectiveness of the filtering techniques relative to existing SAT-based methods, which is the primary focus of this paper. Overall, No-Opponent-Cycle outperforms the SAT-based methods by up to three orders of magnitude. Furthermore, we observe that the Memoization approach (NOC-Memo) does not significantly improve performance over the Eager approach (NOC-Eager). Moreover, NOC-Eager achieves performance comparable to the best non-CP-based algorithms in practice: it is at most one order of magnitude slower on the hardest games (from the PGSolver Collection) and no more than a factor of two slower on all other instances. Additional comparative results with other CP methods are provided in the online appendix [20].

Table 2. Time Performance of Algorithms on Large-Scale Instances, Average measured in seconds using PAR2 Scoring. (Max vertices: $\approx$ 35 Million).

Benchmark	Oink			ZRA	NOC	
	PP	PP+	PAR	ImpAttr	Eager	Memo
EqC	24.775	24.876	24.159	**14.490**	16.169	82.902
MoC	9.547	9.557	8.718	**3.929**	8.838	8.912
PGS	44.949	44.922	182.643	201.804	**32.189**	55.464
RaG	0.016	0.015	**0.015**	0.053	0.078	0.099

Table 2 presents the results for large games, where we evaluate the scalability of the No-Opponent-Cycle approach in comparison to ZRA and the Oink family of algorithms. SAT-based approaches are omitted, as they were unable to solve these larger instances within reasonable resource limits and therefore do not contribute to a meaningful scalability analysis. Overall, No-Opponent-Cycle shows strong and robust performance: in the Equivalence and Model Checking benchmarks, it is within a small constant factor of ZRA, which remains the fastest method, which is consistent with the main results reported in [1]. In the PGSolver Collection, No-Opponent-Cycle outperforms all Oink algorithms and also surpasses ZRA by a large margin. For Random Games, all algorithms solve the instances almost instantly, making performance differences negligible. These results demonstrate that No-Opponent-Cycle scales effectively and remains competitive with the best non-CP-based (dedicated) solvers on large games.

The runtime metrics for the No-Opponent-Cycle and ZRA approaches include both the solver execution time and the internal problem-preparation time (excluding I/O operations such as file reading). In contrast, for the SAT-based approaches we report only the solving time. This distinction is made because the preparation phase for the SAT-based methods—particularly the encoding generation—is substantially more expensive and was not a primary target of optimization in this work. Additional experiments comparing NOC-Eager, NOC-Memo, NOC-Checker with ZRA, together with their statistical analysis (AVG, SD, WR) presented in the online appendix [20], indicate that the NOC-Memo implementation does not provide a significant performance advantage over the NOC-Eager implementation on these benchmark instances.

4.1 Resource Utilization Analysis

A key motivation for developing the No-Opponent-Cycle propagator was to avoid the resource overhead associated with introducing the Boolean literals and clauses required for the progress measure variables and constraints. We also evaluated the memory consumption of No-Opponent-Cycle compared to other methods.

The results of these additional experiments are summarised in Fig. 3. While No-Opponent-Cycle requires at most 700 MB of memory for games with 13 thousand vertices, and ZRA up to 17 MB, the SAT-based approaches consume significantly more. In particular, we see that CaDiCaL, MiniSat, and Kissat report memory usage exceeding 39 GB for the same instances.

5 Games with Additional Constraints

As we will see in this section, our method flexibly integrates additional constraints—such as quantitative or structural requirements—without any modification to the core algorithm. This inherent adaptability to various constraints is a decisive advantage over non-CP approaches, which necessitate custom adaptations for every additional set of constraints in the problem.

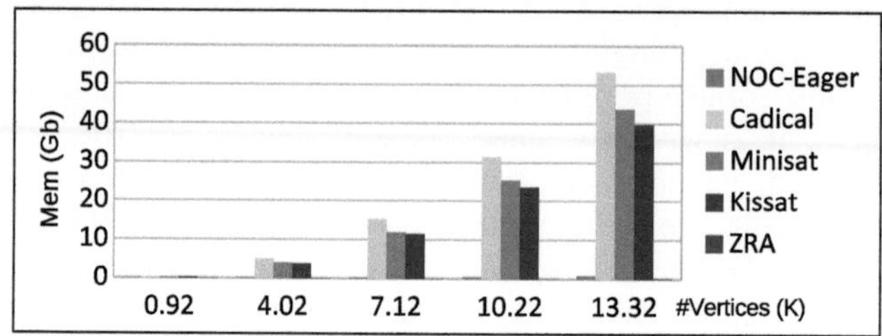

Fig. 3. Memory usage on Gb for solving Jurdziński parity games (one of the hardest games in the PGSolver collection), constrained to 13 thousand vertices.

5.1 Automated Synthesis of Reactive Systems

In (local) model checking, it suffices to determine the winner from a given vertex (satisfiable if won by Player $\bigcirc$, unsatisfiable if won by Player $\square$). However, applications like automated synthesis [8,16,25] require explicit winning strategies, that is, a description of the decisions to be made by the winner at each vertex.

Our satisfiability encoding can capture multiple winning plays simultaneously. For instance, in Fig. 2b, a solution assigning $\top$ to vertices $\mathcal{V}_0$ to $\mathcal{V}_3$ and edges $\mathcal{E}_0$, $\mathcal{E}_1$, $\mathcal{E}_2$, $\mathcal{E}_4$, and $\mathcal{E}_6$ captures two distinct plays: $\langle E_0 E_2 E_4 E_6 \rangle$ and

$\langle E_1 E_4 E_6 \rangle$. To extract individual winning plays, we extend the CSP with variables and constraints such that each solution corresponds to exactly one play. We define a sequence of integer variables $(\mathcal{P}_s)_{s \in \{0..|V|-1\}}$ to represent the path. More specifically, we also add:

- h: the position where the finite prefix ends;
- t: the start of the cycle (tail).

We then introduce three new constraints:

- $(\mathcal{C}_{path})$: $\mathsf{t} \in \{0..\ |V|-1\} \wedge \bigwedge_{s \in \{1..\mathsf{t}\}} \bigvee_{e=(v,w) \in E}(\mathcal{E}_e \wedge \mathcal{P}_{s-1} = v \wedge \mathcal{P}_s = w)$;
  This ensures each step follows a valid edge;

- $(\mathcal{C}_{cycle})$: $\mathsf{h} \in \{0..\mathsf{t}\} \wedge \mathcal{P}_{\mathsf{t}+1} = \mathcal{P}_{\mathsf{h}}$;
  It ensures the cycle closes at position $\mathsf{t}+1$;

- $(\mathcal{C}_{unique})$: $\bigwedge_{i \neq j \in \{0..\mathsf{t}\}}(\mathcal{P}_i \neq \mathcal{P}_j)$; constraint enforces only one cycle.

These constraints can be added in MiniZinc without modifying the core algorithm of No-Opponent-Cycle. Using Chuffed as the backend, we enumerate all solutions to obtain the set of winning plays, represented as vertex sequences.

5.2   Energy Conditions/Games

We consider games with additional quantitative objectives. In *Energy Games* [9], each edge carries an energy level (a weight), and a play is winning if the running sum of these levels remains non-negative at all times, in addition to satisfying the parity condition. This requirement can be encoded using the $\mathcal{P}$ variables introduced earlier. Given a set of energy levels $(L_e)_{e \in E}$, we define integer variables $(\mathcal{I}_s)_{s \in \{0..\mathsf{t}-\mathsf{h}+1\}}$ to represent the edges in the infinite cycle (tail) of the play:

- $\mathcal{C}_{infEdges} : \forall s \in \{0..\mathsf{t}-\mathsf{h}+1\}(\mathcal{I}_s = E(\mathcal{P}_{s+\mathsf{h}}, \mathcal{P}_{s+\mathsf{h}+1}))$

We then impose the following constraint to ensure the total energy level in the cycle is non-negative:

- $\mathcal{C}_{energy} : \sum_{e \in \mathcal{I}} L_e \geq 0$

While a complete model would also enforce energy constraints over all finite prefixes (up to t), we omit these here for brevity. Together, these constraints suffice to model the energy condition, useful in several verification settings.

Many other extensions—such as multi-dimensional energy bounds and other quantitative payoff constraints—can be modelled similarly using additional constraints or objective functions [6,15,26,31]. Further work is needed to precisely define each requirement and establish the appropriate constraints.

6   Conclusions

We have introduced a novel constraint-based approach for parity games which does not require integer-based progress measures. The experiments show that our method significantly outperforms existing constraint-based approaches, significantly reducing runtime and improving memory consumption by over 95% on large instances, and it is competitive against non-CP-based solvers.

Our key innovation, replacing a SAT encoding with a global constraint, not only improves performance but also provides a flexible framework that can be easily extended to accommodate additional constraints, and therefore more complex requirements. In this paper, we have shown that these improvements can enable practical verification of large-scale systems while maintaining the modularity and expressiveness of modern CP-based methods.

Our work goes beyond parity games by easily enabling new solutions to synthesis and providing the flexibility to adapt our core to solve multiplayer games, including more general games (e.g., ω-regular games [10]). Our results demonstrate that constraint-based approaches with carefully designed global constraints offer a promising direction for practical formal verification tools that balance expressiveness and computational efficiency.

References

1. Aggarwal, S., de la Banda, A.S., Yang, L., Gutierrez, J.: A matrix-based approach to parity games. In: Sankaranarayanan, S., Sharygina, N. (eds.) Tools and Algorithms for the Construction and Analysis of Systems, pp. 666–683. Springer, Cham (2023)
2. Benerecetti, M., Dell'Erba, D., Mogavero, F.: Solving parity games via priority promotion. In: Chaudhuri, S., Farzan, A. (eds.) CAV 2016. LNCS, vol. 9780, pp. 270–290. Springer, Cham (2016). https://doi.org/10.1007/978-3-319-41540-6_15
3. Benerecetti, M., Dell'Erba, D., Mogavero, F., Schewe, S., Wojtczak, D.: Priority promotion with parysian flair. J. Comput. Syst. Sci. **147**, 103580 (2025). https://doi.org/10.1016/j.jcss.2024.103580
4. Biere, A., Faller, T., Fazekas, K., Fleury, M., Froleyks, N., Pollitt, F.: CaDiCaL 2.0. In: Gurfinkel, A., Ganesh, V. (eds.) CAV 2024, Part I. LNCS, vol. 14681, pp. 133–152. Springer, Cham (2024). https://doi.org/10.1007/978-3-031-65627-9_7
5. Biere, A., Faller, T., Fazekas, K., Fleury, M., Froleyks, N., Pollitt, F.: CaDiCaL, Gimsatul, IsaSAT and Kissat entering the SAT Competition 2024. In: Heule, M., Iser, M., Järvisalo, M., Suda, M. (eds.) Proc. of SAT Competition 2024 - Solver, Benchmark and Proof Checker Descriptions. Department of Computer Science Report Series B, vol. B-2024-1, pp. 8–10. University of Helsinki (2024)
6. Bozzelli, L., Murano, A., Perelli, G., Sorrentino, L.: Hierarchical cost-parity games. Theoret. Comput. Sci. **847**, 147–174 (2020). https://doi.org/10.1016/j.tcs.2020.10.002
7. Brihaye, T., Bruyère, V., Goeminne, A., Thomasset, N.: On relevant equilibria in reachability games. J. Comput. Syst. Sci. **119**, 211–230 (2021). https://doi.org/10.1016/j.jcss.2021.02.009

8. Bruyère, V.: A game-theoretic approach for the synthesis of complex systems. In: Revolutions and Revelations in Computability. LNCS, vol. 13359, pp. 52–63. Springer, Cham (2022)

9. Chatterjee, K., Doyen, L.: Energy parity games. Theoret. Comput. Sci. **458**, 49–60 (2012). https://doi.org/10.1016/j.tcs.2012.07.038

10. Chatterjee, K., Henzinger, T.A.: A survey of stochastic ω-regular games. J. Comput. Syst. Sci. **78**(2), 394–413 (2012)

11. Chu, G., Stuckey, P.J.: Chuffed: a lazy clause generation constraint solver. GitHub repository (2018). https://github.com/chuffed/chuffed

12. Cotton, S., Asarin, E., Maler, O., Niebert, P., Yovine, S., Lakhnech, Y.: Some progress in satisfiability checking for difference logic. In: Formal Techniques, Modelling and Analysis of Timed and Fault-Tolerant Systems, pp. 263–276 (2004)

13. Dijk, T.: Oink: an implementation and evaluation of modern parity game solvers. In: Beyer, D., Huisman, M. (eds.) TACAS 2018. LNCS, vol. 10805, pp. 291–308. Springer, Cham (2018). https://doi.org/10.1007/978-3-319-89960-2_16

14. Eén, N., Sörensson, N.: An extensible SAT-solver. In: Giunchiglia, E., Tacchella, A. (eds.) SAT 2003. LNCS, vol. 2919, pp. 502–518. Springer, Heidelberg (2004). https://doi.org/10.1007/978-3-540-24605-3_37

15. Fijalkow, N., Zimmermann, M.: Parity and Streett games with costs. Log. Methods Comput. Sci. **10**(2) (2014). https://doi.org/10.2168/lmcs-10(2:14)2014

16. Finkbeiner, B., Hahn, C., Stenger, M., Tentrup, L.: Synthesizing reactive systems from hyperproperties. In: CAV 2018. LNCS, vol. 10981, pp. 158–175. Springer, Cham (2018). https://doi.org/10.1007/978-3-319-96145-3_9

17. Friedmann, O., Lange, M.: The PGSolver Collection of Parity Game Solvers, Version 4.1 (2017). https://github.com/tcsprojects/pgsolver/blob/master/doc/pgsolver.pdf. Accessed 13 Mar 2025

18. Heljanko, K., Keinänen, M., Lange, M., Niemelä, I.: Solving parity games by a reduction to SAT. J. Comput. Syst. Sci. **78**(2), 430–440 (2012). https://doi.org/10.1016/j.jcss.2011.05.004. Games in Verification

19. Hernandez, G., Garcia, J., Gutierrez, J., Tack, G.: Implementation and evaluation of no-opponent-cycle propagators for solving parity games (2025). https://github.com/GonzaloHernandez/NoOpponentCycle. Version 1.0

20. Hernandez, G., Garcia, J., Gutierrez, J., Tack, G.: NOC propagators for solving parity games - appendix. Figshare (2025). https://doi.org/10.6084/m9.figshare.30842819. Online supplement

21. Jurdziński, M.: Small progress measures for solving parity games. In: Reichel, H., Tison, S. (eds.) STACS 2000. LNCS, vol. 1770, pp. 290–301. Springer, Heidelberg (2000). https://doi.org/10.1007/3-540-46541-3_24

22. Jurdziński, M.: Deciding the winner in parity games is in UP ∩ co-UP. Inf. Process. Lett. **68**(3), 119–124 (1998)

23. Jurdziński, M., Morvan, R., Thejaswini, K.S.: Universal algorithms for parity games and nested fixpoints. In: LNCS, pp. 252–271. Springer, Cham (2022)

24. Keiren, J. J. A.: Benchmarks for parity games (extended version) (2015). https://arxiv.org/abs/1407.3121

25. Luttenberger, M., Meyer, P.J., Sickert, S.: Practical synthesis of reactive systems from LTL specifications via parity games: you can teach an old dog new tricks: making a classic approach structured, forward-explorative, and incremental. Acta Informatica **57**(1–2), 3–36 (2020)

26. Mogavero, F., Murano, A., Sorrentino, L.: On promptness in parity games. Fund. Inform. **139**(3), 277–305 (2015). https://doi.org/10.3233/FI-2015-1235

27. Moskewicz, M.W., Madigan, C.F., Zhao, Y., Zhang, L., Malik, S.: Chaff: engineering an efficient SAT solver. In: Proceedings of the 38th Annual Design Automation Conference (DAC 2001), pp. 530–535. ACM (2001). https://doi.org/10.1145/378239.379017
28. Nethercote, N., Stuckey, P.J., Becket, R., Brand, S., Duck, G.J., Tack, G.: MiniZinc: towards a standard CP modelling language. In: Bessière, C. (ed.) Principles and Practice of Constraint Programming – CP 2007. LNCS, vol. 4741, pp. 529–543. Springer, Heidelberg (2007). https://doi.org/10.1007/978-3-540-74970-7_38
29. Ohrimenko, O., Stuckey, P.J., Codish, M.: Propagation via lazy clause generation. Constraints An Int. J. 14(3), 357–391 (2009). https://doi.org/10.1007/S10601-008-9064-X
30. Tarjan, R.: Depth-first search and linear graph algorithms. SIAM J. Comput. 1(2), 146–160 (1972)
31. Weinert, A., Zimmermann, M.: Easy to win, hard to master: optimal strategies in parity games with costs. Log. Methods Comput. Sci. (2016). https://api.semanticscholar.org/CorpusID:1611141
32. Zielonka, W.: Infinite games on finitely coloured graphs with applications to automata on infinite trees. Theoret. Comput. Sci. 200(1), 135–183 (1998)

Contextual Preference Distribution Learning

Benjamin Hudson[1,2(✉)] ⓘ, Laurent Charlin[1,2,3] ⓘ, and Emma Frejinger[1,2] ⓘ

[1] Mila – Quebec Artificial Intelligence Institute, Montreal, QC, Canada
`ben.hudson@mila.quebec`
[2] Département d'informatique et de recherche opérationnelle (DIRO), Université de Montréal, Montreal, QC, Canada
[3] HEC Montréal, Montreal, QC, Canada

Abstract. Decision-making problems often feature uncertainty stemming from heterogeneous and context-dependent human preferences. To address this, we propose a sequential learning-and-optimization pipeline to learn preference distributions and leverage them to solve downstream problems, for example risk-averse formulations. We focus on human choice settings that can be formulated as (integer) linear programs. In such settings, existing inverse optimization and choice modelling methods infer preferences from observed choices but typically produce point estimates or fail to capture contextual shifts, making them unsuitable for risk-averse decision-making. Using a bounded-variance score function gradient estimator, we train a predictive model mapping contextual features to a rich class of parameterizable distributions. This approach yields a maximum likelihood estimate. The model generates scenarios for unseen contexts in the subsequent optimization phase. In a synthetic ridesharing environment, our approach reduces average post-decision surprise by up to 114× compared to a risk-neutral approach with perfect predictions and up to 25× compared to leading risk-averse baselines.

Keywords: Inverse optimization · Preference learning · Sequential learning-and-optimization · Estimate-then-optimize

1 Introduction

Human preferences vary across individuals and contexts [6], often causing *post-decision surprise* [20] in decision-making problems that depend on human choices. We take driver-rider assignment in ridesharing as an illustrative example: while platforms recommend routes using real-time traffic [35], drivers often choose alternate routes to optimize their own objectives, based on personal preferences or tacit knowledge of the road network [29]. These discrepancies introduce uncertainty into downstream problems; for example, if a driver takes an alternate route, their realized arrival time may differ from the platform's estimate, leading to the perception of unreliability. To address this issue, we propose a sequential learning-and-optimization approach (illustrated in Fig. 1): we

T. Guns (Ed.): CPAIOR 2026, LNCS 16595, pp. 243–261, 2026.
https://doi.org/10.1007/978-3-032-27242-3_15

first predict human preference distributions given some contextual features (i.e. covariates), and then leverage them to explicitly minimize the resulting downstream risk. Ultimately, we aim to reduce post-decision surprise in decision-making problems where uncertainty stems from human choices.

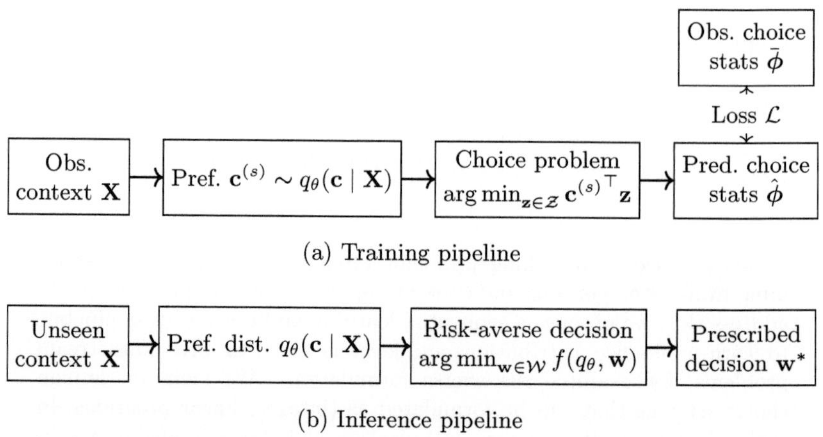

(a) Training pipeline

(b) Inference pipeline

Fig. 1. Our sequential learning-and-optimization approach. (a) During training, we learn a model mapping contextual features to preference (objective function coefficient) distributions in the choice problem (an ILP). **(b)** At inference, we leverage the model to generate scenarios in a risk-averse estimate-then-optimize problem, allowing us to minimize risk introduced by uncertain human preferences.

Existing approaches modelling human choice behaviour with mathematical programs, such as inverse optimization [4,9,31,49] and perturbed utility models [1,17,28], are typically limited by predicting point estimates rather than distributions or by failing to account for how preferences shift with context. This renders them unsuitable for risk-averse decision-making: distributional information is necessary to quantify risk, while conditioning on context helps to avoid overly conservative decisions.

Contribution Statement. Our contributions are summarized as follows:

1. We propose a method to learn context-dependent preference (i.e. objective function coefficient) distributions in integer linear programs (ILPs) from paired context and aggregated solution statistic observations. This is ideal when directly observing the uncertain quantities is impossible or undesirable.
2. Our approach yields a maximum likelihood estimate (MLE), inheriting the desirable statistical properties of consistency and efficiency.
3. Our score function gradient estimator has bounded variance and did not require post hoc variance reduction in our experiments.

4. We demonstrate our approach as part of a sequential learning-and-optimization pipeline for decision-making under human preference uncertainty. In a synthetic ridesharing environment, it reduces average post-decision surprise by 2.4–114× compared to a risk-neutral approach with perfect predictions and by 1.6–25× compared to leading risk-averse baselines.

2   Related Work

We frame our work as inverse optimization (IO), as we recover parameters of an optimization formulation from observed solutions. It relates closely to perturbed utility models (PUMs) and inverse reinforcement learning (IRL), which model choice behaviour as optimization problems or Markov decision processes (MDPs). We leverage integrated learning-and-optimization (ILO) techniques during training and deploy our model in a sequential learning-and-optimization (SLO) pipeline [40].

Inverse Optimization. Recent research in IO addresses settings where observations are corrupted by noise or bounded rationality [9,31] yet the underlying problem formulation is deterministic. The learning task is to recover parameters of this formulation. Contextual IO addresses settings in which these vary via known mappings [4,49]. In contrast, we tackle a setting where the latent parameters are random and the feature mapping function is unknown. The closest IO approach to ours is that of Lin et al. [24], who recover uncertainty sets over problem parameters, but do not consider a contextual setting.

Perturbed Utility Models. PUMs generalize the classical discrete choice setting [1,27,46] and specialized variants like bundle [5,10,28] and route choice [16,17]. These models express choice probabilities as solutions to convex, constrained optimization problems. Most current approaches model a single individual representative of the population [1]: that is, they learn a function mapping from observed features to a point prediction of expected utility (negative cost). Birge et al. [5] explore learning Dirac-Uniform and von Mises-distributed utilities as special cases. A distributional approach enables simulating choices in unseen contexts, something we explore in our experiments.

Inverse Reinforcement Learning. In contrast to IO and PUMs, IRL frameworks model complex choice problems as MDPs, aiming to recover their parameters from observed decisions [7,8,23,50,51]. The applicability of IRL heavily relies on whether a problem fits the MDP paradigm. In contrast, our approach accepts any ILP formulation and is agnostic to problem substructure and the underlying solution method.

Integrated Learning-and-Optimization. ILO, or decision-focused learning (DFL) [26], trains a model by considering the impact of its predictions on downstream decisions (typically measured by regret). Our training pipeline can be seen as an optimal action imitation ILO task, where the goal is to reproduce observed actions given context [40]. While many methods demonstrated on regret minimization can be readily applied to action imitation [2,30,36,41,45,47], we contribute a novel approach for back-propagating through ILPs using a bounded-variance score function gradient estimator.

Sequential Learning-and-Optimization. We take an SLO approach, using predictions from our trained model in a distinct optimization problem. While estimate-then-optimize (ETO) [3,13,14,19,22,43] and conditional robust optimization (CRO) [11,12,37–39,44] predict conditional distributions and uncertainty sets for risk-averse decision-making, both rely on direct observations of the uncertain parameters. In contrast, our approach requires only aggregated solutions to a related problem (the choice problem), making it ideal when direct observation is impossible or undesirable, e.g., for privacy concerns.

3   Contextual Preference Distribution Learning

In this section, we introduce Contextual Preference Distribution Learning (CPDL), demonstrate that it yields an MLE of the model parameters, and discuss the variance of its stochastic gradient estimator.

3.1   Learning Problem Formulation

We model human choice behaviour as an ILP with binary variables (the "choice problem"). This formulation is compact, yet can represent a wide variety of choice settings where an individual selects from a discrete set of alternatives, such as shortest paths [2,17], k-subset selection [5,28,36], and discrete choice [1,27,46]. Given a vector of preferences $\mathbf{c}$, the choice problem yields optimal solutions

$$\mathcal{Z}^*(\mathbf{c}, \mathbf{u}) := \arg\min_{\mathbf{z}} \{\mathbf{c}^\top \mathbf{z} \mid \mathbf{z} \in \mathcal{Z}(\mathbf{u}), \mathbf{z} \in \{0,1\}^m\},$$

where $\mathcal{Z}(\mathbf{u})$ defines the feasible region given some exogenous variables $\mathbf{u}$. We assume an individual selects an optimal solution uniformly at random, denoted by $\mathbf{z}^* \in_R \mathcal{Z}^*(\mathbf{c}, \mathbf{u})$. This defines the conditional choice distribution $p(\mathbf{z}|\mathbf{c}, \mathbf{u})$ as having equal mass over all elements in the optimal set. To reduce notational clutter, we omit conditioning on $\mathbf{u}$ where the context is unambiguous. Human preferences are varied, context-dependent, and can be represented by a distribution $p(\mathbf{c}|\mathbf{X})$, where $\mathbf{X}$ is a $m \times n$ contextual feature matrix, containing an n-dimensional feature vector for each element of $\mathbf{c}$. Thus, we can express the population choice distribution as

$$p(\mathbf{z}|\mathbf{X}) = \int p(\mathbf{z}|\mathbf{c})p(\mathbf{c}|\mathbf{X})dc.$$

We now describe the learning problem. Our goal is to recover the distribution $p(\mathbf{c}|\mathbf{X})$ in a way that can generalize to previously unobserved contexts. We leverage amortized inference to approximate this distribution using a parameterized model $q_\theta(\mathbf{c}|\mathbf{X})$ shared across instances. We model the target and predicted choice distributions $p(\mathbf{z}|\mathbf{X})$ and $q_\theta(\mathbf{z}|\mathbf{X}) = \int p(\mathbf{z}|\mathbf{c})q_\theta(\mathbf{c}|\mathbf{X})d\mathbf{c}$ as discrete exponential-family distributions, allowing us to obtain an MLE of the model parameters θ by optimizing them subject to a moment matching condition [34, 36, 51]. For our model, this condition is given by

$$\mathbb{E}_{p(\mathbf{z}|\mathbf{X})}[\phi(\mathbf{z})] - \mathbb{E}_{q_\theta(\mathbf{c}|\mathbf{X})}\left[\mathbb{E}_{p(\mathbf{z}|\mathbf{c})}[\phi(\mathbf{z})]\right] = 0,$$

where $\phi(\mathbf{z})$ are sufficient statistics of the distribution of $\mathbf{z}$ (target or predicted). We define the inverse problem as

$$\theta^* \in \underset{\theta \in \Theta}{\arg\min} \left\|\mathbb{E}_{p(\mathbf{z}|\mathbf{X})}[\phi(\mathbf{z})] - \mathbb{E}_{q_\theta(\mathbf{c}|\mathbf{X})}\left[\mathbb{E}_{p(\mathbf{z}|\mathbf{c})}[\phi(\mathbf{z})]\right]\right\|_2^2.$$

Unlike previous work [36], we optimize this objective directly using stochastic gradient descent.

3.2 Solution Approach

Let $\mathbf{X}$ be an observed feature matrix and $\bar{\phi} := \mathbb{E}_{p(\mathbf{z}|\mathbf{X})}[\phi(\mathbf{z})]$ be the sufficient statistics of the observed choice distribution. Let the expected sufficient statistics under the predicted distribution be $\hat{\phi} := \mathbb{E}_{q_\theta(\mathbf{c}|\mathbf{X})}\left[\mathbb{E}_{p(\mathbf{z}|\mathbf{c})}[\phi(\mathbf{z})]\right]$. In this paper, we define $\phi(\mathbf{z}) = \mathbf{z}$, corresponding to first-moment matching, although we experiment with matching higher-order moments in App. B.6. The loss of the inverse problem and its gradient are given by

$$\mathcal{L}(\mathbf{X}, \bar{\phi}, \theta) = \left\|\bar{\phi} - \hat{\phi}\right\|_2^2, \tag{1}$$

$$\nabla_\theta \mathcal{L}(\mathbf{X}, \bar{\phi}, \theta) = \left(\bar{\phi} - \hat{\phi}\right) \nabla_\theta \mathbb{E}_{q_\theta(\mathbf{c}|\mathbf{X})}\left[\mathbb{E}_{p(\mathbf{z}|\mathbf{c})}[\phi(\mathbf{z})]\right]$$

$$= \left(\bar{\phi} - \hat{\phi}\right) \mathbb{E}_{q_\theta(\mathbf{c}|\mathbf{X})}\left[\mathbb{E}_{p(\mathbf{z}|\mathbf{c})}[\phi(\mathbf{z})\nabla_\theta \log q_\theta(\mathbf{c}|\mathbf{X})]\right]. \tag{2}$$

We approximate $\hat{\phi}$ with the Monte Carlo estimate:

$$\hat{\phi} \approx \frac{1}{KL} \sum_{k=1}^{K} \sum_{l=1}^{L} \phi\left(\mathbf{z}^{*(kl)}\right) \quad \text{where } \mathbf{z}^{*(kl)} \in_R \mathcal{Z}^*(\mathbf{c}^{(k)}, \mathbf{u}) \text{ and } \mathbf{c}^{(k)} \sim q_\theta(\mathbf{c}|\mathbf{X}).$$

The hyperparameters K and L are the number of samples drawn from the learned cost distribution and the conditional choice distribution, respectively. We find $L = 1$ is sufficient to approximate $\mathbb{E}_{p(\mathbf{z}|\mathbf{c})}[\phi(\mathbf{z})]$, as the choice problem typically has a single solution for a given cost vector. While this estimate involves solving the choice problem KL times, the instances in our setting are not computationally challenging. Computation can be accelerated with warm-starting, solution caching [33], or GPU parallelization [25]. We compute a stochastic gradient estimate by re-weighting each solution by the score of the generating sample from $q_\theta(\mathbf{c}|\mathbf{X})$:

$$\nabla_\theta \mathcal{L}(\mathbf{X}, \bar{\phi}, \theta) \approx \left(\bar{\phi} - \hat{\phi}\right) \left(\frac{1}{K} \sum_{k=1}^{K} \left(\frac{1}{L} \sum_{l=1}^{L} \phi\left(\mathbf{z}^{*(kl)}\right)\right) \nabla_\theta \log q_\theta(\mathbf{c}^{(k)}|\mathbf{X})\right).$$

While the solution distribution is restricted to the discrete exponential family, $q_\theta(\mathbf{c}|\mathbf{X})$ can be any parametrizable distribution with a tractable score function (e.g., normalizing flows). Despite being unconstrained, our experiments show we recover the ground-truth distributional parameters up to an invariant transformation when the distribution is correctly specified.

Variance of the Gradient Estimator. Mohamed et al. [32] show the variance of a score function gradient estimate of a function f under a smoothing distribution $p_\theta(x)$ is given by

$$\mathrm{Var}(\hat{\nabla}_\theta f) = \mathbb{E}_{p_\theta(x)} \left[(f(x)\nabla_\theta \log p_\theta(x))^2\right] - (\hat{\nabla}_\theta f)^2.$$

Large magnitudes of f cause this variance to explode, leading to training instability. In our approach, $f := \mathbb{E}_{p(\mathbf{z}|\mathbf{c})}[\phi(\mathbf{z})]$ represents the moments of the choice distribution. Because these are bounded in $[0, 1]$, the variance of our gradient estimate is bounded from above by $\mathbb{E}_{p_\theta(x)} \left[(\nabla_\theta \log p_\theta(x))^2\right]$, which corresponds to the diagonal elements of the Fisher information matrix. Empirically, this allows CPDL to use higher learning rates than methods with unbounded f (e.g., REINFORCE [42]) without post hoc variance reduction. We compare CPDL and REINFORCE in detail in Appendix A.1.

4   Numerical Experiments

We benchmark CPDL against other methods capable of generating scenarios in a synthetic ridesharing environment.

4.1   Experimental Setting

Our environment comprises a synthetic data-generating process representing driver route choice for the "estimate" phase and a risk-averse driver-rider assignment problem for the "optimize" phase.

Route Choice Problem. Our data-generating process is similar to previous work [2,24,36,47], but features stochastic edge costs whose distributions are feature-dependent. We generate a set of choices (the *samples* $\mathcal{S}$) for each context (the *instance*, $i \in \mathcal{I}$). This is standard in settings with categorical features, where multiple observations per category are expected [7,18].

We begin by uniformly sampling two global n-dimensional vectors $\boldsymbol{\beta}_\mu$ and $\boldsymbol{\beta}_\sigma$. We generate instances representing grid graphs, each having a $|\mathcal{E}| \times n$ contextual feature matrix $\mathbf{X}_i$ sampled from a standard normal distribution and edge cost distributions $\text{LogNormal}(\mathbf{X}_i\boldsymbol{\beta}_\mu, \mathbf{X}_i\boldsymbol{\beta}_\sigma)$. We use a log-normal distribution for its positive support and use in travel-time estimation literature [15, 21, 48]. For each instance, we generate samples by drawing edge costs and solving the resulting shortest path problems $\{\mathbf{z}^{*(s)}_i \mid \forall s \in \mathcal{S}\}$. We define $\phi(\mathbf{z}) = \mathbf{z}$, corresponding to matching the first choice distribution moment: the expected edge selection probability $\bar{\mathbf{p}}_i = \frac{1}{|\mathcal{S}|} \sum_\mathcal{S} \mathbf{z}^{*(s)}_i$. The learner observes only $\mathbf{X}_i$ and $\bar{\mathbf{p}}_i$. The expression of the data-generating process is shown in Appendix B.1.

Risk-Averse Assignment Problem. We leverage the trained model to solve a risk-averse variant of the driver-rider assignment problem, directly addressing risk introduced by uncertain human preferences. Given unseen features $\mathbf{X}$, we sample preferences from the predicted distribution $q_\theta(\mathbf{c}|\mathbf{X})$. For each scenario $k \in \mathcal{K}$, we determine the perceived-shortest paths from all drivers to all riders and compute their cost according to a decision cost vector $\mathbf{g}$—the first column of $\mathbf{X}$ in our experiments. We assemble these into a cost matrix $\mathbf{G}^{(k)}$, representing a scenario. Finally, we obtain an assignment $\mathbf{W}^*$ by solving the conditional value-at-risk (CVaR) minimization problem formulation in Appendix B.2 with $\alpha = 0.95$.

4.2   Evaluation Metrics

Perception of unreliability is caused by both early and late arrivals. Thus, we define post-decision surprise as the squared difference between the expected decision cost under the ground-truth distribution $p(\mathbf{c}|\mathbf{X})$ and the predicted distribution $q_\theta(\mathbf{c}|\mathbf{X})$. Mathematically,

$$\mathcal{L}_{PDS}(\mathbf{W}^*, q_\theta) = \left(\mathbb{E}_{\boldsymbol{\xi} \sim p(\mathbf{c}|\mathbf{X})} \left[g(\mathbf{W}^*, \boldsymbol{\xi}) \right] - \mathbb{E}_{\boldsymbol{\zeta} \sim q_\theta(\mathbf{c}|\mathbf{X})} \left[g(\mathbf{W}^*, \boldsymbol{\zeta}) \right] \right)^2, \qquad (3)$$

where g is the decision cost in a given scenario (i.e. given a realization of uncertainty $\boldsymbol{\xi}$ or $\boldsymbol{\zeta}$). We approximate both expectations with Monte Carlo estimates (Appendix B.2). Because our risk-averse assignment formulation effectively penalizes only late arrivals, we also report this one-sided post-decision *disappointment* metric in Appendix B.4. Finally, to measure the models' capacity to learn the underlying preference distributions, we report the test loss and the squared correlation coefficient (R^2) between the predicted and ground-truth parameters (Appendix B.5).

4.3   Baselines

We compare against four baselines mapping context to distributions: DPO [2], AIMLE [30], REINFORCE [42], and MaxEnt IRL [51]. We omit IO and PUM [17] baselines because they make point predictions, and do not allow sampling scenarios. While all baselines share the same encoder (an MLP with two 32-dimensional hidden layers) and loss (1), they estimate gradients through the choice problem differently. The encoder parameters are shared across edges. Additionally, we evaluate two baselines with perfect predictions: GTD-RN solves a risk-neutral assignment using ground-truth expected preferences, while GTD-RA solves the risk-averse assignment with the ground-truth preference distributions. Further baseline details are in Appendix B.3.

4.4 Results

Figure 2 demonstrates that our method reconstructs the choice distribution in unseen contexts with the highest accuracy (measured by test loss) and generates realistic scenarios, yielding prescribed assignments with low post-decision surprise. Interestingly, while baselines can produce assignments with post-decision high surprise, they rarely lead to post-decision disappointment (Fig. 3); we compare these metrics and discuss this phenomenon further in Appendix B.4. Finally, Appendix B.5 details the models' ability to recover the underlying preference distributions, and Appendix B.6 explores the effects of matching higher-order moments.

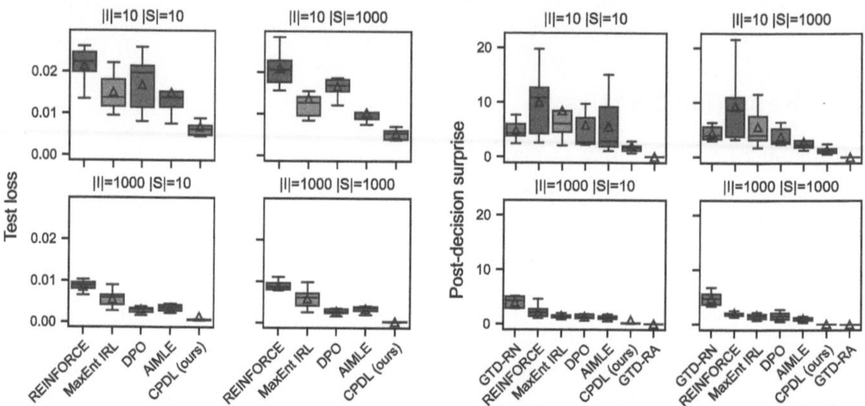

Fig. 2. Model performance across dataset scale and fidelity. Test loss (left) and post-decision surprise (right) as a function of training set size (rows) and observation fidelity (columns), for nine random seeds. Increasing the number of instances ($|\mathcal{I}|$) helps the model learn the context-to-distribution mapping, while increasing the number of samples per instance ($|\mathcal{S}|$) reduces noise in the choice distribution statistics ($\bar{\phi}$).

5   Conclusion

We propose a method to learn contextual preference distributions from observed contexts and aggregated solution statistics. By leveraging a bounded-variance stochastic gradient estimator, our approach yields an MLE of the model parameters. We demonstrate its ability to generate realistic scenarios for unseen contexts in a synthetic ridesharing environment, where we prescribe driver-rider assignments accounting for the uncertainty stemming from human choices. Our approach reduces average post-decision surprise by 2.4–114× compared to a risk-neutral approach with perfect predictions, and by 1.6–25× compared to leading risk-averse baselines. Future work will explore normalizing flows and broader choice settings, like discrete choice and tour choice (VRP) problems.

A   Additional Theoretical Results

A.1   Comparison to REINFORCE

CPDL computes the loss (1) for each datapoint using a batch of samples from the learned distribution $q_\theta(\mathbf{c}|\mathbf{X})$. REINFORCE does the opposite: it computes the loss for a batch of datapoints for each sample from the learned distribution. This approach effectively treats the entire distribution-to-loss pipeline as a black box and yields a different gradient from (2) [42]. Mathematically,

$$\mathcal{L}_{\text{REINFORCE}}(\mathbf{X}, \bar{\phi}, \theta) = \mathbb{E}_{q_\theta(\mathbf{c}|\mathbf{X})} \left[\| \bar{\phi} - \mathbb{E}_{p(\mathbf{z}|\mathbf{c})} \left[\phi(\mathbf{z}) \right] \|_2^2 \right],$$

$$\nabla_\theta \mathcal{L}_{\text{REINFORCE}}(\mathbf{X}, \bar{\phi}, \theta) = \mathbb{E}_{q_\theta(\mathbf{c}|\mathbf{X})} \left[\| \bar{\phi} - \mathbb{E}_{p(\mathbf{z}|\mathbf{c})} \left[\phi(\mathbf{z}) \right] \|_2^2 \nabla_\theta \log q_\theta(\mathbf{c}|\mathbf{X}) \right].$$

As discussed in Sect. 3.2, this gradient estimator can suffer from high variance because the loss for a given batch is unbounded. To mitigate this, our implementation of REINFORCE employs a baseline [32].

B   Additional Experimental Details

B.1   Data-Generating Process

We can express our synthetic data-generating process mathematically as

$$\boldsymbol{\beta}_\mu \sim \text{Uniform}([0, 1]^n)$$
$$\boldsymbol{\beta}_\sigma \sim \text{Uniform}([0, 1]^n)$$
$$\mathbf{X}_i \sim \text{Normal}(\mathbf{0}, \mathbf{I}_{|\mathcal{E}| \times n}) \qquad \forall i \in \mathcal{I}$$
$$\mathbf{c}_i^{(s)} \sim \text{LogNormal}(\mathbf{X}_i \boldsymbol{\beta}_\mu, \mathbf{X}_i \boldsymbol{\beta}_\sigma) \qquad \forall i \in \mathcal{I}, s \in \mathcal{S}$$
$$\mathbf{z}^{*}{}_i^{(s)} \in_R \mathcal{Z}^*(\mathbf{c}_i^{(s)}, \mathbf{u}) \qquad \forall i \in \mathcal{I}, s \in \mathcal{S}$$
$$\bar{\mathbf{p}}_i = \frac{1}{|\mathcal{S}|} \sum_{s \in S} \mathbf{z}^{*}{}_i^{(s)} \qquad \forall i \in \mathcal{I},$$

where $\boldsymbol{\beta}_\mu$ and $\boldsymbol{\beta}_\sigma$ are global n-dimensional vectors. $\mathbf{X}_i$ is a $|\mathcal{E}| \times n$ matrix, and $\mathbf{c}_i^{(s)}$, $\mathbf{z}^{*}{}_i^{(s)}$, and $\bar{\mathbf{p}}_i$ are $|\mathcal{E}|$-dimensional vectors. $\mathcal{I}$ is the set of instances and $\mathcal{S}$ is the set of samples for a given instance.

B.2   Risk-Averse Assignment Formulation

The risk-averse assignment problem is formulated as

$$
\min_{\mathbf{u}\in\mathbb{R}^{|\mathcal{K}|},v\in\mathbb{R},\mathbf{W}} \quad v + \frac{1}{1-\alpha}\frac{1}{|\mathcal{K}|}\sum_{k\in\mathcal{K}} u_k
$$

$$
\begin{aligned}
\text{s.t.} \quad & u_k \geq \sum_{d\in\mathcal{D}}\sum_{r\in\mathcal{R}} G_{dr}^{(k)} W_{dr} - v && \forall k \in \mathcal{K} \\
& u_k \geq 0 && \forall k \in \mathcal{K} \\
& \sum_{\mathcal{D}} W_{dr} = 1 && \forall r \in \mathcal{R} \\
& \sum_{\mathcal{R}} W_{dr} = 1 && \forall d \in \mathcal{D} \\
& W_{dr} \in \{0,1\} && \forall d \in \mathcal{D}, r \in \mathcal{R},
\end{aligned}
\tag{4}
$$

where $\mathbf{u}$ is a $|\mathcal{K}|$-dimensional vector with u_k representing the expected shortfall of scenario k, v is the value-at-risk, $\mathbf{W}$ is a $|\mathcal{D}| \times |\mathcal{R}|$ assignment matrix with $W_{dr} = 1$ indicating driver d is assigned to rider r, and α is the risk-level (0.95 in our experiments). Given a known cost vector $\mathbf{g}$ and exogenous variables $\mathbf{u}_{dr}$ $\forall d \in \mathcal{D}, r \in \mathcal{R}$, we generate a cost scenario $\mathbf{G}^{(k)}$ according to the following process

$$
\begin{aligned}
\mathbf{c}^{(d)} &\sim q_\theta(\mathbf{c}|\mathbf{X}) && \forall d \in \mathcal{D} \\
\mathbf{z}^{*(dr)} &\in_R \mathcal{Z}^*(\mathbf{c}^{(d)}, \mathbf{u}_{dr}) && \forall d \in \mathcal{D}, r \in \mathcal{R} \\
G_{dr}^{(k)} &= \mathbf{g}^\top \mathbf{z}^{*(dr)} && \forall d \in \mathcal{D}, r \in \mathcal{R},
\end{aligned}
$$

where G_{dr}^k is an element of the scenario cost matrix $\mathbf{G}^{(k)}$. The expected solution cost under the predicted distribution is

$$
\mathbb{E}_{\zeta \sim q_\theta(\mathbf{c}|\mathbf{X})}\left[g(\mathbf{W}^*, \zeta)\right] = \frac{1}{|\mathcal{K}|}\sum_{k\in\mathcal{K}}\sum_{d\in\mathcal{D}}\sum_{r\in\mathcal{R}} G_{dr}^{(k)} W_{dr}^*.
$$

B.3   Baselines

Our PyTorch implementation including all baselines is available at https://github.com/ben-hudson/contextual-preference-distribution-learning.

We use the black-box hyperparameter optimization tool in Weights & Biases to select hyperparameters for each method: beginning with 25 randomly sampled experiments, we acquire 25 more by minimizing validation loss with Bayesian optimization. Each experiment trains the given model on a randomly generated dataset with $|\mathcal{I}| = 100$ and $|\mathcal{S}| = 1000$, with an 80–20 train-validation split. The hyperparameter sets with the lowest validation loss are shown in Table 1.

Table 1. Hyperparameters used in our experiments. Perturbation scale defines the variance of the perturbing distribution; a hyperparameter in some methods and a learned quantity in others. Dashes (–) indicate not applicable to given baseline.

Parameter	AIMLE	DPO	MaxEnt IRL	REINFORCE	CPDL
Starting LR	1.06e-6	5.61e-2	8e-4	3e-4	2.5e-3
LR patience (epochs)	20	15	5	7	15
LR patience rel. tol.	6.92%	0.2%	3.64%	4.08%	3.31%
Perturbations (K)	400	300	–	500	300
Perturbation scale	0.6	0.9	–	Learned	Learned
Perturbation sides	2	–	–	–	–
Fixed-point max iters.	–	–	600	–	–
Fixed-point rel. tol.	–	–	3e-8	–	–
Batch size	128	128	5	128	128
Training epochs	200	500	200	200	200

B.4 Post-Decision Surprise and Disappointment

Fig. 3 shows post-decision surprise (computed according to (3)), which penalizes both early and late arrivals, alongside post-decision disappointment, which penalizes only late arrivals. We observe that the baselines consistently achieve much lower disappointment than surprise. This suggests they make overly conservative decisions: the assignments rarely result in late drivers (low disappointment) but frequently result in early ones (high surprise).

Counterintuitively, baseline performance degrades with more data. We examined specific predictions produced by these models—especially REINFORCE, given its similarity to our approach—and compared them to ours. We hypothesize is that this phenomenon is a result of aspects of the data-generating process and training process. The synthetic data-generating process (App. B.1) tends to generate correlated edge costs and edge scales because the coefficients β_μ and β_σ are strictly positive. With small training sets, the model can learn the correct relationship for the distribution location, but an inverse relationship for the distribution scale (i.e., a positive R^2 score but a negative correlation coefficient). Consequently, the model predicts edges with high cost or high variance, when in reality they tend to have both. This results in overly conservative decisions.

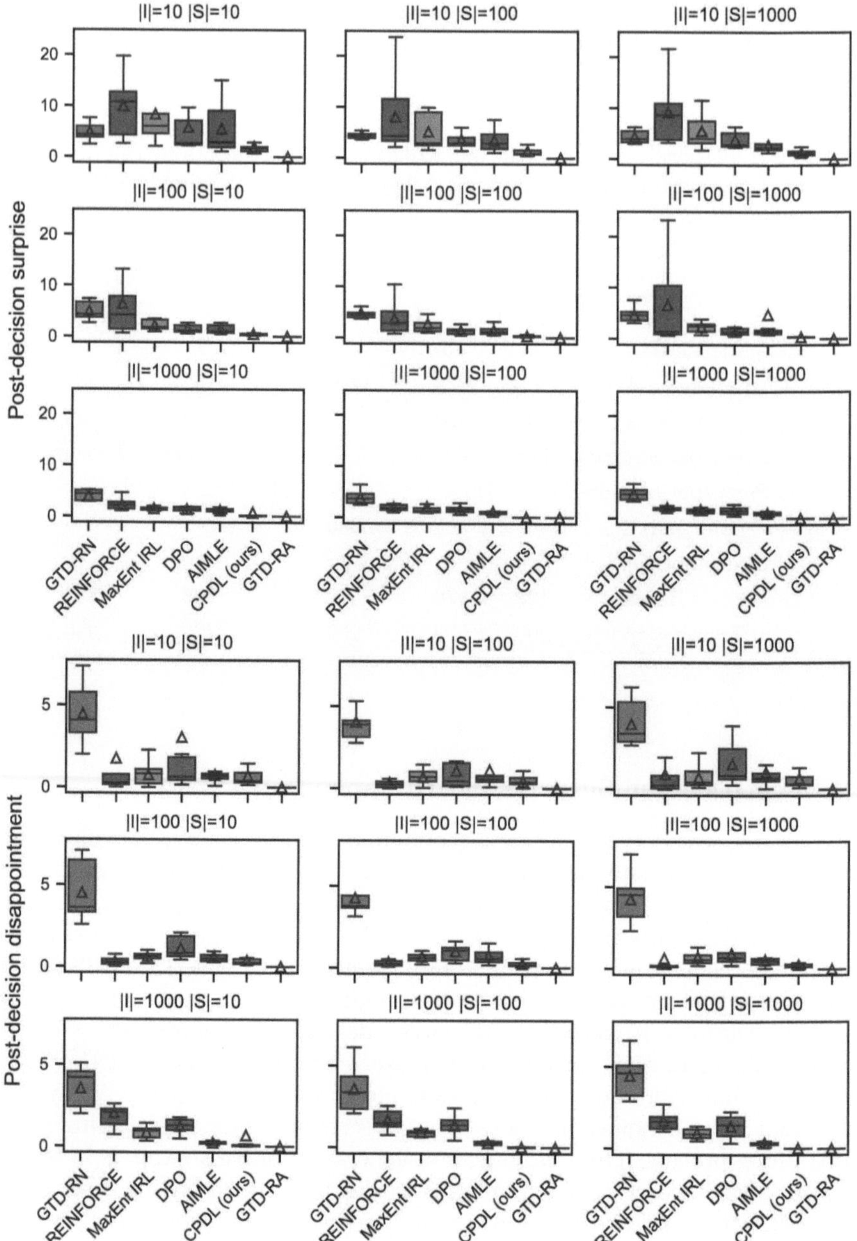

Fig. 3. Post-decision surprise (top) and disappointment (bottom) as a function of training set size (rows) and observation fidelity (columns), aggregated over nine random seeds. While surprise penalizes early and late arrivals, disappointment only penalizes late arrivals. The combination of high surprise and low disappointment suggests overly conservative assignments.

While our model is susceptible to this same issue, it corrects itself with less data, which explains why its performance closely tracks GTD-RA, the risk-averse baseline with access to the ground-truth distributions. This merits further study.

B.5   Preference Distribution Recovery

Fig. 4 shows test loss for additional combinations of training set size and fidelity, which measures the models' ability to reconstruct the observed choice distributions. Figure 5 shows the squared correlation coefficient (R^2) between predicted and ground-truth preference distribution parameters, which measures their recovery up to an affine transformation.

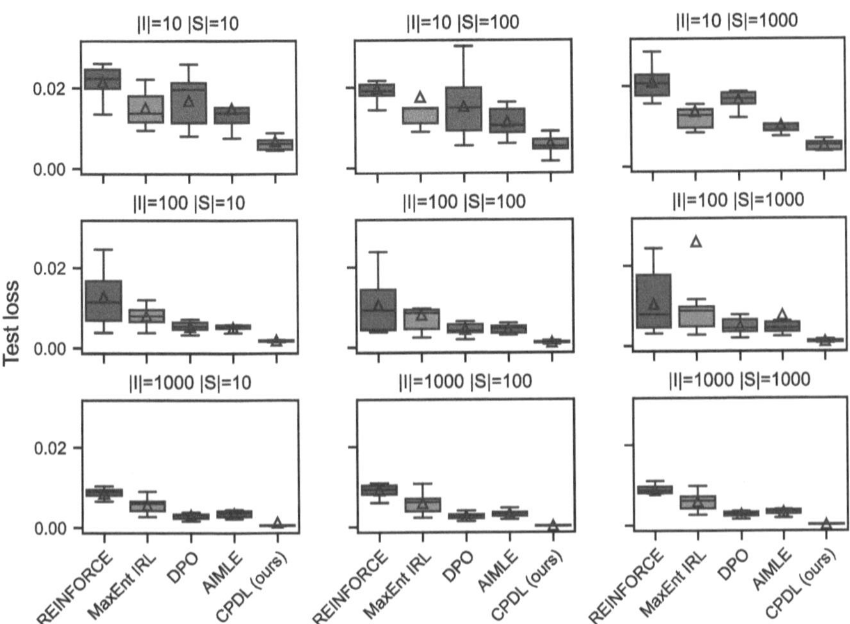

Fig. 4. Test loss as a function of training set size (rows) and observation fidelity (columns), aggregated over nine random seeds.

B.6   Matching Moments Beyond the Mean

Fig. 6 shows the impact of matching the first and second raw moments of the choice distribution, i.e. $\phi(\mathbf{z}) = [\mathbf{z}, \mathbf{z}\mathbf{z}^\top]$, instead of just the first.

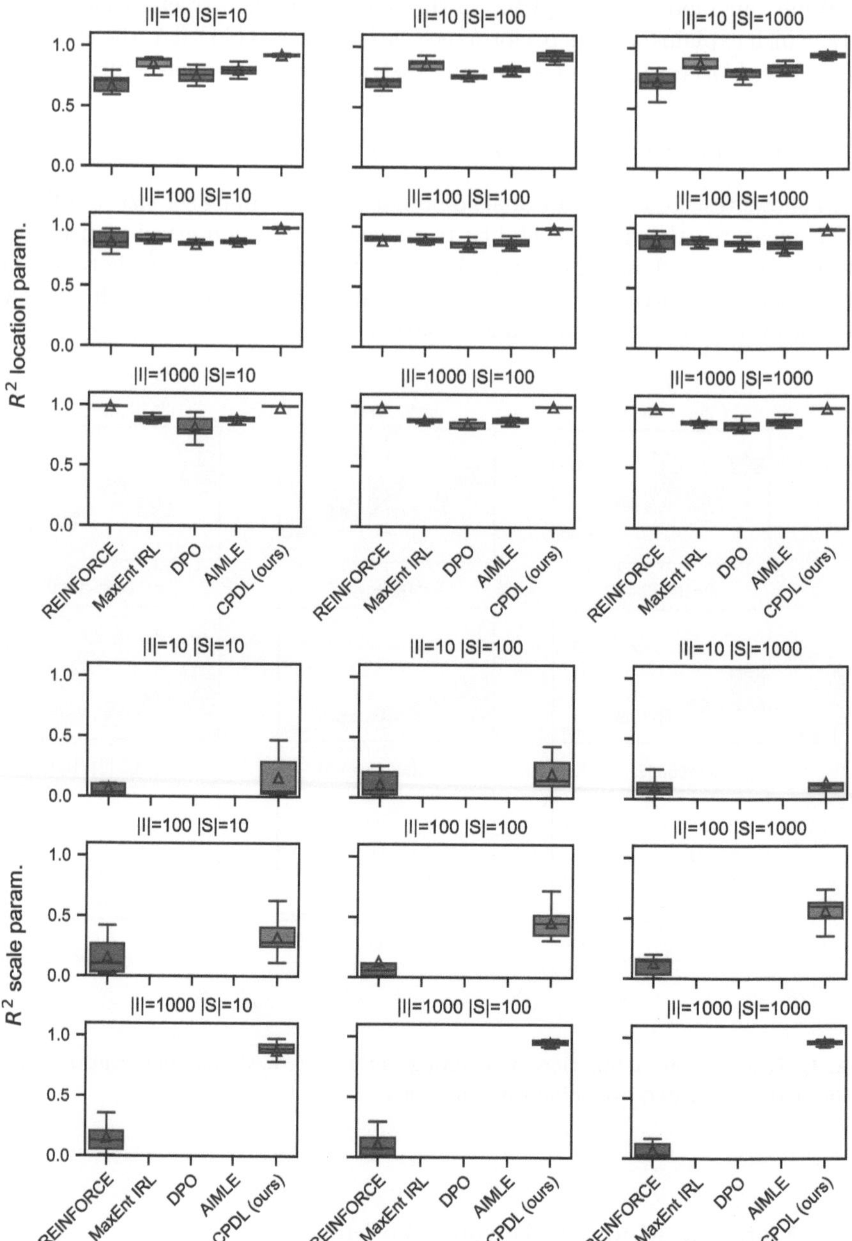

Fig. 5. Recovery of the ground-truth preference distribution location (top) and scale (bottom) up to an affine transformation as a function of training set size (rows) and observation fidelity (columns), aggregated over nine random seeds. The maximum score is one and the minimum score is zero. Only REINFORCE and CPDL are able to learn parameters beyond the location.

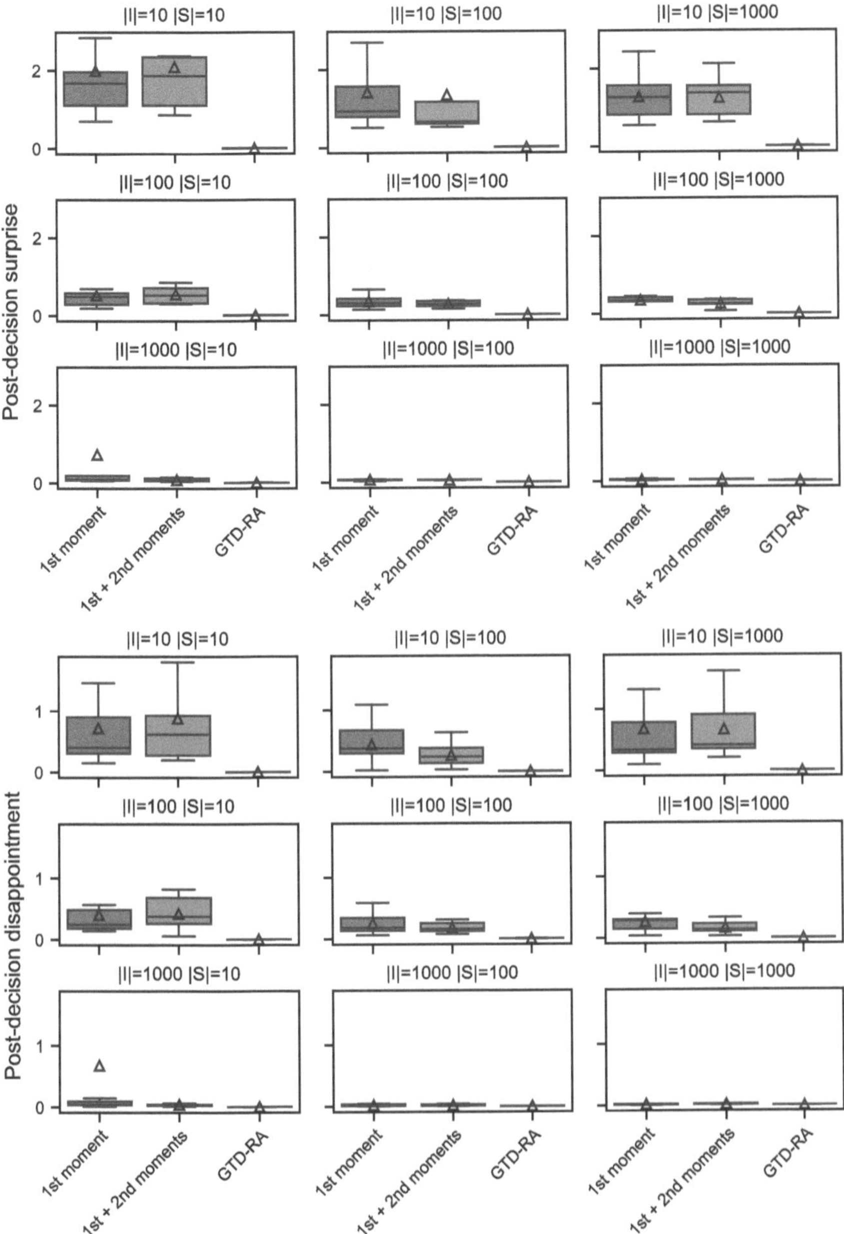

Fig. 6. Post-decision surprise (top) and disappointment (bottom) for CPDL matching only the first moment compared to the first and second moments. On small (low $|\mathcal{I}|$) and low-fidelity (low $|\mathcal{S}|$) datasets, matching only the first moment tends to yield better performance. We hypothesize that the first moment is a less informative learning signal but is more robust to noise. Thus, the tradeoff shifts in favour of higher moments as dataset size and fidelity increase.

References

1. Allen, R., Rehbeck, J.: Identification With Additively Separable Heterogeneity. Econometrica **87**(3), 1021–1054 (2019). https://onlinelibrary.wiley.com/doi/abs/10.3982/ECTA15867, https://doi.org/10.3982/ECTA15867
2. Berthet, Q., Blondel, M., Teboul, O., Cuturi, M., Vert, J.P., Bach, F.: Learning with differentiable perturbed optimizers. In: Advances in Neural Information Processing Systems, vol. 33, pp. 9508–9519. Curran Associates, Inc. (2020), https://papers.nips.cc/paper/2020/hash/6bb56208f672af0dd65451f869fedfd9-Abstract.html
3. Bertsimas, D., McCord, C., Sturt, B.: Dynamic optimization with side information. Eur. J. Oper. Res. **304**(2), 634–651 (2023). https://doi.org/10.1016/j.ejor.2022.03.030
4. Besbes, O., Fonseca, Y., Lobel, I.: Contextual inverse optimization: offline and online learning. Oper. Res. **73**(1), 424–443 (2025). https://doi.org/10.1287/opre.2021.0369
5. Birge, J., Li, X., Sun, C.: Learning from Stochastically Revealed Preference. Adv. Neural. Inf. Process. Syst. **35**, 35061–35071 (2022). https://proceedings.neurips.cc/paper_files/paper/2022/hash/e3c3473812173643147170188ef2b141-Abstract-Conference.html
6. Burton, J.W., Stein, M.K., Jensen, T.B.: A systematic review of algorithm aversion in augmented decision making. J. Behav. Decis. Mak. **33**(2), 220–239 (2020). https://doi.org/10.1002/bdm.2155, https://onlinelibrary.wiley.com/doi/abs/10.1002/bdm.2155
7. Canoy, R., Bucarey, V., Mandi, J., Guns, T.: Learn and route: learning implicit preferences for vehicle routing. Constraints **28**(3), 363–396 (2023). https://doi.org/10.1007/s10601-023-09363-2
8. Canoy, R., Guns, T.: Vehicle routing by learning from historical solutions. In: Schiex, T., de Givry, S. (eds.) CP 2019. LNCS, vol. 11802, pp. 54–70. Springer, Cham (2019). https://doi.org/10.1007/978-3-030-30048-7_4
9. Chan, T.C.Y., Mahmood, R., Zhu, I.Y.: Inverse optimization: theory and applications. Oper. Res. **73**(2), 1046–1074 (2025). https://doi.org/10.1287/opre.2022.0382
10. Chen, N., Farajollahzadeh, S., Wang, G.: Learning Consumer Preferences from Bundle Sales Data (Sep 2022). https://doi.org/10.48550/arXiv.2209.04942
11. Chenreddy, A., Delage, E.: End-to-end Conditional Robust Optimization (Mar 2024). https://doi.org/10.48550/arXiv.2403.04670
12. Chenreddy, A.R., Bandi, N., Delage, E.: Data-Driven Conditional Robust Optimization. Adv. Neural. Inf. Process. Syst. **35**, 9525–9537 (2022). https://papers.nips.cc/paper_files/paper/2022/hash/3df874367ce2c43891aab1ab23ae6959-Abstract-Conference.html
13. Deng, Y., Sen, S.: Predictive stochastic programming. CMS **19**(1), 65–98 (2022). https://doi.org/10.1007/s10287-021-00400-0
14. Elmachtoub, A.N., Lam, H., Zhang, H., Zhao, Y.: Estimate-Then-Optimize versus Integrated-Estimation-Optimization versus Sample Average Approximation: A Stochastic Dominance Perspective (May 2025). https://doi.org/10.48550/arXiv.2304.06833
15. Elmasri, M., Labbe, A., Larocque, D., Charlin, L.: Predictive inference for travel time on transportation networks (Mar 2023). https://doi.org/10.48550/arXiv.2004.11292

16. Fosgerau, M., Frejinger, E., Karlstrom, A.: A link based network route choice model with unrestricted choice set. Trans. Res. Part B: Methodol. **56**, 70–80 (2013). https://doi.org/10.1016/j.trb.2013.07.012

17. Fosgerau, M., Paulsen, M., Rasmussen, T.K.: A perturbed utility route choice model. Trans. Res. Part C: Emerging Technol. **136**, 103514 (2022). https://doi.org/10.1016/j.trc.2021.103514, https://www.sciencedirect.com/science/article/pii/S0968090X21004976

18. Guo, C., Yang, B., Hu, J., Jensen, C.S., Chen, L.: Context-aware, preference-based vehicle routing. VLDB J. **29**(5), 1149–1170 (2020). https://doi.org/10.1007/s00778-020-00608-7

19. Hannah, L., Powell, W., Blei, D.: Nonparametric density estimation for stochastic optimization with an observable state variable. In: Advances in Neural Information Processing Systems. vol. 23. Curran Associates, Inc. (2010). https://proceedings.neurips.cc/paper_files/paper/2010/hash/e1e32e235eee1f970470a3a6658dfdd5-Abstract.html

20. Harrison, J.R., March, J.G.: Decision making and postdecision surprises. Adm. Sci. Q. **29**(1), 26–42 (1984). https://doi.org/10.2307/2393078

21. Hunter, T., Herring, R., Abbeel, P., Bayen, A.: Path and travel time inference from GPS probe vehicle data. NIPS Analy. Netw. Learn. Graphs **12**, 1–8 (2009). https://snap.stanford.edu/nipsgraphs2009/papers/hunter-paper.pdf

22. Kannan, R., Bayraksan, G., Luedtke, J.R.: Residuals-based distributionally robust optimization with covariate information. Math. Program. **207**(1), 369–425 (2024). https://doi.org/10.1007/s10107-023-02014-7

23. Kristensen, D.: On inverse reinforcement learning and dynamic discrete choice for predicting path choices (Nov 2021). http://hdl.handle.net/1866/26530

24. Lin, B., Delage, E., Chan, T.C.: Conformal Inverse Optimization. Adv. Neural. Inf. Process. Syst. **37**, 63534–63564 (2024). https://proceedings.neurips.cc/paper_files/paper/2024/hash/7423902b5534e2b267438c85444a54b1-Abstract-Conference.html

25. Lu, H., Peng, Z., Yang, J.: MPAX: Mathematical Programming in JAX (Feb 2025). https://doi.org/10.48550/arXiv.2412.09734

26. Mandi, J., et al.: Decision-focused learning: foundations, state of the art, benchmark and future opportunities. J. Artifi. Intell. Res. **80**, 1623–1701 (2024). https://doi.org/10.1613/jair.1.15320, http://arxiv.org/abs/2307.13565

27. McFadden, D.: Econometric models of probabilistic choice. Structural analysis of discrete data with econometric applications 198272 (1981)

28. McFadden, D., Fosgerau, M.: A theory of the perturbed consumer with general budgets. Tech. rep., National Bureau of Economic Research (2012). https://www.nber.org/papers/w17953

29. Merchán, D., et al.: 2021 amazon last mile routing research challenge: data set. Transp. Sci. **58**(1), 8–11 (2024). https://doi.org/10.1287/trsc.2022.1173, https://pubsonline.informs.org/doi/full/10.1287/trsc.2022.1173

30. Minervini, P., Franceschi, L., Niepert, M.: Adaptive perturbation-based gradient estimation for discrete latent variable models. In: Proceedings of the AAAI Conference on Artificial Intelligence , vol. 37(8), pp. 9200–9208 (2023). https://doi.org/10.1609/aaai.v37i8.26103

31. Mohajerin Esfahani, P., Shafieezadeh-Abadeh, S., Hanasusanto, G.A., Kuhn, D.: Data-driven inverse optimization with imperfect information. Math. Program. **167**(1), 191–234 (2018). https://doi.org/10.1007/s10107-017-1216-6

32. Mohamed, S., Rosca, M., Figurnov, M., Mnih, A.: Monte Carlo Gradient Estimation in Machine Learning (Sep 2020). https://doi.org/10.48550/arXiv.1906.10652

33. Mulamba, M., Mandi, J., Diligenti, M., Lombardi, M., Bucarey, V., Guns, T.: Contrastive Losses and Solution Caching for Predict-and-Optimize (Jul 2021). https://doi.org/10.48550/arXiv.2011.05354
34. Murphy, K.P.: Probabilistic Machine Learning: Advanced Topics. The MIT Press, Cambridge, Massachusetts (2023)
35. Nguyen, T.: ETA Phone Home: How Uber Engineers an Efficient Route (Nov 2015). https://www.uber.com/en-EG/blog/engineering-routing-engine/
36. Niepert, M., Minervini, P., Franceschi, L.: Implicit MLE: Backpropagating through discrete exponential family distributions. In: Advances in Neural Information Processing Systems, vol. 34, pp. 14567–14579. Curran Associates, Inc. (2021). https://proceedings.neurips.cc/paper_files/paper/2021/hash/7a430339c10c642c4b2251756fd1b484-Abstract.html
37. Ohmori, S.: A predictive prescription using minimum volume k-nearest neighbor enclosing ellipsoid and robust optimization. Mathematics 9(2), 119 (2021). https://doi.org/10.3390/math9020119
38. Patel, Y., Rayan, S., Tewari, A.: Conformal Contextual Robust Optimization (Oct 2023). https://doi.org/10.48550/arXiv.2310.10003
39. Peršak, E., Anjos, M.F.: Integration of Constraint Programming, Artificial Intelligence, and Operations Research. Presented at the (2023). https://doi.org/10.1007/978-3-031-33271-5_9 Contextual Robust Optimisation with Uncertainty Quantification
40. Sadana, U., Chenreddy, A., Delage, E., Forel, A., Frejinger, E., Vidal, T.: A survey of contextual optimization methods for decision-making under uncertainty. Eur. J. Oper. Res. 320(2), 271–289 (2025). https://doi.org/10.1016/j.ejor.2024.03.020. https://www.sciencedirect.com/science/article/pii/S0377221724002200
41. Sahoo, S.S., Vlastelica, M., Paulus, A., Musil, V., Kuleshov, V., Martius, G.: Gradient Backpropagation Through Combinatorial Algorithms: Identity with Projection Works (May 2022). https://doi.org/10.48550/arXiv.2205.15213
42. Silvestri, M., et al.: Score Function Gradient Estimation to Widen the Applicability of Decision-Focused Learning (Jun 2024). https://doi.org/10.48550/arXiv.2307.05213
43. Srivastava, P.R., Wang, Y., Hanasusanto, G.A., Ho, C.P.: On Data-Driven Prescriptive Analytics with Side Information: A Regularized Nadaraya-Watson Approach (Oct 2021). https://doi.org/10.48550/arXiv.2110.04855
44. Sun, C., Liu, L., Li, X.: Predict-then-calibrate: a new perspective of robust contextual LP. Adv. Neural. Inf. Process. Syst. 36, 17713–17741 (2023). https://proceedings.neurips.cc/paper_files/paper/2023/hash/397271e11322fae8ba7f827c50ca8d9b-Abstract-Conference.html
45. Sun, C., Liu, S., Li, X.: Maximum Optimality Margin: A Unified Approach for Contextual Linear Programming and Inverse Linear Programming (May 2023). https://doi.org/10.48550/arXiv.2301.11260
46. Train, K.E.: Discrete Choice Methods with Simulation. Cambridge University Press, Cambridge, 2 edn. (2009). https://doi.org/10.1017/CBO9780511805271, https://www.cambridge.org/core/books/discrete-choice-methods-with-simulation/49CABD00F3DDDA088A8FBFAAAD7E9546
47. Vlastelica, M., Paulus, A., Musil, V., Martius, G., Rolínek, M.: Differentiation of Blackbox Combinatorial Solvers (Feb 2020). https://doi.org/10.48550/arXiv.1912.02175
48. Westgate, B.S., Woodard, D.B., Matteson, D.S., Henderson, S.G.: Large-network travel time distribution estimation for ambulances. Eur. J. Oper. Res. 252(1), 322–333 (2016). https://doi.org/10.1016/j.ejor.2016.01.004

49. Zattoni Scroccaro, P., Atasoy, B., Mohajerin Esfahani, P.: Learning in inverse optimization: incenter cost, augmented suboptimality loss, and algorithms. Oper. Res. (2024). https://doi.org/10.1287/opre.2023.0254
50. Zhao, Z., Liang, Y.: A deep inverse reinforcement learning approach to route choice modeling with context-dependent rewards. Trans. Res. Part C: Emerging Technol. **149**, 104079 (2023). https://doi.org/10.1016/j.trc.2023.104079
51. Ziebart, B.D., Maas, A., Bagnell, J.A., Dey, A.K.: Maximum entropy inverse reinforcement learning. In: Proceedings of the 23rd National Conference on Artificial Intelligence, AAAI 2008, vol. 3, pp. 1433–1438. AAAI Press, Chicago, Illinois (Jul 2008). https://cdn.aaai.org/AAAI/2008/AAAI08-227.pdf

Multi-objective Maximum Satisfiability by Single-Objective Implicit Hitting Set Optimization

Christoph Jabs$^{(\boxtimes)}$ ⓘ, Jeremias Berg ⓘ, and Matti Järvisalo ⓘ

Department of Computer Science, University of Helsinki, Helsinki, Finland
`{christoph.jabs,jeremias.berg,matti.jarvisalo}@helsinki.fi`

Abstract. Significant advances in practical approaches to maximum satisfiability (MaxSAT) solving have made MaxSAT a viable approach to solving complex NP-hard combinatorial optimization problems. Several recent works have extended single-objective MaxSAT algorithms to the multi-objective setting, enabling the enumeration of Pareto-optimal solutions for problems expressed as multi-objective MaxSAT (MO-MaxSAT). We propose and instantiate an alternative approach to MO-MaxSAT solving. Phrased as an implicit hitting set (IHS) approach, our algorithm works by iteratively invoking a single-objective IHS oracle on a scalarization of the multi-objective instance at hand. Our open-source implementation significantly outperforms an earlier-proposed IHS-style approach and complements the current state of the art in algorithmic approaches to MO-MaxSAT.

Keywords: Maximum satisfiability · multi-objective optimization · implicit hitting set approach

1 Introduction

Building on the success of Boolean satisfiability (SAT) [32] solvers, and especially their ability to provide explanations of unsatisfiability, maximum satisfiability (MaxSAT) [2] is today a viable option—alongside other constraint optimization formalisms such as integer-linear programming [7] and finite-domain constraint optimization [35]—for solving NP-hard combinatorial optimization problems. This holds especially true for optimization problems which can be naturally encoded as logical constraints.

The most successful algorithmic approaches employed in current state-of-the-art MaxSAT solvers make use of SAT solvers in different ways. The solution-improving [27,34] and core-guided [18,33] approaches use a SAT solver as the key reasoning engine and extend the working instance with optimization-specific clauses. The clause-learning branch-and-bound approaches [8] extend a SAT solver with optimization-specific reasoning. Finally, the implicit hitting set (IHS)

T. Guns (Ed.): CPAIOR 2026, LNCS 16595, pp. 262–279, 2026.
https://doi.org/10.1007/978-3-032-27242-3_16

approach [10–12,36], as a specific type of an automatic logic-based Benders decomposition [17] approach, makes use of both explanations of unsatisfiability (typically computed with a SAT solver) and hitting set optimization (typically instantiated with an integer programming solver).

While MaxSAT is standardly phrased in terms of single-objective optimization, various recent works have successfully extended single-objective MaxSAT algorithms to the multi-objective setting to address the arguably computationally more involved problem of Pareto optimization [15] under multiple objectives, i.e., multi-objective MaxSAT (MO-MaxSAT). A range of practical algorithmic approaches have been proposed, based on employing or combining solution-improving [22,37] and core-guided search [9,20,22] for MO-MaxSAT as well as an extension of IHS [9] to the multi-objective setting.

We propose and instantiate a novel alternative approach to MO-MaxSAT solving. In contrast to the purely SAT-based MO-MaxSAT algorithms, with the so-called P-Minimal [37] approach as a key representative of the current state of the art, our approach constitutes an IHS approach that is based on iteratively computing lower bounds for a non-dominated (Pareto-optimal) point in the solution space and employing hitting set computations to rule out non-solutions until a representative Pareto-optimal solution for a non-dominated point is found.

Our approach significantly differs and essentially simplifies the only IHS-style MO-MaxSAT algorithm, CLMIHS, proposed earlier [9]. Starting from an unconstrained solution space as the initial "hitting set" relaxation of the search space, CLMIHS computes at each iteration a set of points that can be seen as a lower bound for the whole Pareto frontier of non-dominated points of the original instance at hand. This requires invoking an MO-MaxSAT solver at each iteration, taking the role of what is traditionally called the "hitting set" optimizer in IHS. After obtaining a lower-bounding frontier from the MO-MaxSAT solver, the algorithm proceeds by computing constraints (called cores) for each of its points as explanations of why the lower-bounding frontier is not the actual Pareto frontier sought for. The cores are then added to the current hitting set relaxation, refining the lower-bounding frontier towards the actual Pareto frontier, after which the MO-MaxSAT solver is called again on the refined relaxation in the next iteration, until the MO-MaxSAT solver reports the actual Pareto frontier.

In contrast, instead of employing an MO-MaxSAT solver as a lower-bounding oracle for the whole Pareto frontier at each iteration, our approach, PARETOIHS, consists of an outer loop in which at each iteration exactly one Pareto-optimal solution witnessing a non-dominated point is computed. This is achieved by invoking a *single-objective* IHS oracle—which constitutes the inner loop of the algorithm—on a single-objective scalarization of the multi-objective instance at hand. This has three intuitive benefits: (i) The individual calls to the optimizer are single-objective rather than multi-objective solver calls, with intuitively expected lower computational cost; (ii) using a single-objective IHS oracle allows for directly making use of various search techniques designed for single-objective IHS; and (iii) each time a Pareto-optimal solution is found, a so-called Pareto dominance (PD) cut constraint can be enforced to prune out from the remaining solution space all solutions dominated by the just-computed Pareto-optimal solution. By different choices of the scalarization coefficients, PARETOIHS search can also be ordered in terms of e.g. lexicographic optimal or diverse solutions. Regardless of the choice of scalarization coefficients, search terminates after the so-far introduced PD cuts make the remaining feasible region empty.

Overall, we show empirically that our approach significantly outperforms the earlier-proposed CLMIHS IHS-based approach to MO-MaxSAT, and complements in performance the current state-of-the-art P-Minimal MO-MaxSAT algorithm. Furthermore, we identify a connection of our approach to the MIPPD algorithm proposed in the context of multi-objective integer-linear programming solving [38], based on enumerating non-dominated points via iteratively solving single-objective scalarizations of the multi-objective problem instance. We also instantiate MIPPD—to the best of our knowledge for the first time—in the context of MO-MaxSAT, and show that it has complementary performance against both PARETOIHS and P-Minimal. Finally, we provide open-source implementations of both PARETOIHS and MIPPD for MO-MaxSAT.

2   Preliminaries

For a Boolean variable x there are two literals, x and $\neg x$. A clause $C = (\ell_1 \vee \cdots \vee \ell_n)$ is a disjunction of literals, and a CNF formula $F = (C_1 \wedge \cdots \wedge C_m)$ a conjuction of clauses. When convenient, we view clauses as sets of literals and formulas as sets of clauses. An assignment α maps each variable x to a value in $\{0, 1\}$. The semantics of assignments are extended to literals, clauses and formulas by $\alpha(\neg x) = 1 - \alpha(x)$, $\alpha(C) = \max\{\alpha(\ell) \mid \ell \in C\}$, and $\alpha(F) = \min\{\alpha(C) \mid C \in F\}$. An assignment α for which $\alpha(F) = 1$ is a solution to F; if a solution to F exists, F is satisfiable and otherwise unsatisfiable. Given a formula F, the Boolean satisfiability (SAT) problem [6] is to determine whether F is satisfiable.

An objective $O = \sum_{j=1}^{k} c_j \ell_j$ consists of terms $c_j \ell_j$. Each term is the product of a positive integer constant c_j and a literal ℓ_j. The coefficient of a literal ℓ in O is denoted by $\textsc{Coeff}(\ell, O)$ and the sum of all coefficients in O is $\textsc{CoeffSum}(O)$. The value $O \restriction_\alpha$ of O under an assignment α is $O \restriction_\alpha = \sum_{j=1}^{k} c_j \alpha(\ell_j)$. An instance $\mathcal{I} = (F, (O_1, \ldots, O_p))$ of the multi-objective maximum satisfiability (MO-MaxSAT) problem consists of a CNF formula F and p objectives under minimization. Thereby single-objective MaxSAT instances are of form (F, O). This definition of MaxSAT is equivalent to the more classical definition using hard and weighted soft clauses; see [22]. The solutions of $\mathcal{I}$ are the solutions of F. We assume that MO-MaxSAT instances have at least one solution. Given two solutions α and β of $\mathcal{I}$, α *dominates* β if $O_i \restriction_\alpha \leq O_i \restriction_\beta$ for all $i = 1, \ldots, p$ and $O_i \restriction_\alpha < O_i \restriction_\beta$ for some i. A solution α is *Pareto-optimal* if it is not dominated by any solution. A p-tuple $\mathbf{d} = (v_1, \ldots, v_p)$ of non-negative integers is a *non-dominated point* of $\mathcal{I}$ if there is a Pareto-optimal solution α of $\mathcal{I}$ for which $O_i \restriction_\alpha = v_i$ for all $i = 1, \ldots, p$. Such an α is a representative solution (or witness) for the non-dominated point $\mathbf{d}$. We consider the generic problem of computing the set of all non-dominated points of instance $\mathcal{I}$, i.e., the *non-dominated set* $\{(O_1\restriction_\alpha, \ldots, O_p\restriction_\alpha) \mid \alpha \text{ is Pareto-optimal for } \mathcal{I}\}$, by finding one witness for each non-dominated point.

In the language of integer linear programming (ILP) [7], constraints are expressed as linear inequalities of form $\sum_{i=1}^{k} c_i x_i \leq b$, where c_i and b are integer constants, and x_i are integer decision variables. 0–1 (binary) IPs are a central restriction of general ILP in which all variables are Boolean.

3   The PARETOIHS Algorithm

We propose the PARETOIHS algorithm as a natural lifting of the implicit hitting set approach from single-objective MaxSAT to the multi-objective setting.

Before an overview (Sect. 3.1) and technical details on PARETOIHS (Sects. 3.2–3.3), we start with a central observation underlying PARETOIHS: all k non-dominated points of a given MO-MaxSAT instance $(F, (O_1, \ldots, O_p))$ can be enumerated by computing optimal solutions to a sequence of k monotonically growing sets of constraints under a *fixed scalarization* of the objectives $(O_1, \ldots, O_p)$.

Specifically, consider an instance $(F, (O_1, \ldots, O_p))$, a tuple $\Lambda = (\lambda_1, \ldots, \lambda_p)$ (where $\lambda_i > 0$) of positive scalarization constants and the scalarized objective $O^\Lambda = \sum_{i=1}^{p} \lambda_i O_i$. An optimal solution α^Λ to the single objective optimization problem (F, O^Λ) is a Pareto-optimal solution of $(F, (O_1, \ldots, O_p))$; see e.g. [15, Chapter 3]. In this sense, the non-dominated point $(O_1\restriction_{\alpha^\Lambda}, \ldots, O_p\restriction_{\alpha^\Lambda})$ of $(F, (O_1, \ldots, O_p))$ witnessed by α^Λ corresponds to the tuple Λ of scalarization constants. Different tuples of scalarization constants can correspond to different non-dominated points. However, it is in general not possible to enumerate all non-dominated points of $(F, (O_1, \ldots, O_p))$ by solving single-objective optimization problems arising from different scalarizations, as there may be *unsupported* non-dominated points for which there are no corresponding scalarization constants. Such unsupported non-dominated points do not lie on the convex hull of the non-dominated set [15, Chapter 3]. However, as observed in e.g. [38], all non-dominated points can be enumerated by an incremental procedure that, for a fixed tuple Λ of scalarization constants,

(i)   computes an optimal solution α^Λ to (F, O^Λ), and

(ii) adds constraints to F that block all solutions that are dominated by α^Λ or have equal objective values to it, including α^Λ itself,

until no more solutions can be found.

3.1   Overview of PARETOIHS

Figure 1 gives a high-level overview of our PARETOIHS algorithm for enumerating the non-dominated points of an MO-MaxSAT instance $(F, (O_1, \ldots, O_p))$. The algorithm iterates between two subroutines, the *oracle* and the *optimizer*.

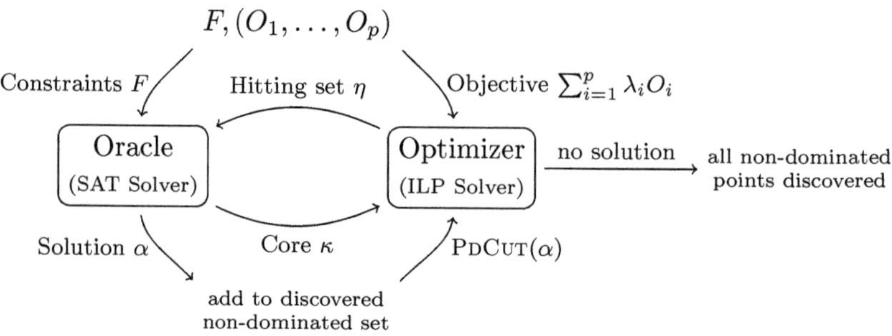

Fig. 1. Illustration of the PARETOIHS algorithm.

On input $(F, (O_1, \ldots, O_p))$ the oracle is initialized with a decision problem consisting of the constraints in F, and the optimizer with a single-objective optimization problem consisting of a scalarization $O^\Lambda = \sum_{i=1}^{p} \lambda_i O_i$ of the input objectives as the objective under minimization. In its inner loop, PARETOIHS iterates between the optimizer and the oracle to compute an optimal solution to the single-objective problem (F, O^Λ) as follows: (i) The optimizer computes an optimal solution η to its current optimization problem under O^Λ. (We call η a hitting set although in practice the constraints of the optimizer are not restricted to hitting-set constraints.) (ii) The oracle checks if F admits a solution α that agrees with η. In practice, the checks performed by the oracle are implemented via the incremental assumption interface provided by SAT solvers [14], invoking the SAT solver with η as the assumptions. If there is no such α, the oracle returns a clausal constraint $\kappa \subseteq \{\neg\ell \mid \ell \in \eta\}$ implied by F, referred to as a core which it adds to the optimizer. As the core is falsified by η, it explains why the oracle rejected η.

If the oracle instead finds a solution α of F that agrees with η, the solution is optimal for (F, O^Λ), since any set of cores is a relaxation of F [11, Theorem 1], and thus also a Pareto-optimal solution of $(F, (O_1, \ldots, O_p))$. The non-dominated point $(O_1\!\restriction_\alpha, \ldots, O_p\!\restriction_\alpha)$ witnessed by α is next added to the so-far found collection of non-dominated points. A constraint called a Pareto dominance (PD) cut PdCut(α) is then formed from α and added to the optimizer. A PD cut blocks all solutions dominated by α, all solutions that have equal objective values, and α itself. The PD cut PdCut(α) associated with α is the constraint $\bigvee_{i=1}^{p} O_i \leq (O_i\!\restriction_\alpha -1)$, encoded into linear constraints with the big-M method [38] using one auxiliary Boolean indicator variable r_i per objective: $O_i - \text{COEFFSUM}(O_i)(1 - r_i) \leq O_i\!\restriction_\alpha -1$ and $\sum_{i=1}^{p} r_i \geq 1$.

The algorithm iterates between steps (i) and (ii) until the constraints in the optimizer become infeasible, at which point search terminates since all non-dominated points have been discovered. The next proposition formalizes the correctness of PARETOIHS.

Proposition 1 (Correctness of PARETOIHS). *Given an input MO-MaxSAT instance $\mathcal{I} = (F, (O_1, \ldots, O_p))$ PARETOIHS eventually terminates after enumerating exactly the non-dominated points of $\mathcal{I}$.*

Proposition 1 follows from the following earlier observations. Firstly, the "inner" loop of the algorithm correctly computes optimal solutions to the scalarization.

Lemma 1 (Theorem 1 in [11]). *Let $\mathcal{C}$ be a set of PD cuts added to the optimizer and η a hitting set for the optimizer's constraints that is minimum cost under O^Λ. Let α be a solution to F that agrees with η. Then α is a minimum cost solution to $(F \wedge \mathcal{C}, O^\Lambda)$.*

Secondly, all non-dominated points are eventually found.

Lemma 2 (Proposition 1 in [38]). *Consider an MO-MaxSAT instance $\mathcal{I} = (F, (O_1, \ldots, O_p))$ and $\mathcal{C} \subseteq \{\text{PDCUT}(\alpha) \mid \alpha \text{ is Pareto-optimal for } \mathcal{I}\}$ a set of PD cuts generated from Pareto-optimal solutions to $\mathcal{I}$. Then the following hold: (i) If α is a Pareto-optimal solution to $(F \wedge \mathcal{C}, (O_1, \ldots, O_p))$ it is also a Pareto-optimal solution to $\mathcal{I}$. (ii) If $F \wedge \mathcal{C}$ is infeasible, $\mathcal{C}$ contains a PD cut for every non-dominated point of $\mathcal{I}$.*

These lemmas yield Proposition 1.

Proof (of Proposition 1 (sketch)). Termination: Each call to the oracle subroutine results in either a core or a PD cut added to the constraint set of the optimizer, and each such constraint blocks at least one of the finitely many solutions of the optimizer's instance.

Exactly enumerating non-dominated points: By Lemma 1 any solution α accepted by the oracle is a minimum-cost solution to $(F \wedge \mathcal{C}, O^\Lambda)$ and thus a Pareto-optimal solution to $\mathcal{I}_\mathcal{C} = (F \wedge \mathcal{C}, (O_1, \ldots, O_p))$ [15, Chapter 3]; hence α is a Pareto-optimal solution to $\mathcal{I}$ by Lemma 2. As α satisfies all PD cuts added so far, it witnesses a previously unseen non-dominated point. All non-dominated points are enumerated based on the set $\mathcal{C}_f$ of PD cuts added to the optimizer by the time of termination. As the conjunction $\mathcal{K}$ of these PD cuts and a set of core constraints implied by F is infeasible, so is $F \wedge \mathcal{C}_f$. Hence by part (ii) of Lemma 2 $\mathcal{C}_f$ contains a PD cut for a witness for each non-dominated point of $\mathcal{I}$.
$\square$

The "inner loop" iterating between the optimizer and the oracle until the oracle finds a next solution α is essentially the IHS algorithm for single-objective MaxSAT. PARETOIHS lifts the IHS approach to the multi-objective setting by wrapping around this inner loop the construction of PD cuts based on the solutions the oracle finds, whereas in the single-objective setting the first solution found by the oracle leads immediately to termination. As a natural generalization of single-objective IHS, PARETOIHS effectively behaves as single-objective IHS when the input instance has a single objective function.

Algorithm 1: Full-fledged PARETOIHS.

Input: MO-MaxSAT instance $(F, (O_1, \ldots, O_p))$
Output: One Pareto-optimal solution per non-dominated point

1 $O^\Lambda \leftarrow \sum_{i=1}^{p} \lambda_i O_i$ // with chosen scalarization coefficients
2 $\mathcal{K} \leftarrow$ SEEDABLE(F) // initialize optimizer constraints via seeding
3 $L \leftarrow 0, \quad \mathcal{A} \leftarrow$ NONDOMARCHIVE$(O_1, \ldots, O_p)$ // lower/upper bounds
4 Precomp$(k, F, \mathcal{K}, (O_1, \ldots, O_p), \mathcal{A})$
5 **repeat**   // once for each non-dominated point
6       $\alpha \leftarrow$ Ihs$(F, \mathcal{K}, O^\Lambda, L, \mathcal{A})$
7   **yield** α // α is Pareto-optimal
8       $\mathcal{K} \leftarrow \mathcal{K} \cup$ PDCUT(α) // introduce PD cut
9   SIMPLIFYARCHIVE$(\mathcal{A}, $PDCUT$(\alpha))$
10 **until** $\mathcal{K}$ *unsat*   // checked by optimizer

3.2 Full-Fledged PARETOIHS

Algorithm 1 presents the PARETOIHS algorithm in full fledge, with various additional search techniques and heuristic refinements. These include both techniques lifted from single-objective IHS and new techniques specific to the multi-objective setting. In the pseudocodes, lines corresponding to such optional algorithmic refinements are highlighted with a gray background.

Many of the heuristics and search techniques either make use of or refine an upper U or lower L bound on the optimal (minimum) cost of the single-objective instance $(F \wedge \mathcal{K}, O^\Lambda)$ consisting of the constraints F, extracted cores and PD cuts $\mathcal{K}$, and a scalarized objective $O^\Lambda = \sum_{i=1}^{p} \lambda_i O_i$ solved during PARETOIHS search. As PD cuts and cores are only ever added during search, the optimal cost of $(F \wedge \mathcal{K}, O^\Lambda)$ is non-decreasing between the computations of non-dominated points. Thus any previously computed lower bound L remains valid throughout the execution of PARETOIHS. In contrast, any upper bound U obtained from an intermediate (non-optimal) solution α to $(F \wedge \mathcal{K}, O^\Lambda)$ is in general only valid until the next PD cut is added, since α might violate future PD cuts. As intermediate solutions might still be valid after the addition of PD cuts, we maintain an *upper bound archive* $\mathcal{A}$ that contains all found intermediate solutions that (a) are not dominated by other solutions in the archive, and (b) do not violate any PD cuts that have been added to $\mathcal{K}$. The archive is updated whenever a new solution α is found (by verifying that α is not dominated by some other solution in the archive, and removing solutions that α dominates) or a new PD cut is added (by removing solutions that violate it). Essentially, $\mathcal{A}$ contains solutions from which an upper bound for $(F \wedge \mathcal{K}, O^\Lambda)$ can be obtained whenever one is needed.

Given an input instance $(F, (O_1, \ldots, O_p))$ Algorithm 1 starts by forming a scalarization O^Λ of the objectives $(O_1, \ldots, O_p)$. The algorithm is correct for any choice Λ of positive scalarization constants; we discuss strategies for choosing the coefficients in Sect. 3.3. Next, the set $\mathcal{K}$ of constraints for the optimizer is initialized on Line 1. We utilize a refinement known as *constraint seeding* [12] that initializes $\mathcal{K}$ to contain all clauses of F that only contain literals in O^Λ. Constraint seeding "kickstarts" the IHS main loop by initializing the optimizer

---

Algorithm 2: The IHS subroutine of the PARETOIHS algorithm.

---

1 **Procedure** Ihs$(F, \mathcal{K}, O^\Lambda, L, \mathcal{A})$ // single-objective IHS optimization
2 $\quad\alpha \leftarrow \arg\min_{\alpha \in \mathcal{A}} O^\Lambda|_\alpha, \qquad U \leftarrow O^\Lambda|_\alpha$ // initial upper bound
3 $\quad$**repeat**   // the IHS loop
4 $\quad\quad$**if** α *defined* **then**   // non-trivial upper bound
5 $\quad\quad\quad \zeta \leftarrow$ REDUCEDCOSTFIXING$(\mathcal{K}, U, \alpha)$
6 $\quad\quad$**else** $\zeta \leftarrow \emptyset$
7 $\quad\quad$opt?$, \eta \leftarrow$ OPTIMIZER$(\mathcal{K} \wedge \zeta, O^\Lambda, U)$
8 $\quad\quad$**if** opt? **then** $L \leftarrow O^\Lambda|_\eta$ // update lower bound
9 $\quad\quad$sat?$, \alpha, \kappa \leftarrow$ ORACLEEXTEND(F, η)
10 $\quad\quad O_{\mathrm{WCE}} \leftarrow O^\Lambda$ // objective to simulate WCE with
11 $\quad\quad$**while** *not* sat? **do**   // the WCE loop
12 $\quad\quad\quad \kappa \leftarrow$ SHRINK(F, κ) // minimize the core
13 $\quad\quad\quad \mathcal{K} \leftarrow \mathcal{K} \wedge \kappa$
14 $\quad\quad\quad O_{\mathrm{WCE}} \leftarrow O_{\mathrm{WCE}} - \min\{\text{COEFF}(\ell, O_{\mathrm{WCE}}) \mid \ell \in \kappa\} \sum_{\ell \in \kappa} \ell$
15 $\quad\quad\quad \eta \leftarrow \eta \setminus \{\neg\ell \mid \text{COEFF}(\ell, O_{\mathrm{WCE}}) = 0\}$
16 $\quad\quad\quad$sat?$, \alpha, \kappa \leftarrow$ ORACLEEXTEND(F, η)
17 $\quad\quad$ADDTOARCHIVE$(\mathcal{A}, \alpha)$ // removes dominated elements
18 $\quad\quad \alpha \leftarrow \arg\min_{\alpha \in \mathcal{A}} O^\Lambda|_\alpha, \qquad U \leftarrow O^\Lambda|_\alpha$ // update upper bound
19 $\quad$**until** $L = U$
20 $\quad$**return** α

---

with cores that are *explicitly* present in the input formula, instead of initializing $\mathcal{K}$ as the empty set. A lowerbound L on the optimal cost of $(F \wedge \mathcal{K}, O^\Lambda)$ is initialized to 0 and an upper bound archive $\mathcal{A}$ is initialized to contain no solutions. (We assume that the set $\mathcal{K}$, the bound L and the archive $\mathcal{A}$ are modified in place so that state changes are preserved between invocations of any subroutine.) The initialization phase ends with heuristic precomputation of up to k lexicographically optimal solutions of $(F, (O_1, \ldots, O_p))$ that can be found simply by varying the scalarization constants. We detail this `Precomp` procedure in Sect. 3.3.

The main loop of Algorithm 1 (Lines 5–10) iterates once per non-dominated point. Each iteration begins with the implicit hitting set subroutine (`Ihs`) computing an optimal (minimum-cost) solution α to $(F \wedge \mathcal{K}, O^\Lambda)$ on Line 6, making use of the lower bound L and upper bound archive $\mathcal{A}$ in its search heuristics. After each invocation of `Ihs`, the solution α is Pareto-optimal and a witness for a previously unseen non-dominated point, and hence it is yielded on Line 7. The iteration ends with adding a new PD cut based on α to the constraints $\mathcal{K}$ of the optimizer and updating the archive $\mathcal{A}$ on Line 8.

Algorithm 2 details the `Ihs` subroutine of PARETOIHS that, in terms of Fig. 1, handles the back and forth between the optimizer (invoked on Line 7) and the oracle (Line 9) to compute an optimal (minimum-cost) solution to its input instance $(F \wedge \mathcal{K}, O^\Lambda)$. Here we make use of weight-aware core extraction [4] that generalizes the so-called disjoint cores technique [28] and extracts more than one core from each hitting set, which is why the oracle is also invoked on Line 16. The algorithm also maintains an upper bound U and lower bound L on the optimal cost of $(F \wedge \mathcal{K}, O^A)$ and terminates when the bounds match, at which point a solution α for which $O^A\!\restriction_\alpha = L$ has also been found.

Each iteration of Algorithm 2 starts with the upper bound U obtained as the smallest scalarized objective value of any solution in the upper bound archive $\mathcal{A}$ (Lines 2 and 18). If $\mathcal{A}$ is empty, α and U are undefined. If α is defined, a multi-objective lifting of *reduced cost fixing* [1] is employed to compute a set ζ of variables that can be fixed based on the reduced costs of variables in an optimal solution of the linear programming relaxation of the optimizer's constraints. To preserve the state of the oracle throughout the algorithm—in contrast to previous work on single-objective IHS—the fixings ζ are only added to the optimizer and not the oracle (cf. Lines 4–7).

Next, the optimizer is invoked on Line 7. Instead of requiring the optimizer to always compute an optimal solution to its constraints we allow it to return any solution η for which $O^A\!\restriction_\eta < U$. This is because computing optimal hitting sets can be expensive, and hitting sets of cost lower than U already ensure that the algorithm makes progress. If η cannot be extended to a solution, a new core is obtained; if it can, the extended solution has the same cost as η and thus constitutes a tighter upper bound. However, if U is undefined, we require the call to return an optimal solution. In addition to the hitting set η, the call to the optimizer returns an indicator opt? for optimality of η. If optimal, the lower bound L is updated on Line 8. The cost of any optimal hitting set being a lower bound on the minimum cost of $(F \wedge \mathcal{K}, O^A)$ follows from of the correctness of single-objective IHS [11].

On Line 9, the oracle is invoked to extend η into a solution to F. If η can be extended into a solution, sat? is true and α contains the extended solution. If η cannot be extended into a solution, sat? is false, α is not modified, and κ contains a core that is added to the optimizer (Line 13). Solutions provided by the oracle are added to the upper bound archive $\mathcal{A}$ (Line 17). The inner loop terminates once the lower bound L and the upper bound U match (Line 19).

Lines 10–16 of Algorithm 2 constitute weight-aware core extraction (WCE) and core minimization [31], aiming at computing multiple cores per each hitting set, typically without much overhead compared to extracting a single core. Concretely, on Line 10, we initialize the auxiliary objective O_{WCE} used by WCE as the scalarized objective O^A. In WCE, the coefficient of each literal in the core is lowered in O_{WCE} by the smallest coefficient over all literals in the core (Line 14). All literals that then have coefficient 0 in O_{WCE} are removed from the hitting set (Line 15) before checking for another solution (Line 16), potentially leading to multiple cores. As a byproduct, an extended solution to add to $\mathcal{A}$ is always obtained. Core minimization is interleaved with WCE: after extracting a core, core minimization heuristically shrinks the core by attempting to remove literals and checking—in practice under resource limits—whether it is still implied by the formula (Line 12).

Algorithm 3: The subroutine for precomputing lexicographically optimal solutions.

---

1 **Procedure** Precomp$(k, F, \mathcal{K}, (O_1, \ldots, O_p), \mathcal{A})$
2      $(\mathcal{S}, \mathcal{D}) \leftarrow (\emptyset, \emptyset)$ // solutions & corresponding non-dominated points
3   **foreach** $P \in \{P_1, \ldots, P_k \mid P_i \text{ is a permutation of } (1, \ldots, p)\}$ **do**
4          $\lambda \leftarrow 1$
5   **foreach** $i \in P$ **do** $\lambda_i \leftarrow \lambda$, $\lambda \leftarrow 1 + \lambda \cdot \textsc{CoeffSum}(O_i)$
6          $O_P^{\Lambda} \leftarrow \sum_{i=1}^{p} \lambda_i O_i$
7          $\alpha \leftarrow \text{Ihs}(F, \mathcal{K}, \emptyset, O_P^{\Lambda}, 0, \mathcal{A})$
8          $d^{\alpha} \leftarrow (O_1|_{\alpha}, \ldots, O_p|_{\alpha})$
9   **if** $d^{\alpha} \notin \mathcal{D}$ **then** $\mathcal{S} \leftarrow \mathcal{S} \cup \{\alpha\}$, $\mathcal{D} \leftarrow \mathcal{D} \cup \{d^{\alpha}\}$ **yield** α
10   **foreach** $\alpha \in \mathcal{S}$ **do**
11          $\mathcal{K} \leftarrow \mathcal{K} \cup \textsc{PdCut}(\alpha)$, $\textsc{SimplifyArchive}(\mathcal{A}, \textsc{PdCut}(\alpha))$

3.3   On Choosing Scalarization Coefficients

For initializing the optimizer in PARETOIHS, the coefficients Λ for the scalarized objective function O^{Λ} need to be chosen. Intuitively, the coefficients ascribe different "importance" to the objectives in terms of the order in which the solutions will be enumerated. If the user has domain knowledge on the relative importance of the objectives, this knowledge can be used for weighting the objectives, especially in cases when not enough time might be available for enumerating *all* non-dominated points. A straightforward option in practice is to choose $\Lambda^1 = (1, \ldots, 1)$, not influencing the relative size of coefficients in the different objectives. As an example of a different choice, consider $\Lambda^{\text{lex}} = (\lambda_1^{\text{lex}}, \ldots, \lambda_p^{\text{lex}})$ where $\lambda_1^{\text{lex}} = 1$ and $\lambda_i^{\text{lex}} = 1 + \sum_{j=1}^{i-1} \lambda_j^{\text{lex}} \textsc{CoeffSum}(O_i)$. Intuitively, Λ^{lex} corresponds to the strictest possible preference in which the ith objective is considered more important than the jth objective for all $i > j$. Specifically, when using Λ^{lex}, PARETOIHS will enumerate the non-dominated points in lexicographic order wrt the order $(O_1, \ldots, O_p)$ of the objectives.

Beyond choosing static scalarization coefficients at the start of PARETOIHS, the following strategy for varying the scalarization coefficients allows for precomputing so-called, lexicographic optima [15, Chapter 5], i.e., Pareto-optimal solutions for which the objective values are lexicographically minimal under some ordering of the objectives. To precompute up to k lexicographic optima before the main loop of the algorithm (see Algorithm 1 Line 4), iterate over k permutations P of the objective indices, and perform the following steps, detailed in Algorithm 3: (i) Compute Λ^{lex} under the order specified by P (Line 5) and set O_P^{Λ} accordingly (Line 6). (ii) Find the optimal solution under O_P^{Λ} (which will be the lexicographic optimum under the order specified by P) and yield it as Pareto-optimal if it has not yet been seen (Line 9). The check on Line 9 is needed to avoid separate PD cuts being generated for two solutions that are optimal under different O_P^{Λ}, but witness the same non-dominated point.

4   Empirical Evaluation

We empirically evaluate the performance of PARETOIHS, comparing its performance to state-of-the-art MO-MaxSAT algorithms.

Competing Approaches. We consider three competing approaches: (i) The P-Minimal algorithm [23,26,37] as implemented in Scuttle [19], representing the state of the art in MO-MaxSAT solving; (ii) CLMIHS as a previously-proposed implicit hitting set approach to MO-MaxSAT [9]; and (iii) the multi-objective ILP algorithm proposed in [38], which we refer to as MIPPD. To the best of our knowledge, this is the first time that MIPPD is employed for MO-MaxSAT. The MIPPD algorithm enumerates non-dominated points by solving a single-objective scalarization with the help of an ILP optimizer, while iteratively blocking dominated areas of the solution space with PD cuts. PARETOIHS can therefore be seen as a variant of MIPPD, where the single-objective optimization is realized with a single-objective implicit hitting set approach. Our instantiation of MIPPD uses Λ^1 as scalarization coefficients.

Implementation Details. We implemented PARETOIHS and MIPPD in the Scuttle MO-MaxSAT solver [19] (version 0.5.0), using Gurobi 12.0.1 as the ILP solver and CaDiCaL 2.1.3 as the SAT solver. The implementations are available at https://bitbucket.org/coreo-group/scuttle. In the subdirectory `cpaior26/`, the repository also contains a digital supplement including all benchmarks used, and more detailed empirical data. Our implementation leaves out objectives in PD cuts whenever the cut, together with a known lower bound, cannot be satisfied via the given objective. In PD cuts over two objectives, we reuse indicator variables to use auxiliary variables sparingly. Our implementation realizes the reified constraints required for PD cuts with Gurobi's native support for indicator constraints, as using native indicator constraints resulted on average in faster solving times compared to the manual addition of big-M constraints. (Using Gurobi's native support for minimizing scalarizations of multiple objectives via its multi-objective API turned out less effective.) In the default configuration of PARETOIHS used in the comparative evaluation, Λ^1 are used as the scalarization coefficients, and reduced cost fixing, weight-aware core extraction, core minimization, and constraint seeding are employed. Up to 8 lexicographic optima are computed before the main loop. (Section 4.3 provides empirical data on the impact of these choices.)

Benchmarks. As benchmark instances, we use a set of 300 instances from 6 domains, as earlier used in [20]: 80 package upgradability (Packup) [24] instances, split into four sets of 20 that have 2–5 objectives, respectively; 20 learning decision rules (LIDR) [29] instances with 2 objectives; 20 satellite photograph scheduling (Spot 5) [16] instances with 2 objectives; 20 flying tourist problem (FTP) [30] instances with 2 objectives; and two sets of 80 random set covering instances (Set-Cover EP, Set-Cover SC) generated in different ways [22] both further split into four sets that have 2–5 objectives, respectively. Additionally,

Table 1. Number of solved instances (#) and VBS contribution per solver and benchmark domain. The P-Minimal CB algorithm that employs core boosting as a preprocessing technique is not considered in the VBS computation or in the comparison of the best-performing solvers marked in bold.

Solver	DAL		Packup		LIDR		Spot 5		FTP		SC EP		SC SC	
	#	VBS	#	VBS	#	VBS	#	VBS	#	VBS	#	VBS	#	VBS
PARETOIHS	2	0	**23**	**20**	3	**3**	2	1	8	2	31	11	**25**	9
CLMIHS	0	0	2	0	0	0	0	0	4	0	22	0	6	0
MIPPD	1	0	20	1	2	0	**7**	**6**	**12**	**10**	31	11	**25**	14
P-Minimal	4	**4**	12	3	**6**	**3**	2	0	9	0	**32**	**15**	13	4
P-Minimal CB	5	–	13	–	6	–	13	–	7	–	38	–	33	–

we use 20 development assurance level (DAL) [5,13] instances with 7 objectives randomly picked from the instances used in [21] and encoded to CNF. For the total benchmark set, the number of clauses ranges from 20 to 1.3M with a mean of 51k and the number of variables from 69 to 98k with a mean of 8k.

Experimental Setup. The experiments reported on were run on 2.50-GHz Intel Xeon Gold 6248 machines with 381-GB RAM in RHEL under a per-instance 1-h time and 32-GB memory limit.

4.1 Performance Comparison to the State of the Art

We start by comparing the performance of PARETOIHS and the competing approaches in terms of the number of solved instances and their virtual best solver (VBS) contributions (i.e., the number of instances on which a specific solver was the fastest).

Table 1 lists the numbers of solved instances per benchmark family (#) and contributions to the VBS. PARETOIHS significantly outperforms the previously proposed IHS-style algorithm for MO-MaxSAT (CLMIHS) on all benchmark families and MIPPD on the DAL, Packup, and LIDR families, justifying the increased complexity of PARETOIHS compared to MIPPD. It should be noted that in the Set-Cover EP and Set-Cover SC benchmark domains, all clauses in the input formula are seeded into the optimizer, which means that PARETOIHS and MIPPD perform essentially the same search. Figure 2(left) provides a more-detailed per-instance runtime comparison of PARETOIHS and MIPPD. We observe that, as expected, PARETOIHS and MIPPD perform essentially identically on the two set covering domains; while their VBS contributions differ, the two VBS contributions would be the same under a 5-s tolerance. In contrast, PARETOIHS performs better on Packup and DAL. For the FTP and Spot 5 domains, on which MIPPD performs better, the runtime logs suggest that core extraction appears to be time-consuming for the oracle.

Lastly, we turn to P-Minimal as arguably the current state of the art in MO-MaxSAT solving [20,22]. PARETOIHS significantly outperforms P-Minimal on

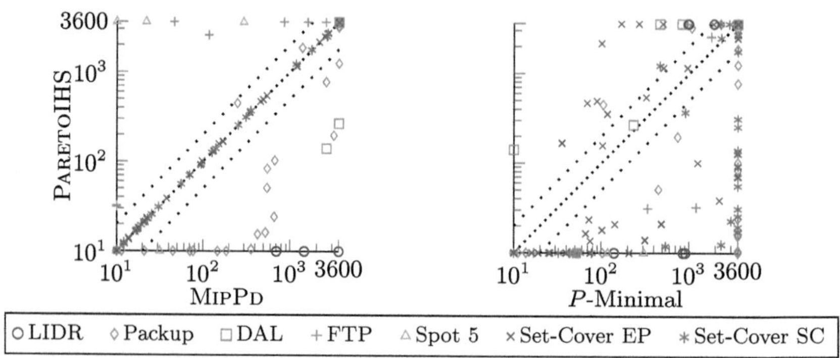

Fig. 2. Per-instance runtime comparison of PARETOIHS, P-Minimal, and MIPPD.

the Packup and Set-Cover SC domains, while more narrowly losing to P-Minimal on FTP and Set-Cover EP (while still contributing to the VBS). A more fine-grained per-instance runtime comparison of PARETOIHS and P-Minimal (Fig. 2 right) highlights the complementary performance of the two solvers.

In addition to the base P-Minimal algorithm, we present the number of solved instances for P-Minimal with the recently-proposed preprocessing technique of *core boosting* [20] (P-Minimal CB in Table 1), which was found to perform well with P-Minimal. We observe that for the Spot 5 and Set-Cover SC benchmark domains, core boosting allows P-Minimal to solve significantly more instances than either PARETOIHS or MIPPD. However, on other domains such as Packup, core boosting provides only a minor boost for P-Minimal so P-Minimal CB is still outperformed by PARETOIHS.

4.2 Search Progress Comparison to the State of the Art

We turn to evaluating the relative search progress of the best-performing algorithms PARETOIHS, P-Minimal, and MIPPD over time. For an instance $\mathcal{I}$, we report after different number of seconds t of runtime the number of non-dominated points enumerated by each algorithm divided by the (when necessary, estimated) total number of non-dominated points of $\mathcal{I}$. This results in progress values in the range $[0, 1]$; the higher the value the better in terms of search progress. For instances on which none of the solvers managed to terminate in 1-h with the total number of non-dominated points, we use as an estimate for the total number the number of non-dominated points the solvers managed to find together. Figure 3 shows the median, 25th, and 75th percentile progress over all instances of a given benchmark domain over time for each algorithm. On the DAL and Packup domains, PARETOIHS and MIPPD make steady progress, while P-Minimal makes almost no progress at all. On LIDR, P-Minimal makes the fastest progress, while on Spot 5 and FTP MIPPD performs the best. On the set covering domains, the performance of PARETOIHS and MIPPD are matched, both clearly dominating P-Minimal. It is also worth noting that progress for

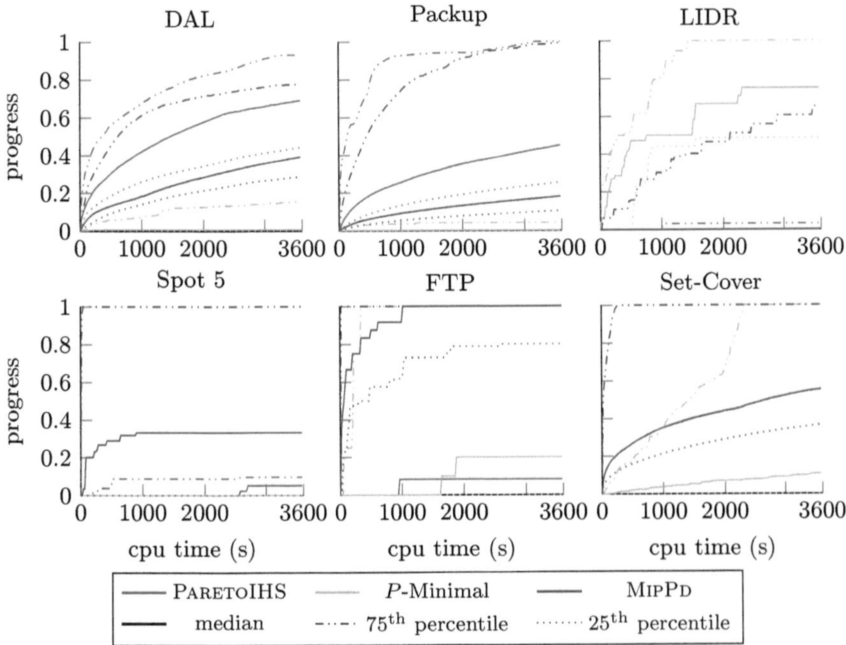

Fig. 3. Search progress for each algorithm in terms of median, 25th, and 75th percentile of progress for all instances of a given family over time.

PARETOIHS and MIPPD seems to be more steady; P-Minimal appears to more likely to get stuck for periods of time without making progress.

4.3   Marginal Contributions of Search Techniques in PARETOIHS

Finally, we evaluate the marginal contributions of the various search techniques employed in PARETOIHS, shown in Fig. 4. Here "default" refers to the default configuration used also in the just-detailed comparative results. (The digital supplement includes more detailed pairwise configuration comparisons.) Using the Λ^1 constants (as in the default configuration) leads to significantly better performance than using the Λ^{lex} constants detailed in Sect. 3.3. Figure 4 includes configurations that differ from the default configuration only in turning off one of: precomputing lexicographic optima, reduced cost fixing, weight-aware core extraction, core minimization, or constraint seeding, respectively. We observe that all of these techniques contribute positively to the overall performance of PARETOIHS as removing any one results in fewer instances solved overall. On these benchmarks, core minimization and constraint seeding have the strongest impact, as removing any one of them leads to 5 fewer instances solved. Removing weight-aware core extraction results in 3 fewer instances solved, while removing either the precomputation of lexicographic optima or reduced cost fixing (RCF) results in 2 fewer instances solved.

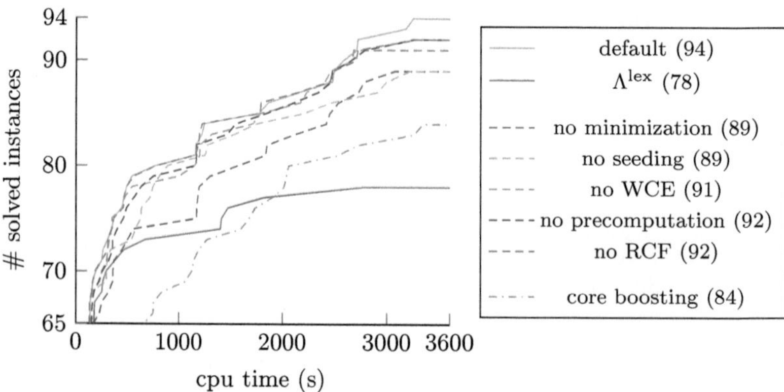

Fig. 4. Number of solved instances over time for PARETOIHS in various configurations.

As a final side-remark, we discuss challenges related to integrating core boosting [20] to PARETOIHS. Roughly, core boosting a multi-objective instance $(F, (O_1, \ldots, O_p))$ results in a reformulated instance $(F \wedge \mathcal{E}, (O_1^R, \ldots, O_p^R))$. where $\mathcal{E}$ is a set of additional reformulation constraints. The design space of integrating core boosting with PARETOIHS is large and includes questions like whether the constraints in $\mathcal{E}$ should be added to the optimizer or the oracle, and whether the optimizer should work on a scalarization of the reformulated objectives or the original ones. We experimented with several different configurations, but were not successful in achieving an overall performance improvement. The core boosting configuration that performed the best on these benchmarks (the line "core boosting" in Fig. 4) runs the so-called single-objective OLL core-guided algorithm [33] on each objective separately to form $\mathcal{E}$ and $(O_1^R, \ldots, O_p^R)$, adds $\mathcal{E}$ into the oracle, encodes $\mathcal{E}$ into the optimizer as described in [25], and builds O^A based on the reformulated objectives. The digital supplement includes a per-instance runtime plot demonstrating that, while core boosting degrades performance of PARETOIHS on most instances, we do see some benefits on instances from the Spot 5 domain. Overall we were so-far unable to benefit from core boosting to improve PARETOIHS.

5   Conclusions

We proposed a novel IHS-style approach to MO-MaxSAT solving, with the key insight that it suffices to iteratively run single-objective IHS on a scalarization of a multi-objective instance to one-by-one compute representative Pareto-optimal solutions and employ PD-cuts to iteratively refine the search space still to be explored. We showed empirically that the approach significantly outperforms an earlier-proposed IHS-style approach and complements in performance the current state of the art in MO-MaxSAT solving. We also identified a connection between our approach and the previously proposed MIPPD algorithm, instantiated for the first time here for MO-MaxSAT. Potential directions for future work are applying the PARETOIHS algorithm to settings such as multiobjective pseudo-Boolean optimization by using a different oracle, and investigating whether performance can be improved by altering the scalarization constants during search.

Acknowledgments. This work was funded by Research Council of Finland under grants 356046 and 362987. The authors thank the Finnish Computing Competence Infrastructure for computational and data storage resources.

References

1. Bacchus, F., Hyttinen, A., Järvisalo, M., Saikko, P.: Reduced cost fixing for maximum satisfiability. In: Lang, J. (ed.) Proceedings of the Twenty-Seventh International Joint Conference on Artificial Intelligence, IJCAI 2018, Stockholm, Sweden, 13–19 July 2018, pp. 5209–5213. ijcai.org (2018). https://doi.org/10.24963/ijcai.2018/723
2. Bacchus, F., Järvisalo, M., Martins, R.: Maximum satisfiability. In: Biere et al. [6], pp. 929–99. https://doi.org/10.3233/FAIA201008
3. Beck, J.C. (ed.): CP 2017. LNCS, vol. 10416. Springer, Cham (2017). https://doi.org/10.1007/978-3-319-66158-2
4. Berg, J., Järvisalo, M.: Weight-aware core extraction in SAT-based MaxSAT solving. In: Beck [3], pp. 652–670. https://doi.org/10.1007/978-3-319-66158-2_42
5. Bieber, P., Delmas, R., Seguin, C.: DALculus – theory and tool for development assurance level allocation. In: Flammini, F., Bologna, S., Vittorini, V. (eds.) SAFECOMP 2011. LNCS, vol. 6894, pp. 43–56. Springer, Heidelberg (2011). https://doi.org/10.1007/978-3-642-24270-0_4
6. Biere, A., Heule, M., van Maaren, H., Walsh, T. (eds.): Handbook of Satisfiability-Second Edition, Frontiers in Artificial Intelligence and Applications, vol. 336. IOS Press (2021). https://doi.org/10.3233/FAIA336
7. Chen, D., Batson, R.G., Dang, Y.: Applied Integer Programming: Modeling and Solution. Wiley (2009). https://doi.org/10.1002/9781118166000
8. Coll, J., Li, C.M., Li, S., Habet, D., Manyà, F.: Solving weighted maximum satisfiability with branch and bound and clause learning. Comput. Oper. Res. **183**, 107195 (2025). https://doi.org/10.1016/j.cor.2025.107195
9. Cortes, J., Lynce, I., Manquinho, V.: New core-guided and hitting set algorithms for multi-objective combinatorial optimization. In: Sankaranarayanan, S., Sharygina, N. (eds.) TACAS 2023, Part II. LNCS, vol. 13994, pp. 55–73. Springer, Cham (2023). https://doi.org/10.1007/978-3-031-30820-8_7
10. Davies, J.: Solving MAXSAT by decoupling optimization and satisfaction. Ph.D. thesis. University of Toronto (2013). https://hdl.handle.net/1807/43539
11. Davies, J., Bacchus, F.: Solving MAXSAT by solving a sequence of simpler SAT instances. In: Lee, J. (ed.) CP 2011. LNCS, vol. 6876, pp. 225–239. Springer, Heidelberg (2011). https://doi.org/10.1007/978-3-642-23786-7_19
12. Davies, J., Bacchus, F.: Exploiting the power of MIP solvers in MAXSAT. In: Järvisalo, M., Van Gelder, A. (eds.) SAT 2013. LNCS, vol. 7962, pp. 166–181. Springer, Heidelberg (2013). https://doi.org/10.1007/978-3-642-39071-5_13

13. Delmas, K., Chambert, L., Frazza, C., Seguin, C.: Optimization of development assurance level allocation. In: 2023 IEEE/AIAA 42nd Digital Avionics Systems Conference (DASC), pp. 1–10 (2023). https://doi.org/10.1109/DASC58513.2023.10311260
14. Eén, N., Sörensson, N.: Temporal induction by incremental SAT solving. In: Strichman, O., Biere, A. (eds.) First International Workshop on Bounded Model Checking, BMC@CAV 2003, Boulder, Colorado, USA, 13 July 2003. Electronic Notes in Theoretical Computer Science, vol. 89, pp. 543–560. Elsevier (2003). https://doi.org/10.1016/S1571-0661(05)82542--3
15. Ehrgott, M.: Multicriteria Optimization, 2nd edn. Springer (2005). https://doi.org/10.1007/3-540-27659-9
16. Heras, F., Larrosa, J., de Givry, S., Schiex, T.: 2006 and 2007 Max-SAT evaluations: contributed instances. J. Satisf. Boolean Model. Comput. **4**, 239–250 (2008). https://doi.org/10.3233/sat190046
17. Hooker, J.: Logic-Based Benders Decomposition: Theory and Applications. Springer (2024). https://doi.org/10.1007/978-3-031-45039-6
18. Ihalainen, H., Berg, J., Järvisalo, M.: Refined core relaxation for core-guided MaxSAT solving. In: Michel, L.D. (ed.) 27th International Conference on Principles and Practice of Constraint Programming, CP 2021, Montpellier, France (Virtual Conference), 25–29 October 2021. LIPIcs, vol. 210, pp. 28:1–28:19. Schloss Dagstuhl - Leibniz-Zentrum für Informatik (2021). https://doi.org/10.4230/LIPIcs.CP.2021.28
19. Jabs, C.: Scuttle: A multi-objective MaxSAT solver. https://bitbucket.org/coreo-group/scuttle
20. Jabs, C., Berg, J., Järvisalo, M.: Core boosting in SAT-based multi-objective optimization. In: Dilkina, B. (ed.) CPAIOR 2024, Part II. LNCS, vol. 14743, pp. 1–19. Springer, Cham (2024). https://doi.org/10.1007/978-3-031-60599-4_1
21. Jabs, C., Berg, J., Järvisalo, M.: Engineering and evaluating multi-objective pseudo-Boolean optimizers. In: Casini, G., Dundua, B., Kutsia, T. (eds.) JELIA 2025, Part I. LNCS, vol. 16093, pp. 115–134. Springer, Cham (2025). https://doi.org/10.1007/978-3-032-04587-4_8
22. Jabs, C., Berg, J., Niskanen, A., Järvisalo, M.: From single-objective to bi-objective maximum satisfiability solving. J. Artif. Intell. Res. **80**, 1223–1269 (2024). https://doi.org/10.1613/jair.1.15333
23. Jackson, D., Estler, H.C., Rayside, D.: The guided improvement algorithm for exact, general-purpose, many-objective combinatorial optimization. Technical report, Compute Science and Artificial Intelligence Laboratory (2009). http://hdl.handle.net/1721.1/46322
24. Janota, M., Lynce, I., Manquinho, V., Marques-Silva, J.: PackUp: tools for package upgradability solving. J. Satisf. Boolean Model. Comput. **8**, 89–94 (2012). https://doi.org/10.3233/sat190090
25. Katsirelos, G.: Core-guided linear programming-based maximum satisfiability. In: Berg, J., Nordström, J. (eds.) 28th International Conference on Theory and Applications of Satisfiability Testing, SAT 2025, Glasgow, Scotland, 12–15 August 2025. LIPIcs, vol. 341, pp. 17:1–17:17. Schloss Dagstuhl - Leibniz-Zentrum für Informatik (2025). https://doi.org/10.4230/LIPIcs.SAT.2025.17
26. Koshimura, M., Nabeshima, H., Fujita, H., Hasegawa, R.: Minimal model generation with respect to an atom set. In: Peltier, N., Sofronie-Stokkermans, V. (eds.) Proceedings of the 7th International Workshop on First-Order Theorem Proving, FTP 2009, Oslo, Norway, 6–7 July 2009. CEUR Workshop Proceedings, vol. 556. CEUR-WS.org (2009). https://ceur-ws.org/Vol-556/paper06.pdf

27. Koshimura, M., Zhang, T., Fujita, H., Hasegawa, R.: QMaxSAT: a partial Max-SAT solver. J. Satisf. Boolean Model. Comput. **8**, 95–100 (2012). https://doi.org/10.3233/sat190091

28. Li, C.M., Manyà, F., Planes, J.: Detecting disjoint inconsistent subformulas for computing lower bounds for Max-SAT. In: Proceedings, The Twenty-First National Conference on Artificial Intelligence and the Eighteenth Innovative Applications of Artificial Intelligence Conference, Boston, Massachusetts, USA, 16–20 July 2006, pp. 86–91. AAAI Press (2006). http://www.aaai.org/Library/AAAI/2006/aaai06-014.php

29. Maliotov, D., Meel, K.S.: MLIC: a MaxSAT-based framework for learning interpretable classification rules. In: Hooker, J. (ed.) CP 2018. LNCS, vol. 11008, pp. 312–327. Springer, Cham (2018). https://doi.org/10.1007/978-3-319-98334-9_21

30. Marques, R., Russo, L.M.S., Roma, N.: Flying tourist problem: flight time and cost minimization in complex routes. Expert Syst. Appl. **130**, 172–187 (2019). https://doi.org/10.1016/j.eswa.2019.04.024

31. Marques-Silva, J., Lynce, I.: On improving MUS extraction algorithms. In: Sakallah, K.A., Simon, L. (eds.) SAT 2011. LNCS, vol. 6695, pp. 159–173. Springer, Heidelberg (2011). https://doi.org/10.1007/978-3-642-21581-0_14

32. Marques-Silva, J., Lynce, I., Malik, S.: Conflict-driven clause learning SAT solvers. In: Biere et al. [6], pp. 133–182. https://doi.org/10.3233/FAIA200987

33. Morgado, A., Dodaro, C., Marques-Silva, J.: Core-guided MaxSAT with soft cardinality constraints. In: O'Sullivan, B. (ed.) CP 2014. LNCS, vol. 8656, pp. 564–573. Springer, Cham (2014). https://doi.org/10.1007/978-3-319-10428-7_41

34. Paxian, T., Reimer, S., Becker, B.: Dynamic polynomial watchdog encoding for solving weighted MaxSAT. In: Beyersdorff, O., Wintersteiger, C.M. (eds.) SAT 2018. LNCS, vol. 10929, pp. 37–53. Springer, Cham (2018). https://doi.org/10.1007/978-3-319-94144-8_3

35. Rossi, F., van Beek, P., Walsh, T. (eds.): Handbook of Constraint Programming, Foundations of Artificial Intelligence, vol. 2. Elsevier (2006). https://www.sciencedirect.com/science/bookseries/15746526/2

36. Saikko, P., Berg, J., Järvisalo, M.: LMHS: a SAT-IP hybrid MaxSAT solver. In: Creignou, N., Le Berre, D. (eds.) SAT 2016. LNCS, vol. 9710, pp. 539–546. Springer, Cham (2016). https://doi.org/10.1007/978-3-319-40970-2_34

37. Soh, T., Banbara, M., Tamura, N., Le Berre, D.: Solving multiobjective discrete optimization problems with propositional minimal model generation. In: Beck [3], pp. 596–614. https://doi.org/10.1007/978-3-319-66158-2_38

38. Sylva, J., Crema, A.: A method for finding the set of non-dominated vectors for multiple objective integer linear programs. Eur. J. Oper. Res. **158**, 46–55 (2004). https://doi.org/10.1016/S0377-2217(03)00255--8

Cost-Minimal Parameter Correction Subsets for Unsatisfiable Constraint Problems

Antoine Laviolette[1], Claude-Guy Quimper[1]([✉]), and Raphaël Boudreault[2]

[1] Université Laval, Québec, Canada
antoine.laviolette.1@ulaval.ca, claude-guy.quimper@ift.ulaval.ca
[2] Thales, Québec, Canada
raphael.boudreault@thalesgroup.com

Abstract. Decision support systems developed for real-world applications often rely on solving complex combinatorial problems using carefully crafted constraint models. However, users of such systems may input data that leads to a problem without any solution, rendering it unsatisfiable. The responsibility of understanding and correctly modifying the data generally falls on the user. Previous research on unsatisfiability mainly focused on reporting subsets of problematic constraints, which requires a good understanding of the underlying model. We introduce the concept of *Cost-Minimal Parameter Correction Subset* (CMPCS) as a subset of data parameters and their correction to make an instance satisfiable at a minimal cost. We propose a problem-independent approach, called *CMPCS-Finder*, which aims at reporting a CMPCS given an unsatisfiable instance. We empirically evaluate this approach on existing combinatorial benchmark problems and demonstrate that, in most cases, it successfully identifies the changes in the data that initially caused the instances to be unsatisfiable.

Keywords: Combinatorial Optimization · Unsatisfiability · Constraints · Correction · Modeling

1 Introduction

Decision support systems developed for real-world applications often rely on solving complex combinatorial problems and are usually designed as follows. A carefully crafted constraint model is used in the background and depends on specific parameters as input. Each time a user interacts with the system, new parameters are provided, but the underlying model remains unchanged. For instance, a scheduling system would have constraints in the model to ensure that resources are not overloaded and that tasks are given sufficient time to execute while meeting their deadlines. Here, the user data would contain the list

---

This project has received financial support from the Mitacs Accelerate program.

T. Guns (Ed.): CPAIOR 2026, LNCS 16595, pp. 280–296, 2026.
https://doi.org/10.1007/978-3-032-27242-3_17

of resources, as well as the tasks and their durations, deadlines, and resource consumption. However, during the use of such a system, it may happen that the combination model/data does not allow for any solution. Since the model is known to have been extensively tested and used for a long period of time with various datasets, it becomes clear that the unsatisfiability comes from the user data and not from the model.

The responsibility of understanding and correctly modifying the data generally falls on the user. To facilitate their task, it is possible to automatize some sanity checks, such as assertions, on the data. For example, one can easily detect if a task has a release time after its deadline. However, in other cases, the source of unsatisfiability can be much more complex due to the incompatibility between a large set of constraints and their associated parameters.

Previous research on explaining the unsatisfiability of such systems mainly focuses on reporting subsets of problematic constraints, which requires a good understanding of the underlying model [7,9,13]. Some approaches address only specific problems [1,22], while others target particular types of constraints [10, 27,28].

Thus, this paper aims to provide a new problem-independent approach, *CMPCS-Finder*, to detect inconsistencies in the data of a presumed correct constraint model given an unsatisfiable instance. For this, we introduce the concept of *Cost-Minimal Parameter Correction Subset* (CMPCS) as a subset of data parameters and their corrections to make an instance satisfiable at minimal cost. The cost of this subset of corrections can be measured by the number of parameters that need to be changed in the input or by any function that expresses the amount of effort required to change the input. Our proposed approach has the advantage of reporting a CMPCS interpretable by a user of such a system (e.g., a project manager using a scheduling system) without requiring any knowledge of the model.

The key contributions of our approach are:

- To propose a data-driven problem-independent method;
- To provide minimal corrections that are understandable to a user with little knowledge of the constraint model;
- To provide a system supporting a wide range of non-linear constraint types.

The paper is structured as follows. We first present the required background and related work on explaining the unsatisfiability of constraint problems. Then, we propose a methodology, *CMPCS-Finder*, to find a CMPCS, while detailing techniques to reduce its computational complexity. We empirically evaluate *CMPCS-Finder* on three existing combinatorial benchmark problems: the *Resource-Constrained Project Scheduling Problem* (RCPSP), the *Tank Allocation Problem* (TAP), and the *Roster Shift Bool Problem* (RSBP). Finally, we conclude and suggest directions for future research.

2   Background and Related Work

A *Constraint Satisfaction Problem* (CSP) is defined by a set of *variables* $\mathcal{X}$, where each variable $X \in \mathcal{X}$ is associated with a set of potential values $\mathrm{dom}(X)$ that form its *domain*. The variables are subject to a set of *constraints* $\mathcal{C}$ that encode the relationships between the variables and restrict their possible assignments. A subset of constraints $S \subseteq \mathcal{C}$ is said to be *satisfiable* if there exists a possible assignment of variables to values in their domains that satisfies all constraints $C \in S$. We say that S is *unsatisfiable* if no such assignment exists. A *solution* to a CSP is any assignment that makes $\mathcal{C}$ satisfiable. A *Constraint Optimization Problem* (COP) is a CSP where a special variable O, called the *objective*, is to be minimized. An *optimal* solution of a COP is a solution that minimizes O. In the following, we refer more generally to CSPs and COPs as *constraint problems*.

The *model* of a problem is the parameterized definition of a constraint problem in a logical (or mathematical) formulation. The parameters $\mathcal{P}$, which are usually numbers or sets, may be used in the definition of the constraints, e.g., as the coefficients of a linear equation. A constraint C is associated with a (possibly empty) parameter set $P(C) \subseteq \mathcal{P}$ such that changing the values of the parameters in $P(C)$ changes the solution space of the constraint C. A parameter may be associated with multiple constraints. The assignment of the model parameters to specific values defines an *instance* of the problem. In practice, the parameters are computed from the data provided by the user as input. High-level modeling languages, such as MiniZinc [19], allow models to be carefully designed using solver-independent declarative elements. Once the parameters are specified, the instances are compiled and given to the chosen solver.

The explanation of unsatisfiability has been extensively studied since the 1980s through the identification of *Minimal Unsatisfiable Subsets* (MUS) and *Minimal Correction Subsets* (MCS) [16]. A subset of constraints $M \subseteq \mathcal{C}$ is a MUS if and only if M is unsatisfiable and M' is satisfiable $\forall M' \subset M$. In other words, a MUS contains only the *core* constraints explaining the unsatisfiability. It is minimal in the sense of set inclusion. In operations research, these sets are sometimes called *Irreducibly/Minimally Inconsistent/Infeasible Subsystems* [31], while in Boolean satisfiability, they are referred to *Minimal Unsatisfiable Cores* [18]. Note that an unsatisfiable problem may admit more than one MUS. A subset of constraints $M \subseteq \mathcal{C}$ is an MCS if and only if $\mathcal{C} \setminus M$ is satisfiable and $\mathcal{C} \setminus M'$ is unsatisfiable $\forall M' \subset M$. In other words, an MCS forms a set of constraints to remove from $\mathcal{C}$ to *correct* it; that is, to make it satisfiable. It is also minimal in the sense of inclusion. Among the most common MUS/MCS extraction and enumeration techniques are QuickXplain [9] and MARCO [15,16]. Detailed surveys of other techniques were made by Gupta *et al.* [6] and Marques-Silva *et al.* [17].

The work of Leo *et al.* [14] led to the development and integration of findMUS into MiniZinc, a debugging tool based on the extraction and enumeration of MUSes. Based on the previous ideas of Jussien *et al.* [10] on grouping constraints and hierarchical modeling, findMUS can identify a MUS regardless of the chosen solver while leveraging the hierarchical structure of MiniZinc constraints to speed up the process. This acceleration is particularly significant for the computation of a single MUS, which is sufficient in practice to independently correct the model for each conflict. This approach thus allows for effectively identifying conflicts in the constraint model, but requires knowledge of the model to be interpretable by a user. Another related tool is MiniBrass [26], a soft constraint modeling language that allows solving problems with constraint *preferences*. This can be useful in always guaranteeing a solution to the problem for the user while requiring only small changes to the original model. On the other hand, this approach may ignore the constraint conflicts that come from the user-provided data. Furthermore, MiniBrass is limited to specific constraint types.

Senthooran *et al.* [27,28] designed an industrial optimization system that can interactively report constraint conflicts and suggest corrections based on the computation of a MUS and an MCS. Their approach uses MiniZinc and find-MUS, where annotations in the model describe constraints in order to link them to the user interface. The proposed corrections presented to the users relax a subset of constraints forming an MCS. Additionally, they add slack variables to relax the constraints in the MCS instead of removing the entire constraints. This helps make the corrections of the unsatisfiability less disruptive. Each constraint uses their own slack variable. This means that two constraints using the same parameter will have independent slack variables. This can be useful for obtaining fine-grained corrections, but it might not be applicable in some real-world scenarios. A limitation of their system is that it assumes the model is directly known and understandable by the users. Furthermore, only linear inequalities are supported in the models. Therefore, if a global constraint is used, Senthooran *et al.*'s approach will decompose it into linear inequalities and might return a subset of constraints from the decomposition that is unknown to the user.

Lauffer and Topcu [13] designed an unsatisfiability explanation system dedicated to resource-constrained scheduling problems. This system computes MUSes which can be represented either as a set of constraints or as a set of related tasks. For frequently occurring MUSes, developers can also craft specific messages to help users understand the unsatisfiability. Similarly, Poveda *et al.* [22] propose an interactive system tailored for workforce allocation, where users can select the right set of tasks to omit to make the problem satisfiable.

Bogaerts *et al.* [2] developed a generic approach to generate explanations as sequences of simple steps. Their method constructs explanations iteratively by leveraging the inference steps performed by the solver during the resolution process, while a cost function evaluates the simplicity of the explanations. The main challenge lies in effectively using the solver's information to provide clear and accessible explanations to the users. Subsequent improvements were proposed by Bleukx *et al.* [1] and Gamba *et al.* [7] to simplify the sequences, notably by filtering out unnecessary steps. While this approach has the advantage of explaining the unsatisfiability to the user, it does not provide a correction to make the model satisfiable. Its use is further restricted to instances for which the unsatisfiability is detected in the root node.

Korikov and Beck [12] introduced a quadratic programming algorithm for computing counterfactual explanations based on the previous work of Wang [33]. The algorithm takes any user's question that can be represented by linear or quadratic constraints. Counterfactual explanations [32] answer questions of the form "Why this solution and not some other solution q?" They then show to the user that with some changes to the input, the optimal solution can be q.

3   Methodology

In the following, we propose *CMPCS-Finder* to provide corrections for an unsatisfiable instance that are understandable by a user. The corrections are modifications to the parameters of the instance that can restore feasibility. A CMPCS consists of a subset of parameters and their corrections to make the instance satisfiable at minimal cost. We represent a CMPCS as a set of equations of the form `<parameter>' = <original value>[+|-]<correction>`. The cost can be measured by the number of parameters to change or by any function expressing the amount of effort required to change the input. Our approach needs not only to solve the underlying NP-Hard CSP, but also to find a *Cost-Minimal Parameter Correction Subset* (CMPCS) to repair the unsatisfiability and provide insight into the underlying cause. Correcting the unsatisfiability is, in itself, computationally harder than solving the original problem. Therefore, we present a general approach that uses MCS reductions and is later refined in the hope of reducing the total computation time. *CMPCS-Finder* is summarized in Fig. 1. It starts with an unsatisfiable instance, the existing hard model, and the instance (user) data. The *Soft Model* is created by the expert with, optionally, an MCS reduction model. The correctable parameters subset is extracted from the unsatisfiable instance and injected into the *Soft Model*. Optionally, the subset can pass through the *MCS Reduction* to reduce its size. Then, using some settings for the objective function, the *Soft Model* outputs a CMPCS.

3.1   Softening the Parameters

When using *CMPCS-Finder*, the modeling expert defines a subset of *correctable parameters* $\mathcal{P}^{cor} \subseteq \mathcal{P}$ that can have their values corrected. The expert chooses the correctable parameters based on their understanding of the real-world application. These can be parameters of constraints, including unary constraints that define the domain of a variable. Parameters that are not defined as correctable can impact unsatisfiability in general, but since they are not selected by the modeling expert, they should not contribute to unsatisfiability in isolation. This approach is particularly useful for data-driven problems, which are common in various industries. Problems that are not data-driven (e.g., the N-Queen Problem) would not benefit from *CMPCS-Finder*.

We are interested in unsatisfiable instances where the values of correctable parameters are the cause of the unsatisfiability. Moreover, we are interested in non-trivial unsatisfiable instances. An instance is trivially unsatisfiable if its

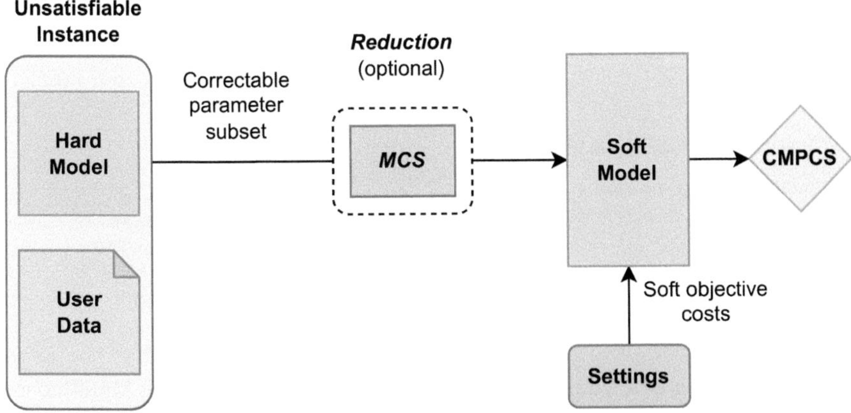

Fig. 1. High-level summary of *CMPCS-Finder*.

unsatisfiability can be detected by a simple algorithm known to the modeling expert, for example, by an assertion on the parameters.

In order to compute a CMPCS, *CMPCS-Finder* transforms the preexisting hard model by replacing the correctable parameters with new variables called *parameter variables*. The resulting model is a *soft* model with the same constraints as the preexisting model. In the soft model, the objective function is replaced (or added, if the original problem is a satisfaction problem) by a cost function that minimizes the corrections needed to restore satisfiability. In practice, four forms of objective functions are of interest:

1. Minimizing the absolute differences between the parameter variables and their original parameter values;
2. Minimizing the number of parameter variables assigned to a different value than their original parameter value;
3. Minimizing objective 1, then minimizing objective 2;
4. Minimizing objective 2, then minimizing objective 1.

The first objective minimizes the amount of change to the parameters in order to obtain a satisfiable instance, while the second objective minimizes the number of parameters to change. It is possible to adapt the first objective by choosing a different distance measure between the parameter variable and its original value, such as the square of the difference. For the first and second objectives, it is also possible to choose a different aggregator, such as the weighted sum.

An example of a preexisting model, written in the MiniZinc modeling language [19], is shown in Fig. 2a while the *soft model* is shown in Fig. 2b. In the MiniZinc language, variables are declared with `var` and parameters with `par`. After the keyword, the integer type is declared with `int` or with a specific range of integer values (e.g., `0..100`). This corresponds to domain for variables or allowed values for parameters. MiniZinc will assert that the parameter's value for a given instance is strictly inside the given range. The `constraint` and

```
1 include "all_different.mzn";
2 var int: y;
3 par 0..100: a; par -20..20: b;
4
5 constraint y >= a;
6 constraint y < b;
7 constraint all_different([y, a, b]);
8
9 solve minimize y;
10 a = 10; b = 8; % User parameters
```

(a) Preexisting simple CSP model.

```
1 include "all_different.mzn";
2 var int: y;
3 par 0..100: a ; par -20..20: b;
4 var 0..100: Xa; var -20..20: Xb; % New parameter variables
5
6 constraint y >= Xa;
7 constraint y < Xb;
8 constraint all_different([y, Xa, Xb]);
9
10 solve minimize Wa * abs(Xa - a) + Wb * abs(Xb - b);
11 a = 10; b = 8;                    % User parameters
12 par int: Wa = 2; par int: Wb = 1; % Objective weights
```

(b) Soft model of the CSP.

Fig. 2. Comparison of a simple CSP model and its soft version, in the MiniZinc modeling language, as created by *CMPCS-Finder*.

`solve minimize` keywords are respectively used for declaring constraints and the objective function. In this example, the `all_different` function is declared to use the solver-specific global constraint or a decomposition alternative if none exists. The correctable parameters a and b are defined on line 3 with allowed values $[0, 100]$ and $[-20, 20]$. In the preexisting model, these ranges serve as a check to ensure that they are assigned to correct values while importing the data. In the soft model, variables X_a and X_b replace the parameters used in the constraints. Their domains are set to the allowed values of the parameters, but one can sometimes restrict these variables further. The CMPCS found by *CMPCS-Finder* is $\{b' = 8 + 4\}$ which means the parameter b had a value of 8 and was corrected to $8 + 4 = 12$ to make the instance satisfiable. This CMPCS has a cost of 4. In our example, we can see that the smaller X_b is, the more constrained X_a is. In order for the solution to be relaxed, one necessarily needs to assign X_b to a value strictly greater than X_a can be. For that reason, the domain of X_b could be set directly to $[\min(\text{dom}(a)) + 1, 20]$. These domain adjustments are problem-specific and need to be programmed by the expert, but they allow to reduce the search space.

Although, most parameters are easily converted into variables, there are situations where the conversion cannot be automatic. Some global constraints only accept constant parameters. Such constraints require to be encoded differently in the model to allow the correctable parameter to be turned into a variable. A parameter can define the domain of a variable, but a variable cannot define the domain of another variable. In such a situation, the domain of the variable can be encoded using parameterized unary constraints (e.g. $p_1 \leq X \leq p_2$). A parameter can define the length of an array of variables. To make this parameter correctable, the whole model might have to be revised. The domain of a variable associated with a correctable parameter can be copied from the given parameter's allowed values. In industry, it is good practice to also restrict the values of parameters to avoid invalid data to reach the solver.

We implemented *CMPCS-Finder* using a custom compiler to create the soft model. The soft model is made from the original constraint model and some settings set by the modeling expert, such as the subset of *correctable parameters* or weights for the cost function. With these as inputs, our compiler can generate most parts of the soft model, accelerating the expert's job of providing an unsatisfiability resolving tool to the users.

Unlike traditional approaches in soft constraint literature [27,28,34], *CMPCS-Finder* allows constraints to be softened in multiple ways, which has two advantages. First, it eliminates the need to rewrite constraints to achieve a one-to-one relationship between a constraint and a parameter. This improves the readability of the model and simplifies the modeling expert's task. Second, this unlocks the use of many global constraints, which often have multiple input parameters that can be replaced by variables. Thus, common global constraints (e.g. GCC, CUMULATIVE, NVALUE, KNAPSACK, etc.) can be used directly instead of decomposing them, which allows the use of their powerful filtering algorithms.

3.2   Reducing the Parameter Variables

Solving the soft model to produce a CMPCS can take a long computation time due to the complexity of the problem. For many industrial problems, there are numerous parameters. This makes it hard for the solver to find a solution in a reasonable time, since the parameters are changed into variables. To reduce computation complexity at the expense of an optimal solution, we present the *MCS Reduction* technique that can be executed before solving the soft model. This technique reduces the number of parameters that can be softened by the solver, which we call a *reduced correctable parameter subset*. In other words, parameters that are not in this subset will not be replaced by parameter variables. The reduction approach can be executed in parallel while, in addition, solving the soft model without reduction. The system can report to the user the first CMPCS found by any technique.

MCS Reduction. This reduction is inspired by the computation of an MCS. However, instead of finding a minimal subset of constraints to correct, we find a set of constraints such that the union of their associated correctable parameters is minimal. In other words, we want to find the set of constraints to correct that involves the least number of parameters. To do so, we use the following model.

$$\min \sum_{p \in \mathcal{P}} \beta_p$$

$$\alpha_C \iff C \qquad\qquad \forall C \in \mathcal{C} \ (1)$$

$$\alpha_C \vee \beta_p \qquad\qquad \forall C \in \mathcal{C}, \forall p \in P(C) \ (2)$$

Constraints (1) reify every constraint $C \in \mathcal{C}$ with a Boolean variable α_C, indicating whether the constraint is satisfied or not. We associate a Boolean variable β_p with each parameter $p \in \mathcal{P}$, indicating whether the parameter can belong to the correction set. Constraints (2) ensure that unsatisfied constraints have their parameters selected. The objective is to minimize the number of selected parameters. The parameters whose variable β_p is set to true are selected to be correctable parameters. Once the parameters are selected, we create the soft model as usual, but only the selected correctable parameters are replaced by parameter variables. Parameters that are not selected remain constant in the model.

According to the definition of an MCS, removing these constraints corrects the unsatisfiability. Since *CMPCS-Finder* focuses on the user data, we instead correct the parameters that are related to these constraints. This makes the correction more flexible than simply removing all constraints in the MCS. In the example from Fig. 2, a possible MCS might be $\{y \geq a\}$, and the reduced subset of correctable parameters would be $\{a\}$. Using this reduced subset of correctable parameters in the soft model leads to a faster computation time and the new CMPCS $\{a' = 10 - 4\}$. The cost of this solution is 8, which is a higher cost than the CMPCS found without reduction (4).

This parameter reduction technique allows solving a soft model with fewer variables, reducing the instance complexity. However, this comes at the price that the CMPCS might not be optimal with respect to the whole set of correctable parameters. In fact, there is even the possibility that too few parameters are selected for correction, leading to a soft model that remains unsatisfiable.

To address this issue, there exist conditions that guarantee the reduction technique selects correctable parameters that make the instance satisfiable. One such condition is the existence of correctable parameter values that turn constraints into tautologies. This is the case for a constraint $X \geq p$, where the parameter p can be fixed to the smallest value in the domain of X.

Constraints that contain no correctable parameters are put in the background and cannot be deactivated. If the unsatisfiability is caused by background constraints, then the data has no error, but the model itself is incorrect.

4 Experiments

We implemented *CMPCS-Finder* in Python 3.12. The first component takes as input a model written in MiniZinc 2.8.5 [19] and outputs the soft model and the MCS reduction model. The second component uses the models to compute a CMPCS. *CMPCS-Finder* can use the MCS reduction model to compute the reduced set of correctable parameters. With the reduced set (`MCS-Reduction`) or the complete set (`No-Reduction`), *CMPCS-Finder* uses the soft model to compute a CMPCS.

In our experiments, *CMPCS-Finder* either used the solver Chuffed 0.13.2 [4] or OR-Tools CP-SAT 9.10.4067 [21], depending on the problem. All experiments were run on a 32-core Intel Xeon 4110 CPU@2.10 GHz and 32 GB of memory, on Windows. We set a 100-second time limit for the full approach, i.e. reducing the parameter set and finding the CMPCS. We also set specific timing rules for finding the reduced set of correctable parameters: we used the best reduced set found within 50s, and if no set was found in 50s, we let the solver run until it has found one or timed out.

We considered three combinatorial problems, the *Resource-Constrained Project Scheduling Problem* (RCPSP) [8,23], the *Tank Allocation Problem* (TAP) [25], and the *Roster Shift Bool Problem* (RSBP) [30]. These problems were chosen because they have standard benchmarks involving multiple parameters. In the following, we provide a description of the problems and how we altered the instances to make them unsatisfiable.

4.1 Resource-Constrained Project Scheduling Problem

The RCPSP [8,23] requires to schedule a set of tasks $\mathcal{T}$ executing on a set of cumulative resources $\mathcal{R}$. Each task $i \in \mathcal{T}$ has a duration p_i and a consumption $h_{i,r}$ over a resource $r \in \mathcal{R}$. Each resource $r \in \mathcal{R}$ has a constant capacity c_r. At any time, the sum of the resource consumption of the tasks that execute on the resource r must not exceed the resource capacity c_r. The tasks are subject to precedence constraints, where some tasks must complete before other tasks start. The objective is to minimize the makespan, i.e. the time to complete all the tasks. We implemented the model given in CSPLib [20] in MiniZinc.

The correctable parameters are the durations p_i, the resource consumptions $h_{i,r}$, and the resource capacities c_r. In real life, a task could be shortened by allowing resources to work overtime. For example, a 5-day task can take 4 d through nighttime labor [3,5,29]. Resource consumptions or resource capacities can be altered by temporarily hiring additional resources. We therefore have $\mathcal{P}^{\mathrm{cor}} = \{p_i \mid i \in \mathcal{T}\} \cup \{h_{i,r} \mid i \in \mathcal{T}, r \in \mathcal{R}\}$. We assume that the precedences cannot be corrected. The *CMPCS-Finder*'s soft model creates a variable X for each correctable parameter. The objective function is:

$$\min 2\sum_{i \in \mathcal{T}} |X_{p_i} - p_i| + \sum_{i \in \mathcal{T}, r \in \mathcal{R}} |X_{h_{i,r}} - h_{i,r}|$$
$$+ 5\sum_{r \in \mathcal{R}} |X_{c_r} - c_r|$$

We randomly selected 10 instances from PSPLib [11] with 30, 60, 90, and 120 tasks. As a first step to make these instances unsatisfiable, we added a constraint for the makespan to be no greater than the pre-calculated optimal makespan. This preserves the feasibility of the instances, but reduces the solution space to optimal solutions. In Sect. 4.4, we explain how the correctable parameters are modified to make the instance unsatisfiable.

4.2 Tank Allocation Problem

The TAP [25] is a problem of allocation between a set of cargoes $\mathcal{F}$ and a set of tanks $\mathcal{T}$. Each tank $t \in \mathcal{T}$ has a capacity v_t and a set of neighboring tanks $\mathcal{N}_t \subset \mathcal{T}$. Each cargo $f \in \mathcal{F}$ has a load q_f, a set of incompatible tanks $\mathcal{U}_f \subset \mathcal{T}$, and a set of incompatible cargoes $\mathcal{O}_f \subset \mathcal{F}$. One needs to assign each cargo to one or multiple compatible tanks. We simplify the problem by maximizing the number of unallocated tanks. We use the model provided in CSPLib [24] that we implemented in MiniZinc.

The correctable parameters are the cargo loads and the tank capacities, i.e. $\mathcal{P}^{\text{cor}} = \{q_f \mid f \in \mathcal{F}\} \cup \{v_t \mid t \in \mathcal{T}\}$. We do not consider tank and cargo compatibility as a correctable parameter nor the neighboring relations. The *CMPCS-Finder*'s soft model creates a variable X for each correctable parameter. The objective function is:

$$\min \sum_{t \in \mathcal{T}} |X_{v_t} - v_t| + 2 \sum_{f \in \mathcal{F}} |X_{q_f} - q_f|$$

We generated 10 instances using the instance generator available in CSPLib. We ran the scripts using $|\mathcal{F}| \in \{5, 10, 15\}$, $|\mathcal{T}| \in \{10, 15, 25\}$. We added the objective function as a constraint so the instances remain satisfiable, but its solution space is restricted to optimal solutions. Finally, we altered some correctable parameters as described in Sect. 4.4.

4.3   Roster Shift Bool Problem

The RSBP is an assignment problem that is part of the 2023 MiniZinc Challenge [30]. There is a set of employees $\mathcal{E}$ that need to be assigned to a set of shifts $\mathcal{S}$. Each employee $e \in \mathcal{E}$ has a number of hours c_e associated to the length of their contract. An employee e also has a set of skill levels $\mathcal{R}_e = \{1, 2, \ldots, l_e\}$. Each shift $s \in \mathcal{S}$ has a start time start_s, an end time end_s, and a required skill level req_s. A binary matrix variable $A_{e,s}$ is used to choose if an employee e works the shift s.

The correctable parameters are the start times, the end times, the skills required for a shift, and the contracts, i.e. $\mathcal{P}^{\text{cor}} = \{\text{start}_s \mid s \in \mathcal{S}\} \cup \{\text{end}_s \mid s \in \mathcal{S}\} \cup \{\text{req}_s \mid s \in \mathcal{S}\} \cup \{c_e \mid e \in \mathcal{E}\}$. The employees' skills are not correctable. The *CMPCS-Finder*'s soft model has a variable X for each correctable parameter and the following objective function:

$$\min 2 \sum_{e \in \mathcal{E}} |X_{c_e} - c_e| + \sum_{s \in \mathcal{S}} |X_{\text{start}_s} - \text{start}_s|$$
$$+ \sum_{s \in \mathcal{S}} |X_{\text{end}_s} - \text{end}_s| + 3 \sum_{s \in \mathcal{S}} |X_{\text{req}_s} - \text{req}_s|$$

We used the "large.dzn" instance provided by the MiniZinc Challenge to craft 10 smaller satisfiable instances. For each instance, we selected a subset of employees and shifts from the large instance. The original instance has $|\mathcal{S}| = 297$ shifts and $|\mathcal{E}| = 25$ employees. Our instances have 10, 15, or 25 shifts with 5, 8, or 10 employees. To make the instances unsatisfiable, we first solved the original instance and let d be the objective value, i.e. the cumulative difference of hours between the employees' contracts and their assigned hours. We replaced the objective function with the constraint:

$$\sum_{e \in \mathcal{E}} \left| c_e - \sum_{s \in \mathcal{S}} (\text{end}_s - \text{start}_s) A_{e,s} \right| \leq d.$$

The instance remains satisfiable, but the solution space becomes restricted. We then altered some correctable parameters as explained in the next section.

4.4   Creating Unsatisfiable Instances

We want to create a benchmark of unsatisfiable instances, which are not "trivial" to explain. For example, with the RCPSP, a polynomial-time algorithm could detect whether the precedences form a cycle forcing, by transitivity, a task to be its own predecessor. We want to construct unsatisfiable instances that are close to feasible ones, which is useful for comparing our corrections.

We created a generator of unsatisfiable instances that takes as input a satisfiable instance. It randomly selects three correctable parameters. It modifies each selected parameter p according to the formula $p' :- p + \Delta_p$, where Δ_p is drawn uniformly from a given range. The modified instance is submitted to the solver for a satisfiability test. If the instance is still satisfiable (or is not proven to be unsatisfiable after five minutes), the generator restarts from the original instance by choosing three new parameters and three new modifications. In order to help this process converge to an infeasible instance, both bounds of the range are updated according to the formula $x' :- x^{\left(1+\frac{i}{8.25}\right)}$ where x is the original range bound and i is the iteration number, starting with $i = 1$. In other words, the generator allows modifications to the original parameters to be larger every failed iteration.

As explained in previous sections, each problem has 10 original satisfiable instances. Using our generator, we generated five unsatisfiable instances for every original instance, for a total of 50 unsatisfiable instances for each problem[1].

For each instance, we saved the three parameters that were altered by the generator. These parameters may not form a CMPCS, as their cost might not be minimal. Indeed, maybe only one of the three parameter changes made the instance unsatisfiable, or maybe a parameter was increased by 10 units, while only 8 units were sufficient. For experimental comparison, we define `Oracle` as a reduction technique that considers these three parameters as the reduced correctable parameter subset. Thus, this oracle uses the soft model to compute a CMPCS based on knowledge from the unsatisfiable instance generation.

4.5   Results

Qualitative Evaluation. We first proceed to a qualitative evaluation of our approach by showing some CMPCSs computed by the *CMPCS-Finder* with `No-Reduction` for the RCPSP. Recall that CMPCS is a set of equations of the form `<parameter>' = <original value>[+|-]<correction>`. For instance, the CMPCS returned for instance 2_j3010_7 is $\{p'_8 = 3 - 2\}$, which means that the duration of task 8 was 3 units of duration and was corrected to $3 - 2 = 1$ to make the instance feasible. This CMPCS has a cost of 4, since the duration parameter has a weight of 2 in the objective function, and $2 \cdot |-2| = 4$. The format we use makes it relatively easy for a program to output the CMPCS in natural language, for example as "the duration of task 8 should be decreased by 2."

---

[1] Generator values, models, and instances are available at https://github.com/ AntoineLaviolette1/cmpcs-finder.

Computing a single CMPCS might not be sufficient for a human to fully understand why an instance is unsatisfiable. In our example, other tasks than task 8 could have their duration reduced to fix the instance. Here are two other CMPCS whose cost 4 is also optimal: $\{p'_{29} = 6 - 1, h'_{19,3} = 3 - 2\}$ and $\{p'_{29} = 6 - 2, h'_{16,3} = 2 - 1, h'_{19,3} = 3 - 1\}$. We can see that the infeasibility arises from a shortage of resource 3 with the combination of task precedences, packing the tasks in a tight time window. The three CMPCSs propose to reduce the consumption of resource 3, either by reducing the duration of the tasks that require this resource, or by reducing the consumption of the tasks of this resource. In this case, the user can choose any proposed CMPCS or decide to make other corrections to solve the problem with resource 3. Here are the corrections that would revert the instance to its original form: $\{p'_{24} = 11 - 4, h'_{26,3} = 12 - 3, c'_2 = 22 + 2\}$ The cost of this correction is $2 \cdot |-4| + |-3| + 5 \cdot 2 = 21$. Using the `Oracle` technique to reduce the correctable parameters to $\{p_{24}, h_{26,3}, c_2\}$, *CMPCS-Finder* finds the CMPCS $\{h'_{26,3} = 12 - 7\}$ that has a cost of 7, slightly worse than the three CMPCSs found by `No-Reduction`. In this case, *CMPCS-Finder* obtains a lower cost, since it does not use a reduced correctable parameter subset. The original instance alterations are consistent with our explanations: the consumption of resource 3 needs to be reduced and the duration of task 24, that consumes resource 3, also needs to be reduced. Finding multiple CMPCSs or *Near-Cost-Minimal* Parameter Correction Subsets can provide different valid explanations that a user, without any knowledge of the underlying model, can understand and draw conclusions from.

MCS Comparison. We compare our CMPCS approach against an MCS, since both approaches find corrections to restore feasibility. We extract an MCS from the same unsatisfiable instance (2_j3010_7) as in our previous evaluation. The MCS returned is CUMULATIVE(3). Therefore, removing the CUMULATIVE constraint associated with resource 3 makes the instance satisfiable. This is a worse correction than what *CMPCS-Finder* finds. In both cases, resource 3 is identified as problematic. But the MCS method only provides one disruptive correction that is, in many cases, impractical. The *CMPCS-Finder* finds a smaller and less disruptive correction, i.e., to reduce the duration of task 8 by 2 units.

Quantitative Evaluation. Figure 3 compares the solving times (log scale) of `No-Reduction` and `MCS-Reduction`. The instances are sorted by the computation time of the `MCS-Reduction`. For the RCPSP, the `MCS-Reduction` does not improve the computation time. The solver is able to solve the soft model with a full set of correctable parameters. The `MCS-Reduction` provides computation times that are still reasonable. For the TAP, there is no clear faster technique. For the RSBP, the `MCS-Reduction` is the fastest way to solve this problem, by far. This is not an anomaly, since for every instance of RSBP, the `MCS-Reduction` was much faster.

Figure 4 shows the cost of the CMPCS for the two techniques and `Oracle`. The instances are sorted by the `MCS-Reduction`'s cost. `Oracle` globally obtains

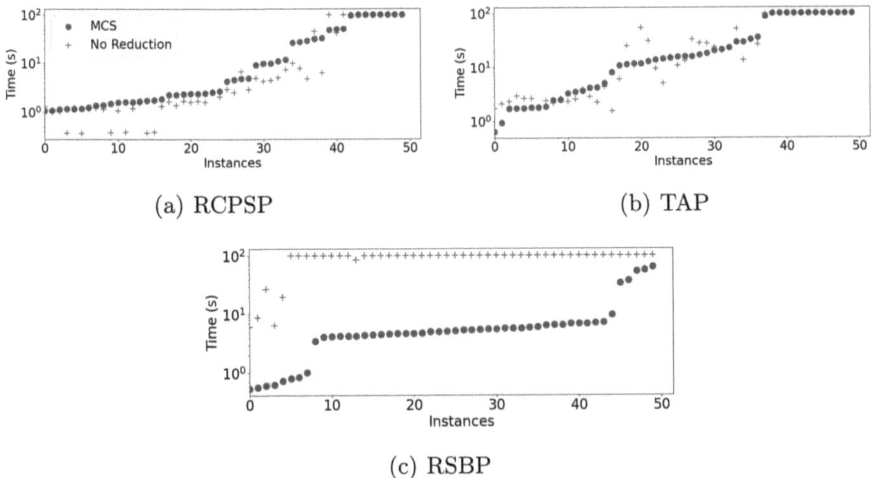

(a) RCPSP

(b) TAP

(c) RSBP

Fig. 3. Graphs comparing the computation time for MCS-Reduction and No-Reduction.

CMPCS with the highest cost. We also observe that the MCS-Reduction increases sometimes by a small amount the cost of the CMPCS, compared to No-Reduction. This is more frequent with RSBP, the harder problem. The No-Reduction technique finds the CMPCS with the lowest cost, but this is not always the case because of the timeout. The MCS-Reduction found a lower cost for 3 instances on the RCPSP, 2 on the TAP and 1 on the RSBP.

Considering both figures, for problems that take longer to compute a CMPCS, such as RSBP, the MCS-Reduction technique can be useful for its quick computation time. The No-Reduction technique appears to generally take the same amount of time and can produce better CMPCS for easier problems.

Table 1. Comparison of the techniques by problem.

Problem	Technique	Optimal	Cor par.
RCPSP	No-Reduction	40	314.00
	MCS-Reduction	43	59.04
TAP	No-Reduction	38	31.00
	MCS-Reduction	38	21.58
RSBP	No-Reduction	6	57.60
	MCS-Reduction	50	2.46

In Table 1, column **Optimal** indicates the number of CMPCS that were proven optimal. For all instances, the solver found at least a solution (a CMPCS), but not all proven optimal. The column **Cor par.** gives the average number of

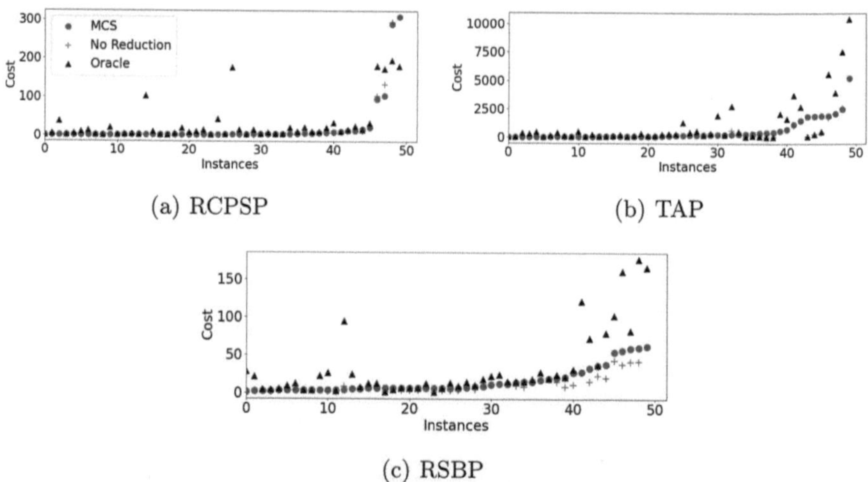

(a) RCPSP

(b) TAP

(c) RSBP

Fig. 4. Graphs comparing the cost of the CMPCS for MCS-Reduction, No-Reduction, and Oracle.

correctable parameters used. The MCS-Reduction significantly reduces the number of correctable parameters. Our results show only small differences in average computation times with smaller correctable parameter subsets for RCPSP and TAP. An average of 20 to 60 correctable parameters yields too many combinations for *CMPCS-Finder* to find a solution quickly. The time taken to compute the reduction is equivalent to the time saved when solving a smaller soft model. For the RSBP, there are few parameters, which accelerates the computation.

5 Conclusion

We introduced *CMPCS-Finder*, a data-oriented method for restoring feasibility in unsatisfiable constraint problems by identifying minimal changes to user-provided parameters. By formalizing *Cost-Minimal Parameter Correction Subsets* (CMPCS) and generating soft models in which parameter values become optimization variables, the approach focuses directly on suggesting concrete corrections that users can apply to their data. We proposed a reduction technique to limit the number of parameters considered, which can accelerate solving depending on the problem structure. Experiments on RCPSP, TAP, and RSBP instances showed the relevance of this approach and demonstrated that, in most cases, it successfully identifies low cost changes in the data.

Future research can use the concept of a parameter-driven approach instead of the traditional constraint-driven approach to create not only corrections but also explanations that users can understand without a computer science background.

References

1. Bleukx, I., Devriendt, J., Gamba, E., Bogaerts, B., Guns, T.: Simplifying step-wise explanation sequences. In: Proceedings of the 29th International Conference on Principles and Practice of Constraint Programming (CP 2023), pp. 11:1–11:20 (2023)
2. Bogaerts, B., Gamba, E., Guns, T.: A framework for step-wise explaining how to solve constraint satisfaction problems. Artif. Intell. **300**, 103550 (2021)
3. Boudreault, R., Simard, V., Lafond, D., Quimper, C.G.: A constraint programming approach to ship refit project scheduling. In: Proceedings of the 28th International Conference on Principles and Practice of Constraint Programming (CP 2022), pp. 10:1–10:16 (2022)
4. Chu, G.G.: Improving Combinatorial Optimization. Ph.D. thesis. The University of Melbourne (2011)
5. Cloutier, S., Quimper, C.G.: Cumulative scheduling with calendars and overtime. In: Proceedings of the 30th International Conference on Principles and Practice of Constraint Programming (CP 2024), pp. 7:1–7:16 (2024)
6. Dev Gupta, S., Genc, B., O'Sullivan, B.: Explanation in constraint satisfaction: a survey. In: Proceedings of the 30th International Joint Conference on Artificial Intelligence, pp. 4400–4407 (2021)
7. Gamba, E., Bogaerts, B., Guns, T.: Efficiently explaining CSPs with unsatisfiable subset optimization. In: 29th International Joint Conference on Artificial Intelligence, pp. 1381–1388. (2021)
8. Hartmann, S., Briskorn, D.: An updated survey of variants and extensions of the resource-constrained project scheduling problem. Eur. J. Oper. Res. **297**(1), 1–14 (2022)
9. Junker, U.: QUICKXPLAIN: Preferred explanations and relaxations for over-constrained problems. In: Proceedings of the 19th National Conference on Artifical Intelligence (AAAI), pp. 167–172 (2004)
10. Jussien, N., Ouis, S.: User-friendly explanations for constraint programming. In: Proceedings of the 11th Workshop on Logic Programming Environments, WLPE 2001 (2001)
11. Kolisch, R., Sprecher, A.: PSPLIB - a project scheduling problem library. Eur. J. Oper. Res. **96**(1), 205–216 (1997)
12. Korikov, A., Beck, J.C.: Objective-based counterfactual explanations for linear discrete optimization. In: Proceedings of the 20th International Conference on Integration of Constraint Programming, Artificial Intelligence, and Operations Research (CPAIOR 2023), pp. 18–34 (2023)
13. Lauffer, N., Topcu, U.: Human-understandable explanations of infeasibility for resource-constrained scheduling problems. In: ICAPS 2019 Workshop XAIP (April 2019)
14. Leo, K., Tack, G.: Debugging unsatisfiable constraint models. In: Salvagnin, D., Lombardi, M. (eds.) CPAIOR 2017. LNCS, vol. 10335, pp. 77–93. Springer, Cham (2017). https://doi.org/10.1007/978-3-319-59776-8_7
15. Liffiton, M.H., Malik, A.: Enumerating infeasibility: finding multiple muses quickly. In: Gomes, C., Sellmann, M. (eds.) CPAIOR 2013. LNCS, vol. 7874, pp. 160–175. Springer, Heidelberg (2013). https://doi.org/10.1007/978-3-642-38171-3_11
16. Liffiton, M.H., Previti, A., Malik, A., Marques-Silva, J.: Fast, flexible MUS enumeration. Constraints **21**(2), 223–250 (2016)

17. Marques-Silva, J., Mencía, C.: Reasoning about inconsistent formulas. In: Proceedings of the 29th International Joint Conference on Artificial Intelligence, pp. 4899–4906 (2020)
18. Nadel, A., Ryvchin, V., Strichman, O.: Accelerated deletion-based extraction of minimal unsatisfiable cores. J. Satisfiabil. Boolean Model. Comput. **9**, 27–51 (2014)
19. Nethercote, N., Stuckey, P.J., Becket, R., Brand, S., Duck, G.J., Tack, G.: MiniZinc: towards a standard cp modelling language. In: Bessière, C. (ed.) CP 2007. LNCS, vol. 4741, pp. 529–543. Springer, Heidelberg (2007). https://doi.org/10.1007/978-3-540-74970-7_38
20. Nightingale, P., Demirović, E.: CSPLib problem 061: Resource-constrained project scheduling problem (RCPSP) (1999)
21. Perron, L., Furnon, V.: OR-Tools. Google (2024). https://developers.google.com/optimization/
22. Povéda, G., et al.: Trustworthy and explainable decision-making for workforce allocation (2024). https://arxiv.org/abs/2412.10272
23. Pritsker, A.A.B., Waiters, L.J., Wolfe, P.M.: Multiproject scheduling with limited resources: a zero-one programming approach. Manage. Sci. (1969)
24. Schaus, P.: CSPLib problem 051: Tank allocation (1999)
25. Schaus, P., Régin, J.C., Van Schaeren, R., Dullaert, W., Raa, B.: Cardinality reasoning for bin-packing constraint: application to a tank allocation problem. In: Proceedings of the 18th International Conference on Principles and Practice of Constraint Programming (CP 2012), pp. 815–822 (2012)
26. Schiendorfer, A., Knapp, A., Anders, G., Reif, W.: MiniBrass: soft constraints for MiniZinc. Constraints **23**(4), 403–450 (2018)
27. Senthooran, I., et al.: Human-centred feasibility restoration. In: Proceedings of the 27th International Conference on Principles and Practice of Constraint Programming (CP2021), pp. 203–243 (2021)
28. Senthooran, I., et al.: Human-centred feasibility restoration in practice. Constraints **28**(2), 203–243 (2023)
29. Smith, S.F., Fox, M.S., Ow, P.S.: Constructing and maintaining detailed production plans: investigations into the development of knowledge-based factory scheduling systems. AI Mag. **7**(4), 45–61 (1986)
30. Stuckey, P.J., Feydy, T., Schutt, A., Tack, G., Fischer, J.: The MiniZinc Challenge 2008–2013. AI Mag. **35**(2), 55–60 (2014)
31. van Loon, J.N.M.: Irreducibly inconsistent systems of linear inequalities. Eur. J. Oper. Res. **8**(3), 283–288 (1981)
32. Wachter, S., Mittelstadt, B., Russell, C.: Counterfactual explanations without opening the black box: automated decisions and the gdpr. Harvard J. Law Technol. **31**(2), 841–887 (2017)
33. Wang, L.: Cutting plane algorithms for the inverse mixed integer linear programming problem. Operat. Res. Lett. **37**(2) (2009)
34. Yang, J.: Infeasibility resolution based on goal programming. Comput. Operat. Res. **35**(5), 1483–1493 (2008)

Towards Solving Polynomial-Objective Integer Programming with Hypergraph Neural Networks

Minshuo Li[1,2], Yaoxin Wu[2,3]([✉]), Pavel Troubil[4], Yingqian Zhang[2,3], and Wim P. M. Nuijten[1,2]

[1] Department of Mathematics and Computer Science, Eindhoven University of Technology, Eindhoven, The Netherlands
`{m.li7,W.P.M.Nuijten}@tue.nl`
[2] Eindhoven Artificial Intelligence Systems Institute, Eindhoven University of Technology, Eindhoven, The Netherlands
`yqzhang@tue.nl`
[3] Department of Industrial Engineering and Innovation Sciences, Eindhoven University of Technology, Eindhoven, The Netherlands
`y.wu2@tue.nl`
[4] Delmia R&D, Dassault Systèmes, 's-Hertogenbosch, The Netherlands
`pavel.troubil@3ds.com`

Abstract. Complex real-world optimization problems often involve both discrete decisions and nonlinear relationships between variables. Many such problems can be modeled as polynomial-objective integer programs, encompassing cases with quadratic and higher-degree variable interactions. Nonlinearity makes them more challenging than their linear counterparts. In this paper, we propose a hypergraph neural network (HNN) based method to solve polynomial-objective integer programming (POIP). Besides presenting a high-degree-term-aware hypergraph representation to capture both high-degree information and variable-constraint interdependencies, we also propose a hypergraph neural network, which integrates convolution between variables and high-degree terms alongside convolution between variables and constraints, to predict solution values. Finally, a search process initialized from the predicted solutions is performed to further refine the results. Comprehensive experiments across a range of benchmarks demonstrate that our method consistently outperforms both existing learning-based approaches and state-of-the-art solvers, delivering superior solution quality with favorable efficiency. Note that our experiments involve both polynomial objectives and constraints, demonstrating our HNN's versatility for general POIP problems and highlighting its advancement over the existing literature.

Keywords: Polynomial-objective integer programming · Hypergraph neural network · Nonlinear integer programming

T. Guns (Ed.): CPAIOR 2026, LNCS 16595, pp. 297–313, 2026.
https://doi.org/10.1007/978-3-032-27242-3_18

1   Introduction

Integer programming has been widely applied to real-world applications involving discrete decisions, such as photolithography scheduling [10] and supply chain optimization [2]. Many integer programming problems are NP-hard, meaning that the computational time required to reach optimality grows exponentially with problem size. In particular, nonlinear integer programming (NLIP) frequently arises due to physical laws, statistical measures, nonlinear regression, and other complex relationships [1]. The nonlinearity makes these problems even more challenging to solve than their linear counterparts, highlighting the need for efficient solution methods that go beyond traditional techniques.

Over the past few decades, different algorithms have been proposed to address the challenges of NLIP, typically following two main approaches. Local approaches rely on gradient information to find locally optimal solutions [3], but often struggle with complex structures containing multiple local optima. Global approaches follow a divide-and-conquer strategy, partitioning the solution space and then searching within each partition to identify the optimal solution [25,38]. This often incurs prohibitive computational times for instances with high nonlinearity or intricate constraint structures. Furthermore, algorithms for these approaches are typically closed-source or tailored to specific NLIP problems, which restricts their broader application [27]. These limitations motivate the exploration of alternative paradigms.

A promising alternative paradigm is machine learning, which has recently made major advances in integer linear programming (ILP) The advances include learning better policies within specific solvers, such as branching [16,34,35] and presolving [28,31], as well as learning general guidance for ILP solvers, such as solution prediction [12,17] and neighborhood selection [21,46]. On the one hand, due to the fundamental differences between linear and nonlinear formulations, these methods are not directly applicable to NLIP. On the other hand, research on learning-based methods for NLIP remains relatively limited, with only a handful of works [14,18]. These methods are mainly built on specific problem structures or algorithms, thus restricting their broader applicability. Therefore, there is a need for learning-based methods that can effectively address a wide range of NLIP problems and work seamlessly across different solvers.

To address these limitations, this paper aims to push the frontier of learning-for-NLIP towards solving polynomial-objective integer programming (POIP), an important subclass of NLIP. By Taylor's formula [36], POIP captures many practical nonlinearities and is representative of NLIP challenges that cannot be efficiently solved by current solvers. More specifically, we innovate a learning-based approach to NLIP and propose a hypergraph neural network (HNN) based model that predicts variable values in optimal solutions based on a hypergraph representation of problem instances. The predicted solution serves as an effective initial solution, which can be further refined by any solver or complementary search algorithm. Our major contributions are summarized as follows.

- A high-degree-term-aware hypergraph representation for general POIP, which encodes interactions among variables within high-degree terms and relations between variables and constraints to reflect problem structure.
- A hypergraph neural network that learns the mapping between instances' hypergraph representations and their corresponding optimal solutions. It applies a convolution across variables and high-degree terms to capture variable representations, together with a convolution between variables and constraints to further involve variable-constraint interdependencies.
- Experimental results on diverse benchmark datasets demonstrate that our method significantly improves the efficiency of Gurobi and SCIP on quadratic and quintic integer programming problems, consistently outperforming existing learning-based approaches.

2   Related Work

2.1   Learning-Based Methods for ILP

Learning-based methods for ILP can be broadly categorized into two classes [47]. The first class concerns learning key policies within solvers. Among these, most studies learn variable selection [13,16,20,30,35] and node selection [29,34] in branch and bound (B&B) tree search. Some works learn other policies including cutting plane selection [11,40], primal heuristic selection [8], and presolving settings [28,31]. The second class focuses on learning general policies that are applicable across different solvers. One approach involves predicting solutions, either as initial solution values for further refinement [12,24] or as direct feasible solutions [17,33]. Another research direction involves learning strategies in general heuristics, including neighborhood selection strategies for large neighborhood search [21,32,42,46] and search strategies for diving heuristics [35].

2.2   Learning-Based Methods for NLIP

Learning-based methods for NLIP remain underexplored in the literature. A few noteworthy contributions have developed learning-for-NLIP methods for specific problems. Bonami et al. [5] train a classifier to decide whether linearizing quadratic integer programs leads to better solver performance. Ferber et al. [14] propose to learn surrogate linear objective functions for nonlinear programs with linear constraints. Tang et al. [39] introduce differentiable correction layers for end-to-end learning on parametric nonlinear programming with fixed problem structure. Chen et al. [7] and Wu et al. [41] prove the theoretical expressive power of graph neural networks for quadratic terms. While these works represent valuable progress, they are all tailored to specific problem settings and do not generalize to polynomial-objective integer programming (POIP).

Two recent studies are more directly related to our work. Xiong et al. [45] develop a graph neural network to predict solutions for quadratic programming. However, both their graph representation and neural network are restricted to quadratic terms, whereas our framework is designed to handle POIP with arbitrary degrees. Ghaddar et al. [18] propose quantile regression to learn instance-specific branching rules inside a closed-source solver RAPOSa [19] for solving polynomial programming. This approach requires access to internal solver modifications and is limited to a specific solver. Instead, our approach predicts solutions and serves as an external module to provide initial solution values for any solver or search algorithm, without the requirement of internal changes.

The literature review highlights the limitations of current learning-based methods for NLIP. Our work pushes the frontier of this domain by proposing a hypergraph neural network framework, which addresses integer programming with high-degree objective terms and integrates seamlessly with existing solvers.

3   Preliminaries

3.1   Definition of Polynomial-Objective Integer Programming

Without loss of generality, polynomial-objective integer programming (POIP) refers to a class of optimization problems that maximizes a polynomial objective function defined over a set of integer variables while satisfying a set of constraints. A generic POIP formulation can be written as:

$$\max_{x} \quad f(x_1, x_2, \cdots, x_n), \tag{1}$$

$$\text{s.t.} \quad \sum_{i=1,\cdots,n} a_{ij} x_i \leq b_j, \quad j = 1, 2, \ldots, m, \tag{2}$$

$$l_i \leq x_i \leq u_i, x_i \in \mathbb{Z}, \quad i = 1, 2, \ldots, n, \tag{3}$$

where n is the number of variables and m is the number of constraints; f is a polynomial expression over the variables $x_1, x_2, \cdots, x_n$; a_{ij} and b_j are, respectively, the coefficient of the variable x_i and the right-hand side scalar in the j-th constraint; l_i and u_i are the lower and upper bounds for x_i.

3.2   Graph and Hypergraph Neural Networks

A graph is defined as $G = (V, E)$, where V is the set of vertices and $E \subseteq V \times V$ is the set of pairwise edges. Graph Neural Networks (GNNs) learn node or edge representations by iteratively aggregating and transforming information from local neighborhoods [44]. Owing to their ability to exploit graph structure, GNNs have been applied to a variety of IP problems that have natural graph representations [4,6,37,43].

A hypergraph extends a graph by allowing edges, called hyperedges, to connect any number of vertices. This enables the modeling of higher-order relationships that cannot be captured by pairwise edges alone. Hypergraph Neural Networks (HNNs) extend GNN principles to learning representations of hypergraphs. They have shown promising performance across diverse hypergraph-based applications [26]. A representative framework is UniGNN [23], which introduced a two-stage message passing mechanism:

$$(\text{UniGNN}) \begin{cases} h_\epsilon = \phi_1(\{h_v\}_{v \in \mathcal{N}_\epsilon}), \\ \bar{h}_v = \phi_2(h_v, \{h_\epsilon\}_{\epsilon \in \mathcal{N}_v}), \end{cases} \tag{4}$$

where h_ϵ and h_v denote the representations for hyperedge ϵ and vertex v, respectively. The sets $\mathcal{N}_\epsilon$ and $\mathcal{N}_v$ represent all the vertices contained in the hyperedge ϵ and all the hyperedges containing the vertex v, respectively. The functions ϕ_1 and ϕ_2 are permutation-invariant aggregation functions, and $\bar{h}_v$ represents the updated representation of vertex v. The two-stage message passing mechanism in Eq. 4 can be iteratively executed to progressively refine the vertex and hyperedge embeddings.

3.3   Graph Representations for Integer Programming

Graph representations are commonly used to transform integer programming (IP) instances into structures suitable for graph neural network processing. Gasse et al. [16] introduce a bipartite graph, in which one set of vertices represents variables and the other set represents constraints, with edges encoding variable-constraint incidences, i.e., a variable appearing in a constraint. Ding et al. [12] extend the representation to a tripartite graph by adding a vertex to represent the objective function, thereby enriching the structural information. Subsequent graph representations of IP are often built on bipartite and tripartite graphs, because of their effectiveness and simplicity.

While effectively capturing variable-constraint relationships, current graph representations are restricted to pairwise interactions and thus struggle to model nonlinear or higher-order structures that frequently arise in practical IP problems. To overcome this limitation, Heydaribeni et al. [22] use hyperedges to connect variables appearing in the same constraint, and Xiong et al. [45] employ hyperedges to represent quadratic terms involving both variables and constraints. Both graph representations remain limited, for example to IP with linear or quadratic terms, and cannot generalize to IP instances with arbitrary high-degree terms. In this paper, we address this gap by proposing a hypergraph neural network tailored for learning over graph representations of POIP.

4   Methodology

This section presents our hypergraph neural network framework for tackling POIP problems, including hypergraph representation, solution prediction via hypergraph neural network, and solution repair and refinement. The overview of our framework is illustrated in Fig. 1 and detailed below.

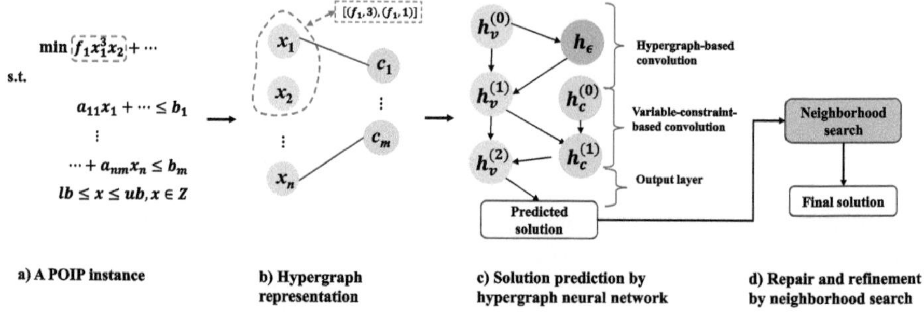

Fig. 1. The framework of the proposed method. For a POIP instance shown in a), our method first transforms it into a hypergraph shown in b), where orange dashed circles denote hyperedges representing high-degree terms. Their raw features (term coefficients and variable degrees) are illustrated in the orange dashed boxes. This hypergraph is then processed by a hypergraph neural network shown in c) for representation learning and solution prediction, where h_ϵ, $h_v^{(i)}$, and $h_c^{(i)}$ represent hyperedge embeddings, variable embeddings after the i-th update, and constraint embeddings after the i-th update, respectively. Finally, a neighborhood-search based repair-and-refinement process, shown in d), turns the predicted results into a high-quality feasible solution.

4.1   High-Degree-Term-Aware Hypergraph Representation

Representing general POIP poses two unique challenges: (i) high-degree terms induce multi-variable interactions that cannot be captured by standard pairwise connections [16], and (ii) the satisfaction of constraints depends intricately on variable assignments, forming the essential variable-constraint relationship. To address these challenges, we encode a POIP instance as a hypergraph.

Formally, our hypergraph is defined as $\mathcal{G} = (\mathcal{V}, \mathcal{C}, \mathcal{H}, \mathcal{E})$, where $\mathcal{V}$ denotes variable vertices, $\mathcal{C}$ denotes constraint vertices, $\mathcal{H}$ denotes hyperedges for high-degree objective terms, and $\mathcal{E}$ are edges encoding variable–constraint incidence. Specifically, each variable in a POIP instance is represented by a vertex $v \in \mathcal{V}$, and each constraint is represented by a vertex $c \in \mathcal{C}$. For every high-degree term $c_\alpha \prod_{i' \in \{1, \cdots, |\mathcal{V}|\}} x_{i'}^{\alpha_{i'}}$ appearing in the objective function, we create a hyperedge $\epsilon \in \mathcal{H}$ that connects all variables contained in the term. A raw feature $[\omega_{v_{i'}\epsilon} = (c_\alpha, \alpha_{i'})]_{v_{i'} \in \mathcal{N}_\epsilon}$ is assigned to the hyperedge ϵ, where $\mathcal{N}_\epsilon$ denotes the variables contained in ϵ. For variable-constraint relationships, we add an edge $e_{vc} \in \mathcal{E}$ between variable vertex v and constraint vertex c whenever the corresponding variable appears with a nonzero coefficient in the corresponding constraint. The associated coefficient is assigned to the edge's feature, ensuring that the numerical dependency between variable and constraint is preserved.

The high-degree-term-aware hypergraph representation integrates both structural and parametric information of a POIP instance, and provides a foundation for the hypergraph neural network introduced in the next subsection. A complete specification of the raw features is provided in Appendix A.1.

For example, Fig. 1(b) illustrates the high-degree-term-aware hypergraph representation for the POIP instance in Fig. 1(a). Variable vertices (left) and constraint vertices (right) represent variables and constraints separately; edges (straight lines) represent variable-constraint relationships, connecting a variable to each constraint in which it has a nonzero coefficient (e.g., x_1 is connected to c_1); hyperedges (circles) capture the relationships of variables in high-degree terms in the objective. For the term $f_1 x_1^3 x_2$, the hyperedge feature set is $\{(f_1, 3), (f_1, 1)\}$, associating the shared coefficient with each variable's exponent.

4.2 Solution Prediction via Hypergraph Neural Network

Based on the hypergraph representation of POIP instances, our neural network is designed to capture two complementary types of relationships: (i) higher-order interactions among variables induced by high-degree terms, and (ii) interdependencies between variables and constraints. To achieve this, we draw on concepts from Hypergraph Neural Networks (HNNs) [26], which support message passing between vertices and hyperedges. Building on this framework, we first introduce a hyperedge-based convolution that aggregates information from hyperedges into variable embeddings, enabling the model to represent higher-order structures. We then complement this with a variable–constraint convolution that propagates information along standard edges, thereby modeling the dependencies between variables and constraints. Finally, the variable embeddings are passed through an output layer to generate the predictions of variable values in high-quality solutions. The architecture of our model is illustrated in Fig. 1(c).

Hyperedge-Based Convolution. Our HNN begins with a hyperedge-based convolution applied to hyperedges and their associated variable vertices in order to effectively extract higher-order information arising from the high-degree terms. Inspired by UniGNN [23] (see Sect. 3.2), we formulate the convolution as presented in Eq. 5 and Eq. 6. Each iteration updates hyperedge embeddings by aggregating incident variable information (Eq. 5), then updates variable embeddings by aggregating the connected hyperedge information (Eq. 6). In this way, information from high-degree terms is integrated into the embeddings of the variables.

$$h_\epsilon \leftarrow \sum_{v \in \mathcal{N}_\epsilon} h_v h_{v\epsilon}, \forall \epsilon \in \mathcal{H}, \tag{5}$$

$$h_v \leftarrow \phi_{\mathcal{H}}(h_v, \mathrm{mean}(\{h_\epsilon h_{v\epsilon}\}_{\epsilon \in \mathcal{N}_v^H})) + h_v, \forall v \in \mathcal{V}, \tag{6}$$

Formally, h_v and $h_{v\epsilon}$ denote the embeddings for variable vertex v and feature vector $\omega_{v\epsilon}$ (see Sect. 4.1), respectively. These embeddings are initialized from raw features using two-layer MLPs, and the subsequent iterations adopt the embeddings from the previous iteration (where Eq. 5 and Eq. 6 are applied) as input. h_ϵ denotes the embeddings for hyperedge ϵ, and both its initialization and subsequent updates are calculated by Eq. 5. $\mathcal{N}_v^H$ denotes the set of hyperedges containing the variable v; $\mathcal{N}_\epsilon$ denotes the set of variable vertices contained in the hyperedge ϵ. The function $\phi_{\mathcal{H}}$ is parameterized by another two-layer MLP activated by LeakyReLU. The hyperedge-based convolution (i.e., Eq. 5 and Eq. 6) is repeated for L iterations to fully capture information from high-degree terms.

Variable-Constraint-Based Convolution. After the hyperedge-based convolution embeds higher-order relationships into variable vertices, the model still needs to account for variable-constraint interdependencies. To this end, we apply a variable-constraint-based convolution that explicitly processes message passing along the edges connecting variable and constraint vertices.

This convolution operates through bidirectional message passing: variable embeddings (that already aggregate higher-order information from the hyperedge-based convolution) are first propagated to constraint vertices (Eq. 7), and then, the updated constraint representations are passed back to the variable vertices (Eq. 8). Through the two-step convolutions, the variable embeddings are enriched with information of the constraints they are required to satisfy.

$$h_c \leftarrow f_{\mathcal{C}}(h_c, \sum_{v \in \mathcal{N}_c} \phi_{\mathcal{C}}(h_c, h_v, h_{vc})) + h_c, \forall c \in \mathcal{C}, \tag{7}$$

$$h_v \leftarrow f_{\mathcal{V}}(h_v, \sum_{c \in \mathcal{N}_v^C} \phi_{\mathcal{V}}(h_c, h_v, h_{vc})) + h_v, \forall v \in \mathcal{V}, \tag{8}$$

Formally, h_c and h_{vc} denote the embeddings for the constraint vertex c and the edge e_{vc} connecting v and c, separately, and they are initialized from the raw features of constraint vertices and edges via two-layer MLPs. h_v denotes the embedding for variable vertex v and is passed from the hyperedge-based convolution. The set $\mathcal{N}_c$ contains all variable vertices connected to constraint c, while $\mathcal{N}_v^C$ contains all constraint vertices connected to variable v. Finally, $\phi_{\mathcal{C}}$, $\phi_{\mathcal{V}}$, $f_{\mathcal{C}}$, and $f_{\mathcal{V}}$ are implemented as two-layer MLPs activated by LeakyReLU. The variable-constraint-based convolution is executed once.

Solution Prediction and Refinement. After hyperedge-based and variable-constraint-based convolutions, the variable embeddings involve both high-order interactions from high-degree terms and the interdependencies between variables and constraints. We finally feed these embeddings into a two-layer MLP, which outputs a scalar value for each variable as the solution prediction.

Without loss of generality, we focus on POIP with binary variables. This is reasonable because any bounded integer variable can be binarized, enabling a bounded POIP to be reformulated as an equivalent binary POIP [9]. The entire HNN is trained in a supervised manner using the binary cross-entropy loss:

$$\mathcal{L}_{\mathrm{BCE}} = -\frac{1}{N} \sum_{i=1}^{N} [y_i \log(\sigma(\hat{y}_i)) + (1 - y_i) \log(1 - \sigma(\hat{y}_i))],$$

where N is the number of logits from the HNN, $y \in \{0,1\}^N$ and $\hat{y} \in \mathbb{R}^N$ represent the ground truth labels and the predicted logits, and $\sigma(\cdot)$ is the sigmoid function.

The predictions produced by our model provide initial solution values for downstream algorithms and solvers. To make use of them, we adopt the parallel neighborhood optimization framework in [45, 46] that embeds an off-the-shelf solver to refine the predicted solution during inference. Specifically, the framework employs adaptive large neighborhood search. It first repairs the predictions into feasible solutions by fixing promising variables to their predicted values while allowing the remaining variables to be reoptimized by a solver, and then further refines these feasible solutions through additional neighborhood search to achieve better objective values. We refer interested readers to [45, 46] for more details.

5 Experimental Results

In this section, we conduct comprehensive experiments to demonstrate the effectiveness of our proposed method in solving POIP instances. Section 5.1 describes the experimental setup. Section 5.2 reports comparative results on quadratic and quintic POIP benchmarks. Section 5.3 investigates generalization to public datasets and quadratically constrained benchmarks. Finally, Sect. 5.4 presents an ablation study on the model architecture and high-degree term handling.

5.1 Setup

Benchmark Instances. Our experiments were conducted on three POIP benchmark problem instances. The first one is a synthetic quadratic integer programming (QIP) benchmark introduced by Xiong et al. [45], derived from the Quadratic Multiple Knapsack Problem (QMKP). It contains instances at five scales (Mini, 1k, 2k, 5k, 10k), where the first three scales are used for training and all except Mini are included in testing. The second benchmark contains hard quadratic instances from the public library QPLIB [15], evaluated also by Xiong et al. [45]. Third, we introduce a new synthetic quintic integer programming benchmark based on the Capacitated Facility Location Problem with Traffic Congestion (CFLPTC). This dataset includes five instance scales (50×10, 50×20, 150×30, 200×30, 500×100), with the first four scales used for training and the last three for testing. Detailed formulations and descriptions for all benchmarks are provided in Appendix B.

Baselines. We compare our method against three recent learning-based methods tailored for quadratic programming (QP): (1) NeuralQP [45] that introduces a hypergraph neural network to predict solutions for quadratically constrained QPs, (2) GNNQP [7] and (3) TriGNN [41], both investigating the theoretical expressive power of graph neural networks for quadratic terms. Similar to our HNN method, these models serve as solution predictors and are combined with exact solvers for repair and refinement (see Sect. 4.2). We adopt SCIP and Gurobi as the exact solvers in this experiment, and we also include them as standalone baselines to provide a comprehensive comparison.

Metrics. We evaluate performance using the relative primal gap (in percentage), defined as $\text{gap}_\% = |\text{OBJ} - \text{BKS}|/(|\text{BKS}| + 10^{-10}) \times 100$, where OBJ is the objective value obtained by a method and BKS is the best-known solution of the instance. BKS values of QPLIB instances are publicly available, while those for synthetic datasets are the best objective value found across our experiments. Under the same time limit, a lower $\text{gap}_\%$ indicates solutions closer to BKS and thus stronger performance. Wilcoxon signed-rank test at the 95% confidence level is applied to assess statistical significance of the difference among methods.

Implementations. All learning-based methods follow the same evaluation procedure. Each trained model first generates a solution prediction, which is then improved using the repair-and-refinement strategy (Sect. 4.2) under a fixed time limit. The time limits vary by benchmark and instance scale: 100, 600, 1800, and 3600 s for QMKP instances at scales 1k, 2k, 5k, and 10k, respectively; 100 s for QPLIB instances; and 60, 60, and 1000 s for CFLPTC instances at scales 150×30, 200×30, and 500×100. The exact-solver baselines (SCIP and Gurobi) are given the same time budgets to solve each instance from scratch. Each method was run five times per instance.

For inference, we matched training and testing benchmarks where possible. On the QMKP and CFLPTC datasets, models trained on the same benchmark with identical or smaller instance scales were used for testing. For QPLIB, the models were trained on QMKP-1k instances. More details on experimental implementations can be found in Appendix C.

Table 1. Comparison on QMKP datasets in terms of mean and standard deviation of $\text{gap}_\%$. The best results are highlighted in bold and $*$ indicates the overall results are statistically different from the best one. "–" denotes that the corresponding test was omitted, as models were only evaluated on instances with identical or larger scales than the training instances, following the setting in [45].

Method	Train	Base solver: Gurobi					Base solver: SCIP				
		1k	2k	5k	10k	Overall	1k	2k	5k	10k	Overall
Exact solver	–	0.38	0.23	29.08	26.69	$14.10^{*}_{\pm 13.89}$	5.15	6.10	35.73	27.61	$18.65^{*}_{\pm 13.38}$
Neural QP	Mini	0.35	0.09	0.03	0.03	$0.11_{\pm 0.13}$	0.54	0.11	29.38	17.90	$12.99_{\pm 13.21}$
	1k	0.35	0.09	0.04	0.04		0.38	0.11	28.68	19.44	
	2k	–	0.09	0.05	0.04		–	**0.10**	29.54	17.41	
GNNQP	Mini	0.45	0.09	0.05	0.03	$7.22^{*}_{\pm 15.00}$	0.54	0.11	29.38	17.90	$16.08_{\pm 15.52}$
	1k	0.49	0.10	0.06	38.95		0.52	0.11	29.48	38.09	
	2k	–	0.09	0.05	39.05		–	0.11	29.53	31.12	
TriGNN	Mini	0.43	0.09	0.04	0.04	$0.13_{\pm 0.17}$	0.52	0.12	28.70	**17.25**	$12.87_{\pm 12.95}$
	1k	0.46	0.09	**0.03**	0.04		0.47	0.10	**28.63**	18.77	
	2k	–	0.10	0.03	0.04		–	0.12	28.66	18.26	
Ours	Mini	**0.16**	0.08	0.03	**0.03**	$\mathbf{0.09}_{\pm 0.10}$	**0.18**	0.10	28.67	17.93	$\mathbf{12.76}_{\pm 13.21}$
	1k	0.31	0.10	0.05	0.04		0.33	0.11	29.33	16.79	
	2k	–	**0.07**	0.05	0.03		–	0.10	29.41	17.45	

5.2 Performance on Polynomial-Objective Integer Programs

We first evaluate all methods on synthetic benchmarks: the QMKP benchmark and the CFLPTC benchmark. Results are shown in Table 1 and Table 2.

On QMKP, we compare our method against all learning-based baselines and the two exact solvers. Table 1 reports the mean and standard deviation of $gap_\%$ across instance scales. It shows that the learning-based methods reach at least comparable solution qualities compared to exact solvers on 1k-scale instances and achieve better performance on larger-scale instances, highlighting the promise of learning-based approaches. Our method achieves performance that is at least comparable to, and in several cases better than, the learning-based baselines on individual scales, and it attains superior overall results across the QMKP dataset. Notably, while NeuralQP, TriGNN, and GNNQP are specifically designed for quadratic problems, our approach targeting more general POIP problems with high-degree terms can still outperform these specialized models.

Table 2 shows the results on the quintic CFLPTC benchmark. The compared baselines include Gurobi and SCIP, as well as NeuralQP. Note that NeuralQP, TriGNN, and GNNQP were implemented only for quadratic POIP. We compare with NeuralQP because it is a practically oriented method and shows strong performance on QMKP datasets. In contrast, GNNQP and TriGNN offer valuable insights into expressive power but are primarily theoretically oriented. The model of NeuralQP was trained on a quadratic reformulation of CFLPTC obtained by introducing auxiliary variables and constraints, as detailed in Appendix B. Across all evaluated scales, our method produces significantly smaller primal gaps than the exact solvers and NeuralQP, demonstrating its ability to effectively exploit the quintic objective structure. Moreover, the performance of our method remains strong on the largest CFLPTC instances of 500×100 scale, which are considerably larger than those seen during training, indicating good scalability and generalization across instance sizes.

Table 2. Comparison on CFLPTC datasets in terms of mean and standard deviation of $gap_\%$. The best results are highlighted in bold and * indicates the overall results are statistically different from the best one.

Method	Train	Base solver: Gurobi				Base solver: SCIP			
		150×30	200×30	500×100	Overall	150×30	200×30	500×100	Overall
Exact Solver	–	51.42	58.20	38.66	$48.89^*_{\pm 11.11}$	66.17	73.11	72.31	$69.14^*_{\pm 4.68}$
Neural QP	50×10 & 50×20	28.03	42.92	23.82	$37.04^*_{\pm 13.95}$	54.63	62.65	71.52	$62.09^*_{\pm 8.09}$
	150×30 & 200×30	42.20	62.18	26.47		58.08	67.11	69.90	
Ours	50×10 & 50×20	9.95	8.96	2.78	$\mathbf{6.59}_{\pm 6.39}$	31.52	**21.05**	62.62	$\mathbf{42.01}_{\pm 16.67}$
	150×30 & 200×30	**5.23**	**7.08**	**2.65**		**29.97**	46.31	**62.36**	

5.3   Generalization to Other Problem Structures

We next investigate how well the learned models generalize beyond the synthetic POIP benchmarks to problems with different structures. We consider the public QPLIB subset mentioned in Sect. 5.1 and compare our method with all baselines. As shown in Table 3, our method attains the lowest (bold) or second lowest (italics) average gap on most instances, and the lowest overall average gap, demonstrating its capability to transfer well to structurally distinct problems.

We also investigate problems with nonlinear constraints. We used RandQCP datasets built by [45], which are collections of quadratically constrained independent set problems. The nonlinearity in constraints is also represented by hyperedges connecting their variable vertices with other components in the hypergraph, while the HNN and repair-and-refinement process are unchanged. This indicates that our HNN has the potential to be easily extended to IP variants with polynomial terms in either objective functions or constraints. The results, summarized in Table 4, show that our method still achieves consistently superior performance compared to the baselines.

Table 3. Comparison on QPLIB instances by mean value and standard deviation of gap%. The best and second best results are highlighted in bold and italics, respectively. * indicates statistically significant difference to the best result.

Instance ID	Gurobi	NeuralQP	GNNQP	TriGNN	Ours
2315	59.75	15.92	*13.10*	**12.54**	13.49
2733	1.38	0.28	*0.18*	**0.10**	0.38
2957	2.11	2.07	1.88	*1.13*	**0.71**
3347	0.52	**0.20**	9.28	5.93	*0.34*
3402	**2.80**	7.10	4.95	*4.12*	6.64
3584	66.75	15.04	**7.75**	15.59	*15.03*
3752	14.09	*1.70*	3.55	5.04	**1.30**
3841	26.69	9.90	8.57	**0.53**	*6.13*
3860	48.95	15.92	*15.56*	16.06	**14.69**
3883	8.25	**0.36**	*0.43*	0.74	0.63
5962	16.89	*15.04*	19.49	15.76	**7.06**
overall	22.56*$_{\pm23.55}$	7.58*$_{\pm6.60}$	7.67*$_{\pm6.00}$	7.05$_{\pm6.32}$	**6.04**$_{\pm5.70}$

5.4   Ablation Study

We perform an ablation study on QMKP to evaluate the contribution of each architectural component. Two variants disable one of the convolution modules: **w/o-HyConv**, which removes the hyperedge-based convolution, and **w/o-VCConv**, which removes the variable–constraint-based convolution. For fairness, both variants use minimally modified graph representations that preserve

Table 4. Results on RandQCP datasets by mean and standard deviation of gap%. The best results are highlighted in bold and * indicates the overall results have statistically significant difference to the best one.

Method	Train	Base solver: Gurobi					Base solver: SCIP				
		1k	2k	5k	10k	Overall	1k	2k	5k	10k	Overall
Exact solver	–	3.21	6.45	5.83	9.88	$6.34^{*}_{\pm 9.84}$	38.82	42.08	53.12	53.07	$46.77^{*}_{\pm 6.60}$
Neural QP	Mini	0.47	0.43	0.25	0.22	$0.33^{*}_{\pm 0.22}$	0.60	0.51	0.37	0.31	$0.43^{*}_{\pm 0.23}$
	1000	0.54	0.42	0.22	0.22		0.62	0.50	0.36	0.32	
	2000	–	0.39	0.25	0.19		–	0.47	0.34	0.29	
GNNQP	Mini	3.53	2.29	3.23	3.36	$3.07^{*}_{\pm 0.68}$	3.60	2.40	3.29	3.42	$3.14^{*}_{\pm 0.66}$
	1000	3.48	2.30	3.25	3.36		3.62	2.40	3.31	3.40	
	2000	–	2.32	3.23	3.38		–	2.37	3.29	3.41	
TriGNN	Mini	0.55	0.45	0.28	0.27	$0.37^{*}_{\pm 0.23}$	0.67	0.49	0.37	0.29	$0.48^{*}_{\pm 0.88}$
	1000	0.57	0.47	0.27	0.25		0.72	0.53	0.38	0.32	
	2000	–	0.50	0.28	0.23		–	0.47	0.36	0.70	
Ours	Mini	**0.32**	**0.31**	**0.19**	**0.14**	$\mathbf{0.26}_{\pm 0.19}$	**0.46**	**0.37**	**0.27**	**0.17**	$\mathbf{0.34}_{\pm 0.19}$
	1000	0.43	0.32	0.23	0.17		0.50	0.42	0.30	0.21	
	2000	–	0.38	0.20	0.18		–	0.40	0.32	0.26	

all necessary relationships (Appendix C), while all other settings remain identical to our full model. To further examine the role of our hyperedge-based convolution in modeling high-order variable interactions, we introduce two additional variants that replace it with high-degree aggregation mechanisms from prior methods: **NeuralQP-HD**, which adopts NeuralQP's high-degree handling block, and **GNNQP-HD**, which uses the corresponding module from GNNQP. In both cases, only the high-order processing component is altered.

All models are trained on QMKP-1k and evaluated on QMKP-1k and QMKP-2k, using Gurobi for repair-and-refinement and measuring solving quality via gap% under identical time limits. As shown in Table 5, removing either convolution module reduces performance, and substituting our high-order mechanism with that from NeuralQP or GNNQP also leads to worse results.

Table 5. Comparison of our model and ablation baselines on QMKP instances in terms of gap%. The best result of each testing dataset is highlighted in bold.

	NeuralQP-HD	GNNQP-HD	w/o-VCConv	w/o-HyConv	Ours
1k	0.33	0.36	0.42	**0.30**	0.31
2k	0.90	2.16	1.92	0.81	**0.10**
Average	0.61	1.26	1.17	0.56	**0.20**

6   Conclusion

This paper introduces a novel hypergraph neural network (HNN) framework for polynomial-objective integer programming problems (POIP). Our approach contributes two key innovations: a high-degree-term-aware hypergraph representation that captures variable interactions in high-degree terms and variable-constraint interdependencies inherent in POIP problems, and a hypergraph neural network architecture that integrates hypergraph-based and bipartite-graph-based convolutions to enable accurate solution prediction. Comprehensive experimental evaluations across quadratic and quintic programming problems demonstrate that our method outperforms both state-of-the-art exact solvers and specialized learning-based approaches, establishing its effectiveness and practical value for challenging POIP applications.

While promising, our work represents just one step towards addressing the broader challenges of nonlinear integer programming. Future research directions include 1) designing more comprehensive representations and neural network structures for general nonlinear instances such as those with trigonometric and logarithmic functions and 2) exploring end-to-end frameworks that directly output feasible solutions without requiring repair mechanisms.

Supplementary Materials

Code and appendix are available at https://github.com/PineappleLiMs/Solving_POIP_with_HNN.git. In appendix, we provide more details of our HNN-based framework (Appendix A), the benchmarks (Appendix B), and our implementation (Appendix C). We also report additional results, including the predictive performance (Appendix D.1), CFLPTC comparisons between the native and quadratic formulations (Appendix D.2), shifted geometric mean (Appendix D.3), and comparisons across different solver random seeds (Appendix D.4), to further substantiate the superiority of our HNN. Finally, Appendix E presents the complexity analysis of our method.

Acknowledgments. This research is part of the LEO (Learning and Explaining Optimization) project which is co-funded by Holland High Tech | TKI HSTM via the PPS allowance scheme for public-private partnerships, and by Dassault Systèmes. This work used the Dutch national e-infrastructure with the support of the SURF Cooperative using grant no. EINF-11638.

References

1. Ahmadi, A.A., Majumdar, A.: Some applications of polynomial optimization in operations research and real-time decision making. Optim. Lett. **10**, 709–729 (2016)
2. Bai, Y., Hwang, T., Kang, S., Ouyang, Y.: Biofuel refinery location and supply chain planning under traffic congestion. Transp. Res. Part B: Methodol. **45**(1), 162–175 (2011)

3. Bazaraa, M.S., Sherali, H.D., Shetty, C.M.: Nonlinear programming: theory and algorithms. John Wiley & Sons (2006)
4. Bengio, Y., Lodi, A., Prouvost, A.: Machine learning for combinatorial optimization: a methodological tour d'horizon. Eur. J. Oper. Res. **290**(2), 405–421 (2021)
5. Bonami, P., Lodi, A., Zarpellon, G.: A classifier to decide on the linearization of mixed-integer quadratic problems in CPLEX. Oper. Res. **70**(6), 3303–3320 (2022)
6. Cappart, Q., et al.: Combinatorial optimization and reasoning with graph neural networks. J. Mach. Learn. Res. **24**(130), 1–61 (2023)
7. Chen, Z., Chen, X., Liu, J., Wang, X., Yin, W.: Expressive power of graph neural networks for (mixed-integer) quadratic programs. In: Proceedings of the 42nd International Conference on Machine Learning. Proceedings of Machine Learning Research, vol. 267, pp. 7880–7911 (2025)
8. Chmiela, A., Khalil, E.B., Gleixner, A.M., Lodi, A., Pokutta, S.: Learning to schedule heuristics in branch and bound. In: Advances in Neural Information Processing Systems. vol. 34, pp. 24235–24246 (2021)
9. Dantzig, G.B.: Linear Programming and Extensions. Princeton Landmarks in Mathematics and Physics (2016)
10. Deenen, P., Nuijten, W., Akcay, A.: Scheduling a real-world photolithography area with constraint programming. IEEE Trans. Semicond. Manuf. **36**(4), 590–598 (2023)
11. Deza, A., Khalil, E.B.: Machine learning for cutting planes in integer programming: a survey. In: Proceedings of the Thirty-Second International Joint Conference on Artificial Intelligence, IJCAI-23, pp. 6592–6600 (2023)
12. Ding, J., et al.: Accelerating primal solution findings for mixed integer programs based on solution prediction. In: Proceedings of the AAAI Conference on Artificial Intelligence. vol. 34, pp. 1452–1459 (2020)
13. Feng, S., Yang, Y.: SORREL: suboptimal-demonstration-guided reinforcement learning for learning to branch. In: Proceedings of the AAAI Conference on Artificial Intelligence. vol. 39, pp. 11212–11220 (2025)
14. Ferber, A.M., et al.: SurCo: learning linear surrogates for combinatorial nonlinear optimization problems. In: International Conference on Machine Learning. vol. 202, pp. 10034–10052 (2023)
15. Furini, F., et al.: QPLIB: a library of quadratic programming instances. Math. Program. Comput. **11**, 237–265 (2019)
16. Gasse, M., Chételat, D., Ferroni, N., Charlin, L., Lodi, A.: Exact combinatorial optimization with graph convolutional neural networks. In: Advances in Neural Information Processing Systems. vol. 32, pp. 15554–15566 (2019)
17. Geng, Z., et al.: Differentiable integer linear programming. In: The Thirteenth International Conference on Learning Representations (2025)
18. Ghaddar, B., Gómez-Casares, I., González-Díaz, J., González-Rodríguez, B., Pateiro-López, B., Rodríguez-Ballesteros, S.: Learning for spatial branching: an algorithm selection approach. INFORMS J. Comput. **35**(5), 1024–1043 (2023)
19. González-Rodríguez, B., Ossorio-Castillo, J., González-Díaz, J., González-Rueda, Á.M., Penas, D.R., Rodríguez-Martínez, D.: Computational advances in polynomial optimization: RAPOSa, a freely available global solver. J. Global Optim. **85**(3), 541–568 (2023)
20. Gupta, P., Gasse, M., Khalil, E.B., Mudigonda, P.K., Lodi, A., Bengio, Y.: Hybrid models for learning to branch. In: Advances in Neural Information Processing Systems (2020)

21. Han, Q., et al.: A GNN-guided predict-and-search framework for mixed-integer linear programming. In: The Eleventh International Conference on Learning Representations (2023)
22. Heydaribeni, N., Zhan, X., Zhang, R., Eliassi-Rad, T., Koushanfar, F.: Distributed constrained combinatorial optimization leveraging hypergraph neural networks. Nat. Mach. Intell. **6**(6), 664–672 (2024)
23. Huang, J., Yang, J.: UniGNN: a unified framework for graph and hypergraph neural networks. In: Proceedings of the Thirtieth International Joint Conference on Artificial Intelligence, IJCAI-21, pp. 2563–2569 (2021)
24. Huang, T., Ferber, A.M., Zharmagambetov, A., Tian, Y., Dilkina, B.: Contrastive predict-and-search for mixed integer linear programs. In: Proceedings of the 41st International Conference on Machine Learning. Proceedings of Machine Learning Research, vol. 235, pp. 19757–19771 (2024)
25. Kesavan, P., Allgor, R.J., Gatzke, E.P., Barton, P.I.: Outer approximation algorithms for separable nonconvex mixed-integer nonlinear programs. Math. Program. **100**, 517–535 (2004)
26. Kim, S., Lee, S.Y., Gao, Y., Antelmi, A., Polato, M., Shin, K.: A survey on hypergraph neural networks: an in-depth and step-by-step guide. In: Proceedings of the 30th ACM SIGKDD Conference on Knowledge Discovery and Data Mining, pp. 6534–6544 (2024)
27. Kronqvist, J., Bernal, D.E., Lundell, A., Grossmann, I.E.: A review and comparison of solvers for convex MINLP. Optim. Eng. **20**(2), 397–455 (2019)
28. Kuang, Y., et al.: Accelerate Presolve in large-scale linear programming via reinforcement learning. IEEE Trans. Pattern Anal. Mach. Intell. **47**(8), 6660–6672 (2025)
29. Labassi, A.G., Chételat, D., Lodi, A.: Learning to compare nodes in branch and bound with graph neural networks. In: Advances in Neural Information Processing Systems. vol. 35, pp. 32000–32010 (2022)
30. Li, S., Kulkarni, J., Menache, I., Wu, C., Li, B.: Towards foundation models for mixed integer linear programming. In: The Thirteenth International Conference on Learning Representations (2025)
31. Liu, C., et al.: L2P-MIP: learning to Presolve for mixed integer programming. In: The Twelfth International Conference on Learning Representations (2024)
32. Liu, D., Fischetti, M., Lodi, A.: Learning to search in local branching. In: Proceedings of the AAAI Conference on Artificial Intelligence. vol. 36, pp. 3796–3803 (2022)
33. Liu, H., et al.: Apollo-MILP: an alternating prediction-correction neural solving framework for mixed-integer linear programming. In: The Thirteenth International Conference on Learning Representations (2025)
34. Maudet, G., Danoy, G.: Search strategy generation for branch and bound using genetic programming. Proc. AAAI Conf. Artif. Intell. **39**(11), 11299–11308 (2025)
35. Nair, V., et al.: Solving mixed integer programs using neural networks (2021)
36. Rudin, W.: Real and complex analysis (1987)
37. Smit, I.G., et al.: Graph neural networks for job shop scheduling problems: a survey. Comput. Oper. Res. **176**, 106914 (2025)
38. Smith, E.M., Pantelides, C.C.: A symbolic reformulation/spatial branch-and-bound algorithm for the global optimisation of nonconvex MINLPs. Comput. Chem. Eng. **23**(4–5), 457–478 (1999)
39. Tang, B., Khalil, E.B., Drgoňa, J.: Learning to optimize for mixed-integer nonlinear programming. arXiv preprint arXiv:2410.11061 (2024)

40. Wang, J., et al.: Learning to cut via hierarchical sequence/set model for efficient mixed-integer programming. IEEE Trans. Pattern Anal. Mach. Intell. **46**(12), 9697–9713 (2024)
41. Wu, C., et al.: On representing convex quadratically constrained quadratic programs via graph neural networks. Transactions on Machine Learning Research (2025)
42. Wu, Y., Song, W., Cao, Z., Zhang, J.: Learning large neighborhood search policy for integer programming. Adv. Neural. Inf. Process. Syst. **34**, 30075–30087 (2021)
43. Wu, Y., et al.: Graph learning assisted multi-objective integer programming. Adv. Neural. Inf. Process. Syst. **35**, 17774–17787 (2022)
44. Wu, Z., Pan, S., Chen, F., Long, G., Zhang, C., Yu, P.S.: A comprehensive survey on graph neural networks. IEEE Trans. Neural Netw. Learn. Syst. **32**, 4–24 (2020)
45. Xiong, Z., Zong, F., Ye, H., Xu, H.: NeuralQP: A general hypergraph-based optimization framework for large-scale QCQPs. arXiv preprint arXiv:2410.03720 (2024)
46. Ye, H., Xu, H., Wang, H., Wang, C., Jiang, Y.: GNN&GBDT-guided fast optimizing framework for large-scale integer programming. In: International Conference on Machine Learning. Proceedings of Machine Learning Research, vol. 202, pp. 39864–39878 (2023)
47. Zhang, J., et al.: A survey for solving mixed integer programming via machine learning. Neurocomputing **519**, 205–217 (2023)

Integrating Bayesian Optimization and Combinatorial Optimization for LEO Satellite Constellation Design

Linhao Luo$^{(\boxtimes)}$, Trudy Lam, and Vicky Mak

Deakin University, Waurn Ponds, Geelong, VIC 3216, Australia
`{s224261834,trudy.lam,vicky.mak}@deakin.edu.au`

Abstract. We study the problem of selecting optimal design parameters for Low-Earth-Orbit (LEO) satellite constellations, where the performance of a design is determined by the maximum Satellite–Ground Link (SGL) duration it enables. The design space spans altitude, inclination, plane count, satellites per plane, and phasing, rendering a full mathematical programming model intractable.

We propose a two-layer optimization framework: Bayesian Optimization (BO) explores the constellation design space, while each candidate design is evaluated by solving a large Satellite–Ground Link Scheduling Problem (SGLSP). The SGLSP is solved using PINCH, our fast primal heuristic based on LP-relaxation-guided fixing and clustering. PINCH produces high-quality schedules significantly faster than Gurobi, one of the most advanced commercial mixed-integer linear programming solvers.

This acceleration allows BO to evaluate constellation designs for 40 ground stations within hours, making large-scale exploration of the design space computationally feasible. The framework scales to large instances and offers a practical approach for complex space-system design problems with combinatorial structure. Moreover, by modifying the weights in the cost–quality trade-off, the same computational pipeline can be steered toward different design priorities (e.g., longer SGL duration, lower deployment cost, or reduced latency), enabling users to obtain constellation configurations that match their specific requirements and preferences.

1 Introduction

Low Earth orbit (LEO) satellite constellations have become a cornerstone of next-generation global connectivity. Satellite networks are valued for their global coverage and reduced launch costs due to advancements in aerospace technology [23]. Deploying satellite constellations has become foundational in modern telecommunications, remote sensing, and global connectivity. These constellations consist of numerous satellites orbiting Earth in well-designed formations, providing a wide range of services, from internet access to Earth observation.

T. Guns (Ed.): CPAIOR 2026, LNCS 16595, pp. 314–330, 2026.
https://doi.org/10.1007/978-3-032-27242-3_19

Advances in reusable launch vehicles and miniaturized satellite platforms have sharply reduced the cost to orbit, allowing operators to deploy thousands of LEO spacecraft in carefully phased orbital planes. Operating between 500 km and 1,200 km above Earth, LEO satellites deliver round-trip latency below 50 ms and substantially lower free space path loss than geostationary (GEO) and medium Earth orbit (MEO) counterparts.

The effectiveness of a LEO network depends on thoughtful constellation design. Key architectural parameters include the number of orbital planes, satellites per plane, altitude, inclination angle, and the phasing factor. An orbital plane is a path around the Earth where satellites in the same plane follow the same ground track but space out along the track. The inclination angle is the tilt of the orbital plane relative to the Earth's equatorial plane. The phasing factor is how the satellites within an orbital plane are evenly spaced. Together, they govern spatial diversity, ground track repeatability, revisit time, and handover cadence. Designers must balance global service availability against launch mass, in-orbit propulsion, and spectrum reuse while respecting uneven user distributions and regulatory filing limits.

Recent studies have moved beyond isolated link-budget calculations towards a multi-perspective optimization of constellation geometry. This shift significantly expands the design space, rendering exhaustive search methods impractical. To address this complexity, Bayesian Optimization (BO), genetic algorithms, and surrogate-assisted search methods have been used to explore potential configurations [8,14]. In this paper, we take a leap forward by integrating satellite-ground link duration as a meaningful metric for solution quality within a BO framework so as to effectively reduce the size of search space and efficiently guide the selection of promising design parameters.

Since constellation performance is reflected in multiple aspects, its evaluation is inherently challenging. Against the backdrop of increased adoption of LEO satellites, Shi et al. [16] estimated satellite coverage by considering the latitude and longitude ranges of the evaluation sites. However the "coverage" considered in their paper does not necessarily mean that communication is feasible. For a more precise measurement of quality of LEO constellation design, we use the optimal solution to the Satellite–Ground Link Scheduling Problem (SGLSP) as a metric to assess total coverage that communication is feasible. The SGLSP is an allocation optimisation problem for scheduling satellite-ground links (SGLs) for low-orbit communication satellite constellations [11] and is NP-hard. Given a fixed LEO constellation and a set of ground stations, each satellite pass offers a few minute's link opportunity, during which data can be transferred between the satellite and ground antennas. The link duration is established when the beams of both the satellite and ground antenna overlap. The combined span of preparation time, link duration, and release time is referred to as the tracking duration. See Fig. 1 for a visual overview of this process.

The total SGL duration is quantified as a key metric of service quality and incorporated into the cost model used in BO. This model also accounts for multiple cost streams, including satellite manufacturing, launch expenses, operations,

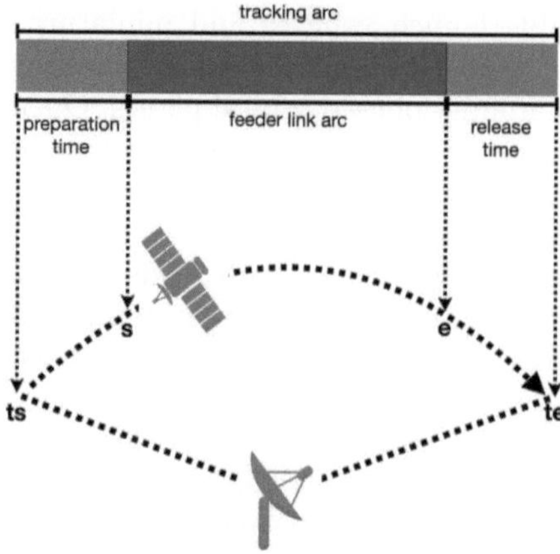

Fig. 1. The process of SGL establishment. This figure is adapted based on Liu et al. [11]. t_s and t_e denote the start and end times of the tracking arc, respectively, while s and e denote the start and end times of the feeder link arc. Communication is possible within the time covered by the feeder link arc only.

and maintenance, and captures the economic effects of latency on communication performance. The proposed approach identifies plane count, satellites per plane, inclination angle, and phasing factor that enhance downlink opportunities while avoiding excessive resource consumption.

2 Related Work

The design of Low Earth Orbit (LEO) constellations typically aims to maximize global coverage and service continuity while minimizing deployment and operational costs. Traditional studies approach this problem by modelling geometric parameters—such as the number of planes, satellites per plane, altitude, and inclination—and evaluating designs mainly through coverage probability. Within this paradigm, surrogate-assisted multi-objective optimization and BO have been widely used to explore the design space, showing that suitable orbital configurations can reduce conflict density and improve scheduling slack [8,14]. However, these works generally assume a fixed scheduling strategy and do not optimize the link-scheduling component itself.

More recent studies adopt a broader multidisciplinary design optimization framework that incorporates orbital mechanics, payload power, atmospheric drag, link budgets, and launch logistics. Examples include the sequential RBF surrogate with SVM classification proposed by Shi et al. [16] for mixed discrete–continuous variables, and the Kriging-based reconfiguration method of Zuo et al. [24], both of which achieve substantial reductions in computational time for solution evaluation.

In contrast to these studies, which rely on coverage as the primary performance metric, our work shifts the focus to optimizing the total feasible communication time—formulated as the SGLSP. Incorporating SGLSP directly into the design process allows us to capture how scheduling decisions influence constellation performance and enables the joint optimization of orbital geometry and link-scheduling strategy, offering a more comprehensive perspective than prior coverage-centric approaches.

The relevance of this integration is further underscored by the increasing attention devoted to SGLSP in recent years. The rapid deployment of large-scale LEO constellations has heightened the need for reliable link scheduling to support applications such as navigation augmentation, broadband connectivity, and Earth observation. Unlike GEO or MEO systems, LEO satellites provide only short visibility windows, and the increasing traffic demand requires far more frequent and complex scheduling decisions [6].

In early small-scale LEO constellations, Spangelo et al. [17] designed a mixed-integer programming model to schedule communications between one satellite and 20 ground stations using a commercial solver. Metaheuristic methods explored by Xhafa et al. [20] and Petelin et al. [13] were applied to comparably small instances involving up to 15 ground stations and 20 satellites. For larger constellations with hundreds of satellites, Liu et al. [11] proposed a knowledge-assisted adaptive large neighborhood search that alternates heuristic destroy-and-repair moves with a CPLEX-based searching method. Although this hybrid approach improves solution quality, its runtime still relies heavily on the commercial solver to achieve convergence.

To enable faster solutions to the SGLSP, we reformulate it as a Maximum Weighted Independent Set (MWIS) problem on a conflict graph. This transformation allows the use of well-studied graph optimization techniques. Metaheuristics such as Tabu search with quadratic penalties [1] and efficient local search methods tailored for sparse graphs [10] have demonstrated competitive performance on benchmark instances. Meanwhile, exact branch-and-reduce algorithms have achieved theoretical progress on bounded-degree graphs [21,22]. However, the exponential complexity makes them impractical for constellations involving tens of thousands of tracking arcs. These observations suggest that neither standalone metaheuristics nor exact MWIS algorithms are sufficient for large-scale constellation design scenarios that require repeated evaluations within a higher-level optimization loop. To address this gap, we develop a hybrid AI-OR framework that combines relaxation-guided decomposition and mixed-integer programming with learning-based global exploration via Bayesian Optimization.

3   Contributions

Our main contributions are summarized as follows:

- **Introduction of SGL Duration as a Key Metric** – We introduce SGL duration as an essential criterion for evaluating constellation quality, directly reflecting service continuity and coverage performance.

- **Hybrid AI–OR Framework for SGLSP and Constellation Design** – We provide a mathematical formulation of the SGLSP and propose PINCH within a Bayesian Optimization framework. PINCH delivers schedules comparable in quality to those generated by commercial MILP solvers. Our efficient framework integrates relaxation-guided decomposition and integer programming with learning-driven global search.
- **Development of a Cost Model for BO** – We design a comprehensive cost model that incorporates budgetary constraints and latency penalties, which is embedded into the BO framework to guide constellation parameter selection.
- **Periodic Surrogate Rebuild for Stable BO Optimization** – We introduce a periodic surrogate rebuild mechanism that recalibrates the normalization statistics and refits the BO model, ensuring consistent objective scaling and substantially improving the stability of the search process.
- **Extensive Computational Experiments on Instances** – We conduct extensive experiments on simulated test instances of varying sizes to demonstrate the scalability and effectiveness of our approach.
- **Flexible Framework for Integrated Optimization** – We propose a flexible framework capable of integrating various cost models and objective functions, enabling broader applicability to future constellation design and scheduling problems.

4 Satellite-Ground Link Scheduling Optimisation

4.1 Satellite-Ground Link Duration Calculation

The SGL durations are computed using our Python-based simulator, which estimates the visibility windows for each satellite–ground station pair and generates realistic instances for the scheduling problem.

The simulation begins with the generation of Sub-Satellite Point (SSP) trajectories, representing the ground tracks traced on Earth's surface directly beneath the satellites. This process requires key orbital parameters, including altitude, inclination angle, eccentricity, right ascension of the ascending node (RAAN), and the time of passage of the ascending node (tAN). The mathematical derivations for calculating these ground tracks can be found in [3].

In this study, we adopt the Walker-delta constellation configuration. This configuration consists of P orbital planes, each separated by an equal right-ascension increment, containing S satellites that are evenly phased by pattern unit F [2,19]. Owing to its symmetry and efficient global coverage, the Walker-delta configuration remains the industry standard for LEO broadband systems such as Starlink and OneWeb. Our simulator incorporates line-of-sight and minimum-elevation constraints with respect to each ground station, while also accounting for the inclination angle of the constellation. The outputs are start and end epochs of every feasible SGL (ts and te in Fig. 1), which serve as the essential input for the subsequent scheduling optimization.

4.2   Binary Integer Programming Model

After enumerating all visible SGL windows, the SGLSP can be reformulated as a MWIS problem. Using $G(V, E)$ to represent the MWIS, each SGL tracking arc corresponds to a vertex $v \in V$ and an edge $e \in E$ corresponds to a conflict between two tracking arcs. The weight of a vertex is its link duration. The conflict rules—covering ground-antenna transition limits, satellite switching/transition requirements, and containment constraints—are taken from Liu et al. [11], and we refer the reader to their work for the detailed formulation explanation.

The pairwise exclusive tracking arcs can be precomputed. Specifically, if a pair of tracking arcs (a_1, a_2) violates any given constraint, they are classified as conflict tracking arc pairs. Leveraging this property, the system can be represented as an undirected graph.

Further, each clique in this graph encapsulates all pairwise conflicts among its vertices, allowing us to replace the set of pairwise conflict constraints with a much smaller set of clique-packing constraints. Given an undirected graph $\mathcal{G}(\mathcal{V}, \mathcal{E})$, where $\mathcal{V}$ denotes the set of tracking arcs and $\mathcal{E}$ is the set of conflict tracking arc pairs. Let $\mathcal{C}$ denote the set of cliques in $\mathcal{G}$ and let $\mathcal{V}(c)$ denote all tracking arcs in clique $c \in \mathcal{C}$. With these definitions, the new problem can be defined as follows:

(**The MWIS Model based on cliques**)

$$\text{maximize} \sum_{v \in \mathcal{V}} x_v (e_v - s_v) \tag{1a}$$

$$\text{subject to} \sum_{v \in V(c)} x_v \leq 1, \qquad \forall c \in \mathcal{C} \tag{1b}$$

$$x_v \in \{0, 1\}, \quad \forall v \in \mathcal{V} \tag{1c}$$

4.3   Solution Algorithm: Primal INteger Clustering Heuristic (PINCH)

Model 2 can be solved to optimality using commercial solvers such as Gurobi, but only for relatively small problem instances. Real-world SGLSP instances, however, are significantly larger, spanning long planning horizons and involving numerous ground stations and satellites, resulting in a vast, highly structured search space. To address this complexity, we introduce a Primal INteger Clustering Heuristic (PINCH), which decomposes the original problem into smaller, manageable subproblems. These are then recombined and refined through a merge-and-repair process, enabling the generation of high-quality solutions with improved efficiency.

The detailed procedure of the PINCH algorithm is presented in Algorithm 1. The process begins by applying a fast linear programming relaxation to fix a subset of decision variables, thereby reducing the problem size. The resulting conflict graph is then decomposed into smaller, tractable subproblems using the Leiden algorithm, a high-performance community detection method that efficiently

---

Algorithm 1. PINCH

Input: Conflict graph $\mathcal{G} = (\mathcal{V}, \mathcal{E})$;
Output: Near-optimal schedule $x^\star$

1: Initialize incumbent schedule $x^\star \leftarrow \emptyset$ *// start with empty solution*
2: Solve LP relaxation of MWIS on $\mathcal{G}$ to obtain x^{LP} *// fractional solution*
3: Fix variables with $x^{\mathrm{LP}} = 1$ to 1 and remove their neighbors   *// reduce problem size*
4: Partition residual graph into clusters $\mathcal{C} = \{C_1, ..., C_k\}$ using Leiden algorithm   *// clustering*
5: **for** each cluster $C \in \mathcal{C}$ **in parallel do**
6:     $x_C \leftarrow \mathrm{IP\_SOLVE}(C)$ *// solve cluster MWIS*
7: **end for**
8: Merge local solutions $\{x_C\}$ into global solution x *// combine partial solutions*
9: $x^\star \leftarrow \mathrm{REPAIRCONFLICTS}(x, \mathcal{G})$ *// enforce feasibility*
10: **return** $x^\star$ *// final near-optimal schedule*

---

refines partitions [18]. Each subproblem is subsequently solved in parallel using an integer programming solver. Finally, a merge-and-repair procedure, described in Algorithm 2, is employed to reconcile the locally optimal solutions and ensure global feasibility. This workflow achieves near-optimal schedules within a fraction of the time required by simply using a commercial IP solver while preserving overall solution quality, and the solution time advantage is critical to our general BO framework for the main LEO constellation design optimisation problem.

Algorithm 2. REPAIRCONFLICTS Procedure

Input: Infeasible solution vector x on graph $\mathcal{G} = (\mathcal{V}, \mathcal{E})$
Output: Repaired solution $x^\star$

1: $P \leftarrow$ vertices incident to edges $(u, v) \in \mathcal{E}$ where $x(u) = x(v) = 1$ *// find conflicts*
2: $Q \leftarrow$ vertices $v \in \mathcal{V}$ with $x(v) = 1$ and $v \notin P$
3: $S \leftarrow$ vertices remaining after removing Q and their immediate neighbors   *// conflict region*
4: $\mathcal{G}_S \leftarrow$ subgraph of $\mathcal{G}$ induced by S *// build conflict subgraph*
5: $x_S \leftarrow \mathrm{IP\_SOLVE}(\mathcal{G}_S)$ *// solve MWIS locally*
6: $x^\star \leftarrow$ replace x entries for S with x_S *// update global solution*
7: **return** $x^\star$ *// repaired schedule now feasible*

Efficiency and Runtime Analysis with Gurobi as Baseline

Figure 2 compares the runtime distributions of PINCH and Gurobi [5]. The experiments were conducted on 81 single-constellation and 60 dual-constellation designs, each consisting of 36 satellites and 40 ground stations. Due to page limit, detailed dataset statistics and full experimental results over all 141 instances are available in the online supplementary material[1]. The results show that PINCH solves more than 90% of the instances within one minute, whereas Gurobi often requires over ten minutes for a large portion of the cases.

Moreover, across all 141 instances, the relative objective gap between PINCH and Gurobi is at most 0.43%, with a mean of 0.14% (median 0.12%, standard deviation 0.12%) and a minimum essentially 0.00%. Moreover, 71.6% of instances fall within 0.20%.

These results demonstrate that PINCH achieves substantial runtime savings with minimal sacrifice in objective quality

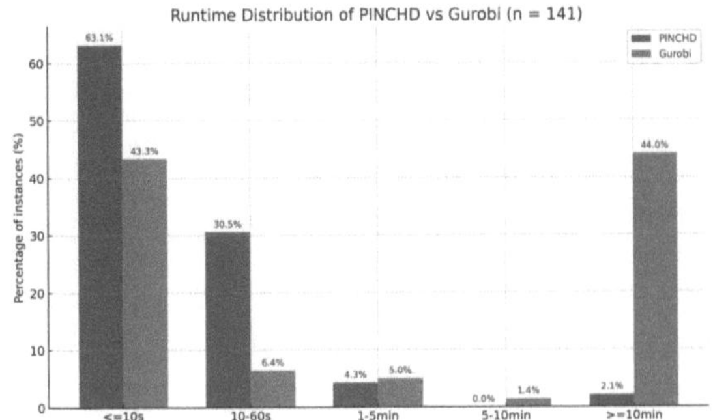

Fig. 2. Runtime distribution of PINCH and Gurobi on 141 satellite-scheduling instances. Bars report the percentage of instances solved within five time intervals ($\leq 10\,\mathrm{s}$, 10–60 s, 1–5 min, 5–10 min, $\geq 10\,\mathrm{min}$).

5   Satellite Constellation Cost Model

The cost model is parameterized to capture realistic scaling behavior and structural relationships between constellation design variables and economic impact. It reflects qualitative trends—such as increasing launch cost with altitude and constellation size, and latency dependence on orbital configuration—while remaining analytically tractable for large-scale design exploration. The optimization framework is independent of any specific cost instantiation, and practitioners may incorporate their own high-fidelity or industry-calibrated cost data without modifying the underlying optimization methodology.

Under this modeling perspective, a satellite constellation incurs three principal cost streams: manufacturing, launch, and multi-year operations, while its revenue potential depends on the aggregate duration of SGLs, moderated by a latency penalty that rises with orbital altitude. Let N, h, and i denote the number of satellites, orbital altitude, and inclination angle, respectively. The cost model employed in our optimization problem is summarized below.

---

[1] https://github.com/Alto00o/bo-pinch-leo-cpaior26.

- **Manufacturing cost** (C_{build}): assumed proportional to the fleet size, $C_{\text{build}} = c_{\text{build}} N$, where c_{build} is the unit build cost per spacecraft.
- **Launch cost** (C_{launch}): modeled relative to a reference mission that places mass m into a baseline orbit of altitude h_0 and inclination i_0: $C_{\text{launch}} = c_0\, m\,(1 + \alpha_h + \alpha_i)N$, where $\alpha_h = k_H \dfrac{\max(0,\, h-h_0)}{\Delta h}$, $\alpha_i = k_I \dfrac{\max(0,\, i-i_0)}{\Delta i}$. c_0 is the baseline cost per kilogram, m is the mass of the satellite, Δh and Δi define the surcharge step sizes, and k_H, k_I are dimensionless penalty factors that account for the additional propellant and trajectory-correction effort required by higher or more inclined orbits.
- **Operations cost** (C_{ops}): covers telemetry, tracking and command, insurance, and data back-haul over a service life of Y years, $C_{\text{ops}} = c_{\text{ops}} NY$, with c_{ops} denoting the annual operating cost per satellite.
- **Latency penalty** (P_{lat}): imposed to reflect the degradation in service quality as altitude increases, $P_{\text{lat}} = h\, w_L$, where w_L is a weighting factor that converts each millisecond of delay into a cost.
- **Link–time benefit** (B_{link}): the economic return from connectivity, assumed to follow a concave power-law in the total scheduled link time (SGL, in seconds) obtained from the SGLSP, $B_{\text{link}} = w_D \left(\frac{SGL}{10^6}\right)^{\beta}$, where w_D is the monetization weight (e.g. USD per million seconds), and $\beta > 0$ is the Link Benefit Index, which characterizes the sensitivity of economic return to the SGL duration. A higher value of β indicates a stronger emphasis on link availability in the evaluation.

For a given design, the net present cost is therefore:

$$\text{NetCost} = C + P_{\text{lat}} - B_{\text{link}}. \tag{2}$$

where $C = C_{\text{build}} + C_{\text{launch}} + C_{\text{ops}}$, as these three terms collectively represent the *monetary costs* of manufacturing, launching, and operating the satellites. Our optimization framework searches the design space for constellations that minimize (2) while satisfying budget ceilings and service-level requirements.

6 Bayesian Optimization

The design space of a LEO constellation is inherently high-dimensional, and evaluating its performance requires solving the computationally expensive SGLSP. To efficiently identify near-optimal designs under a limited evaluation budget, we employ BO [4], which avoids exhaustive exploration of the entire search space. We define the design variable vector as $x = \{p, s, h, i, F\}$, where p denotes the number of orbital planes, s the number of satellites per plane, h the constellation altitude, i the inclination angle, and F the phasing factor. These parameters jointly determine the constellation geometry evaluated during the BO process.

6.1   Surrogate Model

In BO, the surrogate model provides a probabilistic approximation of the expensive objective $f(x)$. For each candidate point x, the surrogate outputs a predictive distribution characterized by its mean and variance: $f(x) \mid \mathcal{D} \sim \mathcal{N}(\mu(x),\, \sigma^2(x))$. These statistics serve as inputs to acquisition functions. In this work, we employ two surrogate families:

Gaussian Process (GP). A GP [15] defines a prior over functions

$$f \sim \mathcal{GP}(m(x),\, k(x, x')),$$

with the mean function $m(x)$ and covariance kernel $k(x, x')$. Given observations $\mathcal{D} = \{(x_i, y_i)\}_{i=1}^{n}$, and covariance matrix K such that $K_{ij} = k(x_i, x_j)$, the GP posterior at a candidate point x is Gaussian with closed-form predictive mean and variance:

$$\mu_{\mathrm{GP}}(x) = k(x, X)\, [K + \sigma_n^2 I]^{-1} y, \tag{3}$$

$$\sigma_{\mathrm{GP}}^2(x) = k(x, x) - k(x, X)\, [K + \sigma_n^2 I]^{-1} k(X, x), \tag{4}$$

where σ_n^2 is the noise variance. The analytical tractability of the GP posterior makes it the canonical surrogate choice for BO, enabling exact computation of acquisition functions.

Random Forests (RF). Although RF are not Bayesian in the strict sense, their bootstrap-aggregated ensemble induces a natural predictive distribution. Let $h_1(x), \ldots, h_M(x)$ denote the predictions of M regression trees. The surrogate mean and uncertainty are then obtained from the empirical mean and variance of these tree predictions [7]. This Monte-Carlo–style estimate can be viewed as a form of Bayesian model averaging under the infinitesimal jackknife framework, providing practical uncertainty quantification without requiring any smoothness assumptions.

6.2   Acquisition Function

Given the surrogate predictive distribution, the acquisition function guides the selection of the next evaluation point. In this work, we employ the Expected Improvement (EI) criterion due to its strong empirical performance and analytical tractability [9].

Let f^+ denote the best objective value observed so far. The improvement at a candidate point x is defined as $I(x) = \max\{f(x) - f^+, 0\}$. The EI acquisition function is the expectation of $I(x)$ under the surrogate's predictive distribution: $EI(x) = \mathbb{E}[\, I(x) \mid \mathcal{D}\,]$.

Assuming a Gaussian predictive model, the integral admits a closed-form solution:

$$EI(x) = (\mu(x) - f^+)\, \Phi(z) + \sigma(x)\, \phi(z), \tag{5}$$

where $z = \frac{\mu(x)-f^+}{\sigma(x)}$, and $\Phi(\cdot)$ and $\phi(\cdot)$ denote the CDF and PDF of the standard normal distribution, respectively. The first term in Eq. (5) promotes exploitation of regions with high predicted mean, while the second term encourages exploration of regions with high uncertainty. Since both GP and RF surrogates provide a predictive mean $\mu(x)$ and an uncertainty estimate $\sigma(x)$, EI can be applied uniformly to either model.

Each evaluation of $f(x)$ calls the PINCH structure to compute the SGL schedule and hence the net cost in (2). The search terminates when the budget of function calls is reached. Algorithm 3 summarizes the procedure.

Algorithm 3. Bayesian Optimization framework

Input: Search ranges for $x = (p, s, h, i, F)$
Output: Best design $x^\star$

 1: Generate initial dataset $\mathcal{D}_0$ with n_0 samples   // *initial design evaluations*
 2: Set $t \leftarrow 0$ // *initialize iteration counter*
 3: **while** evaluation budget not exhausted **do**
 4: Fit Surrogate Model to dataset $\mathcal{D}_t$ // *update model*
 5: $x_{t+1} \leftarrow \arg\max_x \mathrm{EI}(x|\mathcal{D}_t)$ // *acquisition step*
 6: $d(x_{t+1}) \leftarrow \mathrm{PINCH}(x_{t+1})$ // *evaluate SGLSP*
 7: $f(x_{t+1}) \leftarrow \textsc{NetCost}(x_{t+1}, d(x_{t+1}))$ // *compute objective*
 8: $\mathcal{D}_{t+1} \leftarrow \mathcal{D}_t \cup \{(x_{t+1}, f(x_{t+1}))\}$ // *update dataset*
 9: $t \leftarrow t+1$ // *increment iteration*
10: **if** stopping criterion satisfied **then**
11: **break**   // *stop if convergence reached*
12: **end if**
13: **end while**
14: **return** $x^\star = \arg\min_{x \in \mathcal{D}_t} f(x)$ // *best found design*

---

7 Results

In this study, we initialise BO with 50 samples and allow up to 200 evaluations. The design parameters are searched within $x = (p, s, h, i, F) \in [0, 18] \times [3, 36] \times [500, 1200] \times [30, 87.5] \times [0, 6]$. The total number of satellites is restricted to between 12 and 36 in our experiments, however users who adopt our algorithm may select different lower and upper bounds. The ground station network is fixed at 40 locations, referenced from the worldwide ground station distribution provided by Liu et al. [11], and is consistently applied throughout all experiments.

The cost model parameters are set as follows: $c_{\text{build}} = 0.8$ M\$/satellite, $m = 0.26$ t, $c_0 = 2.7$ k\$/kg, $k_H = 0.15$, $h_0 = 500$ km, $k_I = 0.35$, $i_0 = 28.5°$, $c_{\text{ops}} = 0.04$ M\$/year, $Y = 5$ years, $w_L = 0.1$ M\$/km, $w_D = 20$ M\$/10^6s, and $\beta = 1.2$.

All experiments are performed on a 13th Generation Intel®Core™i7-13650HX CPU at 2.60 GHz, running Windows11 with 64 GB of RAM, and implemented in Python 3.10.16. The BO framework is implemented using the `skopt` package from `scikit-optimize` [12]. The PINCH structure is based on Gurobi 11 [5] with default settings, and the Leiden Algorithm is implemented using the `leidalg` library.

7.1   Single and Dual Constellation Design

In LEO satellite constellation design, a single-constellation architecture uses one homogeneous orbital shell for all coverage, whereas a dual-constellation architecture combines two distinct orbital shells (e.g., different altitudes or inclinations) to better trade off coverage, capacity, and latency across regions. To investigate how constellation design influences the cost–performance trade-off, we first compare single and dual-constellation configurations under the same BO framework. Dual constellations, due to their orbital diversity, are expected to enhance coverage continuity and reduce latency penalties; however, they require higher deployment and operational costs. Conversely, single constellations feature simpler architectures and lower costs, though they may offer less scheduling flexibility. This comparison allows us to assess whether the additional complexity of dual constellations leads to meaningful performance improvements.

Figure 3 shows the distribution of all candidate designs evaluated by BO with axes representing cost, SGL duration, and latency penalty. Out of 200 evaluations, the surrogate model selected single-constellation configurations 172 times (orange triangles) and dual-constellation configurations only 28 times (blue circles), indicating a clear preference for single-constellation designs in choosing next candidate points. The 3-D scatter indicates that single-constellation designs dominate the lower-cost and lower-latency regions, while achieving SGL durations comparable to, or exceeding, those of most dual-constellation designs. Only a few dual configurations penetrate the cloud of single-constellation points, and none establish a superior Pareto frontier. These results suggest that, under the current objective formulation and with the fixed network of 40 ground stations, introducing a second sub-constellation rarely provides a Pareto improvement. Therefore, a single-constellation architecture appears to be sufficient in this experimental setting.

7.2   Effect of Normalization and Surrogate Rebuilding on BO Performance

Having verified that single-constellation architectures suffice for the present study, we now analyze the ablation experiment on normalized BO variants. The goal is to understand how objective normalization and periodic surrogate rebuilding affect the stability and efficiency of the search process.

Normalization. As the three components of the net cost differ significantly in scale, the BO surrogate may overfit the dominant term and consequently produce poorly calibrated acquisition scores. To mitigate this issue, we apply a component-wise standardization to each cost component using the empirical

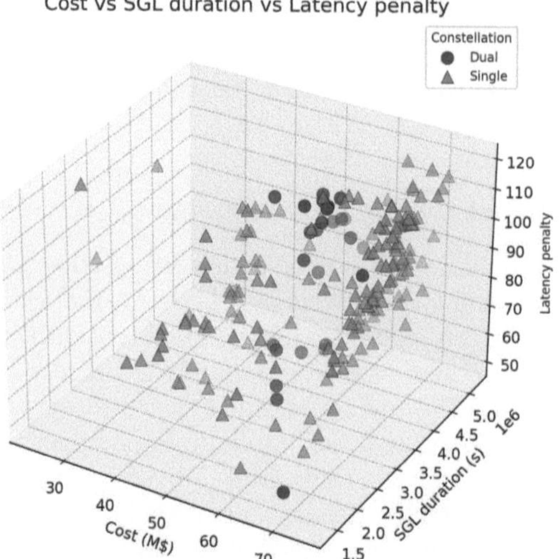

Fig. 3. Three-dimensional scatter plot of Bayesian-optimization samples illustrating the trade-offs between Cost (million USD), total SGL duration (seconds), and latency penalty. Blue circles represent dual-constellation designs, while orange triangles represent single-constellation designs. (Color figure online)

mean and standard deviation computed from all previously evaluated designs: $\hat{X} = \frac{X - \mu_X}{\sigma_X}$, where X denotes any of the components C, P_{lat}, or B_{link}. The normalized scalar objective used by BO is then defined as a weighted linear combination of $\hat{X}$. This normalization step corrects the severe scale imbalance among the components, resulting in a smoother and better-conditioned response surface for both GP and RF surrogates, thereby improving the stability and reliability of the BO search process.

Periodic Rebuild. Although normalization stabilizes the objective scale locally, the empirical mean and variance of $C, P_{\text{lat}},$ and B_{link} naturally evolve as BO explores new constellation designs. Consequently, the distribution of the scalarized objective becomes progressively non-stationary. To maintain consistency between the surrogate and the current objective distribution, we introduce a *Periodic Rebuild* mechanism. Every K iterations, we recompute (μ_X, σ_X) using all evaluated designs, then renormalize all historical observations under the updated statistics and refit a fresh surrogate model on the complete BO history. This ensures that all surrogate predictions and acquisition evaluations are based on a consistent and up-to-date objective scaling, preventing drift caused by incremental learning on differently scaled targets.

Ablation Study. To understand the effectiveness of normalization and the periodic rebuilding mechanism, we conduct an ablation study using both GP and RF surrogates. For each surrogate type, we evaluate three variants: the baseline model without normalization, a normalized version using the component-wise standardization, and a normalized version augmented with the periodic rebuild procedure. We then evaluate each best design in the multi-objective space defined by cost, latency and SGL duration, and compute its Euclidean distance to the global 3D Pareto frontier composed of all feasible designs observed across the six runs.

Table 1. Comparison of the best designs obtained by each BO variant.

Method	Cost(M\$)	Latency	SGL duration($\times 10^6$s)	Distance to Pareto	#Pareto points
GP	43.10	50.00	2.03	0.0	12
GP^{norm}	26.08	60.49	1.59	$3.55 \times 10^{-15} \approx 0$	35
$\text{GP}^{norm}_{rebuild}$	25.62	50.00	1.26	0.0	40
RF	20.75	51.64	0.84	0.0	6
RF^{norm}	26.11	54.86	1.39	0.0	26
$\text{RF}^{norm}_{rebuild}$	20.56	52.29	1.09	0.0	34

Table 1 shows that the best solution of each BO variant lies exactly on the 3D Pareto frontier. This means that all six configurations are capable of reaching non-dominated constellation designs, and the main difference between them is where they land on the frontier rather than whether they reach it at all.

A clearer separation emerges when considering how many nondominated solutions each method contributes. $\text{GP}^{norm}_{rebuild}$ yields the largest number of Pareto-optimal designs, followed by GP^{norm} and $\text{RF}^{norm}_{rebuild}$. Together with RF^{norm}, these variants collectively dominate most of the discovered Pareto set. The remaining models contribute far fewer nondominated designs: GP and RF without normalization contribute only 12 and 6 points, respectively.

These results indicate that GP-based surrogates generally offer better guidance than RF-based models, and that normalization substantially improves BO's ability to explore high-quality trade-off regions, especially when combined with periodic model rebuilding.

The 2D Pareto projections in Figs. 4a and 4b make the trade-offs more interpretable. In the cost–SGL view (Fig. 4a), RF and $\text{RF}^{norm}_{rebuild}$ occupy the low-cost end of the frontier, providing the cheapest designs with low SGL duration. At the other extreme, the GP configuration attains the largest SGL duration among all methods, but at a significantly higher cost. Among the normalized variants, GP^{norm} and RF^{norm} populate the middle part of the frontier and offer stronger SGL performance than RF and $\text{RF}^{norm}_{rebuild}$, while keeping the cost at a moderate level. $\text{GP}^{norm}_{rebuild}$ also lies in this intermediate region, but slightly shifts towards lower cost and lower SGL.

A similar pattern can be observed in the latency–SGL projection (Fig. 4b). GP and $\text{GP}^{norm}_{rebuild}$ lie on the low-latency edge of the frontier, corresponding to constellations with smaller altitudes and thus shorter signal propagation delay. GP^{norm} moves towards the high-SGL region with slightly larger latency, while

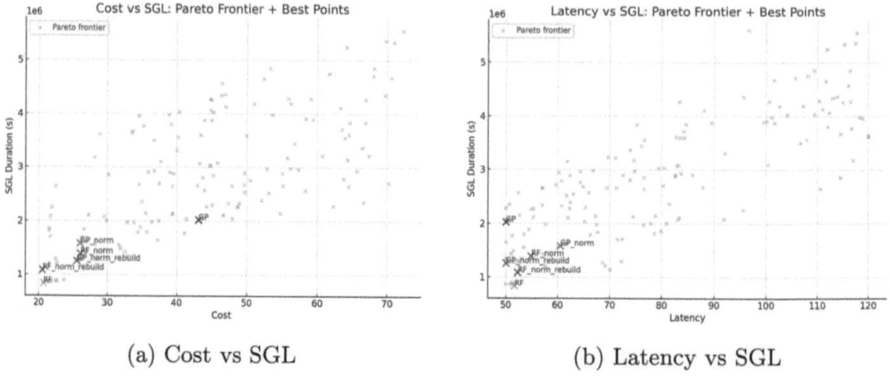

(a) Cost vs SGL   (b) Latency vs SGL

Fig. 4. Comparison of Pareto projections in 2D: Cost–SGL and Latency–SGL.

the RF-based variants remain on the lower-SGL part of the frontier compared to the GP-based configurations. Overall, the combination of Table 1 and Figs. 4a and 4b indicates that BO consistently discovers Pareto-optimal designs under all configurations, while normalization and model rebuilding reshape the search trajectory so that RF-based runs tend to favor low-cost Pareto solutions, the plain GP configuration pushes the search towards the extreme high-SGL region, and the normalized GP variants produced more balanced compromises between cost, latency, and SGL duration.

8 Conclusions

In this paper, we address the LEO constellation design problem through a two-layer approach. The outer layer employs BO to identify a suitable set of design parameters, while the inner layer utilizes the PINCH framework to solve the SGLSP, thereby determining the maximum SGL duration for each constellation design. PINCH took a fraction of Gurobi's computation time with comparable solution quality.

The future work of this study points to two main directions. First, extending the framework to include additional satellite subsystems, such as battery management, payload characteristics, thermal control, and antenna pointing, would yield a more realistic and comprehensive design model. These factors directly influence cost, scheduling priorities, and service quality.

Second, enhancing the BO framework could further improve performance. Multi-fidelity BO, which combines low- and high-fidelity evaluations, can accelerate convergence without sacrificing accuracy. Another promising direction is the design of scale-invariant or problem-specific surrogate kernels that better capture heterogeneous objective components, which may further improve numerical stability and reduce sensitivity to objective scaling. Such advancements would enhance scalability and robustness, making the framework more effective for large-scale constellation design.

References

1. Alidaee, B., Kochenberger, G., Lewis, K., Lewis, M., Wang, H.: A new approach for modeling and solving set packing problems. Eur. J. Oper. Res. **186**(2), 504–512 (2008). https://doi.org/10.1016/j.ejor.2006.12.068, https://www.sciencedirect.com/science/article/pii/S0377221707001889
2. Ballard, A.H.: Rosette constellations of earth satellites. IEEE Trans. Aerosp. Electron. Syst. **16**(5), 656–673 (1980). https://doi.org/10.1109/TAES.1980.308918
3. Capderou, M.: Satellites: Orbits and missions (2005). https://doi.org/10.1007/b139118
4. Frazier, P.I.: A tutorial on bayesian optimization (2018). https://arxiv.org/abs/1807.02811
5. Gurobi Optimization, L.: Gurobi Optimizer Reference Manual (2024). https://www.gurobi.com
6. Hou, Z., Yi, X., Zhang, Y., Kuang, Y., Zhao, Y.: Satellite-ground link planning for LEO satellite navigation augmentation networks. IEEE Access **7**, 98715–98724 (2019). https://doi.org/10.1109/ACCESS.2019.2930626
7. Hutter, F., Hoos, H.H., Leyton-Brown, K.: Sequential model-based optimization for general algorithm configuration. In: International Conference on Learning and Intelligent Optimization, pp. 507–523. Springer (2011)
8. Jiang, C., Luo, Z., Guan, M., Zhu, H.: Low orbit regional enhanced navigation constellation for BDS-3 design based on bayesian optimization algorithm. Geodesy and Geodynamics (2024). https://doi.org/10.1016/j.geog.2023.12.003, in press
9. Jones, D.R., Schonlau, M., Welch, W.J.: Efficient global optimization of expensive black-box functions. J. Global Optim. **13**(4), 455–492 (1998)
10. Li, R., Wang, Y., Hu, S., Jiang, J., Ouyang, D., Yin, M.: Solving the set packing problem via a maximum weighted independent set heuristic. Math. Probl. Eng. **2020**(1), 3050714 (2020). https://doi.org/10.1155/2020/3050714, https://onlinelibrary.wiley.com/doi/abs/10.1155/2020/3050714
11. Liu, Z., Liu, J., Liu, X., Yang, W., Wu, J., Chen, Y.: Knowledge-assisted adaptive large neighbourhood search algorithm for the satellite-ground link scheduling problem. Comput. Ind. Eng. 110219 (2024). https://doi.org/10.1016/j.cie.2024.110219, https://www.sciencedirect.com/science/article/pii/S0360835224003401
12. Pedregosa, F., et al.: Scikit-learn: machine learning in python. J. Mach. Learn. Res. **12**, 2825–2830 (2011)
13. Petelin, G., Antoniou, M., Papa, G.: Multi-objective approaches to ground station scheduling for optimization of communication with satellites. Optim. Eng. **24**(1), 147–184 (2023)
14. Portillo, I.D., Cameron, B.G., Crawley, E.F.: A technical comparison of three low earth orbit satellite constellation systems to provide global broadband. In: Proceedings of the 69th International Astronautical Congress (IAC). International Astronautical Federation, Bremen, Germany, paper IAC-18-B2.1.7 (2018)
15. Rasmussen, C.E., Williams, C.K.I.: Gaussian Processes for Machine Learning. The MIT Press (2005). https://doi.org/10.7551/mitpress/3206.001.0001, https://doi.org/10.7551/mitpress/3206.001.0001
16. Shi, R., Liu, L., Long, T., Wu, Y., Wang, G.G.: Multidisciplinary modeling and surrogate assisted optimization for satellite constellation systems. Struct. Multidiscip. Optim. **58**(5), 2173–2188 (2018). https://doi.org/10.1007/s00158-018-2032-1

17. Spangelo, S., Cutler, J., Gilson, K., Cohn, A.: Optimizationbased scheduling for the single-satellite, multi-ground station communication problem. Comput. Oper. Res. **57**, 1–16 (2015). https://doi.org/10.1016/j.cor.2014.11.004, https://linkinghub.elsevier.com/retrieve/pii/S0305054814002809
18. Traag, V.A., Waltman, L., van Eck, N.J.: From Louvain to Leiden: guaranteeing well-connected communities. Sci. Rep. **9**(1) (2019). https://doi.org/10.1038/s41598-019-41695-z
19. Walker, J.G.: Some circular orbit satellite constellations. J. Br. Interplanetary Soc. **31**, 299–311 (1970)
20. Xhafa, F., Herrero, X., Barolli, A., Barolli, L., Takizawa, M.: Evaluation of struggle strategy in genetic algorithms for ground stations scheduling problem. J. Comput. Syst. Sci. **79**(7), 1086–1100 (2013)
21. Xiao, M., Nagamochi, H.: An exact algorithm for maximum independent set in degree-5 graphs. Discrete Appl. Math. **199**, 137–155 (2016). https://doi.org/10.1016/j.dam.2014.07.009, https://www.sciencedirect.com/science/article/pii/S0166218X14003102, sixth Workshop on Graph Classes, Optimization, and Width Parameters, Santorini, Greece, October 2013
22. Xiao, M., Nagamochi, H.: Exact algorithms for maximum independent set. Inf. Comput. **255**, 126–146 (2017). https://doi.org/10.1016/j.ic.2017.06.001, http://dx.doi.org/10.1016/j.ic.2017.06.001
23. Zhu, X., Jiang, C.: Integrated satellite-terrestrial networks toward 6G: architectures, applications, and challenges. IEEE Internet Things J. **9**(1), 437–461 (2022). https://doi.org/10.1109/JIOT.2021.3126825
24. Zuo, X., et al.: Satellite constellation reconfiguration using surrogate-based optimization. J. Aerosp. Eng. **35**(4), 04022043 (2022). https://doi.org/10.1061/(ASCE)AS.1943-5525.0001438

Explainability Results for the Rotating Workforce Scheduling Problem

Esther Mugdan[1]([✉])[iD], Lucas Kletzander[2][iD], and Nysret Musliu[2][iD]

[1] AI, University of Basel, Basel, Switzerland
esther.mugdan@unibas.ch
[2] DBAI, TU Wien, Vienna, Austria
{lucas.kletzander,nysret.musliu}@tuwien.ac.at

Abstract. Scheduling is a highly relevant aspect of industrial work. From assigning jobs to machines to creating shift plans for employees; schedules must be created to ensure efficiency, cover requirements and adhere to working regulations. As creating schedules while keeping track of all constraints is often a notoriously difficult job, automated methods can be employed. However, sometimes no solution can be found due to conflicting problem specifications. In this case, it is important to explain which constraints contribute to infeasibility and how the problem can be relaxed. We study the Rotating Workforce Scheduling Problem, for which we develop a framework that generates explanations for instances with incompatible constraints. We show how Minimal Correction Sets can be used to provide detailed explanations for infeasible problems, caused by hard constraint violations or by conflicting optimisation goals. We perform a case study and experiments on (real-life) instances that reveal that explanations can be efficiently generated.

Keywords: Scheduling · Explainability · Constraint Programming

1 Introduction

In many industries, shift schedules are used to regulate the working times of employees. When creating such shift plans, one usually has to take into consideration many constraints and preferences from different parties, resulting in a highly constrained problem.

In workforce scheduling, adhering to workplace regulations and reducing costs are often employers' primary interests. However, the resulting schedules provide the structure for a significant part of the employee's lives. Many different versions of such problems have been addressed over the years [6,12,16]. While traditionally the focus has been on reducing cost, a growing body of work on employee well-being [11] shows the importance of incorporating these aspects. It is known that effects like fatigue due to bad workplace conditions impact not only economic costs of companies [41] but also lead to severe consequences in both psychological and physiological health of individuals [20,32], including a

T. Guns (Ed.): CPAIOR 2026, LNCS 16595, pp. 331–349, 2026.
https://doi.org/10.1007/978-3-032-27242-3_20

higher risk for certain diseases and disorders [34], and reduced social contacts with family and friends [2]. Therefore, practical guidelines [24] and metrics have been proposed to evaluate risk-related characteristics [17,18]. Case studies on personnel planning focusing on the impact of including well-being aspects in the objective function for heuristic optimisation approaches have been presented [40]. Shifts are created by minimising a weighted sum of violations of undesired characteristics, like working weekends, long night shift stints, and others.

Scheduling problems with many constraints and optimisation goals are often solved using different methods or solvers. Many of these approaches have successfully been implemented in practice. However, sometimes a method is unable to find a solution. There are two reasons why this can occur. Either the solver could not find a solution within the given computation limits, or a solution does not exist, as the specified instance is over-constrained. In any case, when a stakeholder wants a solution to their problem instance, it is generally not enough to tell them that *there is no solution*. Instead, it is important to explain why no solution could be found and how the problem can be resolved.

Our aim is to develop a framework which can be used to solve and explain the Rotating Workforce Scheduling (RWS) Problem under different configurations, including new well-being objectives. For this purpose, we use Constraint Programming (CP). The goal is to obtain a solution to a problem instance and its configuration. If the specified input is unsatisfiable, detailed explanations should be provided which allow the user to understand the cause of infeasibility and resolve it. Using this framework, we can not only tell a stakeholder that there is no solution but also explain *why* this is the case and *how* it can be resolved.

We test our CP implementation using different solvers. We compare the individual results and find that the choice of the solvers for finding explanations in the form of Minimal Correction Sets (MCSs) is crucial. Using the best solver combinations, we were able to obtain positive results. We show that explanations in the form of MCSs can efficiently be found for unsatisfiable RWS instances under the satisfaction version of the problem. Additionally, we present how different preferences on constraints can be considered when providing explanations in the form of MCSs. We also consider feasible RWS instances with optimisation goals. Here, we test the feasibility of threshold combinations on the optimisation goals. We find that we can efficiently identify conflicting goals, especially for smaller instances. For larger instances, we do not always obtain optimal explanations; however, for most instances, we can get some explanation.

2   Related Work

Explainability for infeasible problems is an essential field of study, which is gaining prominence in many areas. In general, two primary forms of Explainability are of interest, namely (a) finding a maximal part of the problem that is satisfiable, i.e. finding Maximal Satisfiable Subsets (MSS). In this case, the complement of an MSS, namely the Minimal Correction Set (MCS), can be provided as well, which consists of the constraints that need to be removed to regain feasibility. On the other hand, (b) we are interested in the cause of infeasibility, i.e.

finding a Minimal Unsatisfiable Subset (MUS). Some algorithms return a single MSS or MUS. These often build on the general idea of growing (shrinking) mechanisms that start with a satisfiable (unsatisfiable) subset of the problem and test whether the addition (removal) of constraints still yields a satisfiable (unsatisfiable) subset. Doing these tests iteratively and adding (removing) the corresponding constraints eventually yields an MSS (MUS). However, few algorithms can efficiently enumerate all or at least more than one MSS/MUS.

There are tailored approaches for Linear Programming [21,43] and Numerical Constraint Satisfaction Problems [33], but they can not easily be generalised. An early general approach was CAMUS [29–31], which makes use of the hitting set relation between MUSs and MCSs and aims to return all MCSs at once. While this approach shows promising results in some cases, the number of MCSs can be exponential in the size of a problem instance, making the algorithm intractable. To circumvent this problem and report MUSs incrementally, the Dualize and Advance Algorithm (DAA) was developed [3]. This algorithm also uses the hitting set relation between the constraint sets. However, DAA performs worse than CAMUS.

Since it is often not feasible to enumerate all MUSs, and as CAMUS and DAA are often unable to return at least a single MUS, work has been done to efficiently enumerate *some* MUSs [28]. This work proposed the MARCO algorithm, which should provide the first MUS in roughly the same time t as the best algorithm designed for computing a single MUS. In addition, any following MUS should be returned as quickly as possible (roughly after a period of t). MARCO stands for Mapping Regions of Constraint Sets, as the algorithm builds a Boolean formula *map* which maps each subset of the constraints to 1 iff the subset has not yet been evaluated in terms of satisfiability. The algorithm starts with an empty *map* formula. In each iteration, a model of *map* is calculated, from which the set of constraints that are true in the model are extracted. If this set of constraints is satisfiable, it is grown to an MSS; otherwise, it is shrunk to an MUS. In each case the corresponding set is returned and the *map* formula is updated. The MARCO algorithm uses two solvers: one solver for growing and shrinking and a second solver (*map*-solver) to generate models of the *map* formula. A computational study showed that compared to CAMUS and DAA, MARCO can provide at least one MUS/MSS in a timely manner and often finds more MUSs/MSSs in general [28].

We employ CP to model our problem and provide explanations. MiniZinc was introduced as a standard CP modelling language that allows the integration of multiple different solvers [37]. Later the FindMUS solver was introduced for MiniZinc which builds on the MARCO algorithm and provides MUSs [37].

We use the CPMpy [22] Python library to model our problem. In comparison to MiniZinc, the constraints can be directly modelled in Python, and the model as well as results can be directly processed further. Additionally, CPMpy allows the use of a range of different solvers and provides tools to directly generate explanations within the framework. The library provides an implementation of MARCO, which can be used with the desired solvers, and implementations of algorithms for the generation of a single MUS/MCS/MSS, which support finding an optimal MUS/MCS/MSS according to preferences given for the constraints.

We study Explainability for the RWS Problem, which can be classified as a single-activity tour scheduling problem with non-overlapping shifts and rotation constraints [4]. Over the years, several approaches have been used to solve the problem, including use in commercial software for almost 25 years [35]. Initially, only the satisfaction version of the problem was studied, where only hard constraints need to be fulfilled. A complete method based on CP was introduced [36]. It was further extended [26], in particular by introducing several optimisation goals used in practice, turning the satisfaction problem into an optimisation problem. The CP method was used in an instance space analysis, leading to the creation of new instances to better cover the transition between feasible and infeasible instances [27]. The current state-of-the-art approach is a branch&cut (b&c) framework, which reduces the runtime substantially compared to previous methods for the satisfaction version and a limited set of optimisation goals [5].

A general approach based on Abstract Argumentation has been proposed for the explanation of scheduling problems [10]. This work was extended by a concrete implementation called *Schedule Explainer* for makespan scheduling which allows for interactive explanations [9]. *Crosscheck* was developed as an explainable scheduling tool tested for the Mars 2020 Rover Mission [1]. The methods used in this paper are related to work by Senthooran et al. [42]. However, we investigate different toolchains and MCS computation methods on a different real-life application with a cyclic structure, which makes resolving conflicts potentially very challenging. We include novel ideas of using redundant bounds for the MCS computation and incorporating a-priori preferences of the users. Bleukx et al. [7] apply MUS and MCS computation to deal with disruptions in a workforce allocation and scheduling problem and define objectives for feasibility restoration. Other recent methods include couterfactual explanation through constraint relaxation [23], the use of oversubscription planning [14], or using stepwise explanations in efficient MUS computation [19].

3   Problem Definition

In the Rotating Workforce Scheduling (RWS) Problem, pre-defined shift types are considered. These have to be assigned to employees according to several constraints so as to form a complete schedule over multiple weeks. The assignment of shift types to certain employees and days must adhere to the required demand. Additional restrictions are given in the form of minimal and maximal lengths for sequences of shifts and limitations to consecutive shift assignments. In many applications, a rotating schedule, where each employee rotates through the same sequence of shifts with different offsets, is a preferred way of scheduling. However, the rotating structure can create dependencies of assignments across large parts of a schedule that can make it more difficult to understand why an instance does not allow a solution. We distinguish between hard and soft constraints, where hard constraints must be met and soft constraints are optional objectives that optimise the solution. We provide a formal definition of the problem and introduce new soft constraints that focus on employee well-being.

3.1   Formal Definition

For the general problem definition of RWS we build on definitions and notation that have been introduced previously [35,36]. An instance of the RWS Problem is given by the following formulation.

- n: Number of employees.
- d: Length of the schedule. The total length of the planning period is $n \cdot d$, as each employee rotates through all n rows. We set $d = 7$, corresponding to the number of days in a week.
- $\mathbf{A}$: Set of work shifts (activities), enumerated from 1 to m. We consider instances with 2 or 3 shifts where shift 3 is a night shift. A day off is denoted by a special activity off with numerical value 0, $\mathbf{A}^+ = \mathbf{A} \cup \{\text{off}\}$.
- T: Temporal requirements matrix, describing the demand. An $m \times d$ matrix where each element $T_{i,j}$ corresponds to the number of employees that need to be assigned shift $i \in \mathbf{A}$ at day j.
- $\ell_{\text{work}}, u_{\text{work}}$: Minimal and maximal length of blocks of consecutive work shifts.
- ℓ_a, u_a: Minimal and maximal lengths of blocks of consecutive assignments of shift a for each $a \in \mathbf{A}^+$.
- $\mathbf{F}_2, \mathbf{F}_3$: Sequences of shifts of length 2 and 3 that are forbidden in the schedule (e.g. N D, a night shift followed by a day shift). This is typically required due to legal or safety concerns.

The task is to construct a cyclic schedule S, represented as an $n \times d$-matrix, where each $S_{i,j} \in \mathbf{A}^+$ denotes the shift or day off that employee i is assigned during day j in the first period of the cycle.

3.2   Hard Constraints

In total, we consider ten hard constraints that are defined as follows:

- h^{demand}: The assigned shifts in S must meet the demand T exactly for each shift and day.
- h^{overlap}: A day must have at most one shift $a \in \mathbf{A}^+$ assigned.
- h^{minon}: A sequence of working days must cover at least ℓ_{work} days in a row.
- h^{maxon}: A sequence of working days must cover at most u_{work} days in a row.
- h^{minoff}: A sequence of off days must cover at least ℓ_{off} days in a row.
- h^{maxoff}: A sequence of off days must cover at most u_{off} days in a row.
- h^{minshift}: A sequence of working days consisting only of shift a must cover at least ℓ_a days in a row, for each shift $a \in \mathbf{A}$.
- h^{maxshift}: A sequence of working days consisting only of shift a must cover at most u_a days in a row, for each shift $a \in \mathbf{A}$.
- $h^{\text{forbidden}}$: Any sequences in $\mathbf{F}_2$ must not occur in the schedule S.
- $h^{\text{forbidden3}}$: Any sequences in $\mathbf{F}_3$ must not occur in the schedule S.

3.3 Soft Constraints

We also consider optimisation variants of RWS and extend existing soft constraints [26] and propose novel objectives to be minimised. These aim to increase the well-being of employees and to meet applicable best practices of Britain's national regulator for workplace health and safety (HSE) shift work guidelines [24]. We address the reduction of consecutive night shifts by the constraint $s^{N>3}$. A focus on consistent working block lengths is set by $s^{\ell_{\text{dev}}}$. Lastly, ensuring regular free weekends is addressed by three already introduced constraints, namely $s^{\text{ww}}, s^{d_{\text{max}}}$ and $s^{d_{\text{rms}}}$ [26]. Additionally, we propose $s^{N\text{ww}}$ to broaden the definition of free weekends. The resulting six soft constraints to be minimised are defined as follows.

- $s^{N>3}$: Minimise the number of consecutive night shifts exceeding 3.
- $s^{\ell_{\text{dev}}}$: Minimise the squared deviation of working sequence lengths to 5.
- s^{ww}: Minimise the number of working weekends. A weekend is free if Saturday and Sunday are off. Otherwise, it is a working weekend.
- $s^{d_{\text{max}}}$: Minimise the maximum distance d_{max} between consecutive free weekends ($d_{\text{max}} = n + 1$ if no weekend is free).
- $s^{d_{\text{rms}}}$: Minimise root mean squared distances between weekends $\sqrt{1/n \sum_{i=1}^{n} \hat{d}_i^2}$, where $\hat{d}_i$ is the distance minus 1 to the next free weekend if weekend i is free, and n otherwise.
- $s^{N\text{ww}}$: Minimise the number of working weekends, where a weekend is only considered free if there is additionally no night shift on Friday.

3.4 Problem Solutions

A solution x to an RWS instance consists of a cyclic schedule S of dimensions $n \times d$ that assigns shifts from $\mathbf{A}^+$ to each employee and day. Given an instance I of the RWS Problem we can set the (potentially empty) set of soft constraints S to be considered. This configuration determines which soft constraints are optimised. Additionally, we have the option of using thresholds to set bounds on the desired soft constraints. A solution x to a problem instance, given its configuration, is evaluated over the set of hard constraints $\mathcal{H}$ and its set of soft constraints S using the violation vector $v(x)$. With each constraint $c \in \mathcal{H} \cup S$ we associate a violation $v_c(x)$ in the solution x that indicates for a hard constraint how often it is violated (e.g. in days) and for a soft constraint its value to be minimised. Using this, we can define the following.

Definition 1 (Feasibility). *Let x be a solution of an instance I with hard constraints $\mathcal{H}$. Then x is called feasible if $v_h(x) = 0$ for all $h \in \mathcal{H}$.*

Definition 2 (Optimalility). *Let x be a solution of an instance I with soft constraints S. Then x is called optimal if x is feasible and $\sum_{s \in S} v_s(x)$ is minimal.*

When working with thresholds on soft constraints, we additionally define a constant threshold vector $t(x)$ which must be taken into account. Thresholds can be defined for any soft constraint $s \in \mathcal{S}$ in the current configuration. For a solution x of an instance configuration with thresholds, we define the following.

Definition 3 (Acceptability). *Let x be a solution of an instance I with soft constraints $\mathcal{S}$ and thresholds t. Then x is called acceptable if x is feasible and $v_s(x) \leq t_s(x)$ for all $s \in \mathcal{S}$ where a threshold is set.*

4   Constraint Modelling

The RWS Problem has been solved using a constraint model in MiniZinc[1] using only hard constraints [36] and some soft constraints [26]. We build on the previous formulations and adapt them using the Python library CPMpy[2] from [22]. We additionally include the novel soft constraints proposed above. In the following, we give a high-level description of how the hard and soft constraints are implemented.

We represented a solution (i.e. a schedule) as a one-dimensional list S of length $n \cdot d$, where each entry S_i is an element of $\mathbf{A}^+$ for all $i \in \{1, \ldots, n \cdot d\}$. As the schedule is cyclic, we can define the first entry S_1 as a working day and the last entry $\mathsf{S}_{n \cdot d}$ as a day off. This allows for easier computation of subsequent constraints. However, we need to define an offset $o \in [0, d)$ that indicates at which position in S the solution schedule S starts. This offset is always fixed in preprocessing. For the sake of simplicity, we omit offset and modulo calculations when describing the constraints.

We distinguish three different types of constraints: definitions, hard bounds and soft bounds. Definitions amount to the way the constraints are defined and implemented. The hard and soft bound constraints are used to provide explanations for infeasible configurations. For hard constraints we introduce decision variables representing the number of workers per day and shift, as well as minimal and maximal block lengths. These decision variables have restricted domains. For demand and minimal block lengths, the domain is $[0, \ldots, val]$, where val corresponds to the required value (i.e. T, $\ell_{\mathrm{work}}, \ell_{\mathrm{off}}, \ell_1, \ell_2$, etc.) given in the input instance. For maximal block lengths, the domain is $[val, \ldots, n \cdot d]$, where val is again the value of the corresponding data input (i.e. $u_{\mathrm{work}}, u_{\mathrm{off}}, u_1, u_2$, etc.). To enforce that the decision variables correspond exactly to the value they represent, we introduce what we call hard bounds. For each variable representing demand or minimal block lengths, we introduce up to 5 separate bounds as shown below.

$$\mathsf{v}_i \geq j \quad \forall j \in \{max(0, i - 5), i\} \tag{1}$$

where i represents the demand of a day and shift or a minimal block length. Analogously, for maximal block lengths up to 5 separate bounds are introduced.

$$\mathsf{v}_i \leq j \quad \forall j \in \{i, min(i + 5, n \cdot d)\} \tag{2}$$

---

[1] https://www.minizinc.org/.
[2] https://github.com/CPMpy/cpmpy.

where i represents a maximal block length. This allows us to give more precise information on how an infeasible configuration must be altered to regain feasibility, i.e. we know by *how much* we need to relax a certain numerical input, not only *that* it needs to be altered.

The soft bounds correspond to the thresholds set in the configuration. For each soft constraint s we again use a decision variable v_s that represents its value. The soft bounds for constraints with specified thresholds then correspond to the following, where t_s is the threshold value set on soft constraint s.

$$\mathrm{v}_s \leq t_s \tag{3}$$

To ensure that the demand is met by our schedule S we introduce the following constraints, that force the equality of the cover variable $v_{a,i}$ of each shift a and day i and the sum of corresponding entries in S. $v_{a,i}$ has upper bound $T_{a,i}$ and up to 5 lower bounds as explained above. The constraints are given below.

$$\forall i \in \{1, \ldots, d\} \left(\forall a \in \mathbf{A} \left(\sum_{j=0}^{n-1} \mathrm{S}[7 * j + i] = v_{a,i} \right) \right) \tag{4}$$

5 Explaining Infeasibility

Especially when working with stakeholders, simply stating that "there is no solution" is not enough. Therefore, we are interested in *explaining* infeasibility. We can do so by providing explanations in different formats that isolate the reasons why a problem cannot be solved. We distinguish three types of Constraint Sets that provide explanations for infeasibility. Given a problem that is defined over a set of constraints $\mathcal{C}$, we define the following.

Definition 4 (MUS). *Let $M \subseteq \mathcal{C}$. Then M is a Minimal Unsatisfiable Set (MUS) if M is unsatisfiable and for each $c \in M$ it holds that $M \setminus \{c\}$ is satisfiable.*

Definition 5 (MSS). *Let $M \subseteq \mathcal{C}$. Then M is a Maximal Satisfiable Set (MSS) if M is satisfiable and for each $c \in \mathcal{C} \setminus M$ it holds that $M \cup \{c\}$ is unsatisfiable.*

Definition 6 (MCS). *Let $M \subseteq \mathcal{C}$. Then M is a Minimal Correction Set (MCS) if $M = \mathcal{C} \setminus M'$, where M' is a MSS.*

Minimal Correction Sets give us the set of constraints that need to be removed from $\mathcal{C}$ in order to restore feasibility, which is the focus in our framework. Removing only parts of an MCS from $\mathcal{C}$ will not restore feasibility. Using MCSs we can directly provide information on how the problem must be modified and relaxed to be solvable. The resulting problem from such a modification is an MSS of the original set of constraints.

Since we are faced with a large number of MCSs, choosing one is not straightforward. Choosing an MCS should be done according to the preferences of stakeholders. While some might be interested in the smallest amount of changes, others may want to keep the demand intact. Using such preferences, we can use weights to give priorities to different constraints. Given a total weight function $w : \mathcal{C} \mapsto \mathcal{N}$ that maps every constraint to a natural number we can find optimal MCSs. An MCS is optimal if the sum of the weights of its constraints is minimal.

Definition 7 (Minimal Size). *To minimise the number of constraints in the MCS, we define* $w(c) = 1 \ \forall c \in \mathcal{C}$.

Definition 8 (No Demand). *To get the smallest MCS where the demand should not be changed, we define* $w(c) = 10 \ \forall c \in \mathcal{C}_{\mathrm{demand}}$ *and* $w(c) = 1$ *for all other constraints.*

Consider the example using $n{=}4$ employees, $d{=}7$ days and three shift types $\mathbf{A} = \{1, 2, 3\}$. The instance is defined in Table 1.

Table 1. Demand T and constraint bounds for the example

Shift	Mon	Tue	Wed	Thu	Fri	Sat	Sun
1	2	1	1	1	1	2	2
2	1	1	1	1	1	1	1
3	1	1	1	1	2	0	0

$[\ell_{\mathrm{work}}, u_{\mathrm{work}}] = [4, 7]$
$[\ell_{\mathrm{off}}, u_{\mathrm{off}}] = [1, 4]$
$[\ell_1, u_1] = [2, 7]$
$[\ell_2, u_2] = [2, 7]$
$[\ell_3, u_3] = [1, 3]$
$\mathbf{F}_2 = \{\langle 2, 1\rangle,$
$\langle 3, 1\rangle, \langle 3, 2\rangle\}$
$\mathbf{F}_3 = \{\}$

This instance is infeasible, but even at this size it has 462 MCSs. Using weights, we can find a minimal MCS $\{T_{3,Fri} \geq 2, \ u_{\mathrm{work}} \leq 7\}$. This means that reducing the demand on Fridays for shift 3 to 1 shift and relaxing the upper bound on working sequences to 8 restores feasibility. Computing an MCS with the weight function No Demand still returns an MCS containing a demand constraint, indicating it is not possible to restore feasibility without reducing the demand.

6   Evaluation

We test our CP framework using the CPMpy algorithm framework on the RWS Problem to show that we can provide explanations in form of Minimal Correction Sets for infeasible configurations of RWS instances. We focus on two different problem configurations. First, we study the satisfaction version of the problem, without optimisation goals, on infeasible instances. Secondly, we evaluate (over constrained) soft constraint threshold configurations on feasible instances.

6.1 Satisfaction Configuration

We used FairSubset [38] in the online tool provided[3] to extract a representative set of 50 instances from a total of 1843 infeasible instances stemming from existing benchmarks [27]. As features for the selection we used the coordinates introduced to present instances in a 2D grid indicating hardness of an instance [27].

We tested our model on these instances by comparing approaches using different solvers. First, we evaluated generating (all) MCSs with MARCO. Secondly, we compared different approaches to provide single MCSs. In both cases, we employ the implementations of the algorithms provided in the CPMpy framework.

We performed our experiments on a cluster with Ubuntu 22.04.2 LTS with $2\times$ Intel Xeon CPU E5-2650 v4 (2.2GHz, 12 physical cores, no hyperthreading). We ran our experiments single-threaded using Python 3.11. We employed the solvers OR-Tools CP-Sat (v9.11, [39]), z3 (v4.13, [13]), Exact (v2.1, [15]) and pysat (v1.8, [25]). Each run was given a 1h timeout to prove the unsatisfiability of the instance using OR-Tools. We additionally used a 1h timeout for providing explanations. Each run additionally had a memory limit of 16'384 megabytes.

Table 2. Comparison of running MARCO with the respective configuration. Percentages of instances in relation to all 50 instances are given for failed proofs of infeasibility, memory limit exhaustion and cases where CSs could be found. Additionally, the total number of CSs found is indicated.

primary-map solvers	failed proof	memory limit	MCS found	#MCS found
z3-OR-Tools	24%	0%	**20%**	**100**
z3-pysat	24%	8%	18%	51
Exact-OR-Tools	24%	0%	16%	62
Exact-pysat	24%	0%	16%	64

Generating all MCSs with MARCO. We first evaluated how many different MCS can be found. For this, we employ MARCO using z3 or Exact as primary solvers and OR-Tools or pysat as map solvers. This yields a total of four solver combinations. A summary of our findings is depicted in Table 2. We observe that 24% of instances can not be proven unsatisfiable within the one hour time limit. Even if no proof of unsatisfiability was found, we performed the MARCO algorithm. However, no solver combination could provide MCSs using MARCO for these instances within the time limit. For the combination z3-pysat we observe that the runs for 4 instances ran out of memory. No other solver combination had this issue. When we compare the percentage of instances for which MCSs could be found, the configuration z3-OR-Tools was able to find MCSs for a total of 10 out of 50 instances. In total, using this configuration yielded 100 MCSs over

---

[3] https://delaney.shinyapps.io/FairSubset/.

all instances. For 7 out of the 10 instances z3-OR-Tools found at least as many MCSs as all other solvers. Additionally, only one instance that was solved by another solver (z3-pysat) was not solved by z3-OR-Tools.

In comparison to z3-OR-Tools, Exact performed worse with two instances less for both OR-Tools and pysat. However, while the Exact approaches could find MCSs for fewer instances than z3-pysat, Exact was able to provide more MCSs in total. Therefore, it is hard to compare these remaining three configurations especially because each configuration was able to solve at least one instance that none of the other two configurations could solve. Additionally, for some instances, the number of MCSs found varied a lot between the approaches. For example, using exact-ortools provided 17 MCSs on an instance, for which none of the other approaches found any MCS. Using Exact-pysat yielded 27 MCSs on an instance for which Exact-ortools could not find any MCS and z3-pysat only 3. Lastly, for one instance, Exact-pysat could find only 3 MCSs, Exact-OR-Tools 9 and z3-pysat 19 MCSs. These examples show that while the configurations performed worse than z3-OR-Tools, it is hard to compare them.

Finding a Single MCS. We could not enumerate MCSs using MARCO for a majority of the instances. However, we are usually only interested in a single MCS and do not need to enumerate all of them especially as this is computationally more expensive. We, therefore, focus on finding a single MCS using different weight configurations as introduced in Sect. 5. Employing a timeout on finding a single MCS might result in non-minimal Correction Sets (CS), meaning that the resulting set of constraints is not necessarily subset minimal. However, its size was optimized until timeout, so it is likely to be quite small. If a CS was provided before the timeout, minimality is guaranteed (MCS).

A summary of the results can be found in Table 3. We compare results for finding any CS (base), i.e. no weight function, weight-minimal CS (minimal), i.e. $w(c) = 1 \; \forall c$, and weight-minimal CS without demand (no demand), for which we set a weight of 100 on demand constraints and a weight of 1 on all remaining hard bounds. We observe that OR-Tools can find a CS for each instance over all configurations, even for the 24% of instances where no proof of infeasibility could initially be generated. We also observe that the mean time used by OR-Tools is roughly 30 min, which is half of the 1h timeout. In fact, only 40% of instances ran for the whole hour, indicating that an MCS could be found for 60% of the instances. Additionally, for about 40% of the instances that ran for the whole hour an MCS of size one could be found, resulting in a total of 76% where an MCS was found. In contrast, Exact always reports MCSs, but only for 9 out of 50 instances. All MCSs were found before the timeout of one hour. However, compared to OR-Tools, Exact took roughly 10 times longer to provide an MCS. In the best case, Exact was about 4 times slower and in the worst case around 25 times slower than OR-Tools. This shows that OR-Tools is not only able to provide (M)CSs for all instances but is also considerably faster than Exact.

We compared the sizes of CSs found by OR-Tools for the three configurations. We observed that there is no big difference in the sizes of CS. However, when comparing the base and minimal configurations, we observed that the sizes of

Table 3. Comparison for finding any CS (base), weight-minimal CS (minimal), or weight-minimal CS without demand (no demand) within 1h using OR-Tools or Exact. Percentage of instances for which a CS/MCS was found, mean and standard deviation (σ) of CS size and run times are indicated.

	solver	CS/MCS	size	σ(size)	time	σ(time)
base	OR-Tools	**100% / 76%**	**1.86**	1.23	1934	1582
	Exact	18% / 18%	2.78	1.40	3162	1428
minimal	OR-Tools	**100% / 76%**	1.88	1.24	**1898**	1590
	Exact	18% / 18%	2.78	1.48	3172	1434
no demand	OR-Tools	**100% / 76%**	1.92	1.38	1983	1611
	Exact	18% / 18%	3.11	1.85	3168	1423

CSs are the same for each instance except instance 3966. For this instance, a CS of size 3 was found for the minimal configuration, but a true subset of size 2 for the base and no demand case. For all configurations, the instance was run until a timeout, indicating that under the weight-minimal configuration, a non-subset-minimal CS was returned. For the no demand case, we observed that for two instances an MCS of size 6 was provided (instead of sizes 4 and 5). In both cases, these were found in about 10–15 seconds, indicating their optimality. When comparing with the MCS found for the base and minimal case, we found that the MCSs contained a demand constraint. Additionally, CSs for four other instances contained demand bounds for the base and minimal configuration. In each case, a CS of the same size without demand bounds could be found.

6.2   Optimisation Configuration

We now want to analyse configurations using thresholds on soft constraints for feasible instances. We want to test whether threshold combinations still allow for acceptable solutions. If not, we provide CSs as explanations for infeasibility. To this end, we consider 20 feasible real-life instances of different sizes.

In order to determine suitable thresholds for each dimension we first determined the $opt(s)$ and $avg(s)$ value for each soft constraint s and instance i. The $opt(s)$ value corresponds to the expected violation of s for instance i if s is the only optimisation goal. On the other hand, $avg(s)$ corresponds to the expected violation of s if the sum of all soft constraint violations is set as the optimisation goal. We determined these values by running each of the 20 instances once optimising the sum of all soft constraints and once for each of the soft constraints, considering only a single soft constraint as the optimisation goal. We performed 5 runs per configuration using OR-Tools. For the avg values, we took results from the runs with the least summed objective.

Optimization Performance. We compared our optimal values opt with objective values obtained in previous work for the same benchmark set [26], which used a

MiniZinc model of the problem and solved the instances with the Chuffed solver [8], also using a timeout of one hour, but on a more performant machine with 2.7 GHz. We compared the objective values of the four soft constraints that are used by the authors as well. For the working weekends constraint s^{ww}, we could prove optimal objectives for 5 additional instances and found a better objective value for the only instance for which our approach could not prove optimality. Similarly, for working weekends including Friday night s^{Nww}, we proved optimality for 4 additional instances and again found a better objective for the only instance where we could not prove optimality. For the maximal weekend distance s^{dmax}, our approach could not prove optimality for one instance, while [26] could. However, we obtained the same objective value, indicating that the optimal value was found, but optimality could not be proven within the time limits. While we could not prove optimality for any additional instances, we were able to find solutions within the time limit for two instances, which Chuffed failed to do. Lastly, for s^{drms}, we could also not prove optimality on any new instances. However, we found better values for three instances and were again able to find a solution within the time limit for an instance that Chuffed could not solve.

Threshold Definition. Using the *opt* and *avg* values we can implement thresholds of the following form: *Reduce constraint s_1 by $x\%$ compared to the average value and increase constraint s_2 by at most 10% .* This is implemented as follows. Given a factor $f(s) \in [-1, 1]$ by which the value of the soft constraint s should be decreased or increased, we can calculate the threshold $t(s)$:

$$t(s) = max\Big(opt(s), avg(s) + \lceil d(s) \cdot |f(s)| \rceil \cdot sgn\big(f(s)\big)\Big)$$

where $sgn(\ )$ is the sign function and $d(s)$ is the distance between $avg(s)$ and $opt(s)$ (and 1 if $avg(s) = opt(s)$). We decided on this formulation for rounding for multiple reasons. First of all, rounding is necessary as we deal with integer valued soft constrains. Secondly, traditional rounding always rounds up/down in the same direction for positive and negative numbers, i.e. 0.6 will be rounded to 1 and -0.6 will be rounded to 0. However as we use positive and negative factors to set a threshold relative to the average value, we want thresholds with the same absolute distance to the average value (mirrored rounding), i.e. for $avg = 0$ and $d = 1$ a factor +/-0.6 should result in +/-1. Lastly, we use ceiling rounding as we want a difference of at least 1 to the average value for thresholds with a factor other than 0. We are interested in finding soft constraint thresholds that will lead to infeasibility. This allows us to evaluate the CSs provided for these infeasible configurations and to learn which combinations of soft constraints are most conflicting. For this purpose, we look at six different threshold configurations. We test for each soft constraint reducing its violation by 100% (i.e. setting the threshold to *opt*) while increasing each of the remaining 5 soft constraints by at most 10%. In the following, we identify the configuration by the constraint that is reduced to the optimum. I.e. *reduced* s^{ww} refers to the configuration where $f(s^{\mathrm{ww}}) = -1$ and $f(s) = 0.1$ for all other soft constraints. For each of these configurations, there are two MCSs, namely one consisting of the reduced constraint

s and one which is a subset of the remaining constraints. Intuitively, there is a solution with $v(s) = opt(s) = t(s)$, therefore removing at most all of the other thresholds must yield a solution. On the other hand, we know that there exists a solution with $v(s') = avg(s') \leq t(s')$ for all $s' \neq s$, since we have $f(s') = 0.1$. Therefore, removing the threshold on s must also yield a solution.

Finding a Single MCS. As we are interested in which constraints conflict with the reduced constraint, we generate CSs containing only soft constraints with a weight function $w(s) = 10$ if s is being reduced and $w(s') = 1$ for all $s' \neq s$. We use OR-Tools to solve our instances within 1h. If the configuration of the instance was proven unsatisfiable or its status is not known (i.e. neither a solution was found nor a proof of unsatisfiability), we proceed to finding a CS with the specified weight function, comparing OR-Tools and Exact. We discuss the results in Table 4. For each configuration of reduced soft constraint, we depict the percentage of instances that could be solved feasibly or optimally (sat/opt), instances for which an unsatisfiability proof was found (unsat), as well as the percentage of remaining instances (unknown). For unsatisfiable and unknown instances, we proceed by generating CSs using OR-Tools and Exact. We observe that with OR-Tools we are able to find CSs even for the unknown instances. For each of the configurations, OR-Tools can find a CS for all instances that are either unsatisfiable or unknown except one. This one instance is usually an instance which has been shown to be hard to solve, and in one case, it is the largest instance and therefore inherently difficult to solve. While OR-Tools shows promising results, we observe that not all found CSs are subset-minimal. In general, for roughly 35% of the cases where a CS is found, we can not guarantee that it is an MCS. In comparison, Exact mostly fails to find a CS even for instances that are proven unsatisfiable. However, all found CSs are MCSs.

Table 4. Results for finding CSs with thresholds using OR-Tools or Exact. Percentages of instances where solutions were found (sat/opt) and where no solution was found (unsat/unknown) are indicated. Percentages of instances where the solvers could find a CS under 1h are also given.

reduced	sat	opt	unsat	unknown	OR-Tools CS/MCS	Exact CS/MCS
$s^{N>3}$	40%	30%	5%	25%	25%/15%	0%/0%
$s^{\ell_{\mathrm{dev}}}$	5%	20%	25%	50%	70%/45%	10%/10%
s^{ww}	35%	35%	0%	30%	25%/15%	0%/0%
$s^{d_{\mathrm{max}}}$	5%	30%	15%	50%	60%/40%	10%/10%
$s^{d_{\mathrm{rms}}}$	15%	30%	5%	50%	50%/30%	5%/5%
$s^{N_{\mathrm{ww}}}$	30%	35%	0%	35%	30%/20%	0%/0%

Soft Constraint Interactions. To evaluate the influence of the soft constraints on each other, we evaluate how often a constraint s' is mentioned in a CS for the configuration reducing s. A visualisation of this can be found in Fig. 1. We depict the constraint to be reduced on the x-axis and the relative occurrences of constraints in CSs on the y-axis. We count the occurrences in a CS in percentages relative to the total number of CSs found for this configuration. As multiple constraints can occur in one CS, the summed percentages in each column do not necessarily sum up to 100%. We observe that the night shift constraint ($s^{N>3}$) occurs the least often in a conflict set, while the length deviation constraint ($s^{\ell_{dev}}$) occurs very often in a CS. In fact, almost every CS for reducing distances between weekends ($s^{d_{max}}$, $s^{d_{rms}}$) contains the length deviation constraint. However, if we compare how often the weekend distance constraints occur in a CS for reducing length deviation, we find that while $s^{d_{max}}$ occurs in roughly 70% of the CSs, the $s^{d_{rms}}$ constraint is only problematic in about 50% of CSs. In general, we observe that there is no direct correspondence between the percentage with which constraint s' occurs in CSs when reducing s and the percentage with which constraint s occurs in CSs when reducing s'.

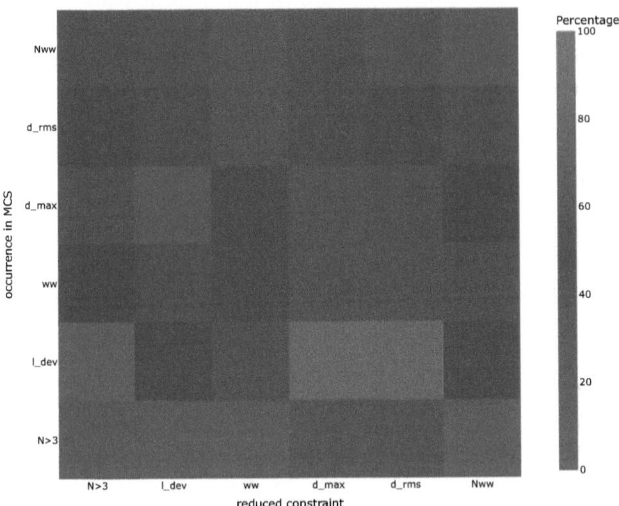

Fig. 1. Heat map indicating the percentage of occurrences of a constraint in an CS for a configuration using thresholds.

7   Conclusion

We have proposed an Explainability framework for the extended Rotating Workforce Scheduling (RWS) Problem that includes several novel soft constraints. We created a constraint model of the RWS Problem using CPMpy. We showed how soft and hard constraints as well as thresholds can be modelled. This model was then used for providing detailed explanations. Using a small example, we showed how (Minimal) Correction Sets (MCS/CS) can be generated and how preferences can be specified to obtain preferred CSs. The experiments showed that the proposed granular bound constraints give precise explanations showing how much constraints need to be changed. We first conducted experiments on infeasible RWS instances. While using MARCO to enumerate all MCSs did not yield the best results, we could employ OR-Tools to provide CSs in under an hour for all infeasible instances. In a second step, we tested threshold combinations for soft constraints on feasible instances. While we could not always find MCSs, especially for larger instances, we were generally able to provide CSs in cases where no solution could be found. These CSs additionally allowed us to determine conflicting soft constraints. Our results provide valuable insights for the different stakeholders involved in the generation of employee schedules with many constraints, and is planned to be incorporated into software of our industry partner. In the future, it would be interesting to further evaluate our constraint model. Testing the performances of other solvers, as well as running experiments with higher time-outs, could yield new insights. We also want to apply the granular bound constraints on other optimization problems.

Acknowledgments. The financial support by the Austrian Federal Ministry for Labour and Economy and the National Foundation for Research, Technology and Development and the Christian Doppler Research Association is gratefully acknowledged. In addition, this research was funded in part by the Austrian Science Fund (FWF) [10.55776/COE12].

Disclosure of Interests. The authors have no competing interests to declare that are relevant to the content of this article.

References

1. Agrawal, J., Yelamanchili, A., Chien, S.: Using explainable scheduling for the mars 2020 rover mission. arXiv preprint arXiv:2011.08733 (2020)
2. Arlinghaus, A., Bohle, P., Iskra-Golec, I., Jansen, N., Jay, S., Rotenberg, L.: Working time society consensus statements: evidence-based effects of shift work and non-standard working hours on workers, family and community. Ind. Health **57**(2), 184–200 (2019)
3. Bailey, J., Stuckey, P.J.: Discovery of minimal unsatisfiable subsets of constraints using hitting set dualization. In: Proceedings of the 7th International Symposium on Practical Aspects of Declarative Languages (PADL), pp. 174–186 (2005)
4. Baker, K.R.: Workforce allocation in cyclical scheduling problems: a survey. J. Oper. Res. Soc. **27**(1), 155–167 (1976)
5. Becker, T., Schiffer, M., Walther, G.: A general branch-and-cut framework for rotating workforce scheduling. INFORMS J. Comput. **34**(3), 1548–1564 (2022)
6. Van den Bergh, J., Beliën, J., De Bruecker, P., Demeulemeester, E., De Boeck, L.: Personnel scheduling: a literature review. Eur. J. Oper. Res. **226**(3), 367–385 (2013)
7. Bleukx, I., Boumazouza, R., Guns, T., Laage, N., Poveda, G.: Modeling and explaining an industrial workforce allocation and scheduling problem. Leibniz Int. Proc. Inform. **340** (2025)

8. Chu, G., Stuckey, P.J., Schutt, A., Ehlers, T., Gange, G., Francis, K.: Chuffed, a lazy clause generation solver (2018). https://github.com/chuffed/chuffed

9. Čyras, K., Lee, M., Letsios, D.: Schedule explainer: an argumentation-supported tool for interactive explanations in makespan scheduling. In: Proceedings of the 3rd International Workshop on Explainable, Transparent Autonomous Agents and Multi-Agent Systems (EXTRAAMAS), pp. 243–259. Springer (2021)

10. Čyras, K., Letsios, D., Misener, R., Toni, F.: Argumentation for explainable scheduling. In: Proceedings of the 33rd AAAI Conference on Artificial Intelligence, pp. 2752–2759 (2019)

11. Dall'Ora, C., Ball, J., Recio-Saucedo, A., Griffiths, P.: Characteristics of shift work and their impact on employee performance and wellbeing: a literature review. Int. J. Nurs. Stud. **57**, 12–27 (2016)

12. De Bruecker, P., Van den Bergh, J., Beliën, J., Demeulemeester, E.: Workforce planning incorporating skills: state of the art. Eur. J. Oper. Res. **243**(1), 1–16 (2015)

13. De Moura, L., Bjørner, N.: Z3: an efficient SMT solver. In: Proceedings of the 14th International Conference on Tools and Algorithms for the Construction and Analysis of Systems (TACAS), pp. 337–340 (2008)

14. Eifler, R.: Explaining goal conflicts in oversubscription planning. Phd thesis, Saarländische Universitäts-und Landesbibliothek (2025). https://doi.org/10.22028/D291-46295

15. Elffers, J., Nordström, J.: Divide and conquer: towards faster Pseudo-Boolean solving. In: Proceedings of the 18th International Joint Conference on Artificial Intelligence (IJCAI), pp. 1291–1299 (2018)

16. Ernst, A.T., Jiang, H., Krishnamoorthy, M., Sier, D.: Staff scheduling and rostering: a review of applications, methods and models. Eur. J. Oper. Res. **153**(1), 3–27 (2004)

17. Folkard, S., Lombardi, D.A.: Modeling the impact of the components of long work hours on injuries and "accidents." Am. J. Ind. Med. **49**(11), 953–963 (2006)

18. Folkard, S., Robertson, K.A., Spencer, M.B.: A fatigue/risk index to assess work schedules. Somnologie **11**(3), 177–185 (2007)

19. Gamba, E., Bogaerts, B., Guns, T.: Efficiently explaining CSPs with unsatisfiable subset optimization. J. Artif. Intell. Res. **78**, 709–746 (2023)

20. Gärtner, J., Bohle, P., Arlinghaus, A., Schafhauser, W., Krennwallner, T., Widl, M.: Scheduling matters-some potential requirements for future rostering competitions from a practitioner's view. In: Proceedings of the 12th International Conference on the Practice and Theory of Automated Timetabling (PATAT), pp. 33–42 (2018)

21. Gleeson, J., Ryan, J.: Identifying minimally infeasible subsystems of inequalities. ORSA J. Comput. **2**(1), 61–63 (1990)

22. Guns, T.: Increasing modeling language convenience with a universal n-dimensional array, CPpy as python-embedded example. In: Proceedings of the 18th workshop on Constraint Modelling and Reformulation (ModRef) at CP 2019 (2019)

23. Gupta, S.D., O'Sullivan, B., Quesada, L.: Counterfactual explanation through constraint relaxation. In: 2024 IEEE 36th International Conference on Tools with Artificial Intelligence (ICTAI), pp. 396–403 (2024). https://doi.org/10.1109/ICTAI62512.2024.00064

24. Health and Safety Executive: Managing shiftwork-health and safety guidance (2006)

25. Ignatiev, A., Morgado, A., Marques-Silva, J.: PySAT: a Python toolkit for prototyping with SAT oracles. In: Proceedings of the 21st International Conference on Theory and Applications of Satisfiability Testing (SAT), pp. 428–437 (2018)

26. Kletzander, L., Musliu, N., Gärtner, J., Krennwallner, T., Schafhauser, W.: Exact methods for extended rotating workforce scheduling problems. In: Proceedings of the 29th International Conference on Automated Planning and Scheduling (ICAPS), pp. 519–527 (2019)

27. Kletzander, L., Musliu, N., Smith-Miles, K.: Instance space analysis for a personnel scheduling problem. Ann. Math. Artif. Intell. **89**, 617–637 (2021)

28. Liffiton, M.H., Malik, A.: Enumerating infeasibility: finding multiple muses quickly. In: Proceedings of the 10th International Conference on Integration of AI and OR Techniques in Constraint Programming for Combinatorial Optimization Problems (CPAIOR), pp. 160–175 (2013)

29. Liffiton, M.H., Sakallah, K.A.: On finding all minimally unsatisfiable subformulas. In: Proceedings of the 8th International Conference on Theory and Applications of Satisfiability Testing (SAT), pp. 173–186 (2005)

30. Liffiton, M.H., Sakallah, K.A.: Algorithms for computing minimal unsatisfiable subsets of constraints. J. Autom. Reason. **40**, 1–33 (2008)

31. Liffiton, M.H., Sakallah, K.A.: Generalizing core-guided max-sat. In: Proceedings of the 12th International Conference on Theory and Applications of Satisfiability Testing (SAT), pp. 481–494 (2009)

32. Lock, A.M., Bonetti, D.L., Campbell, A.: The psychological and physiological health effects of fatigue. Occup. Med. **68**(8), 502–511 (2018)

33. Martínez Gasca, R., Valle Sevillano, C.d., Gómez López, M.T., Ceballos Guerrero, R.: NMUS: structural analysis for improving the derivation of all muses in over-constrained numeric CSPs. In: Proceedings of the 12th Conference of the Spanish Association for Artificial Intelligence (CAEPIA), pp. 160–169 (2007)

34. Moreno, C.R., et al.: Working time society consensus statements: evidence-based effects of shift work on physical and mental health. Ind. Health **57**(2), 139–157 (2019)

35. Musliu, N., Gärtner, J., Slany, W.: Efficient generation of rotating workforce schedules. Discret. Appl. Math. **118**(1–2), 85–98 (2002)

36. Musliu, N., Schutt, A., Stuckey, P.J.: Solver independent rotating workforce scheduling. In: Proceedings of the 15th International Conference on International Conference on Integration of AI and OR Techniques in Constraint Programming for Combinatorial Optimization Problems (CPAIOR), pp. 429–445 (2018)

37. Nethercote, N., Stuckey, P.J., Becket, R., Brand, S., Duck, G.J., Tack, G.: MiniZinc: towards a standard cp modelling language. In: proceedings of the 13th International Conference on Principles and Practice of Constraint Programming (CP), pp. 529–543 (2007)

38. Ortell, K.K., Switonski, P.M., Delaney, J.R.: FairSubset: a tool to choose representative subsets of data for use with replicates or groups of different sample sizes. J. Biol. Methods **6**(3) (2019)

39. Perron, L., Didier, F.: CP-SAT. https://developers.google.com/optimization/cp/cp_solver/

40. Petrovic, S., Parkin, J., Wrigley, D.: Personnel scheduling considering employee well-being: insights from case studies. In: Proceedings of the 13th International Conference on the Practice and Theory of Automated Timetabling (PATAT). vol. 1, pp. 10–23 (2021)

41. Rosekind, M., Gregory, K., Mallis, M., Brandt, S., Seal, B., Lerner, D.: The cost of poor sleep: workplace productivity loss and associated costs. J. Occup. Environ. Med./Am. Coll. Occup. Environ. Med. **52**, 91–98 (2010)
42. Senthooran, I., et al.: Human-centred feasibility restoration in practice. Constraints **28**(2), 203–243 (2023)
43. Van Loon, J.: Irreducibly inconsistent systems of linear inequalities. Eur. J. Oper. Res. **8**(3), 283–288 (1981)

Generalised Arc Consistency
via the Synchronised Product of Finite
Automata with Respect to a Constraint

Nicolas Beldiceanu$^{(\boxtimes)}$ (iD)

IMT Atlantique, LS2N, UMR CNRS 6004, F-44307 Nantes, France
`nicolas.beldiceanu@imt-atlantique.fr`

Abstract. Given an m by n matrix V of domain variables $v_{i,j}$ (with i from 1 to m and j from 1 to n), where each row i must be accepted by a specified Deterministic Finite Automaton (DFA) $\mathcal{A}_i$ and each column j must satisfy the same constraint `ctr`, we show how to use the *synchronised product of DFAs wrt constraint* `ctr` to obtain a Berge-acyclic decomposition ensuring Generalised Arc Consistency (GAC). Such decomposition consists of one `regular` and n `table` constraints. We illustrate the effectiveness of this method by solving a hydrogen distribution problem, finding optimal solutions and proving optimality quickly.

1 Introduction

The synchronised product of automata is a key technique, well established in formal methods since the 1960s [3], for modelling concurrent systems. However, its potential as an effective propagation mechanism in constraint programming for achieving GAC filtering [2] for specific matrix models was not investigated.

Motivation. Motivated by a problem of generating cyclic schedules for continuous hydrogen supply to customers via containers,[1] we exploit the concept of synchronised product of automata wrt a constraint to obtain GAC on a subset of the constraints, specifically relaxing container capacities and customer demands. In this context, the matrix V's elements, $v_{i,j}$ (with $i \in [1, m]$, m = number of containers; $j \in [1, n]$, n = number of time slots), denote the specific location (refill site or customer) visited by container i at time j.

- Rows (container i's itinerary) are governed by a `regular` constraint [5], which enforces a specific cyclic sequence of customer and refill sites.
- Columns (all containers at time slot j) are restricted by an `alldifferent` [6], ensuring no two containers are at the same site simultaneously.

To achieve GAC on this conjunction (C) of m `regular` constraints and n `alldifferent` constraints, we propose a technique based on the synchronised

---

[1] Companies producing and distributing green hydrogen like Lhyfe, see https://www.lhyfe.com/, were the source of our hydrogen distribution problem.

© The Author(s), under exclusive license to Springer Nature Switzerland AG 2026
T. Guns (Ed.): CPAIOR 2026, LNCS 16595, pp. 350–358, 2026.
https://doi.org/10.1007/978-3-032-27242-3_21

product construction. This product is formed over the automata corresponding to the m rows, and is explicitly restricted by the `alldifferent` constraints of the columns. The resulting automaton captures the interaction between all row and column constraints, providing a powerful mechanism for filtering out inconsistent values and for directly proving the infeasibility of conjunction (C) even before search: the right automaton-based abstraction eliminates backtracking.

Contributions Our contributions include:

1. For a matrix model where each row must satisfy a given DFA and each column fulfils a same given constraint, a method that captures such a conjunction (C) by generating a single synchronised product automaton.
2. The demonstration that the resulting synchronised product automaton can be reformulated as a Berge-acyclic [1] conjunction of one `regular` [5] and n `table` [4] constraints, which permits obtaining GAC.
3. To address the original problem by considering the containers' capacities and customers' demands, we extract minimal size solutions from the synchronised product automaton and solve a tiny MIP model for each extracted solution.

Paper Organisation. (A) We begin with the *necessary theoretical framework*. In Sect. 2 we recall the notion of a synchronised product of automata wrt a constraint. Building upon this, in Sect. 3 we give a Berge-acyclic reformulation of a synchronised automata as a conjunction of a `regular` and `table` constraints. (B) In Sect. 4 we *introduce our motivating application*, the Hydrogen Distribution Problem (HDP). (C) Finally, we *present the integrated solution*. In Sect. 5 we demonstrate the core contribution, showing how the reformulated synchronised product automaton solves the cyclic and disjunctive aspects of the HDP without backtracking. This leads to Sect. 6, where we detail a tiny MIP model to handle the remaining capacity and demand constraints.

2 Background

In this section, we recall the necessary definitions for the rest of the article.

Definition 1 (synchronised product of automata). *Given m DFA $\mathcal{A}_i = (Q_i, \Sigma_i, \delta_i, q_{0i}, F_i)$ (with $i \in [1, m]$), where each automaton $\mathcal{A}_i$ is defined by its set of states Q_i, its local input alphabet Σ_i, its transition function δ_i, its initial state q_{0i}, and its set of final states F_i, the synchronised product $\mathcal{A}_1 \times \mathcal{A}_2 \times \cdots \times \mathcal{A}_m$, denoted $\mathcal{A}_{prod} = (Q_{prod}, \Sigma_{prod}, \delta_{prod}, q_{0prod}, F_{prod})$, is defined as:*

- $Q_{prod} = Q_1 \times Q_2 \times \cdots \times Q_m,$
- $\Sigma_{prod} = \Sigma_1 \times \Sigma_2 \times \cdots \times \Sigma_m,$
- $q_{0prod} = (q_{0,1}, q_{0,2}, \ldots, q_{0,m}),$
- $F_{prod} = F_1 \times F_2 \times \cdots \times F_m,$
- $\delta_{prod}((q_1, q_2, \ldots, q_m), (\ell_1, \ell_2, \ldots, \ell_m)) = (\delta_1(q_1, \ell_1), \delta_2(q_2, \ell_2)), \ldots, \delta_m(q_m, \ell_m)).$

The definition of a *synchronised product of automata wrt a constraint* extends Definition 1 by introducing a specific synchronisation constraint on the product of the local alphabets $\Sigma_1, \Sigma_2, \ldots, \Sigma_m$ of automata $\mathcal{A}_1, \mathcal{A}_2, \ldots, \mathcal{A}_m$.

Definition 2 (synchronised product of automata wrt a constraint). *Given m DFA $\mathcal{A}_i = (Q_i, \Sigma_i, \delta_i, q_{0i}, F_i)$ (with $i \in [1, m]$) and a constraint $\mathcal{C}$ on m variables, the* synchronised product of automata $\mathcal{A}_1, \mathcal{A}_2, \ldots, \mathcal{A}_m$ wrt $\mathcal{C}$, *denoted $\mathcal{A}^{\mathcal{C}}_{prod}$, is an automaton induced by $\mathcal{A}_{prod}$ on the same alphabet Σ_{prod}, where the transition function is restricted to $\delta^{\mathcal{C}}_{prod}((q_1, q_2, \ldots, q_m), (\ell_1, \ell_2, \ldots, \ell_m)) = (\delta_1(q_1, \ell_1), \delta_2(q_2, \ell_2)), \ldots, \delta_m(q_m, \ell_m))$ where constraint $\mathcal{C}(\ell_1, \ell_2, \ldots, \ell_m)$ holds.*

Notation: $\min(\mathcal{A}^{\mathcal{C}}_{prod})$ denotes the minimal automaton associated with $\mathcal{A}^{\mathcal{C}}_{prod}$. We recall the notion of equivalent letters in a DFA. Two letters are equivalent if the DFA does not distinguish them in terms of state changes, regardless of position. This will help reduce the alphabet size of the synchronised product.

Definition 3 (equivalent letters). *Given a DFA $\mathcal{A} = (Q, \Sigma, \delta, q_0, F)$, two distinct letters ℓ_1 and ℓ_2 of Σ are equivalent, iff $\forall q \in Q, \delta(q, \ell_1) = \delta(q, \ell_2)$.*

Definition 4 (Berge-acyclic constraint network). *A constraint network is Berge-acyclic if any two constraints share at most one variable and if there are no circular sequences of distinct variables and constraints.*

3   Berge-Acyclic Reformulation via Synchronised Product

The challenge lies in propagating GAC over the conjunction of m row-wise `regular` constraints and n column-wise constraints $\mathcal{C}$. Building the synchronised product automaton $\mathcal{A}^{\mathcal{C}}_{prod}$ results in a Berge-acyclic [1] reformulation that *completely avoids search tree exploration*. Achieving GAC on the resulting subconstraints is equivalent to achieving GAC on the original system. We rewrite

$$\bigwedge_{i \in [1,m]} \texttt{regular}(\mathcal{A}_i, \langle v_{i,1}, v_{i,2}, \ldots, v_{i,n} \rangle) \;\wedge\; \bigwedge_{j \in [1,n]} \mathcal{C}(v_{1,j}, v_{2,j}, \ldots, v_{m,j}) \quad (1)$$

into the following conjunction of one `regular` and n `table` constraints:

$$\texttt{regular}(\overline{\min(\mathcal{A}^{\mathcal{C}}_{prod})}, \langle w_1, w_2, \ldots, w_n \rangle) \;\wedge$$

$$\bigwedge_{j \in [1,n]} \texttt{table}\left(\langle v_{1,j}\; v_{2,j} \ldots v_{m,j}\; w_j \rangle, \left\langle \begin{matrix} \ell_{1,1} & \ell_{2,1} & \ldots & \ell_{m,1} & | & \ell_{m+1,1} \\ \ell_{1,2} & \ell_{2,2} & \ldots & \ell_{m,2} & | & \ell_{m+1,2} \\ \vdots & \vdots & \ddots & \vdots & | & \vdots \\ \ell_{1,p} & \ell_{2,p} & \ldots & \ell_{m,p} & | & \ell_{m+1,p} \end{matrix} \right\rangle \right), \quad (2)$$

I The `regular` constraint, i.e. the first term of (2):

- $\overline{\min(\mathcal{A}^{\mathcal{C}}_{prod})}$ is the minimal synchronised product automaton, $\min(\mathcal{A}^{\mathcal{C}}_{prod})$, whose input alphabet is a reduced set of unique integer labels. To reduce the size of the input alphabet of $\overline{\min(\mathcal{A}^{\mathcal{C}}_{prod})}$, each label represents an equivalence class of one or more feasible transition tuples from $\min(\mathcal{A}^{\mathcal{C}}_{prod})$ (as defined in Definition 3). These labels are the initial domain of the variables $w_1, w_2, \ldots, w_n$ as specified by the `table` constraints in II.
- The sequence of variables $w_1, w_2, \ldots, w_n$ represents the sequence of global column labels read by the automaton $\min(\mathcal{A}^{\mathcal{C}}_{prod})$.

II The `table` constraints, i.e. the second term of (2):
- Each `table` constraint functionally determines the auxiliary variable w_j from the original column variables $v_{1,j}, v_{2,j}, \ldots, v_{m,j}$.
- The table itself lists all feasible tuples where each row, $\ell_{1,k}, \ell_{2,k}, \ldots, \ell_{m,k}, \ell_{m+1,k}$, consists of: (i) A valid assignment of values to the original column variables, that is, one of the feasible tuples associated with the transitions of the minimal synchronised product automaton $\min(\mathcal{A}^{\mathcal{C}}_{prod})$; (ii) The corresponding unique global label $\ell_{m+1,k}$ from the reduced global alphabet that the automaton $\overline{\min(\mathcal{A}^{\mathcal{C}}_{prod})}$ reads for this column.

The constraint network of the reformulation given by (2) is Berge-acyclic [1], since (i) there is at most one shared variable between any two constraints, and (ii) the hypergraph of the constraint network contains no cycles.

Limits of the Approach. The primary limitation lies in the potential state-space complexity of the resulting automaton, $\min(\mathcal{A}^{\mathcal{C}}_{prod})$. In the worst case, the number of states, $|Q^{\mathcal{C}}_{prod}|$, is bounded by the product $\prod_{i=1}^{m} |Q_i|$, and the size of the input alphabet is bounded by the number of feasible assignments to $\mathcal{C}$. Crucially, the size of $\min(\mathcal{A}^{\mathcal{C}}_{prod})$ is independent of n (the number of columns). Since the complexity only depends on the number of rows (m) and the tightness of constraint $\mathcal{C}$, the approach is well suited for our motivating problem, where the number of containers (m) is limited, but the time horizon (n) is large. Furthermore, the size of the automaton is often mitigated by two factors:

- Constraint Pruning and State Reachability: Both a tighter column constraint $\mathcal{C}$ and more restrictive row automata $\mathcal{A}_i$ prune transition possibilities, making some states in $\mathcal{A}_{prod}$ unreachable, leading to a smaller $\min(\mathcal{A}^{\mathcal{C}}_{prod})$.
- Alphabet Minimisation: If the automata associated with the matrix's rows contain equivalent letters, the alphabet substitution process may significantly reduce the size of the input alphabet of the automaton $\overline{\min(\mathcal{A}^{\mathcal{C}}_{prod})}$.

4 The Hydrogen Distribution Problem (HDP)

We consider a set of locations, $1, 2, \ldots, loc$, where '1' denotes the production site where each container must be fully refilled for a minimum duration of R time units. The locations $2, 3, \ldots, loc$ denote the customer sites, each with a constant instantaneous demand D_k, with k in $[2, loc]$. We also consider a set of m containers, $1, 2, \ldots, m$, with individual capacities C_i, i ranging from 1 to m.

Continuous Service and Scheduling Constraints. Since all customers require uninterrupted supply, a strict, instantaneous container swap is enforced: as soon as one container is removed from any site (customer or production), a new container must immediately take its place. Consequently, exactly one container is present at each location at any given moment, meaning $m = loc$. The following constraints apply:

- [*Cyclic Constraint*] Each container follows a cyclic sequence of stops (customers & production sites) from a predefined set of cyclic sequences, where the starting location is arbitrary.
- [*Disjunctive Constraint*] At any instant, each customer site $k \in [2, loc]$ must be served by one container.
- [*Demand and Capacity Constraints*] At any instant, each customer site $k \in [2, loc]$ must be served by a non-empty container that meets its demand D_k.

Objective: The objective is to minimise container movements. To achieve this, we break down the initial problem into sub-problems by clustering nearby customers during a preprocessing phase. Consequently inter-site travel time is considered negligible. Next, we create cyclic schedules for each sub-problem that maximise cycle duration. This minimises the number of reloads required, assuming an infinite time frame in which the cyclic schedule is repeated.

5   Managing Cyclic and Disjunctive Constraints in the HDP

To handle the cyclic and disjunctive constraints, we generate a synchronised product of automata with respect to a constraint: (i) the cyclic restriction of each container is encoded as a regular expression that is converted into a finite automaton; (ii) the disjunctive constraint, ensuring that each customer is served by only one container at a time, is modelled using an **alldifferent** constraint. This synchronised product automaton wrt **alldifferent** accepts certain words. A word specifies, for each container, the precise order of visited sites that adheres to both the cyclic and disjunctive requirements. These words represent the layout of feasible solutions for the HDP when relaxing the demand and capacity constraints.

We present two examples of synchronised automata that were generated from our HDP. Then, we provide statistics on the sizes of the automata that were generated using this approach on a test set of 118 instances.

Example 1 (infeasible system of cyclic constraints). Consider the four expressions R_1, R_2, R_3 and R_4, which respectively restrict the four rows (container itineraries) of a matrix V. The alphabet $\Sigma = \{1, 2, 3, 4\}$ represents the possible locations (refill site/customers). Rows R_2, R_3, and R_4 are constrained by a cyclic sequence requirement. To break symmetry, R_1 is not a cyclic sequence but a fixed sequence. The constraints are defined as follows:

- $R_1 = 2^+1^+3^+1^+4^+1^+$, a fixed sequence for symmetry breaking,
- $R_2 =$ the cyclic sequence associated with $3^+4^+1^+$. This is formally defined by the regular expression $1^*3^+4^+1^+|3^*4^+1^+3^+|4^*1^+3^+4^+$,
- $R_3 =$ the cyclic sequence associated with $4^+1^+2^+3^+1^+$,
- $R_4 =$ the cyclic sequence associated with $4^+2^+1^+$.

The row automata associated with R_1–R_4 enforce these constraints, and their synchronised product wrt the `alldifferent` constraint $\mathcal{C}$ yields an empty automaton. This shows, without using any search, that there is no solution to this conjunction, regardless of the number of columns in the matrix V.

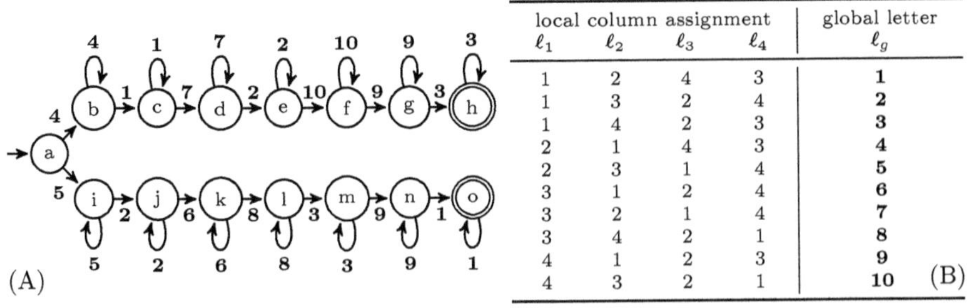

| local column assignment | | | | global letter |
ℓ_1	ℓ_2	ℓ_3	ℓ_4	ℓ_g
1	2	4	3	1
1	3	2	4	2
1	4	2	3	3
2	1	4	3	4
2	3	1	4	5
3	1	2	4	6
3	2	1	4	7
3	4	2	1	8
4	1	2	3	9
4	3	2	1	10

(B)

Fig. 1. (A) Minimal synchronised product automaton $\overline{\min(\mathcal{A}^{\mathcal{C}}_{prod})}$ with an input alphabet that matches the table's column ℓ_g in Table (B); (B) Tuples of the `table` constraints used in Reformulation (2).

Example 2 (feasible system of cyclic constraints). Now consider a modification of Example 1 where R_1 is kept unchanged, but R_2, R_3, and R_4 respectively correspond to the cyclic sequences $4^+1^+2^+3^+1^+$, $2^+4^+1^+$, and $3^+4^+1^+$. Part (A) of Fig. 1 shows the corresponding synchronised product of the automata of R_1, R_2, R_3 and R_4 wrt `alldifferent`. Part (B) provides the tuples for the `table` constraints in Reformulation (2). This table maps the feasible column assignments $(\ell_1, \ell_2, \ell_3, \ell_4)$, which satisfy the `alldifferent` constraint and are valid transitions for the row automata's current states, to a unique global letter ℓ_g, a potential value of the auxiliary variables w_j (with $j \in [1, n]$). We observe that:

- The minimal DFAs for R_1, R_2, R_3, R_4 have 7, 29, 13, and 13 states.
- The synchronised product $\min(\mathcal{A}^{\mathcal{C}}_{prod})$ has only 15 states (a–o), much less than the worst-case product of 34307 states. This reduction is due to the tightness of `alldifferent`, which makes most combined states unreachable.
- Of the $4! = 24$ possible `alldifferent` assignments for the column variables $\{v_{1,j}, v_{2,j}, v_{3,j}, v_{4,j}\}$, only 10 are feasible transitions, as shown in Table (B) of Fig. 1. Such pruning of the global alphabet contributes to efficiency.

Table 1 presents observed statistics from 118 test instances of the hydrogen distribution problem, where the number of containers m is fixed per table, leading to evaluating 161 synchronised automata, as there may be several possible cyclic sequences that a container can choose from. The column constraint $\mathcal{C}$ is `alldifferent`. The results in Table 1 confirm that the resulting synchronised product automata are small relative to the theoretical worst-case product of states in our motivating problem. For $m = 4$, the average number of output states is only 6.4, which demonstrates the practical viability of the approach for models with small m and a tight column constraint $\mathcal{C}$.

Table 1. Observed size of the synchronised automaton $(\min(\mathcal{A}_{prod}^{\mathcal{C}}))$ for $m = 3$ and $m = 4$ containers; Statistics are given for the minimum (min), maximum (max), sample mean $(\bar{x})$, and sample standard deviation (s); "In" refers to the initial DFAs, "Out" to the resulting synchronised product automaton $\min(\mathcal{A}_{prod}^{\mathcal{C}})$.

(82 products with $m = 3$ from which 2 were infeasible)	min	max	$\bar{x}$	s
In states product $(\prod_{i=1}^{m} |Q_i|)$	196	1805	437.7	435.5
Out states (# states of $\min(\mathcal{A}_{prod}^{\mathcal{C}})$)	0	13	5.6	2.6
Out alphabet (# global letters)	0	6	3.5	1.3
(179 products with $m = 4$ from which 81 were infeasible)	min	max	$\bar{x}$	s
In states product $(\prod_{i=1}^{m} |Q_i|)$	2058	229593	29920.3	37359.4
Out states (# states of $\min(\mathcal{A}_{prod}^{\mathcal{C}})$)	0	61	6.4	9.0
Out alphabet (# global letters)	0	24	4.2	4.7

6 Handling Demand and Capacity Constraints in the HDP

To integrate demand and capacity constraints while maximising the schedule length, we first introduce a property of the synchronised product of automata associated with the HDP to identify a reduced set of essential paths, which we call minimal solutions. We then formulate a tiny MIP problem for each minimal solution to determine the optimal duration of each stage.

Identifying Minimal Solutions. We present a property of the synchronised product of automata used to model the cyclic and disjunctive constraints of the HDP.

Property 1. Given automata $\mathcal{A}_1, \mathcal{A}_2, \ldots, \mathcal{A}_m$ that do not contain any cycles of length greater than one, the synchronised product of $\mathcal{A}_1, \mathcal{A}_2, \ldots, \mathcal{A}_m$ wrt any constraint does not contain any cycles of length greater than one.

Property 1 can be proved by contradiction. Assume that the synchronised product automaton contains a cycle of length greater than one. It would then accept an infinite number of words. This would imply that all automata $\mathcal{A}_i$ (with $i \in [1, m]$) must also contain such cycles, a contradiction. Since each cyclic constraint of the HDP corresponds to the union of a fixed number of automata with no cycles of length greater than one, this ensures that the synchronised automaton, once self-loops are removed, has a finite number of solutions. They correspond to the possible starting points in the cyclic sequence associated with each container.

Definition 5. *We call the* minimal set of solutions *of an automaton $\mathcal{A}$ with no cycle of length greater than one, the finite set of solutions of automaton $\mathcal{A}$ from which we discard all self-loops.*

Formulating a Compact MIP Model. The MIP determines the optimal duration of each stage in the minimal solution, ensuring demand fulfilment and capacity limits are respected, while maximising the total duration. Given a minimal solution of length n, we create n integer variables $p_1, p_2, \ldots, p_n$, where p_k (with $k \in [1, n]$) is the duration of the k-th stage. We have the constraints: [Refill] At each stage k (with $k \in [1, n]$), one of the containers is always at the production site for refill/loading. Thus, the minimum duration of any stage p_k is set to the refill time R, i.e. $p_k \geq R$. [Demand] For each container $i \in [1, m]$ a critical subsequence $\mathcal{S}$ of visited customer locations is defined as the locations visited by container i between two successive visits to the production site. For each such subsequence $\mathcal{S}$, we post a linear capacity constraint: $\sum_{k \in \mathcal{S}} D_k \cdot p_k \leq C_i$. C_i is the capacity of container i, and D_k is the demand of the customer location being visited at stage k. [Objective Function] The objective is to maximise $\sum_{k=1}^{n} p_k$.

Example 3. The synchronised product automaton depicted in Fig. 1 has the minimal solutions '**4 1 7 2 10 9 3**', '**5 2 6 8 3 9 1**'. They represent valid sequences of global production stages (locations visited by each container) that satisfies the cyclic and disjunctive constraints. The 1st solution corresponds to the sequences visited by containers 1 to 4: '2 1 3 1 4 4 1', '1 2 2 3 3 1 4', '4 4 1 2 2 2 2', and '3 3 4 4 1 3 3'. The 1st location in each sequence is the starting location for that container. Figure 2 illustrates the linear constraints that are created for the first minimal solution of '**4 1 7 2 10 9 3**' of the automaton depicted in Fig. 1.

The total time it took to solve the 118 instances of our test case, including generating cyclic automata and their synchronised product, searching for optimal solutions, and proving optimality, was 5.2 s on an M2 Ultra using SICStus 4.8.0. For reproducibility purposes, instances are available in the report arXiv:2512.09975. We do not report results on an other CP model because it was inefficient, as the number of variables was proportional to the maximum duration of the schedule, and as it did not decompose the problem into two phases.

$$\begin{cases} R = 6 \\ C_1 = 900 \\ C_2 = 900 \\ C_3 = 900 \\ C_4 = 900 \\ D_2 = 8 \\ D_3 = 15 \\ D_4 = 34 \end{cases}$$

(A) matrix rows and constraints:

Row (①,②,③): $2\ 1\ 3\ 1\ 4\ 4\ 1$ — ①: $8 \cdot p_1 \le 900$, ②: $15 \cdot p_3 \le 900$, ③: $34 \cdot (p_5 + p_6) \le 900$

Row (④,⑤): $1\ 2\ 2\ 3\ 3\ 1\ 4$ — ④: $8 \cdot (p_2 + p_3) + 15 \cdot (p_4 + p_5) \le 900$, ⑤: $34 \cdot p_7 \le 900$

Row (⑥): $4\ 4\ 1\ 2\ 2\ 2\ 2$ — ⑥: $34 \cdot (p_1 + p_2) + 8 \cdot (p_4 + p_5 + p_6 + p_7) \le 900$

Row (⑦): $3\ 3\ 4\ 4\ 1\ 3\ 3$ — ⑦: $15 \cdot (p_1 + p_2 + p_6 + p_7) + 34 \cdot (p_3 + p_4) \le 900$

Columns: $p_1\ p_2\ p_3\ p_4\ p_5\ p_6\ p_7$ — $p_1, p_2, \cdots, p_7 \ge 6$, maximise $p_1 + p_2 + \cdots + p_7$

(B) corresponding optimal solution:

p_1	p_2	p_3	p_4	p_5	p_6	p_7
2	1	3	1	4	4	1
1	2	2	3	3	1	4
4	4	1	2	2	2	2
3	3	4	4	1	3	3
$p_1 = 6$	$p_2 = 6$	$p_3 = 6$	$p_4 = 6$	$p_5 = 20$	$p_6 = 6$	$p_7 = 14$

Fig. 2. (A) MIP model stated for the first minimal solution of the synchronised product automaton of Fig. 2, and (B) the corresponding optimal solution that maximises the total duration. Grey areas indicate the refill site.

7 Conclusion

We used a well-known construction from concurrent system modelling to obtain a reformulation performing GAC on a conjunction of constraints on a matrix of variables, where each row verifies a **regular** constraint and each column verifies the same constraint.

Acknowledgments. Athanaël Jousselin and Victor Spitzer explained the hydrogen distribution problem to me and created the experimental instances used in this paper. Olivier Peton provided feedback on an early draft of the paper, and Justin Pearson found the relevant reference for the synchronized product of automata.

References

1. Beeri, C., Fagin, R., Maier, D., Yannakakis, M.: On the desirability of acyclic database schemes. J. ACM **30**(3), 479–513 (1983)
2. Bessière, C.: Chapter 3: constraint propagation. In: Rossi, F., van Beek, P., Walsh, T. (eds.) Handbook of Constraint Programming, pp. 29–83. Elsevier (2006)
3. Hartmanis, J.: Symbolic analysis of a decomposition of information processing machines. Inf. Control **3**(2), 154–178 (1960)
4. Lhomme, O., Régin, J.-C.: A fast arc consistency algorithm for N-ARY constraints. In: 20th National Conference on Artificial Intelligence (AAAI-05), pp. 405–410 (2005)
5. Pesant, G.: A regular language membership constraint for finite sequences of variables. In: Wallace, M.G. (ed.) Principles and Practice of Constraint Programming (CP'2004). LNCS, vol. 3258, pp. 482–495. Springer-Verlag (2004)
6. Régin, J.-C.: A Filtering algorithm for constraints of difference in CSP. In: 12th National Conference on Artificial Intelligence (AAAI-94), pp. 362–367 (1994)

Open-Source Implementation of Slack Induction by String Removals for Routing and Orienteering Problems

Martin Pajerský, Václav Sobotka, and Hana Rudová

Faculty of Informatics, Masaryk University, Botanická 68a, Brno, Czech Republic
`sobotka@mail.muni.cz, hanka@fi.muni.cz`

Abstract. Vehicle routing problems aim to efficiently serve customer requests while respecting constraints such as vehicle capacities and delivery time windows. When it is not possible to fulfill all requests, orienteering problems are considered, where only a subset of locations is visited to maximize the total collected scores. This paper presents an open-source implementation of the slack induction by string removals (SISRs) heuristic [9] for three representative variants of the vehicle routing problem. Our implementations match the solution quality of the original solver, and their runtimes scale better with increasing instance sizes. We propose an SISRs extension to solve three variants of the orienteering problem, with specific modifications to address their unique requirements. The developed solver achieves strong performance compared to the current state of the art, computing 49 new best-known solutions and delivering about 85 % of the best-known solutions [38] across all considered variants of the orienteering problem. Overall, we provide an efficient open-source solver for a variety of routing problems [39], as the availability of such implementations is very limited.

Keywords: Vehicle routing · Orienteering problem · Slack induction by string removals · Optimization · Heuristics · Open-source solver · C++

1 Introduction

Vehicle routing problems (VRP) are complex, and inherent limitations of exact methods prevent their wider adoption for real-world and large-scale problem instances. This led to the use of heuristics. Over the long history of vehicle routing research, increasingly complicated methods have been developed to solve specific VRP variants [11]. It is then difficult to reuse even parts of these methods or to adapt them to different constraints. Simpler and more flexible methods are easier to adapt to practical constraints [25]. Slack induction by string removals (SISRs), introduced by Christiaens and Vanden Berghe in 2020 [9], has achieved highly competitive results, especially on large-scale problems. It stands out not just for its performance but also for its design simplicity and ability to handle a wide range of problem variants with only minor algorithmic modifications.

T. Guns (Ed.): CPAIOR 2026, LNCS 16595, pp. 359–377, 2026.
https://doi.org/10.1007/978-3-032-27242-3_22

Despite its success, there is no publicly available implementation of SISRs supporting multiple VRP variants. Generally, the extensive range of increasingly complex problem variants and heuristic methods strongly contrasts with the limited availability of high-quality, scalable, open-source solver implementations in the VRP area [57]. This, in turn, makes reproducibility and further development unnecessarily difficult. To fill this gap, we contribute efficient open-source C++ implementations of the SISRs heuristic for the most common VRP variants [39]. Compared to the original paper, our implementations scale better with increasing instance size, resulting in up to an order of magnitude better runtimes for the largest instances. Furthermore, we introduce and provide SISRs for the family of team orienteering problems (TOP). Our implementation achieves state-of-the-art results across three variants of TOP, finding the best-known solutions for 85 % of standard benchmark instances, including 49 new best-known solutions [38].

2   Problem Overview

Motivated directly by real-world applications, numerous VRP and orienteering problem (OP) variants have been introduced and extensively researched. These variants modify problem attributes, objectives, or constraints to fit various operational settings [4,17,60]. Here, we briefly describe the most studied variants considered in our paper, three for VRPs and three for OPs.

We note that for the capacitated vehicle routing problem and the team orienteering problem, the concepts of time/duration and traveled distance are identical, as is standard in the literature. Despite this, we commonly distinguish between time/duration and traveled distance, as it is important in problems that include additional time-related characteristics.

2.1   Vehicle Routing Problems

Capacitated VRP (CVRP) is the most fundamental and extensively studied variant [7,55]. The vehicles have a predefined capacity. The total load of all customers served by a vehicle must not exceed the vehicle capacity. In the simplest version of CVRP, all vehicles have homogeneous capacity.

VRP with Time Windows (VRPTW) assigns to each customer a time window in which the service must start [52]. All customers must be visited within their designated time windows, and vehicles must wait if they arrive before the window opens. Along with the start and end of the time window, a service time is usually specified, representing the time required to complete delivery at the customer's location. The depot itself can also have a time window, limiting the duration of routes. In addition, VRPTW is often associated with a hierarchical objective. First, the goal is to minimize the number of vehicles used, and then to minimize the total distance traveled.

Pickup and Delivery Problem with Time Windows (PDPTW) is a direct extension of VRPTW. Instead of transporting goods from the depot to customers, requests are generalized to start and end at any pair of locations [40, 41]. A vehicle serving a request loads goods in a pickup location and transfers them to the corresponding delivery location. This order must be respected, and each pickup-delivery pair must be served by the same vehicle. Vehicles can start either empty or partially loaded from the depot. As the vehicle visits pickup and delivery locations, its load changes non-monotonically, depending on the order in which pickup/delivery locations are visited.

2.2   Team Orienteering Problems

With a limited time budget, serving all customers may become impossible. This motivates the *orienteering problems* (OP), where an optimal route is found only for a subset of requests, leaving the remainder *absent*. Customers have a *score* (reward/profit), which is earned for serving them. The measure of collected scores defines the objective in OP, i.e., the maximization of the total sum of scores from visited customers. Numerous versions of OP have been formulated [17]. We focus on three classical variants where several vehicles form a *team*.

Team OP (TOP) defines a fixed number of vehicles greater than one. The objective is to maximize the total score collected by the whole fleet/team. The durations of the vehicle routes are limited. Note that the fleet size is fixed because an unlimited fleet allows serving all customers trivially.

TOP with Time Windows (TOPTW) extends TOP by incorporating time window constraints identical to those described for VRPTW, where customers must be served within the specified time intervals. Also, each customer defines their service time as in VRPTW.

Capacitated TOP (CTOP) extends TOP by introducing customer demands and service times. As in CVRP, the vehicle capacity constraint must be respected. We note that TOP with both time windows and capacity constraints is referred to as *selective vehicle routing problem with time windows* (SVRPTW) [6].

3   Related Work

VRP [4, 54] and OP [17] families share core solution techniques focused on searching for optimal routes. While OP introduces an additional decision on customer selection, method overlap is significant. Due to the computational complexity of these NP-hard problems [26, 27], approaches range from exact mathematical modeling to flexible heuristics and metaheuristics.

The most effective exact algorithms nowadays rely on branch-and-bound [12]. Solvers like VRPSolver [42] and the learning-based RouteOpt [63] define the state of the art. Exact methods for OP additionally incorporate mechanisms for customer selection [6]. However, scaling remains problematic; systematically solving VRP instances with over 200 customers [42] and OP instances with over 100 customers [29, 35] remains a significant challenge. Consequently, over 80 % of VRP research between 2009 and 2015 relied on heuristics or metaheuristics [7].

Constructive heuristics, such as merging routes to save cost [10,45], provide initial feasible solutions. These can be refined by improvement heuristics, e.g., relocate [56] moves customers to a different position in a route. More advanced methods, such as the bi-level filter-and-fan [53], extend the improvement process through hierarchical search. Single-solution approaches navigate neighborhoods of a single candidate. Iterated local search [33] was applied to TOPTW [16], including multi-start versions [3]. Simulated annealing has also proven competitive for TOP [31] and TOPTW [32], achieving effective convergence and efficient memory usage. Population-based approaches such as genetic algorithms, ant colony optimization, and particle swarm optimization have been used for both VRP [37,44] and OP [13,22,23] families. Those methods are often combined with other strategies, such as greedy randomized construction and local search for CTOP [5].

Large neighborhood search (LNS) [8,50], adaptive (A)LNS [47], and slack induction by string removals (SISRs) [9] explore large neighborhoods using destroy and repair operators in contrast to small neighborhoods of relocate or k-opt [30]. Both ALNS [43] and SISRs are so-called *unified* algorithms, providing competitive performance across a wider range of VRP variants. Hybrid genetic search (HGS) [58] is a hybrid of a population-based genetic algorithm, combining crossover search operations with classical improvement moves. Generalized into the unified HGS [59], this method is state of the art for over 29 VRP variants, though it struggles with pickup-and-delivery constraints. Hybrid (H)ALNS [20] combines ALNS with exact integer programming (using CPLEX) to solve sub-route optimization and set packing problems. This approach has improved the best-known solutions for large TOP instances (up to 400 requests) and was later adapted for CTOP [19].

Competitions like DIMACS[1] and EURO Meets NeurIPS 2022[2] encourage the development of open-source, scalable solvers, but the resulting codes are often not published. For the canonical CVRP, we are aware of implementations of the HGS [57], ALNS [48], adaptive iterative local search (AILS-II) [36], and recent FILO [1] and FILO2 [2] solvers, which target extremely large problem instances. Alternatives covering a wider range of problem variants include the Python-based library PyVRP [62] and generic solvers such as OR-Tools [14] or the LKH3 solver [21], which, however, offer limited options for adaptation or modification to more specific use cases. Overall, open-source implementations of promising heuristic algorithms that cover multiple problem variants and scale well, such as SISRs, are generally scarce.

4 Overview of Slack Induction by String Removals

We will provide an overview of SISRs, focusing on features relevant to our extension for orienteering problems. For further reading, we refer to the original work Christiaens and Vanden Berghe [9].

---

[1] http://dimacs.rutgers.edu/programs/challenge/vrp/.
[2] https://euro-neurips-vrp-2022.challenges.ortec.com/.

1: **procedure** LARGENEIGHBORHOODSEARCH(s)
2:      $s_{\text{best}} \leftarrow s$
3:      $T \leftarrow T_0$
4:      **while** not reached *maxIteration* **do**
5:          $s' \leftarrow s$
6:          DESTROY(s')
7:          REPAIR(s')
8:          **if** $f(s') < f(s) - T\ln(\mathcal{U}(0,1))$ **then** $s \leftarrow s'$
9:          **if** $f(s') < f(s_{\text{best}})$ **then** $s_{\text{best}} \leftarrow s'$
10:         $T \leftarrow \alpha \cdot T$

Fig. 1. Pseudo-code of the main LNS procedure in the SISRs algorithm.

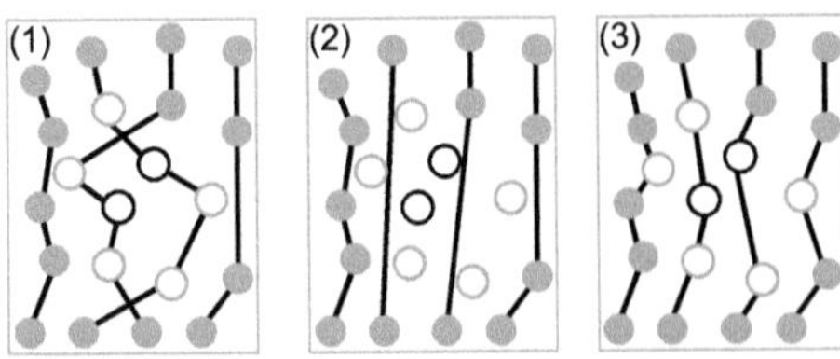

Fig. 2. Example of two adjacent strings with white nodes (1), where black emphasizes pivots. In (2), strings are destroyed. In (3), customers are greedily added after repair.

SISRs, as a method from the LNS family, explores a large portion of the solution space by iteratively destroying and repairing solutions (see Fig. 1). It employs only a single pair of destroy and repair methods on the current solution s', omitting the adaptive mechanism of ALNS [47]. However, the simulated annealing metaheuristic [24] is still used to guide the search and generate neighboring solutions (see its Metropolis criterion on line 8, with T as the temperature and a cooling constant α on line 10). The initial solution is constructed naively by assigning each customer to a separate route.

Destroy Heuristic. The purpose of the destroy heuristic is to remove strings of customers from a solution, leaving some customers absent. The *string* is a consecutive sequence of customer requests in a route (see Fig. 2 (1)). This prepares space for the following repair.

Moreover, the DESTROY procedure is based on the removal of *adjacent strings* (see example in Fig. 2 (2)) located around close *pivot* nodes. Such removals are intended to introduce the required slacks while allowing customers to be exchanged among the affected routes. The *capacity slack* considers the unutilized capacity of a vehicle, and the *spatial slack* takes into account the reachable geographical area of a vehicle.

To simplify the search for adjacent strings, SISRs' destroy uses an *adjacency table*. For each customer, the table contains a list of all customers, sorted in ascending order by distance. This table is precomputed before the search begins and remains unchanged throughout execution.

Repair Heuristic. The repair heuristic in SISRs extends the widely used greedy insertion (see Fig. 2 (3)). A purely greedy insertion heuristic considers all feasible positions for all absent customers. A customer is placed in a position with the smallest increase in cost, and this process is repeated until all customers are served. However, consistently inserting at the overall best place has a significant weakness. It tends to postpone inserting the most expensive requests, ending with very few poor options when most routes are already full [47]. To overcome this problem, SISRs first orders absent customers using a predefined sorting rule. Then, the absent customers are greedily inserted into the solution in this fixed order rather than relying solely on the most cost-effective positions. For the base CVRP, the sorting method is chosen randomly from four predefined options (each selected based on its weight). The customers may be shuffled randomly (RANDOM), ordered by descending demand (DEMAND), or in decreasing/increasing distance from the depot (FAR/CLOSE-ORIGIN). More details about the REPAIR procedure are also provided in Sect. 5.3 to contrast it with its OPs proposal.

Fleet Minimization. The fleet minimization procedure is another version of the LNS heuristic, which is processed prior to the cost minimization to handle the primary objective. It subsequently lowers the number of vehicles used in solutions and allows for the possibility of not serving all customers. To explore the neighborhood, the procedure uses previously introduced DESTROY and REPAIR methods. An additional acceptance criterion prioritizes solutions serving customers who were previously often absent. Finally, finding a solution that serves all customers reduces the number of routes.

Adaptations to Problem Variants. For the VRPTW, the REPAIR procedure needs to be extended by a feasibility check emerging from the time windows. In more detail, vehicles are assumed to depart from customers as soon as possible, which is the best choice from a feasibility standpoint. Then, for each position in the route, the largest possible increase in the departure time (*time slack*) is computed in linear time, proportional to the route size. Using these precomputed values, the insertion feasibility itself is checked in constant time [49]. The time slack values are recomputed only after a customer is inserted or removed from the route. Thanks to the additional time-window variables, new possibilities for customer sorting in REPAIR emerge. The three additional orderings are based on the customer time windows and their ascending length (TW-WIDTH), ascending start (TW-START), and descending end (TW-END).

Further extensions to the PDPTW are introduced. In the DESTROY procedure, it may happen that after removing requests, their corresponding pickup-delivery pair remains in the route. The method needs to be modified so that any unpaired requests are removed after the removal operator is applied. Furthermore, the REPAIR procedure must account for non-monotonic changes in vehicle capacity along the route. Similarly to time constraints, the capacity constraint can be checked again in constant time using precalculated statistics. The REPAIR is also modified to consider the pickup-delivery node pair. After finding a feasible place for the pickup node, the positions for its corresponding delivery must also be explored. This introduces an additional nested loop before evaluating the insertion cost. Three more sorting methods are used in REPAIR. Requests are ordered by decreasing distance between the pickup and delivery nodes (PD-DIST) and by increasing distance of the pickup/delivery location from the depot (CLOSE-PICKUP/DELIVERY). Finally, a time window feasibility check must be performed for each pickup-delivery pair. Each time a pickup location is fixed, the time slacks within the route change, and they must be recalculated before finding a suitable spot for the delivery.

5   SISRs for Team Orienteering Problems

5.1   Large Neighborhood Search

The objective dramatically changes in the case of OPs. The quality of the solution strictly depends on its total score, and the goal of maximizing this value drives the search. The large neighborhood search from Sect. 4 can be applied to OPs, even though it was originally defined for distance minimization tasks. This is realized by redefining the objective function to $f(s) = -score(s)$.

Optimizing the score brings a new challenge compared to classical VRPs. Different solutions frequently share the same objective value, complicating navigation in the solution space. For example, simply reordering customer visits in a solution might change its available slack, but the objective function does not capture the change. Vidal et al. [61] named this a *staircase aspect*, and to counter this issue in their solver, they used a weighted total solution cost as an auxiliary objective. On the other hand, state-of-the-art hybrid solver HALNs [20] performs effectively without such an addition. Inspired by the work of Vidal et al., we introduce a parameter ω that penalizes the distance in the objective function. The final form of the objective function for OP is $f(s) = -score(s) + \omega \cdot distance(s)$, where $0 \leq \omega \ll 1$. Finally, note that the LNS for fleet size minimization is not used as the number of vehicles is fixed. Next, we introduce SISRs for the base TOP and later describe its modifications for TOPTW and CTOP.

Initial Solution Construction. Since the number of vehicles is fixed in TOP, the initial solution generation method used for VRP with an unlimited fleet (described in Sect. 4) makes little sense as inserting a single customer per vehicle would leave most customers absent. On the other hand, TOP allows feasible solutions that include empty routes. Therefore, we adopt a straightforward strategy and initialize all vehicles as empty at the beginning of the search. This approach remains consistent with the LNS procedure, and the construction of the first non-trivial solution occurs in its first iteration (see Fig. 1), where the REPAIR method initiates the search.

```
1: procedure REPAIRTOP(s)
2:    pick sorting algorithm
3:    SORT(A)
4:    A_new = ∅
5:    for each c ∈ A, c selected using Shaw shuffle
6:       p ← none
7:       for each r ∈ R such that r has enough capacity for c
8:          for each p_r ∈ r
9:             if 1 − γ < U(0,1) or duration(r with c at p_r) > D_max then continue
10:            if p = none or distance(c, (r, p_r)) < distance(c, p) then p ← (r, p_r)
11:       if p ≠ none then insert c at p
12:       else if allowAbsence then
13:             A_new ← A_new ∪ {c}
14:       else R ← R ∪ {new route with c}
15:    A ← A_new
```

Fig. 3. Pseudo-code of the REPAIRTOP procedure.

5.2   Destroy Heuristic

We use the same DESTROY procedure as for the VRP. The original adjacent string removal can be applied to TOP without modification. The heuristic is naturally compatible with solutions where only part of the customers are visited, which is ideal for TOP. The slack introduced by this procedure brings enough diversity and prepares the ground for the critical search phase, which lies in the repair step. While one might consider enhancements such as score-aware seeding, where nodes with lower scores have a higher probability of being selected as pivots for removal, such modifications would go against the original simplicity of the method.

5.3   Repair Heuristic

The repair step for TOP is based on the original REPAIR procedure. Changes to the procedure are made in multiple places; a pseudo-code of REPAIRTOP is presented in Fig. 3. To highlight the differences from the original procedure, we cross out the unused lines and underline all added and modified parts. We now describe the most important aspects of the original algorithm and the changes made to it. Detailed discussions are provided in the following subsections.

The method starts identically as for VRP, with the sorting methods used to define the order in which absent customers A are inserted into the solution s. However, the accompanying sort methods (used in lines 2 and 3) are redesigned to fit the problem as discussed below. Additionally, the absent customers are not inserted in the strict order specified by the sorting method; instead, randomness is introduced using the *Shaw shuffle* (line 5).

Blink rate $\gamma \in \langle 0, 1 \rangle$ brings additional randomness into the search. It is a probability (typically close to zero) of ignoring a candidate insertion position

p_r into the route $r \in R$ (line 9). Each p_r must also be checked against the maximum route duration constraint: the duration of route r together with the candidate customer c inserted at p_r must not exceed the maximum duration $D_{\max}$. In the case of TOP, the duration is identical to the traveled distance.

The selection and assignment of the best insertion position remain unchanged (lines 10 – 11), continuing to rely on the lowest insertion cost (based on the traveled distance) combined with blinks. If no feasible position is found, the fixed fleet size must be kept. REPAIRTOP exclusively operates in a mode that allows customer absences (the original SISRs also operates with forbidden absences). Therefore, the route creation (line 14) is omitted.

Sorting Methods. The new objective and the differences in problem formulation create possibilities for new sorting methods. Some of the common orderings for VRP are not applicable. Namely, ordering customers by decreasing distance from the depot is not ideal, as visiting distant nodes significantly reduces the route slack. Given the limited time and need to select a subset of customers to serve, it is generally better to prioritize nodes located closer to the depot. The following sorting methods are used for TOP (and later for TOPTW and CTOP).

RANDOM shuffles the customers.
SCORE orders customers by decreasing score.
CLOSE-ORIGIN orders customers by increasing distance from the origin depot.
CLOSE-DESTINATION orders customers in increasing distance from the destination depot.
SCORE-TO-COST-KNN orders customers by the ratio between their score and the sum of distances to their k geographically nearest nodes. The precomputed adjacency table introduced on page 363 can be used for efficient computation, as it contains nodes sorted by distance. The number of k neighbors is adjustable; we use two variants with $k = 2$ and $k = 5$.

An example of the sorting methods' effect is demonstrated in Fig. 4. First, an input solution to the REPAIRTOP is shown on the left. The solution serves just one route, with a black square depicting both the origin and destination depot. White-colored nodes are numbered from 1 to 4, and each number corresponds to its score. The following three graphs depict repaired solutions, where the ordering SCORE-TO-COST-KNN with $k = 2$ resulted in the highest score increase of 5 compared to the other methods, which increased the score by 4 or 1.

Shaw Shuffle. To introduce more randomness and enhance search diversification, we designed a stochastic rank-based selection of nodes. The sorting methods (except for random ordering) rely on static values and maintain a rigid ordering. As a result, the list of absent customers tends to remain unchanged across iterations, making it highly unlikely that customers ranked worse in the ordering will be selected. To overcome this issue, we adopt a technique introduced by Paul Shaw in [51], therefore the name *Shaw shuffle*.

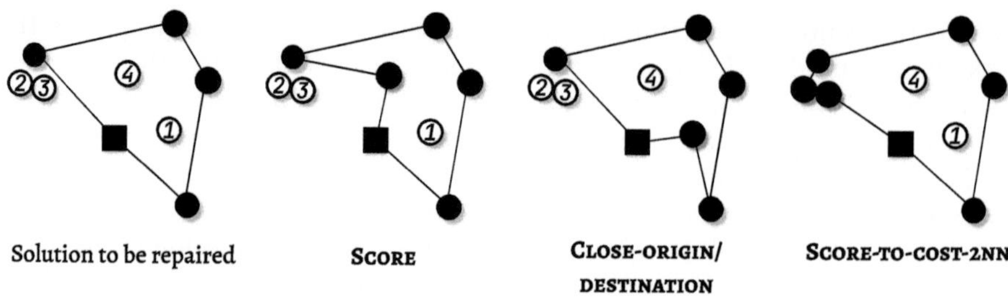

Solution to be repaired   **SCORE**   **CLOSE-ORIGIN/ DESTINATION**   **SCORE-TO-COST-2NN**

Fig. 4. Example effect of different sorting methods in REPAIRTOP procedure.

Instead of simply iterating over the sorted set of absent customers A, the next customer $c \in A$ is selected stochastically. The selection is guided by a parameter $q \in \langle 0, 1 \rangle$, controlling how much the original ordering should be respected. In each iteration, the next index i from the ordered set A is sampled as follows $i = \lfloor |A| \cdot (1 - \mathcal{U}(0,1)^q) \rfloor$, where $\mathcal{U}(0,1)$ is an uniform distribution. The customer $c = A[i]$ is subsequently selected for reinsertion and removed from A. The probability of selecting worse-ranked customers exponentially decreases. When $q = 0$, the best-ranked customers are always selected, and in the case of $q = 1$, the choice is completely random. Regarding the actual value of q, we select it uniformly at random from the range $(0, \sigma\rangle$ in each iteration of SISRs, and σ is a configurable hyperparameter. We found that such a random choice better adapts across a wider range of problem types and intances, than a fixed q. Lastly, we note that if the initially picked sorting algorithm is RANDOM, the Shaw shuffle can be omitted for further efficiency.

5.4 Adaptations to Problem Variants

Our approach to solving TOP maintains high adaptability, and the methods can be extended to new problem variants similar to those described in SISRs for VRP. Adding time window constraints directly leads to the TOPTW variant. The algorithmic changes are the same as described for VRPTW. Specifically, in the repair method, time slack calculations are used to verify feasibility of insertions, and three sorting methods based on time window properties (TW-WIDTH, TW-START, and TW-END) are employed. The calculations of route durations must also take service times and possible time window waiting times into account.

CTOP variant differs in capacity constraints, which affect only the repair method. Similar to CVRP, these constraints require an additional feasibility check when evaluating insertion positions within a route. Moreover, route duration calculations involve service times. A new sorting method is developed to leverage the vehicles' capabilities. The trade-off between customers' scores and demand may provide a helpful guide for reinsertion, so we order customers by decreasing their score-to-demand ratio. This idea was previously used, for example, by Gunawan et al. [18] to generate initial solutions for CTOP. We call this sorting SCORE-TO-DEMAND.

6 Experimental Evaluation

The experiments were conducted within the MetaCentrum[3] distributed computing infrastructure in the Czech Republic. Computations were run on nodes with Debian GNU/Linux 12 (x86_64), using AMD EPYC 7543 CPUs at 2.80 GHz. The solver was implemented in C++ (see source code with datasets at [39]). The code was compiled with GCC 12.2.0, targeting the C++20 standard and enabling -O3 optimization. Detailed tabular results are provided in [38].

6.1 Vehicle Routing Problems

Datasets. The CVRP instances used in this study are available through the CVRPLIB website[4]. The data for VRPTW and PDPTW were obtained from the Sintef portal[5]. A widely used benchmark set by Uchoa et al. [55] of 100 instances is used for CVRP, with the customer counts ranging from 100 to 1000. For VRPTW, we experimented with the benchmark set introduced by Gehring and Homberger [15]. The set contains 200 to 1000 customers, totaling 300 instances. For the PDPTW variant, a standard benchmark set by Li and Lim [28] is employed. Problem sizes include 200, 400, 600, 800, and 1000 pickup or delivery requests, totaling 298 instances.

Comparison with the Original Solver. The implementation was experimentally evaluated using the same benchmark instances as those in the original paper. We followed the parameter settings and replicated the experiments using our implementation for comparison. Although runtime is not directly comparable across hardware, we report it to validate the behavior of our implementation. The number of iterations varies between problem variants and is typically scaled with the instance size, allowing for more iterations for larger problems. For the comparison between our result and the original, we use the gap equal to $(our - original)/original$ as a percentage.

For the CVRP, Fig. 5 reports the best distance, average distance, and average computation time for a given number of customers. Each problem instance was run 50 times, as in the paper. The percentage gap on the y-axis is calculated per instance for each metric. The red lines represent equal performance (zero percentage gap). For the best/average distance, 35/3 instances yield the same result, 40/49 instances yield a better result for our solver, and 25/48 instances have a worse result, i.e., the results are balanced. While the hardware differs, our implementation scales better as the instance size increases. For the 100-customer instance, our solver is only slightly faster; for the 1000-customer instance, our solver runs more than 10 times faster.

For the VRPTW benchmark, we report the best results from 5 runs per instance as in the SISRs paper. Table 1 shows the average fleet size for six subsets

---

[3] https://metavo.metacentrum.cz.
[4] http://vrp.atd-lab.inf.puc-rio.br.
[5] https://www.sintef.no/projectweb/top.

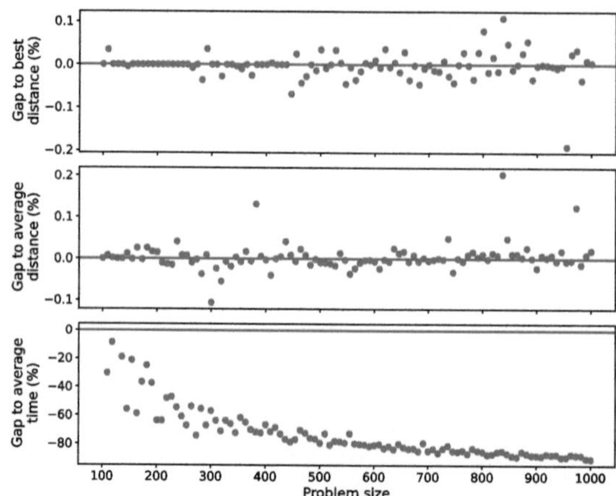

Fig. 5. Comparison with the original SISRs paper on the CVRP.

Table 1. Comparison on the VRPTW benchmark with the SISRs paper, presenting the average number of vehicles per subset and the average computational time in min.

Size	200		400		600		800		1000	
	Orig	Our	Orig	Our	Orig	Our	Orig	Our	Orig	Our
R1	18.2	18.2	36.4	36.4	54.5	54.5	72.8	72.8	91.9	91.9
R2	4.0	4.0	8.0	8.0	11.0	11.0	15.0	15.0	19.0	19.0
C1	18.9	18.9	37.6	37.6	57.4	57.4	**74.9**	**75.0**	**93.9**	**94.0**
C2	6.0	6.0	**11.8**	**12.0**	**17.4**	**17.5**	**23.4**	**23.5**	**29.1**	**29.2**
RC1	18.0	18.0	36.0	36.0	55.0	55.0	**72.8**	**72.0**	90.0	90.0
RC2	4.3	4.3	8.4	8.4	11.4	11.4	15.4	15.4	18.2	18.2
Time	11.0	2.5	42.0	8.2	95.3	14.1	216.6	25.8	324.2	38.9

(e.g., R1, R2...) per problem size, each containing 10 instances. First, the best result per instance is selected and then averaged across the subset. Multiplying by 10 yields the total number of vehicles across a subset. In 23 subsets, the results are identical. The differences amount to only a few vehicles, but grow with the problem size and are highlighted in bold. In the case of RC1, with 800 customers, the difference is the most evident. The likely explanation is that eight instances differed by only one vehicle, although this cannot be confirmed with certainty. For completeness, we also compared distances across subsets with equal fleet sizes. The largest percentage gap was 0.027%. Notably, the ratio between our runtimes and the original behaves similarly to CVRP. We have also executed 30 runs on each instance and provided the results in [38].

For PDPTW, the SISRs paper contains only a table listing the cumulative number of vehicles, the cumulative total distance, and the average computational time for each problem size. The solutions averaged are the best of 5 runs.

Table 2. Comparison on the PDPTW benchmark with the SISRs paper with the cumulative number of vehicles, cumulative total distances, and average time in minutes.

Size	Original			Our		
	CNV	CTD	Time	CNV	CTD	Time
200	600	185257	6.3	601	183390	1.5
400	1133	444247	23.7	1136	432906	5.0
600	1629	906717	66.0	1629	889148	11.3
800	2119	1526559	140.1	2118	1501673	17.7
1000	2577	2276009	226.6	2576	2217682	31.2

Compared to the original, our implementation found solutions with four additional vehicles across the groups, and required one less vehicle for two groups (see Table 2). In terms of runtime, our implementation consistently outperforms the original, and the margin again increases with the instance size. We have again executed 30 runs on each instance (see [38]).

6.2 Team Orienteering Problems

Datasets. For TOP, we use 82 instances with 2 to 4 vehicles and 102 to 401 requests, as reported in the literature and originating from Dang et al. [13]. For TOPTW, we have 76 instances with 48 to 288 customers, each in a variant with 1 to 4 vehicles (304 instances in total). The set originates from Montemanni et al. [34] who extended benchmarks with one vehicle [46]. For CTOP, we use datasets 2 and 3 introduced by Tarantilis et al. [53] (dataset 1 has instances where all customers can be served). There are 90 and 30 instances, respectively, with 337 to 577 customers, all of which have fleets of 6 to 8 vehicles.

Parameter Setting. For all three OP variants, we set the number of iterations to $|N| \times 50000$, where N denotes the number of nodes in the given instance. For the core SISRs parameters, we keep the same values as in the original work. The only (yet important) exception is the change in the initial and final temperatures during simulated annealing, reflecting the vastly different objective structures in OPs compared to standard VRPs. To accommodate these differences, we set the starting temperature to 50.0 (down from 100.0) and the final temperature to 0.01 (down from 1.0). Next, the previously discussed parameters ω and σ are common to all TOP, TOPTW, and CTOP. For all three OP variants, we set these core parameters to the common values $\omega = 1^{-5}$ and $\sigma = 0.3$. Regarding the sorting method weights, we follow the scheme from the original SISRs paper. For RANDOM and SCORE as the main sortings, we set the weights to 4. For the remaining weights, we set 1 for sortings directly inherited from SISRs, i.e., CLOSE-ORIGIN[6], TW-WIDTH, TW-START, and TW-END.

---

[6] We omit the CLOSE-DESTINATION sorting as origin and destination depots overlap.

For the newly introduced OP-specific sortings, we performed an ablation analysis to either include (weight 1) or exclude (weight 0) the given sorting method. To avoid overfitting, the analysis was performed on tuning instances derived from the benchmarks by adding Gaussian noise to the key characteristics (coordinates, scores, demands, service times, and time windows) of the original instances. For all three OPs, we compared the base method, which introduces no new sorting, with methods that include individual sortings. Sortings that improved over the baseline performance are included. The final setting of sorting weights is in Table 3; the weights subject to ablation analysis are in bold.

Table 3. Weight of all sortings in TOP, TOPTW, and CTOP.

	Random	Score	Close origin	Tw width	Tw start	Tw end	Score-to cost-knn	Score-to demand
TOP	4	4	1	-	-	-	1	-
TOPTW	4	4	1	1	1	1	1	-
CTOP	4	4	1	-	-	-	0	1

Results. In all experiments, the reference best-known solutions (BKS) used to calculate BKS gaps are based on *all* referenced papers and on *our* results. On each instance of the TOP benchmark, we performed 20 independent runs as is common in the literature. The compared methods are particle swarm optimization inspired algorithm (PSOiA) by Dang et al. [13], Pareto mimic algorithm (PMA) by Ke et al. [23], hybrid adaptive large neighborhood search (HALNS) by Hammami et al. [20], and large neighborhood search (LNS2) by Chaigneau et al. [8]. Note that PSOiA and PMA report results based on just 10 runs per instance, making the mean performance more important in interpreting their results. Reported computation time is the average per run, except only for HALNS, where the lowest time required to identify the best solution is shown. The aggregated results are presented in Table 4. SISRs-OP shows the best average performance among all compared methods while also finding 3 new BKSs.

Table 4. Results for TOP benchmark.

	PSOiA	PMA	HALNS	LNS2	SISRs-OP
# BKS	68	74	**75**	53	66
# Highest means	2	8	25	11	**48**
Avg. gap-best (%)	0.063	**0.024**	**0.024**	0.157	0.042
Avg. gap-mean (%)	0.522	0.445	0.318	0.701	**0.281**
Avg. time (s)	11031.0	1004.1	783.4	**20.6**	455.6

For TOPTW, we compare our approach with the evolution strategy (ES) by Karabulut et al. [22] and the multi-start iterated local search (MSILS) by

Amarouche et al. [3]. These methods report the best and average results based on 5 runs. While many more papers have benchmarked this dataset, their performance is lower and omitted here for brevity (see [22] for additional comparisons). Table 5 summarizes the aggregated results split by the instance variants with a different number of vehicles. Overall, the performance of SISRs-OP and MSILS is on par, while ES exhibits greater variability and significantly larger gaps. SISRs-OP shows better average performance than MSILS, while the latter performs better in terms of average gaps to the best solutions. Interestingly, SISRs-OP performance clearly improves with increasing vehicle count. In summary, SISRs-OP exhibits strong performance, finding the same number of BKSs as MSILS (257 of 304), of which 7 are new BKSs.

Table 5. Results for TOPTW benchmark for ES (E), MSILS (M), and SISRs-OP (S).

# vehicles	1			2			3			4		
	E	M	S	E	M	S	E	M	S	E	M	S
# BKS	48	**71**	68	34	**62**	**62**	34	**65**	64	44	59	**63**
# Highest mean	41	**71**	64	23	54	**59**	28	60	**62**	33	50	**72**
Avg. gap-best (%)	0.524	**0.112**	0.127	0.570	**0.101**	0.139	0.433	**0.038**	0.049	0.506	0.070	**0.045**
Avg. gap-mean (%)	0.802	**0.145**	0.175	0.972	**0.215**	0.222	0.684	0.176	**0.145**	0.871	0.244	**0.123**
Avg. time (s)	**22.9**	100.8	98.3	**52.4**	113.6	114.2	**43.3**	86.6	123.3	**60.8**	100.6	127.6

For CTOP, we provide comparison with variable space search with tabu search (VSS-Tabu) and variable space search with simulated annealing (VSS-SA) by Ben-Said et al. [5], bi-level filter-and-fan method (BiF&F-s) by Tarantilis et al. [53], and hybrid adaptive large neighborhood search (HALNS) by Hammami [19]. They all report results based on the best of 10 runs. Following historical conventions, only the computational time required to reach the best solution is provided. The CTOP dataset is the most challenging among the studied orienteering problem benchmarks due to the larger instance and fleet sizes. Table 6 shows subset-wise aggregated results. Clearly, SISRs-OP dominates all competing methods in all measured aspects, including runtime. Overall, SISRs-OP sets a new state of the art for CTOP, computing 105 of 120 BKSs, including 39 new ones.

Table 6. Results for CTOP benchmark.

subset	BiF&F-s		VSS-Tabu		VSS-SA		HALNS		SISRs-OP	
	2	3	2	3	2	3	2	3	2	3
# BKS	12	19	47	13	32	16	51	17	**77**	**28**
Avg. gap (%)	0.431	0.044	0.067	0.037	0.091	0.024	0.056	0.047	**0.022**	**0.003**
Avg.time to best (s)	3061	4686	1487	5763	893	4667	1175	3085	**547**	**686**

7   Conclusion

The paper presents our open-source implementations of the popular algorithm SISRs for common variants of VRPs and TOPs [39]. Our experiments show that the VRP solvers match the performance reported in the original paper and the runtimes scale better as instance sizes increase. For the OPs, we design and introduce SISRs-OP, closely following the original idea of algorithm simplicity. SISRs-OP sets the new state of the art for CTOP and finds BKSs in about 85 % of instances across all benchmarks, including 49 new BKSs [38]. To conclude, recognizing that high-quality, simple, flexible, scalable, and publicly available solvers are scarce, we contribute our implementations of SISRs and SISRs-OP.

Acknowledgments. Computational resources were provided by the e-INFRA CZ project (ID:90254), supported by the Ministry of Education, Youth and Sports of the Czech Republic.

References

1. Accorsi, L., Vigo, D.: A fast and scalable heuristic for the solution of large-scale capacitated vehicle routing problems. Transp. Sci. **55**(4), 832–856 (2021)
2. Accorsi, L., Vigo, D.: Routing one million customers in a handful of minutes. Comput. Oper. Res. **164**, 106562 (2024)
3. Amarouche, Y., Guibadj, R.N., Chaalal, E., Moukrim, A.: Effective neighborhood search with optimal splitting and adaptive memory for the team orienteering problem with time windows. Comput. Oper. Res. **123**, 105039 (2020)
4. Archetti, C., Coelho, L.C., Speranza, M.G., Vansteenwegen, P.: Beyond fifty years of vehicle routing: insights into the history and the future. Eur. J. Oper. Res. **330**(2), 355–372 (2026)
5. Ben-Said, A., El-Hajj, R., Moukrim, A.: A variable space search heuristic for the capacitated team orienteering problem. J. Heuristics **25**(2), 273–303 (2019)
6. Boussier, S., Feillet, D., Gendreau, M.: An exact algorithm for team orienteering problems. 4OR Quart. J. Belgian, French Italian Oper. Res. Soc. **5**, 211–230 (2007)
7. Braekers, K., Ramaekers, K., Van Nieuwenhuyse, I.: The vehicle routing problem: state of the art classification and review. Comput. Indus. Eng. **99**, 300–313 (2016)
8. Chaigneau, C., Bostel, N., Grimault, A.: A large neighborhood search-based approach to tackle the very large scale team orienteering problem in industrial context. Comput. Oper. Res. **176**, 106954 (2025)
9. Christiaens, J., Vanden Berghe, G.: Slack induction by string removals for vehicle routing problems. Transp. Sci. **54**(2), 417–433 (2020)
10. Clarke, G., Wright, J.W.: Scheduling of vehicles from a central depot to a number of delivery points. Oper. Res. **12**(4), 568–581 (1964)
11. Cordeau, J.F., Gendreau, M., Laporte, G., Potvin, J.Y., Semet, F.: A guide to vehicle routing heuristics. J. Oper. Res. Soc. **53**(5), 512–522 (2002)
12. Costa, L., Contardo, C., Desaulniers, G.: Exact branch-price-and-cut algorithms for vehicle routing. Transp. Sci. **53**(4), 946–985 (2019)
13. Dang, D.C., Guibadj, R.N., Moukrim, A.: An effective PSO-inspired algorithm for the team orienteering problem. Eur. J. Oper. Res. **229**(2), 332–344 (2013)

14. Furnon, V., Perron, L.: OR-Tools routing library. https://developers.google.com/optimization/routing/
15. Gehring, H., Homberger, J.: Parallelization of a two-phase metaheuristic for routing problems with time windows. J. Heuristics **8**, 251–276 (2002)
16. Gunawan, A., Lau, H., Vansteenwegen, P., Lu, K.: Well-tuned algorithms for the team orienteering problem with time windows. J. Oper. Res. Soc. **68**, 861–876 (2017)
17. Gunawan, A., Lau, H.C., Vansteenwegen, P.: Orienteering problem: a survey of recent variants, solution approaches and applications. Eur. J. Oper. Res. **255**(2), 315–332 (2016)
18. Gunawan, A., Zhu, J., Ng, K.M.: The capacitated team orienteering problem: a hybrid simulated annealing and iterated local search approach. In: Proceedings of the 13th International Conference on the Practice and Theory of Automated Timetabling – PATAT 2021: Volume II, pp. 299–303 (2021)
19. Hammami, F.: An efficient hybrid adaptive large neighborhood search method for the capacitated team orienteering problem. Expert Syst. Appl. **249**, 123561 (2024)
20. Hammami, F., Rekik, M., Coelho, L.C.: A hybrid adaptive large neighborhood search heuristic for the team orienteering problem. Comput. Oper. Res. **123**, 105034 (2020)
21. Helsgaun, K.: An extension of the Lin-Kernighan-Helsgaun TSP solver for constrained traveling salesman and vehicle routing problems: Technical report. Technical report, Roskilde University, Denmark (2017). http://webhotel4.ruc.dk/~keld/research/LKH-3/.
22. Karabulut, K., Tasgetiren, M.F.: An evolution strategy approach to the team orienteering problem with time windows. Comput. Indus. Eng. **139**, 106–109 (2020)
23. Ke, L., Zhai, L., Li, J., Chan, F.T.: Pareto mimic algorithm: an approach to the team orienteering problem. Omega **61**, 155–166 (2016)
24. Kirkpatrick, S., Gelatt, C.D., Vecchi, M.P.: Optimization by simulated annealing. Science **220**(4598), 671–680 (1983)
25. Laporte, G.: Fifty years of vehicle routing. Transp. Sci. **43**, 408–416 (2009)
26. Laporte, G., Martello, S.: The selective travelling salesman problem. Discret. Appl. Math. **26**(2), 193–207 (1990)
27. Lenstra, J.K., Kan, A.H.G.R.: Complexity of vehicle routing and scheduling problems. Networks **11**(2), 221–227 (1981)
28. Li, H., Lim, A.: A metaheuristic for the pickup and delivery problem with time windows. In: Proceedings 13th IEEE International Conference on Tools with Artificial Intelligence. ICTAI 2001, pp. 160–167 (2001)
29. Li, J., Zhu, J., Peng, G., Wang, J., Zhen, L., Demeulemeester, E.: Branch-price-and-cut algorithms for the team orienteering problem with interval-varying profits. Eur. J. Oper. Res. **319**(3), 793–807 (2024)
30. Lin, S.: Computer solutions of the traveling salesman problem. Bell Syst. Tech. J. **44**(10), 2245–2269 (1965)
31. Lin, S.W.: Solving the team orienteering problem using effective multi-start simulated annealing. Appl. Soft Comput. **13**(2), 1064–1073 (2013)
32. Lin, S.W., Yu, V.F.: A simulated annealing heuristic for the team orienteering problem with time windows. Eur. J. Oper. Res. **217**(1), 94–107 (2012)
33. Lourenço, H.R., Martin, O.C., Stützle, T.: Iterated local search: framework and applications. In: Gendreau, M., Potvin, J.Y. (eds.) Handbook of Metaheuristics, pp. 363–397. Springer, US (2010)
34. Montemanni, R., Gambardella, L.M.: Ant colony system for team orienteering problems with time windows. Found. Comput. Decision Sci. **34**, 287–306 (2009)

35. Keshtkaran, M., Ziarati, K., Vigo, A.B., D.: Enhanced exact solution methods for the team orienteering problem. Int. J. Prod. Res. **54**(2), 591–601 (2016)
36. Máximo, V.R., Courdeau, J.F., Nascimento, M.C.V.: AILS-II: An adaptive iterated local search heuristic for the large-scale capacitated routing problem. INFORMS J. Comput. **36**(4), 939–1146 (2024)
37. Nalepa, J., Blocho, M.: Adaptive memetic algorithm for minimizing distance in the vehicle routing problem with time windows. Soft. Comput. **20**, 2309–2327 (2015)
38. Pajerský, M., Sobotka, V., Rudová, H.: Appendix for Open-source implementation of slack induction by string removals for routing and orienteering problems. In: 23rd International Conference on the Integration of Constraint Programming, Artificial Intelligence, and Operations Research (CPAIOR) (2026). available from https:// is.muni.cz/go/cpaior-26-appendix
39. Pajerský, M., Sobotka, V., Rudová, H.: Open-source implementation of slack induction by string removals for vehicle routing and team orienteering problems (2026). Available from https://github.com/hankarudova/open-source-sisr-routing
40. Parragh, S.N., Doerner, K.F., Hartl, R.F.: A survey on pickup and delivery problems: Part I: transportation between customers and depot. J. für Betriebswirtschaft **58**, 21–51 (2008)
41. Parragh, S.N., Doerner, K.F., Hartl, R.F.: A survey on pickup and delivery problems: Part II: transportation between pickup and delivery locations. J. für Betriebswirtschaft **58**, 81–117 (2008)
42. Pessoa, A., Sadykov, R., Uchoa, E., et al.: A generic exact solver for vehicle routing and related problems. Math. Program. **183**, 483–523 (2020)
43. Pisinger, D., Ropke, S.: A general heuristic for vehicle routing problems. Comput. Oper. Res. **34**(8), 2403–2435 (2007)
44. Puljić, K., Manger, R.: Comparison of eight evolutionary crossover operators for the vehicle routing problem. Math. Commun. **18**, 359–375 (2013)
45. Reyes-Rubiano, L., Juan, A., Bayliss, C., Panadero, J., Faulin, J., Copado, P.: A biased-randomized learnheuristic for solving the team orienteering problem with dynamic rewards. Trans. Res. Proc. **47**, 680–687 (2020)
46. Righini, G., Salani, M.: Decremental state space relaxation strategies and initialization heuristics for solving the orienteering problem with time windows with dynamic programming. Comput. Oper. Res. **36**(4), 1191–1203 (2009)
47. Ropke, S., Pisinger, D.: An adaptive large neighborhood search heuristic for the pickup and delivery problem with time windows. Transp. Sci. **40**(4), 455–472 (2006)
48. Santini, A., Schneider, M., Vidal, T., Vigo, D.: Decomposition strategies for vehicle routing heuristics. INFORMS J. Comput. **35**(3), 543–559 (2023)
49. Savelsbergh, M.: The vehicle routing problem with time windows: minimizing route duration. INFORMS J. Comput. **4**, 146–154 (1992)
50. Shaw, P.: A new local search algorithm providing high quality solutions to vehicle routing problems. APES Group, Dept of Computer Science, University of Strathclyde, Glasgow, Scotland, UK 46 (1997)
51. Shaw, P.: Using constraint programming and local search methods to solve vehicle routing problems. In: Principles and Practice of Constraint Programming – CP 1998, pp. 417–431. LNCS Springer (1998)
52. Solomon, M.M.: Algorithms for the vehicle routing and scheduling problems with time window constraints. Oper. Res. Int. J. **35**, 254–265 (1987)
53. Tarantilis, C., Stavropoulou, F., Repoussis, P.: The capacitated team orienteering problem: a bi-level filter-and-fan method. Eur. J. Oper. Res. **224**(1), 65–78 (2013)
54. Toth, P., Vigo, D.: Vehicle Routing: Problems, Methods, and Applications, 2nd edn. Society for Industrial and Applied Mathematics, USA (2014)

55. Uchoa, E., Pecin, D., Pessoa, A., Poggi, M., Vidal, T., Subramanian, A.: New benchmark instances for the capacitated vehicle routing problem. Eur. J. Oper. Res. **257**(3), 845–858 (2017)
56. Van Breedam, A.: Improvement heuristics for the vehicle routing problem based on simulated annealing. Eur. J. Oper. Res. **86**(3), 480–490 (1995)
57. Vidal, T.: Hybrid genetic search for the CVRP: Open-source implementation and SWAP* neighborhood. Comput. Oper. Res. **140**, 105643 (2022)
58. Vidal, T., Crainic, T.G., Gendreau, M., Lahrichi, N., Rei, W.: A hybrid genetic algorithm for multidepot and periodic vehicle routing problems. Oper. Res. **60**, 611–624 (2012)
59. Vidal, T., Crainic, T.G., Gendreau, M., Prins, C.: A unified solution framework for multi-attribute vehicle routing problems. Eur. J. Oper. Res. **234**(3), 658–673 (2014)
60. Vidal, T., Laporte, G., Matl, P.: A concise guide to existing and emerging vehicle routing problem variants. Eur. J. Oper. Res. **286**(2), 401–416 (2020)
61. Vidal, T., Maculan, N., Ochi, L.S., Penna, P.H.V.: Large neighborhoods with implicit customer selection for vehicle routing problems with profits. Transp. Sci. **50**(2), 720–734 (2016)
62. Wouda, N.A., Lan, L., Kool, W.: PyVRP: a high-performance VRP solver package. INFORMS J. Comput. **36**(4), 943–955 (2024)
63. You, Z., Yang, Y., Wang, X., Yin, W.: Two-stage learning to branch in branch-price-and-cut algorithms for solving vehicle routing problems exactly (2023). available at SSRN: https://doi.org/10.2139/ssrn.4630549

Heuristic Multiobjective Discrete Optimization Using Restricted Decision Diagrams

Rahul Patel[1]([✉]) [iD], Elias B. Khalil[1] [iD], and David Bergman[2] [iD]

[1] University of Toronto, Toronto, ON, Canada
`rm.patel@mail.utoronto.ca`, `elias.khalil@utoronto.ca`
[2] University of Connecticut, Mansfield, USA
`david.bergman@uconn.edu`

Abstract. Decision diagrams (DDs) have emerged as a state-of-the-art method for exact multiobjective integer linear programming. When the DD is too large to fit into memory or the decision-maker prefers a fast approximation to the Pareto frontier, the complete DD must be restricted to a subset of its states (or nodes). We introduce new node-selection heuristics for constructing restricted DDs that produce a high-quality approximation of the Pareto frontier. Depending on the structure of the problem, our heuristics are based on either simple rules, machine learning with feature engineering, or end-to-end deep learning. Experiments on multiobjective knapsack, set packing, and traveling salesperson problems show that our approach is highly effective, recovering over 85% of the Pareto frontier while achieving 2.5× speedups compared to exact DD on average, with very few non-Pareto solutions. The code is available at https://github.com/rahulptel/HMORDD.

Keywords: Multiobjective Optimization · Decision Diagrams · Machine Learning

1 Introduction

Real-world decision-making rarely involves a single criterion. Instead, one must frequently trade off conflicting goals, such as maximizing profit while minimizing risk, or maximizing service quality while minimizing cost. These scenarios are modeled with *multiobjective optimization* (MOO). Unlike single-objective optimization, where the goal is to find a single optimal solution, the goal in MOO is to identify the *Pareto frontier*—the set of feasible solutions for which no objective can be improved without worsening another. In this work, we focus on *multiobjective integer linear programming* (MOILP), i.e., MOO problems with integer variables and linear constraints and objectives. Just as the framework of integer linear programming is widely used to model single-objective decision-making tasks, its multiobjective counterpart has found various applications in multicriteria decision-making [17,18].

In recent years, *decision diagrams* (DDs) have emerged as a powerful alternative to traditional enumerative methods. Initially introduced for representing

T. Guns (Ed.): CPAIOR 2026, LNCS 16595, pp. 378–403, 2026.
https://doi.org/10.1007/978-3-032-27242-3_23

switching circuits [25] and later popularized for formal verification [12], DDs were adapted for discrete optimization problems by compactly encoding their feasible set [6]. A DD is a directed acyclic graph where paths from the root to the terminal node correspond to feasible solutions. Building on this success, recent works have extended DDs to the multiobjective setting [3, 4] and achieved state-of-the-art results for problems amenable to dynamic programming formulations. By representing the solution space compactly, exact DDs often outperform traditional algorithms for problems with 2–10 objectives and hundreds of variables.

The major bottleneck for exact methods such as DDs is that the size of the Pareto frontier can grow exponentially with instance size, making exact enumeration computationally prohibitive and often overwhelming for the decision-maker. In many practical scenarios, a user requires that a high-quality approximation of the frontier be delivered quickly, rather than a delayed exact set. To address this issue, several heuristics have been proposed in the literature [2, 16, 30, 31, 36, 39].

Within the context of DDs, one can *restrict* the diagram to obtain an approximate Pareto frontier. A restricted DD limits the width of the graph (the number of nodes per layer), thereby discarding some feasible solutions to maintain a manageable size. While restricted DDs are widely used to obtain bounds [5, 7, 13, 27] in single-objective problems, their application to multiobjective optimization remains largely unexplored. To the best of our knowledge, this work is the first to address the challenge of constructing a restricted DD that quickly recovers a large fraction of the true Pareto frontier, thereby extending the use of DDs to heuristic MOO.

Illustrative Example. Consider the knapsack problem in Example 1 and the corresponding exact DD in Fig. 1.

Example 1 (A bi-objective knapsack problem).

$$\min_{x \in \{0,1\}^3} \; (x_1 + 10x_2 + 3x_3, \; 2x_1 + 3x_2 + x_3) \text{ such that } 3x_1 + x_2 + 2x_3 \leq 5.$$

The state in this case, represented by a square node, is equal to the weight of the items selected in the knapsack. We start with an empty knapsack, represented by the root node **r** with weight 0. From there on, in each layer we decide whether to select an item or not. For example, selecting (not selecting) item x_1 is represented by a solid (dashed) arc from the root node, resulting in a state with a value 3 (or 0). Note that only the red nodes are used by the two Pareto solutions. A restricted DD comprising only of the red nodes preserves the Pareto frontier. Additionally, the red nodes are a small fraction of the total nodes in the DD. Hence, if one can construct a restricted DD with only the red nodes, the Pareto frontier can be recovered quickly.

Empirical Justification. This phenomenon is not accidental: in random instances of the multiobjective knapsack problem (MOKP), multiobjective set packing problem (MOSPP), and multiobjective traveling salesperson problem (MOTSP), only a very small fraction of DD nodes (between 1% and 12%) lead to Pareto-optimal solutions, as detailed in Appendix A. We refer to these nodes as *Pareto*

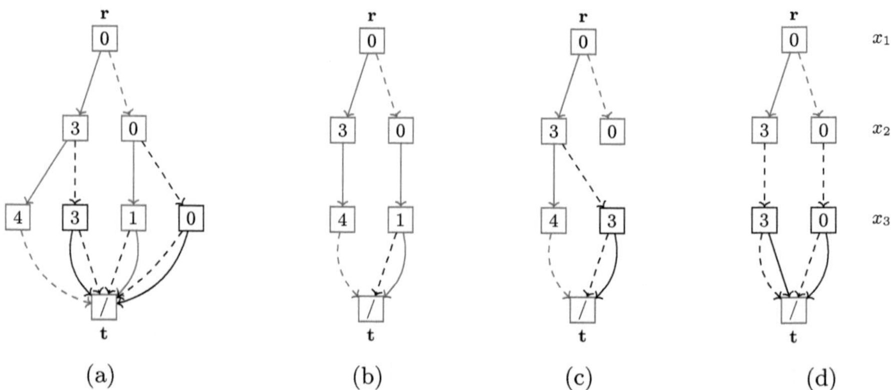

Fig. 1. The exact and restricted DDs for Example 1. The paths corresponding to the Pareto-optimal solutions $(1, 1, 0)$ and $(0, 1, 1)$, along with their associated Pareto nodes, are highlighted in red. (a) Exact DD. (b) Restricted DD with complete recovery of the Pareto frontier. (c) Restricted DD with partial recovery of the Pareto frontier. (d) Restricted DD with no recovery of the Pareto frontier. (Color figure online)

nodes. If one could identify these nodes in advance and restrict the DD to them, the enumeration of the Pareto frontier would be substantially faster, as it dramatically reduces the number of nodes that must be explored.

Contributions. Building on these conceptual and empirical insights, we tackle the challenge of rapidly generating a high-quality approximation of the Pareto frontier by constructing restricted DDs that retain most Pareto nodes. We develop two families of node-selection heuristics (NOSHs) to guide the restriction process: rule-based and learning-based. Figure 2 illustrates our methods.

Rule-based heuristics do not require prior knowledge about the distribution of problem instances and operate on each instance independently. For example, in the MOKP, the DD state is equal to the knapsack weight. A simple rule is to sort the nodes in each layer by decreasing weight and select only the top-ranked nodes up to the width limit.

In contrast, learning-based heuristics assume access to a sufficiently large collection of training instances, along with their Pareto frontiers, from which a binary classifier can be learned. Nodes that are predicted to be Pareto are then prioritized over the ones that are not. Once trained, the classifier is invoked on each node of the DDs of previously unseen instances that are similar in structure to those seen in training.

We demonstrate that NOSHs effectively identify Pareto nodes, leading to much faster solution enumeration compared to exact DDs and recovering a large fraction of the true Pareto frontier on three representative multiobjective problems: MOKP, MOSPP, and MOTSP.

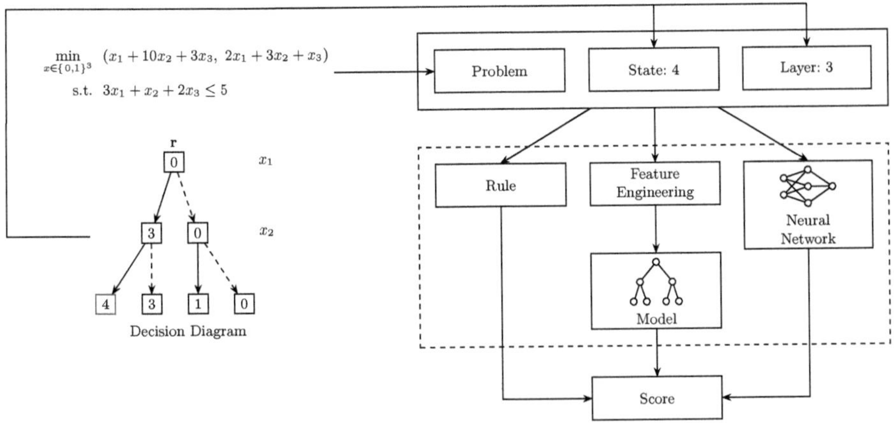

Fig. 2. A schematic illustration of the node-selection pipeline in restricted DD construction with width 2. A problem instance and a DD node—represented by its state and layer—are provided as input to a node-selection heuristic. Heuristics may be rule-based or learning-based; the latter include feature-engineered models and neural networks. The heuristic outputs a score for the node, which guides the construction of the width-restricted DD. The heuristic components depicted in the figure are highlighted with a dashed border.

2   Preliminaries

2.1   Multiobjective Integer Linear Programming

We define an MOILP with K objectives as $\mathcal{P} \equiv \min_x \{Cx : x \in \mathcal{X} \subseteq \mathbb{Z}^N\}$, where x is the decision vector, $\mathcal{X}$ is the feasible set, $C \in \mathbb{R}^{K \times N}$ is the objective coefficient matrix. Let $\mathcal{Z} = \{Cx : x \in \mathcal{X}\}$ denote the set of feasible objective vectors. For two vectors $z^1, z^2 \in \mathbb{R}^K$, we say z^1 *dominates* z^2 (denoted $z^1 \prec z^2$) if $z_k^1 \leq z_k^2$ for all $k \in \{1, \cdots, K\}$ and $z^1 \neq z^2$. A vector $z \in \mathcal{Z}$ is *nondominated* if there exists no $z' \in \mathcal{Z}$ such that $z' \prec z$. Solving the MOILP amounts to finding the *Pareto frontier* $\mathcal{Z}^\star = \{z \in \mathcal{Z} \mid \nexists z' \in \mathcal{Z} \text{ such that } z' \prec z\}$, which is the set of all nondominated vectors. The set of feasible decision vectors that map to the Pareto frontier is defined as the *efficient set* $\mathcal{X}^\star = \{x \in \mathcal{X} \mid Cx \in \mathcal{Z}^\star\}$.

2.2   Decision Diagrams

A DD for problem $\mathcal{P}$ is a layered directed acyclic graph $\mathcal{B} = (\mathcal{N}, \mathcal{A})$ with node set $\mathcal{N}$ and arc set $\mathcal{A}$. The node set is partitioned into $N+1$ layers, $\mathcal{N} = \mathcal{L}_1 \cup \cdots \cup \mathcal{L}_{N+1}$. Layers $\mathcal{L}_1$ and $\mathcal{L}_{N+1}$ contain only the single *root node* **r** and *terminal node* **t**, respectively. The *width* of the DD is the maximum size of any layer, denoted $\omega(\mathcal{B}) = \max_j |\mathcal{L}_j|$. For example, the exact DD in Fig. 1 has 4 layers and a width of 4.

Arcs are directed from layer j to $j+1$ for $j \in \{1, \ldots, N\}$. For each arc $a \in \mathcal{A}$ originating from a node in $\mathcal{L}_j$, we define two attributes. An *assignment value* $d(a) \in \mathbb{Z}$, representing the value assigned to decision variable x_j. An *objective weight* $v(a) \in \mathbb{R}^K$, representing the contribution of this assignment to the objective function. A path $p = (a_1, \ldots, a_N)$ from $\mathbf{r}$ to $\mathbf{t}$ defines a complete solution $x(p) = (d(a_1), \ldots, d(a_N)) \in \mathbb{Z}^N$ with an associated objective vector $v(p) = \sum_{i=1}^N v(a_i)$. For instance, the path $(\mathbf{r} - 3, 3 - 4, 4 - \mathbf{t})$ in the exact DD of Fig. 1 corresponds to the decision vector $(x_1, x_2, x_3) = (1, 1, 0)$ and achieves an objective of $(11, 5)$.

Let $\mathcal{X}_\mathcal{B} = \{x(p) : p \text{ is an } \mathbf{r}\text{-}\mathbf{t} \text{ path in } \mathcal{B}\}$ be the set of solutions encoded by the DD. The DD is *exact* for problem $\mathcal{P}$ if $\mathcal{X}_\mathcal{B} = \mathcal{X}$ and the arc weights satisfy $v(p) = Cx(p)$ for all paths. Under these conditions, the image of $\mathcal{X}_\mathcal{B}$ in objective space is $\mathcal{Z}_\mathcal{B} = \{v(p) : p \text{ is an } \mathbf{r}\text{-}\mathbf{t} \text{ path}\}$. The nondominated vectors in $\mathcal{Z}_\mathcal{B}$ are denoted by $\mathcal{Z}_\mathcal{B}^\star$ and is equal to the Pareto frontier $\mathcal{Z}^\star$ of $\mathcal{P}$.

A decision diagram $\mathcal{B}'$ is said to be *restricted* for problem $\mathcal{P}$ if it encodes a subset of the feasible solutions, i.e., $\mathcal{X}_{\mathcal{B}'} \subseteq \mathcal{X}$. In practice, restricted DDs are often maintained by imposing a *maximum width* W on the graph. During construction, if the number of nodes in a layer $\mathcal{L}_j$ exceeds W, a heuristic filtering procedure is triggered. This procedure retains a subset of W nodes and removes the rest, ensuring that $\omega(\mathcal{B}') \leq W$. Consequently, the image of the feasible set $\mathcal{Z}_{\mathcal{B}'}$ is a subset of the objective space $\mathcal{Z}$. The nondominated vectors in $\mathcal{Z}_{\mathcal{B}'}$ are represented by $\mathcal{Z}_{\mathcal{B}'}^\star$, which is an approximate Pareto frontier of $\mathcal{P}$.

To construct a DD we leverage the dynamic programming formulation of the problem. As a result, each node $u \in \mathcal{N}$ is associated with a *state* $s(u)$ from a state space $\mathcal{S}$ given by the formulation. The state plays a key role in the design of NOSH for constructing restricted DDs. We refer the reader to [3, 4] for additional details on dynamic programming formulation and Pareto frontier enumeration using the multicriteria shortest path algorithm. The notation used to define MOILP and DDs is summarized in Appendix B.

3 Methodology

The central challenge in solving MOILPs via DDs is the exponential growth of the state space $\mathcal{S}$ with problem size. To address this, we propose NOSHs, a framework for constructing restricted DDs that prioritize the retention of Pareto nodes, which typically form a small fraction of the total nodes as highlighted in Appendix A. Given a maximum width W, our goal is to construct a restricted DD $\mathcal{B}'$ such that $\omega(\mathcal{B}') \leq W$ while maximizing the approximation quality of the Pareto frontier.

We formalize this using a generic node scoring framework described in Sect. 3.1, and subsequently detail how this framework specializes into rule-based heuristics, classical machine learning with feature engineering, and end-to-end deep learning.

3.1 Node Selection Heuristics

When the width of a layer $\mathcal{L}_j$ exceeds the maximum width W, a filtering procedure must select a size-W subset of nodes to retain. We unify this selection process under a single scoring entity, denoted as the *scorer* S_Θ. The scorer is a function $S_\Theta : \mathcal{S} \setminus \{\mathbf{s(r)}, \mathbf{s(t)}\} \times \{2, \ldots, N\} \to [0, 1]$ parameterized by Θ. This function maps the state and layer index of a node $u \in \mathcal{N} \setminus \{\mathbf{r}, \mathbf{t}\}$'s, $(s(u), l(u))$ to a score representing the likelihood of u being a Pareto node.

Our objective is to learn (or design) parameters Θ that maximize the expected recovery of the exact Pareto frontier $\mathcal{Z}^\star$. Let $\mathcal{Z}^\star_{\mathcal{B}'(S_\Theta)}$ denote the approximate Pareto frontier obtained from a restricted DD constructed using scorer S_Θ. The optimization objective is:

$$\max_{\Theta} \mathbb{E}_{\mathcal{P}} \left[\frac{|\mathcal{Z}^\star_{\mathcal{B}'(S_\Theta)} \cap \mathcal{Z}^\star|}{|\mathcal{Z}^\star|} \right], \tag{1}$$

where the expectation is taken over a distribution of training instances. An ideal scoring function achieves an objective of 1; scoring functions that eliminate too many Pareto nodes achieve a low score, as many of the returned solutions would be dominated. Based on the structure of S_Θ and the nature of Θ, we categorize NOSHs into three distinct approaches.

Rule-Based Scoring. In this setting, S_Θ is a fixed, deterministic function where $\Theta = \emptyset$ (i.e., there are no learnable parameters). These heuristics rely on domain knowledge and intrinsic properties of the state, as illustrated below for example.

- *Scalar States:* If $s(u) \in \mathbb{R}$, natural candidates for the scorer are the state value itself ($S(u) = s(u)$, denoted `Scal+`) or its negation. For example, the state in MOKP is the total weight of the items that are selected in the state. Since the objectives are in the maximization direction, states that include many items can be thought to be conducive to high-quality solutions, leading to a natural scoring rule.
- *Set-based States:* If $s(u)$ is a set, the scorer may utilize the cardinality of the set ($S(u) = |s(u)|$, denoted `Card+`) or its negation. For example, the state in MOSPP is the set of items that can be added to already-selected items without violating any of the packing constraints. The larger this set, the more items one can potentially add down the line, the larger the objective values.

While computationally inexpensive, these rules are rigid and their performance sensitive to instance structure.

Learning-Based Scoring with Feature Engineering. Here, the scorer is a composite function $S_\Theta(u) = f_\Theta(\psi(s(u), l(u)))$. A fixed extraction function ψ maps the raw state and layer to a hand-crafted feature vector in $\mathbb{R}^d$. These features capture our understanding of the problem. A parameterized binary classification model $f_\Theta : \mathbb{R}^d \to [0, 1]$ (e.g., Logistic Regressor, Random Forest) maps these features to a probability score. In this context, Θ represents the parameters of the classifier f. This approach adapts to data but is limited by the expressiveness of the feature map ψ.

End-to-End Learning-Based Scoring. To overcome the limitations of manual feature engineering and capture the complex dependencies between objectives, constraints, and decision variables, we propose a generic end-to-end deep

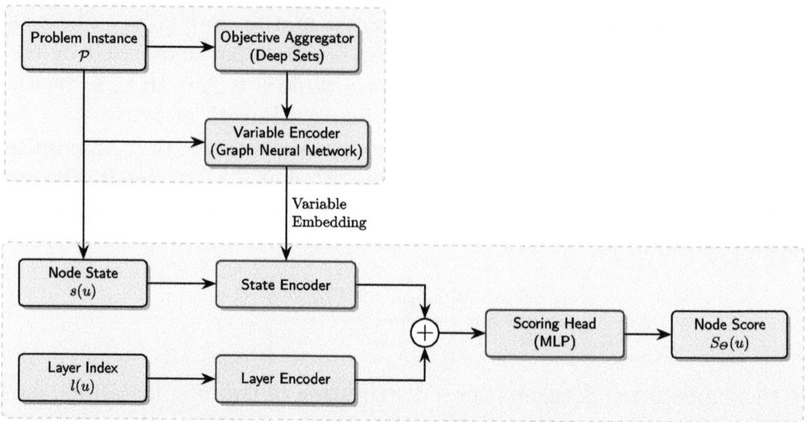

Fig. 3. The generic end-to-end node selection heuristic architecture.

learning architecture. The scorer S_Θ maps the raw state and layer definition $(s(u), l(u))$ directly to a score. As illustrated in Fig. 3, this architecture operates in two distinct phases: a computationally intensive *Instance Embedding Phase* performed once per problem instance, and a lightweight *Scoring Phase* performed for each node in the decision diagram.

The instance embedding phase begins with *Objective Aggregator*. Since the order of objectives is arbitrary, it combines information across objectives into a permutation-invariant representation using Deep Sets [38]. The key result in [38] establishes that any permutation-invariant function f mapping a set O to a representation can be approximated in the form $f(O) = \rho\left(\sum_{x \in O} \gamma(x)\right)$, where γ and ρ are flexible functions (e.g., neural networks).

The *Variable Encoder*, initialized with the aggregated embeddings from the previous step, captures the intricate dependencies among various components of the problem instance. We exploit the fact that the MOILP can be represented as a graph and use a Graph Neural Network (GNN) to process it. This results in *variable embeddings* that encapsulate the instance's structural information. Crucially, the variable embeddings are computed only once per instance, avoiding redundant computations per DD node.

Later, these pre-computed variable embeddings are combined with dynamic state information to obtain a *State Embedding*. Finally, the state embedding is combined with a *Layer Embedding* (derived from the layer index) to obtain a comprehensive *DD Node Embedding*. This embedding is used by the *Scoring Head* to output a scalar score for a DD node.

3.2   Training Formulation

For the learning-based approaches, we employ supervised learning. We generate a node-wise training dataset $\mathcal{D}$ derived from m problem instances. For each instance i, we use the ground-truth efficient set $\mathcal{X}_i^\star$ to label the nodes in the DD $\mathcal{N}_i$. The dataset is defined as the union of these labeled nodes:

$$\mathcal{D} = \{(\mathcal{P}_i, s(u), l(u), y_u) \mid i \in \{1, \ldots, m\}, u \in \mathcal{N}_i\}, \tag{2}$$

where $y_u \in \{0, 1\}$ is the binary label indicating whether u is a Pareto node.

We seek parameters $\Theta^\star$ that minimize the empirical risk over $\mathcal{D}$ using the weighted binary cross-entropy loss:

$$\Theta^\star \in \arg\min_\Theta \frac{1}{|\mathcal{D}|} \sum_{(\cdot, y_u) \in \mathcal{D}} \mathcal{L}_{\mathrm{WBCE}}(S_\Theta(s(u), l(u)), y_u), \tag{3}$$

where $\mathcal{L}_{\mathrm{WBCE}}(p, y) = -\alpha \cdot y \log p - (1 - y) \log(1 - p)$ and $\alpha \in \mathbb{R}_+$. By minimizing this loss, S_Θ learns to assign higher probabilities to Pareto nodes.

3.3 Case Study: Multiobjective Traveling Salesperson Problem

In the dynamic programming formulation for the MOTSP [8], a node u at layer j represents a partial tour of length $j - 1$. The state $s(u)$ is defined as the tuple $s(u) = (V_u, i)$, where $V_u \subseteq V$ is the set of vertices visited on the path from the root to u, and $i \in V_u$ is the index of the last vertex added to the path (i.e., the current city). At the root node $\mathbf{r}$ (layer 1), the state is $(\{1\}, 1)$, assuming the tour always starts at vertex 1. For a node u with state (V_u, i), a transition to vertex j is valid if $j \notin V_u$. The succeeding state is $(V_u \cup \{j\}, j)$. The travel costs associated with different edges are represented using a matrix c, where c_{ij}^k denotes the cost of edge (i, j) for the k-th objective.

Rule-Based Heuristics. Our rule-based heuristics leverage the edge cost matrices c^k to estimate the quality of a state. Due to the conflicting nature of the K objectives, we first compute the *rank* of each edge for every objective. Let $R \in \mathbb{R}^{N \times N \times K}$ be a tensor where R_{ijk} is the rank of cost c_{ij}^k among all edges for objective k. We derive a scalar score for node u by aggregating these ranks. First, we aggregate across objectives to obtain a single "cost" matrix $R^{\mathrm{agg}} \in \mathbb{R}^{N \times N}$. The entry corresponding to edge (i, j) is given by $R_{ij}^{\mathrm{agg}} = \bigoplus_1 \{R_{ijk} : k \in \{1, \ldots, K\}\}$, where $\bigoplus_1$ is an aggregation function (e.g., mean, maximum or minimum).

The score for a node u with state (V_u, i) is determined by the potential extensions to the set of unvisited cities $U_u = V \setminus V_u$. Specifically, the score is computed as $S(u) = \bigoplus_2 \{R_{ij}^{\mathrm{agg}} : j \in U_u\}$, where $\bigoplus_2 \in \{\mathrm{high}, \mathrm{low}\}$ determines whether we prioritize the best or worst immediate extension[1]. We denote these heuristics using the prefix "Ord" followed by the aggregation method and the prioritization direction. For instance, OrdMeanHigh aggregates ranks using the mean ($\bigoplus_1$) and scores the node based on the most favorable (higher rank) extension ($\bigoplus_2$). Conversely, a suffix of "Low" implies prioritizing edges with lower aggregated ranks (worst extensions).

---

[1] For two ranks $r_1, r_2 \in \mathbb{R}_+$, rank r_1 is higher than r_2 if $r_1 < r_2$.

Learning-Based Heuristic. For the learning-based approach, we instantiate the end-to-end architecture proposed in Fig. 3. In the instance embedding phase, the input to the objective aggregator are the node features $h_i^{v,0} \in \mathbb{R}^{N \times d_{node}}$ edge features $h_{ij}^{e,0} \in \mathbb{R}^{N \times N \times K}$. The node features are initialized with the 2D coordinates of the cities and may optionally include hand-crafted features. The edge features are initialized using the cost matrices across the K objectives, where $h_{ijk}^{e,0} = c_{ij}^k$. The objective aggregator transforms these features $h^{v,0}$ and $h^{e,0}$ into objective-invariant embeddings $h^{v,1} \in \mathbb{R}^{N \times d_{\mathrm{emb}}}$ and $h^{e,1} \in \mathbb{R}^{N \times N \times d_{\mathrm{emb}}}$

To capture the complex structural dependencies of the TSP, we utilize an Edge-augmented Graph Transformer (EGT) [20]. The EGT extends the standard Transformer by introducing *edge channels* that evolve structurally aware edge embeddings alongside node embeddings. Let $h^{v,\ell-1} \in \mathbb{R}^{N \times d_{\mathrm{emb}}}$ and $h^{e,\ell-1} \in \mathbb{R}^{N \times N \times d_{\mathrm{emb}}}$ denote the node and edge embeddings input to layer ℓ. We first linearly transform these embeddings using learnable weight matrices. For a single attention head with dimension d_k, we compute the query $Q^\ell = h^{v,\ell-1}W_Q^\ell$, key $K^\ell = h^{v,\ell-1}W_K^\ell$, value $V^\ell = h^{v,\ell-1}W_V^\ell$, structural bias $E^\ell = h^{e,\ell-1}W_E^\ell$, and gate $G^\ell = h^{e,\ell-1}W_G^\ell$ matrices, where $W_Q^\ell, W_K^\ell, W_V^\ell \in \mathbb{R}^{d_{\mathrm{emb}} \times d_k}$ are the node projection matrices and $W_E^\ell, W_G^\ell \in \mathbb{R}^{d_{\mathrm{emb}} \times 1}$ are the edge projection matrices. Consequently, the resulting matrices have dimensions $Q^\ell, K^\ell, V^\ell \in \mathbb{R}^{N \times d_k}$ and $E^\ell, G^\ell \in \mathbb{R}^{N \times N}$. The edge-augmented attention $A^\ell \in \mathbb{R}^{N \times N}$ is computed by modulating the standard dot-product similarity with the edge-derived bias and gating terms as follows:

$$A^\ell = \mathrm{softmax}\left(\hat{H}^\ell\right) \odot \sigma(G^\ell), \tag{4}$$

$$\hat{H}^\ell = \mathrm{clip}\left(\frac{Q^\ell(K^\ell)^T}{\sqrt{d_k}}\right) + E^\ell, \tag{5}$$

where $\sigma(\cdot)$ is the sigmoid function and $\odot$ denotes element-wise multiplication. Here, E^ℓ acts as a structural bias added to the node affinities, while G^ℓ gates the attention values, controlling the information flow based on edge attributes.

The node embeddings are updated by aggregating V^ℓ weighted by A^ℓ. In practice, this operation is performed by multiple heads in parallel, and their outputs are combined and projected to form the input for the next layer. We stack L such layers to obtain the final instance-specific embeddings $h^{v,L}$ which acts as the variable embedding $h_{emb}^v \in \mathbb{R}^{N \times d_{emb}}$.

The primary challenge in the node scoring phase is to efficiently map the dynamic DD state $s(u) = (V_u, i)$ to a vector representation using the precomputed embeddings h_{emb}^v. We represent the set of visited nodes V_u by modifying the variable embeddings. We define a state mask $M \in \mathbb{R}^{N \times d_{\mathrm{emb}}}$ with learnable parameters θ_{vis} and θ_{unvis}. The mask for vertex j is set to θ_{vis} if $j \in V_u$ and θ_{unvis} otherwise. This mask is added element-wise to the variable embeddings: $h_{\mathrm{mask}}^v = h_{\mathrm{emb}}^v + M$. The resulting set of embeddings is then aggregated into a single vector $h_{\mathrm{vis}} \in \mathbb{R}^{d_{\mathrm{emb}}}$ using a Deep Sets-based network [38], capturing the composition of the partial tour regardless of visit order. To explicitly capture the current position in the graph, we extract the embedding corresponding to the last visited vertex i: $h_{\text{last}} = h^v_{\text{emb}}[i]$.

Finally, the current layer index $l(u)$ is projected to an embedding $h_{\text{layer}} \in \mathbb{R}^{d_{\text{emb}}}$ via a *Layer Encoder*, a multi-layered perceptron (MLP). The DD node representation is the sum of these embeddings $h_{\text{node}} = h_{\text{vis}} + h_{\text{last}} + h_{\text{layer}}$. This vector is passed to the scoring head (an MLP) to predict the likelihood of node u leading to a Pareto-optimal solution.

4   Computational Setup

We test the proposed approach on MOSPP, MOKP and MOTSP. We use a computing cluster with Intel Xeon CPU E5-2683 CPUs, a memory limit of 16GB, and a time limit of 1800 s. The DD manipulation code responsible for generating exact, restricted, or reduced DDs is based on [3]. We create a Python binding for this C++ code using pybind11. The problem definition and instance-generation scheme are detailed in Appendix C.

For the learning-based NOSH, each problem class and instance size has 1000 training, 100 validation, and 100 testing instances. We compute the Pareto frontier using the exact DD and extract all Pareto nodes. As shown in Appendix A, Pareto nodes constitute only a small fraction of total nodes, leading to a significant class imbalance. To address this during dataset construction, we apply undersampling to the negative class: for each Pareto (positive) node, we randomly sample one non-Pareto (negative) node, resulting in a 1:1 ratio between positive and negative samples. This not only creates a balanced dataset but only also limits its size, making the training tractable.

Node Selection Heuristics and Baselines. As highlighted in Sect. 3, NOSHs can be categorized into three distinct types. We refer to the rule-based heuristic as `NOSH-Rule`, the learning-based approach that relies on feature engineering as `NOSH-ML-FE` and the approach that learns the scoring function end-to-end using deep learning as `NOSH-ML-E2E`. The `NOSH-ML-E2E` is the most extensible method to other problem classes among the proposed NOSHs. By modeling the MOILP as a graph, modern deep learning architectures can be leveraged to learn rich representations for scoring DD nodes. However, this expressivity incurs the computational cost of training deep learning models with multiple hyperparameters. Consequently, we adopt a complexity-aware strategy: we prioritize the lightweight `NOSH-Rule`, employing learning-based variants only when the problem complexity necessitates it. The state definition along with the implementation details of various NOSHs is provided in Appendix D. The approach for selecting the width for restricted DDs is given in Appendix E.1.

The learning-based NOSH approach for the MOKP leverages XGBoost 2.0.1 [14], a highly efficient implementation of gradient-boosted decision trees. The primary motivation for choosing XGBoost lies in its ability to deliver strong performance with minimal tuning when meaningful, hand-crafted node features are available. Compared to neural network-based methods, this allows for more interpretable and computationally efficient models. We perform a grid search to select the best-performing model based on validation accuracy, which is then used for evaluation on the test set. The selected models achieve classification accuracies between 85% and 88% using a threshold of $\tau = 0.5$ on the predicted scores. The corresponding mean absolute errors fall within the range of 0.18 to 0.23. In contrast, designing hand-crafted features for the MOTSP is considerably more challenging. Therefore, to develop an end-to-end node scoring model for this setting, we employ a graph transformer implemented using PyTorch [32]. Details of the hyperparameters used for the node scorers are provided in Appendix E.2.

Our primary baseline is the state-of-the-art exact DD method (referenced as Exact) based on coupled enumeration [3]. We also benchmarked against NSGA-II. Despite its widespread use, NSGA-II failed to produce good approximations of the Pareto frontier, even with increased runtime. When the population size was scaled to match that of the Exact method, it struggled to find feasible solutions. With smaller, more typical population sizes, the resulting frontier was consistently of lower quality than those obtained by NOSH-based approaches. Consequently, we focus our analysis on the DD-based comparisons. Detailed experimental settings and the supplementary NSGA-II analysis are available in Appendix F.

Metrics. We report the following metrics to evaluate and compare the performance of different methods.

1. **Width:** The width of the constructed DDs. Ideally, smaller restricted DDs are preferred, provided they still capture most of the Pareto frontier.
2. **Time:** The total time (in seconds) required to construct the DD and enumerate the Pareto frontier. Faster methods are desirable, especially when approximating the frontier for large or complex instances. Note that the time required to compile the DD is a small fraction of the total time, even for the learning-based NOSH. Therefore, we report only the total time for constructing the DD and enumerating the Pareto frontier.
3. **Cardinality:** The fraction of true Pareto-optimal solutions captured by a method, relative to the total number of Pareto-optimal solutions. Let $\hat{\mathcal{Z}}^\star$ denote the set of solutions obtained by a given method. Then, cardinality is computed as $(|\hat{\mathcal{Z}}^\star \cap \mathcal{Z}^\star|/|\mathcal{Z}^\star|) \times 100$. A higher cardinality indicates that a larger portion of the true Pareto frontier is recovered.
4. **Precision:** The fraction of solutions identified by the method that are truly Pareto-optimal. This is calculated as $(|\hat{\mathcal{Z}}^\star \cap \mathcal{Z}^\star|/|\hat{\mathcal{Z}}^\star|) \times 100$. A higher precision suggests that the approximate Pareto frontier consists predominantly of true Pareto-optimal solutions.
5. **Inverted Generational Distance (IGD)** [15]: The average distance from each solution in $\mathcal{Z}^\star$ to its closest solution in $\hat{\mathcal{Z}}^\star$. Lower IGD values indicate a closer and more comprehensive approximation of the true Pareto frontier. To ensure scale invariance across objectives, we normalize each objective using the minimum and maximum values observed in $\mathcal{Z}^\star$ before computing IGD.

Table 1. MOSPP results averaged over "Inst." test instances. Each column corresponds to a specific instance size (N, K), with rows giving the metrics for the Exact and `NOSH-Rule` methods. Refer to Sect. 4 for column description.

Metric	Method	$N = 100$					$N = 150$				
		$K=3$	$K=4$	$K=5$	$K=6$	$K=7$	$K=3$	$K=4$	$K=5$	$K=6$	$K=7$
Width	Exact	5,766	6,034	5,936	5,976	5,707	471,602	518,556	464,330	468,787	590,908
	NOSH-Rule	50	50	50	50	50	5,000	5,000	5,000	5,000	5,000
Time ↓	Exact	1	1	2	5	31	11	51	261	567	783
	NOSH-Rule	1	1	1	2	13	1	7	77	183	314
Cardinality ↑	Exact	100	100	100	100	100	100	100	100	100	100
	NOSH-Rule	85	84	89	87	87	99	99	99	99	99
Precision ↑	Exact	100	100	100	100	100	100	100	100	100	100
	NOSH-Rule	89	90	94	94	93	100	100	99	100	100
IGD ↓	Exact	0.000	0.000	0.000	0.000	0.000	0.000	0.000	0.000	0.000	0.000
	NOSH-Rule	0.012	0.016	0.013	0.017	0.019	0.000	0.000	0.001	0.001	0.001
$\|\hat{\mathcal{Z}}^\star\|$	Exact	238	1,117	4,765	9,117	25,457	787	6,099	29,061	59,951	103,489
	NOSH-Rule	226	1,051	4,591	8,550	24,534	786	6,089	28,953	59,724	103,275
Inst.		100	100	100	100	100	100	100	92	54	27

Table 2. MOKP results averaged across 100 test instances. Refer to Sect. 4 for column description.

N	K	Method	Width	Time ↓	Cardinality ↑	Precision ↑	IGD ↓	$\|\hat{\mathcal{Z}}^\star\|$
40	7	Exact	9,709	84	100	100	0.000	25,098
		NOSH-Rule	2,000	**2**	19	64	0.128	4,131
			3,000	19	61	89	0.047	12,972
		NOSH-ML-FE	2,000	16	60	74	0.042	20,482
			3,000	36	**88**	**96**	**0.012**	22,267
50	4	Exact	12,359	7	100	100	0.000	3,564
		NOSH-Rule	2,500	**1**	17	36	0.085	1,132
			3,500	2	52	70	0.032	2,256
		NOSH-ML-FE	2,500	3	61	70	0.020	3,062
			3,500	4	**88**	**92**	**0.006**	3,367
80	3	Exact	20,097	27	100	100	0.000	2,442
		NOSH-Rule	4,000	**3**	10	19	0.064	1,015
			6,000	7	62	71	0.013	1,954
		NOSH-ML-FE	4,000	6	46	53	0.012	2,039
			6,000	12	**93**	**95**	**0.002**	2,363

5   Results

Evidently, some restricted DDs are better than others in that they approximate the true Pareto frontier more completely. We will show how NOSHs finds such "accurate" restricted DDs for MOSPP, MOKP and MOTSP in Sects. 5.1, 5.2 and 5.3, respectively. The metrics Width, Cardinality, Precision and $\|\hat{\mathcal{Z}}^\star\|$ are rounded to the nearest integer, whereas Time is rounded up. In all tables, the

Exact method is shown in gray as a reference baseline and is not included in the comparative evaluation, as it represents the ground-truth Pareto frontier.

5.1 Multiobjective Set Packing Problem

Table 1 reports the performance of Exact and NOSH-Rule for the MOSPP. The method NOSH-Rule is based on the Card+ heuristic detailed in Sect. 3.1. The NOSH-Rule has Cardinality and Precision in the range of $\sim 85\%$ to 99% and $\sim 90\%$ to 99%, respectively. NOSH-Rule achieves up to $11\times$ speedup over Exact, with most instances exhibiting $2\times$–$7\times$ improvements.

For larger instances ($N = 150$ and $K \geq 5$), the reported metrics are averaged over fewer than 100 test instances. This is because the Exact method often exceeds the memory limit as instance size increases, making it unable to compute the Pareto frontier for all cases. Consequently, we apply NOSH-Rule only to those instances where Exact completes within the time and memory constraints, which explains the value of "Inst." being less than 100. Note that applying learning-based NOSH would be a challenge in this setting as it would depend on labeled training data, which limits its applicability to larger instances as we do not have access to the exact Pareto frontier. In summary, NOSH-Rule achieved a good trade-off between solution quality and enumeration time, without the need for expensive data labeling.

Table 3. MOTSP results averaged across 100 test instances. Refer to Sect. 4 for column description.

N	K	Method	Width	Time ↓	Cardinality ↑	Precision ↑	IGD ↓	$|\hat{\mathcal{Z}}^{\star}|$
15	3	Exact	24,024	3	100	100	0.000	868
		NOSH-Rule	4,804	**2**	1	2	0.085	439
		NOSH-ML-E2E	4,804	**2**	**91**	**95**	**0.003**	832
	4	Exact	24,024	28	100	100	0.000	9,210
		NOSH-Rule	4,804	**3**	1	4	0.082	3,254
		NOSH-ML-E2E	4,804	13	**89**	**95**	**0.004**	8,546

5.2 Multiobjective Knapsack Problem

The performance of different approaches for the MOKP is presented in Table 2. The NOSH-Rule method, derived from the Scal+ heuristic (Sect. 3.1), is sensitive to the width of the restricted DD, with Cardinality and Precision ranging from 52%–62% and 69%–89%, respectively. In contrast, NOSH-ML-FE consistently outperforms NOSH-Rule, achieving Cardinality between 87%–92% and Precision between 92%–95%, along with lower IGD values. Moreover, NOSH-ML-FE attains speedups of $1.75\times$–$2.33\times$ over Exact in most settings. Overall, these results highlight the advantage of NOSH-ML-FE in achieving superior trade-offs between solution quality and runtime. Additionally, the use of XGBoost with handcrafted features enhances interpretability by enabling identification of the most influential features.

5.3   Multiobjective Traveling Salesperson Problem

The results for MOTSP are presented in Table 3. The NOSH-Rule method, based on the OrdMeanLow heuristic, achieves significant speedups but performs poorly in terms of solution quality, with Cardinality and Precision close to zero and relatively high IGD values, indicating a weak approximation of the Pareto frontier. In contrast, NOSH-ML-E2E substantially improves solution quality, achieving Cardinality between 89%–91% and Precision around 95%, along with very low IGD values. It also attains speedups of $1.5\times$–$2.15\times$ compared to the Exact method. These results demonstrate that NOSH-ML-E2E is effective in learning useful node representations from raw data and accurately identifying Pareto-optimal nodes.

6   Related Work

As surveyed in [17,18], the literature on exact approaches to MOILPs is vast and generally partitioned into decision space methods [1] and objective space methods [10,11]. The former searches in the space of feasible solutions whereas the latter search in the space of objective vectors. Many of these approaches are confined to biobjective and triobjective problems. Notable exceptions include the KSA algorithm [23] and the DD approach [4]. The latter has been shown to outperform KSA by a substantial margin on combinatorial problems that are amenable to a dynamic programming formulation, which is why we focus on this promising algorithmic paradigm in this work. Specifically, our approach allows efficient state-space exploration, similar to [24], by eliminating states less likely to contribute to a Pareto optimal solution. Note that the authors of [4] enhance a basic decision diagram approach in [3] by introducing a series of Pareto frontier preserving operations. Those would directly apply here, but we utilize the basic decision diagram approach for transparency and ease of implementation.

Evolutionary or genetic algorithms (GAs) have long been used for multiobjective optimization. The Pymoo paper and software package [9] summarize and implement state-of-the-art GAs such as NSGA-II. Note that the NSGA-II [16] is widely used (45,000+ citations) so outperforming it is a good sanity check of the promise of any new method. However, it is known that GAs typically struggle with integer variables and hard constraints, a limitation that is not exhibited by the DD approach. Another class of heuristics based on integer and linear programming appear in [2,30,31]. They extended the single-objective "Feasibility Pump" [19] heuristic to the multiobjective case. The method in [2] is evaluated only on triobjective problems with binary variables whereas NOSH will be applied to problems with more objective and discrete variables, a substantial generalization. With the expansive literature on heuristic approaches to MOILP, we opted to compare only with NSGA-II as it is the best known and has been a strong competitor and benchmark comparison algorithm for over two decades.

Relative to single-objective optimization, there has been much less work on ML for multiobjective discrete optimization. ML-based methods have been proposed for unconstrained continuous multiobjective problems that arise in deep learning applications such as multi-task learning [28] or molecule generation [21,26]. Because they are not equipped to deal with hard constraints, these methods do not apply to combinatorial optimization. More relevant to this paper is the work of [37] who train a *graph neural network* (GNN) to guide the exact algorithm of [35]. This GNN-based method is evaluated on knapsack problems with 3–5 objectives only and requires a much larger amount of time than NOSH. For example, on instances with 4 objectives and 50 variables, NOSH runs in about 11 s on average (Table 2) whereas the method in [37] (Table 1) runs for 1,000 s on a faster CPU than ours. Additionally, no publicly available code was provided for this rather sophisticated GNN architecture, making a direct comparison challenging.

7 Conclusion and Future Work

We demonstrated the use of restricted DDs as a heuristic to solve multiobjective integer linear programming problems. In fact, to the best of our knowledge, we are the first to invoke restricted DDs in the context of multiobjective optimization as they have only appeared in the single objective case. We presented two types of node selection heuristics, rule-based and learning-based, to construct the restricted DDs for the MOKP, MOSPP and MOTSP. The results demonstrate that node selection heuristics provide a high-quality approximation of the true Pareto frontier and are significantly faster than the exact DDs. Specifically, NOSH-Rule, NOSH-ML-FE and NOSH-ML-E2E performed exceedingly well for the MOSPP, MOKP and MOTSP, respectively.

Furthermore, instead of classifying a node in isolation, one can formulate the problem of constructing the restricted DDs as a structured output prediction task. Specifically, given an exact DD the goal is to predict a subgraph consisting only of nodes and arcs used by the Pareto optimal solutions. Predicting a subgraph is a combinatorial task which has received significant attention from the machine learning community [22] but has not been applied to optimization applications such as ours here. One limitation of our approach is its reliance on the complete Pareto frontier to generate the training dataset for the learning-based node selection heuristics. In the future, we aim to relax this requirement by utilizing partial Pareto frontiers for training data generation. This relaxation may necessitate an increase in the number of training instances to maintain performance.

A Fraction of Pareto Nodes in DDs

Table 4 reports the percentage of Pareto nodes observed across three benchmark problem classes—MOSPP, MOKP, and MOTSP—for varying numbers of objectives and decision variables, averaged over ten instances per configuration.

Table 4. Percentage (rounded) of Pareto nodes across different sizes (number of objectives, number of variables) for 10 instances per size.

MOSPP	Pareto Node (%)	MOKP	Pareto Node (%)	MOTSP	Pareto Node (%)
(3, 100)	1	(3, 80)	3	(3, 15)	2
(5, 100)	1	(5, 40)	7	(4, 15)	6
(7, 100)	2	(7, 40)	12		

Overall, the results indicate that Pareto nodes constitute only a small fraction of the total nodes in a DD.

B   Mathematical Notation

The notation used to describe MOILP and DDs is presented in Tables 5 and 6, respectively.

Table 5. Mathematical notation for MOILP

Symbol	Description
$\mathcal{P}$	A multiobjective integer linear programming problem
N	Number of decision variables
K	Number of objectives
$x \in \mathbb{Z}^N$	Integer-valued decision vector
$\mathcal{X} \subseteq \mathbb{Z}^N$	Feasible solution set
$C \in \mathbb{R}^{K \times N}$	Objective function coefficient matrix
$\mathcal{Z}$	Set of feasible objective vectors (Image of ($\mathcal{X}$) in the objective space)
$\prec$	Dominance relation: $z^1 \prec z^2$ implies z^1 dominates z^2
$\mathcal{Z}^* \subseteq \mathcal{Z}$	Pareto frontier (set of nondominated objective vectors)

C   Problem Definition and Instance Generation

In this section, we provide the formulation of the MOILP problems considered in this work and corresponding instance generation scheme.

C.1   MOTSP

Problem Definition. The MOTSP extends the classical traveling salesperson problem by incorporating multiple, potentially conflicting cost metrics associated with traversing arcs. We consider a complete directed graph $\mathcal{G} = (\mathcal{V}, \mathcal{E})$ where $\mathcal{V} = \{1, \ldots, N\}$ is the set of vertices (cities) and $\mathcal{E} = \{(i, j) : i, j \in \mathcal{V}, i \neq j\}$ is the set of edges. There are K cost matrices, where c_{ij}^k denotes the cost of edge (i, j) for the k-th objective. The problem consists of finding a Hamiltonian cycle (a permutation of the vertices) that simultaneously minimizes the objective vectors. In the context of the MOILP formulation, the decision variables $x = (x_1, \ldots, x_N) \in \mathcal{V}^N$ represent the permutation of vertices visited, such that x_j is the index of the vertex visited at step j.

Table 6. Mathematical notation for DDs

Symbol	Description
$\mathcal{B} = (\mathcal{N}, \mathcal{A})$	A decision diagram for problem $\mathcal{P}$
$\mathcal{N}$	Set of nodes in a DD
$\mathcal{A}$	Set of arcs (directed edges) in a DD
$\mathcal{L}_j$	Set of nodes in layer j, for $j \in \{1, \ldots, N+1\}$
$\mathbf{r}, \mathbf{t}$	Root node (layer 1) and terminal node (layer $N+1$)
$\omega(\mathcal{B})$	Width of the DD: $\max_j |\mathcal{L}_j|$
$a \in \mathcal{A}$	An arc in the DD
$d(a)$	Assignment value of arc a (value for variable x_j)
$v(a)$	Objective weight of arc a (vector in $\mathbb{R}^K$)
p	A path from $\mathbf{r}$ to $\mathbf{t}$
$x(p)$	Solution vector encoded by path p
$v(p)$	Objective vector accumulated along path p
$\mathcal{X}_\mathcal{B}$	Set of feasible integer solutions encoded by $\mathcal{B}$
$\mathcal{Z}_\mathcal{B}$	Set of objective vectors encoded by $\mathcal{B}$
$\mathcal{Z}_\mathcal{B}^\star$	Set of nondominated vectors in $\mathcal{Z}_\mathcal{B}$ (Pareto frontier)
Restricted DDs and Construction	
$\mathcal{B}'$	Restricted DD where $\mathcal{X}_{\mathcal{B}'} \subseteq \mathcal{X}$
W	Maximum width parameter for restricted DDs
$\mathcal{Z}_{\mathcal{B}'}^\star$	Approximate Pareto frontier derived from $\mathcal{B}'$
$\mathcal{S}$	State space defined by the DP formulation of $\mathcal{P}$
$s(u) \in \mathcal{S}$	State associated with node u

Instance Generation. The instances are generated as in [29]. For each K, we generated integer coordinates for N cities on a 1000×1000 square (uniformly at random) and used Euclidean distances to create the distance matrix.

C.2   MOKP

Problem Definition. The MOKP involves selecting a subset of items to maximize multiple profit objectives subject to a single weight capacity constraint. Given N items, let $w \in \mathbb{Z}_+^N$ be the vector of item weights and $B \in \mathbb{Z}_+$ be the knapsack capacity. The profit of item j for the k-th objective is given by the entry C_{kj}. The problem is formulated as:

$$\max_{x \in \{0,1\}^N} \left\{ Cx \mid \sum_{j=1}^{N} w_j x_j \leq B \right\}.$$

Here, the decision variable $x_j = 1$ implies item j is selected, and $x_j = 0$ otherwise.

Instance Generation. We follow the instance generation scheme used in [23], where each profit C_{kj} and weight w_j were drawn uniformly at random from the integer interval $[1, \ldots, 1000]$. The capacity of the knapsack was set to $B :=$ $\lceil 0.5 \sum_{j=1}^{N} w_j \rceil$.

C.3 MOTSP

Problem Definition. The MOTSP seeks to select a subset of variables to maximize objectives subject to pairwise conflict constraints (or packing constraints). Let $A \in \{0,1\}^{M \times N}$ be a binary constraint matrix with M constraints, and $\mathbf{1}$ be a vector of ones of size M. The problem is defined as:

$$\max_{x \in \{0,1\}^N} \{Cx : Ax \leq \mathbf{1}\}.$$

The constraint $Ax \leq \mathbf{1}$ implies that for any row m of A, at most one variable j with $A_{mj} = 1$ can be set to 1. This is equivalent to finding a maximum weight independent set on the intersection graph defined by A.

Instance Generation. The instance generation for the MOSPP is based on the previous work by [34]. Specifically, we fix the number of constraints $M = N/5$. Then for each constraint, we select the number of variables that participate in it from a random uniform distribution over $[2, 20]$, resulting in an average of 10 variables per constraint.

We identified a minor issue in this instance generation process. Specifically, certain variables were not involved in any constraints. To address this, we randomly assign such variables to any one of the existing constraints, ensuring all variables participate meaningfully in the problem formulation.

D Implementation Details of Node Selection Heuristics

In this section, we describe the implementation of the node selection heuristics for the MOKP and MOSPP. The corresponding details for the MOTSP are presented in Sect. 3.3.

D.1 Multiobjective Knapsack Problem

State Definition. The state $s(u)$ at layer j (where the decision for item $j-1$ has just been made) tracks the resource consumption. It is defined as a scalar:

$$s(u) = q, \tag{6}$$

where $q \in \mathbb{Z}_{\geq 0}$ represents the accumulated weight of items selected in the path from the root to node u. Formally, if u is reached by a path p with assignments $x_1, \ldots, x_{j-1}$, then $s(u) = \sum_{t=1}^{j-1} w_t x_t$. A transition $x_j = 1$ is feasible only if $s(u) + w_j \leq B$. If $x_j = 1$, the next state is $s(u) + w_j$; if $x_j = 0$, the next state is $s(u)$.

Rule-Based Heuristic. As the state in MOKP is scalar-valued, the NOSH-Rule method uses Scal+ to select the nodes along with minWeight variable ordering [33]. In minWeight variable ordering, we sort the items based on the ascending order of their weight before constructing the DD.

Learning-Based Heuristic. The features used by the NOSH-ML method are presented in Table 7. Note that these features rely on the fact that the problem has only one constraint. Extending this approach to problems with multiple constraints can be non-trivial.

Table 7. Features associated with a DD node for the MOKP.

Feature scope	Feature	Count
Instance features	The number of objectives	1
	The number of items (or variables)	1
	The capacity of the knapsack	1
	The mean, min., max., and std. of the weights	4
	The mean, min., max., and std. of the values	12
Layer-variable features	The weight associated with variable	1
	Average value across objectives	1
	Maximum value across objectives	1
	Minimum value across objectives	1
	Standard deviation across objectives	1
	Ratio of average value across objectives to weight	1
	Ratio of maximum value across objectives to weight	1
	Ratio of minimum value across objectives to weight	1
Layer-index features	Normalized layer index	1
State features	Normalized state (weight of the knapsack at the current node)	1
	Ratio of state by capacity	1
Total		44

D.2 Multiobjective Set Packing Problem

State Definition. The state $s(u)$ at layer j must capture the availability of the remaining variables $x_j, \ldots, x_N$ given the decisions made so far. The state is defined as the set of *eligible* variables:

$$s(u) = \mathcal{V}_{eligible} \subseteq \{j, \ldots, N\}, \tag{7}$$

where $k \in \mathcal{V}_{eligible}$ implies that selecting variable k (setting $x_k = 1$) does not violate any constraints given the variables already selected in the path to u. At the root node, $s(\mathbf{r}) = \{1, \ldots, N\}$. Given a node u at layer j with state $s(u)$:

- If $x_j = 0$, the constraint set remains unchanged for the remaining variables. The next state is $s(u) \setminus \{j\}$.
- If $x_j = 1$ (valid only if $j \in s(u)$), we must remove j and all future variables $k > j$ that conflict with j (i.e., variables k such that $\exists m, A_{mj} = 1 \wedge A_{mk} = 1$). The next state is $\{k \in s(u) \setminus \{j\} \mid$ variable k does not conflict with $j\}$.

Rule-Based Heuristic. The NOSH-Rule method uses the Card+ heuristic to select states as they are represented as a set. Ties among nodes with the same cardinality are broken by randomly selecting a subset of them to fit the width limit. To build the DD, we use the minState [6] variable ordering, where we select the variable appearing in the minimum number of states in the current layer to construct the next layer.

E Additional Computational Details

E.1 Width Selection for Restricted DDs

One of the key considerations in constructing restricted DDs is determining its width. If the width is too small, the DD may be overly restrictive, making it difficult to recover the true Pareto frontier. Conversely, if the width is too large, the performance of the restricted DD may closely resemble that of the exact DD, offering little computational advantage.

In the DD literature, the width is typically chosen based on empirical validation, tuned to the downstream task performance using the restricted DD. In this work, we begin by setting the initial width to approximately 20% of the average width of the exact DDs for a given problem class and instance size. If the method achieves both cardinality and precision above 80%, we retain the current width. Otherwise, we incrementally increase the width budget by 10%. Conversely, if the method performs exceptionally well with the initial width, we systematically reduce it to identify a minimal effective width.

As demonstrated in Sect. 5, effective width budgets for the MOSPP, MOKP, and MOTSP are found to be 1%, 30%, and 20% of the average exact DD width, respectively.

E.2 Hyperparameter Configuration

Appendix E.2 provides the architectural and training details for the learning-based heuristic for the MOTSP. Appendix E.2 outlines the configuration of the learning-based heuristic for the MOKP, which is based on XGBoost. Finally, Appendix E.2 describes the parameters of the NSGA-II-based evolutionary baseline.

MOTSP Scorer Configuration. We enrich the raw vertex features of the MOTSP by computing additional features based on the distances per objective. Specifically, for each vertex, we compute the mean, minimum, maximum, standard deviation, and interquartile range (75th percentile minus 25th percentile)

Table 8. Hyperparameter configuration for the XGBoost-based scorers used in MOKP

Parameter	Size		
	(7, 40)	(4, 50)	(3, 80)
max_depth	9	7	5
min_child_weight	10000	10000	1000
eta	0.3		
objective	binary:logistic		
num_round	250		
early_stopping_rounds	20		
eval_metric	logloss		

of the distances to other nodes for each objective. These five statistical features are concatenated with the original 2D coordinates of the vertex, resulting in a 7-dimensional feature vector for each vertex. This enhanced representation provides the model with richer contextual information about each node's relative position and importance in the instance.

The Objective Aggregator processes this input using a Deep Sets architecture, which is permutation invariant over objectives. It transforms the enriched node features $h_0^v \in \mathbb{R}^{N \times K \times 7}$ and edge features $h_0^e \in \mathbb{R}^{N \times N \times K}$ into fixed-size embeddings $h_{\text{agg}}^v \in \mathbb{R}^{N \times 32}$ and $h_{\text{agg}}^e \in \mathbb{R}^{N \times N \times 32}$, respectively, with an embedding dimension of 32. These embeddings are then passed into the downstream Graph Transformer network.

Concretely, we define feature encoders and aggregators using functions:

$$\gamma_v : \mathbb{R}^7 \to \mathbb{R}^{32}, \quad \rho_v : \mathbb{R}^{64} \to \mathbb{R}^{32}, \quad \gamma_e : \mathbb{R} \to \mathbb{R}^{32}, \quad \rho_e : \mathbb{R}^{32} \to \mathbb{R}^{32}$$

Here, γ_v and γ_e are per-objective encoders applied independently to each node and edge for a given objective, while ρ_v and ρ_e are permutation-invariant aggregators that combine the representations across objectives. The output of this aggregation serves as a unified embedding that captures information across all objectives. These encoders and aggregators are implemented as a single linear layer with ReLU activation in the output.

The EGT is configured with 4 layers and 8 heads per layer and no dropout. The architecture uses ReLU as activation and omits biases in both multi-head attention and linear layers. A hidden-to-input dimension ratio of 2 is used in the MLP blocks. The model is trained for 100 epochs using the Adam optimizer. The learning rate is linearly warmed up from 5×10^{-5} to 5×10^{-4}, after which it follows a cosine decay schedule down to a minimum of 5×10^{-5}.

MOKP Scorer Configuration. Table 8 describes the configuration details for the XGBoost-based node scorer for MOKP, including the number of trees, learning rate, and maximum depth.

Table 9. NSGA-II Configuration

Config	MOKP & MOSPP	MOTSP
Crossover	TwoPointCrossover	OrderCrossover
Mutation	BitflipMutation	InversionMutation
Sampling	BinaryRandomSampling	PermutationRandomSampling

NSGA II Configuration. The crossover, mutation, and sampling strategies for the NSGA-II algorithm applied to different problems is provided in Table 9. For each problem type and size, we set the time budget for MOSPP, MOKP, and MOTSP to match the average runtime of `NOSH-Rule`, `NOSH-ML-FE`, and `NOSH-ML-E2E`, respectively. Initially, we set the population size to the average number of nondominated solutions observed for each problem size. However, this led to out-of-memory errors for some sizes, prompting us to reduce the population size accordingly.

F   Additional Results

Tables 10, 11, and 12 present the complete NSGA-II results for MOSPP, MOKP, and MOTSP, respectively. In addition, Table 10 includes the performance of various rule-based NOSH variants explored in our experiments.

Table 10. MOTSP results averaged across test instances. Methods prefixed with `Ord` correspond to different rule-based NOSHs. NSGA-II-p denotes NSGA-II with population size p. Refer to Sect. 4 for column description.

N	K	Method	Width	Time ↓	Cardinality ↑	Precision ↑	IGD ↓	$|\hat{Z}^*|$
15	3	Exact	24,024	3	100	100	0.000	868
		NSGA-II-100	–	3	4	31	3.558	100
		NSGA-II-500	–	3	7	20	3.531	278
		OrdMeanHigh	4,804	2	0	1	0.116	376
		OrdMeanLow	4,804	2	1	2	0.085	439
		OrdMaxHigh	4,804	2	0	1	0.116	366
		OrdMaxLow	4,804	2	1	3	0.090	425
		OrdMinHigh	4,804	2	0	1	0.105	369
		OrdMinLow	4,804	1	1	2	0.087	440
		NOSH-ML-E2E	4,804	2	**91**	**95**	**0.003**	832
	4	Exact	24,024	28	100	100	0.000	9,210
		NSGA-II-100	–	25	0	11	4.056	100
		NSGA-II-500	–	25	2	28	4.008	500
		OrdMeanHigh	4,804	2	0	2	0.096	2,737
		OrdMeanLow	4,804	2	1	4	0.082	3,254
		OrdMaxHigh	4,804	2	1	2	0.094	2,721
		OrdMaxLow	4,804	2	1	3	0.081	3,281
		OrdMinHigh	4,804	2	0	1	0.095	2,790
		OrdMinLow	4,804	2	1	2	0.084	3,229
		NOSH-ML-E2E	4,804	7	**89**	**95**	**0.004**	8,546

Table 11. MOSPP results averaged over test instances. Each column corresponds to a specific instance size (N, K). NSGA-II-p denotes NSGA-II with population size p. Refer to Sect. 4 for column description.

Metric	Method	$N = 100$					$N = 150$				
		$K = 3$	$K = 4$	$K = 5$	$K = 6$	$K = 7$	$K = 3$	$K = 4$	$K = 5$	$K = 6$	$K = 7$
Width	Exact	5,766	6,034	5,936	5,976	5,707	471,602	518,556	464,330	468,787	590,908
	NSGA-II-100	–	–	–	–	–	–	–	–	–	–
	NSGA-II-500	–	–	–	–	–	–	–	–	–	–
	NOSH-Rule	50	50	50	50	50	5,000	5,000	5,000	5,000	5,000
Time ↓	Exact	1	1	2	5	31	11	51	261	567	783
	NSGA-II-100	1	1	1	2	13	1	7	77	182	313
	NSGA-II-500	1	1	1	2	13	1	7	77	182	313
	NOSH-Rule	1	1	1	2	13	1	7	77	183	312
Cardinality ↑	Exact	100	100	100	100	100	100	100	100	100	100
	NSGA-II-100	2	1	1	1	1	0	0	0	0	0
	NSGA-II-500	0	0	0	0	4	0	0	1	1	0
	NOSH-Rule	**85**	**84**	**89**	**87**	**87**	**99**	**99**	**99**	**99**	**99**
Precision ↑	Exact	100	100	100	100	100	100	100	100	100	100
	NSGA-II-100	12	15	14	30	50	0	8	24	31	38
	NSGA-II-500	0	0	0	1	56	0	1	28	36	44
	NOSH-Rule	**89**	**90**	**94**	**94**	**93**	**100**	**100**	**99**	**100**	**100**
IGD ↓	Exact	0.000	0.000	0.000	0.000	0.000	0.000	0.000	0.000	0.000	0.000
	NSGA-II-100	0.234	0.252	0.281	0.260	0.289	0.437	0.205	0.232	0.293	0.309
	NSGA-II-500	1.413	1.234	1.111	0.487	0.198	2.350	0.282	0.142	0.189	0.222
	NOSH-Rule	**0.012**	**0.016**	**0.013**	**0.017**	**0.019**	**0.000**	**0.000**	**0.001**	**0.001**	**0.001**
$|\hat{\mathcal{Z}}^{\star}|$	Exact	238	1,117	4,765	9,117	25,457	787	6,099	29,061	59,951	103,489
	NSGA-II-100	26.7	45.8	63.8	96.6	100.0	19.6	97.7	100.0	100.0	100.0
	NSGA-II-500	6.016	9.684	15.862	45.206	445.142	0.078	74.084	499.028	499.981	499.956
	NOSH-Rule	226	1,051	4,591	8,550	24,534	786	6,089	28,953	59,724	103,275
Inst.		100	100	100	100	100	100	100	92	54	27

Table 12. MOKP results averaged across test instances. NSGA-II-p denotes NSGA-II with population size p. Refer to Sect. 4 for column description.

N	K	Method	Width	Time ↓	Cardinality ↑	Precision ↑	IGD ↓	$|\hat{Z}^*|$
40	7	Exact	9,709	84	100	100	0.000	25,098
		NSGA-II-100	–	60	0	0	0.282	100
		NSGA-II-500	–	60	0	0	0.199	500
		NOSH-Rule	2,000	**2**	19	64	0.128	4,131
			3,000	19	61	89	0.047	12,972
		NOSH-ML-FE	2,000	16	60	74	0.042	20,482
			3,000	36	**88**	**96**	**0.012**	22,267
50	4	Exact	12,359	7	100	100	0.000	3,564
		NSGA-II-100	–	12	0	0	0.137	100
		NSGA-II-500	–	12	0	0	0.059	497
		NOSH-Rule	2,500	**1**	17	36	0.085	1,132
			3,500	2	52	70	0.032	2,256
		NOSH-ML-FE	2,500	3	61	70	0.020	3,062
			3,500	4	**88**	**92**	**0.006**	3,367
80	3	Exact	20,097	27	100	100	0.000	2,442
		NSGA-II-100	–	58	0	0	0.068	100
		NSGA-II-500	–	58	0	0	0.023	499
		NOSH-Rule	4,000	**3**	10	19	0.064	1,015
			6,000	7	62	71	0.013	1,954
		NOSH-ML-FE	4,000	6	46	53	0.012	2,039
			6,000	12	**93**	**95**	**0.002**	2,363

References

1. Adelgren, N., Gupte, A.: Branch-and-bound for biobjective mixed-integer linear programming. INFORMS J. Comput. **34**(2), 909–933 (2022)
2. An, D., Parragh, S.N., Sinnl, M., Tricoire, F.: A matheuristic for tri-objective binary integer linear programming. Comput. Oper. Res. **161**, 106397 (2024)
3. Bergman, D., Bodur, M., Cardonha, C., Cire, A.A.: Network models for multiobjective discrete optimization. INFORMS J. Comput. (2021)
4. Bergman, D., Cire, A.A.: Multiobjective optimization by decision diagrams. In: Rueher, M. (ed.) CP 2016. LNCS, vol. 9892, pp. 86–95. Springer, Cham (2016). https://doi.org/10.1007/978-3-319-44953-1_6
5. Bergman, D., Cire, A.A., Hoeve, W.J.v., Hooker, J.N.: Optimization bounds from binary decision diagrams. INFORMS J. Comput. **26**(2), 253–268 (2014)
6. Bergman, D., Cire, A.A., Van Hoeve, W.J., Hooker, J.: Decision Diagrams for Optimization, vol. 1. Springer (2016). https://doi.org/10.1007/978-3-319-42849-9
7. Bergman, D., van Hoeve, W.-J., Hooker, J.N.: Manipulating MDD relaxations for combinatorial optimization. In: Achterberg, T., Beck, J.C. (eds.) CPAIOR 2011. LNCS, vol. 6697, pp. 20–35. Springer, Heidelberg (2011). https://doi.org/10.1007/978-3-642-21311-3_5
8. Bertsekas, D.: Dynamic Programming and Optimal Control: Volume I, vol. 4. Athena Scientific (2012)
9. Blank, J., Deb, K.: Pymoo: multi-objective optimization in python. IEEE Access **8**, 89497–89509 (2020)
10. Boland, N., Charkhgard, H., Savelsbergh, M.: The triangle splitting method for biobjective mixed integer programming. In: Lee, J., Vygen, J. (eds.) IPCO 2014.

LNCS, vol. 8494, pp. 162–173. Springer, Cham (2014). https://doi.org/10.1007/978-3-319-07557-0_14

11. Boland, N., Charkhgard, H., Savelsbergh, M.: A criterion space search algorithm for biobjective integer programming: the balanced box method. INFORMS J. Comput. **27**(4), 735–754 (2015)

12. Bryant, R.E.: Graph-based algorithms for boolean function manipulation. Comput. IEEE Trans. **100**(8), 677–691 (1986)

13. Cappart, Q., Goutierre, E., Bergman, D., Rousseau, L.M.: Improving optimization bounds using machine learning: decision diagrams meet deep reinforcement learning. In: Proceedings of the AAAI Conference on Artificial Intelligence, vol. 33, pp. 1443–1451 (2019)

14. Chen, T., Guestrin, C.: XGBoost: a scalable tree boosting system. In: Proceedings of the 22nd ACM SIGKDD International Conference on Knowledge Discovery and Data Mining, KDD '16, pp. 785–794. ACM, New York, NY, USA (2016). https://doi.org/10.1145/2939672.2939785

15. Coello Coello, C.A., Reyes Sierra, M.: A study of the parallelization of a coevolutionary multi-objective evolutionary algorithm. In: Monroy, R., Arroyo-Figueroa, G., Sucar, L.E., Sossa, H. (eds.) MICAI 2004. LNCS (LNAI), vol. 2972, pp. 688–697. Springer, Heidelberg (2004). https://doi.org/10.1007/978-3-540-24694-7_71

16. Deb, K., Pratap, A., Agarwal, S., Meyarivan, T.: A fast and elitist multiobjective genetic algorithm: NSGA-II. IEEE Trans. Evol. Comput. **6**(2), 182–197 (2002)

17. Ehrgott, M.: Multicriteria Optimization. Springer Science & Business Media (2006). https://doi.org/10.1007/3-540-27659-9_9

18. Ehrgott, M., Gandibleux, X., Przybylski, A.: Exact methods for multi-objective combinatorial optimisation. In: Greco, S., Ehrgott, M., Figueira, J.R. (eds.) Multiple Criteria Decision Analysis. ISORMS, vol. 233, pp. 817–850. Springer, New York (2016). https://doi.org/10.1007/978-1-4939-3094-4_19

19. Fischetti, M., Glover, F., Lodi, A.: The feasibility pump. Math. Program. **104**, 91–104 (2005)

20. Hussain, M.S., Zaki, M.J., Subramanian, D.: Global self-attention as a replacement for graph convolution. In: Proceedings of the 28th ACM SIGKDD Conference on Knowledge Discovery and Data Mining, pp. 655–665 (2022)

21. Jain, M., et al.: Multi-objective GflowNets. In: International Conference on Machine Learning, pp. 14631–14653. PMLR (2023)

22. Joachims, T., Hofmann, T., Yue, Y., Yu, C.N.: Predicting structured objects with support vector machines. Commun. ACM **52**(11), 97–104 (2009)

23. Kirlik, G., Sayın, S.: A new algorithm for generating all nondominated solutions of multiobjective discrete optimization problems. Eur. J. Oper. Res. **232**(3), 479–488 (2014)

24. Kuroiwa, R., Beck, J.C.: Large neighborhood beam search for domain-independent dynamic programming. In: 29th International Conference on Principles and Practice of Constraint Programming (CP 2023). Schloss-Dagstuhl-Leibniz Zentrum für Informatik (2023)

25. Lee, C.Y.: Representation of switching circuits by binary-decision programs. Bell Syst. Tech. J. **38**(4), 985–999 (1959)

26. Lin, X., Yang, Z., Zhang, X., Zhang, Q.: Pareto set learning for expensive multi-objective optimization. In: Advances in Neural Information Processing Systems, vol. 35, pp. 19231–19247 (2022)

27. Nafar, M., Römer, M.: Using clustering to strengthen decision diagram bounds for discrete optimization. In: Proceedings of the AAAI Conference on Artificial Intelligence, vol. 38, pp. 8082–8089 (2024)

28. Navon, A., Shamsian, A., Fetaya, E., Chechik, G.: Learning the pareto front with hypernetworks. In: International Conference on Learning Representations (2020)
29. Özpeynirci, Ö., Köksalan, M.: An exact algorithm for finding extreme supported nondominated points of multiobjective mixed integer programs. Manage. Sci. **56**(12), 2302–2315 (2010)
30. Pal, A., Charkhgard, H.: A feasibility pump and local search based heuristic for bi-objective pure integer linear programming. INFORMS J. Comput. **31**(1), 115–133 (2019)
31. Pal, A., Charkhgard, H.: FPBH: a feasibility pump based heuristic for multi-objective mixed integer linear programming. Comput. Oper. Res. **112**, 104760 (2019)
32. Paszke, A., et al.: PyTorch: an imperative style, high-performance deep learning library. In: Advances in Neural Information Processing Systems, vol. 32 (2019)
33. Patel, R., Khalil, E.B.: LEO: learning efficient orderings for multiobjective binary decision diagrams. In: International Conference on the Integration of Constraint Programming, Artificial Intelligence, and Operations Research, pp. 83–110. Springer (2024). https://doi.org/10.1007/978-3-031-60599-4_6
34. Stidsen, T., Andersen, K.A., Dammann, B.: A branch and bound algorithm for a class of biobjective mixed integer programs. Manage. Sci. **60**(4), 1009–1032 (2014)
35. Tamby, S., Vanderpooten, D.: Enumeration of the nondominated set of multiobjective discrete optimization problems. INFORMS J. Comput. **33**(1), 72–85 (2021)
36. Tricoire, F.: Multi-directional local search. Comput. Oper. Res. **39**(12), 3089–3101 (2012)
37. Wu, Y., Song, W., Cao, Z., Zhang, J., Gupta, A., Lin, M.: Graph learning assisted multi-objective integer programming. In: Koyejo, S., Mohamed, S., Agarwal, A., Belgrave, D., Cho, K., Oh, A. (eds.) Advances in Neural Information Processing Systems, vol. 35, pp. 17774–17787. Curran Associates, Inc. (2022)
38. Zaheer, M., Kottur, S., Ravanbakhsh, S., Poczos, B., Salakhutdinov, R.R., Smola, A.J.: Deep sets. In: Guyon, I., et al. (eds.) Advances in Neural Information Processing Systems, vol. 30. Curran Associates, Inc. (2017)
39. Zhang, Q., Li, H.: MOEA/D: a multiobjective evolutionary algorithm based on decomposition. IEEE Trans. Evol. Comput. **11**(6), 712–731 (2007)

Exact Synthetic Populations for Scalable Societal and Market Modeling

Thierry Petit[1,2(✉)] and Arnault Pachot[1,2]

[1] Emotia, Paris 75008, France
apachot@pollitics.com
[2] STATION F, 5 Parvis Alan Turing, 75013 Paris, France
tpetit@pollitics.com

Abstract. We introduce a constraint-programming framework for generating synthetic populations that reproduce target statistics with high precision while enforcing full individual consistency. Unlike data-driven approaches that infer distributions from samples, our method directly encodes aggregated statistics and structural relations, enabling exact control of demographic profiles without requiring any microdata. We validate the approach on official demographic sources and study the impact of distributional deviations on downstream analyses. This work is conducted within the Pollitics project developed by Emotia, where synthetic populations can be queried through large language models to model societal behaviors, explore market and policy scenarios, and provide reproducible decision-grade insights without personal data.

Keywords: Constraint Programming · Synthetic Populations · Polls

1 Introduction

This article presents an original method for Synthetic Population Generation (SPG) based on Constraint Programming (CP), designed to enforce both global distributional targets and individual-level coherence.

SPG is generally organized into two main families of methods [5], Synthetic Reconstruction (SR), and Heuristic Combinatorial Optimization (HCO).

SR methods generate individuals by sampling attributes from marginal or reconstructed joint distributions. Classical SR includes iterative proportional fitting and iterative proportional updating schemes, and Monte Carlo based approaches that infer a joint distribution before sampling [4]. Markov models and probabilistic graphical models have also been used, including Bayesian networks [22]. Copula based methods reconstruct dependence structures [13,14]. Recent work applies deep generative models, such as autoencoders and variational autoencoders for rare feature combinations [8], and generative adversarial networks (GANs) for tabular synthesis surveyed in [7]. Hybrid neural approaches with differentiable constraints have also been explored [21].

HCO methods construct synthetic populations by selecting or recombining individuals from a microdata sample so that the resulting distribution approximates known marginals. Foundational contributions formulate the task as an optimisation problem measured by a discrepancy between synthetic and target distributions [23,24]. Variants mainly differ in their search strategies, including hill climbing, simulated annealing, and genetic algorithms.

SR and HCO methods typically require either microdata samples or training sets from which joint distributions can be learned, or alternatively detailed multiway contingency tables that are rarely fully available in public statistics. Rare contributions explore sample free optimisation starting from artificial individuals [2] but as far as we now, all rely on fitness functions defined with respect to target distributions and do not offer guarantee of internal coherence, since individual consistency emerges indirectly from optimisation or sampling rather than from explicit constraints. Therefore, existing methods are not well-suited to the operational setting of our system (named Pollitics), where we construct digital twins of individuals, companies, or training centers to support direct querying or economic simulations. First, we rarely have access to reliable, openly available real microdata. Second, the information we do have consists of precise statistics expressed as percentages that must be matched exactly. Third, we require a declarative approach to enforce strict individual level coherence (for example, no retired minors). Our synthetic agents are subsequently queried by Large Language Models to provide justifications and behavioural explanations, rather than being used solely through nomenclature mapping tables or analytical models.

CP offers a declarative framework in which structural constraints are enforced explicitly and systematically, independently of any optimisation objective. A CP based generator must nevertheless resolve several modelling challenges: (i) The enforcement of marginal or joint statistical distributions such as age pyramids, gender ratios, or income brackets. (ii) The encoding of internal coherence rules for each individual and the optional promotion of value diversity in the population. (iii) The ability to scale to hundreds of individuals defined by multiple categorical attributes, which requires a model consistent with a batched solving strategy. In this paper, we introduce and evaluate a novel CP driven method that addresses these challenges. Our code has been used in various contexts, including a polling MVP (Minimal Viable Product) and an application that required generating more than 55000 synthetic individuals across 570 municipalities for fine grained territorial economic modelling.

2   Constraint-Based Population Generation

Our approach formulates population generation as a constraint programming (CP) optimization problem. We construct a set of individuals from a declarative constraint model. These individuals jointly satisfy three types of constraints:

1. **Exact compliance with target distributions.** A global distribution constraint ensures that the generated population matches the given statistical targets exactly.

2. **Local logical coherence of individuals.** Each individual must satisfy a set of local constraints that encode admissible combinations of characteristic values, e.g., no retired child.
3. **Optional structural constraints.** Additional constraints (such as diversity requirements or structural restrictions on subgroups).

We impose no restriction on the constraints attached to each individual, in particular we do not assume any tractable structural property of the underlying constraint hypergraph, such as Berge-acyclicity [3]. The central difficulty is to scale. Our solution is to generate individuals by batches, which requires defining constraints that satisfy a suitable *monotonicity* property so that feasibility is preserved as the population grows. We use the following terminology: we consider a set of categorical *features* (or *characteristics*) $\mathcal{F}$ indexed by $f \in \{1, \ldots, F\}$. Each feature f is associated with a finite domain D_f representing its possible categories. For a population of size N, the value of characteristic f for individual i is represented by a CP variable x_f^i, and each such variable takes values in D_f. The set of variables $X_f = \{x_f^1, \ldots, x_f^N\}$ thus represents all instantiations of characteristic f across the population, while an individual is the tuple $x^i = (x_1^i, \ldots, x_F^i)$.

2.1   Matching Target Distributions: A Motivating Example

In human surveys, the variance of an estimated proportion $\hat{p}$ (where $\hat{p}$ denotes the empirical frequency of a category, p its true population proportion, and n the number of respondents) follows the classical binomial form $v(\hat{p}) = p(1 - p)/n$. Structural biases must often be corrected because many characteristics of the sample cannot be controlled. In contrast, synthetic populations can be constructed to match known distributions directly, rather than relying on large sample sizes n to reduce variance. To illustrate this property, consider two categorical characteristics: age (X_1 with associated domain $D_1 = \{0, 1, 2, 3\}$) and location (X_2, domain $D_2 = \{0, 1, 2, 3\}$), with the allocation shown in Table 1 for $N = 100{,}000$ individuals.

Table 1. Example of a categorical population distribution.

	D_2			
	0	1	2	3
D_1 0	7000	15750	1750	10500
1	7000	15750	1750	10500
2	4000	9000	1000	6000
3	2000	4500	500	3000

We consider a simple voting model where each individual chooses among A, B, or DK, with probabilities conditioned on age and location (Table 2).

Table 2. Vote probabilities conditioned on age (left) and location (right).

X_1 (age)	0	1	2	3
vote_A	0.45	0.25	0.25	0.05
vote_B	0.25	0.35	0.55	0.85
vote_DK	0.30	0.40	0.20	0.10

X_2 (location)	0	1	2	3
vote_A	0.25	0.25	0.45	0.35
vote_B	0.65	0.15	0.15	0.45
vote_DK	0.10	0.60	0.40	0.20

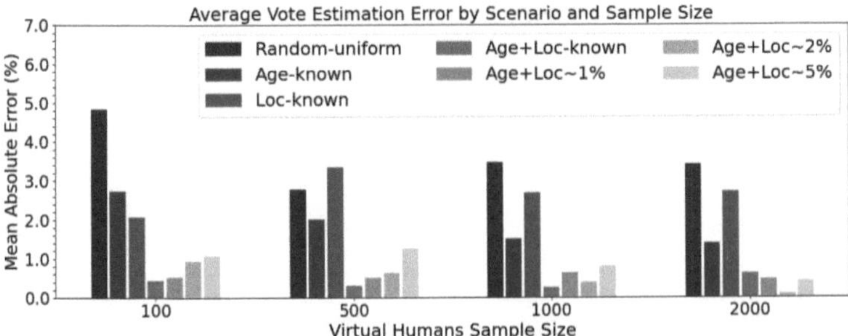

Fig. 1. Absolute error between estimated and true vote proportions (A, B, DK), as a function of sample size. Each group of three bars represents the error for vote A, B, and DK in a specific scenario, compared with the true results over $N = 100{,}000$ individuals.

These ground-truth distributions yield reference proportions A $= 29.5\%$, B $= 37.27\%$, DK $= 33.25\%$. We then simulate synthetic populations under different scenarios: (i) random assignment of x_1^i, x_2^i; (ii) matching only age marginals; (iii) matching only location marginals; (iv) matching both; (v) matching perturbed marginals within $\pm k\%$.

We assume a perfect ground-truth voting model: each individual casts a vote by sampling exactly from the conditional probabilities in Table 2. We compute each individual's voting probabilities as the arithmetic mean of the two values. Figure 1 shows that estimates are strongly biased when x_1^i and x_2^i do not follow their target distributions. Matching both marginals eliminates almost all error, and even slightly perturbed marginals yield highly accurate estimates.

2.2 Batch-Based Generation

Independent Characteristics. We distinguish a special class of characteristics, denoted $\mathcal{F}_{\text{indep}} \subseteq \mathcal{F}$, called *independent features*. An independent feature is defined as one that is not subject to any constraints, neither globally nor through individual-level restrictions, such as persona first names. Its values can be generated randomly from its domain before solving the constraint model, using a sampling without replacement strategy to promote diversity.

Generation Algorithm. We design a batch-based generation process where individual-level constraints are restricted to the variables of the current batch, whereas global feature constraints are stated over all known values of the corresponding features, thus incorporating the constants derived from previously generated batches. Recall that all individuals in a given population take their characteristic values from the same set of domains.

Definition 1 (Abstract Constraints). *We denote by $\mathcal{C}^*$ the set of* abstract constraints *stated at the schema level such that:* $\mathcal{C}^* = \mathcal{C}^*_{\text{feat}} \cup \mathcal{C}^*_{\text{ind}}, \mathcal{C}^*_{\text{feat}} \cap \mathcal{C}^*_{\text{ind}} = \emptyset.$

- *Feature-level (vertical) constraints: each $C^* \in \mathcal{C}^*_{\text{feat}}$ is instantiated on all variables of the features in its scope.*
- *Individual (horizontal) constraints: each $C^* \in \mathcal{C}^*_{\text{ind}}$ is instantiated on a subset of variables of every individual.*

Algorithm 1: Batch-Based Generation

Input: Nb. of batches B, batch size n, features $\mathcal{F}$, $\mathcal{C}^*$

$\mathcal{F} \leftarrow \mathcal{F} \setminus \mathcal{F}_{\text{indep}}$;

$\mathcal{P} \leftarrow \emptyset$;

for $b = 1$ *to* B **do**

    $\mathcal{M} \leftarrow$ new constraint model;

    for *each individual* $i = 1$ *to* n **do**

        State $X = \{x_i^1, x_i^2, \ldots, x_i^f\}$ in $\mathcal{M}$;

        for *each* $C^* \in \mathcal{C}^*_{\text{ind}}$ **do**

            Instantiate $C(Y)$ with $Y \subseteq \{x_i^1, \ldots, x_i^F\}$;

            Add $C(Y)$ to $\mathcal{M}$;

        Add constraint $[(x_i^1, \ldots, x_i^F) \notin \mathcal{P}]$ to $\mathcal{M}$;

    for *each* $C^* \in \mathcal{C}^*_{\text{feat}}$ **do**

        $Y \leftarrow \emptyset$;

        for *each feature* f *involved in* C^* **do**

            $X_f \leftarrow \{x_1^f, \ldots, x_n^f\} \cup \{x_k^f \mid k \in \mathcal{P}\}$;

            $Y \leftarrow Y \cup X_f$;

        Instantiate $C(Y)$;

        Add $C(Y)$ to $\mathcal{M}$;

    $S \leftarrow$ Solve $\mathcal{M}$;

    $\mathcal{P} \leftarrow \mathcal{P} \cup S$;

Return $\mathcal{P}$;

2.3 Feature-Level Constraints

The essential constraint for accurately simulating a population is the enforcement of target distributions on each feature. We call this set of global constraints the *distribution constraints*. In our framework, other feature-level constraints can be regarded either as facilities provided to the user or as constraints for improving the solving process (implied constraints that do not remove solutions while they improve propagation), as in most cases they can be handled in post-processing. For instance, a diversity constraint can be enforced after the generation, by randomly selecting individuals once the categorical distribution has been satisfied. We therefore refer to these additional constraints as *optional feature-level constraints*.

Distribution Constraint. The *distribution constraint* enforces the alignment of each feature with a target categorical distribution. This constraint is related to global cardinality constraints, which are widely studied in the literature [19]. However, our approach requires constraints that are enforced as best as possible through an optimization objective rather than satisfaction, similarly to cost function networks and soft constraints [1,6,11].

In our setting, each target percentage associated with a category can be viewed as a bin with a desired fill level. During batched generation, every individual assigned to a category contributes one unit to the corresponding bin. Each bin has a hard upper bound equal to the batch size, but a soft target defined by the prescribed distribution. As a result, the solver naturally tends to allocate individuals so as to reduce the deviation from the target fill levels whenever underfilled bins remain, or to incur at most an additional unit of overflow per individual otherwise. As shown in the experimental section, a decomposition into primitive constraints is the most effective formulation for this purpose. This approach benefits directly from solver-level explanation and learning mechanisms, such as those implemented in OR-Tools CP-SAT [15].

Proposition 1 (Largest Remainder Rounding). *Let $p_1, \ldots, p_q \in [0, 100]$ be target percentages with $\sum_{i=1}^{q} p_i = 100$ and let $N \in \mathbb{N}$ be the total number of individuals. Define the (real-valued) ideal allocations*

$$f_i = \frac{p_i}{100} N \quad (i = 1, \ldots, q),$$

their integer parts $t_i^{(0)} = \lfloor f_i \rfloor$, and fractional parts $r_i = f_i - t_i^{(0)}$. Let

$$R = N - \sum_{i=1}^{q} t_i^{(0)} = \sum_{i=1}^{q} r_i, \quad \text{so that} \quad 0 \le R < q.$$

Let $S \subseteq \{1, \ldots, q\}$ be the indices of the R largest r_i (break ties arbitrarily), and set $t_i = t_i^{(0)} + \mathbf{1}_{\{i \in S\}}$. Then $\sum_{i=1}^{q} t_i = N$ and $|t_i - f_i| < 1$ for all i.

Proof. By construction, $\sum_i t_i^{(0)} \le N$ and $R = N - \sum_i t_i^{(0)}$ is an integer with $0 \le R < q$. Adding 1 to exactly the R indices with largest r_i yields

$$\sum_{i=1}^{q} t_i = \sum_{i=1}^{q} t_i^{(0)} + R = N.$$

Moreover, for each i, either $t_i = t_i^{(0)}$ giving $|t_i - f_i| = r_i < 1$, or $t_i = t_i^{(0)} + 1$ giving $|t_i - f_i| = 1 - r_i < 1$. $\qquad\square$

Definition 2 (Distribution Constraint). *Let obj_f be a variable and $X_f \in \mathcal{F}$ be the variable set of a feature discretized into q disjoint bins $B_1, \ldots, B_q$. Each B_j is associated with a target percentage p_j $(1 \le j \le q)$, from which we compute t_j as the global target number of individuals to allocate in B_j after generating all batches, using the largest remainder method.*

Let e_j be the number of individuals already generated in B_j. For the current batch of size n, the constraint is expressed as:

$$t_j^{batch} = \max\left(0,\, t_j - e_j\right), \tag{1}$$

$$b_{ij} = \begin{cases} 1 & if \ x_i^f \in B_j, \\ 0 & otherwise, \end{cases} \tag{2}$$

$$\sum_{j=1}^{q} b_{ij} = 1 \quad \forall i \in \{1, \ldots, n\}, \tag{3}$$

$$c_j = \sum_{i=1}^{n} b_{ij}, \tag{4}$$

$$\delta_j = |\, c_j - t_j^{batch}\,|, \tag{5}$$

$$obj_f \ge \sum_{j=1}^{q} \delta_j. \tag{6}$$

Explanations.

(1) t_j^{batch} adjusts the global target t_j to account for e_j.
(2) b_{ij} is a Boolean indicator: it equals 1 if individual i is placed in bin B_j, where x_i^f denotes the value of feature X_f for individual i.
(3) Each individual must belong to exactly one bin.
(4) c_j counts the number of individuals assigned to bin B_j in the current batch.
(5) δ_j is the absolute deviation from the batch target t_j^{batch}.
(6) Variable obj_f is to be minimized: total deviation across all bins.

The constraint of Definition 2 satisfies a property that allows to use it in a batch-based solving process. For an assignment S on X and $Y \subseteq X$, we write $S[Y]$ for the projection of S onto Y, defined by $S[Y](x) = S(x)$ for all $x \in Y$.

Definition 3 (Extension-Preserving Optimality). *Let X be a set of variables and obj a variable, and the problem [min obj subject to: $C(X)$ and $obj \ge C_{obj}(X)$,] where C_{obj} is an expression over the variables in X (e.g. $C_{obj}(X) = \sum_{j=1}^{q} \delta_j(X)$). Given $Y \subseteq X$, the problem is* extension-preserving optimal *on Y if every optimal solution S_1 of the restricted problem on Y can be extended to an optimal solution S_2 on X such that $S_1[Y] = S_2[Y]$.*

Proposition 2. *Let X be the set of variables involved in the distribution model of Definition 2 for a batch of N individuals, and let $Y \subseteq X$ be the subset corresponding to a sub-batch of $M < N$ individuals. The decomposition of the distribution constraint is extension-preserving optimal on Y in the sense of Definition 3.*

Proof. For Y, the $\delta_j(Y)$'s are computed with respect to the same global targets t_j as in the full problem on X with N individuals. An optimal solution on Y minimizes $\sum_j |c_j(Y) - t_j|$. Since $\sum_{j=1}^{q} b_{ij} = 1$ (Definition 2), any global improvement of the objective must come from a strictly better combination of bin counts with respect to the same targets t_j. Therefore, any global solution with strictly smaller objective would induce a strictly smaller value of $\sum_j |c_j(Y) - t_j|$ on Y, contradicting the optimality of the solution on Y. $\qquad\square$

The extension-preserving property holds for the distribution constraint in isolation, but individual coherence constraints and additional global constraints (such as diversity) may alter this behaviour. We measure this impact empirically in the experimental section.

Optional Feature-Level Constraints. Although our system does not limit the global constraints that can be implemented on features, we implemented a declarative API in which all constraints are expressed through a JSON specification. This format supports several *optional feature-level constraints*, including the classical *AllDifferent* constraint [18], and a dedicated *diversity constraint*. We focus on describing this diversity constraint, derived from CP existing approaches [9,10,12,17]. This constraint is useful when no specific distribution is known for a feature, such as first names, or preferences from which no statistical data is available.

Definition 4 (Diversity Constraint). *Let X_f be a feature, and let I_{exist} and I_{batch} be the index sets of already generated individuals and of the current batch, respectively. For any distinct $i_1, i_2 \in I_{\text{exist}} \cup I_{\text{batch}}$ with $i_1 < i_2$, define*

$$b_{i_1 i_2} = \begin{cases} 0 & \text{if } x_{i_1}^f \neq x_{i_2}^f, \\ 1 & \text{if } x_{i_1}^f = x_{i_2}^f. \end{cases}$$

The diversity objective is expressed as

$$obj \geq \sum_{\substack{i_1 < i_2 \\ i_1, i_2 \in I_{\text{exist}} \cup I_{\text{batch}}}} b_{i_1 i_2}.$$

Explanation. Minimizing obj penalizes identical feature values.

Proposition 3. *Let X be the set of variables involved in the diversity constraint of Definition 4 for a batch of N individuals, and let $Y \subseteq X$ be the subset corresponding to a sub-batch of $M < N$ individuals. The diversity constraint is extension-preserving optimal on Y in the sense of Definition 3.*

Proof. The diversity objective is $obj_f \geq \sum_{i_1 < i_2} b_{i_1 i_2}$, where each $b_{i_1 i_2}$ depends only on the pair of individuals (i_1, i_2) considered. The variables $b_{i_1 i_2}$ involving indices in $X \setminus Y$ are independent of all variables indexed in Y. As all variables in X_f have the same initial domain, any optimal assignment on Y can be extended by optimally choosing the $b_{i_1 i_2}$ for pairs in $X \setminus Y$, without affecting the objective contribution of Y. Such an extension attains the global optimum. $\qquad\square$

Handling Interdependent Distributions. In many applications, features exhibit statistical dependencies. For example, voting intentions may vary across age categories, so that the target distribution of a "vote" feature X_g depends on the category of an "age" feature X_f. Two modelling strategies can be used.

- *Two-phase generation (sequential case).* If the dependency structure between features is acyclic and known (e.g. the distribution of X_g depends only on X_f, or more generally if a topological ordering exists), we first generate all individuals by enforcing the distribution constraint on the parent features. For each category v of a parent feature X_f, we then generate the dependent feature X_g in the subpopulation $\{i : x_f^i = v\}$ using its own distribution constraint.
- *Joint generation (cyclic or mutually dependent case).* When several features mutually constrain each other, we introduce a *joint feature* whose domain is the Cartesian product of their domains:

$$D_{\text{joint}} = \{(v_{f_1}, \ldots, v_{f_F}) \mid v_f \in D_f\}.$$

A single distribution constraint is then applied to this joint feature.

When detailed joint statistics for D_{joint} are available (e.g. cross-tabulated survey or census data), each joint category $(v_{f_1}, \ldots, v_{f_F})$ is assigned its target percentage. Otherwise, only the known joint statistics are assigned, and an approximate distribution for the remaining combinations can be constructed from the available marginals using a domain-specific rule or an independence assumption, which must be explicitly stated. In all cases, target counts for the joint categories are obtained using the largest remainder method, and the standard distribution constraint of Definition 2 is applied to the joint feature.

2.4 Individual Constraints

The constraints ensuring the internal consistency of each generated individual typically express dependencies or interactions between characteristics, such as a

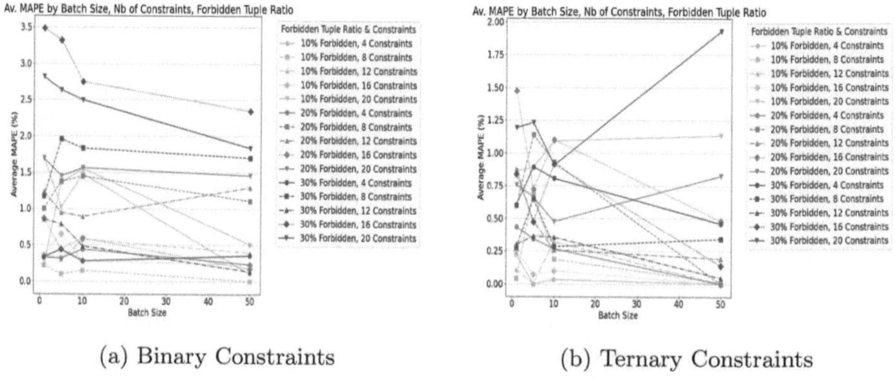

(a) Binary Constraints   (b) Ternary Constraints

Fig. 2. Average MAPE by batch size, number of constraints and tightness.

relation between age and professional activity, or between a city and its administrative department. Formally, they can be modeled using standard logical, Boolean, and table constraints for specifying the set of allowed or forbidden tuples.

3   Applications

3.1   Synthetic Population Generation

The experiments were run on an Apple M2 Pro (16 GB RAM, macOS Ventura 13.5) with Python 3.11 and OR-Tools 9.14. To interpret these experiments properly, it is important to consider the intended use case of the framework, designed for SPG on demand, through JSON specifications possibly dynamically refreshed. We fixed a 30-second solving time limit to ensure responsive user-interaction.

Trade-offs in Constrained Generation. This experiment evaluates how the presence and structure of individual constraints affects the system's ability to match global targets. The MAPE (Mean Absolute Percentage Error) is calculated as follows: for each feature f with a target distribution defined by q bins, the target percentages p_i (where $i = 1, \ldots, q$ and $\sum p_i = 100\%$) are compared against the actual percentages k_i derived from the generated population. The MAPE is the mean of the absolute errors $|p_i - k_i|$, expressed as a percentage. To handle cases where target percentages might be zero (avoiding division by zero), we use absolute difference normalized by the total count, averaged across bins: $\text{MAPE}_f = \frac{100}{n} \sum_{i=1}^{n} |p_i - k_i|$. The MAPE for an instance is the average of MAPE_f across all features.

The experiment generates synthetic instances of 100 individuals with 15 features per individual, each with a distribution constraint on a number of bins ranging from 2 to 15, randomly assigned per instance and feature. The number of individual constraints ranges from 4 to 20, incremented by 4 (i.e., [4, 8, 12, 16, 20]). The proportion of forbidden tuples in each constraint, set to [10%, 20%, 30%]. While it might be possible for one constraint to lie near the phase transition, considering multiple compatibility constraints all positioned at this critical hardness is not representative of practical cases of human populations. Figure 2 shows two graphs that plot average MAPE against batch size (in [1, 5, 10, 50]) in the generation process, one for binary individual constraints (two features) and one for ternary individual constraints (three features).

Both plots reveal the trade-off between constraint enforcement and distribution accuracy: higher constraint counts logically tend to increase MAPE, particularly at smaller batch sizes, indicating greater difficulty in matching global targets. Increasing values for the largest batch size in the ternary case are due to the imposed 30-second resolution limit (with a greater time limit we observed the same trend as for other instances).

To complete the experiment, we build on the setup of Trade-offs in Constrained Generation to assess scalability and the effect of batch size on performance. Table 3 shows the execution time per batch for the median case of 12 constraints and 20% of forbidden tuples.

Table 3. Average runtime per batch and percentage of locally optimal batches, for different population and batch sizes.

Pop. Size	Batch Size	Av. Time (s)	% Optimal batches
100	1	0.60	100.0
100	5	0.73	100.0
100	10	0.80	100.0
1000	1	4.59	100.0
1000	5	4.76	100.0
1000	10	4.85	100.0
5000	1	12.55	100.0
5000	5	15.12	99.94
5000	10	19.85	98.02

We observe a moderate constant multiplicative increase in runtime as the population size grows. 5000 individuals is the limit for solving all batches to their local optimum in less than 30 s when the batch size is in $[1, 10]$.

District Demographic Simulation. We evaluate our method on five interdependent demographic features for the Mulhouse district (France): age, area, gender, employment, and political orientation. Target distributions come from INSEE (age, gender, employment, area) and electoral sources (ideology). The categorical domains are as follows: age has seven groups (0–18, 19–30, 31–40, 41–50, 51–65, 66–75, 76+); area contains seven municipalities (Mulhouse, Riedisheim, Illzach, Brunstatt-Didenheim, Pfastatt, Sausheim, Lutterbach); gender has three categories (male, female, other); employment has five (student, unemployed, retired, employed, self-employed/director); ideology has five (left, center-left, center-right, right, unknown).

Two individual constraints are enforced: individuals aged 0–18 cannot be retired, and their political orientation must be "unknown". Inter-feature dependencies are represented using composite features treated as new variables with their own target distributions: (X_2, X_1) (age by area), (X_1, X_4) (age by employment), and (X_3, X_5) (gender by ideology). Consistency between base and composite variables is enforced through allowed-tuple relations (Table 4).

All distributions are matched exactly except for age (X_1), which carries the strongest set of constraints. The generated population remains close to the real district structure. Deviations between target and generated distributions may also reveal inconsistencies in the input statistics, suggesting a secondary use of our framework for validating cross-distribution coherence.

Table 4. Mean Absolute Percentage Error (MAPE) for each feature.

MAPE (%)	X_1	X_2	X_3	X_4	X_5	(X_1, X_4)	(X_2, X_1)	(X_3, X_5)
Value	4.9	0.0	0.0	0.0	0.0	0.0	0.0	0.0

3.2 Application Usage

Virtual Polling. We developped Pollitics (`pollitics.com`), a virtual polling platform that replaces human survey panels with synthetic populations generated from official demographic statistics.

Each virtual individual is queried through a Large Language Model, producing aggregated results that mimic traditional opinion polls. National populations for dozens of countries have been generated using publicly available demographic data (age, gender, etc.) as well as psychological statistics. The two major advantages of using synthetic populations are the absence of privacy concerns and the ability to iterate as many times as required, with new questions or specific populations.

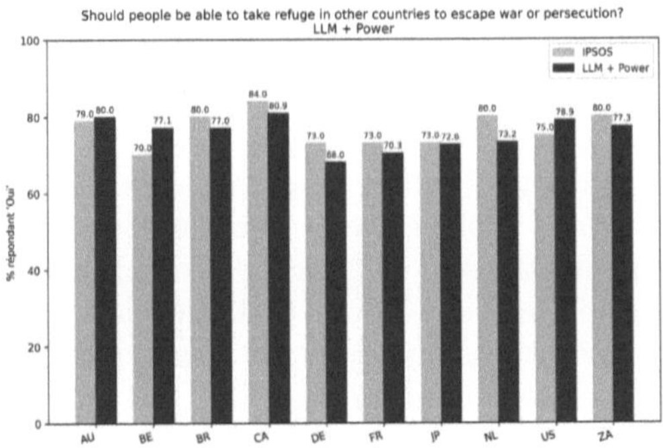

Fig. 3. Comparison between human IPSOS polling results and virtual polling results obtained with synthetic populations queried via LLMs ("LLM + Power") on the question: *"Should people be able to take refuge in other countries to escape war or persecution?".*

Once a population is instantiated, Pollitics asks each synthetic individual the userâĂŹs question and aggregates their answers in real time. In this application, questions are binary. For each respondent, an LLM first produces three probabilities $p_{\mathrm{yes}}, p_{\mathrm{no}}, p_{\mathrm{dk}}$; a vote is then sampled accordingly. Different aggregation modes can be applied. (1) *Raw mode* simulates each ballot directly from p_{yes}, with an optional rule counting cases with $p_{\mathrm{yes}} > 0.5$ as "yes." (2) *Gumbel mode* injects Gumbel noise into the logits of the three probabilities, modelling the stochastic variability observed in human respondents, who do not always act deterministically given their latent preferences.

$$g_i \sim \text{Gumbel}(0,1), \qquad \tilde{p}_i = \sigma\left(\log\frac{p_i}{1-p_i} + g_i\right)$$

(3) *Power mode* sharpens the distribution by applying an exponent $\alpha > 1$ to the probabilities below and above a given threshold, reinforcing decisive preferences and mitigating the dampening of strong opinions induced by LLM safety filters, an effect also noted in recommendation settings [20]. These transformations may be selected or combined depending on the desired simulation behaviour.

In addition, a context generation process step builds a concise background tailored to the persona and question using up-to-date information organized chronologically when relevant ; sources are selected according to the target population to ensure cultural relevance (e.g., prioritizing Spanish media for a Spanish population) and classified as *facts* (neutral, verifiable information such as statistics or documented events) and *viewpoints* (commonly expressed perspectives associated with the group without asserting factual validity).

For broad, nonâĂŞtime-sensitive questions, the system has empirically shown results close to human polls, suggesting that well-structured synthetic populations combined with LLM-based reasoning can approximate real survey outcomes. An illustrative example is shown in Fig. 3 (see https://pollitics.com/foundations for other examples).

Territorial Economic Intelligence. Beyond aggregate national polling, we built a scalable intelligence tool for French territorial and industrial monitoring based on synthetic populations. It is designed to continuously ingest new data sources and incorporate up-to-date contextual information in real time.

In this setting, synthetic populations are generated for 570 communes in the Lyon–Saint-Étienne–Roanne area (about 100,000 digital twins), using public INSEE statistics, for total of one hundred thousands individuals. The process combines the *dossier_ complet* INSEE dataset (see https://www.insee.fr/fr/statistiques/zones/2011101), communal reference files, electoral data (https://data.gouv.fr), labour-market indicators, household and housing statistics, and cross-tabulated demographic characteristics (age × gender, formation × gender, activity, employment categories). These distributions are automatically extracted, normalised, with a particular care on interdependent data to generate faithful synthetic inhabitants. A similar strategy is used to model *legal entities*, where synthetic companies are instantiated from aggregated economic and employment indicators, including datasets provided by France Travail, enabling simulations of industrial dynamics (Fig. 4).

These territorial populations enable applications in local economic analysis, including the estimation of a company's local image, perceived industrial attractiveness, and attitudes toward heritage preservation or circular-economy initiatives. Each virtual inhabitant represents a statistically grounded profile; the system can therefore simulate local sentiment or behavioural responses by interrogating individuals one by one through an LLM, just as in virtual polling.

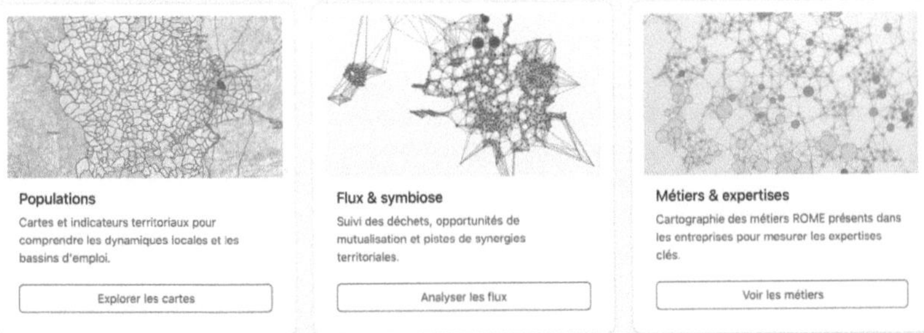

Fig. 4. French territorial economic intelligence tool: synthetic populations of Lyon–Saint-Étienne–Roanne districts mapped to local economic indicators.

The overall approach yields a granular, traceable, and updatable territorial intelligence layer, capable of supporting market studies, public-policy diagnostics, and strategic planning across hundreds of communes.

AI-Driven Text Evaluation and Message Optimization. We used synthetic populations for AI-driven evaluation of written communication, enabling users to analyse, compare, and optimise messages for specific audiences. The system leverages large language models to assess clarity, tone, persuasion strength, emotional resonance, and potential misinterpretations. Each message is processed through a multi-criteria evaluation pipeline that provides structured feedback, highlighting strengths, weaknesses, and opportunities for improvement (Fig. 5).

Messages can be evaluated not only in absolute terms, but also *as perceived by selected demographic groups*. By querying virtual individuals one by one, the system estimates how different audiences (e.g. young adults, retirees, industrial workers, local residents) would react to a proposed text. This enables message optimization tuned to specific territories, market segments, or socio-economic profiles. The result is a versatile tool for refining public communication, marketing materials, policy messaging, and stakeholder engagement strategies, with rapid feedback loops and reproducible evaluations.

4   Discussion and Conclusion

This work introduced a Constraint Programming framework for generating exact synthetic populations. Although the model is entirely driven by aggregated statistics and does not require microdata, each individual is locally coherent thanks to explicit logical constraints, yielding structured populations that are simultaneously distribution-accurate and semantically valid. This property constitutes a distinctive property in the SPG literature, allowing us to cover a broad range of new applications. The method scales to thousands of individuals and supports a broad range of applications, including national virtual polling,

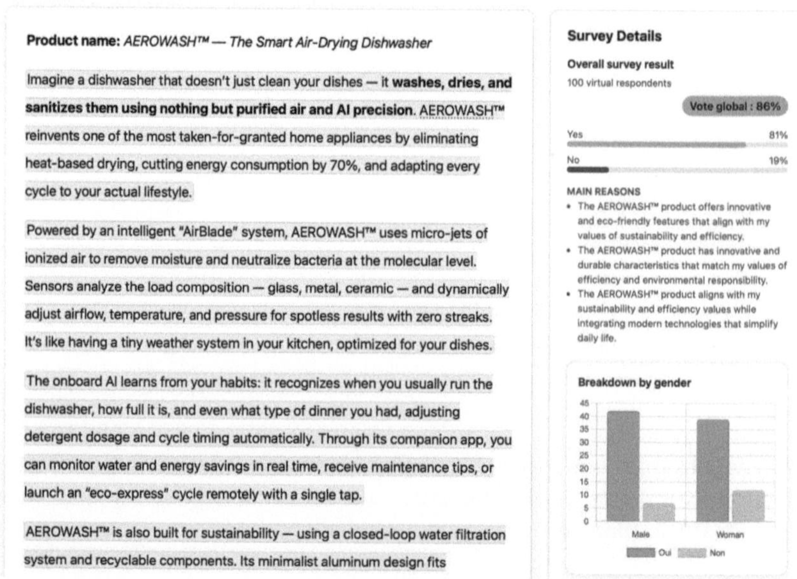

Fig. 5. Illustrative example of AI-driven text evaluation: multi-criteria scoring and optimisation suggestions for a target audience.

territorial economic intelligence, and AI-driven text evaluation for human and corporate digital twins. More broadly, the results illustrate the complementary roles of constraint programming and large language models. LLMs cannot replace optimization approaches capable of tackling problems of combinatorial nature. However, they excel at synthesizing heterogeneous information, extracting weak signals, and enabling natural-language interaction with synthetic agents. This paper demonstrates a productive synergy between leveraging LLMs for the tasks they are genuinely well suited to and relying on the formal guarantees offered by a constraint-reasoning approach.

From an ethical perspective, we emphasize that such systems are decision-support tools and must not be treated as substitutes for human judgment. At the same time, they offer important positive externalities: they enable qualitative and quantitative studies for organizations that could not otherwise afford them, allow large-scale simulations with no privacy risk since no generated individual corresponds to a real person, and make it possible to represent and query minority groups that are statistically important but extremely hard to reach in traditional surveys. This provides a principled way to explore viewpoints that are structurally under-sampled in human polling. Finally, we emphasize that all questions submitted to our system are filtered through a dedicated ethical-safety layer, described in [16], ensuring that inappropriate, discriminatory, or harmful queries cannot be processed.

As future work, we plan to generate populations for measuring deviations in the resulting cross-distributions to highlight inconsistencies in public statistics, providing a principled approach to data validation.

References

1. Allouche, D., et al.: Tractability and decompositions of global cost functions. Artif. Intell. **238**, 14–42 (2016). https://arxiv.org/abs/1502.02414
2. Barthelemy, J., Toint, P.L.: Synthetic population generation without a sample. Transp. Sci. **47**(2), 266–279 (2013). https://doi.org/10.1287/trsc.1120.0408
3. Beldiceanu, N., Carlsson, M., Debruyne, R., Petit, T.: Reformulation of global constraints based on constraints checkers. Constraints An Int. J. **10**(4), 339–362 (2005). https://doi.org/10.1007/S10601-005-2809-X
4. Casati, D., Müller, K., Fourie, P.J., Erath, A., Axhausen, K.W.: Synthetic population generation by combining a hierarchical, simulation-based approach with reweighting by generalized raking. Transp. Res. Record J. Transp. Res. Board. (2493), 107–116 (2015). https://doi.org/10.3141/2493-12
5. Chapuis, K., Taillandier, P., Drogoul, A.: Generation of synthetic populations in social simulations. J. Artif. Soc. Soc. Simul. **25**(2), 6 (2022). https://doi.org/10.18564/jasss.4762
6. Cooper, M.C., de Givry, S., Sánchez-Fibla, M., Schiex, T., Zytnicki, M., Werner, T.: Soft arc consistency revisited. Artif. Intell. **174**(7–8), 449–478 (2010). https://doi.org/10.1016/J.ARTINT.2010.02.001
7. Figueira, A., Vaz, B.: Survey on synthetic data generation, evaluation methods and gans. Mathematics **10**(15), 2733 (2022). https://doi.org/10.3390/math10152733
8. Garrido, S., Borysov, S.S., Pereira, F.C., Rich, J.: Prediction of rare feature combinations in population synthesis: application of deep generative modelling. Transp. Res. Part C: Emerging Technol. **120**, 102787 (2020). https://doi.org/10.1016/j.trc.2020.102787
9. Hebrard, E., Hnich, B., O'Sullivan, B., Walsh, T.: Finding diverse and similar solutions in constraint programming. In: Proceedings of the Twentieth National Conference on Artificial Intelligence (AAAI), pp. 372–377 (2005). https://cdn.aaai.org/AAAI/2005/AAAI05-059.pdf
10. Hebrard, E., O'Sullivan, B., Walsh, T.: Distance constraints in constraint satisfaction. In: Proceedings of the 20th International Joint Conference on Artificial Intelligence (IJCAI), pp. 106–111 (2007). https://www.ijcai.org/Proceedings/07/Papers/015.pdf
11. van Hoeve, W.J.: Over-constrained problems. In: Hentenryck, P.V., Milano, M. (eds.) Hybrid Optimization: The Ten Years of CPAIOR, Lecture Notes in Computer Science, Hybrid Optimization, pp. 191–225. Springer (2011)
12. Ingmar, L., de la Banda, M.G., Stuckey, P.J., Tack, G.: Modelling diversity of solutions. In: Proceedings of the Thirty-Fourth AAAI Conference on Artificial Intelligence (AAAI). vol. 34, pp. 1528–1535 (2020). https://ojs.aaai.org/index.php/AAAI/article/view/5512
13. Jeong, B., Lee, W., Kim, D.S., Shin, H.: Copula-based approach to synthetic population generation. PLoS ONE **11**(8), e0159496 (2016). https://doi.org/10.1371/journal.pone.0159496
14. Jutras-Dubé, P., Al-Khasawneh, M.B., Yang, Z., Bas, J., Bastin, F., Cirillo, C.: Copula-based transferable models for synthetic population generation. Preprint (2024). https://arxiv.org/abs/2302.09193, arXiv:2302.09193v3

15. Perron, L., Furnon, V.: The CP-SAT solver: a conflict-programming solver for satisfiability and optimization. In: Proceedings of the 27th International Conference on Principles and Practice of Constraint Programming (CP 2021), Open Access Series in Informatics vol. 210, pp. 1–14 (2021). https://egon.cheme.cmu.edu/ewo/docs/CP-SAT%20and%20OR-Tools.pdf
16. Petit, T., Pachot, A., Conan-Vrinat, C., Dubarry, A.: A declarative framework for large language model integration: towards a new theory of task-specific AI orchestration. In: Proceedings of the 17th International Conference on Machine Learning and Computing, pp. 474–489. Lecture Notes in Networks and Systems, Springer Nature (2025). https://doi.org/10.1007/978-3-031-94892-3_35
17. Petit, T., Trapp, A.C.: Enriching solutions to combinatorial problems via solution engineering. INFORMS J. Comput. **31**(3), 429–444 (2019). https://doi.org/10.1287/ijoc.2018.0855
18. Régin, J.C.: A filtering algorithm for constraints of difference in CSPs. In: Proceedings of the 12th National Conference on Artificial Intelligence, AAAI pp. 362–367 (1994). https://www.researchgate.net/publication/200034395_A_Filtering_Algorithm_for_Constraints_of_Difference_in_CSPs
19. Schmied, M., Régin, J.: Efficient implementation of the global cardinality constraint with costs. In: 30th International Conference on Principles and Practice of Constraint Programming (CP 2024). Leibniz International Proceedings in Informatics (LIPIcs), vol. 307, pp. 27:1–27:18. Schloss Dagstuhl-Leibniz-Zentrum für Informatik (2024). https://doi.org/10.4230/LIPIcs.CP.2024.27
20. Sinacola, E., Pachot, A., Petit, T.: Llms, virtual users, and bias: predicting any survey question without human data. In: Proceedings of 17th International Conference on Machine Learning and Computing, pp. 396–407. Lecture Notes in Networks and Systems, Springer Nature (2025). https://doi.org/10.1007/978-3-031-94892-3_29
21. Stoian, M.C., Dyrmishi, S., Cordy, M., Lukasiewicz, T., Giunchiglia, E.: How realistic is your synthetic data? constraining deep generative models for tabular data. Proceedings of the International Conference on Learning Representations (ICLR) (2024). 10.48550/arXiv. 2402.04823, arXiv:2402.04823
22. Sun, L., Erath, A.: A bayesian network approach for population synthesis. Transp. Res. Part C: Emerging Technol. **61**, 49–62 (2015). https://doi.org/10.1016/j.trc.2015.10.010
23. Voas, D., Williamson, P.: An evaluation of the combinatorial optimisation approach to the creation of synthetic microdata. Int. J. Popul. Geogr. **6**(5), 349–366 (2000). https://doi.org/10.1002/1099-1220(200009/10)6:5<349::AID-IJPG196>3.0.CO;2-5
24. Williamson, P., Birkin, M., Rees, P.H.: The estimation of population microdata by using data from small area statistics and samples of anonymised records. Environ Plan A **30**(5), 785–816 (1998). https://doi.org/10.1068/a300785

Optimization Over Trained (and Sparse) Neural Networks: A Surrogate Within a Surrogate

Hung Pham[1], Aiden Ren[1], Ibrahim Tahir[1], Jiatai Tong[2(✉)], and Thiago Serra[3]

[1] Bucknell University, Lewisburg, PA, USA
hqp001@bucknell.edu, zr002@bucknell.edu, it005@bucknell.edu
[2] Northwestern University, Evanston, IL, USA
jiataitong2026@u.northwestern.edu
[3] University of Iowa, Iowa City, IA, USA
thiago-serra@uiowa.edu

Abstract. In constraint learning, we use a neural network as a *surrogate* for part of the constraints or of the objective function of an optimization model. However, the tractability of the resulting model is heavily influenced by the size of the neural network used as a surrogate. One way to obtain a more tractable surrogate is by pruning the neural network first. In this work, we consider how to approach the setting in which the neural network is actually a given: how can we solve an optimization model embedding a large and predetermined neural network? We propose surrogating the neural network itself by pruning it, which leads to a sparse and more tractable optimization model, for which we hope to still obtain good solutions with respect to the original neural network. For network verification and function maximization models, that indeed leads to better solutions within a time limit, especially—and surprisingly—if we skip the standard retraining step known as finetuning. Hence, a pruned network with worse inference for lack of finetuning can be a better surrogate.

Keywords: Constraint learning · Neural network pruning · Neural network verification · Piecewise linear approximation · Rectified linear units

1 Introduction

In the last five years, we have seen a growing interest in approximating a constraint or an objective function of an optimization model with a neural network. This approach is often denoted as *constraint learning*. The most compelling circumstance for using constraint learning is when the exact form of some constraints or part of the objective function is unknown, but can be approximated using available data. When the exact form is known, another compelling circumstance is when the formulation is intractable for a combination of factors such as being nonlinear, nonconvex, and very large [87,112,132,145].

Constraint learning has been applied in optimization models for scholarship allocation [8], chemotherapy [83], molecular design [85], power grid operation [18,76,92], and automated control in general [106,114,136,140]. We can also use optimization models involving neural networks to evaluate those neural networks for adversarial perturbations [2,19,23,104,116], compression [30,110,111], counterfactual explanations [66,126], equivalence [67], expressiveness [16,108, 109], monotonicity [75], and reachability [28,77]. Conversely, when the neural network is trained as a reinforcement learning policy, the constraints in the optimization model can be used for limiting the action space [13,24]. Many frameworks have been proposed to embed neural networks and other machine learning models as part of optimization models, such as JANOS [8], relu-MIP [78], OMLT [17], OCL [31], OptiCL [83], Gurobi Machine Learning [42], and PySCIPOpt-ML [128].

Some neural network architectures are more convenient to embed in optimization models than others. In particular, we typically use the Rectified Linear Unit (ReLU) activation function [44,93] for a couple of reasons. First, this activation function became widely popular in the 2010s due to its good classification performance and relatively lower training cost [38,70,93]. Similar to what was already known for other neural network architectures much earlier [22,36,60], we now know that ReLU networks are also universal function approximators [47,141]. More recently, Gaussian Error Linear Units (GELUs) [52] have been used from the very beginning in many foundation models based on the transformer architecture [129], such as GPT [99], BERT [26], and ViT [27]. Nevertheless, a ReLU is a piecewise linear approximation of a GELU, which brings us to the next point. Second, and perhaps most importantly for the continued use of ReLUs nowadays, each neuron with ReLU activation represents a piecewise linear function—and by consequence a neural network with only ReLUs is also a piecewise linear function [3]. In the context of nonlinear optimization, substantial work has been devoted to using piecewise linear approximations in surrogate models [5,21,32,63,81,82,84,86,103,131,134] because we can resort to linear optimization for iterative improvements over such approximations. In fact, there are many active lines of research on what piecewise linear representations can be obtained from different neural network architectures, most of which focused on ReLUs [3,4,16,40,43,45,46,53–55,62,88–90,95,97,100,102,105,108,109,119,125,133].

With a piecewise linear representation, we can embed the neural network in the optimization model using a Mixed-Integer Linear Programming (MILP) formulation. However, such formulations range from being small but having a weak linear relaxation to having a stronger relaxation but being prohibitively large [61], which makes it difficult to use an off-the-self MILP solver. That motivated studies on how to tackle such models, ranging from reformulation [61,76,107,124] to methods for obtaining smaller big M coefficients through activation bounds [6,19,33,41,57,74,114,124,144] and parameter rescaling [98], activation inferences for fixing variables and limiting search space [11,108,120,139], cutting planes [2], and heuristic solutions [96,121,122].

Another—perhaps overlooked—approach is using neural networks that are smaller [15] and sparser [106,139].

In this work, we explore this latter approach of using sparser neural networks. On the one hand, we know that (i) sparser optimization models tend to be solved faster; and (ii) we can sparsify a neural network, which is known as *network pruning*, and still recover its performance by retraining for a few steps, which is known as *finetuning*. In fact, it is already reasonably well established that pruning a neural network in advance leads to a more tractable constraint learning model [106,139]. On the other hand, finetuning requires access to training data and requires a non-negligible runtime overhead. Moreover, if the purpose of solving an optimization model that embeds a neural network is to verify properties of the neural network itself, then it is not immediate if—and how—to use a pruned counterpart of the given neural network as a proxy. Therefore, we evaluate how much we can prune, and how much we should finetune, a neural network to still obtain effective and efficient surrogate optimization models.

2   From Notation to Formulation

In this paper, we consider feedforward networks with fully-connected layers of neurons having ReLU activation. Note that convolutional layers can be represented as fully-connected layers with a block-diagonal weight matrix, for which reason we abstract that possibility. We also abstract that fully-connected layers are often followed by a softmax layer [12], since the largest output of softmax matches the largest input of softmax, for which reason it is not necessary to include the softmax layer in applications such as network verification.

Each neural network has an input $\boldsymbol{x} = [x_1 \ x_2 \ \ldots \ x_{n_0}]^\top$ from a bounded domain $\mathbb{X}$ and corresponding output $\boldsymbol{y} = [y_1 \ y_2 \ \ldots \ y_m]^\top$, and each layer $l \in \mathbb{L} = \{1, 2, \ldots, L\}$ has output $\boldsymbol{h}^l = [h_1^l \ h_2^l \ldots h_{n_l}^l]^\top$ from neurons indexed by $i \in \mathbb{N}_l = \{1, 2, \ldots, n_l\}$. Let $\boldsymbol{W}^l$ be the $n_l \times n_{l-1}$ matrix where each row corresponds to the weights of a neuron of layer l, $\boldsymbol{W}_i^l$ the i-th row of $\boldsymbol{W}^l$, and $\boldsymbol{b}^l$ the vector of biases associated with the units in layer l. With $\boldsymbol{h}^0$ for $\boldsymbol{x}$ and $\boldsymbol{h}^L$ for $\boldsymbol{y}$, the output of each unit i in layer l consists of an affine function $g_i^l = \boldsymbol{W}_i^l \boldsymbol{h}^{l-1} + b_i^l$ followed by the ReLU activation $h_i^l = \max\{0, g_i^l\}$. When training the neural network, we vary the paramenters in $\{(\boldsymbol{W}^l, \boldsymbol{b}^l)\}_{l \in \mathbb{L}}$ to better fit the values given for the set of input–output pairs $\{(\boldsymbol{x}^{(i)}, \boldsymbol{y}^{(i)})\}_{i \in \mathbb{S}}$ representing the training set.

When optimizing over the trained neural network, we flip what is variable and what is constant in relation to training: we vary the input $\boldsymbol{x} = \boldsymbol{h}^0$ and the outputs at each layer $\{(\boldsymbol{g}^l, \boldsymbol{h}^l)\}_{l \in \mathbb{L}}$ for the fixed parameters $\{(\boldsymbol{W}^l, \boldsymbol{b}^l)\}_{l \in \mathbb{L}}$. For each layer $l \in \mathbb{L}$ and neuron $i \in \mathbb{N}_l$, we also use a binary variable z_i^l representing the ReLU activation to map inputs to outputs using MILP:

$$\boldsymbol{W}_i^l \boldsymbol{h}^{l-1} + \boldsymbol{b}_i^l = \boldsymbol{g}_i^l \tag{1}$$

$$(\boldsymbol{z}_i^l = 1) \rightarrow (\boldsymbol{h}_i^l = \boldsymbol{g}_i^l) \tag{2}$$

$$(\boldsymbol{z}_i^l = 0) \rightarrow (\boldsymbol{g}_i^l \leq 0 \wedge \boldsymbol{h}_i^l = 0) \tag{3}$$

$$\boldsymbol{h}_i^l \geq 0 \tag{4}$$

$$\boldsymbol{z}_i^l \in \{0, 1\} \tag{5}$$

The indicator constraints (2)–(3) can be modeled with big M constraints [10].

In the absence of other decision variables, one general form of representing a linear optimization model embedding a neural network is as follows:

$$\max \boldsymbol{c}^T \boldsymbol{x} + \boldsymbol{d}^T \boldsymbol{y} \tag{6a}$$

$$\text{s.t.} \boldsymbol{A}\boldsymbol{x} + \boldsymbol{B}\boldsymbol{y} \leq b \tag{6b}$$

$$\boldsymbol{y} = \text{NN}(\boldsymbol{x}) \tag{6c}$$

We use $\boldsymbol{y} = \text{NN}(\boldsymbol{x})$ as a shorthand for the input–output mapping defined by the set of constraints (1)–(5) $\forall l \in \mathbb{L}, i \in \mathbb{N}_l$ across the entire neural network.

Next, we describe two applications based on special cases of this formulation. They are both used for evaluating our approach.

2.1 Network Verification

For neural networks used for classification, one application of embedding them in optimization models is to determine if there is an adversarial perturbation for a chosen input $\boldsymbol{x}^{(i)}, i \in \mathbb{S}$ from the training set. If $\boldsymbol{x}^{(i)}$ is correctly classified as class $j \in \{1, \ldots, m\}$ by having an output $\boldsymbol{y}$ such that $y_j^{(i)} > y_k^{(i)}$ $\forall k \in \{1, \ldots, m\} \setminus \{j\}$, then we can try to determine if there is a similar input $\boldsymbol{x}$ with a different classification. By varying the inputs within $\{\boldsymbol{x} : \|\boldsymbol{x} - \boldsymbol{x}^{(i)}\|_1 \leq \varepsilon\}$ for a chosen ε, we try to find an input that is better classified with a chosen class $j' \neq j$, i.e., $y_{j'} > y_j$.[1] That leads to the following MILP formulation:

$$\max y_{j'} - y_j \tag{7a}$$

$$\text{s.t.} \sum_{k \in \{1, \ldots, n_0\}} |\boldsymbol{x}_k - \boldsymbol{x}_k^{(i)}| \leq \varepsilon \tag{7b}$$

$$\boldsymbol{y} = \text{NN}(\boldsymbol{x}) \tag{7c}$$

In the model above, any solution with a positive objective function value entails an adversarial perturbation; whereas a nonpositive optimal value implies that no such perturbation exists for the given choices of i, j', and ε.

---

[1] With j' fixed, we are only ensuring that j' would be a better classification than j, since there might be another class j'' dominating both, i.e., $y_{j''} > y_{j'} > y_j$.

2.2 Function Maximization

For neural networks used for regression, one application of embedding them in optimization models is to optimize over the function surrogated by the neural network. The case of finding the maximum for a neural network with a single output—i.e., $m = 1$—leads to the following MILP formulation:

$$\max y_1 \tag{8a}$$
$$\text{s.t.} \, \boldsymbol{y} = \text{NN}(\boldsymbol{x}) \tag{8b}$$
$$\boldsymbol{x} \in \mathbb{X} \tag{8c}$$

3 From Network Pruning to Sparse Surrogates

Neural networks have a very peculiar trait: assuming that two neural networks can be trained to achieve the same level of accuracy, it is often easier to train the largest one than it is to train the smallest one. But after training a neural network that is larger than it needs to be, we can simplify the network by removing neurons or connections and then still recover a similar accuracy by carefully adjusting the remaining parameters. In fact, this is becoming mainstream knowledge with the constant discussion about number of parameters in large language models and the application of pruning techniques for obtaining comparable variants that are smaller and faster for inference [35,80,117,137].

3.1 Background

Neural networks are pruned by either (i) removing connections, which is equivalent to zeroing out specific parameters—known as *unstructured* pruning; or (ii) removing units, such as neurons, convolutional filters, or layers—known as *structured* pruning. The latter has greater appeal for performance, since it goes beyond reducing storage to using smaller hardware and running the model faster, but then we need to prune less to still recover the original performance [20].

But Why Do We Need Network Pruning? A larger neural network has a smoother loss landscape [72,118], which facilitates training convergence; and the larger size may also prevent layers from becoming inactive [101], which is a common cause for unsuccessful training. **Why Does Network Pruning Work?** In larger networks, there is redundancy among the parameters [25], and zeroing out parameters leads to a loss landscape from which finetuning the pruned network for recovering the original performance converges considerably faster than the original training [65]. From a model flexibility perspective, unstructured pruning at moderate rates has little effect on the expressiveness of the neural network architecture [16]. **How Much Can We Prune?** In sufficiently large networks, we can remove as much as half of the parameters and still recover the original performance—or even improve upon it [56]. However, that varies with the task for which the network is trained [73]. Moreover, pruning may have a disparate effect across classes [58,59,94], which may lead to pruned networks that exacerbate existing performance differences [123]. On the bright side, a smaller amount of pruning may actually correct such distortions [39]. **What Should We Prune?** The two main philosophies [9] are (1) to remove parameters with the smallest absolute value—dating back to [48,64,91]; and (2) to remove parameters with the smallest expected impact on the output—dating back to [49,50,69], and including the special case of exact compression [37,110,111,115]. **And When Should We Prune?** Most studies have focused on pruning once (*one-shot*) and after training, but recent work has shown that it might be beneficial to prune iteratively and during training [34], or even before training [71].

Mathematical optimization has been extensively used in more sophisticated pruning methods [1,7,14,16,29,30,49–51,69,79,110,111,113,130,135,142,143]. Those methods tend to perform better than simpler heuristics at higher pruning rates. However, they also come at a greater computational cost. For moderate pruning, something as simple as removing the weights with the smallest absolute values, known as *Magnitude Pruning* (MP) [48,64,91], remains a very competitive choice [143]. We take the liberty of calling it "unreasonably effective".

The use of pruned neural networks in mathematical optimization, however, has been less explored. Among the first studies on embedding neural networks, Say et al. [106] observed that a modest amount of unstructured pruning—removing about 20% of the parameters—significantly reduced the runtime for solving the optimization model. Moreover, Xiao et al. [139] and Cacciola et al. [15] observed that structured pruning—having fewer neurons and therefore fewer binary decision variables mapping the activation state of each neuron—leads to comparable neural networks that are more easily verified.

3.2 The Sparse Surrogate Approach

Suppose that we have a (dense) neural network $\mathcal{D}$ to which $\mathcal{S}$ is a sparse counterpart obtained by network pruning, with $\boldsymbol{y}^D = \mathcal{D}(\boldsymbol{x})$ and $\boldsymbol{y}^S = \mathcal{S}(\boldsymbol{x})$ as the corresponding outputs from those two neural networks for a same input $\boldsymbol{x}$.

We will succinctly describe how to use $\mathcal{S}$ for obtaining solutions for constraint learning models on $\mathcal{S}$. We will use the models from Sects. 2.1 and 2.2.

Network Verification. First, we consider the case of a network verification problem, in which we validate if an adversarial input to the pruned network is an adversarial input to the original network. Let $\text{VNN}(\mathcal{N}, \boldsymbol{x}^{(i)}, \varepsilon, j, j')$ be the MILP formulation (7a)–(7c) for verifying neural network $\mathcal{N} \in \{\mathcal{D}, \mathcal{S}\}$ starting from input $\boldsymbol{x}^{(i)}$ and with maximum $L1$-norm distance ε for obtaining another input $\boldsymbol{x}$ in which the output for class j' is as large as possible in comparison to that of class j, i.e., we want to find an input $\boldsymbol{x}$ maximizing $y_{j'}^N - y_j^N$ for $\boldsymbol{y}^N = \mathcal{N}(\boldsymbol{x})$. For the purpose of verification, it suffices to find an $\boldsymbol{x}$ such that $y_{j'}^N > y_j^N$. To find a solution with positive value for dense model $\textbf{VNN}(\mathcal{D}, \varepsilon, x^{(i)}, j, j')$, we resort to solving sparse model $\textbf{VNN}(\mathcal{S}, \varepsilon, x^{(i)}, j, j')$ as outlined in Algorithm 1.

Function Maximization. Next, we consider the case of a function maximization problem, in which we check if an input to the pruned network yields a better value than the inputs previously tried on the dense network. Let $\mathbf{FM}(\mathcal{N})$ be the MILP formulation (8a)–(8c) for maximizing the output of the function modeled by network $\mathcal{N} \in \{\mathcal{D}, \mathcal{S}\}$ over domain $\mathbb{X}$. Unlike the network verification case, there is no criterion for prematurely stopping the search in this application. To find a solution with better value for dense model $\mathbf{FM}(\mathcal{D})$, we resort to solving sparse model $\mathbf{FM}(\mathcal{S})$ as outlined in Algorithm 2.

In both cases, we assume that the algorithms will be run up to a reasonable time limit, so that we would not be done solving the sparse model but would have obtained and evaluated a sufficient number of solutions in the dense model.

4 Experiments for Network Verification

We evaluated the time for finding an adversarial input for a (dense) neural network $\mathcal{D}$ by directly solving model $\text{VNN}(\mathcal{D}, x^{(i)}, \varepsilon, j, j')$, which we denote as *Dense Runtime*, in comparison to indirectly solving model $\text{VNN}(\mathcal{S}, x^{(i)}, \varepsilon, j, j')$ while resorting to Algorithm 1, which we denote as *Pruned Runtime*. Our goal is to find if, and when, Pruned Runtime is shorter than Dense Runtime.

4.1 Technical Details

We used the source code of SurrogateLIB [127] as the basis for training neural networks and producing network verification problems, starting with the MNIST dataset [68] and then extending the code to also work with the Fashion-MNIST dataset [138]. For each of those datasets, we tried all combined variations of inputs sizes $n_0 = 18^2$ (compressed images) and $n_0 = 28^2$ (images at original size), number of ReLU layers $L \in \{2, 4\}$, and uniform layer width $n_i \in \{32, 64\} \; \forall i \in \{1, \dots, L\}$. For each dataset and choice of hyperparameters, we used 10 randomization seeds for training neural networks, associating them with verification problems on distinct samples from the training set, and choosing some $\varepsilon \in [4.5, 5.5]$. In total, we have 160 verification problems.

Algorithm 1 Heuristic for obtaining an adversarial input to a (dense) neural network $\mathcal{D}$ by trying to solve the same problem on its sparse counterpart $\mathcal{S}$

1: **while** trying to solve $\mathbf{VNN}(\mathcal{S}, \varepsilon, x^{(i)}, j, j')$ **do**	▷ MILP solver call
2: **if** solution (x, y^S) found **then**	▷ Feasible solution callback
3: $y^D = \mathcal{D}(x)$	▷ Output of the dense model $\mathcal{D}$
4: **if** $y^D_{j'} > y^D_j$ **then**	▷ Check if x is adversarial to $\mathcal{D}$
5: **return** x	▷ Adversarial input found
6: **end if**	
7: **end if**	
8: **end while**	
9: **return** $\emptyset$	▷ No adversarial input found

Algorithm 2 Heuristic for maximizing the output of a (dense) neural network $\mathcal{D}$ by trying to solve the same problem on its sparse counterpart $\mathcal{S}$

1: $x^* \leftarrow \emptyset$ ▷ Initialize best solution as none
2: $y^* \leftarrow -\infty$ ▷ Placeholder for best solution value
3: **while** trying to solve $\mathbf{FM}(\mathcal{S})$ **do**   ▷ MILP solver call
4:     **if** solution (x, y^S) found **then**   ▷ Feasible solution callback
5:         $y^D = \mathcal{D}(x)$ ▷ Output of the dense model $\mathcal{D}$
6:         **if** $x^* = \emptyset$ or $y^D > y^*$ **then**   ▷ Check if x is the first or a better solution
7:             $x^* \leftarrow x$ ▷ Update best solution
8:             $y^* \leftarrow y^D$ ▷ Update best solution value
9:         **end if**
10:     **end if**
11: **end while**
12: **return** x^* ▷ Provide best solution found

For producing pruned versions of those networks, we used the PyTorch library. On the number of parameters removed, we applied layerwise pruning rates of 0.3, 0.5, 0.8, 0.9, and 0.95. On what to prune, we applied both Magnitude Pruning (MP), which corresponds to pruning the parameters with the smallest absolute value; and Random Pruning (RP), which corresponds to pruning parameters randomly. In addition, we also evaluated *unstructured pruning*, in which case the parameters across the layer are pruned indiscriminately; and *structured pruning*, in which case we consider all the parameters associated with a neuron and decide about pruning whole neurons instead. Finally, as a last step we also opted between finetuning the pruned network or not finetuning it and keeping it as it was after being pruned. For each of the 160 original verification problems, the combination of all the pruning choices above resulted in solving variants in 40 pruned versions of each neural network. In total, we use 6,560 models.

We used the BisonNet cluster. The steps involving the solution of MILP models with Gurobi were run on Intel Xeon Gold 6442Y CPUs. The steps involving network training and pruning were run on AMD EPYC 7252 CPUs with NVIDIA RTX A5000 GPUs. The neural networks were trained and pruned using Torch 2.0.0. Each model on MNIST was trained for 5 epochs, and each model on Fashion-MNIST was trained for 40 epochs. Without finetuning, there was a single round of pruning. With finetuning, there were 5 pruning rounds and each round had 5 epochs of retraining. The MILP models were solved using Gurobi Optimizer 10.0.1 with a time limit of 300 s for each of the 6,560 models.

4.2   Results and Analysis

Our best results were obtained by using Algorithm 1 with unstructured MP instead of solving the verification model directly. We compare the runtimes (in seconds) to solve the verification model directly and indirectly for MNIST in Fig. 1 and Fashion-MNIST in Fig. 2. We vary pruning rate, whether finetuning is used, and whether finetuning time is counted as part of the runtime. We report the percentage of instances above and below the identity line, corresponding to the direct and the indirect approaches being faster. Those percentages do not add up to 100% if there are ties, which includes when both methods time out.

We draw the following conclusions and provide a rationale to each conclusion by listing observations about the plots in those figures:

(I) Using our approach in network verification is advantageous in terms of (i) individual runtimes as well as (ii) number of instances solved:
  (i) We found adversarial inputs faster for most instances regardless of pruning rate and of finetuning the pruned neural networks or not.
  (ii) When there are timeouts (i.e., not finding an adversarial input) from solving the verification problem directly (as with MNIST), then our approach with a small pruning rate reduces the number of timeouts.
(II) The pruning rate can be adjusted for different purposes, from finding more adversarial inputs up to a time limit at lower rates (ii again) to finding (iii) most adversarial inputs faster at higher rates and (iv) fewer adversarial inputs but in a much shorter amount of time at the highest rate:
  (iii) The number of runtime improvements increases with pruning rate up to 90% for both datasets, finetuned or not, and then decreases.
  (iv) The differences between runtimes become more extreme in our favor with greater pruning rates, but then the timeouts also increase.
(III) The pruned neural network does not need to be a good classifier to help us find adversarial inputs (v). In fact, it is not helpful to finetune the network to improve accuracy after pruning at lower pruning rates (vi). For higher pruning rates, finetuning the neural network can be helpful (vii), but the cost of finetuning would have to be amortized over solving multiple verification problems on the same neural network (viii):
  (v) The lower accuracy in the case without finetuning, almost approaching random guessing (10% on either dataset), did not prevent us from using the pruned neural networks for obtaining adversarial inputs.
  (vi) The results were better without finetuning for the lowest pruning rates (one for MNIST and three for Fashion-MNIST).
  (vii) The percentage difference of instances solved faster with finetuning is only on the second most significant digit for the higher pruning rate, except the highest (e.g., 98.8% instead of 93.5% on MNIST and 93.8% instead of 91.2% on Fashion-MNIST for pruning rate 90%).
  (viii) If we account for the finetuning cost, then it is generally faster to solve the verification problem directly.

Given the cost–benefit advantage of not using finetuning, we conducted the following ablations restricted to the results without finetuning.

First, we considered the impact of other network pruning choices. Figure 3 compares the joint results on MNIST and Fashion-MNIST by using unstructured MP as before, then by only replacing unstructured with structured pruning, and then by only replacing MP with RP. Considering possible questions due to the

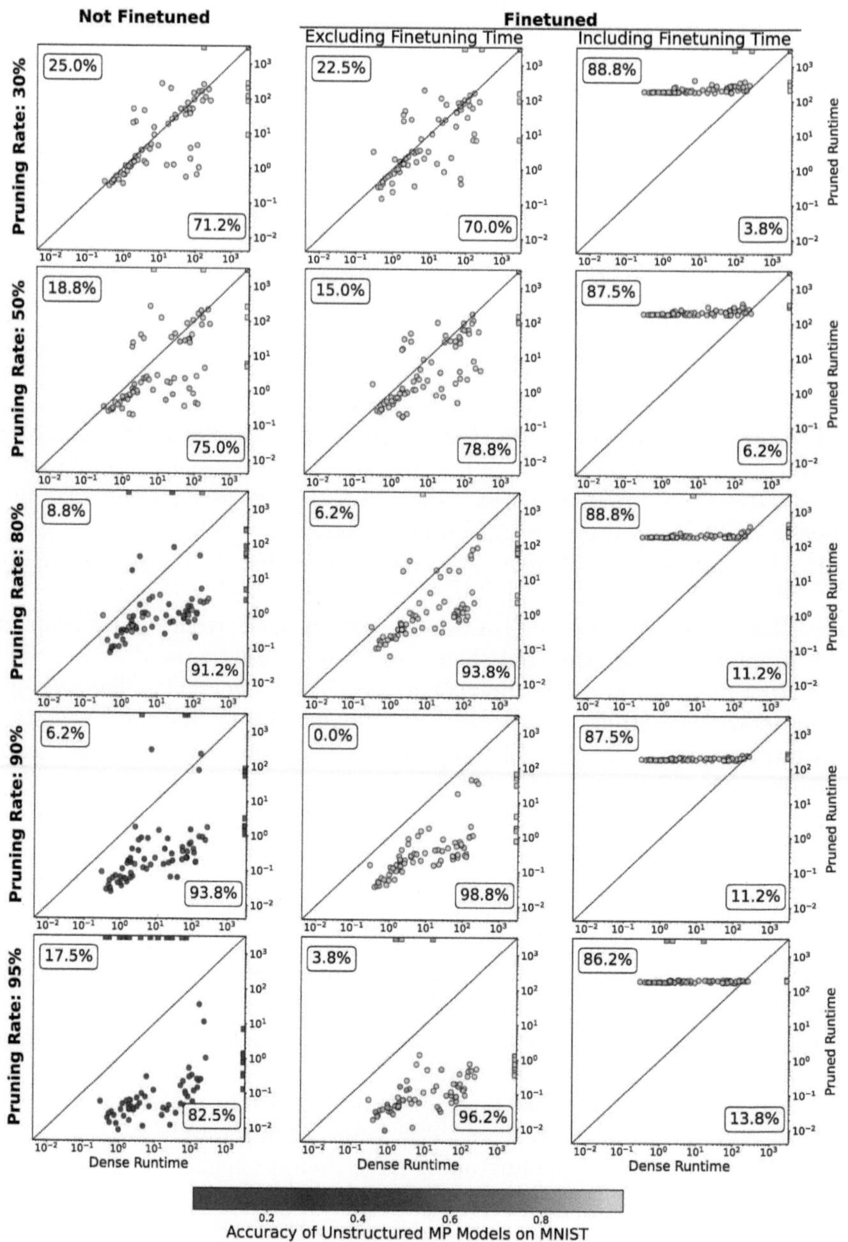

Fig. 1. Time to find adversarial input to networks trained on MNIST by solving the verification problem directly (x axis) or indirectly with Algorithm 1 (y axis) per pruning rate, use of finetuning, and inclusion of finetuning in runtime. Squares on top or (and) right sides indicate no adversarial input found for either (both). Ties are not counted.

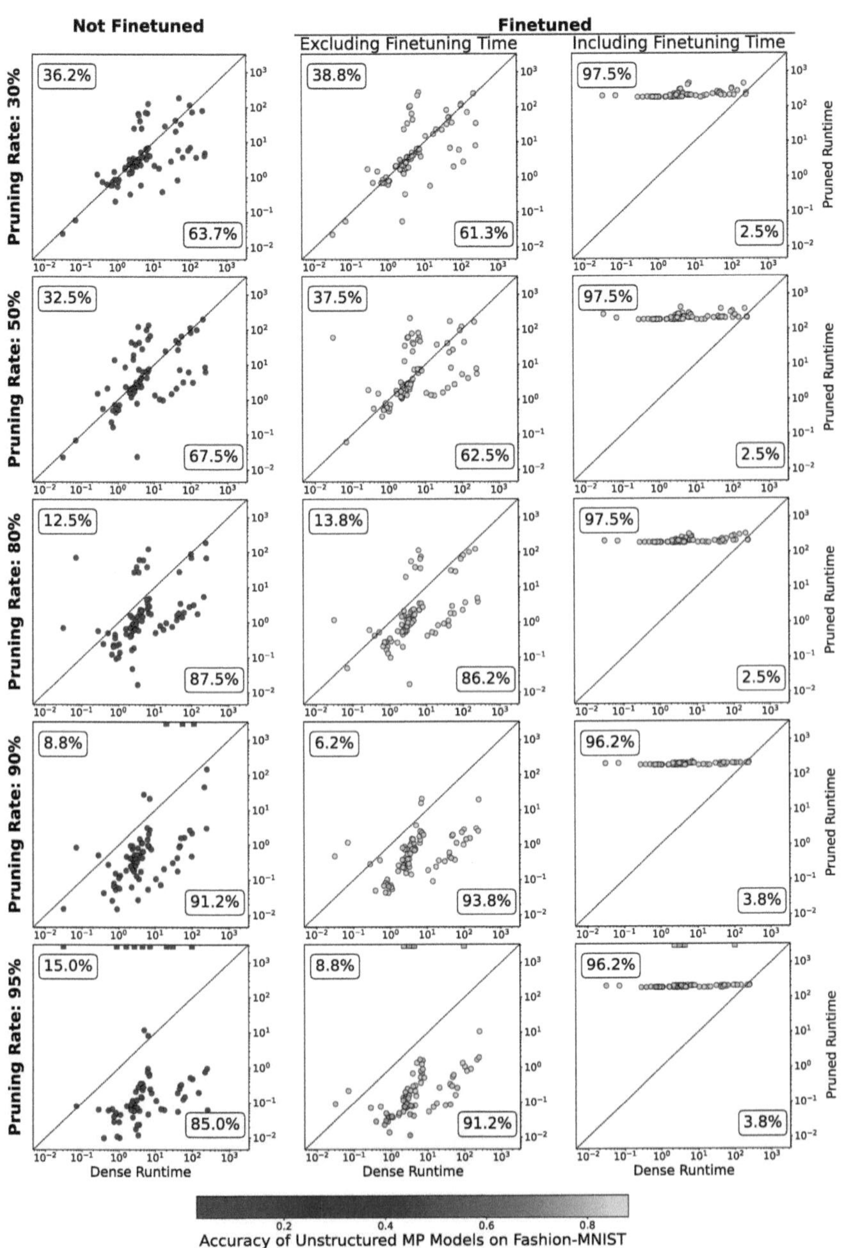

Fig. 2. Time to find adversarial input to networks on Fashion-MNIST by solving the verification problem directly (x axis) or indirectly with Algorithm 1 (y axis) per pruning rate, use of finetuning, and inclusion of finetuning in runtime. Squares on top or (and) right sides indicate no adversarial input found for either (both). Ties are not counted.

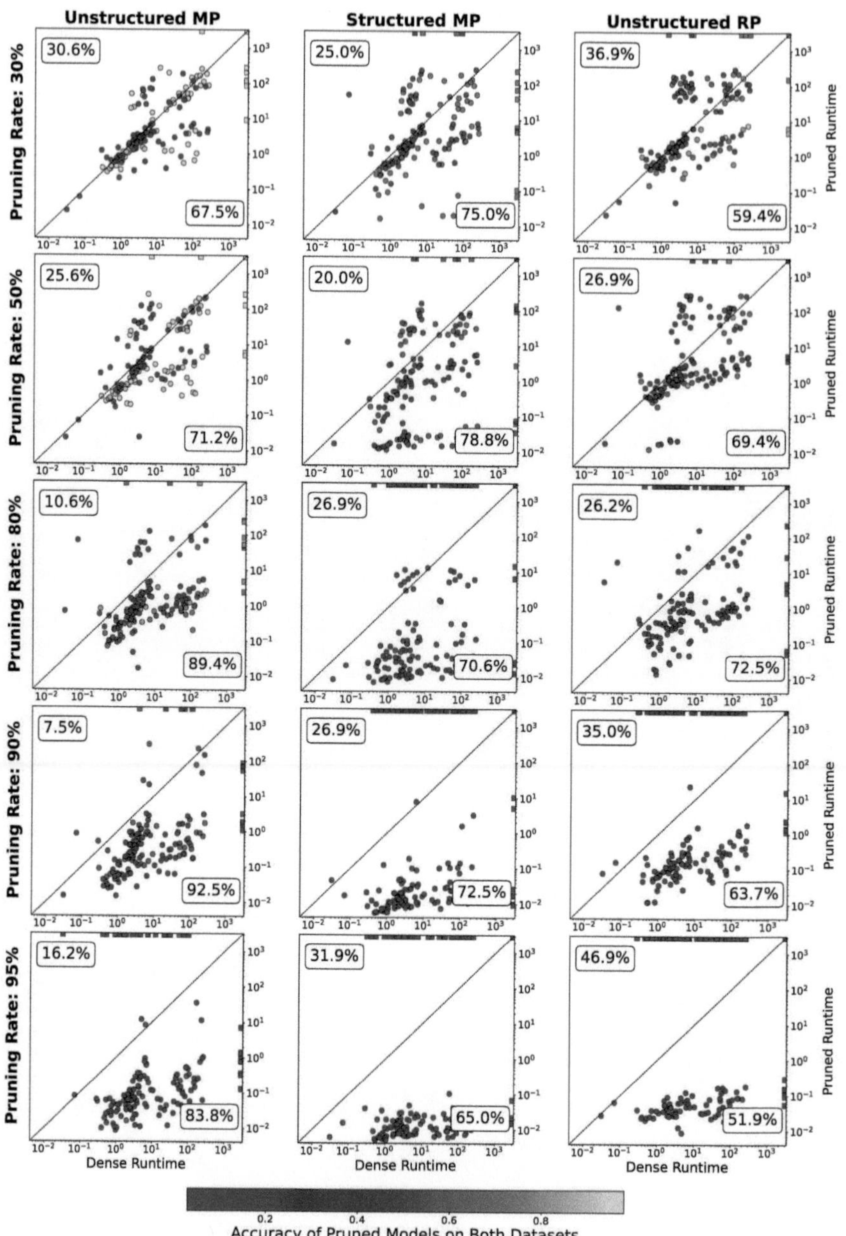

Fig. 3. Time to find adversarial input to networks on MNIST and Fashion-MNIST by solving the verification problem directly (x axis) or indirectly with Algorithm 1 (y axis) per pruning rate for different forms of network pruning. Squares on top or (and) right sides indicate no adversarial input found for either (both). Ties are not counted.

Table 1. Percentage of instances for which solving the verification problem indirectly on a pruned network is faster by pruning rate (columns); using unstructured and structured pruning (top and bottom rows); and not finetuning (NF) or finetuning (F). Unfavorable figures (below 50%) are reported in italics for greater emphasis.

		Pruning Rate				
		0.3	0.5	0.8	0.9	0.95
Unstructured	NF	67.5%	71.3%	89.4%	92.5%	83.8%
	F	65.6%	70.6%	90.0%	96.3%	93.8%
Structured	NF	75.0%	78.8%	70.6%	72.5%	65.0%
	F	59.4%	57.5%	56.3%	*49.4%*	*46.3%*

favorable results for structured pruning in some of the plots, Table 1 summarizes the number of instances on both datasets in which an adversarial input would have been found faster than by directly solving the verification problem according to the structural type of pruning and the use of finetuning.

The purpose of considering RP, which is not an appealing choice intuitively, was to evaluate if a deliberate choice on how to prune would influence the results. We can see that it does—and that the only deliberate choice that we considered was MP. We opted for MP due to its small computational cost and widely regarded effectiveness in practice. Hence, we leave is open to future work if it would be beneficial to replace MP with a more sophisticated network pruning algorithm. That improvement seems reasonable to expect, but we believe that it would be beyond the scope of this particular study.

Second, we considered how the results vary based on the dimensions of the neural networks considered. Table 2 summarizes the number of instances on both datasets in which an adversarial input would have been found faster than by directly solving the verification problem according to input size, number of layers (network depth), and neurons per layer (layer width).

Table 2. Percentage of instances for which solving the verification problem indirectly on a pruned network is faster by pruning rate (columns); disaggregated in terms of input size (top rows), number of layers (middle), and neurons per layer (bottom rows).

		Pruning Rate				
		0.3	0.5	0.8	0.9	0.95
Input Size	18^2	75.0%	80.6%	86.9%	84.4%	76.3%
	28^2	67.5%	69.4%	73.1%	80.6%	72.5%
Network Depth	2	80.6%	84.4%	86.3%	86.3%	76.3%
	4	61.9%	65.6%	73.8%	78.8%	72.5%
Layer Width	32	69.4%	78.8%	78.1%	81.9%	75.6%
	64	73.1%	71.3%	81.9%	83.1%	73.1%

We draw more conclusions and justify them based on the above ablations:

(IV) Structured pruning without finetuning is also potentially applicable at lower pruning rates (ix), with the caveat that more timeouts may occur (x). However, finetuning after structured pruning appears to make the pruned networks significantly different from the original neural network, since their adversarial inputs are less compatible (xi):

(ix) For the two lowest pruning rates (up to 50%), more instances are solved faster with structured MP than with unstructured MP.

(x) Across all pruning rates, structured pruning leads to more timeouts.

(xi) Finetuning has a positive effect when applied to neural networks that were subject to unstructured pruning, and that positive effect grows with pruning rate. The opposite happens with structured pruning.

(V) Pruning criteria (such as MP vs. RP) affect the results significantly (xii), and may help tackling networks with larger inputs and more layers (xiii):

(xii) The results for unstructured MP are better than those for unstructured RP, and the difference grows with the pruning rate.

(xiii) The benefit is more significant in neural networks with smaller input size or smaller depth, but less affected by layer width.

5   Experiments for Function Maximization

We evaluated the best solution obtained for a (dense) neural network $\mathcal{D}$ by directly solving model $\mathbf{FM}(\mathcal{D})$, which we denote as *Dense Maximum Value*, in comparison to indirectly solving model $\mathbf{FM}(\mathcal{S})$ while resorting to Algorithm 2, which we denote as *Pruned Maximum Value*. Our goal is to find if, and when, Pruned Maximum Value is greater than Dense Maximum Value.

5.1   Technical Details

We used randomly initialized networks with 5 seeds for all combinations of input sizes $n_0 \in \{100, 1000, 10000\}$, number of layers $L \in \{2, 3, 4, 5\}$, uniform layer width $n_i \in \{50, 100, 200\} \ \forall i \in \{1, \ldots, L\}$, and with an extra one-neuron output layer. That setting is similar to other papers using this problem [96, 121]. Based on the results from Sect. 4, we used only unstructured MP without finetuning. When using Algorithm 2, we solved $\mathbf{FM}(\mathcal{S})$ by setting parameters MIPFocus, PoolSearchMode to 1 and PoolSolutions to 1000, ensuring that a greater number of solutions is obtained. We report the baseline of directly solving $\mathbf{FM}(\mathcal{S})$ with default parameter values, since it gave better results in this case. We set the time limit to 600 s for all models. All other settings are as in Sect. 4.

5.2   Results and Analysis

Our approach based on Algorithm 2 generally produced better results for the instances that would be typically regarded as harder to solve for having a larger number of linear regions [121]. These are modest results in comparison with Sect. 4, but with noteworthy similarities and differences. Table 3 summarizes the number of networks for which we found better values with Algorithm 2 by input size, number of layers (network depth), and neurons per layer (layer width).

We draw the following conclusions and justifications based on the results:

(A) The results from Algorithm 2 are generally better if at least one of the dimensions of the neural network has a larger value (a), and consistently more so in the case of layer width (b).

(a) In 32 out of 35 cases where any dimension is not the smallest, at least half of the instances have a better solution. In contrast, larger input sizes were less favorable to our approach in the case of network verification.
(b) Unlike input size and network depth, any larger value for network width corresponds to a favorable case for Algorithm 2. In contrast, network width was the least relevant dimension in network verification.

(B) Increasing along a dimension or along the pruning rate while fixing the other does not necessarily lead to better results (c), but very large networks yield considerably better results when using the highest pruning rate (d).

(c) We see at least non-monotonic variation of percentages if we increase any dimension along the same pruning rate, or if we increase the pruning rate along the same dimension. In contrast, we generally observe a monotonic improvement up to 90% pruning rate in network verification.
(d) In all cases in which one of the dimensions is the largest, the best results are obtained for the highest pruning rate, unlike in network verification.

6   Conclusion

With the goal of solving optimization problems embedding a dense neural network, we tackled those problems indirectly through drastically sparsified neural networks serving as surrogates. Those problems become very difficult to solve as the neural networks grow larger in size, and the surrogate is a pruned version of the same neural network. By making the models sparser, we naturally expect to find solutions faster, and in some cases we do not expect or do not need to necessarily find an optimal solution. We believe that this work contributes to understanding how to tackle constraint learning models more effectively.

We have found that a cost-effective approach is applying unstructured pruning while carefully choosing which connections to prune but not finetuning

Table 3. Percentage of instances for which solving the function maximization problem indirectly on a pruned network yields a better solution by pruning rate (columns); disaggregated in terms of input size (top rows), number of layers (middle), and neurons per layer (bottom rows). Favorable figures (above 50%) are reported in bold.

		Pruning Rate				
		0.3	0.5	0.8	0.9	0.95
Input Size	100	33.3%	40.0%	48.3%	50.0%	45.0%
	1,000	**63.3%**	48.3%	**63.3%**	50.0%	50.0%
	10,000	**65.0%**	**58.3%**	**56.7%**	**55.0%**	**78.3%**
Network Depth	2	48.9%	40.0%	24.4%	24.4%	24.4%
	3	**57.8%**	44.4%	**66.7%**	**55.6%**	**51.1%**
	4	**62.2%**	**55.6%**	**66.7%**	**64.4%**	**73.3%**
	5	46.7%	**55.6%**	**66.7%**	**62.2%**	**82.2%**
Layer Width	50	**51.7%**	43.3%	35.0%	31.7%	33.3%
	100	**58.3%**	**51.7%**	**65.0%**	**55.0%**	**60.0%**
	200	**51.7%**	**51.7%**	**68.3%**	**68.3%**	**80.0%**

the pruned network afterwards. We obtained consistently strong results in network verification, even if the surrogate network had very low accuracy. We also obtained encouraging, albeit modest, results in function maximization, but under different conditions than we found them for network verification.

Acknowledgments. The authors were supported by the National Science Foundation grant 2104583, including Jiatai Tong while at Bucknell University.

Disclosure of Interests. The authors have no competing interests to declare that are relevant to the content of this article.

References

1. Aghasi, A., Abdi, A., Nguyen, N., Romberg, J.: Net-Trim: convex pruning of deep neural networks with performance guarantee. In: NeurIPS (2017)
2. Anderson, R., Huchette, J., Ma, W., Tjandraatmadja, C., Vielma, J.P.: Strong mixed-integer programming formulations for trained neural networks. Math. Program. (2020)
3. Arora, R., Basu, A., Mianjy, P., Mukherjee, A.: Understanding deep neural networks with rectified linear units. In: ICLR (2018)
4. Averkov, G., Hojny, C., Merkert, M.: On the expressiveness of rational ReLU neural networks with bounded depth. In: ICLR (2025)
5. Babayev, D.A.: Piece-wise linear approximation of functions of two variables. J. Heuristics (1997)
6. Badilla, F., Goycoolea, M., Muñoz, G., Serra, T.: Computational tradeoffs of optimization-based bound tightening in ReLU networks. arXiv:2312.16699 (2023)

7. Benbaki, R., et al.: Fast as CHITA: neural network pruning with combinatorial optimization. In: ICML (2023)
8. Bergman, D., Huang, T., Brooks, P., Lodi, A., Raghunathan, A.U.: JANOS: an integrated predictive and prescriptive modeling framework. INFORMS J. Comput. (2022)
9. Blalock, D., Ortiz, J., Frankle, J., Guttag, J.: What is the state of neural network pruning? In: MLSys (2020)
10. Bonami, P., Lodi, A., Tramontani, A., Wiese, S.: On mathematical programming with indicator constraints. Math. Program. (2015)
11. Botoeva, E., Kouvaros, P., Kronqvist, J., Lomuscio, A., Misener, R.: Efficient verification of (RELU)-based neural networks via dependency analysis. In: AAAI (2020)
12. Bridle, J.S.: Probabilistic interpretation of feedforward classification network outputs, with relationships to statistical pattern recognition (1990)
13. Burtea, R.A., Tsay, C.: Safe deployment of reinforcement learning using deterministic optimization over neural networks. Comput. Aided Chemic. Eng. (2023)
14. Cacciola, M., Frangioni, A., Li, X., Lodi, A.: Deep neural networks pruning via the structured perspective regularization. SIAM J. Math. Data Sci. (2023)
15. Cacciola, M., Frangioni, A., Lodi, A.: Structured pruning of neural networks for constraints learning. Oper. Res. Lett. (2024)
16. Cai, J., et al.: Getting away with more network pruning: from sparsity to geometry and linear regions. In: CPAIOR (2023)
17. Ceccon, F., et al.: Omlt: optimization & machine learning toolkit. J. Mach. Learn. Res. **23**(349), 1–8 (2022)
18. Chen, Y., Shi, Y., Zhang, B.: Data-driven optimal voltage regulation using input convex neural networks. Electric Power Syst. Res. (2020)
19. Cheng, C., Nührenberg, G., Ruess, H.: Maximum resilience of artificial neural networks. In: ATVA (2017)
20. Cheng, Y., Wang, D., Zhou, P., Zhang, T.: Model compression and acceleration for deep neural networks: the principles, progress, and challenges. IEEE Sign. Process. Magazine (2018)
21. Chien, M.J., Kuh, E.: Solving nonlinear resistive networks using piecewise-linear analysis and simplicial subdivision. IEEE Trans. Circuits Syst. (1977)
22. Cybenko, G.: Approximation by superpositions of a sigmoidal function. Math. Control Sign. Syst. (1989)
23. De Palma, A., Behl, H., Bunel, R.R., Torr, P., Kumar, M.P.: Scaling the convex barrier with active sets. In: ICLR (2021)
24. Delarue, A., Anderson, R., Tjandraatmadja, C.: Reinforcement learning with combinatorial actions: an application to vehicle routing. In: NeurIPS (2020)
25. Denil, M., Shakibi, B., Dinh, L., Ranzato, M., Freitas, N.: Predicting parameters in deep learning. In: NeurIPS (2013)
26. Devlin, J., Chang, M.W., Lee, K., Toutanova, K.: BERT: pre-training of deep bidirectional transformers for language understanding. In: NAACL (2019)
27. Dosovitskiy, A., et al.: An image is worth 16x16 words: transformers for image recognition at scale. In: ICLR (2021)
28. Dutta, S., Jha, S., Sankaranarayanan, S., Tiwari, A.: Output range analysis for deep feedforward networks. In: NFM (2018)
29. Ebrahimi, A., Klabjan, D.: Neuron-based pruning of deep neural networks with better generalization using kronecker factored curvature approximation. In: IJCNN (2023)

30. ElAraby, M., Wolf, G., Carvalho, M.: OAMIP: optimizing ANN architectures using mixed-integer programming. In: CPAIOR (2023)
31. Fajemisin, A., Maragno, D., den Hertog, D.: Optimization with constraint learning: a framework and survey. European J. Operational Res. (2023)
32. Feijoo, B., Meyer, R.R.: Piecewise-linear approximation methods for nonseparable convex optimization. Manage. Sci. (1988)
33. Fischetti, M., Jo, J.: Deep neural networks and mixed integer linear optimization. Constraints (2018)
34. Frankle, J., Carbin, M.: The lottery ticket hypothesis: finding sparse, trainable neural networks. In: ICLR (2019)
35. Frantar, E., Alistarh, D.: SparseGPT: massive language models can be accurately pruned in one-shot. In: ICML (2023)
36. Funahashi, K.I.: On the approximate realization of continuous mappings by neural networks. Neural Netw. (1989)
37. Ganev, I., Walters, R.: Model compression via symmetries of the parameter space (2022). https://openreview.net/forum?id=8MN_GH4Ckp4
38. Glorot, X., Bordes, A., Bengio, Y.: Deep sparse rectifier neural networks. In: AISTATS (2011)
39. Good, A., et al.: Recall distortion in neural network pruning and the undecayed pruning algorithm. In: NeurIPS (2022)
40. Grigsby, J.E., Lindsey, K.: On transversality of bent hyperplane arrangements and the topological expressiveness of ReLU neural networks. SIAM J. Appl. Algebra Geometry (2022)
41. Grimstad, B., Andersson, H.: ReLU networks as surrogate models in mixed-integer linear programs. Comput. Chemic. Eng. (2019)
42. Gurobi Optimization: Gurobi Machine Learning. https://github.com/Gurobi/gurobi-machinelearning (2025). Accessed 09 Feb 2025
43. Haase, C.A., Hertrich, C., Loho, G.: Lower bounds on the depth of integral ReLU neural networks via lattice polytopes. In: ICLR (2023)
44. Hahnloser, R., Sarpeshkar, R., Mahowald, M., Douglas, R., Seung, S.: Digital selection and analogue amplification coexist in a cortex-inspired silicon circuit. Nature (2000)
45. Hanin, B., Rolnick, D.: Complexity of linear regions in deep networks. In: ICML (2019)
46. Hanin, B., Rolnick, D.: Deep ReLU networks have surprisingly few activation patterns. In: NeurIPS (2019)
47. Hanin, B., Sellke, M.: Approximating continuous functions by ReLU nets of minimal width. arXiv:1710.11278 (2017)
48. Hanson, S., Pratt, L.: Comparing biases for minimal network construction with back-propagation. In: NeurIPS (1988)
49. Hassibi, B., Stork, D.: Second order derivatives for network pruning: optimal Brain Surgeon. In: NeurIPS (1992)
50. Hassibi, B., Stork, D., Wolff, G.: Optimal brain surgeon and general network pruning. In: IEEE International Conference on Neural Networks (1993)
51. He, Y., Zhang, X., Sun, J.: Channel pruning for accelerating very deep neural networks. In: ICCV (2017)
52. Hendrycks, D., Gimpel, K.: Gaussian error linear units (GELUs). arXiv:1606.08415 (2016)
53. Hertrich, C., Basu, A., Summa, M.D., Skutella, M.: Towards lower bounds on the depth of ReLU neural networks. In: NeurIPS (2021)

54. Hertrich, C., Loho, G.: Neural networks and (virtual) extended formulations, (2024). arXiv:2411.03006
55. Hinz, P., van de Geer, S.: A framework for the construction of upper bounds on the number of affine linear regions of ReLU feed-forward neural networks. IEEE Trans. Inf. Theory (2019)
56. Hoefler, T., Alistarh, D., Ben-Nun, T., Dryden, N., Peste, A.: Sparsity in deep learning: pruning and growth for efficient inference and training in neural networks. J. Mach. Learn. Res. (2021)
57. Hojny, C., Zhang, S., Campos, J.S., Misener, R.: Verifying message-passing neural networks via topology-based bounds tightening. In: ICML (2024)
58. Hooker, S., Courville, A., Clark, G., Dauphin, Y., Frome, A.: What do compressed deep neural networks forget? arXiv:1911.05248 (2019)
59. Hooker, S., Moorosi, N., Clark, G., Bengio, S., Denton, E.: Characterising bias in compressed models. arXiv:2010.03058 (2020)
60. Hornik, K., Stinchcombe, M., White, H.: Multilayer feedforward networks are universal approximators. Neural Networks (1989)
61. Huchette, J.: Advanced mixed-integer programming formulations: Methodology, computation, and application. Ph.D. thesis, Massachusetts Institute of Technology (2018)
62. Huchette, J., Muñoz, G., Serra, T., Tsay, C.: When deep learning meets polyhedral theory: a survey. INFORMS J. Comput. (2026)
63. Huchette, J., Vielma, J.P.: Nonconvex piecewise linear functions: advanced formulations and simple modeling tools. Oper. Res. (2023)
64. Janowsky, S.: Pruning versus clipping in neural networks. Phys. Rev. A (1989)
65. Jin, T., Roy, D., Carbin, M., Frankle, J., Dziugaite, G.: On neural network pruning's effect on generalization. In: NeurIPS (2022)
66. Kanamori, K., Takagi, T., Kobayashi, K., Ike, Y., Uemura, K., Arimura, H.: Ordered counterfactual explanation by mixed-integer linear optimization. In: AAAI (2021)
67. Kumar, A., Serra, T., Ramalingam, S.: Equivalent and approximate transformations of deep neural networks. arXiv:1905.11428 (2019)
68. LeCun, Y., Bottou, L., Bengio, Y., Haffner, P.: Gradient-based learning applied to document recognition. Proc. IEEE (1998)
69. LeCun, Y., Denker, J., Solla, S.: Optimal brain damage. In: NeurIPS (1989)
70. LeCun, Y., Bengio, Y., Hinton, G.: Deep learning. Nature (2015)
71. Lee, N., Ajanthan, T., Torr, P.: SNIP: Single-shot network pruning based on connection sensitivity. In: ICLR (2019)
72. Li, H., Xu, Z., Taylor, G., Studer, C., Goldstein, T.: Visualizing the loss landscape of neural nets. In: NeurIPS (2018)
73. Liebenwein, L., Baykal, C., Carter, B., Gifford, D., Rus, D.: Lost in pruning: the effects of pruning neural networks beyond test accuracy. In: MLSys (2021)
74. Liu, C., Arnon, T., Lazarus, C., Strong, C., Barrett, C., Kochenderfer, M.J., et al.: Algorithms for verifying deep neural networks. Found. Trends® Optimization (2021)
75. Liu, X., Han, X., Zhang, N., Liu, Q.: Certified monotonic neural networks. In: NeurIPS. vol. 33 (2020)
76. Liu, X., Dvorkin, V.: Optimization over trained neural networks: difference-ofconvex algorithm and application to data center scheduling. IEEE Control Syst. Lett. (2025)
77. Lomuscio, A., Maganti, L.: An approach to reachability analysis for feed-forward ReLU neural networks. arXiv:1706.07351 (2017)

78. Lueg, L., Grimstad, B., Mitsos, A., Schweidtmann, A.M.: reluMIP: open source tool for MILP optimization of relu neural networks (2021). https://doi.org/10.5281/zenodo.5601907, https://github.com/ChemEngAI/ReLU_ANN_MILP

79. Luo, J., Wu, J., Lin, W.: ThiNet: a filter level pruning method for deep neural network compression. In: ICCV (2017)

80. Ma, X., Fang, G., Wang, X.: LLM-Pruner: on the structural pruning of large language models. In: NeurIPS (2023)

81. Magnani, A., Boyd, S.P.: Convex piecewise-linear fitting. Optimization Eng. (2009)

82. Mangasarian, O.L., Rosen, J.B., Thompson, M.E.: Global minimization via piecewise-linear underestimation. J. Global Optimization (2005)

83. Maragno, D., Wiberg, H., Bertsimas, D., Birbil, S.I., Hertog, D.d., Fajemisin, A.: Mixed-integer optimization with constraint learning. Oper. Res. (2023)

84. McCormick, G.P.: Computability of global solutions to factorable nonconvex programs: part I — convex underestimating problems. Mathematical Programming (1976)

85. McDonald, T., Tsay, C., Schweidtmann, A.M., Yorke-Smith, N.: Mixed-integer optimisation of graph neural networks for computer-aided molecular design. Comput. Chemic. Eng. (2024)

86. Misener, R., Floudas, C.A.: Piecewise-linear approximations of multidimensional functions. J. Optimiz. Theory Appl. (2010)

87. Misener, R., Biegler, L.: Formulating data-driven surrogate models for process optimization. Comput. Chemic. Eng. (2023)

88. Montúfar, G.: Notes on the number of linear regions of deep neural networks. In: Sampling Theory and Applications (SampTA) (2017)

89. Montúfar, G., Pascanu, R., Cho, K., Bengio, Y.: On the number of linear regions of deep neural networks. In: NeurIPS. vol. 27 (2014)

90. Montúfar, G., Ren, Y., Zhang, L.: Sharp bounds for the number of regions of maxout networks and vertices of Minkowski sums. SIAM J. Appl. Algebra Geometry (2022)

91. Mozer, M., Smolensky, P.: Using relevance to reduce network size automatically. Connection Sci. (1989)

92. Murzakhanov, I., Venzke, A., Misyris, G.S., Chatzivasileiadis, S.: Neural networks for encoding dynamic security-constrained optimal power flow. In: Bulk Power Systems Dynamics and Control Sympositum (2022)

93. Nair, V., Hinton, G.: Rectified linear units improve restricted boltzmann machines. In: ICML (2010)

94. Paganini, M.: Prune responsibly. arXiv:2009.09936 (2020)

95. Pascanu, R., Montúfar, G., Bengio, Y.: On the number of response regions of deep feedforward networks with piecewise linear activations. In: ICLR (2014)

96. Perakis, G., Tsiourvas, A.: Optimizing objective functions from trained ReLU neural networks via sampling. arXiv:2205.14189 (2022)

97. Phuong, M., Lampert, C.H.: Functional vs. parametric equivalence of ReLU networks. In: ICLR (2020)

98. Plate, C., Hahn, M., Klimek, A., Ganzer, C., Sundmacher, K., Sager, S.: An analysis of optimization problems involving ReLU neural networks. arXiv:2502.03016 (2025)

99. Radford, A., Narasimhan, K., Salimans, T., Sutskever, I.: Improving language understanding by generative pre-training. https://cdn.openai.com/research-covers/language-unsupervised/ language_understanding_paper.pdf (2018). Accessed 04 Mar 2025

100. Raghu, M., Poole, B., Kleinberg, J., Ganguli, S., Dickstein, J.: On the expressive power of deep neural networks. In: ICML (2017)
101. Riera, C., Rey, C., Serra, T., Puertas, E., Pujol, O.: Training thinner and deeper neural networks: jumpstart regularization. In: CPAIOR (2022)
102. Rolnick, D., Kording, K.: Reverse-engineering deep ReLU networks. In: International Conference on Machine Learning (ICML) (2020)
103. Rosen, J.B., Pardalos, P.M.: Global minimization of large-scale constrained concave quadratic problems by separable programming. Math. Program. (1986)
104. Rössig, A., Petkovic, M.: Advances in verification of ReLU neural networks. J. Global Optimization (2021)
105. Safran, I., Reichman, D., Valiant, P.: How many neurons does it take to approximate the maximum? In: SODA (2024)
106. Say, B., Wu, G., Zhou, Y.Q., Sanner, S.: Nonlinear hybrid planning with deep net learned transition models and mixed-integer linear programming. In: IJCAI (2017)
107. Schweidtmann, A.M., Mitsos, A.: Deterministic global optimization with artificial neural networks embedded. J. Optimization Theory Appl. (2019)
108. Serra, T., Ramalingam, S.: Empirical bounds on linear regions of deep rectifier networks. In: AAAI (2020)
109. Serra, T., Tjandraatmadja, C., Ramalingam, S.: Bounding and counting linear regions of deep neural networks. In: ICML (2018)
110. Serra, T., Yu, X., Kumar, A., Ramalingam, S.: Scaling up exact neural network compression by ReLU stability. In: NeurIPS (2021)
111. Serra, T., Kumar, A., Ramalingam, S.: Lossless compression of deep neural networks. In: CPAIOR (2020)
112. Shi, C., Emadikhiav, M., Lozano, L., Bergman, D.: Constraint learning to define trust regions in optimization over pre-trained predictive models. INFORMS J. Comput. (2024)
113. Singh, S.P., Alistarh, D.: WoodFisher: efficient second-order approximation for neural network compression. In: NeurIPS (2020)
114. Sosnin, P., Tsay, C.: Scaling mixed-integer programming for certification of neural network controllers using bounds tightening. In: CDC (2024)
115. Sourek, G., Zelezny, F.: Lossless compression of structured convolutional models via lifting. In: ICLR (2021)
116. Strong, C.A., et al.: Global optimization of objective functions represented by ReLU networks. Mach. Learn. (2021)
117. Sun, M., Liu, Z., Bair, A., Kolter, J.Z.: A simple and effective pruning approach for large language models. In: ICLR (2024)
118. Sun, R., Li, D., Liang, S., Ding, T., Srikant, R.: The global landscape of neural networks: an overview. IEEE Sign. Process. Magazine **37** (2020)
119. Telgarsky, M.: Representation benefits of deep feedforward networks (2015). arXiv:1509.08101
120. Tjeng, V., Xiao, K., Tedrake, R.: Evaluating robustness of neural networks with mixed integer programming. In: ICLR (2019)
121. Tong, J., Cai, J., Serra, T.: Optimization over trained neural networks: Taking a relaxing walk. In: CPAIOR (2024)
122. Tong, J., Zhu, Y., Serra, T., Burer, S.: Optimization over trained neural networks: going large with gradient-based algorithms. In: CPAIOR (2026)
123. Tran, C., Fioretto, F., Kim, J.E., Naidu, R.: Pruning has a disparate impact on model accuracy. In: NeurIPS (2022)

124. Tsay, C., Kronqvist, J., Thebelt, A., Misener, R.: Partition-based formulations for mixed-integer optimization of trained ReLU neural networks. In: NeurIPS (2021)
125. Tseran, H., Montúfar, G.: On the expected complexity of maxout networks. In: NeurIPS (2021)
126. Tsiourvas, A., Sun, W., Perakis, G.: Manifold-aligned counterfactual explanations for neural networks. In: AISTATS (2024)
127. Turner, M., Chmiela, A., Koch, T., Winkler, M.: SurrogateLIB: an extendable library of mixed- integer programs with embedded machine learning predictors (2024). https://doi.org/10.5281/zenodo.11231147
128. Turner, M., Chmiela, A., Koch, T., Winkler, M.: PySCIPOpt-ML: embedding trained machine learning models into mixed-integer programs. In: CPAIOR (2025)
129. Vaswani, A., et al.: Attention is all you need. In: NeurIPS (2017)
130. Verma, S., Pesquet, J.C.: Sparsifying networks via subdifferential inclusion. In: ICML (2021)
131. Vielma, J.P., Ahmed, S., Nemhauser, G.: Mixed-integer models for nonseparable piecewise-linear optimization: unifying framework and extensions. Oper. Res. (2010)
132. Wang, K., Lozano, L., Cardonha, C., Bergman, D.: Optimizing over an ensemble of trained neural networks. INFORMS J. Comput. (2023)
133. Wang, Y.: Estimation and comparison of linear regions for ReLU networks. In: IJCAI (2022)
134. Williams, H.P.: Model Building in Mathematical Programming. Wiley, 5 edn. (2013)
135. Wu, D., Modoranu, I.V., Safaryan, M., Kuznedelev, D., Alistarh, D.: The iterative optimal brain surgeon: faster sparse recovery by leveraging second-order information. In: NeurIPS (2024)
136. Wu, G., Say, B., Sanner, S.: Scalable planning with deep neural network learned transition models. J. Artifi. Intell. Res. (2020)
137. Xia, M., Gao, T., Zeng, Z., Chen, D.: Sheared LLaMA: accelerating language model pre-training via structured pruning. In: ICLR (2024)
138. Xiao, H., Rasul, K., Vollgraf, R.: Fashion-MNIST: a novel image dataset for benchmarking machine learning algorithms. arXiv 1708.07747 (2017)
139. Xiao, K.Y., Tjeng, V., Shafiullah, N.M., Madry, A.: Training for faster adversarial robustness verification via inducing ReLU stability. In: ICLR (2019)
140. Yang, S., Bequette, B.W.: Optimization-based control using input convex neural networks. Comput. Chemic. Eng. (2021)
141. Yarotsky, D.: Error bounds for approximations with deep ReLU networks. Neural Netw. (2017)
142. Ye, M., Gong, C., Nie, L., Zhou, D., Klivans, A., Liu, Q.: Good subnetworks provably exist: pruning via greedy forward selection. In: ICML (2020)
143. Yu, X., Serra, T., Ramalingam, S., Zhe, S.: The combinatorial brain surgeon: pruning weights that cancel one another in neural networks. In: ICML (2022)
144. Zhao, H., Hijazi, H., Jones, H., Moore, J., Tanneau, M., Hentenryck, P.V.: Bound tightening using rolling-horizon decomposition for neural network verification. In: CPAIOR (2024)
145. Zhu, Y., Burer, S.: An extended validity domain for constraint learning. arXiv:2406.10065 (2024)

Introducing Automatically Derived Subproblem Relaxations in a Logic-Based Benders Decomposition Solver

Sachin Rajendran[1]([envelope]) [iD], Stephen Maher[2] [iD], and Elina Rönnberg[1] [iD]

[1] Department of Mathematics, Linköping University, Linköping, Sweden
`sachin.rajendran@liu.se`
[2] GAMS Software GmbH, Frechen, Germany

Abstract. Logic-based Benders decomposition (LBBD) effectively combines mixed integer programming and constraint programming to solve difficult optimisation problems. While the decomposition exploits different solution and modelling paradigms, the separation destroys problem information connecting master and subproblem decisions. Alongside cut generation, a standard way of re-introducing this connection is through the addition of valid inequalities in the master problem. These inequalities typically constitute subproblem relaxations and they are usually problem specific and derived by hand. This paper proposes a general approach for deriving valid inequalities and demonstrates how they can be applied to cumulative and disjunctive constraints. In an effort to develop a general purpose LBBD solver, the general derivation of valid inequalities have been implemented as part of a new cumulative constraint handler for SCIP and the effectiveness is evaluated on two variants of scheduling problems. The computational experiments show the potential of applying these general valid inequalities to the master problem.

Keywords: logic-based Benders decomposition · generic valid inequalities · implication structure · SCIP

1 Introduction

Logic-Based Benders Decomposition (LBBD) has been successfully applied to a variety of applications in transportation, production management, supply chain, computing, and telecommunications [5]. Computationally effective implementations of LBBD commonly rely on problem-specific acceleration strategies such as cut-strengthening techniques, customised cuts, and adding subproblem-based valid inequalities to the master problem. Typically, such valid inequalities constitute a relaxation of a set of subproblem constraints and their variables, or expressed in terms of master problem variables but hand-crafted for a specific subproblem structure [5]. In this paper we introduce a generalised form of subproblem relaxations, expressed in terms of master problem variables. Preliminary computational experiments are performed to assess the potential of including such a subproblem relaxation. Further, we exploit the property of Linear

T. Guns (Ed.): CPAIOR 2026, LNCS 16595, pp. 443–452, 2026.
https://doi.org/10.1007/978-3-032-27242-3_26

Programming (LP) relaxations in Constraint Integer Programming (CIP) and investigate the effect of the same relaxation in the subproblem for cut generation.

1.1 Preliminaries

Our work addresses a relatively general problem structure with implication constraints. In the following, the notion of an item should be interpreted in its most wide sense, e.g. a job, a vehicle, a product, or a patient. Let the set J index items and let the set S index subproblems. A subproblem $s \in S$ corresponds to an independent facility, machine, segment of a timeline or anything that an item can be assigned to. As decision variables, let x_{js} be a binary variable indicating if item j is assigned to subproblem s or not, and let y_{js} be defined such that it can belong to any type of domain $\mathcal{D}^{\mathrm{S}}_{js}$ if the corresponding $x_{js} = 1$, and that it has to be zero otherwise, $j \in J$, $s \in S$. To model the implication structure, we use the notation $(y_{js})_{j \in J | x_{js}=1}$ to include the variables $y_{js} \in \mathcal{D}^{\mathrm{S}}_{js}$ for each subproblem $s \in S$ where $x_{js} = 1$, and exclude them otherwise. Also let $x = (x_{js})_{j \in J, s \in S}$ and $y = (y_{js})_{j \in J, s \in S}$. Before decomposition, the problem is of the form

$$
\begin{aligned}
[\mathrm{P}] \quad \min \quad & h(x) + g(y) & & (1)\\
\mathrm{s.t.} \quad & \mathcal{C}^{\mathrm{M}}(x), & & (2)\\
& \mathcal{C}^{\mathrm{S}}_{s}\big((y_{js})_{j \in J | x_{js}=1}\big), & s \in S, & (3)\\
& y_{js} \in \mathcal{D}^{\mathrm{S}}_{js}, & j \in J,\ s \in S \mid x_{js} = 1, & (4)\\
& y_{js} = 0, & j \in J,\ s \in S \mid x_{js} = 0, & (5)\\
& x_{js} \in \{0,1\}, & j \in J,\ s \in S. & (6)
\end{aligned}
$$

The reason for addressing this particular problem structure is that it captures many of the problems to which LBBD has been successfully applied, see e.g. [2,3,6,7,9]. Thus, effectively handling such problems in a generic LBBD solver, accelerated by general purpose techniques, has wide-reaching benefits. This paper is part of a research project towards implementing such a solver in SCIP [4] and its purpose is to complement the existing initiative with the NUT-MEG solver [10] by addressing this more specific structure. Both NUTMEG and our ongoing work are based on the branch-and-check form of the LBBD algorithm and use a Mixed-Integer Programming (MIP) formulation in the master problem. In the subproblem, NUTMEG elegantly leverages the conflict analysis capabilities of lazy clause generation Constraint Programming (CP) solvers. This means it is designed to handle general problem structures where the objective depends only on master problem variables. One property of the NUTMEG implementation is that the complete model is handled in the subproblem. This makes it less suitable for problems with the type of implication structure introduced above, where there are known benefits from handling them explicitly, see e.g. [9].

1.2   Contributions and Outline

The current section is concluded with a brief review of the use of subproblem relaxations. Section 2 provides the decomposition of [P] and a proposition stating the general valid inequalities that can be used as subproblem relaxation constraints in the master problem. Section 3 presents a general type of scheduling problem structure where the subproblem either contains cumulative or disjunctive constraints. For this structure, we show how to use common relaxations of the subproblem constraints as an input to our general subproblem relaxation in the master problem. We also show how these generically-derived valid inequalities generalises commonly used hand-crafted ones, confirming the potential of our inequalities. Section 4 details how the relaxation of the subproblem constraints are implemented in a new constraint handler that is used with the Benders decomposition framework in SCIP [11]. Further, this constraint handler generates valid inequalities for either the master problem or subproblem. The code is publicly available in https://gitlab.liu.se/eliro15/SCIP_LBBD_cumulative. Section 5 shows and explains the computational results for some specific models and publicly available instances from the literature: cumulative facility scheduling [6] and single-facility scheduling with a segmented timeline [12], both with makespan and fixed costs objectives.

1.3   Literature Review

Without aiming for completeness, this review provides some examples of subproblem relaxations derived from the subproblem constraints. For cumulative facility scheduling with fixed costs, the so-called energy relaxation can be used [6]. In this problem, tasks are assigned to facilities and then scheduled on the respective facilities, complying with task-specific release times and deadlines, and capacity restrictions stated by a cumulative constraint. The energy relaxation is hand-crafted from the cumulative constraint. For each possible interval between a release time-deadline pair, it states that the claimed capacity of tasks in this interval cannot exceed the total available amount of capacity in the interval.

Another important problem structure is single machine scheduling with sequence-dependent setup times and multiple time windows [8], which is a generalisation of single-facility scheduling with segmented time line [1]. In this problem, jobs are assigned to segments of a time line and then scheduled within the segment by a disjunctive constraint, adhering to job-and-segment-dependent release times and deadlines. For this problem, a segment relaxation is hand-crafted from the disjunctive constraint and for each segment, the total processing time of the assigned jobs are restricted to at most the length of the segment. This type of relaxation is also used for similar scheduling problems such as in [7].

Examples of vehicle routing applications include [13], where a condition on the total travel time in each subproblem is used. Alternatively, [2] uses clique inequalities representing when a vehicle cannot service a subset of customers.

2 Decomposition and Subproblem Relaxation

The problem is decomposed such that the master problem and subproblems contain the variables x_{js} and y_{js}, respectively, $j \in J$, $s \in S$. Assuming $g(y)$ is bounded from below by $\hat{g}$, the master problem in a given iteration is

$$
\begin{aligned}
[\text{MP}] \quad \min \quad & h(x) + \varphi \\
\text{s.t.} \quad & \mathcal{C}^{\text{M}}(x), \\
& \mathcal{B}_\omega(x, \varphi), && \omega \in \Omega, \\
& [\text{Subproblem relaxation}] \\
& x_{js} \in \{0,1\}, && j \in J,\ s \in S, \\
& \varphi \geq \hat{g},
\end{aligned}
$$

where Ω is a set of all generated Benders cuts and $\mathcal{B}_\omega(x, \varphi)$, $\omega \in \Omega$, is a bounding function. Given a solution, $\hat{x} = \left(\hat{x}_{js}\right)_{j \in J, s \in S}$, the subproblem is

$$
\begin{aligned}
[\text{SP}(\hat{x})] \quad \min \quad & g(y) \\
\text{s.t.} \quad & \mathcal{C}^{\text{S}}_s\big((y_{js})_{j \in J | \hat{x}_{js}=1}\big), && s \in S, \\
& y_{js} \in \mathcal{D}^{\text{S}}_{js}, && j \in J,\ s \in S \mid \hat{x}_{js} = 1.
\end{aligned}
$$

Note that when $\hat{x}_{js} = 0$ holds for a master problem variable, the corresponding variable $\hat{y}_{js} = 0$ is not present in the subproblem constraints, $j \in J$, $s \in S$.

We introduce subproblem relaxations on a specific form, given by the following proposition. For the proposition to be applicable, a required input is a model for the subproblem, or a relaxation of the subproblem, where at least a part of the formulation consists of linear inequalities. Such a model can be used both in the proposition and to formulate a relaxed subproblem for generating cuts.

Proposition 1. *For each $s \in S$, let the constraints*

$$
\sum_{j \in J} a^{\text{LEQ}}_{jsu} w_{js} \leq b^{\text{LEQ}}_{su},\ u \in U, \qquad (7)
$$

$$
\sum_{j \in J} a^{\text{GEQ}}_{jsv} w_{js} \geq b^{\text{GEQ}}_{sv},\ v \in V, \qquad (8)
$$

$$
\mathcal{C}^{\text{L}}_s\big((w_{js})_{j \in J | x_{js}=1}\big), \qquad (9)
$$

$$
w_{js} \in \mathcal{D}^{\text{L}}_{js},\ j \in J \mid x_{js} = 1, \qquad (10)
$$

$$
w_{js} = 0,\ j \in J \mid x_{js} = 0, \qquad (11)
$$

$$
x_{js} \in \{0,1\},\ j \in J, s \in S, \qquad (12)
$$

provide a relaxation of the feasible set described by constraints (3)–(6) in [P]. *Then,*

$$
\sum_{j \in J} a^{\text{MIN}}_{jsu} x_{js} \leq b^{\text{LEQ}}_{su},\ \text{with } a^{\text{MIN}}_{jsu} = \min_{s.t.} (9) - (11) a^{\text{LEQ}}_{jsu} w_{js},\ u \in U, \qquad (13)
$$

$$
\sum_{j \in J} a^{\text{MAX}}_{jsv} x_{js} \geq b^{\text{GEQ}}_{sv},\ \text{with } a^{\text{MAX}}_{jsv} = \max_{s.t.} (9) - (11) a^{\text{GEQ}}_{jsv} w_{js},\ v \in V, \qquad (14)
$$

are valid inequalities for [P].

Proof. Consider the feasible set described by (7)–(12). Using (11) gives that (7) and (8) can be equivalently stated as

$$\sum_{j \in J} a_{jsu}^{\mathrm{LEQ}} w_{js} x_{js} \leq b_{su}^{\mathrm{LEQ}}, \ u \in U, \ \text{and} \ \sum_{j \in J} a_{jsv}^{\mathrm{GEQ}} w_{js} x_{js} \geq b_{sv}^{\mathrm{GEQ}}, \ v \in V.$$

Since $a_{jsu}^{\mathrm{MIN}} x_{js} \leq a_{jsu}^{\mathrm{LEQ}} w_{js} x_{js}$, $a_{jsv}^{\mathrm{MAX}} x_{js} \geq a_{jsv}^{\mathrm{GEQ}} w_{js} x_{js}$, $j \in J$, $s \in S$, $u \in U$, $v \in V$, the result follows. $\qquad\square$

Note that an equality constraint can be represented by its two corresponding inequality constraints and the results of the proposition can be applied to these.

3 Scheduling Problems with Implication Structure

We study two applications in our preliminary investigations: cumulative facility scheduling and single-facility scheduling with a segmented timeline, using two different objectives, both to minimise assignment cost and to minimise makespan. In the following we state a model that captures both applications, in line with model [P]. Let J index the set of jobs and S index a set of facilities. The variables are defined as in [P]. For $j \in J$, $s \in S$, let the processing time and resource consumption per time unit of job j in facility s be p_{js} and c_{js}, respectively. Let the domain $\mathcal{D}_{js}^{\mathrm{s}} = \{r_{js}, \ldots, d_{js} - p_{js}\}$, where r_{js} is the release time and d_{js} is the deadline for job $j \in J$ in facility $s \in S$. In facility $s \in S$ the maximum allowed resource consumption per time unit is C_s.

The model is given by

$$
\begin{aligned}
\min \quad & h(x) + g(y), \\
\text{s.t.} \quad & \sum_{s \in S} x_{js} = 1, && j \in J, \\
& \textsc{Cumulative}\big((y_{js}, p_{js}, c_{js})_{j \in J \mid x_{js}=1}, C_s\big), && s \in S, \\
& y_{js} \in \{r_{js}, \ldots, d_{js} - p_{js}\}, && j \in J, s \in S \mid x_{js} = 1, \\
& y_{js} = 0, && j \in J, s \in S \mid x_{js} = 0, \\
& x_{js} \in \{0, 1\}, && j \in J, s \in S,
\end{aligned}
$$

where the first and second constraint corresponds to (2) and (3), respectively, in [P]. Application-specific models are obtained as follows: For cumulative facility scheduling, release times and deadlines are subproblem independent and replaced by $r_{js} = r_j$ and $d_{js} = d_j$, $j \in J$, $s \in S$. For single-facility scheduling, $p_{js} = p_j$, $j \in J$, $s \in S$. There is also a disjunctive constraint in the subproblem instead of the cumulative constraint, which is obtained by fixing $c_{js} = 1$, $C_s = 1$, $j \in J$, $s \in S$, in the cumulative constraint. When minimising assignment cost the objectives are $h(x) = \sum_{j \in J, s \in S} f_{js} x_{js}$ and $g(y) = 0$, where f_{js} is the cost of assigning job $j \in J$ to facility $s \in S$, and when minimising makespan they are $h(x) = 0$ and $g(y) = \max_{j \in J, s \in S} y_{js} + p_{js}$. For single-facility scheduling,

$h(x) = \sum_{j \in J, s \in S} 2p_j(1 - x_{js})$ which includes a penalty of $2p_j$ if job $j \in J$ is unscheduled. The decomposition of the problem follows that of [P].

A commonly used MIP formulation for the cumulative constraint is to make a time-indexed formulation and introduce capacity constraints for each time step. From this, a relaxation is obtained by aggregating constraints over time intervals. A derivation of this relaxation and its use in Proposition 1 is as follows.

For $s \in S$, let $r_s^{\mathrm{MIN}} = \min_{j \in J} r_{js}$ and $d_s^{\mathrm{MAX}} = \max_{j \in J} d_{js}$ be the earliest release time and latest deadline, respectively. Let $T_s = \{r_s^{\mathrm{MIN}}, \ldots, d_s^{\mathrm{MAX}}\}$ be the set of integer time steps, $s \in S$. Introduce the variables $w_{js} = (w_{jst})_{t \in \{r_{js}, \ldots, d_{js} - p_{js}\}}$, where w_{jst} indicates if job j starts at time t in subproblem s or not, $j \in J$, $s \in S$. Its domain, for $s \in S$, $t \in \{r_{js}, \ldots, d_{js} - p_{js}\}$ is defined as $w_{jst} \in \{0,1\}$, $j \in J \mid x_{js} = 1$ and $w_{jst} = 0$, $j \in J \mid x_{js} = 0$, corresponding to (10) and (11) in Proposition 1. For $s \in S$, the capacity for time step $t' \in T_s$ is respected if

$$\sum_{j \in J} c_{js} \sum_{t=t'-p_{js}+1}^{t'} w_{jst} \leq C_s, \qquad \sum_{t \in \{r_{js}, \ldots, d_{js} - p_{js}\}} w_{jst} = 1, \; j \in J \mid x_{js} = 1.$$

Let the latter constraint correspond to (9) in Proposition 1 and note that a relaxation of the former constraint can be obtained by summing such constraints over a subset of time steps. Introduce U_s to index a set of time intervals for $s \in S$ and, for $u \in U_s$, let $T'_{su} = \{t_{su}^{\mathrm{START}}, \ldots, t_{su}^{\mathrm{END}}\} \subseteq T_s$ denote a set of consecutive time steps. The linear formulation we use as Constraint (7) in Proposition 1 is

$$\sum_{j \in J} c_{js} \sum_{t'=T'_{su}} \sum_{t=t'-p_{js}+1}^{t'} w_{jst} \leq C_s \left(t_{su}^{\mathrm{END}} - t_{su}^{\mathrm{START}}\right), \; u \in U_s.$$

To compute a_{jsu}^{MIN} in Proposition 1, note that a lower bound on the number of times the sum $\sum_{t=t'-p_{js}+1}^{t'} w_{jst}$ can take the value one is obtained by considering the number of time steps in the interval between r_{js} and d_{js} that are outside of the interval $t_{su}^{\mathrm{START}}, \ldots, t_{su}^{\mathrm{END}}$ and assume that those will be used. It then follows from Proposition 1 that, given a $u \in U_s$, $s \in S$

$$a_{jsu}^{\mathrm{MIN}} = \min_{\substack{\sum_{t \in \{r_{js}, \ldots, d_{js} - p_{js}\}} w_{jst} = 1 \\ w_{jst} \in \{0,1\} \text{ if } x_{js} = 1 \\ w_{jst} = 0, \text{ if } x_{js} = 0}} \sum_{t'=T'_{su}} \sum_{t=t'-p_{js}+1}^{t'} w_{jst}$$

$$= \max\left\{0, p_{js} - \max\left\{0, t_{su}^{\mathrm{START}} - r_{js}\right\} - \max\left\{0, d_{js} - t_{su}^{\mathrm{END}}\right\}\right\}, \tag{15}$$

and

$$\sum_{j \in J} c_{js} a_{jsu}^{\mathrm{MIN}} x_{js} \leq C_s \left(t_{su}^{\mathrm{END}} - t_{su}^{\mathrm{START}}\right)$$

is a valid inequality for our scheduling problem.

The derived inequalities relate to the ones introduced in Sect. 1.3 as follows. First, note that the energy and segment relaxations only include jobs for which it holds that $t_{su}^{\text{START}} \leq r_{js}$ and $d_{js} \leq t_{su}^{\text{END}}$, while (15) can also handle partial contributions of a job. Choosing the set U_s, $s \in S$, so that the time intervals are between all release time and deadline pairs and considering only $j \in J$ such that $t_{su}^{\text{START}} \leq r_{js}$ and $d_{js} \leq t_{su}^{\text{END}}$ hold, our valid inequality coincides with the energy relaxation. Using a similar argument, but with the set U_s chosen so that a time interval is a segment, our valid inequality becomes the segment relaxation.

4   Implementation

The concepts presented in this paper are implemented using the Benders decomposition framework of SCIP [11]. A branch-and-cut version of Benders decomposition is applied using the default Benders decomposition plugin, where the subproblems are solved as SCIP instances. A feature of SCIP is that general implementations of both classical Benders and LBBD-based cuts are available. Classical cuts are generated from the LP relaxation of the subproblem, while the LBBD-based no-good cuts are generated from the solution to the CIP.

Our implementation extends SCIP with a variant of a CUMULATIVE constraint handler. The new constraint handler, termed CUMULATIVEBENDERS, permits jobs to be optionally selected in a cumulative constraint to consume resources. Prior to solving each subproblem—given a master assignment solution— CUMULATIVEBENDERS adds a CUMULATIVE constraint that includes only the jobs assigned to this subproblem.

The CUMULATIVEBENDERS constraint is important regarding the implementation of Constraints (7)–(12). The valid inequalities for both the master and subproblem are implemented as part of the CUMULATIVEBENDERS constraint, which are added during the initialisation stage immediately prior to presolving each problem. As a result, the CUMULATIVEBENDERS constraint handler provides a general method to add subproblem relaxations for the CUMULATIVE and DISJUNCTIVE constraints when applying Benders decomposition. For our preliminary experiments in this paper, we chose the sets U_s, $s \in S$, and the jobs so that we use the inequalities corresponding to the energy and segment relaxations.

5   Computational Results and Conclusions

The computational experiments assess the effect of using the generic relaxation of Proposition 1 in the master or subproblem when applying LBBD to the problems in Sect. 3. The problems are solved by SCIP 10.0 using Gurobi Optimizer 11.0 as the LP solver. The experiments were executed on a compute node comprising two Intel Xeon Gold 6130 processors (32 cores, 2.10 GHz) and 376 GB of RAM. Each instance was solved using 8 CPU cores with a time limit of 20 min. For cumulative facility scheduling, the 335 instances presented in [6] are used. For single-facility scheduling, the 480 instances from [12], based on [1], are used.

The results presented in Fig. 1 show that adding generically-derived valid inequalities to the master problem exhibit similar performance improvements as problem-specific subproblem relaxations. The explanation of this performance is that the generically-derived valid inequalities focus on eliminating master problem solutions that induces infeasible subproblems—thus, reducing the number of LBBD-based cuts that are generated. An exception is Fig. 1b, where the subproblems are always feasible and hence there are no master solutions to eliminate. In this case, objective-based valid inequalities are potentially better suited to improve computational performance. Overall, when subproblems become infeasible, the generically-derived valid inequalities are computationally effective.

Figs. 1c and 1d shows that adding valid inequalities to the subproblem does not greatly improve computational performance. This is due to the valid inequalities being designed to eliminate infeasible subproblems, and not necessarily strengthening the LP relaxation of the subproblem. In most instances, the valid inequalities do not sufficiently strengthen the LP relaxation of the subproblem to generate classical Benders optimality and feasibility cuts—hence, relying only on LBBD-based cuts to eliminate infeasible or suboptimal solutions.

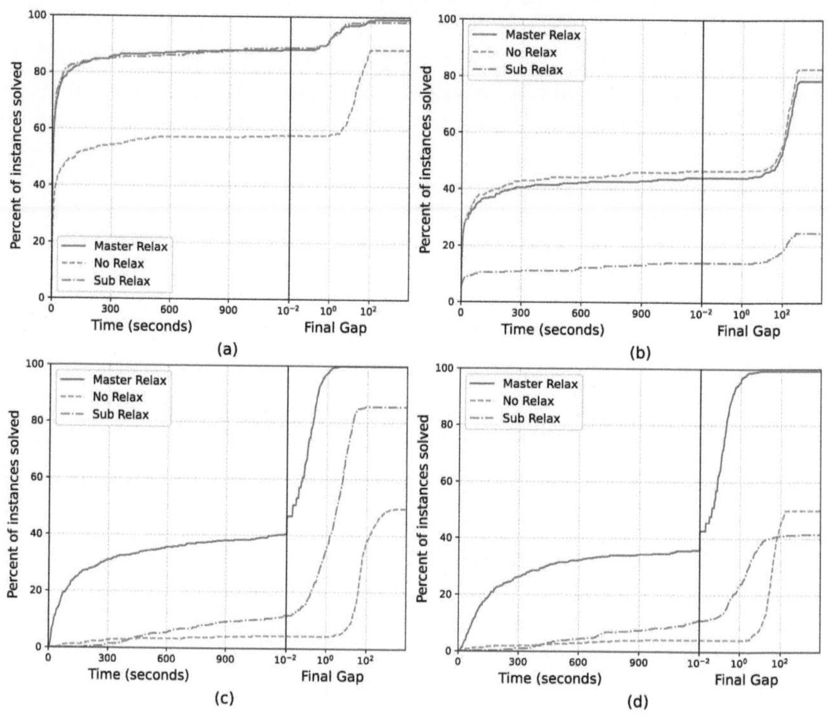

Fig. 1. Percentage of instances solved to optimality within 1200 s and the remaining optimality gap thereafter. Cumulative facility scheduling: (a) total cost, (b) makespan. Single-facility scheduling: (c) total cost, (d) makespan.

In conclusion, a framework for deriving valid inequalities is a valuable contribution to improve the general performance of LBBD. As confirmed by these results and from [10], valid inequalities based on the constraints of the problem are very effective at eliminating master solutions which induce infeasible subproblems. Extending the concepts in this paper to derive objective-based valid inequalities is a necessary step in the development of a general LBBD solver.

Acknowledgements. The work of S. Rajendran and E. Rönnberg was mainly supported by the Excellence Center at Linköping Lund in Information Technology (ELLIIT), and partially supported by the Wallenberg AI, Autonomous Systems and Software Program (WASP). Computational experiments were performed on resources provided by the National Academic Infrastructure for Supercomputing in Sweden (NAISS).

Disclosure of Interests. The authors have no competing interests.

References

1. Coban, E., Hooker, J.N.: Single-facility scheduling by logic-based benders decomposition. Ann. Oper. Res. **210**, 245–272 (2013)
2. Fachini, R.F., Armentano, V.A.: Logic-based benders decomposition for the heterogeneous fixed fleet vehicle routing problem with time windows. Comput. Ind. Eng. **148**, 106641 (2020)
3. Harjunkoski, I., Grossmann, I.E.: Decomposition techniques for multistage scheduling problems using mixed-integer and constraint programming methods. Comput. Chem. Eng. **26**, 1533–1552 (2002)
4. Hojny, C., et al.: The SCIP optimization suite 10.0. Technical report, Optimization Online, November 2025. https://optimization-online.org/2025/11/the-scip-optimization-suite-10-0/
5. Hooker, J.: Logic-Based Benders Decomposition - Theory and Applications. Springer Nature, Switzerland (2024)
6. Hooker, J.N.: Planning and scheduling by logic-based Benders decomposition. Oper. Res. **55**, 588–602 (2007)
7. Jain, V., Grossmann, I.E.: Algorithms for hybrid MILP/CP models for a class of optimization problems. INFORMS J. Comput. **13**(4), 258–276 (2001)
8. Karlsson, E., Rönnberg, E.: Strengthening of feasibility cuts in logic-based benders decomposition. In: Stuckey, P.J. (ed.) Integration of Constraint Programming, Artificial Intelligence, and Operations Research, pp. 45–61. Springer International Publishing, Cham (2021)
9. Karlsson, E., Rönnberg, E.: Logic-based Benders decomposition with a partial assignment acceleration technique for avionics scheduling. Comput. Oper. Res. **146**, 105916 (2022)
10. Lam, E., Gange, G., Stuckey, P.J., Van Hentenryck, P., Dekker, J.J.: Nutmeg: a MIP and CP hybrid solver using branch-and-check. SN Oper. Res. Forum **1**, 22:1–22:27 (2020)
11. Maher, S.J.: Implementing the branch-and-cut approach for a general purpose Benders' decomposition framework. Eur. J. Oper. Res. **290**(2), 479–498 (2021)

12. Saken, A., Karlsson, E., Maher, S., Rönnberg, E.: Computational evaluation of cut-strengthening techniques in logic-based benders' decomposition. Oper. Res. Forum **4** (2023)
13. Thorsteinsson, E.S.: Branch-and-check: a hybrid framework integrating mixed integer programming and constraint logic programming. In: Walsh, T. (ed.) CP 2001. LNCS, vol. 2239, pp. 16–30. Springer, Heidelberg (2001). https://doi.org/10.1007/3-540-45578-7_2

A Hybrid Learning-Based Matheuristic to Solve the Vehicle Routing Problem with Stochastic Demands

Gaël Reynal[1]($\boxtimes$) , Quentin Cappart[1,3,4] , Guy Desaulniers[1,2] ,
and Louis-Martin Rousseau[1,3]

[1] Polytechnique Montréal, Montreal, Canada
gael-simon.reynal@etud.polymtl.ca
[2] GERAD, Montreal, Canada
[3] CIRRELT, Montreal, Canada
[4] UCLouvain, Louvain-la-Neuve, Belgium

Abstract. The vehicle routing problem with stochastic demands is a combinatorial optimization problem that arises in industrial applications such as waste management and facility replenishment. In these applications, one could aim at using a specific number of vehicles to better align with available resources, thereby including a fixed-fleet constraint in the problem. Standard column generation heuristics, that are usually efficient in this context, struggle to handle this additional constraint and cannot quickly produce good feasible solutions, mainly because the labeling algorithm used during the pricing becomes inefficient. We introduce a hybrid pricing heuristic that generates columns by combining a greedy component aiming for a quick generation of good columns, a reinforcement learning module to compute critical routes disregarded by the greedy construction, and a tabu search procedure to explore the search space around the generated routes. We embed our method within an existing restricted master heuristic framework: we first perform a column generation phase using our pricing heuristic to quickly generate a set of high-quality columns, which we then complete with a greedy randomized adaptive search procedure. The resulting restricted master problem is then solved as a mixed-integer program. We evaluate our approach on 40 benchmark instances with up to 60 customers and achieve an average optimality gap of 1% within a 5-min total computation time. Our matheuristic also provides more best average solution cost and optimal solutions than the competing heuristics considered.

Keywords: vehicle routing · stochastic demands · restricted master heuristic · pricing heuristics · reinforcement learning

1 Introduction

Vehicle routing problems (VRPs) are widely applicable in industrial contexts, including freight/parcel delivery, waste management, and facility replenishment.

T. Guns (Ed.): CPAIOR 2026, LNCS 16595, pp. 453–469, 2026.
https://doi.org/10.1007/978-3-032-27242-3_27

For this reason, numerous variants of this problem have been studied. The basic situation involves a fleet of vehicles of known capacity departing from a single depot to serve a set of customers with known demands. Additional constraints can be added to better represent the context of the problem at hand. For instance, customer demands may be unknown in contexts such as waste collection [17]. In this case, the demands are stochastic, and only their probability distribution is known beforehand; their realization is revealed only upon reaching the customer. This still allows for *a priori* optimization: the routes are created with a *recourse strategy* to handle the possibility of overloads, a case in which the realized demand exceeds the remaining capacity in the vehicle. This variant is referred to as the *vehicle routing problem with stochastic demands* (VRPSD) [30].

In this work, we introduce an additional constraint to the VRPSD, namely, a fixed-fleet constraint, which specifies a predetermined fleet size. Although the common practice is to impose a lower bound on the number of vehicles used in a solution [10,14], in industrial settings, solutions employing a fixed, given number of vehicles may be preferred because they better correspond to the resources actually available. For example, a firm might want to use all the vehicles at its disposal or provide work to each of its employees. Consequently, the fixed-fleet constraint remains highly relevant in practice.

Looking at the solution approaches, *path-flow formulations* have proven to be an effective way to address this variant [10,12,20]. In these formulations, each feasible route is represented by a variable, which leads to an exponential number of variables overall. This naturally motivates the use of *column generation* [10–12] to solve the linear relaxation. The algorithm iteratively solves a *restricted master problem* (RMP), a restriction of the complete model limited to a subset of its variables, and a *pricing problem* that identifies variables with negative reduced costs to add to the RMP at each iteration. When no such variables remain, the RMP solution is optimal for the linear relaxation. Because this solution is not necessarily feasible for the original problem, the column generation procedure must be integrated into a broader algorithmic framework. To achieve this, one can use a *branch-and-price* algorithm that integrates column generation within a branch-and-bound procedure, possibly enhanced with cutting planes [10–12]. While this method guarantees finding an optimal solution if the pricing problem is solved exactly, it typically produces only a limited number of feasible solutions during the search and can be computationally demanding. As a result, it is less suitable in contexts where the goal is to obtain a good integer solution quickly.

Another option is to solve the RMP as a mixed-integer program (MIP) once the column generation process has converged, a technique commonly known as the *restricted master heuristic* (RMH, see Sadykov et al. [26]). In theory, it should yield a good feasible solution; however, since column generation is not designed to produce complementary columns that support integrality, solving the RMP as a MIP alone is often insufficient. To address this limitation, completion procedures, such as the one proposed by Reynal et al. [25], focus on leveraging the columns identified during column generation to generate complementary columns through a fast heuristic before solving the RMP as a MIP.

However, the introduction of the fixed-fleet constraint causes classical RMHs, such as that of Reynal et al. [25], to struggle even with solving the linear relaxation, often preventing the final MIP from yielding a feasible solution. This difficulty is particularly pronounced in instances with long routes and a small fleet size, as the underlying labeling algorithm becomes computationally complex When the goal is to obtain a good feasible solution rapidly, alternative pricing strategies must be considered. Metaheuristics are well known for their ability to quickly generate high-quality solutions, which can guide the column generation process toward more feasible outcomes [1,15,24].

On another front, *machine learning* has shown promising results when applied to routing problems, including within column generation frameworks. For example, Morabit et al. [21] introduced a bipartite graph neural network at each column generation iteration to select which columns to add to the RMP from a larger candidate set. Their model was trained to imitate a MIP formulation containing all candidate columns that aims at selecting a minimal set of columns that maximizes the objective function decrease. Later, Chi et al. [6] enhanced this approach by employing *reinforcement learning* (RL) to predict the most promising columns to add to the RMP from the candidate set. Their method outperformed the expert model proposed by Morabit et al. [21] on VRPTW instances with up to 50 customers. However, it required an efficient algorithm to generate candidate columns, a challenging task, particularly when long routes are involved. More recently, Morabit et al. [22] proposed a pricing heuristic based on supervised learning for the VRPTW. In many VRP variants, including the VRPSD, the pricing problem can be formulated as an NP-hard elementary shortest path problem with resource constraints. Their method reduces the network size through arc selection, performed by either a random forest or a deep neural network, achieving a reduction of approximately 40% in computational time compared to traditional column generation. A similar approach was designed in [13] for an electric bus scheduling problem. Beyond column generation, Cappart et al. [4] combined RL with constraint programming to address the *traveling salesman problem with time windows*. This line of research has since been extended to further improve efficiency and generality [5,18,19].

These recent works suggest that incorporating a RL model can help the algorithm identify additional high-quality columns to add to the RMP, which motivates our main contribution. We propose a new pricing heuristic designed for the VRPSD with a fixed-fleet constraint. Our heuristic consists of three complementary components:

1. A *greedy module* that rapidly generates columns during the initial iterations.
2. A *RL module* that enables the discovery of high-quality columns even when their search is computationally demanding, and
3. A *tabu search* that enhances diversification within the batch of generated columns, thereby promoting a better exploration of the search space.

We integrate this pricing heuristic into the framework introduced by Reynal et al. [25]. Specifically, our method follows the RMH structure: a column generation phase, followed by the same GRASP-based completion phase (with the fixed-fleet constraint handled as a soft constraint), and finally, the resolution of the MIP. We experiment on the same set of 40 instances as they did except that we impose the fixed-fleet constraint together with a fixed time limit. Our approach achieves superior performance compared to Reynal et al. [25], which fails to find a feasible solution in 60% to 35% of the instances depending on the available computation time, and also outperforms the GRASP of Mendoza et al. [20] in terms of the number of optimal solutions found across all tested computation time limits.

2   Problem Description

The vehicle routing problem with stochastic demands can be formulated as follows. Let $\mathcal{N}$ be a set of n_C independent customers, each of them having an uncertain demand following a Poisson distribution of parameter μ_j (see, e.g., [7,12,25]). All customers must be served using exactly one of K identical vehicles of capacity Q departing from the depot, denoted as 0. Because the demands are uncertain, we compute the probability $P(OV \mid j, k, n)$ that the nth overload occurs (denoted OV) when reaching customer j with total expected load k :

$$P(OV \mid j, k, n) = 1 - \sum_{i=0}^{nQ-k} \frac{(\mu_j)^i}{i!} e^{-\mu_j}. \tag{1}$$

The overloads are handled using a classical detour-to-depot recourse policy as in [7,12,25], thus designing the routes a priori. In this case, the realization of a customer demand is only revealed upon reaching it. If it results in an overload, the vehicle makes a round trip to the depot to replenish before continuing the route normally.

A vehicle route, denoted by $r = (j_0^r, j_1^r, j_2^r, \ldots, j_{|r|+1}^r)$, starts and ends at the depot $j_0^r = j_{|r|+1}^r = 0$ and visits an ordered list of $|r|$ customers $\mathcal{V}_r = (j_1^r, j_2^r, \ldots, j_{|r|}^r)$ with $|r| \geq 1$. To be feasible, it must respect the constraints:

$$\sum_{l=1}^{|r|} \mu_{j_l^r} \leq Q, \tag{2}$$

$$j_{i_1}^r \neq j_{i_2}^r, \qquad \forall(i_1, i_2) \in \{1, \ldots, |r|\}, i_1 < i_2. \tag{3}$$

Capacity constraint (2) ensures that the total expected demand of a route is less than the vehicle capacity, which limits the number of expensive potential overloads occurring in a route as in several previous works (see, e.g., [12]). The *elementarity constraints* (3) forbid a customer to be visited more than once in the same route. We denote by $\mathcal{R}$ the set of *feasible routes*.

Let $d_{i,j} \in \mathbb{R}^+$ be the Euclidean distance between any pair of nodes i and j, customer or depot. We also introduce a failure penalty $p_{j,k,n} \in \mathbb{R}^+$ accounting for the cost of applying the recourse policy when the nth overload occurs upon reaching customer j with an expected load of k:

$$p_{j,k,n} = 2 \, d_{0,j} \, P(OV \mid j, k, n) \tag{4}$$

with $p_{0,k,n} = 0$ for all possible values of k and n. We are now able to compute the expected cost of a route r, denoted c_r, as its total expected travel distance. Let $\kappa_l^r = \sum_{i=1}^{l-1} \mu_{j_i^r}$ be the expected load of a vehicle following route r when it reaches the lth node j_l^r. Then,

$$c_r = \sum_{l=1}^{|r|+1} \left(d_{j_{l-1}^r, j_l^r} + \sum_{n=1}^{\infty} \left(p_{j_l^r, \kappa_l^r, n} - p_{j_{l-1}^r, \kappa_{l-1}^r, n} \right) \right). \tag{5}$$

In practice, the probability that an overload occurs is only taken into account if it is greater than 10^{-4}, as in previous works [12,25]. The VRPSD with the fixed-fleet constraint consists in finding exactly K feasible routes such that each customer in $\mathcal{N}$ is served by exactly one route, while minimizing the sum of the expected route costs. This problem is illustrated in Fig. 1.

To model the VRPSD, we adopt a path-flow formulation. For each feasible route $r \in \mathcal{R}$ and each customer $j \in \mathcal{N}$, let $\nu_{j,r}$ be a binary parameter equal to 1 if route r serves customer j. Moreover, let x_r be a binary variable that takes value 1 if route r is part of the solution and 0 otherwise. The proposed integer program is

$$\min_{x} \quad \sum_{r \in \mathcal{R}} c_r x_r \tag{6}$$

$$\text{s.t.} \quad \sum_{r \in \mathcal{R}} \nu_{j,r} x_r = 1, \qquad \forall j \in \mathcal{N} \tag{7}$$

$$\sum_{r \in \mathcal{R}} x_r = K \tag{8}$$

$$x_r \in \{0, 1\}, \qquad \forall r \in \mathcal{R}. \tag{9}$$

The objective function (6) minimizes the sum of the route expected costs. Constraints (7) ensure that each customer is served by exactly one vehicle, whereas constraint (8) sets the number of routes to select to K. Finally, constraints (9) ensure the integrality of the variables.

3 Technical Background on Restricted Master Heuristics

RMHs form a class of matheuristics typically designed for large-scale MIPs. They rely on column generation to build a reduced-sized MIP that can be solved more efficiently. Column generation is an iterative process that aims to solve a linear relaxation, called the *master problem*. At each iteration, it begins by solving the RMP, a version of the master problem limited to a subset $\hat{\mathcal{R}} \subset \mathcal{R}$ of its variables. Solving the RMP yields a primal solution and a dual vector $\left(\gamma_0, (\gamma_j)_{j \in \mathcal{N}} \right)$, corresponding to the dual variables associated with constraints (7) and (8), respectively. The current primal solution is optimal for the master problem if no variable $x_r \in \mathcal{R} \setminus \hat{\mathcal{R}}$ has a negative reduced cost $\bar{c}_r$, defined as

$$\bar{c}_r = c_r - \sum_{j \in \mathcal{N}} \nu_{j,r} \gamma_j - \gamma_0. \tag{10}$$

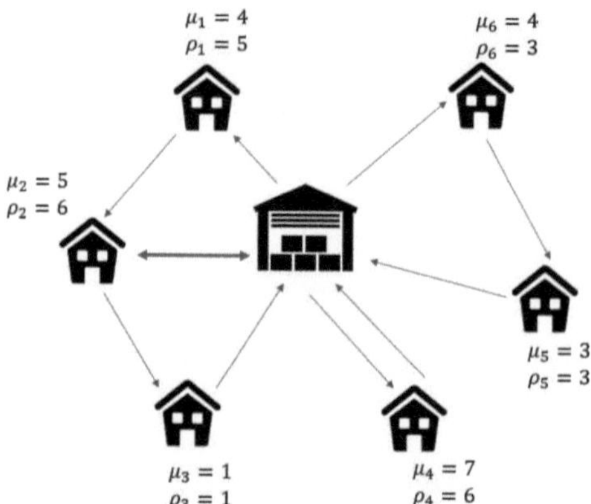

Fig. 1. A toy example of a VRPSD instance with 3 vehicles of capacity 10 and a possible solution. μ_j is the expected demand of customer j while ρ_j is its actual realization, only known upon visiting this customer. An overload occurs at customer 2, triggering a return to the depot before the route is normally completed.

To verify this condition, the algorithm solves the following pricing problem:

$$\min_{r \in \mathcal{R} \setminus \hat{\mathcal{R}}} \bar{c}_r. \tag{11}$$

If variables with a negative reduced cost are found, they are added to the RMP, and a new iteration is performed. Otherwise, the algorithm establishes that the current primal solution is optimal for the master problem and terminates.

For the VRPSD, the pricing problem is formulated as an *elementary shortest path problem with resource constraints* (ESPPRC) [8] on an acyclic network $G = (V, A)$ [7]. The set of vertices V includes a source node and a sink node, denoted as $\langle 0, 0 \rangle$ and $\langle 0, Q \rangle$, respectively, to represent the depot. It also contains nodes of the form $\langle j, k \rangle$, where j denotes a customer and $k \in [0, Q - \mu_j]$ represents a cumulative expected demand when reaching customer j. The arc set A is partitioned into three subsets:

$$
\begin{aligned}
A^1 &= \{\langle 0, 0 \rangle, \langle j, 0 \rangle | j \in \mathcal{N}\} &&\text{(leaving source)} \\
A^2 &= \{(\langle j, k \rangle, \langle i, k + \mu_j \rangle) \mid j, i \in \mathcal{N}, j \neq i, k \in [0, Q - \mu_j - \mu_i]\} &&\text{(linking nodes)} \\
A^3 &= \{(\langle j, k \rangle, \langle 0, Q \rangle) \mid j \in \mathcal{N}, k \in [0, Q - \mu_j]\} &&\text{(entering sink)}
\end{aligned}
\tag{12}
$$

Each arc in the above sets has an adjusted cost computed as

$$
\bar{c}_a = \begin{cases}
d_{0,j} + \sum_{n=1}^{\infty} p_{j,0,n} - \gamma_j & \text{if } a = (\langle 0, 0 \rangle, \langle j, 0 \rangle) \in A^1 \\
d_{j,i} + \sum_{n=1}^{\infty} (p_{i,k+\mu_j,n} - p_{j,k,n}) - \gamma_i & \text{if } a = (\langle j, k \rangle, \langle i, k + \mu_j \rangle) \in A^2 \\
d_{j,0} - \gamma_0 & \text{if } a = (\langle j, k \rangle, \langle 0, Q \rangle) \in A^3,
\end{cases}
\tag{13}
$$

guaranteeing that the cost of a source-to-sink path in G is equal to the reduced cost of the corresponding route.

Because a customer is represented by multiple nodes, a feasible path in G may not correspond to a feasible route in $\mathcal{R}$ due to potential violations of the elementarity constraints (3). To address this, one resource constraint per customer must be added, ensuring that at most one of the nodes representing a given customer can be included in any feasible path.

Most previous works (e.g., [12,25]) solve the pricing problem using a labeling algorithm, often relying on relaxations such as the ng-ESPPRC [2] to accelerate computation. However, for instances involving longer routes, particularly under the fixed-fleet constraint with a tight fleet size, such dynamic programming algorithms tend to lose efficiency, making them irrelevant for a RMH. We thus adapt the RMH from Reynal et al. [25] by replacing their labeling algorithm by a new pricing heuristic that aims to generate high-quality columns efficiently. Once done, the column generation phase is followed by a GRASP completion phase in which the fixed-fleet constraint has been passed as a soft constraint. The resulting RMP is then solved as a MIP.

4   Design of Our Pricing Heuristic

This section presents our core contribution, a hybrid pricing heuristic for the VRPSD. It is formalized in Algorithm 1. First, we generate an initial pool of routes using a greedy procedure (line 2). Then, if there are not enough columns with a negative reduced cost (parametrized with a value Θ in line 3), we complete the generation with a reinforcement learning construction (line 4). At this step, the generated routes may be too similar and we propose to diversify them using a tabu search (line 7). The improved routes are then returned. The next paragraphs describe the three main procedures: the greedy construction, the RL module, and the tabu search.

Algorithm 1. HYBRIDPRICING(.) - Hybrid pricing heuristic

Require: Θ: threshold defining if learning should be used for generating new routes.
Ensure: $\mathcal{R}_{\mathrm{HP}}$: set of generated routes
  1: $\mathcal{R}_{\mathrm{HP}} \leftarrow \emptyset$
  2: $\mathcal{R} \leftarrow \mathrm{GREEDY}()$
  3: **if** $\left|\{r \in \mathcal{R} \mid \bar{c}_r < 0\}\right| < \Theta \cdot |\mathcal{R}|$ **then**
  4:     $\mathcal{R}_{\mathrm{LEARNING}} \leftarrow \mathrm{LEARNING}()$
  5:     $\mathcal{R} \leftarrow \mathcal{R} \cup \mathcal{R}_{\mathrm{LEARNING}}$
  6: **for** $r \in \mathcal{R}$ **do**
  7:     $r' \leftarrow \mathrm{TABUSEARCH}(r)$
  8:     $\mathcal{R}_{\mathrm{HP}} \leftarrow \mathcal{R}_{\mathrm{HP}} \cup \{r'\}$
  9: **return** $\mathcal{R}_{\mathrm{HP}}$

4.1 Greedy Construction GREEDY(.)

The greedy component aims to quickly generate columns using two heuristics, namely, *nearest-neighbor* and *best-insertion*, each of which builds routes by adding one customer at a time. We start by describing how they build one route at a time.

Nearest-Neighbor Heuristic. The procedure begins with a route departing from the depot and iteratively selects the next location (customer or depot) to visit. At each step, the route must remain feasible, i.e., satisfy the capacity and elementarity constraints (2) and (3). The chosen extension is the one yielding the least reduced cost. The route is completed when it includes a return to the depot.

Best-Insertion Heuristic. The construction process starts with an empty route. At each iteration, it evaluates the insertion of each remaining customer into all possible positions along the current route. The selected insertion is the one that results in the feasible route with the least reduced cost. The algorithm terminates when no further feasible insertion can improve the solution.

Improvement with Beam-Search. To enable these two heuristics to generate multiple columns simultaneously, we integrate them into a beam-search procedure. Instead of constructing a single route, each heuristic maintains a pool of n_G routes, where n_G is a user-defined parameter. At each iteration, a candidate pool is formed by considering all possible extensions (for the nearest-neighbor heuristic) or insertions (for the best-insertion heuristic) of the routes currently in the pool. The pool is then updated with the n_G candidates exhibiting the least reduced costs. Routes that are already complete are added to the candidate pool without generating new extensions or insertions. The algorithm terminates when the pool contains only complete routes. The beam-search width n_G is set to $5n_C$, where n_C is the number of customers. We observed that larger n_G values lead to larger RMPs, which are more time-consuming to solve while providing only marginal performance improvements.

4.2 Learning Module - LEARNING(.)

Solving a problem with deep reinforcement learning requires to model it as a *Markov decision process* and to design a neural procedure to solve it. Let us define the set of states (S), the set of actions (A), the transition function $(T : S \times A \to S)$ and the reward function $(R : S \times A \to \mathbb{R})$, as the tuple $\langle S, A, T, R \rangle$. Our model is as follows:

States. We define a state $s \in S$ as the triplet $(j_s, k_s, \mathcal{V}_s)$ where j_s is the last customer visited by the route (i.e. the current location of the agent), k_s is the expected cumulated load gathered in the route up to this point and $\mathcal{V}_s$ is the set of customers already visited by the route. By the problem definition, we know that $k_s = \sum_{j \in \mathcal{V}_s} \mu_j$. A state is terminal if the agent is back to the depot.

Actions. An action corresponds to the selection of the customer $j \in \mathcal{V}$ to visit next. We note that the selection must satisfy the capacity constraint (2) and the elementarity constraints (3). Formally, the set of actions $\mathcal{A}_s$ available at a state $(j_s, k_s, \mathcal{V}_s)$ is as follows:

$$\mathcal{A}_s = \{j \in \mathcal{V} \backslash \mathcal{V}_s \mid k_s + \mu_j \leq Q\}. \tag{14}$$

Transition Function. Executing an action $a \in \mathcal{A}_s$ at state $s = (j_s, k_s, \mathcal{V}_s)$ leads the agent to the state $s' = (a, k_s + \mu_a, \mathcal{V}_s \cup \{a\})$.

Reward Function. Given a state $s = (j_s, k_s, \mathcal{V}_s)$ corresponding to partial route r, choosing to perform action $a \in \mathcal{A}_s$ leading to route $r' = r :: a$ comes with the reward $\bar{c}_{(\langle j_s, k_s \rangle, \langle a, k_s + \mu_a \rangle)}$, which is the adjusted cost of the arc associated with the transition as defined in Eq. (13)

Features. The next question is how to represent a state $s \in S$ in a structure amenable to learning. Following related works in the field of learning for combinatorial optimization, we propose to leverage an architecture based on a graph neural network. Briefly, this architecture operates on inputs having an underlying graph structure. We refer to the work of Cappart et al. [3] for an extended description of this architecture and its use in the context of combinatorial optimization. Let $G = (V, E)$ be a graph representing a current state of the pricing problem. We define the set of vertices V as the set of customers and the set of edges E as the possible connections between customers. Besides, each vertex and edge is decorated with a set of features. They are summarized in Table 1 and mainly corresponds to information either related to the instance (e.g., distances to the depot) or to the pricing state (e.g., the customer resources). The depot is represented by node 0 and corresponds to a customer with a dual variable γ_0 and an expected demand $\mu_0 = 0$.

Neural Architecture. Our architecture is based on an *edge-featured graph attention network* [28] (4 layers of 96 neurons) followed by a multilayer perceptron (3 layers of 96 neurons) with a ReLU activation function. The final activation is a sigmoid that we use to obtain a value comprised between 0 and 1 for each vertex v, which we interpret as the quality of the decision of going to the location related to vertex v, given the current state.

Training Phase. The training data are obtained by solving VRPSD instances from [7]. As done by Reynal et al. [25], we also include variations of these instances with a smaller fleet size K and larger capacity Q to create instances featuring longer routes, yielding a total of 149 instances. Among those, 40 were kept as the dataset for this work and thus not included in the training.

For the remaining 109 instances, we collect dual variable values obtained at each iteration of a column generation algorithm applied to solve the root node of the BPC algorithm, without adding any cutting planes. Each of the 4386 pricing problem instances obtained this way is a training instance for our model, with 60% being used for training, 20% for validation and 20% for testing.

Table 1. Definition of our input graph $G = (V, E)$ representing a state $(j_s, k_s, \mathcal{V}_s)$.

Features on the vertices $j \in V$	
Description	Value
Current location of the agent	1 if $j = j_s$ else 0
Customer resources	1 if $j \in \mathcal{V}_s$ else 0
Possible customers to visit next	$\mathcal{A}_s$ as defined by (14)
Percentage of remaining capacity	$\frac{k_s + \mu_j}{Q}$
Distance to the depot	$d_{0,j}$
Distance from j_s to j	$d_{j_s,j}$
Cost from j_s to j	$d_{j_s,j} + \sum_{n=1}^{\infty} \left(p_{j,k_s+\mu_j,n} - p_{j_s,k_s,n} \right)$
Features on the edges $(j_1, j_2) \in E$	
Description	Value
Distance to the depot from the head node	d_{0,j_2}
Length of (j_1, j_2)	d_{j_1,j_2}
Cost of (j_1, j_2) when leaving with charge k_s	$d_{j_1,j_2} + \sum_{n=1}^{\infty} \left(p_{j_2,k_s+\mu_{j_2},n} - p_{j_1,k_s,n} \right)$
Dual variable at the head node	γ_{j_2}

We train our model using a standard *proximal policy optimization* (PPO) algorithm [27] with a batch size of 32, and a convergence criterion set to a loss reduction below 10^{-6} over 5 iterations. We use Adam optimizer [16] with default parameters, except for the learning rate, which we set at 10^{-5}. The implementation is done using PyTorch and DGL libraries [23,29]. Training ran for 24,587 epochs, representing around a day of computation time.

Once trained, the model is employed to generate new columns. At each step, it provides the k_L most probable customers to visit next, where k_L is a user-defined parameter. To evaluate each potential extension, the resulting partial route is completed using a nearest-neighbor heuristic until it returns to the depot, and the reduced cost of the completed route is used as the evaluation criterion. The extension selected is the one leading to the completed route with the least reduced cost. As before, route feasibility must be maintained at every step. A beam-search strategy is also applied to generate a pool of columns simultaneously. The construction process terminates once n_L complete routes have been generated. We set $n_L = 50$ and $k_L = 3$. These relatively small values, compared to those used in the greedy procedure, stem from the fact that calling the graph neural network is a computationally expensive operation, which requires us to be parsimonious with the number of calls. In line with this rationale, the threshold Θ is set to 0.05, such that the majority of routes are produced by the greedy procedure, while the learning component focuses on identifying critical routes that the greedy construction fails to capture.

4.3 Diversification with a Tabu Search - TABUSEARCH(.)

A recurrent issue with the previous constructions is that the columns tend to be highly similar, which reduces the effectiveness of the pricing heuristic. To address this limitation, we introduce a diversification mechanism based on a tabu search procedure. Specifically, we apply local perturbations to each route $r = (j_0^r, \ldots, j_{k-1}^r, j_k^r, j_{k+1}^r, \ldots, j_{|r|+1}^r)$ produced by the greedy and learning-based procedures. Two types of moves are considered:

1. *Removal moves*, which remove a customer from the route. Removing the kth customer of route r yields the route $(j_0^r, \ldots, j_{k-1}^r, j_{k+1}^r, \ldots, j_{|r|+1}^r)$.
2. *Insertion moves*, which insert a customer into the route at the best possible position while maintaining route feasibility. Inserting customer j at the kth position of route r yields the route $(j_0^r, \ldots, j_{k-1}^r, j, j_k^r, j_{k+1}^r, \ldots, j_{|r|+1}^r)$.

At each iteration, we perform the first non-tabu improving move found, or otherwise the best non-tabu deteriorating move, using the reduced cost $\bar{c}_r$ as the evaluation function. Once a move is executed, it is added to a tabu list τ of fixed size T to prevent it from being undone by its opposite move during the next T iterations. The tabu list is updated in a first-in, first-out manner: when its maximum capacity is reached, the oldest move is removed to make room for the new one. We perform n_T iterations of this process on each route and return the route with the least reduced cost encountered during the search. In our experiments, we set $n_T = 20$ and $T = 10$. Preliminary tests indicated that increasing these values, up to 50 iterations and a tabu list length of 20, does not lead to significant performance improvements.

5 Experimental Results

In this section, we evaluate our new pricing heuristic and analyze the contribution of each component to its overall performance.

5.1 Experimental Protocol

The tests are conducted on the same set of instances as Reynal et al. [25], except that we consider the fixed-fleet constraint. The fleet size is set to the lower bound on the number of vehicles required to serve all customers $\lceil (\sum_{j \in \mathcal{V}} \mu_j) \setminus Q \rceil$. All instances remain feasible under this configuration. The column generation phase is performed using Gencol 4.5, the MIP resolution with CPLEX 22.1.1, and all experiments are executed on an Intel(R) Core(TM) i7-8700 CPU running at 3.20 GHz. The dataset includes 40 instances of the VRPSD, with 39 to 60 customers. These instances cannot be solved to optimality by the branch-price-and-cut (BPC) algorithm of [12] in one hour.

We evaluate our method (referred to as HYBRIDPRICING) using the same computation times as Reynal et al. [25], namely 2, 5, 10, and 15 min, with approximately 50% of the time allocated to column generation and the remaining 50% to the completion phase and MIP solving. We compare our approach with the following baselines:

- *Reynal et al.* [25]. This RMH leverages first a dynamic programming-based pricing to generate routes guided by dual variables from the RMP. It then applies a GRASP metaheuristic to complete partial solutions and introduce new routes that complement those used in the solution of the current RMP. Finally, it solves the resulting model as a MIP.
- *Mendoza et al.* [20]. This method generates routes with a GRASP for 90% of the total computation time and solves a MIP with all generated routes for the remaining 10%.

All reported results are computed by averaging the cost of the solutions obtained by each method over 5 runs.

5.2   Main Result: Comparison with State-of-the-Art Heuristics

The results over the whole dataset are summarized in Table 2 and in Fig. 2 by means of performance profiles [9]. Because most optimal solutions for these instances are unknown, the gaps are computed with respect to the final lower bound returned by the BPC algorithm of [12] after running for 2 h.

As shown by these results, the RMH of Reynal et al. [25] fails to achieve competitive results. It reaches a feasible solution for 40% of the instances for the 2-min timeout. This proportion only increases to 65% with more computation time available as in the 15-min case. Due to the small fleet size, allowing for longer routes these instances are already challenging for the dynamic programming pricing. The addition of the fixed-fleet constraint (8) prevents the column generation from yielding a feasible solution in a reasonable amount of time.

As in Reynal et al. [25], the GRASP of Mendoza et al. [20] performs better with less computation times. For the 2-min case, it achieves a smaller average optimality gap and a larger number of optimal solutions and best average solution cost. As shown on the performance profiles, its best and worst-case scenarios over 5 runs are also better than our algorithm. As available computation time increases, our method HYBRIDPRICING becomes competitive. We achieve an average optimality gap of 1% in the 5 and 10-min cases, outperforming the GRASP. Both methods yield similar worst-case scenarios, but our algorithm provides a better best-case scenario than the GRASP. We account for stochasticity between the 5 runs used to produce these results by conducting two-sided Wilcoxon signed-rank tests, with the null hypothesis H_0 "*both methods yield identical performance*". We obtain p-values of 0.014 and 0.003 respectively for the 5 and 10-min cases, thus rejecting the null in these configurations.

Finally, the GRASP and our method yield almost identical best-case scenarios for the 15-min timeout case. Moreover, the longer computation time makes HYBRIDPRICING much more robust, thus improving its worst-case scenario. The same two-sided Wilcoxon signed-rank test with the same null hypothesis however yields a p-value of 0.055, preventing from confidently rejecting the null in the 15-min case, even if the p-value is only slightly above the 5% threshold.

These results show that our method requires time to be efficient: it heavily relies on the quality of the routes generated during the column generation phase.

Table 2. Average optimality gap and number of instances solved under certain gaps for all baselines and computation times. Average optimality gap is computed on the instances for which a method finds at least one feasible solution. The number of such instances is provided in parentheses. Bold values indicate the best result per row.

		HYBRIDPRICING	MENDOZA ET AL.	REYNAL ET AL.
2 minutes	*Avg optimality gap*	1.43%(40)	**1.34%**(40)	1.46%(16)
	# best avg solution cost	14	**28**	5
	# optimal solutions	2	**3**	1
	# < 1% optimality gap	**18**	**18**	7
	# < 5% optimality gap	**40**	**40**	16
5 minutes	*Avg optimality gap*	**1.01%**(40)	1.06%(40)	0.95%(24)
	# best avg solution cost	**27**	24	10
	# optimal solutions	**6**	5	3
	# < 1% optimality gap	24	**25**	15
	# < 5% optimality gap	**40**	**40**	24
10 minutes	*Avg optimality gap*	**0.95%**(40)	1.02%(40)	1.62%(25)
	# best avg solution cost	**27**	23	12
	# optimal solutions	**6**	5	3
	# < 1% optimality gap	**26**	**26**	17
	# < 5% optimality gap	**40**	**40**	24
15 minutes	*Avg optimality gap*	**0.90%**(40)	0.94%(40)	0.89%(26)
	# best avg solution cost	**26**	13	21
	# optimal solutions	**10**	6	3
	# < 1% optimality gap	27	**27**	18
	# < 5% optimality gap	**40**	**40**	26

In the 2-min case, it is too short to find high-quality routes to complete with the GRASP. However, the results in the 5-, 10-, and 15-min cases show that the column generation phase identifies critical columns that the GRASP disregards, yielding better solutions.

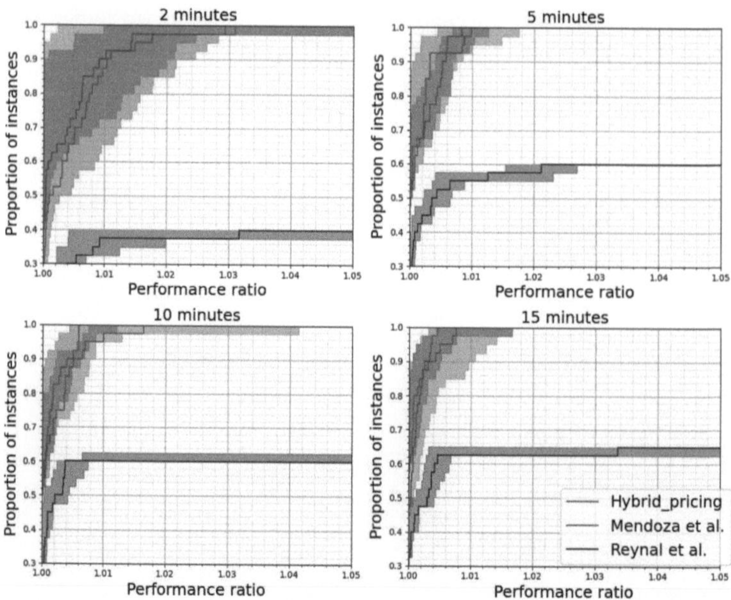

Fig. 2. Performance profiles on the optimality gap for all baselines. Plain line is the average performance over 5 runs while the colored zone around it represents the minimal and maximal performance. Performance ratios are computed with regard to the best solution obtained by all considered methods over all 5 runs.

5.3   Analysis: Ablation Study

To better assess the contribution of each component in our RMH, we conducted an ablation study. Specifically, we independently removed each module of the hybrid pricing framework (i.e. greedy construction, learning component, and tabu search) and computed the number of instances for which they yield the best average solution cost, when benchmarked against the RMH of [25] and the GRASP of [20]. The corresponding results are presented in Table 3.

We first observe from the first three rows that removing any single module leads to a noticeable degradation in performance, indicating that all three components play complementary roles in achieving competitive results. Among them, the tabu search is particularly critical: in all cases, its removal causes a significant drop in performance. This can be explained by the fact that the routes generated by the greedy construction tend to be similar (as is also the case for the routes generated by the RL module), thus emphasizing the importance of diversification mechanisms introduced by the tabu search.

Another noteworthy observation concerns the learning component. When used as the only module to generate columns (variant - GREEDY and TABU), the performance of the RMH is poor for the 2-min timeout (1 best average solution cost), but as more computation time becomes available, it steadily increases to reach 15 best average solution costs for a time of 15 min. This suggests that the learning mechanism requires sufficient time to fully exploit its potential and guide the search process effectively.

Finally, the greedy construction complements the RL module. When used alone (variant - LEARNING and TABU), it quickly reaches its limit in terms of quality, yielding poor performance even when combined with tabu search (variant - LEARNING) for timeouts of less than 15 min. The combination with the learning module (variant - TABU) allows for going through the first easy column generation iterations without wasting time on them, leaving more time to handle the last difficult iterations with the learning module, thus increasing the performance of the algorithm. However, the lack of diversification brought by the tabu search in HYBRIDPRICING prevents it from yielding competitive results.

Table 3. Results of the ablation study. #: number of best avg solution cost; %: number of best avg solution cost in percentage; Δ: loss (in percentage) on the number of best avg solution cost with respect to HYBRIDPRICING.

Variant	2 minutes			5 minutes			10 minutes			15 minutes		
	#	%	Δ	#	%	Δ	#	%	Δ	#	%	Δ
HYBRIDPRICING	14	35	-	27	67.5	-	27	67.5	-	26	65	-
- LEARNING	4	10.0	-71.4	6	15.0	-77.7	9	22.5	-66.7	14	35.0	-46.1
- GREEDY	7	17.5	-50.0	8	20.0	-70.3	12	30.0	-55.6	17	42.5	-34.6
- TABU	4	10.0	-71.4	6	15.0	-77.7	8	20.0	-70.3	12	30.0	-53.8
- LEARNING and TABU	4	10.0	-71.4	4	10.0	-74.1	7	17.5	-79.4	9	22.5	-65.3
- GREEDY and TABU	1	2.5	-92.8	7	20.0	-74.1	13	32.5	-51.9	15	37.5	-42.3

6 Conclusion

This paper introduced a new pricing heuristic for the VRPSD under a fixed-fleet constraint, integrating three complementary components: (1) a greedy construction procedure to rapidly generate initial solutions and stabilize the dual variables in the early iterations of column generation; (2) a deep reinforcement learning agent that exploits problem-specific knowledge acquired during training to produce high-quality columns; and (3) a tabu search mechanism that enhances diversification among the routes generated by the other two components.

Experimental results show that our RMH performs competitively with the recent one of Reynal et al. [25], while achieving a higher ratio of optimal solutions, more best average solution costs, and smaller average optimality gaps than the classical GRASP of Mendoza et al. [20]. The ablation study confirms that each component contributes to the overall performance of the method, highlighting the synergy between learning, constructive, and diversification strategies.

Future works could focus on improving the greedy construction. In its current state, it allows the column generation to quickly go through the early iterations in which the generated columns are less relevant, but the resulting routes are of poor quality and thus rarely contribute to the returned solution. Another option would be to consider a different construction method for the learning module, which only builds the routes by choosing which client to visit next. One could instead consider a clustering algorithm in which the model would take a route as input and return, for each other customer, if it should also be part of the route, inserting the most promising ones at the best possible place.

References

1. Alvelos, F., de Sousa, A., Santos, D.: Combining column generation and metaheuristics. In: Talbi, E. (ed.) Hybrid Metaheuristics. Studies in Computational Intelligence, vol. 434, pp. 285–334. Springer, Berlin (2013)
2. Baldacci, R., Mingozzi, A., Roberti, R.: New route relaxation and pricing strategies for the vehicle routing problem. Oper. Res. **59**(5), 1269–1283 (2011)
3. Cappart, Q., Chételat, D., Khalil, E.B., Lodi, A., Morris, C., Veličković, P.: Combinatorial optimization and reasoning with graph neural networks. J. Mach. Learn. Res. **24**(130), 1–61 (2023)
4. Cappart, Q., Moisan, T., Rousseau, L.M., Prémont-Schwarz, I., Cire, A.A.: Combining reinforcement learning and constraint programming for combinatorial optimization. In: Proceedings of the AAAI Conference on Artificial Intelligence, vol. 35, pp. 3677–3687 (2021)
5. Chalumeau, F., Coulon, I., Cappart, Q., Rousseau, L.-M.: SeaPearl: a constraint programming solver guided by reinforcement learning. In: Stuckey, P.J. (ed.) CPAIOR 2021. LNCS, vol. 12735, pp. 392–409. Springer, Cham (2021). https://doi.org/10.1007/978-3-030-78230-6_25
6. Chi, C., Aboussalah, A., Khalil, E., Wang, J., Sherkat-Masoumi, Z.: A deep reinforcement learning framework for column generation. In: Advances in Neural Information Processing Systems, vol. 35, pp. 9633–9644 (2022)
7. Christiansen, C.H., Lysgaard, J.: A branch-and-price algorithm for the capacitated vehicle routing problem with stochastic demands. Oper. Res. Lett. **35**(6), 773–781 (2007)
8. Costa, L., Contardo, C., Desaulniers, G.: Exact branch-price-and-cut algorithms for vehicle routing. Transp. Sci. **53**(4), 946–985 (2019)
9. Dolan, E.D., Moré, J.J.: Benchmarking optimization software with performance profiles. Math. Program. **91**(2), 201–213 (2002)
10. Florio, A.M., Gendreau, M., Hartl, R.F., Minner, S., Vidal, T.: Recent advances in vehicle routing with stochastic demands: Bayesian learning for correlated demands and elementary branch-price-and-cut. Eur. J. Oper. Res. **306**(3), 1081–1093 (2023)
11. Florio, A.M., Hartl, R.F., Minner, S.: New exact algorithm for the vehicle routing problem with stochastic demands. Transp. Sci. **54**(4), 1073–1090 (2020)
12. Gauvin, C., Desaulniers, G., Gendreau, M.: A branch-cut-and-price algorithm for the vehicle routing problem with stochastic demands. Comput. Oper. Res. **50**, 141–153 (2014)
13. Gerbaux, J., Desaulniers, G., Cappart, Q.: A machine-learning-based column generation heuristic for electric bus scheduling. Comput. Oper. Res. **173**, 106848 (2025)

14. Hoogendoorn, Y., Spliet, R.: An evaluation of common modeling choices for the vehicle routing problem with stochastic demands. Eur. J. Oper. Res. **321**(1), 107–122 (2025)

15. Hu, W., Du, B., Wu, Y., Liang, H., Peng, C., Hu, Q.: A hybrid column generation algorithm based on metaheuristic optimization. Transport **31**(4), 389–407 (2016)

16. Kingma, D.P.: Adam: a method for stochastic optimization. arXiv preprint arXiv:1412.6980 (2014)

17. Liao, S., Xu, Y., Niu, Y., Cao, Z.: Learning-guided bi-objective evolutionary optimization for green municipal waste collection vehicle routing. J. Clean. Prod. **501**, 145316 (2025)

18. Marty, T., et al.: Learning and fine-tuning a generic value-selection heuristic inside a constraint programming solver. Constraints **29**(3), 234–260 (2024)

19. Marty, T., François, T., Tessier, P., Gautier, L., Rousseau, L.M., Cappart, Q.: Learning a generic value-selection heuristic inside a constraint programming solver. In: 29th International Conference on Principles and Practice of Constraint Programming (CP 2023), pp. 25–1. Schloss Dagstuhl–Leibniz-Zentrum für Informatik (2023)

20. Mendoza, J.E., Rousseau, L.M., Villegas, J.G.: A hybrid metaheuristic for the vehicle routing problem with stochastic demand and duration constraints. J. Heuristics **22**, 539–566 (2016)

21. Morabit, M., Desaulniers, G., Lodi, A.: Machine-learning-based column selection for column generation. Transp. Sci. **55**(4), 815–831 (2021)

22. Morabit, M., Desaulniers, G., Lodi, A.: Machine-learning-based arc selection for constrained shortest path problems in column generation. INFORMS J. Optim. **5**(2), 191–210 (2023)

23. Paszke, A., et al.: Automatic differentiation in PyTorch (2017)

24. Prescott-Gagnon, E., Desaulniers, G., Rousseau, L.M.: A branch-and-price-based large neighborhood search algorithm for the vehicle routing problem with time windows. Networks **54**(4), 190–204 (2009)

25. Reynal, G., Cappart, Q., Desaulniers, G., Rousseau, L.M.: Improving column complementarity in a restricted master heuristic with a grasp-guided completion: application to the vehicle routing problem with stochastic demands. Les Cahiers du GERAD G-2025-56, HEC Montréal (2025). https://www.gerad.ca/fr/papers/G-2025-57

26. Sadykov, R., Vanderbeck, F., Pessoa, A., Tahiri, I., Uchoa, E.: Primal heuristics for branch and price: the assets of diving methods. INFORMS J. Comput. **31**(2), 251–267 (2019)

27. Schulman, J., Wolski, F., Dhariwal, P., Radford, A., Klimov, O.: Proximal policy optimization algorithms. arXiv preprint arXiv:1707.06347 (2017)

28. Veličković, P., Cucurull, G., Casanova, A., Romero, A., Lio, P., Bengio, Y.: Graph attention networks. arXiv preprint arXiv:1710.10903 (2017)

29. Wang, M., et al.: Deep graph library: a graph-centric, highly-performant package for graph neural networks. arXiv preprint arXiv:1909.01315 (2019)

30. Yee, J.R., Golden, B.L.: A note on determining operating strategies for probabilistic vehicle routing. Naval Res. Logistics Q. **27**(1), 159–163 (1980)

Scheduling Data Transfers with Priorities for Space Missions

Julien Rouzot[1,2]([✉]), Christian Artigues[1], Clément Carbonnel[3],
Philippe Garnier[2], Emmanuel Hebrard[1], Pierre Lopez[1], and Bertrand Simon[4]

[1] Univ Toulouse, CNRS, LAAS, Toulouse, France
`christian.artigues@laas.fr` , `emmanuel.hebrard@laas.fr` ,
`pierre.lopez@laas.fr`
[2] Univ Toulouse, CNRS, CNES, OMP, IRAP, Toulouse, France
`julien.rouzot@laas.fr` , `philippe.garnier@irap.omp.eu`
[3] University of Montpellier, CNRS, LIRMM, Montpellier, France
`clement.carbonnel@lirmm.fr`
[4] UGA, CNRS, INRIA, Grenoble INP, LIG, Grenoble, France
`bertrand.simon@cnrs.fr`

Abstract. In space missions, scientific data collected by various instruments must be stored onboard before being downlinked to Earth during designated communication windows. For many long range missions, the available bandwidth is shared according to a priority assigned to each memory buffer during such downlink window. The overlapping Memory Dumping Problem (oMDP) consists in finding the priority assignment that minimizes the highest memory peak. This problem has been shown to be weakly NP-hard and has so far only been addressed with heuristic methods. In this paper, we complete the complexity analysis by proving that the problem is strongly NP-hard in the general case, and we propose the first exact method to solve the oMDP. We present a constraint programming approach combining new global constraints and a heuristic branching strategy, and show that our method is competitive with state-of-the-art heuristics while being more generic and able to produce optimality proofs on small instances.

Keywords: Constraint Programming · Scheduling · Data Transfers

1 Introduction

In deep space missions, scientific instruments operate semi-independently, generating valuable scientific data that must be transmitted to Earth through tightly constrained downlink windows. Each instrument typically writes its data in a dedicated onboard buffer, which introduces a key challenge: coordinating data transfers in such a way that buffer overflows are avoided. This paper addresses the scheduling of these data transfers with the goal to avoid data loss.

In the case of the Rosetta mission [3], the data produced by the different instruments is stored on distinct buffers that are dumped only during *downlink*

T. Guns (Ed.): CPAIOR 2026, LNCS 16595, pp. 470–485, 2026.
https://doi.org/10.1007/978-3-032-27242-3_28

windows, where the spacecraft is visible from ground stations of the Deep Space Network (DSN) [8]. Data transfers must be scheduled so that no buffer exceeds its capacity during the mission, which could result in the loss of critical data. More precisely, since actual data production and transfer rates are subject to uncertainties, the objective is to minimize the highest memory level across all buffers over the entire time horizon. This objective is motivated by the need to enhance robustness in the case of operational uncertainties. By minimizing the highest memory level across all instrument buffers, the scheduling plan gains greater tolerance to unexpected fluctuations on the actual data production and transfer rates. A lower memory peak effectively creates a safety margin within each buffer, reducing the risk of overflow. During a downlink window, bandwidth is shared via a priority-based Round-Robin scheme. Buffers with the best priority evenly share the available bandwidth; any unused capacity is then allocated to lower priorities, until all buffers are empty or no bandwidth remains.

We propose an exact method based on constraint programming to solve the problem of assigning priorities that minimize the highest memory peak, under the assumption that the fill rates of each buffer are known along the mission, as they are determined by the science observations plan [3]. This problem is known as the overlapping Memory Dumping Problem [12]. The oMDP was previously shown to be at least weakly NP-hard in [7] when both the number of memory buffers and the number of downlink windows are unbounded. We refine the complexity analysis by proving that the problem is strongly NP-hard in the general case. We also introduce a CP approach, the first exact method for the oMDP, previously tackled only with heuristics in [12] and [7].

We first formally introduce the oMDP in Sect. 2 and present the related works in Sect. 3. We provide an extensive complexity analysis in Sect. 4. We present our CP model with new global constraints and a heuristic branching strategy to address the oMDP in Sect. 5 and we finally highlight the performance of our approaches in Sect. 6, before concluding in Sect. 7.

2   The Overlapping Memory Dumping Problem

We consider the oMDP, illustrated in Fig. 1, where we have a set of buffers $\mathcal{B} = \{1, \ldots, n\}$. For each buffer $i \in \mathcal{B}$, the memory level is $U_i(t) : \mathbb{R}^+ \mapsto \mathbb{R}^+$ with t between 0 and h, where h is the time horizon. Each buffer has a finite capacity C_i, and the memory level must remain below this limit at all times. Data can be transmitted only during a set of downlink windows in $\mathcal{W} = \{1, \ldots, m\}$, that start at w_j^{start} and end at w_j^{end}. The dump rate—i.e. bandwidth—available for each window j is denoted δ_j. The memory peak $r_{i,j}$ for window j is the ratio of the highest memory level of buffer i during the window j and its capacity C_i. The objective is to minimize the highest peak: $\min rmax = \max_{i \in \mathcal{B}, j \in \mathcal{W}} (r_{i,j})$.

We assume that each buffer i has a piecewise constant fill rate function $f_i : \mathbb{R}^+ \mapsto \mathbb{R}^+$ determined by the science observation plan, where each nonzero segment corresponds to a distinct observation. Transfers are controlled through priorities, updated only at the beginning of each downlink window. For each buffer i, priorities $p_{i,j}$ are within the domain $[1, \ldots, n]$, where 1 is the *best* and n the *worst*. The transfer policy follows a simple priority-based Round-Robin scheme for sending atomic data packets. More precisely, whenever bandwidth is available, the system transfers a single packet from one of the non-empty buffers with the highest priority. If all such buffers are empty, it proceeds to the next priority pool, continuing this process until a non-empty buffer is found or all priority pools have been checked. Since buffers can share the same priority, ties are broken using a circular queue, meaning that among buffers with equal priority, preference is given to the one that transferred data least recently.

Although the downlink procedure is discrete, it is shown in [14] that the transfer rate functions resulting from a priority policy p can be tightly approximated as piecewise linear functions, as the size of the data block are extremely small in front of the fill rates and the dump rates. In [7], a procedure based on the work presented in [14] is proposed to compute the transfer rates and the memory level functions resulting from the fill rate functions and a priority assignment, assuming that buffers are filled and dumped continuously.

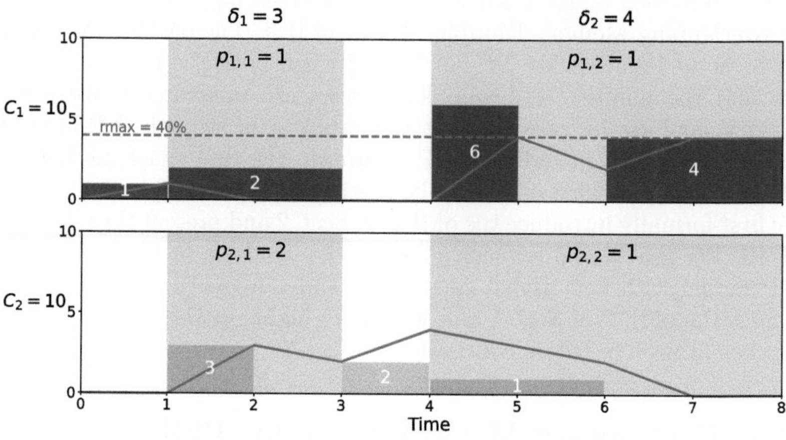

Fig. 1. Evolution of the buffers memory level $U_i(t)$ (red), according to their fill rates (violet and yellow), the dump rates (green), and a priority assignment.

Definition 1 (Transfer Rates [7]). *Given a downlink window j and a priority assignment p, the transfer rate functions $g_i^p(t)$ can be computed recursively by considering the buffers from the best priority rank to the worst. Let $\Omega = \{i_1, \ldots, i_k\}$ be the set of buffers with best priority, ordered by ascending memory level, breaking ties with ascending fill rate at time t. The available bandwidth Δ_j for buffers in Ω is initially set to δ_j. Then, the transfer rate $g_{i_1}^p(t)$ of buffer i_1 at time t is $\Delta_j/|\Omega|$ if its memory is not empty (an equal share of the bandwidth) and $\min(f_{i_1}(t), \Delta_j/|\Omega|)$ otherwise (as fast as it is filled, but no more than an equal share). Then, for $\ell \in [2, k]$, the transfer rate $g_{i_\ell}^p(t)$ is computed similarly, after setting Δ_j to $\Delta_j - g_{i_{\ell-1}}^p(t)$ and Ω to $\Omega \setminus \{i_{\ell-1}\}$. Finally, if*

$\Delta_j > 0$, *transfer rates are similarly computed for lower-priority buffers, rank by rank, until bandwidth is exhausted or all buffers are processed.*

Definition 2 (Memory Level). *Given the fill and transfer rate functions f_i and g_i^p, and the initial memory level for each buffer $U_i(0)$, the memory level $U_i^p(t)$ is the integral over time of the data production minus the data transfer:*

$$U_i^p(t) = U_i(0) + \int_0^t \left(f_i(t') - g_i^p(t')\right) dt'$$

Definition 3 (oMDP). *Given the buffer set $\mathcal{B}$, the downlink windows $\mathcal{W}$, the fill rate functions f_i, and the initial memory level $U_i(0)$ for each buffer i, what is the priority assignment p that minimizes:*

$$rmax = \max_{i \in \mathcal{B}} \left(\max_{t \in [0,h]} \left(U_i^p(t)\right) / C_i \right)$$

3   Related Work

In [12], the authors introduced the oMDP and presented the fast DOWNLINK-COUNT heuristic to solve the problem. This first heuristic approach for oMDP is based on the following assumption: buffers that overflow first if no data dump is performed should be given a better priority. More precisely, for a given downlink window and initial memory levels at the start of this window, DOWNLINKCOUNT *counts* the number of windows before an overflow occurs, when $\delta_j = 0$, $j \in \mathcal{W}$. Then, the priorities are distributed to the buffers, starting with the best priority for the buffer with the lowest downlink count. In case of ties, the same priority is assigned. This process is repeated until all priorities are assigned. This heuristic has a very low computing cost as we only need to sum the fill rate events until the overflow (or the end of the instance) is reached, for each buffer, for a given downlink window. This simple heuristic yields surprisingly good solutions on its own and was used by Rabideau et al. in their ITERATIVELEVELING method, consisting of iteratively reducing the overflow limit. This method was used for the Rosetta science operations planning tool, ASPEN [2].

Even though the Rosetta mission is over, efficient algorithms for scheduling data transfers via priority assignments remain valuable for future missions. The oMDP has been further studied in [7], where the authors introduced a fast and exact SIMULATION procedure to compute memory level functions based on a given priority assignment. They also proposed a polynomial-time algorithm, SINGLEWINDOW, to determine the optimal priority assignment for a single downlink window, as well as an LNS-based heuristic (REPAIRDESCENT) for solving the multiple-window problem. In this paper, we build on this work, so let us present the SIMULATION algorithm. Algorithm 1 computes the function U_i as defined in Definition 2, for a given downlink window j.

This SIMULATION procedure starts by setting the current and the next time point to the start time of the downlink window (lines 1). The function

---

Algorithm 1 SIMULATION: Computes the memory level functions.

Require: A window j, the dump rate δ_j, the start and end dates w_j^{start} and w_j^{end}, $\mathcal{B}$, $f_i(t)_{t \in [w_j^{start}, w_j^{end}]}$, the priorities p_j, the initial memory level $U_i(w_j^{start})$.

1: $t_{now} \leftarrow t_{next} \leftarrow w_j^{start}$ ▷ Current and next breakpoint
2: $events \leftarrow$ **MakeEventList** $\left(f_i(t)_{t \in [w_j^{start}, w_j^{end}]} \right)$ ▷ Ordered by ascending time
3: $k \leftarrow 0$ ▷ Current event index
4: **while** $t_{now} < w_j^{end}$ **do**   ▷ Loop through all events
5:      $t_{now} \leftarrow t_{next}$ ▷ Update current breakpoint
6:      **while** $t_{now} = t_{next}$ **do**
7:          $i \leftarrow events[k].buffer$
8:          $f_i(t_{now}) \leftarrow events[k].value$ ▷ Update current fill rates
9:          $k \leftarrow k + 1$ ▷ Next event
10:         $t_{next} \leftarrow events[k].time$ ▷ Next breakpoint
11:      $g^{p_j}(t_{now}) \leftarrow$ **ComputeTransferRates**$(U(t_{now}), f(t_{now}), \delta_j, p_j)$
12:      **for** $i \in \mathcal{B}$ **do**
13:         $t_{empty} \leftarrow$ **GetEmptyTime**$(t_{now}, U_i(t_{now}), f_i(t_{now}), g^{p_j}(t_{now}))$
14:         **if** $t_{empty} < t_{next}$ **then**
15:           $t_{next} \leftarrow t_{empty}$ ▷ Insert new breakpoint if buffer i becomes empty
16:      **JumpTo**$(t_{next}, U(t_{now}), f(t_{now}), g^{p_j}(t_{now}))$
17: **return** $U(t)_{t \in [w_j^{start}, w_j^{end}]}$

MakeEventList creates a list of events, including the start or the end of each fill rate event, ordered by time (line 2). Each event e is characterized by a *buffer index* (e.buffer), a *time* (e.time) and a *value* (e.value). They correspond to the *breakpoints* that must be evaluated chronologically to update the memory levels. While events correspond to the current breakpoint, the fill rate for the corresponding buffer is updated (line 8). Then, current transfer rates resulting from the memory levels, fill rates, dump rate, and priority assignment p at $t = t_{now}$ are computed in function **ComputeTransferRates** (line 11) as explained in Definition 1. For a given downlink window, these transfer rates do not change unless an observation starts or ends (i.e. a fill rate changes) or a buffer becomes empty (as it will redistribute its share of the bandwidth). While fill rate breakpoints are known, breakpoints due to buffers becoming empty must be computed dynamically. **GetEmptyTime** returns $t_{now} + U_i(t_{now})/(g_i^p(t_{now}) - f_i(t_{now}))$ if $g_i^p(t_{now}) > f_i(t_{now})$, and returns $+\infty$ otherwise. In Algorithm 1, a new event is inserted if a buffer becomes empty before t_{next}. The memory level is then updated at the next time point inside function **JumpTo** that sets $U_i(t_{next})$ to $U_i(t_{now}) + (t_{next} - t_{now}) \times (f_i(t_{now}) - g_i^p(t_{now}))$ for each buffer i. This process is repeated until the end of the current downlink window is reached. With k the number of fill rate events, n the number of buffers, this algorithm simulates a priority assignment in $O(k \log k + n^2 k \log n)$-time.

4   Complexity

D-oMDP (the decision variant of oMDP) was shown to be weakly NP-hard in [7], when both the number of memory buffers and the number of downlink windows are not bounded by a constant, and polynomial for a single downlink window. In this section, we show that the problem is strongly NP-hard.

Theorem 1. *D-oMDP is strongly NP-hard.*

Proof. We use a reduction from the NP-complete problem 3-PARTITION [4], which takes as input a set of $3n$ numbers $A = \{a_1, \ldots, a_{3n}\}$ with $\sum_{i=1}^{3n} a_i = nT$ with $\frac{T}{4} < a_i < \frac{T}{2}$ and asks whether there exists a partition of A into n triplets, each summing to T. The reduction uses $4n$ buffers and $3n$ downlink windows. Window i has duration a_i and dump rate n. For $1 \leq i \leq 3n$, buffer i has capacity a_i, starts at $U_i(0) = 0$ and has a fill rate of 2 during the downlink window i. All remaining buffers have capacity $2T$, start with a memory level of T and have a fill rate of 1 during each downlink window.

Assume we have a solution to the reduced instance. Note that the buffers capacities sum to $3nT$, the maximal data that can be dumped is n^2T and the total data that is produced is $2nT + nT + n^2T$, so all buffers must be full at the end, from which we have the following Lemma:

Lemma 1. *Every buffer must be full at the end of the plan.*

In window i, buffer i must dump exactly a_i data. Otherwise, if it dumps less than a_i, it would exceed its capacity at the end of window i and if it dumps more it would contradict Lemma 1. This means it must be given the best priority tied with exactly $n-1$ non-empty buffers. These buffers must be among buffers $3n+1$ to $4n$, because non-empty buffers from 1 to $3n$ would never re-fill and hence that would contradict Lemma 1. In other words, at window j, the usage of exactly one buffer among $3n + 1$ to $4n$ increases by exactly a_j. Therefore, the reduction is satisfiable if and only if there exists a partition of A into n sets, each summing to T. These sets must contain exactly 3 elements as $\frac{T}{4} < a_i < \frac{T}{2}$.

5   CP Model

We propose to address the oMDP with constraint programming. As presented in Sect. 3, we must calculate the transfer rates for each buffer at each breakpoint (where a fill rate or the dump rate changes) to compute the memory level function according to a priority assignment. We also have to dynamically add new breakpoints when buffers become empty as the bandwidth is redistributed to the other buffers in this case, and thus the transfer rates change. Therefore, modeling the simulation process with classical constraints is extremely challenging and it is very likely that time would have to be discretized in order to obtain an implementable model (by approximating the problem in this way and creating a high number of memory variables). Instead, we encapsulate the simulation in a global constraint, and we only track the memory levels at the beginning of each window, and the maximum memory levels during each window.

The memory variables $mem_{i,j}$ represent the memory level of buffer i at the beginning of a window j. Each window is composed of the non-visibility part where the dump rate is zero and the actual downlink window where the bandwidth is δ_j. Thus, the memory variables serve as indicators of the memory level at both the start of window j and the end of window $j-1$ when $j > 1$. For each buffer, we keep track of the memory peaks $r_{i,j}$ during window j with the peak variables. As the constraint programming solver we use does not handle continuous variables, we have to round memory and peak values. In the case of memory variables, we round to the upper integer. In the case of peaks, since they are a ratio between 0 and 1, we use a parameter r^{ub} to control the rounding, so the actual peak is $r_{i,j} / r^{ub}$. Therefore, increasing r^{ub} improves the precision.

In the oMDP, our only decision variables are the priority variables $p_{i,j}$ for each buffer and window. The objective and constraints are as follows:

$$\min \quad rmax = \left[\max_{i \in \mathcal{B},\, j \in \mathcal{W}} (r_{i,j}) \right] \tag{1}$$

s.t.

$$mem_{i,0} = U_i(0) \qquad\qquad i \in \mathcal{B} \tag{2}$$

$$\textsc{PriorityTransfer}\,(ins_j, mem_j, mem_{j+1}, r_j, p_j) \qquad\qquad j \in \mathcal{W} \tag{3}$$

$$\textsc{DenseRanking}\,(p_j) \qquad\qquad j \in \mathcal{W} \tag{4}$$

$$\textsc{PrioritySymmetry}\,(ins_j, mem_j, p_j) \qquad\qquad j \in \mathcal{W} \tag{5}$$

In our model, the instance is denoted ins and contains all the information about its parameters (observation events, downlink windows, etc.). The dropping of index i (for instance, mem_j or p_j) means that all buffers in window j are taken into account. The instance is split into m single window instances during a pre-processing step, and ins_j is a tuple gathering the parameters of the j-th window. More precisely, $ins_j = \{j, \delta_j, w_j^{start}, w_j^{end}, \mathcal{B}, f(t)_{t \in [w_j^{start}, w_j^{end}]}\}$. The objective is to minimize the highest memory peak (1). We initialize the first memory variables to the initial memory levels (2). We propose a new global constraint $\textsc{PriorityTransfer}$ that maintains and filters the domains of the memory, peak and priority variables (3). We also introduce two new global constraints to eliminate symmetric solutions, one based on the rules of a dense ranking to order the priorities in a unique way (4), and one taking advantage of the oMDP structure (5).

5.1 PriorityTransfer Constraint

$\textsc{PriorityTransfer}$ constraint is responsible for maintaining consistency between the domains of the memory variables $mem_{i,j}$, $mem_{i,j+1}$, the peak variables $r_{i,j}$, and the priority variables $p_{i,j}$.

Definition 4 (PRIORITYTRANSFER).
Let $U(t)_{t \in [w_j^{start}, w_j^{end}]}$ = SIMULATION (ins_j, p_j, mem_j), *then the constraint* PRIORITYTRANSFER $(ins_j, mem_j, mem_{j+1}, r_j, p_j)$ *is satisfied if and only if:*
$$mem_{i,j+1} = U_i\left(w_j^{end}\right) \wedge r_{i,j} = \max_{t \in [w_j^{start}, w_j^{end}]} (U_i(t)) \leq rmax, \forall i \in \mathcal{B}$$

PRIORITYTRANSFER consists of three propagators. Propagator SIMULATION CHECK handles the simplest case, when $mem_{i,j}$ and $p_{i,j}$ are fixed, as we can compute the values of $mem_{i,j+1}$ and $r_{i,j}$ with the SIMULATION algorithm. We know $U(t)$ for $t \in \left[w_j^{start}, w_j^{end}\right]$, by running SIMULATION (ins_j, p_j, mem_j). Thus, the memory level for the next window is $mem_{i,j+1} = U_i(w_j^{end})$ and $r_{i,j} = \max_{t \in [w_j^{start}, w_j^{end}]} (U_i(t))$, for each buffer i. The worst-case time complexity of this propagator is the same as SIMULATION, $O(k \log k + n^2 k \log n)$ with k the number of fill rate events. This propagator is sufficient for the constraint correctness but performs only minimal filtering on the domains of the variables, as it only updates the memory for the next window when all variables are fixed for the current one. To improve filtering, we developed another propagator, SIMULATIONLOWERBOUND, that also uses SIMULATION. This propagator considers a *best-case scenario* for buffer i, to compute a lower bound on $mem_{i,j+1}$ and $r_{i,j}$. After each SIMULATION run, we update the lower bounds of $mem_{i,j+1}$ and $r_{i,j}$. If $\min(r_{k,j}) > rmax$, the solver fails and performs backtracking. The time complexity of this propagator is $O(n(k \log k + n^2 k \log n))$, as we need to run SIMULATION for each buffer.

Theorem 2 (Best-Case Scenario). *From the perspective of a buffer i, given the domains of $mem_{i',j}$ and $p_{i',j}$ for all buffers $i' \in \mathcal{B}$ and a single window j, the minimal memory level peak and memory level at the end of the window will be reached if: memory levels at the start of window j are minimal: $\min (mem_{i',j})$, $\forall i' \in \mathcal{B}$; Priority of buffer i during window j is the best: $\min(p_{i,j})$; Priorities of other buffers during window j are the worst: $\max (p_{i',j})$, $\forall i' \neq i \in \mathcal{B}$.*

Proof. For a given share of the bandwidth allocated to buffer i, increasing the initial memory level $mem_{i,j}$ can only increase its memory level during the window. Increasing the initial memory level of any other buffer $i' \neq i$ can never increase the bandwidth allocated to buffer i, but it can decrease it. In the same manner, giving buffer i a higher (i.e. worse) priority or giving another buffer a lower (i.e. better) priority can only lead to decrease i in the buffers relative priority order and thus decreasing the bandwidth allocated to buffer i. Therefore, the minimal peak and end memory level will be reached if the three conditions of Theorem 2 stand.

Although oMDP is NP-hard, it has been demonstrated that an optimal solution can be computed in polynomial time if we consider a single downlink window [7]. The algorithm to decide whether there exists a priority assignment achieving a given peak usage is called SINGLEWINDOW. We can extend this algorithm to take into account the priority domains to compute this solution and filter some inconsistent values.

Lemma 2. *Given a window j, a buffer i and a value $v \in D(p_{i,j})$, if buffer i exceeds the current rmax when running* SIMULATION *with: $p_{i,j} = v$ and $\forall\ 'i \in \mathcal{B}, i' \neq i, p_{i',j} = \max(p_{i',j}); \forall\ i' \in \mathcal{B}; U_{i'}(w_j^{start}) = \min(mem_{i',j})$, then the assignment $p_{i,j} = v$ is inconsistent.*

Proof. Giving a better priority to any buffer $i' \neq i$ can never result in increasing the bandwidth share allocated to buffer i. In the same way, increasing the memory level of any buffer can only increase the peak usage of buffer i. Therefore, if the scenario in Lemma 2 violates the constraint $r_{i,j} \leq rmax$, the only way to increase the bandwidth allocated to buffer i, and thus reducing its peak, is to assign it a strictly better priority than v.

This leads to the constraint's third propagator, SINGLEWINDOW. In this propagator, memory levels are initialized to $\min(mem_{i,j})$ and the SIMULATION algorithm is run successively with the priority assignment p', that corresponds to the priority variables set to their current upper bound, and the memory levels set to their lower bounds. By Lemma 2, the current priority value of each overflowing buffer is inconsistent and can be removed. We repeat this process until no buffer overflows, or one priority domain is empty. In Rouzot's thesis [13], we provide the pseudo code for this propagator.

5.2 Symmetry Breaking

Dense Ranking. A solution of oMDP is a complete assignment of the priority variables for each downlink window. The priorities correspond to a *ranking* between the different buffers. The use of integer variables to represent the priority group for all buffers is straightforward, but the downside is the existence of equivalent solutions. For instance, for 3 buffers, $p_{1,j} = 1, p_{2,j} = 1, p_{3,j} = 2$ is equivalent to $p_{1,j} = 1, p_{2,j} = 1, p_{3,j} = 3$ (i.e. it represents the same ranking). To break these symmetries, we force the variables to respect the rules of a *dense ranking*: at least one element is equal to 1 (i.e. best rank 1 must be assigned); for any $k \in \{2, \ldots, n\}$, if k is assigned, then rank $k - 1$ is also assigned (i.e. consecutive ranks must be assigned). To enforce these rules, we introduce a new global constraint called DENSERANKING, defined as follows:

Definition 5 (DENSERANKING).   *Let $\mathcal{X} = \{x_1, \ldots, x_n\}$, then the constraint* DENSERANKING $(\mathcal{X})$ *is satisfied if and only if:* $\text{MIN}(\mathcal{X}) = 1 \land \forall i \in \{1, \ldots, n\}, \forall k \in \{2, \ldots, k\}, x_i = k \implies \exists i' \in \{1, \ldots, n\}$ *such that $x_{i'} = k - 1$*

Our propagation algorithm for the DENSERANKING constraint enforces only bound consistency and operates on a set of n variables whose domains are subsets of $\{1, \ldots, n\}$ with consecutive integer values. The pseudo-code for finding a bound support for the DENSERANKING constraint is presented in Algorithm 2, and is used in Algorithm 3, which is the propagation algorithm.

In DENSERANKINGBOUNDSUPPORT, variables are sorted in increasing order of their lower bounds (line 1). Then, we try to build a valid dense ranking—i.e. a support for the current domains of the variables. The variable k tracks the

---

Algorithm 2 DENSERANKINGBOUNDSUPPORT.

Require: $\mathcal{X} = \{x_1, \ldots, x_n\}$
1:   $x_1, \ldots, x_n \leftarrow$ **SortLB**$(\mathcal{X})$ $\triangleright$ Sort the variables by ascending lower bounds
2:   $heap \leftarrow$ **InitBinaryHeap**$(ordering = asc\_ub)$
3:   $k \leftarrow 1$ $\triangleright$ Current rank
4:   $i \leftarrow 1$
5:   **while** $i \leq n$ **do**
6:      **while** $\min(x_i) \leq k$ **do**
7:         $heap.$**Insert**(x_i) $\triangleright$ Insert variables with k in their domain
8:         $i \leftarrow i + 1$
9:      **if** $heap.$**IsEmpty**$()$ **then** $\triangleright$ No variable to cover the current rank
10:        **return false**
11:     **else**
12:        $heap.$**RemoveRoot**$()$ $\triangleright$ Variable with smallest UB takes rank k
13:      **while** $\max(heap.$**Root**$()) = k$ **do**
14:        $heap.$**RemoveRoot**$()$ $\triangleright$ Variables x_i with $\max(x_i) = k$ takes rank k
15:      $k \leftarrow k + 1$
16: **return true**

Algorithm 3 DENSERANKINGBC

Require: $\mathcal{X} = \{x_1, \ldots, x_n\}$
1:   **for** $x \in \mathcal{X}$ **do**
2:     **AdjustLB**$(x, \mathcal{X})$
3:     **AdjustUB**$(x, \mathcal{X})$
4:     **if** $\min(x) > \max(x)$ **then**
5:        **return Fail**

---

current rank being assigned in the construction of the support (line 3). A binary heap is initialized—empty at first—to maintain candidate variables ordered by ascending upper bounds. For each rank k, we add to the heap all variables whose domain includes k (line 7). At least one variable must be selected to cover each rank. Since we assume domains with consecutive values bounded by the lower and upper bounds of the variable, the variable with the smallest upper bound must be selected (line 12). Indeed, variables with larger upper bounds can still cover future ranks, while those with smaller upper bounds may not be able to do so, risking discontinuities in the dense ranking. We prove that this strategy is optimal in Lemma 3. After assigning a variable to rank k, we remove from the heap all variables whose upper bound is equal to k (line 14), as they cannot be used to cover any higher ranks. If no variable in the heap can be assigned to the current rank and some variables are still to be assigned (line 9), it means that the current domains are inconsistent. This process is repeated with rank $k + 1$ until failure or until all variables have been assigned to a rank. In the latter case, a support for the current domains has been successfully found. The time complexity of Algorithm 2 is $O(n \log n)$: sorting the variables initially takes

$O(n \log n)$, and in the main loop, at most n insertions and n root removals are performed sequentially, each in $O(\log n)$.

Lemma 3. *Given the domains of variables $\mathcal{X}$, Algorithm 2 always finds a bound-consistent support for* DENSERANKING, *if such a support exists.*

Proof. When building the support for $\mathcal{X}$ in DENSERANKINGBOUNDSUPPORT, we select at least one variable x that contains k, with the smallest upper bound to cover each rank k. Let us choose any other variable x' that also contains k instead of x. If choosing x' leads to a valid support, it is possible to switch the values of x and x' as $[k, \max(x)]$ is included in $D(x')$ and x can take value k. Thus, choosing the variable with the smallest upper bound at each step is the best strategy. It follows that if no variable can take current rank k, there is no assignment of $\mathcal{X}$ with the current domains such as the DENSERANKING constraint is respected.

In Algorithm 3, the bounds of variable domains are updated by eliminating inconsistent lower and upper bounds values with Algorithm 2. As we are asserting bound consistency only, this can be done in $O(n \log n)$-time by performing a dichotomic search on both bounds with procedures ADJUSTLB and ADJUSTUB. In ADJUSTLB, the domain of variable x is reduced to a subset of $D(x)$ with consecutive values starting from $\min(x)$. The procedure searches for the maximum value of m such that no support can be found for $\mathcal{X}$ with $D(x) = [\min(x), \ldots, m]$. When such m is found, the lower bound of x is set to $m+1$, as all values in $[\min(x), \ldots, m]$ are not bound consistent for DENSERANKING. Similarly, in ADJUSTUB, the procedure search for the minimum value of m such that no support exists for $\mathcal{X}$ with $D(x) = [m, \ldots, \max(x)]$, and the upper bound of the variable is set to $m - 1$. This is repeated for all variables $x \in \mathcal{X}$, so the time complexity of our propagator DENSERANKINGBC is $O(n^2 \log^2 n)$, as DENSERANKINGBOUNDSUPPORT is called $O(n \log n)$ times.

Note that DENSERANKING constraint can be enforced with the global constraint ATLEASTNVALUES$(\mathcal{X}, n)$ [1], based on SOFTALLDIFF [11]. This constraint forces the set of variables $\mathcal{X}$ to take at least n distinct values. By setting n to MAX $(\mathcal{X})$, we ensure that all values in $\{1, \ldots, \text{MAX}(\mathcal{X})\}$ will be assigned to at least one variable. However, this combination of constraints is sometimes not as strong as DENSERANKING. Let us demonstrate it with a simple example: Let $\mathcal{X} = \{x_1, x_2, x_3\}$ with $D(x_1) = \{1, 2, 3\}$, $D(x_2) = \{2, 3\}$, $D(x_3) = \{2, 3\}$. As rank 1 is mandatory for respecting the rules of a DenseRanking, x_1 must take the value 1 and thus $x_1 = 2$ and $x_1 = 3$ are not consistent and those values should be discarded. If we run the arc consistency for ATLEASTNVALUES$(\mathcal{X}, N)$ and $N = \text{MAX}(\mathcal{X})$, $x_1 = 2$ and $x_2 = 3$ will not be filtered even though they are inconsistent for the dense ranking. In the other hand, as we are not providing an algorithm for arc consistency on DENSERANKING, the decomposition may sometimes filter values missed by DENSERANKINGBC.

Priority Symmetry. Symmetry also arises when a group of buffers, having priority over the bandwidth, holds enough data to fully utilize the available dump rate during the downlink window. In this case, the worst-priority buffers receive no bandwidth, regardless of their relative priority order. Let v be the worst priority value among a subset of buffers S with fixed priorities for a given downlink window j. Let us assume that the buffers in S are never all empty during the downlink window j, thus no share of the overall bandwidth will be redistributed to buffers with a priority strictly greater (i.e. worse) than v. In that case, any priority value v' such as $v' > v$ (i.e., a worse priority) is equivalent to the priority value $v + 1$. This observation being made, we can infer a filtering procedure for the priority variables, that is embedded in constraint PRIORITYSYMMETRY. Every time the domain of a priority variable $p_{i,j}$ changes, we run: SIMULATION(ins_j, p_j, mem_j) with $p_j = \{\max(p_{i,j}) \mid \forall i \in \mathcal{B}\}$, $mem_j = \{\min(mem_{i,j}) \mid \forall i \in \mathcal{B}\}$. During the simulation, we keep track of whether each buffer has been allocated bandwidth given the current priority assignment. Let v be the worst priority value among the buffers that received bandwidth. Since any priority assignment strictly greater than v would result in the buffer not receiving bandwidth, we can remove all values strictly greater than $v + 1$ from the priority domains for the current downlink window j, as they represent equivalent priority assignments. This propagator is as costly as SIMULATION that runs in $O(k \log k + n^2 k \log n)$-time, with k the number of observation events (see Sect. 3).

5.3   Search

In NP-hard problems such as the oMDP, the way we explore the search tree may significantly impact the quality of the solution, given a limited time. As PRIORITYTRANSFER strongly relies on the current $rmax$ upper bound to trigger backtracking with SIMULATIONLOWERBOUND and SINGLEWINDOW propagators, we must find good solutions quickly. To that end, we developed a search method based on the DOWNLINKCOUNT heuristic for variable and value selection.

In our search, decisions are made starting from the earliest downlink windows, proceeding to the next only when all priority variables of the current window are fixed. Within a single window, we use DOWNLINKCOUNT (see Sect. 3) for both variable and value selection. Instead of setting the buffer capacity as the overflow limit, we introduce a random limit based on the current $rmax$ upper bound. More precisely, the overflow limit for DOWNLINKCOUNT is drawn uniformly between $0.5 \times \max(rmax)$ and $\max(rmax)$. Adding randomness to the decision process is helpful, so the solver explores new branches of the search tree after each restart, thereby accelerating the resolution. For variable selection, we begin with the variable that has the best priority in the solution produced by DOWNLINKCOUNT. Choosing the buffers with the best priority first is more likely to trigger propagation through the DENSERANKING constraint and the PRIORITYSYMMETRY propagator. For value selection, we follow the priority order given by DOWNLINKCOUNT if the value is present in the domain of the variable. If the priority exceeds the upper bound of the variable, we assign the upper bound; if the priority is lower than the lower bound, we assign the lower bound.

6 Experimental Results

For long-range missions by the ESA (European Space Agency), science observation planning is divided into four steps [5,10]. First, the Long Term Plans (LTP) are established well in advance by the Science Ground Segment in collaboration with the Science Working Team. These high-level plans each cover a few months of the mission. They are then refined into Medium Term Plans (MTP) of a month, Short Term Plans (STP) of a week, and finally Very Short Term Plans (VSTP), which is the smallest and final planning level, spanning half a week. In the case of Rosetta, each new smaller plan corresponds to a new instance to solve, requiring orchestration of data transfers to minimize buffer overflow risks. Proving optimality for non-trivial MTP and LTP instances is extremely challenging, and at these stages, good solutions are often sufficient. However, having an optimality guarantee for STP and VSTP instances is far more valuable.

For their experiments, [7] used 4 real instances, that are MTP of the Rosetta mission with 16 buffers, 64 to 94 downlink windows, and around 10,000 data production events each (see [3] for Rosetta observation planning). They also generated random instances based on these real scenarios. However, these instances often feature long non-visibility periods, during which high memory peaks can occur, making trivial lower bound[1] on $rmax$ easy to match. To address this issue and thus produce instances corresponding to potentially more difficult contexts, we generated new instances from scratch by introducing random events, shortening non-visibility periods, and ensuring all buffers produce substantial data.

Our model is implemented using the OR-Tools original CP solver [9]. For all experiments, we approximate the peaks to a tenth of a percent ($r^{ub} = 10^3$) and always round up peak values. All experiments are run on a Xeon E5-2695 v3 @ 2.30 GHz CPU with 10 GB of RAM, with a time limit of one hour per experiment. Our source code and instances are available in our Git repository[2].

First, we analyze the efficiency of our global constraints in reducing the search space on small randomly generated instances. We then highlight the impact of our search heuristic in quickly finding good solutions for real Rosetta instances. Finally, we present a broader statistical analysis on both Rosetta instances and randomly generated instances of various sizes, comparing our approach against the state-of-the-art: ITERATIVELEVELING [12] and REPAIRDESCENT [7].

Table 1 presents the average search effort required to prove optimality on 20 small instances, each with 4 buffers and 4 downlink windows. We evaluate the impact of disabling constraints or propagators by comparing performance across different model configurations. Specifically, we report the average solution time, the number of branches explored by the solver, and the number of optimal solutions. The first row corresponds to the baseline, where only the SIMULATIONCHECK propagator is enabled. In the subsequent rows, a check mark ($\checkmark$) indicates which additional propagators or constraints are active.

---

[1] A way to find such lower bound is FULLTRANSFER [7], an algorithm that allocates all bandwidth to each buffer during downlink windows.

[2] https://gitlab.laas.fr/roc/julien-rouzot/transfer-scheduling-csp.

Table 1. *Mean resolution time, number of branches explored, and number of optimal solutions for 20 small instances for different combinations of global constraints and propagators enabled. DR: DENSERANKING , PS: PRIORITYSYMMETRY , SW: SINGLEWINDOW , SLB: SIMULATIONLOWERBOUND.*

| Constraints | | | | Resolution time | Number of branches | Optimal solutions |
DR	PS	SW	SLB			
				14 min	168.2 M	16
	✓	✓	✓	3.5 min	10.2 M	19
✓		✓	✓	53.3 s	2.6 M	20
✓	✓		✓	1.7 s	0.14 M	20
✓	✓	✓		1.6 s	0.11 M	20
✓	✓	✓	✓	1.5 s	0.07 M	20

We observe a clear improvement when using our global constraints compared to the baseline, for the resolution time (560 times faster), the number of branches explored (2400 times fewer) and the number of optimal solutions (20 vs. 16). We remark the importance of symmetry breaking with DENSERANKING and PRIORITYSYMMETRY constraints, as the resolution time and the number of branches raise significantly when they are disabled. On the other hand, SIMULATIONLOWERBOUND and SINGLEWINDOW propagators seem to have the most limited impact on the resolution time, but help pruning the search space.

We also compare our CP method against ITERATIVELEVELING heuristic [12] and REPAIRDESCENT heuristic [7] on 480 generated scenarios in Fig. 2.

The synthetic instances include 240 small plans (4 to 16 buffers, 4 to 16 downlinks) and 240 larger plans (8 to 24 buffers, 20 to 80 downlinks). We present the proportion of optimal solutions for 240 small instances with varying numbers of buffers and windows. The darker zone at the bottom of each bar represents the proportion of instances where either ITERATIVELEVELING or REPAIRDESCENT can prove optimality by reaching the lower bound provided by FULLTRANSFER. As intended by our instance generation process, very few instances are trivially proved optimal by the heuristics. In contrast, our CP model succeeds in proving optimality much more frequently, even when the FULLTRANSFER lower bound is unattainable. However, providing non-trivial optimality proofs becomes challenging when the number of buffers and windows increases.

In Rouzot's thesis [13], we provide additional experiments. We first evaluate the impact of our search strategy against two strategies on all four Rosetta real-world instances. The *random* search is used as baseline and the popular generic strategy *min-dom* selects the variable with the smaller domain, following the first-fail philosophy [6]. The first two instances (MTP011, MTP012) are solved to optimality in a few seconds with our search strategy *downlink-count*, while we reach the one hour time limit for the two other strategies. For the two other instances (MTP013, MTP014), we can find much better solutions with *downlink-*

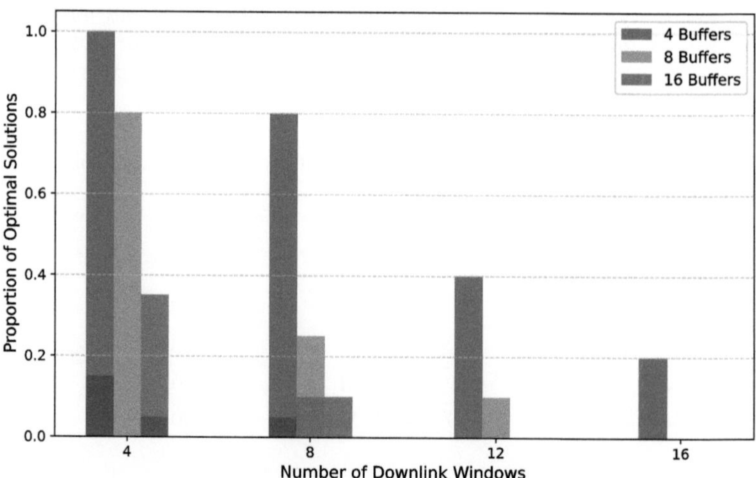

Fig. 2. Proportion of optimal solutions found for 240 synthetic instances. The shaded areas in the bar plot correspond to the proportion of solutions for which the heuristics are able to prove optimality (by reaching a trivial lower bound).

count in a minute than with *random* or *min-dom* in an hour. These results highlight the importance of an efficient branching strategy for solving the oMDP.

We also report the mean objective values for the three methods across instances with varying numbers of buffers and downlink windows. We observe that for few buffers and downlink windows, our CP model is competitive with REPAIRDESCENT, occasionally even outperforming it. In general, ITER-ATIVELEVELING is dominated by the other two methods, except for the largest instances, where the CP model fails to scale.

7   Conclusion

In this paper, we introduced the first exact approach for solving the overlapping Memory Dumping Problem (oMDP) and provided an enhanced complexity analysis, demonstrating that the problem is strongly NP-hard in the general case. We proposed a constraint programming model incorporating novel global constraints and a search strategy to improve its efficiency. The generic DENSERANKING constraint can be applied to any application in which *ranking* is involved. While our approach is slower than the heuristics of [12] and [7] on large instances, it outperforms state-of-the-art methods on smaller instances and is capable of producing non-trivial optimality proofs. Our results suggest that both approaches could be used synergistically for data transfer scheduling: the REPAIRDESCENT heuristic for strategic planning (LTP-MTP) and our CP model for tactical and operational planning (STP-VSTP). Furthermore, by leveraging an expressive framework such as constraint programming, our approach offers greater flexibility, allowing additional operational constraints to be integrated into the model.

Acknowledgements. This work benefited from ANITI AI Cluster funded by the France 2030 program under Grant Agreement No. ANR-23-IACL-0002.

References

1. Bessière, C., Hebrard, E., Hnich, B., Kiziltan, Z., Walsh, T.: Filtering algorithms for the NValue constraint. Constraints **11**, 271–293 (2006)
2. Chien, S., Knight, R., Stechert, A., Sherwood, R., Rabideau, G.: Using iterative repair to increase the responsiveness of planning and scheduling for autonomous spacecraft. In: International Joint Conference on Artificial Intelligence (IJCAI 1999). Stockholm, Sweden (1999)
3. Chien, S., et al.: Activity-based scheduling of science campaigns for the Rosetta orbiter. In: Twenty-Fourth International Joint Conference on Artificial Intelligence, IJCAI-15. Buenos Aires, Argentina (2015)
4. Garey, M.R., Johnson, D.S.: Computers and Intractability; A Guide to the Theory of NP-Completeness. W. H. Freeman & Co., USA (1990)
5. Haddow, C., Whitehead, G., Adamson, K., Sousa, B.: Mission planning-establishing a common concept for esoc's missions. In: SpaceOps 2010 Conference Delivering on the Dream Hosted by NASA Marshall Space Flight Center and Organized by AIAA, p. 1969 (2010)
6. Haralick, R.M., Elliott, G.L.: Increasing tree search efficiency for constraint satisfaction problems. Artif. Intell. **14**(3), 263–313 (1980)
7. Hebrard, E., Artigues, C., Lopez, P., Lusson, A., Chien, S., Maillard, A., Rabideau, G.: An efficient approach to data transfer scheduling for long range space exploration. In: De Raedt, L. (ed.) Proceedings of the Thirty-First International Joint Conference on Artificial Intelligence, IJCAI-22, pp. 4635–4641. International Joint Conferences on Artificial Intelligence Organization (7 2022)
8. Imbriale, W.A.: Large Antennas of the Deep Space Network. John Wiley & Sons (2005)
9. van Omme, N., Perron, L., Furnon, V.: OR-Tools User's Manual. Tech. rep, Google (2014)
10. Pérez-Ayúcar, M., et al.: The Rosetta science operations and planning implementation. Acta Astronaut. **152**, 163–174 (2018)
11. Petit, T., Régin, J.C., Bessière, C.: Specific filtering algorithms for over-constrained problems. In: Principles and Practice of Constraint Programming: 7th International Conference, Paphos, Cyprus, Proceedings 7, pp. 451–463. Springer (2001)
12. Rabideau, G., Chien, S., Nespoli, F., Costa, M.: Managing spacecraft memory buffers with overlapping store and dump operations. In: Workshop on Scheduling and Planning Applications, International Conference on Automated Planning and Scheduling (SPARK, ICAPS 2016). London, UK (2016)
13. Rouzot, J.: Combinatorial Optimization for Space Exploration with Constraint Programming : Data Transfers, Scientific Observations, and Operations Scheduling. Ph.D. thesis, Université de Toulouse (2025). https://www.theses.fr/2025TLSEI026
14. Simonin, G., Artigues, C., Hebrard, E., Lopez, P.: Scheduling scientific experiments on the Rosetta/Philae mission. In: International Conference on Principles and Practice of Constraint Programming., p. 23–37. Springer (2012)

BaB-PoNN: A Bit-Exact Branch-and-Bound Framework for Verified Robustness of Posit Neural Networks

Suleiman Junaidu Sadiq[(✉)] and Martin Mariusz Lester

Department of Computer Science, University of Reading, Reading, UK
s.j.sadiq@pgr.reading.ac.uk, m.lester@reading.ac.uk

Abstract. We present BaB-PoNN, the first formal verification framework for verified robustness of neural networks that use posit arithmetic. BaB-PoNN reasons under the exact posit-8 operational semantics used at inference time, including quire-8 fused accumulation and a single quire-to-posit rounding per affine layer, so every robustness verdict is sound for the deployed implementation. The framework verifies mixed-budget robustness properties in which perturbations are constrained jointly by (i) the number of input coordinates that may change and (ii) the total number of bit flips across those coordinates. We formalize both local robustness, where search is restricted to a data-driven region of interest, and global robustness, where any coordinate may be perturbed within the same hybrid coordinate-bit neighborhood. BaB-PoNN performs an explicit-state search guided by admissible bounds from coordinate swings and per-bit gains, enabling sound pruning while remaining complete: for each input and budget, it either returns a concrete adversarial witness or proves robustness for the specified neighbourhood. Evaluations on posit-8 multilayer perceptron (MLP) and LeNet-5 convolutional neural network (CNN) architectures trained on MNIST show that BaB-PoNN can verify robustness under non-trivial mixed budgets and locate bit-precise adversarial witnesses.

Keywords: Posit · branch-and-bound · neural network verification

1 Introduction

1.1 Motivation and Context

Formal verification provides a principled way to analyse the reliability of neural networks in safety-critical and numerically constrained settings. Even when test accuracy is high, small input perturbations can cause abrupt changes in the predicted class, and such sensitivity cannot be ruled out by testing alone [1,6,14,16]. A central robustness question is therefore: given an input and a constrained perturbation budget, does *every* admissible perturbation preserve the top-1 class, or does there exist a concrete counterexample? [4–6,8,18]

As neural inference increasingly moves to low-precision or domain-specific hardware, this question must be answered with respect to the arithmetic

T. Guns (Ed.): CPAIOR 2026, LNCS 16595, pp. 486–502, 2026.
https://doi.org/10.1007/978-3-032-27242-3_29

semantics of the deployed platform. The implemented network is not a real-valued function, but a finite-state transition system whose behaviour is determined by a particular discrete number system, rounding discipline, and accumulation scheme. Most existing verification frameworks, whether SMT/MILP-based [4,8] or relaxation-based [5,9,18–21], assume real-valued or IEEE floating-point semantics and rely on continuous relaxations or convex abstractions tailored to uniform mantissa-exponent formats. These techniques do not account for the discrete, asymmetric behaviour of modern quantized number systems, and a verification method that is sound for one semantics is *not sound* when the network is executed under a different arithmetic.

The *posit* number system has been proposed as a drop-in alternative to IEEE floating point, offering higher effective precision and improved numerical accuracy at low bit widths by combining tapered precision, a dynamic exponent regime, and a fused accumulator (*quire*) with a single rounding step per affine layer [7]. Posit-based inference and training have been evaluated empirically in deep learning, with frameworks such as *Deep PeNSieve* [13] and the posit-training framework of Lu et al. [12] using tensor-wise scaling and warmup for ResNet-style networks, showing that 8 and 16-bit posit configurations (often with 16-bit master weights) can match float32 accuracy on standard vision and language benchmarks while enabling more area, memory, and energy-efficient accelerators. On the hardware side, Edavoor et al. propose a Posit Quire Processing Engine (PQPE) and show that a posit-16 configuration without quire can match float32 inference accuracy on standard vision workloads, while posit-8 with or without quire loses only a few percentage points of accuracy and reduces LUT and DSP usage by more than 70% on FPGAs [3]. Podobas and Matsuoka implement parameterised posit operators on FPGAs and integrate them into an OpenCL-based linear algebra library, demonstrating that posit units can operate at competitive clock frequencies and achieve orders-of-magnitude speedups over software posit emulation [15]. However, all of these works remain purely empirical: correctness and accuracy are validated by testing and simulation only; the arithmetic units themselves are not accompanied by formally-checked proofs of their semantics; and, crucially, there are no soundness guarantees about the behaviour of posit neural networks under input perturbations. To the best of our knowledge, there is no verification framework that reasons directly about the exact bit-level posit semantics (including quire-based accumulation) and proves robustness properties such as the absence of adversarial examples within a given discrete perturbation budget. This gap motivates a verification approach that works over the exact posit semantics used in deployment.

1.2 Posit Arithmetic and Robustness Sensitivity

In posit arithmetic the distribution of representable real values and the available relative precision are highly nonuniform [7]. Near unit scale, a posit format allocates more bits to the fraction and can offer higher effective precision than an IEEE 754 floating-point format with comparable bit width, while further away it trades precision for dynamic range [2]. As a consequence, the semantic effect

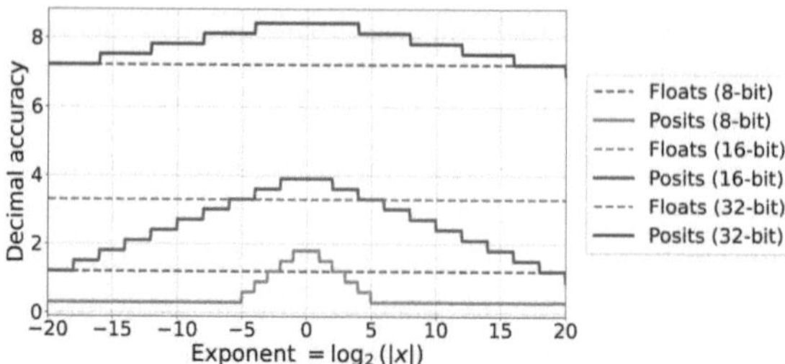

Fig. 1. Illustrative comparison of decimal accuracy across representable magnitudes for 8, 16, and 32-bit posit and IEEE floating-point. Posits allocate higher precision near unit scale ($|x| \approx 1$), while IEEE floats distribute precision more uniformly in log-space.

of changing a posit-encoded value by a single bit flip depends strongly on its magnitude and on the *local density of representable real values*: nominally small changes in the bit pattern can produce large shifts in the decoded real value in some regions and negligible shifts in others. Figure 1 illustrates this nonuniform precision profile for 8, 16, and 32-bit IEEE floating-point formats and the corresponding posit configurations, showing how decimal accuracy is concentrated near unit scale for posits and more evenly spread in log-space for IEEE floats.

Posit neural networks inherit the tapered precision and quire-based fused dot products of the underlying arithmetic, which concentrate representable values around zero and change how dot products are accumulated and rounded [2,7,13]. Consequently, two perturbations with the same Euclidean norm can have different effects on the logits, depending on how they move activations relative to the posit value grid and quire accumulations. From a verification perspective, robustness therefore cannot be characterised solely in terms of continuous ℓ_p norms: the same ℓ_p movement may correspond to very different sequences of bit-level transitions and hence very different classification outcomes. Any meaningful robustness analysis for posit neural networks must explicitly track how perturbations move activations across regions of varying precision, rather than relying on idealised real or IEEE 754 abstractions.

1.3 Contributions

This work introduces BaB-PoNN, a branch-and-bound framework for *neural networks executed under posit arithmetic*. Our main contributions are:

- **Posit-exact verification.** A verification system that operates directly on posit arithmetic, modelling quire-based accumulation and posit rounding at every layer, so that all computations match the deployed posit-8 implementation.
- **Discrete branch-and-bound search.** A finite branch-and-bound procedure over discrete input perturbations, constrained by a joint budget on how many input coordinates may change and how many bits may flip in their posit encodings, and guided by margin-based bounds to prune the search.

- **Robustness and sensitivity for posit networks.** A robustness view tailored to the non-uniform posit value grid, covering both perturbations confined to a small local subset of input coordinates and perturbations allowed anywhere, together with ranked coordinate and bit-level sensitivities that highlight where the network is most vulnerable to discrete input changes.

An open-source implementation of BaB-PoNN, including source code, evaluation data, and extended evaluation results, is available online [17].

2 Related Work

The formal verification of neural networks has followed several distinct methodological paths. Early SMT and MILP-based methods such as Reluplex and related encodings showed that medium-sized ReLU networks can be verified exactly, but at high computational cost [4,8]. Abstract-interpretation and bound-propagation approaches such as AI2 [5], DeepPoly [18], and ERAN [19] trade some precision for much better scalability, using convex relaxations or custom abstract domains to over-approximate reachable sets and certify robustness for larger models. More recent branch-and-bound (BaB) frameworks combine these relaxations with search over subdomains, yielding complete verifiers that are faster and scale to larger benchmarks than earlier LP or SMT-based tools.

Recent BaB-based verifiers push this paradigm further. Wang et al. introduce β-CROWN, a bound-propagation-based verifier that can fully encode neuron split constraints within a branch-and-bound framework by introducing optimisable per-neuron parameters in the primal and dual spaces [20]. Their method tightens ReLU relaxations, jointly optimises intermediate layer bounds, and is implemented to run efficiently on GPUs, yielding a complete BaB verifier that is significantly faster than LP-based approaches and also competitive as an incomplete verifier in terms of verified accuracy and timeout rate. This construction is generalised in the α, β-CROWN framework, which augments β-CROWN with additional dual parameters α associated with neuron splits, providing a unified BaB-compatible formulation of several CROWN-style bound-propagation schemes. Zhou et al. extend the α, β-CROWN framework with BICCOS, which adds branch-and-bound inferred cutting planes to the verifier [21]. From explored subproblems, they derive linear relations between ReLU activation patterns and encode these as additional constraints that tighten later relaxations, reducing the number of BaB subdomains that must be explored and improving verified accuracy and timeout behaviour on large ReLU benchmarks. Complementing these algorithmic advances, Kaulen et al. investigate how to reduce the computational cost of branch-and-bound-based neural network verification by predicting, during a run, whether a robustness query is likely to be solved within a fixed time budget [9]. They define static features (e.g. initial bounds, margins, unstable neuron counts) and dynamic features (e.g. BaB tree depth, number of branches, evolving global bounds), train classifiers to distinguish solvable from unsolvable instances, and use these predictions to terminate likely timeouts early.

Despite these advances, existing BaB-based verifiers assume real or IEEE 754-valued networks and continuous ℓ_p balls, and do not account for posit arithmetic. In contrast, our work develops a BaB framework for posit-quantised networks under their exact arithmetic semantics, reasoning about the discrete perturbation space induced by tapered precision and quire-based accumulation, for which no prior verification framework exists.

3   Preliminaries and Formal Model

3.1   Posit Networks

Let $\mathbb{P}_{n,es}$ denote the finite set of representable posit values for bit width n and exponent size es. We fix $(n, es) = (8, 0)$ and write $\mathbb{P} = \mathbb{P}_{8,0}$, so $\mathbb{P} = \{-64, \ldots, -1, \ldots, -\frac{1}{64}, 0, \frac{1}{64}, \ldots, 1, \ldots, 64\}$.

Posit arithmetic is determined by two core maps: $\mathsf{dec} : \mathbb{P} \to \mathbb{R}$, the decoding map that assigns to each posit value the real number it represents; and $\mathsf{round}_\mathbb{P} : \mathbb{R} \to \mathbb{P}$, the rounding map that sends a real argument to the nearest representable posit, with a fixed tie-breaking rule.

Posit addition and multiplication are the total operations $\oplus, \otimes : \mathbb{P} \times \mathbb{P} \to \mathbb{P}$ defined by:

$$p_1 \oplus p_2 = \mathsf{round}_\mathbb{P}(\mathsf{dec}(p_1) + \mathsf{dec}(p_2)) \qquad p_1 \otimes p_2 = \mathsf{round}_\mathbb{P}(\mathsf{dec}(p_1) \cdot \mathsf{dec}(p_2))$$

Inner products may be accumulated either *stepwise* in $\mathbb{P}$ (rounding after each multiply-accumulate) or using a fused dot product with an exact accumulator (*quire*). In BaB-PoNN we adopt the quire-fused semantics:

$$\mathsf{FDP}(u, v) = \mathsf{round}_\mathbb{P}\Big(\sum_{i=1}^{d} \mathsf{dec}(u_i)\mathsf{dec}(v_i) \Big),$$

where the sum is accumulated exactly in the quire and only the final result is rounded back to $\mathbb{P}$. This contrasts with stepwise accumulation, which rounds after each update: $s_{t+1} = \mathsf{round}_\mathbb{P}(s_t + \mathsf{dec}(u_t)\mathsf{dec}(v_t))$. Because rounding is non-linear, stepwise and quire-fused accumulation can differ even when the same terms are added. More precisely, if $\mathsf{round}_\mathbb{P}(\alpha) = \beta$ with $\beta \neq \alpha$, then stepwise accumulation yields $\alpha \oplus \alpha \oplus \alpha = \beta \oplus \beta \oplus \beta$, whereas quire-fused accumulation preserves 3α until the final rounding, often improving numerical accuracy.

A feed-forward posit network has layers $0, \ldots, L$. Each layer $l \in [0, L]$ has d_l neurons with parameters $W_\ell \in \mathbb{P}^{d_\ell \times d_{\ell-1}}$ (weights) and $b_\ell \in \mathbb{P}^{d_\ell}$ (biases). Given an input $x \in \mathbb{P}^{d_0}$, activations are defined by

$$a_0 = x \qquad z_\ell[j] = \mathsf{FDP}(W_{\ell,j,\cdot}, a_{\ell-1}) \oplus b_{\ell,j} \qquad a_\ell = \sigma_\ell(z_\ell)$$

where σ_ℓ is the ReLU function for $\ell < L$ and the identity on the output layer. All operations stay within $\mathbb{P}$, so the network

$$F = f_L \circ \cdots \circ f_1 \; : \; \mathbb{P}^{d_0} \to \mathbb{P}^C$$

is a deterministic map under this posit-8/quire-8 operational semantics. This semantics is the concrete execution model used throughout the paper.

3.2 Decision Function and Margin

Let $F : \mathbb{P}^{d_0} \to \mathbb{P}^C$ be a feed-forward posit network with posit-valued logits $L(x) = F(x) \in \mathbb{P}^C$ over classes $\{1, \ldots, C\}$. The decoding map $\mathsf{dec} : \mathbb{P} \to \mathbb{R}$ is strictly monotone, so it induces a total order on logits.

We define the (discrete) decision function by comparing decoded logits:

$$\hat{y}(x) = \arg\max_{c \in \{1, \ldots, C\}} \mathsf{dec}\big(L_c(x)\big)$$

where a fixed tie-breaking convention for $\arg\max$ is assumed so that $\hat{y}(x)$ is deterministic.

For a label $y \in \{1, \ldots, C\}$, we define the margin

$$m(x; y) = \max_{c \neq y}\big(\mathsf{dec}(L_c(x)) - \mathsf{dec}(L_y(x))\big)$$

Then $m(x; y) < 0$ exactly means that y has strictly largest decoded logit (and hence $\hat{y}(x) = y$), while $m(x; y) > 0$ means that some $c \neq y$ has a strictly larger decoded logit (and hence $\hat{y}(x) \neq y$). The case $m(x; y) = 0$ corresponds to decoded ties. Thus correctness and robustness properties can be stated as inequalities over $m(x; y)$, while the forward computations remain in $\mathbb{P}$.

3.3 Discrete Perturbations and Hybrid Budgets

Inputs lie in $\mathbb{P}^{d_0}$. Robustness verification restricts perturbations to a scope $S \subseteq \{1, \ldots, d_0\}$ specifying which coordinates may change (for example, a region of interest in an image). For each $i \in S$ we fix a finite alphabet $\Sigma_i \subseteq \mathbb{P}$ of admissible values, determined by the concrete input encoding (e.g. the 256 pixel intensity levels for MNIST); coordinates outside S remain fixed at x_i. The symbolic domain around x is

$$\mathcal{X}(x; S) = \big\{x' \in \mathbb{P}^{d_0} \mid (i \in S \Rightarrow x'_i \in \Sigma_i) \wedge (i \notin S \Rightarrow x'_i = x_i)\big\}$$

which is finite because each Σ_i is finite.

Each $u \in \Sigma_i$ has an n-bit posit encoding $\mathsf{enc}_i(u) \in \{0, 1\}^n$. For $u, v \in \Sigma_i$ we define the bit-level (Hamming) distance

$$\mathrm{Ham}_i(u, v) = \#\big\{j \in \{1, \ldots, n\} \mid \mathsf{enc}_i(u)_j \neq \mathsf{enc}_i(v)_j\big\}$$

i.e. the number of single-bit flips required to transform the encoding of u into that of v.

We impose two budgets: a coordinate budget K (how many input coordinates may change) and a bit budget B (how many bit flips may occur across those coordinates). The mixed-budget neighbourhood is

$$\mathcal{X}(x; S, K, B) = \Big\{x' \in \mathcal{X}(x; S) \ \Big| \ \#\{i : x'_i \neq x_i\} \leq K, \ \sum_{i \in S} \mathrm{Ham}_i(x'_i, x_i) \leq B\Big\}$$

Here K controls sparsity at the feature level, while B measures the bit-level effort in the posit encodings. Since inputs are represented as finite posit codewords, we formulate the perturbation budget in terms of bit flips rather than a Euclidean norm, so that robustness is stated directly over the underlying discrete input encodings.

3.4 Local and Global Robustness

Let (x, y) be a labelled instance. Given scope $S \subseteq \{1, \ldots, d_0\}$ and budgets (K, B), we say that F is *locally (K, B)-robust* at (x, S) if

$$\forall x' \in \mathcal{X}(x; S, K, B). \qquad \hat{y}(x') = y \quad \text{(equivalently } m(x'; y) \leq 0\text{)}$$

Local robustness constrains perturbations to the designated scope S (e.g. a chosen region of interest) under the hybrid budgets.

Global robustness is the special case where every coordinate may change. Writing $S_{\mathrm{all}} = \{1, \ldots, d_0\}$, F is *globally (K, B)-robust* at (x, y) if

$$\forall x' \in \mathcal{X}(x; S_{\mathrm{all}}, K, B). \qquad \hat{y}(x') = y$$

If $S' \subseteq S$ and $K' \leq K$, $B' \leq B$, then $\mathcal{X}(x; S', K', B') \subseteq \mathcal{X}(x; S, K, B)$; hence global robustness implies local robustness for any smaller scope or budget, and any counterexample for (S, K, B) remains valid for any larger scope or budget.

4 BaB-PoNN Verification Framework

We now describe the BaB procedure used in BaB-PoNN. The verifier operates on the discrete transition system induced by the posit/quire semantics and mixed-budget perturbation model of Sect. 3.3, with each transition applying a concrete bit-level perturbation followed by an exact posit-8 forward pass.

4.1 Verification Objective

Let $F : \mathbb{P}^{d_0} \to \mathbb{P}^C$ be a feed-forward posit network with decision function $\hat{y}$ and margin $m(x; y)$ as in Sect. 3. For a reference input x with clean prediction

$$y_{\mathrm{ref}} = \hat{y}(x)$$

a scope $S \subseteq \{1, \ldots, d_0\}$, and budgets $(K_{\mathrm{max}}, B_{\mathrm{max}})$, we consider the mixed-budget perturbation set

$$\mathcal{X}(x; S, K_{\mathrm{max}}, B_{\mathrm{max}})$$

as defined in Sect. 3.3.

The verification problem is to decide whether there exists an admissible perturbation that changes the top-1 class:

$$\exists x' \in \mathcal{X}(x; S, K_{\mathrm{max}}, B_{\mathrm{max}}) \, . \, \hat{y}(x') \neq y_{\mathrm{ref}}$$

Equivalently,

$$\exists x' \in \mathcal{X}(x; S, K_{\max}, B_{\max}) \, . \, m(x'; y_{\text{ref}}) > 0$$

If no such x' exists, then F is $(K_{\max}, B_{\max})$-robust at x with respect to S in the sense of Sect. 3.4; otherwise, any x' with $m(x'; y_{\text{ref}}) > 0$ is a concrete counterexample.

4.2   State Representation

A BaB node encodes a symbolic perturbation state

$$\eta = (x', K, B, M)$$

where x' is the current perturbed input, K and B are the *used* coordinate and bit budgets (number of coordinates whose value differs from x and total number of flipped bits, respectively), and M records all feature-bit pairs that have been exhaustively branched on. The global budgets $K_{\max}$ and $B_{\max}$ are fixed; every reachable node satisfies $0 \leq K \leq K_{\max}$ and $0 \leq B \leq B_{\max}$.

An *apply-transition* flips a single bit of one editable feature. Let $i \in S$ be a feature index and b a bit position. Starting from (x', K, B, M), flipping bit b of feature i yields a new input x'' and updated budgets

$$(x', K, B) \xrightarrow{i,b} (x'', K + \Delta_K(i, b), B + \Delta_B(i, b))$$

where $\Delta_B(i, b) \in \{0, 1\}$ is 1 if the bit actually changes, and $\Delta_K(i, b) \in \{0, 1\}$ is 1 if feature i was equal to its original value in x before the flip and differs afterwards, and 0 otherwise. The apply-transition is only admissible if $K + \Delta_K(i, b) \leq K_{\max}$ and $B + \Delta_B(i, b) \leq B_{\max}$; otherwise it is rejected. A *skip-transition* marks (i, b) as explored by adding it to M without modifying x' or the budgets.

Each node therefore represents all perturbations reachable by applying any sequence of admissible bit flips and skips that lead to the same x', the same used budgets (K, B), and the same set M of exhausted feature-bit pairs.

4.3   Coordinate and Bit-Level Bound Functions

At each BaB node $\eta = (x', K, B, M)$, the verifier recomputes the logits $L(x') = F(x')$ under the exact posit/quire semantics of Sect. 3, and evaluates the decoded margin $m(x'; y_{\text{ref}})$.

For each BaB node with current input x', the verifier computes two sets of local bound functions over the symbolically perturbable features $i \in S$:

- swing(i): the maximal increase in decoded margin obtainable at this node by changing only coordinate i over all admissible values $u \in \Sigma_i$ (for MNIST, each Σ_i has 256 elements);
- gain(i, b): the non-negative margin increase obtained at this node by flipping bit b of feature i once.

Concretely, for each $i \in S$, the implementation enumerates all $u \in \Sigma_i$, substitutes $x'_i \leftarrow u$, and evaluates the exact posit forward pass to record per-class decoded logit extrema and derive $\text{swing}(i)$ as a worst-case contribution to the margin. For each pair (i, b), it evaluates the single-bit flip $x' \mapsto x''$ where $x''_i = x'_i \oplus 2^b$ and sets $\text{gain}(i, b)$ to the resulting non-negative margin change. These quantities are (i) computed using the exact posit forward pass, and (ii) recomputed at each BaB node as needed, so they are always local to the current state x'.

Given a node that has used budgets (K, B), with remaining budgets $K_{\text{left}} = K_{\max} - K$ and $B_{\text{left}} = B_{\max} - B$, the maximal additional margin that any descendant can obtain is over-approximated by

$$\text{UB}_{\text{rem}}(K, B) = \sum_{\text{top } K_{\text{left}}} \text{swing}(i) + \sum_{\text{top } B_{\text{left}}} \text{gain}(i, b)$$

where the first sum ranges over the K_{left} coordinates with largest $\text{swing}(i)$ among those still eligible under the coordinate budget, and the second sum ranges over the B_{left} admissible bit flips with largest $\text{gain}(i, b)$.

4.4   Global Upper Bound and Pruning

At a BaB node $\eta = (x', K, B, M)$, the optimistic upper bound on any descendant margin is

$$\text{UB}(\eta) = m(x'; y_{\text{ref}}) + \text{UB}_{\text{rem}}(K, B),$$

where $m(\cdot; \cdot)$ is the margin from Sect. 3.2 and $\text{UB}_{\text{rem}}(K, B)$ is the residual bound defined in Sect. 4.3.

If $\text{UB}(\eta) \leq 0$, then no descendant $(\text{Desc}(\eta))$ that respects the remaining budgets K_{left} and B_{left} can produce a positive margin, and the entire subtree rooted at η is treated as free of counterexamples. This yields the pruning rule implemented in BaB-PONN: whenever the current margin plus the mixed-budget residual bound is non-positive, the node is closed without further branching.

4.5   Branch Selection

Among all unexplored feature-bit pairs, the verifier selects

$$(i^*, b^*) = \arg \max_{(i,b) \notin M} \text{gain}(i, b),$$

where (i, b) ranges over all admissible feature-bit pairs for the current node.

Two successors are generated:

- *apply*: flip bit b^* of feature i^*, obtaining a child node with updated input x'' and budgets $(K + \Delta_K(i^*, b^*), B + \Delta_B(i^*, b^*))$, provided the global limits $(K_{\max}, B_{\max})$ are not exceeded;
- *skip*: mark (i^*, b^*) as explored by adding it to M and leave x' and (K, B) unchanged.

In the implementation, the *apply* child is explored first whenever it is admissible; if it returns a counterexample, the search terminates, and if it returns unsatisfiable, the *skip* child is then explored. Since each bit is explored at most once along any root-to-leaf path and the budgets $(K_{\max}, B_{\max})$ are finite, the transition graph is finite.

4.6 Branch-and-Bound Procedure

Algorithm 1 summarizes the mixed-budget BaB search: starting from the clean input, it recursively explores admissible perturbations.

Algorithm 1. BaB-PoNN Verification

1: **procedure** $\textsc{Verify}(x, S, K_{\max}, B_{\max})$
2:      construct initial node $\eta_0 \leftarrow (x, 0, 0, \emptyset)$
3:      **return** $\textsc{Search}(x, 0, 0, \emptyset)$
4: **end procedure**
5: **procedure** $\textsc{Search}(x', K, B, M)$
6:      $m \leftarrow m(x'; y_{\mathrm{ref}})$
7:      **if** $m > 0$ **then**
8:          **return** $\textsc{Counterexample}(x')$
9:      **end if**
10:     $\mathrm{UB} \leftarrow \mathrm{UB}(x', K, B, M)$
11:     **if** $\mathrm{UB} \leq 0$ **then**
12:         **return** $\textsc{NoCounterexampleSubtree}$
13:     **end if**
14:     select (i^*, b^*) maximizing $\mathrm{gain}(i, b)$ over $(i, b) \notin M$
15:     **if** flipping (i^*, b^*) respects $(K_{\max}, B_{\max})$ **then**
16:         $found \leftarrow \textsc{Search}(\mathrm{flip}(i^*, b^*, x'), K + \Delta_K(i^*, b^*), B + \Delta_B(i^*, b^*), M)$
17:         **if** $found$ **then**
18:             **return** $\textsc{Counterexample}$
19:         **end if**
20:     **end if**
21:     $M' \leftarrow M \cup \{(i^*, b^*)\}$
22:     **return** $\textsc{Search}(x', K, B, M')$
23: **end procedure**

The procedure is *sound and complete* over the finite neighbourhood $\mathcal{X} := \mathcal{X}(x; S, K_{\max}, B_{\max})$: pruning is admissible (Sect. 4.4), so $\mathrm{UB}(\eta) \leq 0 \Rightarrow \forall x' \in \mathrm{Desc}(\eta) . m(x'; y_{\mathrm{ref}}) \leq 0$, and exhaustive branching (Sect. 4.5) yields $\exists x' \in \mathcal{X} . m(x'; y_{\mathrm{ref}}) > 0$ iff $\textsc{Counterexample}$; otherwise $\forall x' \in \mathcal{X} . m(x'; y_{\mathrm{ref}}) \leq 0$.

5 Implementation

BaB-PoNN is implemented in C using the SoftPosit library [11] for posit and quire arithmetic, with implementation-level optimisations to keep the mixed (K, B)-budget search tractable while preserving exact posit-8 semantics.

Semantics-Preserving Precomputation. To avoid repeated SoftPosit calls, we precompute two finite maps at initialization. For input normalization we tabulate

$$\tau\colon \{0,\ldots,255\} \to \mathbb{P} \qquad \tau(v) \;=\; \mathrm{encode}_{\mathrm{p8}}\left(\frac{v - 127.5}{127.5}\right)$$

and for decoding we tabulate

$$\delta\colon \mathbb{P} \to \mathbb{R} \qquad \delta(p) \;=\; \mathrm{decode}_{\mathrm{p8}}(p)$$

Both tables are filled once using the same SoftPosit routines as the forward pass; all input normalizations and logit/margin computations then use constant-time lookups $\tau(v)$ and $\delta(p)$. This preserves bit-precise agreement with the deployed posit-8 implementation.

Admissible Support, ROI Refinement, and Top-k Restriction. Let $X \subseteq \{1,\ldots,d_0\}$ denote the admissible support on which perturbations are allowed; coordinates in $\{1,\ldots,d_0\} \setminus X$ are kept fixed. Each coordinate i carries a nonnegative influence score $s_i \geq 0$, derived from the per-coordinate gain $k\mathrm{gain}(i)$ by

$$s_i \;=\; \max\{0, k\mathrm{gain}(i)\}$$

where $k\mathrm{gain}(i)$ is the worst-case increase in decision margin obtained by exhaustively enumerating all 256 byte values at coordinate i.

For *global* robustness, there is no geometric constraint on the support: we start from the full index set $\{1,\ldots,d_0\}$, sort the scores $\{s_i\}$ in nonincreasing order to obtain an ordering $i_1,\ldots,i_{d_0}$, and select an influence-guided support

$$X_{\mathrm{glob}} \;=\; \{i_1,\ldots,i_{k^\star}\} \qquad \text{where} \qquad k^\star \;\leq\; \min\{K_{\max}, d_0\}$$

so that BaB only branches on the $k^\star$ most influential pixels consistent with the coordinate budget.

For *local* robustness, we fix a coarse region of interest (ROI) and restrict perturbations to this subset of coordinates. Within this ROI, we optionally apply an *influence-guided ROI heuristic* to refine the search to the most sensitive subregion. For inputs with 2D grid structure, we consider all contiguous $h \times w$ blocks B contained in the ROI, assign each block an aggregate influence

$$S(B) \;=\; \sum_{i \in B} s_i$$

and choose

$$B^\star \;=\; \arg\max_{B} S(B)$$

so that $B^\star$ is the contiguous block whose coordinates have the highest combined influence and are therefore most likely to destabilize the margin under the given budgets. We then set X_{loc} to the index set of $B^\star$. If the user supplies a concrete block instead, we simply take X_{loc} to be the indices of that block.

In both global and local configurations, once X is fixed (X_{glob} or X_{loc}), we may further apply a top-k restriction $X_k = \{i_1, \ldots, i_{\min(k,|X|)}\}$, where $i_1, \ldots, i_{|X|}$ are the indices in X sorted by s_i. The BaB search then explores $\mathcal{X}(x_0; X_k, K_{\max}, B_{\max})$ as in Sect. 3.3, branching only on the most influential coordinates in the admissible support.

Greedy Warm Starts. Before invoking BaB, the verifier can run an optional greedy search that applies a sequence of margin-increasing flips within the $(K_{\max}, B_{\max})$ budgets. At each step it evaluates all admissible perturbations (full-byte changes or single-bit flips) over the support X, chooses the modification that maximally increases the decoded margin against the clean top-1 class y_{ref}, and updates the current input and remaining budgets. This yields a simple discrete hill-climbing warm start: if some iterate x' satisfies $\hat{y}(x') \neq y_{\mathrm{ref}}$, the verifier returns a counterexample immediately; otherwise it proceeds to full BaB search.

Resource Limits and Complexity. To guarantee practical termination, we enforce a wall-clock time limit, a global node cap, a maximum recursion depth, and an "idle" stop that triggers if the best global upper bound fails to improve for a fixed window. In our evaluations, these limits are chosen so they do not bind on the reported runs; all reported "no counterexample" outcomes correspond to exhausting the mixed-budget search.

Let $n = |X|$ be the number of editable coordinates. A single forward pass through the posit-8 network under verification has cost T_{net}, linear in the number of weights. Per node, the swing computation enumerates all 256 values for each of the n coordinates (and 8 bit flips per coordinate), giving worst-case cost $O(n \cdot 256 \cdot T_{\mathrm{net}})$, plus $O(n \log n)$ to sort the coordinate and bit gains for the residual bound. The BaB tree is exponential in the worst case, but in practice the combination of ROI restriction, greedy warm starts, top-k support, mixed-budget pruning, and best-first bit-level branching drastically reduces the number of explored nodes on our benchmarks.

5.1 Runtime Phases

A BaB-PoNN run proceeds in four conceptual phases. For each reference input x_0, the verifier (i) fixes the admissible scope S (global or local) and the influence-guided support X (Sect. 5); (ii) optionally applies a greedy sequence of highest-gain flips within the budgets ($K_{\max}, B_{\max}$), terminating early if it finds a counterexample; (iii) otherwise launches the mixed-budget branch-and-bound search over the admissible perturbation neighbourhood $\mathcal{X}(x_0; S, K_{\max}, B_{\max})$, recomputing logits under the exact posit/quire semantics, applying margin-based pruning, and branching on the best remaining feature-bit pair; and (iv) returns either a concrete counterexample, a *no counterexample within the budgeted neighbourhood* verdict, or an *inconclusive* outcome if time, node, depth, or idle limits are hit first.

6   Evaluation

6.1   Networks and Training

We evaluate BaB-PoNN on two MNIST classifiers quantized to $\mathbb{P}$ with quire-8 accumulation and trained using *Deep PeNSieve* [13]. MNIST consists of 28×28 grayscale images of handwritten digits 0–9, with 60,000 training and 10,000 test images. Both models use quantisation-aware training: float32 shadow weights are cast to posit-8 at each forward pass, and the exported posit-8 parameters are exactly those used by the verifier.

MLP-8. A two-layer multilayer perceptron on flattened 28×28 inputs (784 units) with 8 hidden ReLU units and 10 linear output logits, with $N_\theta = 6{,}290$ parameters. Inputs are normalised by $(x - 127.5)/127.5$. Using Adam (learning rate 10^{-3}, batch size 128) for 30 epochs yields 92.6% test accuracy.

LeNet-5-8. A LeNet-5 convolutional network based on LeCun et al. [10] with 28×28 inputs, two convolutional stages with 5×5 kernels and 2×2 max-pooling (6 and 16 output channels), followed by fully connected layers of size 400, 120, 84, and 10 with ReLU activations after each non-output layer. The model has $N_\theta = 61{,}706$ parameters and achieves 97.0% test accuracy.

6.2   Robustness Verification Setup

We ask whether small mixed-budget perturbations can change the posit-8 prediction. For each network (MLP-8 and LeNet-5-8) and for the first 500 MNIST test inputs x_0, we run BaB with fixed budgets $K_{\max} = 2$ and $B_{\max} = 4$, in both the local and global regime. For any admissible support $X \subseteq \{1, \ldots, 784\}$, the verifier searches $\mathcal{X}(x_0; X, K_{\max}, B_{\max})$ for an input that changes the prediction.

For one image, the global mixed-budget neighbourhood contains $\binom{784}{2} \cdot \binom{16}{4} \approx 5.6 \times 10^8$ distinct perturbed bit patterns, so across 500 test inputs BaB-PoNN reasons about $\approx 2.8 \times 10^{11}$ candidate perturbations. All runs use a recursion depth cap of 5×10^8 nodes and a per-input wall-clock limit of 100,000 seconds.

Local Robustness. Perturbations are restricted to a region of interest: the admissible support X_{loc} is the index set of a single contiguous block selected by the ROI heuristic in Sect. 5, which scans axis-aligned blocks and chooses the one with maximal aggregate influence. Within this block we retain only the two most influential pixels (top-k with $k = 2$), matching $K_{\max} = 2$.

Global Robustness. Perturbations may occur anywhere, so we set $X_{\mathrm{glob}} = \{1, \ldots, 784\}$. Influence scores are computed for all pixels and used only to select the two most influential pixels globally (top-k with $k = 2$) and to define the branching order.

6.3   Execution Environment

Evaluations run on a dual-socket Dell Precision T7810 with two Intel Xeon E5-2630 v4 CPUs (20 cores, 40 hardware threads) and 96 GiB RAM running

Ubuntu 24.04.3 LTS. Each of the 500 inputs is verified by a single-threaded process; a driver schedules these processes up to a concurrency cap of 35.

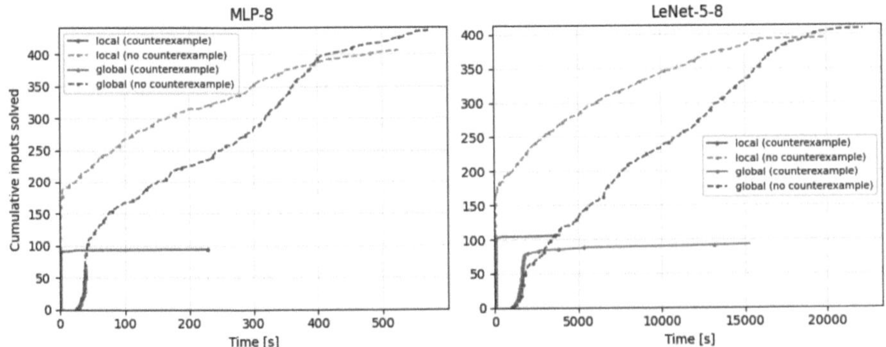

Fig. 2. Input runtime CDFs for MLP-8/LeNet-5-8, split by local/global and outcome.

7   Results

For each configuration c (MLP-8-local, MLP-8-global, LeNet-5-8-local, LeNet-5-8-global) and for each of the first 500 MNIST test inputs, BaB-PoNN returns either *counterexample* or *no counterexample*. Let $N_{\text{no-cex}}(c)$ and $N_{\text{cex}}(c)$ be the corresponding counts, and define the verified robustness rate as $V_{\text{robust}}(c) = 100 \cdot N_{\text{no-cex}}(c)/500$. We also report the median and 95th-percentile per-input runtimes $t_{\text{med}}(c)$ and $t_{95}(c)$.

All four configurations achieve substantial verified robustness, with $V_{\text{robust}}(c)$ between 78.8% and 87.4%. The MLP-8 instances are comparatively easy to verify: median runtimes stay below 1.2 s in the local regime and around 133 s in the global regime, and $t_{95}(c)$ remains below 500 s. LeNet-5-8 has higher medians and much heavier tails, with 95th-percentile runtimes in the 10^4 s range, indicating a small set of hard inputs that force BaB-PoNN to explore a large fraction of the mixed-budget search space.

For each configuration c, we collect the `elapsed_s` values for runs ending in *counterexample* and in *no counterexample*, sort each multiset, and plot the corresponding empirical CDFs in Fig. 2.

Across all configurations, runs that find counterexamples are consistently computational cheaper than runs that prove their absence. For MLP-8, adversarial inputs are almost always found within a few seconds, while *no counterexample* runs typically complete within a few hundred seconds. For LeNet-5-8, both outcomes exhibit heavier tails, and non-existence proofs can reach 10^4 s in the global regime. Together with Table 1, this indicates that adversarial examples, when they exist, are usually discovered early, and that the dominant cost comes from inputs whose entire $\mathcal{X}(x_0; S, K_{\max}, B_{\max})$ neighbourhood must be exhaustively explored before all perturbations can be ruled out.

Empirically, as shown in Fig. 3, every witness satisfies $K(x_0, x') \in \{1, 2\}$, with most mass on $K(x_0, x') = 1$, and $B(x_0, x') \in \{1, 2, 3, 4\}$, with a clear skew towards $B(x_0, x') \in \{3, 4\}$, where $K(x_0, x')$ and $B(x_0, x')$ denote the used coordinate and bit budgets for the perturbation $x_0 \to x'$ (as in Sect. 3.3). Violations of the mixed-budget robustness property are therefore realised by extremely sparse perturbations that modify only one or two pixels and flip at most four bits in their posit-8 encodings. In particular, posit neural networks can admit misclassifying inputs under very small input perturbations (low Hamming cost in the input encoding), and BaB-PoNN effectively locates such witnesses whenever they lie within the admissible neighbourhood $\mathcal{X}(x_0; S, K_{\max}, B_{\max})$.

Table 1. Verified robustness for $(K_{\max}, B_{\max}) = (2, 4)$ over 1st 500 MNIST test inputs.

Config	$V_{\text{robust}}(c)\ [\%]$	$N_{\text{no-cex}}(c)$	$N_{\text{cex}}(c)$	$t_{\text{med}}(c)\ [\text{s}]$	$t_{95}(c)\ [\text{s}]$
MLP-8 (local)	81.2	406	94	1	378
MLP-8 (global)	87.4	437	63	133	463
LeNet-5-8 (local)	78.8	394	106	45	12551
LeNet-5-8 (global)	81.6	408	92	6549	17509

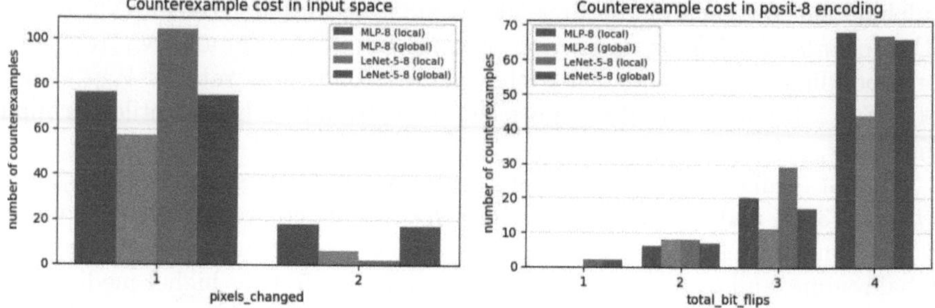

Fig. 3. Counterexample cost in input space (left) and posit-8 encoding (right): distribution of used pixel budget $K(x_0, x')$ and bit budget $B(x_0, x')$ over all counterexamples.

8   Conclusion

We introduced BaB-PoNN, a branch-and-bound framework for mixed-budget robustness verification of posit neural networks. The verifier operates under the exact posit-8 and quire-8 semantics and explores a finite neighbourhood over input coordinates and bit-level perturbations, so reported *counterexample* and *no counterexample* outcomes are sound for the deployed posit implementation.

On posit-8 MLP and LeNet-5 CNN for MNIST, BaB-PoNN certifies robustness, with verified robustness between 78.8% and 87.4%. Adversarial examples, when they exist, are typically found quickly; the dominant computational cost arises on inputs whose mixed-budget neighbourhood must be exhaustively closed. Future work includes tightening bounds and search heuristics, extending the mixed-budget model to other posit formats and datasets, and combining BaB-PoNN with complementary verification techniques for larger networks.

Disclosure of Interests. The authors have no competing interests to declare.

References

1. Carlini, N., Wagner, D.A.: Towards evaluating the robustness of neural networks. In: 2017 IEEE Symposium on Security and Privacy, SP 2017, San Jose, CA, USA, 22–26 May 2017, pp. 39–57. IEEE Computer Society (2017). https://doi.org/10.1109/SP.2017.49
2. Dinechin, F.D., Forget, L., Muller, J., Uguen, Y.: Posits: the good, the bad and the ugly. In: Proceedings of the Conference for Next Generation Arithmetic (CoNGA 2019), pp. 1–10. Association for Computing Machinery (2019). https://doi.org/10.1145/3316279.3316285
3. Edavoor, P.J., Raveendran, A., Selvakumar, D., Desalphine, V., Dharani, S.G., Raut, G.: Design and analysis of posit quire processing engine for neural network applications. In: 36th International Conference on VLSI Design and 2023 22nd International Conference on Embedded Systems, VLSID 2023, Hyderabad, India, 8–12 January 2023, pp. 252–257. IEEE (2023). https://doi.org/10.1109/VLSID57277.2023.00059
4. Ehlers, R.: Formal verification of piece-wise linear feed-forward neural networks. In: D'Souza, D., Narayan Kumar, K. (eds.) ATVA 2017. LNCS, vol. 10482, pp. 269–286. Springer, Cham (2017). https://doi.org/10.1007/978-3-319-68167-2_19
5. Gehr, T., Mirman, M., Drachsler-Cohen, D., Tsankov, P., Chaudhuri, S., Vechev, M.T.: AI2: safety and robustness certification of neural networks with abstract interpretation. In: 2018 IEEE Symposium on Security and Privacy, SP 2018, Proceedings, 21–23 May 2018, San Francisco, California, USA, pp. 3–18. IEEE Computer Society (2018). https://doi.org/10.1109/SP.2018.00058
6. Goodfellow, I.J., Shlens, J., Szegedy, C.: Explaining and harnessing adversarial examples. In: Bengio, Y., LeCun, Y. (eds.) 3rd International Conference on Learning Representations, ICLR 2015, San Diego, CA, USA, 7–9 May 2015, Conference Track Proceedings (2015). http://arxiv.org/abs/1412.6572
7. Gustafson, J.L., Yonemoto, I.T.: Beating floating point at its own game: posit arithmetic. Supercomput. Front. Innov. **4**(2), 71–86 (2017). https://doi.org/10.14529/JSFI170206
8. Katz, G., Barrett, C., Dill, D.L., Julian, K., Kochenderfer, M.J.: Reluplex: an efficient SMT solver for verifying deep neural networks. In: Majumdar, R., Kunčak, V. (eds.) CAV 2017. LNCS, vol. 10426, pp. 97–117. Springer, Cham (2017). https://doi.org/10.1007/978-3-319-63387-9_5
9. Kaulen, K., König, M., Hoos, H.H.: Dynamic algorithm termination for branch-and-bound-based neural network verification. In: Walsh, T., Shah, J., Kolter, Z. (eds.) AAAI-25, Sponsored by the Association for the Advancement of Artificial Intelligence, 25 February–4 March 2025, Philadelphia, PA, USA, pp. 27356–27364. AAAI Press (2025). https://doi.org/10.1609/AAAI.V39I26.34946
10. LeCun, Y., Bottou, L., Bengio, Y., Haffner, P.: Gradient-based learning applied to document recognition. Proc. IEEE **86**(11), 2278–2324 (1998). https://doi.org/10.1109/5.726791

11. Leong, S.H.: Softposit. Zenodo (2020). https://doi.org/10.5281/zenodo.3709035
12. Lu, J., Fang, C., Xu, M., Lin, J., Wang, Z.: Evaluations on deep neural networks training using posit number system. IEEE Trans. Comput. **70**(2), 174–187 (2021). https://doi.org/10.1109/TC.2020.2985971
13. Murillo, R., Barrio, A.A.D., Botella, G.: Deep PeNSieve: a deep learning framework based on the posit number system. Digit. Signal Process. **102**, 102762 (2020). https://doi.org/10.1016/J.DSP.2020.102762
14. Pei, K., Cao, Y., Yang, J., Jana, S.: DeepXplore: automated whitebox testing of deep learning systems. In: Proceedings of the 26th Symposium on Operating Systems Principles, Shanghai, China, 28–31 October 2017, pp. 1–18. ACM (2017). https://doi.org/10.1145/3132747.3132785
15. Podobas, A., Matsuoka, S.: Hardware implementation of POSITS and their application in FPGAs. In: 2018 IEEE International Parallel and Distributed Processing Symposium Workshops, IPDPS Workshops 2018, Vancouver, BC, Canada, 21–25 May 2018, pp. 138–145. IEEE Computer Society (2018). https://doi.org/10.1109/IPDPSW.2018.00029
16. Ruan, W., Huang, X., Kwiatkowska, M.: Reachability analysis of deep neural networks with provable guarantees. In: Lang, J. (ed.) Proceedings of the Twenty-Seventh International Joint Conference on Artificial Intelligence, IJCAI 2018, 13–19 July 2018, Stockholm, Sweden, pp. 2651–2659. ijcai.org (2018). https://doi.org/10.24963/IJCAI.2018/368
17. Sadiq, S.: BaB-ponn: posit neural network branch-and-bound robustness verification (2025). https://github.com/suleimansadiq/BaB-PoNN. gitHub repository
18. Singh, G., Gehr, T., Püschel, M., Vechev, M.T.: An abstract domain for certifying neural networks. Proc. ACM Program. Lang. **3**(POPL), 41:1–41:30 (2019). https://doi.org/10.1145/3290354
19. Singh, G., Gehr, T., Püschel, M., Vechev, M.T.: Boosting robustness certification of neural networks. In: 7th International Conference on Learning Representations, ICLR 2019, New Orleans, LA, USA, 6–9 May 2019. OpenReview.net (2019). https://openreview.net/forum?id=HJgeEh09KQ
20. Wang, S., et al.: Beta-CROWN: efficient bound propagation with per-neuron split constraints for neural network robustness verification. In: Ranzato, M., Beygelzimer, A., Dauphin, Y.N., Liang, P., Vaughan, J.W. (eds.) Advances in Neural Information Processing Systems 34: Annual Conference on Neural Information Processing Systems 2021, NeurIPS 2021, 6–14 December 2021, virtual, pp. 29909–29921 (2021). https://proceedings.neurips.cc/paper/2021/hash/fac7fead96dafceaf80c1daffeae82a4-Abstract.html
21. Zhou, D., Brix, C., Hanasusanto, G.A., Zhang, H.: Scalable neural network verification with branch-and-bound inferred cutting planes. In: Globersons, A., et al. (eds.) Advances in Neural Information Processing Systems 38: Annual Conference on Neural Information Processing Systems 2024, NeurIPS 2024, Vancouver, BC, Canada, 10–15 December 2024 (2024). http://papers.nips.cc/paper_files/paper/2024/hash/33d93e4dc57453e7667b20f62e7c0681-Abstract-Conference.html

Probing Features for Automatic Algorithm Selection for Pseudo-boolean Optimization

Amanda Salinas-Pinto[1], Catalina Pezo[2], Dorit S. Hochbaum[2],
Bistra Dilkina[3], Ricardo Ñanculef[1], and Roberto Asín-Achá[1(✉)]

[1] Universidad Técnica Federico Santa María, Valparaíso, Chile
`amanda.salinas@sansano.usm.cl`, `{ricardo.nanculef,roberto.asin}@usm.cl`
[2] University of California, Berkeley, Berkeley, CA, USA
`{catalina.pezo,dhochbaum}@berkeley.edu`
[3] University of Southern California, Los Angeles, CA, USA
`dilkina@usc.edu`

Abstract. For NP-hard problems like Pseudo-Boolean Optimization (PBO), solver performance varies dramatically across instances, making algorithm selection a critical bottleneck. Current Automatic Algorithm Selection (AAS) systems for PBO depend primarily on static problem features, ignoring crucial dynamic information. In contrast, AAS for Boolean Satisfiability (SAT) has demonstrated that "probing" features, gathered from short solver runs, are essential for making high-quality, time-sensitive predictions. This work improves AAS for PBO by systematically integrating and evaluating these powerful probing features. We investigate their impact across a range of machine learning frameworks, including regression, multiclass/multilabel classification, and a novel hybrid model. This investigation culminates in MetaPB, a new open-source meta-solver that combines both static and probing features for superior solver selection. On benchmarks from the 2024 PBO competition, MetaPB outperforms the best individual solver. Remarkably, it closes the gap with leading commercial solvers, achieving performance competitive with Gurobi while using an entirely open-source portfolio. This study establishes a new state of the art for AAS in PBO and challenges the prevailing view in this domain that algorithm selection should be treated purely as a classification task, highlighting the effectiveness of hybrid, regression-informed approaches.

Keywords: Automatic Algorithm Selection · PBO · Machine learning

1 Introduction

Automatic Algorithm Selection (AAS) is to automatically choose the most effective algorithm from a *portfolio* collection of solvers for a given problem instance. Since its formal inception by Rice in 1976 [24], this field has evolved considerably, demonstrating practical impact across numerous domains including Boolean Satisfiability (SAT) [21,29], Pseudo-Boolean Optimization (PBO) [23], the Traveling

T. Guns (Ed.): CPAIOR 2026, LNCS 16595, pp. 503–519, 2026.
https://doi.org/10.1007/978-3-032-27242-3_30

Salesperson Problem (TSP) [12,18], and various other NP-hard problems [13,28]. The core principle is that no single algorithm performs optimally across all problem instances, which has motivated sophisticated selection strategies that can significantly improve overall performance compared to any single solver. These strategies enable the design of *meta-solvers*, which utilize the solvers in the portfolio by selecting, combining, or configuring them based on the specific input instance *features*.

Most AAS systems use machine learning (ML) techniques to recommend a solver for a given instance [17,27]. These approaches receive *features* that describe the instance and learn the patterns that map these features to solver performance. The quality of the features is essential for guaranteeing good generalization performance on unseen instances. In the domain of PBO, which generalizes SAT by incorporating an objective function and pseudo-Boolean constraints, current algorithm selection approaches have predominantly relied on static features. These features are extracted directly from the problem formulation and offer structural insights but lack information on solver behavior. In contrast, the SAT community has demonstrated that probing features [2,15,17], obtained by briefly running solvers and recording their behavior, offer crucial additional information that significantly improves selection performance, especially in scenarios with tight time budgets. This insight has not been exploited in the PBO domain until now.

This paper explores new learning formulations for AAS in PBO, advancing the state of the art in several directions. First, we introduce a rich set of probing features obtained by briefly running different solvers: a Conflict-Driven Clause Learning (CDCL) PBO solver and a Branch and Bound (BnB) solver. These features are effective in capturing solver behavior characteristics that static features alone cannot, and they improve upon previously proposed probing features taken from the SAT domain.

Second, we conduct an extensive comparison of different ML formulations for algorithm selection, including regression, multiclass classification, multilabel classification, and a hybrid approach that combines regression and multilabel predictions. Our experimental results consistently show that regression models, which predict the performance of each solver and then select the best one, outperform direct classification approaches that attempt to identify the best solver directly. Moreover, a hybrid approach that incorporates information about whether a solver can solve an instance to optimality has achieved the best results overall. Given the prevalence of classification approaches in the literature, these findings suggest re-evaluating common assumptions about the best way to frame AAS in PBO.

Finally, we present MetaPB, an open-source meta-solver that implements the best-performing combination of features and learning formulations identified in our study. MetaPB integrates both static and probing information. In our evaluation on a diverse set of PBO instances, MetaPB significantly outperforms previous AAS approaches and the best individual solver in the portfolio, both in terms of the number of optimally solved instances and overall solution quality.

It also surpasses the best solver from the 2024 PBO competition and achieves performance competitive with commercial tools such as Gurobi, while relying exclusively on open-source solvers.

2 Related Work

2.1 Pseudo-Boolean Optimization Solvers

PBO is an extension of SAT [1] by including pseudo-Boolean constraints and an objective function that we assume, w.l.o.g., to minimize. Let $\mathbb{B} = \{0,1\}$, $\mathbf{x} = (x_1, x_2, ..., x_n) \in \mathbb{B}^n$, and let a literal ℓ_i be x_i or $\bar{x}_i$. A pseudo-Boolean function is a mapping of the form $f(\mathbf{x}) = \sum_{i=1}^{k} a_i \prod_{j \in S_i} \ell_j$ where $a_i \in \mathbb{R}^+$ and S_i is the set of literal indices defining the i-th $term$. A PBO instance can be defined as

$$\min_{\mathbf{x} \in \mathbb{B}^n} f(\mathbf{x}) \text{ s.t. } g_i(\mathbf{x}) \geq a_i \ \forall i \in [m] = \{1, 2, ..., m\},$$

where $f, g_1, g_2 \ldots g_m$ are pseudo-Boolean functions, and $a_i \in \mathbb{R}^+, \forall i \in [m]$. Most PBO solvers linearize the PBO functions or only receive as valid inputs linear PBO instances. Hence, for our work here, we assume the PBO instances to be linear.

In recent years, much effort has been placed on developing new PBO solvers. Below, we describe the PBO solvers that we identified as competitive, complementary, alternatives to be included in a portfolio of solvers.

Gurobi [11] is a Mixed Integer programming (MIP) commercial solver using a branch-and-bound approach based on linear programming, enhanced with presolving, cutting planes, heuristics, and parallel processing techniques.

NaPS [26] is a MaxSAT solver based on MiniSat+ [6] that translates a PBO instance to a MaxSAT instance using Binary Decision Diagrams (BDD), and solves the resulting problem alternating between binary and sequential search strategies.

RoundingSat [5,7] is a PBO solver that combines the CDCL procedure, based on the cutting planes proof system, with LP relaxation and cut generation. Once a feasible solution is found, it is optimized using model-improving and core-guided search.

SCIP [3] is an open-source software tool designed for solving MIP, Mixed Integer Nonlinear Programming (MINLP), and Constraint Programming (CP) problems. To solve PBO, SCIP uses LP relaxations, the branch-and-cut algorithm, and cutting plane techniques.

Mixed-Bag[16] combines three components in sequence: the PaPILO preprocessor [10], SCIP and PB-OLL-RS (a core-guided solver based on RoundingSat) as the final solver.

2.2 Automatic Algorithm Selection for PBO and SAT

AAS [24] is the task of building meta-solvers that select the most appropriate solver, from a portfolio of solvers, to be executed on a specific instance of a problem. Formally, given a set of instances I and a set of solvers A, the problem is to devise a mapping $s^t : I \to A$ such that $s^t(i)$ identifies the most effective algorithm for each instance $i \in I$, given a time-limit t. To evaluate the performance of s^t, the $\hat{m}_{s^t}$ metric introduced in [19], which we define next, has become the standard in the field.

Let $o_a(i, t)$ be the best objective value obtained by a solver $a \in A$ for instance $i \in I$ given a time-limit t. We define $o_{\min}(i, t)$ as the minimum value of $o_a(i, t)$ observed across all solvers $a \in A$ at time t, and $o_{\max}(i, t)$ as the maximum value. The normalized objective value $n_a(i, t)$ of solver $a \in A$, on instance $i \in I$ at t, is defined as:

$$
n_a(i, t) = \begin{cases}
0 & \text{if } o_a(i, t) \text{ is certified optimal,} \\
1 & \text{if } o_a(i, t) = o_{\min}(i, t) = o_{\max}(i, t) \\
3 & \text{if } o_a(i, t) \text{ undefined and } o_{\max}(i, t) \text{ defined,} \\
1 + \dfrac{o_a(i, t) - o_{\min}(i, t)}{o_{\max}(i, t) - o_{\min}(i, t)} & \text{otherwise.}
\end{cases}
\tag{1}
$$

Lower $n_a(i, t)$ values are better, indicating a solver's objective value is closer to the best observed across solvers.

We define the normalized cumulative performance of a solver $a \in A$ at time t, and the normalized cumulative performance of meta-solver $s^t \in \{I \to A\}$ as:

$$
m_{a^t} = \sum_{i \in I} n_a(i, t), \qquad m_{s^t} = \sum_{i \in I} n_{s(i)}(i, t)
\tag{2}
$$

The Single Best Solver at time t (SBS^t) is the solver $a \in A$, such that $\forall a' \in A : m_{a^t} \leq m_{a'^t}$. The Virtual Best Solver, at time t (VBS^t) is a theoretically perfect meta-solver such that $m_{\text{VBS}^t} = \sum_{i \in I} \min_{a \in A} n_a(i, t)$.

In this work, we distinguish between "VBS-all", where A includes all solvers in our portfolio, and "VBS-Open-Source", where A is restricted to the open-source solvers in the portfolio: NaPS, RoundingSat, Mixd-Bag and SCIP, which additionally employ complementary solving techniques. Analogously, we distinguish between "SBS-all," where the best single solver is selected from the full portfolio, and "SBS-Open-Source", where the selection is limited to the open-source solvers.

The $\hat{m}_{s^t}$ metric for meta-solver s^t is defined as:

$$
\hat{m}_{s^t} = \frac{m_{s^t} - m_{\text{VBS}^t}}{m_{\text{SBS}^t} - m_{\text{VBS}^t}}.
\tag{3}
$$

Note that an $\hat{m}_{s^t}$ value of 0 indicates that the algorithm selector is optimal, a value of 1 indicates that the selector performs as well as SBS^t, and values above 1 indicate that the meta-solver is not useful.

In the SAT community, it is highly relevant to understand SAT instances through their features. The SATzilla feature set [15,22] is one of the most commonly used sets of features, with applications in Automatic Algorithm Selection [21], Algorithm Configuration [14], and predicting algorithms' runtime [15]. In [22], the authors present 91 features to characterize a SAT instance. Some of these features are static, easy-to-compute features extracted directly from the SAT formula. Others, called "Probing Features", require running a solver for a limited amount of time, and extracting characteristics from the solver's trajectory. Several works [21,29,30] use ML techniques, along with the SATzilla features, to implement an AAS model for SAT.

For PBO, [23] presented a study of multi-class classification-based AAS methods in an anytime setting, where solvers are selected considering their performance within specified time limits. Their methods used static features extracted from PBO Competition instances [20] (2006-2016) and included competition solvers from 2016 in their portfolio. The best-performing meta-solver of [23] utilized Random Forest classification, and demonstrated a performance that significantly surpassed Gurobi, the SBS, on a testing distribution similar to the training distribution.

3   Methodology

This section presents the solver portfolio, feature sets, and machine learning formulations used in our study.

3.1   Solver Portfolio and PBO Instances

We selected solvers from the 2024 PBO competition's optimization track [25], focusing on the top 10 performers and those with complementary capabilities. Gurobi was included as well due to its strong results in related work [23]. The selected solvers for our portfolio are Gurobi, NaPS, RoundingSat, Mixed-Bag, and SCIP. For the RoundingSat solver, we include both its default hybrid optimization mode, which we will refer to as "RoundingSat", and the core-boosted mode, which we will refer to as "RoundingSat-CB", as these variants demonstrate complementary behavior and competitive performance when evaluated individually.

We collected instances from the 2024 PBO competition website, which included instances from previous competitions held between 2005 and 2016 [25]. Only optimization and linearized instances were considered, yielding an initial dataset of 14,586 instances.

We evaluated solver performance using a 1-hour time limit on an Intel Xeon E5-2670 v3 (64GB RAM). During solver runs, we captured the objective value at five critical time intervals (60, 100, 300, 600, and 3600 s) to analyze performance dynamics under varying time constraints. After filtering out instances solved optimally by every solver in under 60 s (easy) and those unsolved by all solvers (difficult), 6,394 instances remained for analysis. Additionally, we reserved 478 instances used in the 2024 PBO competition's to test solvers on the optimization track [25]. This set was used to evaluate solver performance on challenging instances, selected by the organizers of the competition to test the limits of current PBO solvers.

3.2 Probing Feature Sets

We propose two distinct sets of probing features to characterize solver behavior: one based on CDCL and another based on BnB strategies.

For CDCL, we adapted SAT-based probing features [15] to the context of PBO and introduced complementary features to better capture solver behavior and problem structure. Features were extracted during RoundingSat's CDCL process by running the solver with a 5-second timeout, sampling feature values at conflict points, and averaging them. We refer to this complete set of features as "CDCL", which includes both adapted SAT features and our new features.

Solver Activity Metrics: We include the number of "Decisions", "Propagations", "Conflicts", "Learned Constraints", "Restarts", and "Units", which track the solver's progress.

Clause Quality Metric: The "Learned Clauses Length" measures the average size of learned clauses, where shorter clauses generally increase solver efficiency.

Efficiency Metrics: "Propagation-to-Conflict", "Propagation-to-Decision" and "Decision-to-Conflict", ratios assess solver efficiency in decision-making, propagation, and conflict management. No features of this type have been proposed before.

Clause Quality Metric: The "LBD" (Literal Block Distance) Value evaluates the usefulness of learned clauses, with lower values indicating higher quality. No features of this type have been proposed before.

Conflict Resolution Metrics: "Conflict Resolution Steps" measure the complexity of resolving conflicts, while "Trail Pops" reflect backtracking behavior and recovery efforts. No features of this type have been proposed before.

To capture information related to the Branch and Bound solving process, we propose an entirely new set of probing features, which we denote as "BnB". These features are specifically devised and first presented for this work and extracted using SCIP. The solver was run with a 5-second timeout to gather statistics that reflect search progress, bounding behavior, and solution quality.

Solution Quality indicators: "Primal Bound", "Gap", "Dual Bound".

Variable Metrics: "Counts of Active" (1) and "Inactive" (0) binary variables in the incumbent solution, revealing structural patterns in feasible solutions.

Search Tree Metrics: Number of "Feasible Leaves", "Iterations", and "Solutions Found" capture the exploration extent and search dynamics within the BnB tree.

Solver Activity Metrics: "Average Variable Activity" reflects how actively variables influence branching and decisions during the search.

3.3   Machine Learning Tasks and Models

We define four ML strategies to address the problem of solver selection:

Multiclass Classification: It trains the selector to predict the solver from the portfolio, achieving the lowest normalized objective value within the time limit. In case of ties, the model is trained to choose the fastest one.

Regression: It trains the selector to predict the normalized objective value $n_a(i,t)$ of each solver-instance pair at time t, as defined in Eq. 1. The selector then chooses the solver with the lowest value.

Multilabel Classification: It trains the selector to predict, for each solver in the portfolio, the probability that it fails to solve an instance optimally within the time limit. For each solver, a binary label that indicates success (0) or failure (1) on the instance is predicted. The selector then chooses the solver with the lowest predicted failure probability.

Hybrid: This approach combines the predictions of both regression and multilabel models to compute a score for each solver-instance pair. The final score is defined as:

$$\text{score}_a(i,t) = \alpha \cdot \widehat{n}_a(i,t) + (1 - \alpha) \cdot \widehat{p}_a(i,t), \tag{4}$$

where $\widehat{n}_a(i,t)$ is the predicted normalized objective from the regression model, $\widehat{p}_a(i,t)$ is the predicted failure probability from the multilabel classifier, and $\alpha \in [0,1]$ is a weighting parameter. The selector chooses the solver minimizing this score, thereby balancing expected solution quality and risk of failure at time t.

We evaluated Random Forest [4], Gradient Boosting [9], and K-Nearest Neighbors (KNN) [8] for all ML tasks, as these models are computationally efficient and have shown strong performance in related work [23].

3.4   MetaPB

MetaPB[1] is a meta-solver introduced in this work, designed to predict the most suitable solver for a given PBO instance by combining static and probing features. Figure 1 illustrates the overall pipeline.

Given an input PBO instance, MetaPB first extracts static features that describe the structural properties of the instance. Then, for instances with manageable size, with fewer than $100,000$ constraints, MetaPB adds probing features obtained from short runs of two complementary solvers: SCIP, which provides BnB probing features, and RoundingSAT, which provides CDCL probing features.

For larger instances (with more than $100,000$ constraints), probing becomes computationally expensive, so MetaPB relies only on static features. This ensures it can generate predictions for these larger instances, ensuring coverage of the entire benchmark collection.

---

[1] https://github.com/amanda-salinas-pinto/MetaPB.

Additionally, the best solution found during probing is preserved. Once feature extraction is complete, a pre-trained machine learning model predicts the best solver from the portfolio. Finally, the selected solver processes the PBO instance receiving the best solution found during the computation of the probing features, as a warm start.

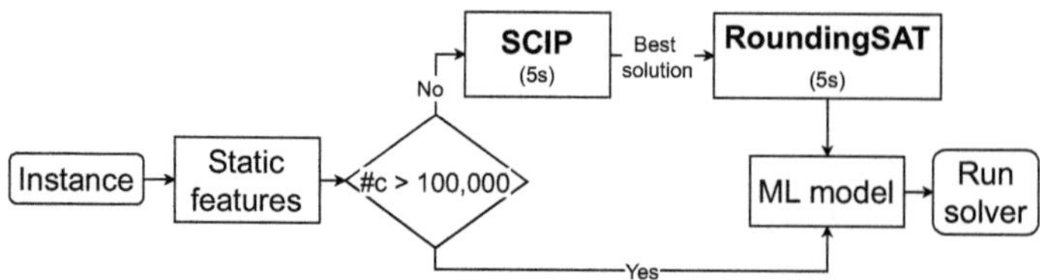

Fig. 1. Overview of the MetaPB meta-solver pipeline. We extract static features for all instances, and dynamically collect Branch and Bound and CDCL probing features for instances with a number of constraints (denoted as #c in the illustration) fewer than 100,000, before predicting the best solver to use.

4 Experimental Setup

This section details our methodology. We first use cross-validation on the training set to select the best configurations. Details on solvers and model settings are in the online supplementary material[2]. We then evaluate the selected best meta-solvers on the test set.

4.1 Validation

From the original set of 6,394 instances from the 2016 PBO competition, we randomly selected 1,500 instances to construct the training set.[3]

Figure 2 (a) provides the normalized objective values (Eq. 1) over 3600 seconds, showing that Gurobi is the Single Best Solver (SBS) across all the five time limits tested, followed by the two configurations of RoundingSAT (see Sect. 3.1), being default RoundingSAT configuration the SBS-Open-Source. We can also observe that the VBS-Open-Source composed only of open-source solvers, outperforms Gurobi from approximately 500 s onward.

Figure 2 (b) shows how often each solver finds the optimal solution. Again, Gurobi performs best overall. Interestingly, the VBS-Open-Source with open-source solvers can surpass Gurobi. Moreover, while Mixed-Bag and SCIP achieve worse normalized objective values on average, they find the optimal solution more frequently.

We train and validate the models independently for each time limit to capture evolving solver performance. Our evaluation employed 10-fold cross-validation.

---

[2] https://github.com/amanda-salinas-pinto/MetaPB/blob/main/appendix/A.pdf.

[3] https://github.com/amanda-salinas-pinto/MetaPB/tree/main/aslib.

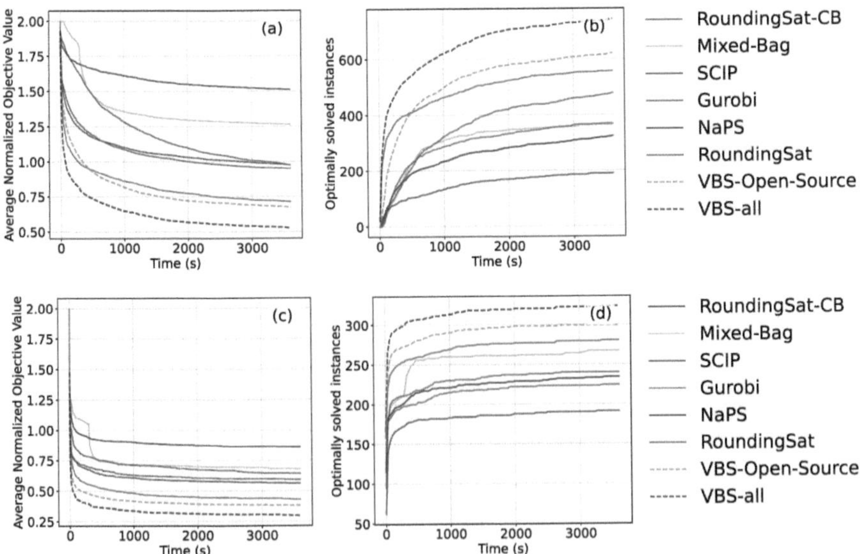

Fig. 2. Preliminary analysis of solver performance on the training set (top figures) and test set (bottom figures). On the left, normalized objective values (Eq. 1) over time. On the right, number of times a solver finds the optimal solution over time.

4.2   Testing

In this section, we evaluate (i) the best-performing metasolvers selected from our validation search and (ii) MetaPB, trained exclusively on open-source solvers using the best configurations. Both are tested on 478 instances from the 2024 PBO Competition, a final evaluation set designed to avoid benchmark imbalance and include challenging instances that was curated by the organizers of the competition.

Figure 2(c) shows that Gurobi achieves the best normalized objective values, while the VBS-Open-Source often surpasses it, highlighting the strength of the open-source portfolio. Figure 2(d) favors Gurobi in terms of the number of optimal solutions found across all time limits. Mixed-Bag approaches Gurobi despite weaker normalized scores, and the VBS-Open-Source again frequently outperforms it. Test instances also present better normalized objective values than training instances, suggesting that solvers could obtain good initial solutions more easily on this set.

4.3   Feature Sets

We adopt the static feature set used for PBO as our baseline [23]. This set comprises 10 features for linear PBO instances, capturing key metrics related to instance size, objective function characteristics, and constraint properties. We also assess the effect of combining these features with the proposed probing features defined in Sect. 3.2.

To denote the different feature sets, we define the following acronyms: **st** for static features only, **st+CDCL** for static with CDCL probing features, **st+BnB** for static with Branch and Bound probing features, and **all** for all the features. We evaluate feature sets' performance across all ML models and tasks described in Sect. 3.

5   Results and Analysis

5.1   Validation Results

Tables 1 and 2 present cross-validated results on $\hat{m}_{st}$ and the average number of optimally solved instances for the best-performing ML model (Random Forest), evaluated on the same feature sets and time limits. An analysis of feature importance is provided in the appendix.[4]

The results reveal several key trends. First, feature sets combining probing and static features consistently achieve better results than just using static features. Most significantly, using all features provides the best results in most cases, suggesting complementarity between CDCL and Branch and Bound probing features.

Second, regression-based approaches systematically outperform multiclass classification on the $\hat{m}$ metric across all experimental conditions. This advantage likely arises from the regression model's ability to capture subtle differences in solver performance, avoiding the oversimplification inherent in discrete classification. The hybrid approach achieves the best overall performance, indicating that incorporating information about whether solvers can solve an instance optimally improves the regression model.

Even though regression achieved lower $\hat{m}$ scores, predicting the normalized objective value does not always lead to the highest number of optimally solved instances. The hybrid approach, which also predicts solver feasibility, overcomes this limitation.

Finally, the $\hat{m}_{st}$ metric tends to worsen for higher time budgets. This behavior aligns with expectations, as tighter time limits impose stronger penalties for unsolved instances, while performance differences between solvers gradually diminish during later stages of execution.

5.2   Impact of the Weighting Parameter α

Figure 3 illustrates how the performance changes in terms of the $\hat{m}$ metric and the number of optimally solved instances as the hybrid model's parameter α is varied for each feature set. When $\alpha = 1$, the model relies exclusively on the regression model, which results in fewer optimally solved instances. Incorporating some information about the number of optimally solved instances improves both metrics compared to using only regression. However, when relying entirely on the multilabel approach ($\alpha = 0$), the $\hat{m}_{st}$ metric worsens despite solving more

---

[4] https://github.com/amanda-salinas-pinto/MetaPB/blob/main/appendix/B.pdf.

Table 1. Cross-validated $\hat{m}_{st}$ (mean $\pm$ SD, lower is better) of best ML models with different feature sets (st = Static, st+CDCL = Static + CDCL, st+BnB = Static + BnB, all = All features).

Task	t	st	st+CDCL	st+BnB	all
Multiclass	60	0.728 $\pm$ 0.034	0.721 $\pm$ 0.042	0.726 $\pm$ 0.033	**0.719 $\pm$ 0.036**
	100	0.782 $\pm$ 0.053	0.750 $\pm$ 0.063	0.778 $\pm$ 0.055	**0.740 $\pm$ 0.058**
	300	0.830 $\pm$ 0.046	0.798 $\pm$ 0.049	0.827 $\pm$ 0.048	**0.793 $\pm$ 0.042**
	600	0.838 $\pm$ 0.060	0.810 $\pm$ 0.053	0.824 $\pm$ 0.058	**0.818 $\pm$ 0.061**
	3600	0.901 $\pm$ 0.063	0.900 $\pm$ 0.063	**0.883 $\pm$ 0.060**	0.891 $\pm$ 0.059
Hybrid (best α)	60	0.283 $\pm$ 0.044	0.284 $\pm$ 0.044	0.278 $\pm$ 0.044	**0.276 $\pm$ 0.043**
	100	0.348 $\pm$ 0.071	0.311 $\pm$ 0.078	0.339 $\pm$ 0.064	**0.309 $\pm$ 0.061**
	300	0.439 $\pm$ 0.082	0.389 $\pm$ 0.097	0.429 $\pm$ 0.087	**0.382 $\pm$ 0.084**
	600	0.455 $\pm$ 0.045	0.444 $\pm$ 0.062	0.443 $\pm$ 0.052	**0.429 $\pm$ 0.050**
	3600	0.814 $\pm$ 0.085	0.790 $\pm$ 0.083	**0.788 $\pm$ 0.076**	0.789 $\pm$ 0.088
Multilabel ($\alpha = 0$)	60	0.830 $\pm$ 0.054	0.798 $\pm$ 0.072	0.830 $\pm$ 0.067	**0.787 $\pm$ 0.092**
	100	0.780 $\pm$ 0.108	**0.768 $\pm$ 0.137**	0.778 $\pm$ 0.110	0.768 $\pm$ 0.115
	300	0.772 $\pm$ 0.130	**0.719 $\pm$ 0.112**	0.773 $\pm$ 0.110	0.743 $\pm$ 0.103
	600	0.903 $\pm$ 0.214	**0.843 $\pm$ 0.151**	0.886 $\pm$ 0.201	0.871 $\pm$ 0.191
	3600	0.827 $\pm$ 0.083	0.831 $\pm$ 0.098	**0.819 $\pm$ 0.085**	0.846 $\pm$ 0.091
Regression ($\alpha = 1$)	60	0.332 $\pm$ 0.040	0.326 $\pm$ 0.043	0.316 $\pm$ 0.045	**0.306 $\pm$ 0.049**
	100	0.384 $\pm$ 0.057	0.365 $\pm$ 0.073	0.380 $\pm$ 0.065	**0.348 $\pm$ 0.073**
	300	0.496 $\pm$ 0.107	0.413 $\pm$ 0.120	0.486 $\pm$ 0.093	**0.406 $\pm$ 0.108**
	600	0.510 $\pm$ 0.051	0.477 $\pm$ 0.085	0.505 $\pm$ 0.047	**0.457 $\pm$ 0.063**
	3600	0.832 $\pm$ 0.074	0.822 $\pm$ 0.076	0.828 $\pm$ 0.068	**0.803 $\pm$ 0.066**

instances. This indicates that there is no clear correlation between the number of optimally solved instances and the $\hat{m}_{st}$ metric. It also suggests that intermediate values of α are more effective for achieving both lower $\hat{m}_{st}$ values and a higher number of optimally solved instances across all feature sets. We can see this behavior in all the feature sets. In this work, we prioritize minimizing the $\hat{m}$ metric and set $\alpha = 0.8$, but if the main objective is to maximize the number of optimally solved instances, a lower α can be chosen.

5.3   Results on the Test Set (2024 PBO Competition)

We first evaluate the best configurations trained on the complete portfolio, ignoring probing overhead. We then present the MetaPB evaluation, which accounts for the full cost of feature extraction and prediction.

Table 2. Cross-validated number of optimally solved instances (mean $\pm$ SD, higher is better) of best ML models with different feature sets (st = Static, st+CDCL = Static + CDCL, st+BnB = Static + BnB, all = All features).

Task	t	st	st+CDCL	st+BnB	all
Multiclass	60	27.70 ± 5.53	$\mathbf{27.90 \pm 5.03}$	27.90 ± 5.37	27.80 ± 5.08
	100	33.00 ± 6.10	$\mathbf{33.70 \pm 6.36}$	33.20 ± 6.08	33.60 ± 5.99
	300	43.00 ± 5.78	42.70 ± 5.51	$\mathbf{43.10 \pm 5.84}$	42.80 ± 5.62
	600	46.20 ± 5.15	46.60 ± 5.24	$\mathbf{46.80 \pm 4.69}$	46.40 ± 5.71
	3600	59.60 ± 5.78	59.10 ± 5.79	$\mathbf{59.90 \pm 5.66}$	59.00 ± 5.73
Hybrid (best α)	60	26.90 ± 5.70	$\mathbf{28.20 \pm 5.38}$	27.00 ± 5.57	27.00 ± 5.35
	100	33.70 ± 5.57	33.70 ± 6.10	33.30 ± 6.10	$\mathbf{34.00 \pm 5.48}$
	300	42.10 ± 5.32	42.70 ± 5.55	41.70 ± 5.35	$\mathbf{43.30 \pm 5.88}$
	600	45.70 ± 4.65	45.60 ± 5.10	46.10 ± 4.32	$\mathbf{46.40 \pm 5.35}$
	3600	67.60 ± 7.27	67.30 ± 7.40	66.60 ± 6.87	$\mathbf{68.40 \pm 6.96}$
Multilabel ($\alpha = 0$)	60	28.20 ± 4.66	$\mathbf{28.70 \pm 4.94}$	28.40 ± 4.88	28.60 ± 5.02
	100	34.20 ± 5.86	35.30 ± 5.93	34.50 ± 5.84	$\mathbf{35.40 \pm 5.68}$
	300	43.90 ± 5.63	44.30 ± 5.62	43.70 ± 5.69	$\mathbf{44.30 \pm 5.57}$
	600	48.20 ± 5.56	48.40 ± 5.75	48.40 ± 5.75	$\mathbf{48.50 \pm 6.00}$
	3600	68.00 ± 6.83	67.80 ± 6.97	68.20 ± 6.94	$\mathbf{68.40 \pm 7.09}$
Regression ($\alpha = 1$)	60	24.20 ± 4.56	$\mathbf{24.40 \pm 4.84}$	23.70 ± 5.42	23.70 ± 4.63
	100	28.80 ± 4.71	$\mathbf{30.10 \pm 4.76}$	29.40 ± 4.36	30.00 ± 4.69
	300	38.30 ± 5.93	39.50 ± 4.94	38.00 ± 4.98	$\mathbf{39.80 \pm 5.23}$
	600	42.70 ± 3.95	43.50 ± 5.14	42.90 ± 3.53	$\mathbf{43.60 \pm 4.82}$
	3600	65.00 ± 6.00	63.80 ± 6.97	$\mathbf{65.30 \pm 5.88}$	64.7 ± 7.64

Performance Using the Complete Portfolio We evaluate the configurations selected at validation time on the test set of the 2024 PBO competition. Results are reported under two metrics: a) the $\hat{m}_{st}$ optimization metric (Table 3) and b) the number of optimally solved instances within the time limit (Table 4).

Table 3 shows that combining static features with probing features consistently improves the $\hat{m}_s^t$ metric compared to using static features alone. Both the **st+CDCL** and **st+BnB** feature sets outperform **st**, and in most cases, using all features provides the best performance, except at the longest timeout, where **st+BnB** performs slightly better. All feature sets also improve upon the SBS (Gurobi), achieving $\hat{m}_s^t$ values below 1.

As shown in Table 4, the meta-solver consistently solves more instances optimally than the SBS (Gurobi) across all time marks. Adding probing features generally increases the number of optimally solved instances compared to using only static features. The benefit of CDCL features is more evident at shorter time marks, whereas BnB features contribute more as the time budget increases.

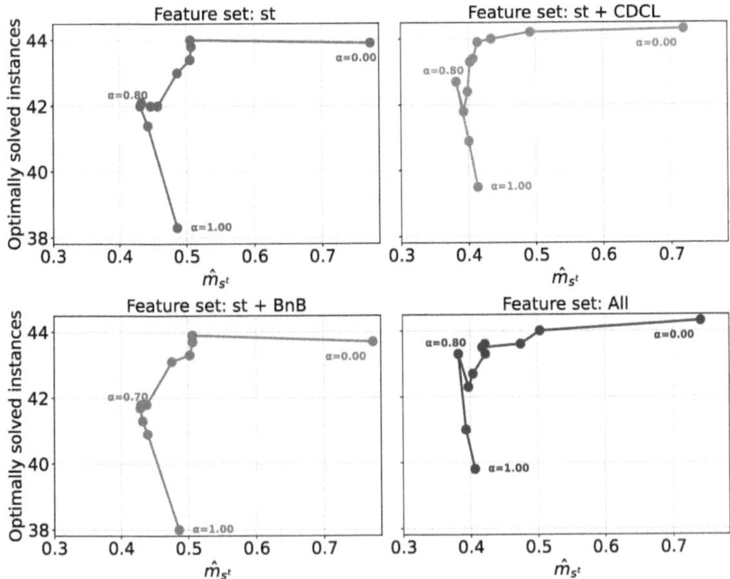

Fig. 3. Effect of varying α in the hybrid approach across the four feature sets under a 300 s time limit. For each, we report the number of optimally solved instances and the $\hat{m}_{st}$ metric. The parameter α controls the trade-off between the regression prediction $\hat{n}_a(i, t)$ and the multilabel prediction $\hat{p}_a(i, t)$. Intermediate values of α lead to better performance.

As a limitation of our study, we observed that the $\hat{m}_{st}$ values generally get worse for increasing time marks for all feature sets, suggesting that AAS for PBO is more challenging as the time limit increases.

MetaPB Evaluation We evaluate MetaPB, our proposed meta-solver that combines static and probing features for solver selection using only open-source solvers. The evaluation uses the test set of the 2024 PBO competition, comparing performance against three baselines: Gurobi (a commercial solver), Mixed-Bag (PBO 2024 winner), and RoundingSat (SBS-Open-Source). Tables 5 and 6 report performance under two metrics: the accumulated normalized objective value m_{st} (Eq. 2) and the total number of optimally solved instances.

The statistical analysis (Table 5) reveals a clear performance hierarchy. The two strongest MetaPB configurations are **all** and **st+BnB**. These variants are statistically equivalent across most time limits ($t = 60, 100, 600$), belonging to the same statistical tier. The **st+BnB** configuration is a more efficient feature set for enhancing the static approach because the cost-to-benefit ratio of **BnB** features is inherently superior to CDCL-based features. This is evidenced by two points: 1) **all** is only statistically superior to **st+BnB** at $t = 300$; and 2) the high probing overhead of **CDCL** features causes the **st+CDCL** variant to be statistically worse than the base **st** variant at short limits ($t = 60, 100$). The

Table 3. $\hat{m}_{st}$ scores (lower is better) for different feature sets, including statistical ranking on the test set. Superscript numbers denote statistical tiers (pairwise Wilcoxon $p < 0.05$). Tier 1 is the best-performing group. Solvers sharing the same tier number are statistically equivalent. All feature sets are statistically better than Gurobi at all timesteps.

t	st	st+CDCL	st+BnB	all
60	0.448^4	0.344^2	0.413^3	$\mathbf{0.316^1}$
100	0.632^4	0.465^2	0.556^3	$\mathbf{0.420^1}$
300	0.688^4	0.606^2	0.633^3	$\mathbf{0.595^1}$
600	0.726^3	0.648^2	0.686^2	$\mathbf{0.627^1}$
3600	0.852^2	0.829^2	$\mathbf{0.789^1}$	0.799^1

Table 4. Number of optimally solved instances per solver and meta-solver (algorithm selector), including *all* solvers in the portfolio.

t	st	st+CDCL	st+BnB	all	Gurobi^t
60	254	254	254	**261**	231
100	262	270	270	**274**	241
300	271	276	**277**	277	254
600	277	278	**280**	280	261
3600	298	295	**302**	300	280

performance of **st+CDCL** only recovers to join the same tier as **st** at longer limits ($t \geq 300$) once the feature cost is amortized.

In comparison with baselines, MetaPB remains highly competitive. The best MetaPB variants (**all** and **st+BnB**) are statistically equivalent to **Gurobi** in most time limits and significantly outperform **st** features alone. While **st+CDCL** is statistically equivalent to the RoundingSat baseline at $t = 60, 100$, all MetaPB variants are ranked superior to RoundingSat from $t = 300$ onward. Importantly, MetaPB consistently achieves results statistically superior to Mixed-Bag, the PBO 2024 winner, which is grouped into the lowest statistical tiers.

Regarding the number of optimally solved instances (Table 6), all MetaPB variants generally outperform both the SBS-Open-Source and Mixed-Bag baselines. The **all** feature set is the best performer, surpassing Gurobi at every timestep except $t = 600$. These results confirm MetaPB's effectiveness for solver selection, demonstrating that it achieves high solution quality and strong performance in solving instances to optimality. This results indicate that MetaPB would have won the Optimization track of the 2024 PBO Competition.

Table 5. Accumulated normalized objective value m_{st} (lower is better) and statistical ranking on the test set for MetaPB variants and baselines. Superscript numbers denote statistical tiers (pairwise Wilcoxon $p < 0.05$). Tier 1 is the best-performing group. Solvers sharing the same tier number are statistically equivalent.

t	st	st+CDCL	st+BnB	all	MB	Gu	RS
60	337^2	359^3	323^1	323^1	738^4	336^1	355^3
100	333^1	340^2	326^1	327^1	732^3	316^1	352^2
300	291^3	297^3	281^2	272^1	656^5	285^1	330^4
600	274^3	269^3	253^2	259^1	453^5	257^1	306^4
3600	236^3	234^3	230^2	234^3	424^5	211^1	288^4

Table 6. Number of optimally solved instances on the test set for MetaPB variants and baselines: Mixed-Bag (MB), Gurobi (Gu), and RoundingSat (RS).

t	st	st+CDCL	st+BnB	all	MB	Gu	RS
60	225	220	234	**239**	194	231	180
100	232	227	239	**243**	202	241	187
300	244	241	250	**254**	216	**254**	202
600	253	253	260	257	256	**261**	217
3600	280	279	281	**284**	267	280	224

6 Conclusions and Future Work

This work demonstrates that combining static and probing features can substantially improve AAS for PBO. Branch and Bound based probing features provided more significant improvements at longer time limits than the CDCL based probing features. However, the best results were achieved by using the full set of static and dynamic features.

Our analysis shows that regression-based and hybrid approaches consistently outperform multiclass classification by more accurately capturing the nuances of solver performance. The hybrid model, which incorporates information on both solution quality and whether a solver solves an instance optimally, achieved the best trade-off between solution quality and the number of optimally solved instances.

Building on these insights, we introduce MetaPB, an open-source meta-solver that relies solely on open-source solvers. MetaPB outperformed all individual solvers in its portfolio and surpassed the best solver in the 2024 PBO competition. It also showed competitive performance against the commercial state-of-the-art solver Gurobi.

Future research should explore dynamic feature representations that evolve along with solver behavior to improve long-term prediction stability. Increasing the diversity of solvers in the portfolio may also enhance generalization, giving

AAS greater adaptability. Moreover, analyzing the time required to extract probing features would help clarify the trade-off between feature extraction cost and predictive accuracy. The observed shifts in solver performance over time suggest strong potential for adaptive scheduling strategies, where solvers are dynamically selected based on runtime progress to maximize efficiency.

Acknowledgments. The research of the 1st, 2nd, 3rd, 4th and 6th authors is supported in part by grant #2112533: "NSF Artificial Intelligence Research Institute for Advances in Optimization (AI4OPT)". The 5th author is supported by ANID CCTVal (CIA250027) grant. The first author is also supported by ANID-Subdirección de Capital Humano/Magíster Nacional/2024-22241061. The 4th author is also supported by grant #2346058: "NRT-AI: Integrating Artificial Intelligence and Operations Research Technologies".

Disclosure of Interests. The authors have no competing interests to declare that are relevant to the content of this article

References

1. Biere, A., Heule, M., van Maaren, H., Walsh, T. (eds.): Handbook of satisfiability - second edition, frontiers in artificial intelligence and applications, vol. 336. IOS Press (2021). https://doi.org/10.3233/FAIA336
2. Bischl, B., et al.: Aslib: a benchmark library for algorithm selection. Artif. Intell. **237**, 41–58 (2016)
3. Bolusani, S., et al.: The SCIP optimization suite 9.0. technical report, optimization online (2024). https://optimization-online.org/2024/02/the-scip-optimization-suite-9-0/
4. Breiman, L.: Random forests. Mach. Learn. **45**, 5–32 (2001)
5. Devriendt, J., Gocht, S., Demirović, E., Nordström, J., Stuckey, P.J.: Cutting to the core of pseudo-boolean optimization: combining core-guided search with cutting planes reasoning. In: Proceedings of the AAAI Conference on Artificial Intelligence vol. 35,no. 5, pp. 3750–3758 (2021). https://doi.org/10.1609/aaai.v35i5.16492
6. Eén, N., Sörensson, N.: Translating pseudo-boolean constraints into sat. J. Satisfiability, Boolean Mod. Comput. **2**(1–4), 1–26 (2006)
7. Divide and conquer: Elffers, J., Nordström. J. Towards faster pseudo-boolean solving. In: IJCAI. **18**, 1291–1299 (2018)
8. Fix, E., Hodges, J.L.: Discriminatory analysis, nonparametric discrimination, consistency properties (1951)
9. Friedman, J.H.: Greedy function approximation: a gradient boosting machine. Ann. Stat. , 1189–1232 (2001)
10. Gleixner, A., Gottwald, L., Hoen, A.: Papilo: a parallel presolving library for integer and linear optimization with multiprecision support. INFORMS J. Comput. **35**(6), 1329–1341 (2023)
11. Gurobi optimization, LLC: gurobi optimizer reference manual (2024). https://www.gurobi.com
12. Huerta, I.I., Neira, D.A., Ortega, D.A., Varas, V., Godoy, J., Asin-Acha, R.: Improving the state-of-the-art in the traveling salesman problem: an anytime automatic algorithm selection. Expert Syst. Appl. **187**, 115948 (2022)

13. Huerta, I.I., Neira, D.A., Ortega, D.A., Varas, V., Godoy, J., As'in Ach'a, R.J.: Anytime automatic algorithm selection for knapsack. Expert Syst. Appl. **158**, 113613 (2020). https://doi.org/10.1016/j.eswa.2020.113613
14. Hutter, F., Hoos, H.H., Leyton-Brown, K.: Sequential model-based optimization for general algorithm configuration. In: Coello, C.A.C. (ed.) LION 2011. LNCS, vol. 6683, pp. 507–523. Springer, Heidelberg (2011). https://doi.org/10.1007/978-3-642-25566-3_40
15. Hutter, F., Xu, L., Hoos, H.H., Leyton-Brown, K.: Algorithm runtime prediction: methods & evaluation. Artif. Intell. **206**, 79–111 (2014)
16. Jabs, C., Berg, J., Järvisalo, M.: PB-OLL-RS and mixed-bag in pseudo-boolean competition, p. 2024 (2024)
17. Kerschke, P., Hoos, H.H., Neumann, F., Trautmann, H.: Automated algorithm selection: survey and perspectives. Evol. Comput. **27**(1), 3–45 (2019)
18. Kotthoff, L., Kerschke, P., Hoos, H., Trautmann, H.: Improving the state of the art in inexact TSP solving using per-instance algorithm selection. In: Dhaenens, C., Jourdan, L., Marmion, M.-E. (eds.) LION 2015. LNCS, vol. 8994, pp. 202–217. Springer, Cham (2015). https://doi.org/10.1007/978-3-319-19084-6_18
19. Lindauer, M., van Rijn, J.N., Kotthoff, L.: The algorithm selection competitions 2015 and 2017. Artif. Intell. **272**, 86–100 (2019)
20. Manquinho, V., Roussel, O., Deters, M.: Pseudo-boolean competition (2010). http://www.cril.univ-artois.fr/PB10
21. Nudelman, E., Leyton-Brown, K., Devkar, A., Shoham, Y., Hoos, H.: Satzilla: an algorithm portfolio for sat. Solver description, SAT competition (2004)
22. Nudelman, E., Leyton-Brown, K., Hoos, H.H., Devkar, A., Shoham, Y.: Understanding random sat: beyond the clauses-to-variables ratio. In: Principles and Practice of Constraint Programming-CP 2004: 10th International Conference, CP 2004, Toronto, Canada, 27 September, 1 October 2004. Proceedings 10, pp. 438–452. Springer (2004)
23. Pezo, C., Hochbaum, D.S., Godoy, J., As'in Ach'a, R.J.: Automatic algorithm selection for pseudo-boolean optimization with given computational time limits. Comput. Oper. Res. **173**, 106836 (2025). https://doi.org/10.1016/j.cor.2024.106836
24. Rice, J.R.: The algorithm selection problem. In: Advances in computers, vol. 15, pp. 65–118. Elsevier (1976)
25. Roussel, O.: Pseudo boolean competition (2024). https://www.cril.univ-artois.fr/PB24/
26. Sakai, M., Nabeshima, H.: Construction of an robdd for a pb-constraint in band form and related techniques for pb-solvers. IEICE Trans. Inf. Syst. **98**(6), 1121–1127 (2015)
27. Shavit, H., Hoos, H.H.: Revisiting satzilla features in 2024. In: 27th International Conference on Theory and Applications of Satisfiability Testing (SAT 2024), pp. 27-1. Schloss Dagstuhl-Leibniz-Zentrum für Informatik (2024)
28. Smith-Miles, K., Baatar, D., Wreford, B., Lewis, R.: Towards objective measures of algorithm performance across instance space. Comput. Oper. Res. **45**, 12–24 (2014)
29. Xu, L., Hutter, F., Hoos, H.H., Leyton-Brown, K.: Satzilla: portfolio-based algorithm selection for sat. J. Artif. Intell. Res. **32**, 565–606 (2008)
30. Xu, L., Hutter, F., Shen, J., Hoos, H.H., Leyton-Brown, K.: Satzilla 2012: improved algorithm selection based on cost-sensitive classification models. In: Proceedings of SAT Challenge, pp. 57–58 (2012)

Singleton Node Consistency for Quadratic Assignment Problems in Cost Function Networks

Guidio Sewa[1,2], David Allouche[1], Simon de Givry[1,2(✉)],
George Katsirelos[2,3], and Thomas Schiex[1,2]

[1] MIAT, INRAE, Université de Toulouse, Toulouse, France
{guidio.sewa,david.allouche,simon.de-givry,thomas.schiex}@inrae.fr
[2] ANITI, Université de Toulouse, Toulouse, France
[3] MIA Paris, INRAE, Paris, France

Abstract. The Quadratic Assignment Problem (QAP) consists of finding a permutation that minimizes a quadratic objective function. Exact methods generally rely on a branch-and-bound procedure, the efficiency of which depends heavily on the quality of its lower bound. In integer linear programming, several bounds have been investigated, exhibiting different trade-offs between speed and quality. The Gilmore-Lawler bound appears to be the most commonly used in practice. It requires solving a linear assignment problem (LAP) for each variable-value pair. We show how to obtain this bound using Singleton Node Consistency (SNC) and LAP. In Cost Function Networks (CFNs), we propose a reformulation that transforms the result of applying LAP to a given variable-value pair into cost functions of arity 1 and 2, which can be added to the original problem. Combined with existing lower bounds for CFNs, including EDAC and a recent CFN propagator for ALLDIFFERENT, this method (SNC-LAP-GLB), used as a preprocessing, significantly increases the initial lower bound and accelerates the search, resulting in competitive results on the QAPLIB benchmark. We then propose an extension of the ALLDIFFERENT propagator for the Global Cardinality Constraint. It allows us to exploit variable symmetries on some challenging QAPLIB instances, thus improving the results.

Keywords: Linear Assignment Problem · ALLDIFFERENT · Global Cardinality Constraint

1 Introduction

The quadratic assignment problem (QAP) was introduced in [26] as a mathematical model for the location of indivisible economical activities. *"We want to assign n facilities to n locations with the cost being proportional to the flow between the facilities multiplied by the distances between the locations plus, eventually, costs for placing the facilities at their respective locations. The objective*

T. Guns (Ed.): CPAIOR 2026, LNCS 16595, pp. 520–537, 2026.
https://doi.org/10.1007/978-3-032-27242-3_31

is to allocate each facility to a location such that the total cost is minimized" [8]. It is a very challenging problem and even MIP solvers struggle to scale beyond the smallest instances with less than thirty variables. The problem has many real-world applications, and has attracted the attention of the community [30].

The QAP is naturally modeled as a cost function network (CFN) [10,11] with a single permutation constraint, and unary and binary cost functions to capture the costs of linear (on facility-location pairs) and quadratic terms (on pairs of facility-location pairs), respectively. Branch-and-bound methods for solving CFNs rely on soft local consistency (SLC) techniques [10]. The idea of SLC is to reformulate a problem into an equivalent one by rewriting the objective function. The rewritten objective function has a constant term which provides an explicit lower bound and may have infinite cost for some values, indicating they are pruned. The reformulation is computed by moving costs from the quadratic/binary level to the linear/unary level, and then to the constant term. This reformulation allows for better communication between the quadratic terms and the permutation (ALLDIFFERENT) constraint, for which a dedicated SLC based on solving the polynomial-time linear assignment problem (LAP) has recently been proposed [38].

This observation was already known in QAP and so-called reduction methods have been developed with exactly the same goal [8]. In particular, the Lagrangian relaxation proposed in [20] can be related to optimal soft arc consistency in CFNs [10]. It dominates the Gilmore-Lawler lower bound (GLB), which is a decomposition approach used in branch-and-bound methods for QAP because of its computation speed.

In crisp constraint networks, several extensions of arc consistency methods have been proposed to get better value pruning [36]. One technique is singleton arc consistency (SAC) [3,5,13,14,29]. In this paper, we propose an adaptation to CFNs of singleton node consistency (SNC), a weaker form of SAC. We show how it can be used to simulate GLB, by providing valid reformulations and even get improved lower bounds when used in conjunction with existing SLC techniques such as *existential directional arc consistency* (EDAC) [10,12] during preprocessing.

In the following, we introduce necessary background on QAP and CFNs. Then, we present our singleton node consistency approach with SLC reformulations. Two iterative algorithms are proposed: SNC-LAP-GLB and a variant SNC-LAP-GREEDY. Next, we exploit symmetries present in some specific QAP instances using an inverted formulation of the problem and a global cardinality constraint (GCC). Following our work on ALLDIFFERENT, and the relationship with the semi-assignment problem (SAP), we develop SLC techniques for the case of GCC in closed form. Replacing LAP by SAP in the previous algorithms, we obtain SNC-SAP-GLB and SNC-SAP-GREEDY. Last, we give detailed experimental results for these four algorithms compared to CP and OR approaches on the QAPLIB benchmark and conclude.

The closest related work in CFNs is *virtual singleton arc consistency via super-reparametrizations* [16,17]. However, a super-reparametrization does not reformulate the problem into an equivalent one. Also, SNC with the greedy strategy applied on binary CFNs (without the ALLDIFFERENT constraint) implies the *existential arc consistent* (EAC) property [12].

2   Background

2.1   Linear Assignment Problem

The Linear Assignment Problem (LAP) of size n is defined by

$$l = \min \sum_{i=1}^{n} \sum_{u=1}^{n} c_{iu} x_{iu} \tag{1}$$

$$\sum_{u=1}^{n} x_{iu} = 1 \quad \forall i \in \{1, \ldots, n\} \tag{2}$$

$$\sum_{i=1}^{n} x_{iu} = 1 \quad \forall u \in \{1, \ldots, n\} \tag{3}$$

$$x_{iu} \in \{0, 1\} \quad \forall i, u \in \{1, \ldots, n\}^2 \tag{4}$$

The constraint matrix Eq. (2)–(3) being totally unimodular, it can be solved by linear programming algorithms in polynomial time, or *e.g.*, using the Hungarian [27] or LAPJV [24] algorithms in $O(n^3)$ time.

2.2   Quadratic Assignment Problem

A Quadratic Assignment Problem (QAP) is defined by a set of n facilities and a set of n locations. For each pair (i, j) of facilities, a flow a_{ij} is specified, and for each pair (u, v) of locations, a distance b_{uv} is specified. Finally, there is a cost c_{iu} of locating facility i at location u. The problem is to assign all facilities to different locations while minimizing the total cost, assuming a unit cost per supply and unit of distance.

Formally, a QAP of size n is defined by

$$q = \min \sum_{i=1}^{n} \sum_{j=1}^{n} \sum_{u=1}^{n} \sum_{v=1}^{n} a_{ij} b_{uv} x_{iu} x_{jv} + \sum_{i=1}^{n} \sum_{u=1}^{n} c_{iu} x_{iu} \tag{5}$$

$$\sum_{u=1}^{n} x_{iu} = 1 \quad \forall i \in \{1, \ldots, n\} \tag{6}$$

$$\sum_{i=1}^{n} x_{iu} = 1 \quad \forall u \in \{1, \ldots, n\} \tag{7}$$

$$x_{iu} \in \{0, 1\} \quad \forall i, u \in \{1, \ldots, n\}^2 \tag{8}$$

where coefficients a, b, and c are assumed to be non-negative integers. W.l.o.g., we assume $a_{ii} = b_{uu} = 0$ for all pairs (i, u). If not, c_{iu} can be updated by

$c_{iu} \leftarrow c_{iu} + a_{ii}b_{uu}$. We have $x_{iu} = 1$ if facility i is assigned to location u, otherwise, $x_{iu} = 0$. The problem is NP-hard, as it contains the Traveling Salesman Problem as a special case [8].

If we ignore the quadratic terms in the objective, we get a LAP. Solving this gives a basic lower bound for QAP.

2.3 Gilmore-Lawler Bound

The Gilmore-Lawler bound (GLB) [21,41] is a strengthened variant of the simple LAP bound. It considers solving first the following n^2 specific LAPs LAP_{GLB}^{iu}, for each pair $(i, u) \in \{1, \ldots, n\}^{2}$:[1]

$$LAP_{GLB}^{iu} : l_{iu} = \min \sum_{j=1, j \neq i}^{n} \sum_{v=1, v \neq u}^{n} a_{ij} b_{uv} x_{jv} \tag{9}$$

$$\sum_{v=1, v \neq u}^{n} x_{jv} = 1 \quad \forall j \in \{1, \ldots, n\} \setminus \{i\} \tag{10}$$

$$\sum_{j=1, j \neq i}^{n} x_{jv} = 1 \quad \forall v \in \{1, \ldots, n\} \setminus \{u\} \tag{11}$$

$$x_{jv} \in \{0, 1\} \quad \forall j \in \{1, \ldots, n\} \setminus \{i\}, v \in \{1, \ldots, n\} \setminus \{u\} \tag{12}$$

Observe that each LAP_{GLB}^{iu} objective function Eq. (9) involves a different part of the a and b coefficients, which means that the individual bounds we compute can be added to the coefficient of the corresponding linear term x_{iu}.

The GLB is obtained by solving this final LAP:

$$glb = \min \sum_{i=1}^{n} \sum_{u=1}^{n} (l_{iu} + c_{iu}) x_{iu} \tag{13}$$

$$\sum_{u=1}^{n} x_{iu} = 1 \quad \forall i \in \{1, \ldots, n\} \tag{14}$$

$$\sum_{i=1}^{n} x_{iu} = 1 \quad \forall u \in \{1, \ldots, n\} \tag{15}$$

$$x_{iu} \in \{0, 1\} \quad \forall i, u \in \{1, \ldots, n\}^{2} \tag{16}$$

2.4 QAP as a Weighted Constraint Satisfaction Problem

A Cost Function Network is a triplet $P = (X, D, F)$ where $X = \{x_1, \ldots, x_n\}$ is a set of n decision variables with finite domain D_i for variable x_i. An assignment I to a set of variables S is a mapping from each variable $x_i \in S$ to a value

---

[1] These LAPs can all be solved together in $O(n^2 \log(n))$ time (see *e.g.*, [8,41]). However, it assumes a decomposition of their objective function as a product of two vectors, which is not preserved by the problem reformulations done in Sect. 3.

in D_i. Let $\ell(S) = \Pi_{i \in S} D_i$ denote the set of all possible assignments for S. An assignment is complete if $S = X$, otherwise it is partial. F is a set of cost functions, each cost function $f_S \in F$ applies to a subset S of X and returns a non-negative integer cost depending on its variable assignment. Returning an infinite cost means a forbidden assignment. A cost function returning only zero or infinite costs defines a hard constraint.

The binary Weighted Constraint Satisfaction Problem (WCSP) is defined by:[2]

$$\min \sum_{i=1}^{n} \sum_{j>i}^{n} f_{ij}(x_i, x_j) + \sum_{i=1}^{n} f_i(x_i) + f_\varnothing \tag{17}$$

$$x_i \in D_i \quad \forall i \in \{1, \ldots, n\} \tag{18}$$

where $f_\varnothing$, f_i, f_{ij} are respectively called empty, unary, and binary cost functions. $f_\varnothing$ is a constant that serves as a trivial problem lower bound. The problem is NP-hard [11]. Complete solving approaches are usually based on branch-and-bound, whose efficiency greatly depends on the quality of its lower bound [10]. In the past, several so-called *Soft Local Consistency* (SLC) methods, including NC, EDAC [10, 12], and $F\varnothing IC$ [32, 38], have been proposed in order to reformulate P into an equivalent problem P' having the same cost for any complete assignment, but a better $f_\varnothing$.

The simplest SLC method is Node Consistency (NC): if all domain values of a variable x_i have a nonzero unary cost, then increase $f_\varnothing \leftarrow f_\varnothing + \min_{x_i \in D_i} f_i(x_i)$ and decrease the unary cost function by this minimum: $\forall x_i \in D_i, f_i(x_i) \leftarrow f_i(x_i) - \min_{x_i \in D_i} f_i(x_i)$. Moreover, if a problem upper bound ub is available, then NC will also prune all domain values $x_i \in D_i$ for which it holds that $f_i(x_i) + f_\varnothing \geq ub$. EDAC is a stronger SLC method which enforces NC and more, but is limited to binary (and ternary) cost functions. $F\varnothing IC$ has been introduced for the propagation of global constraints in CFNs (KNAPSACK [32], ALLDIFFERENT [38]). The aim of $F\varnothing IC$ is to enforce that for each nonunary cost function $f_S, |S| > 1$, it admits a partial assignment $I \in \ell(S)$ such that $f_S(I) + \sum_{x_i \in S} f_i(I[\{x_i\}]) = 0$. EDAC is stronger than $F\varnothing IC$ on binary cost functions.

A QAP of size n can be formulated as a WCSP $P = (X, D, F)$ with:

$$D_i = \{1, \ldots, n\} \qquad \forall i \in \{1, \ldots, n\} \tag{19}$$

$$f_{ij}(x_i = u, x_j = v) = a_{ij}b_{uv} +$$

$$a_{ji}b_{vu} \qquad \forall i, j, u, v \in \{1, \ldots, n\}^4, i < j, u \neq v \tag{20}$$

$$f_{ij}(x_i = u, x_j = u) = \infty \qquad \forall i, j, u \in \{1, \ldots, n\}^3 \tag{21}$$

$$f_i(x_i = u) = c_{iu} \qquad \forall i, u \in \{1, \ldots, n\}^2 \tag{22}$$

$$f_\varnothing = 0 \tag{23}$$

---

[2] Later, we use the term WCSP indifferently to refer to the CFN or its associated minimization problem.

Here, decision variables are associated to facilities and values to locations. So $\{x_i = u\}$ in the WCSP represents $x_{iu} = 1$ in the QAP. We will see in Sect. 4 that associating variables to locations instead of facilities allows to exploit symmetries in some cases.

Using a ALLDIFFERENT(X) hard global constraint instead of Eq. (21) gives better propagation and stronger lower bounds as shown in [38]. We call this QAP model with the ALLDIFFERENT, P_{alldiff}.[3]

Reformulation with the ALLDIFFERENT Constraint. The combination of the ALLDIFFERENT constraint and the unary cost functions $f_i \forall i \in \{1, \ldots, n\}$ is exactly the LAP of Eq. (1–4) where $c_{iu} \equiv f_i(x_i = u)$. We have shown in [38] a reformulation of P_{alldiff} by solving the corresponding LAP, then adding its optimum l to the problem lower bound, $f_\varnothing \leftarrow f_\varnothing + l$, and the reduced costs $rc(i, u)$ in the LAP become the new unary costs: $f_i(x_i = u) \leftarrow rc(i, u)$ for all $i, u \in \{1, \ldots, n\}^2$. This is a valid reformulation because the LAP contains only equality constraints [38, Theorem 1] [31]. Applying this reformulation enforces exactly $F\varnothing IC$ on the ALLDIFFERENT constraint. After the reformulation, P_{alldiff} has a better lower bound $f_\varnothing$, which can be used by the branch-and-bound method to prune the search when $f_\varnothing \geq ub$ or prune domains by NC when $f_i(x_i) + f_\varnothing \geq ub$. We have shown [38] that these pruning rules are stronger than the ones developed for the Weighted ALLDIFFERENT constraint [9,37].

3 Singleton Node Consistency and Linear Assignment Problem

Singleton Arc Consitency (SAC) has been proposed in the context of Constraint Satisfaction Problems. The main idea is to test each partial assignment $\{x_i = u\} \forall x_i \in X, u \in D_i$ and propagate using arc consistency. If a domain wipe-out occurs, then u can be removed from D_i [14]. Refinements of this base algorithm improve practical or worst case performance [3,5,13,29].

Here, our goal is to increase the unary costs by using search and $F\varnothing IC$, a weaker SLC method than soft arc consistency methods such as EDAC.

Based on the previously-defined QAP problem P_{alldiff}, we define the following subproblem P_{GLB}^{iu} where the value u is assigned to the variable x_i and only binary

---

[3] In practice, we add Eq. (21) as a redundant binary constraint to the ALLDIFFERENT constraint only if the function $f_{ij}(x_i = u, x_j = v)$ returns a nonzero cost for some $u \neq v$, otherwise the function f_{ij} can be discarded which makes the WCSP having less binary cost functions.

cost functions (or half quantities of them) related to $\{x_i = u\}$ are kept:[4]

$$P_{GLB}^{iu} : l_{iu}' = \min \sum_{j<i} f_{ji}(x_j, x_i = u) + \frac{1}{2} \sum_{j>i} f_{ij}(x_i = u, x_j) \tag{24}$$

$$\text{ALLDIFFERENT}(X \setminus \{x_i\}) \tag{25}$$

$$x_j \in D_j \setminus \{u\} \quad \forall j \in \{1, \ldots, n\} \setminus \{i\} \tag{26}$$

Our main observation is that we can reformulate the original problem P_{alldiff} using the solution of the modified LAP_{GLB}^{iu}. In P_{alldiff}, the optimum l_{iu}' is added to the unary cost $f_i(x_i = u)$ and the reduced costs $rc(j, v)$ update binary costs:

$$f_i(x_i = u) \leftarrow f_i(x_i = u) + l_{iu}' \tag{27}$$

$$f_{ji}(x_j = v, x_i = u) \leftarrow rc(j, v) \qquad \qquad \forall j < i, v \in D_j \tag{28}$$

$$f_{ij}(x_i = u, x_j = v) \leftarrow \frac{f_{ij}(x_i = u, x_j = v)}{2} + rc(j, v) \quad \forall j > i, v \in D_j \tag{29}$$

Theorem 1. *The updates in Equations (27)–(29) preserve equivalence.*

Proof (Sketch). The proof is the same as for the ALLDIFFERENT constraint [38] but now the modified costs are conditioned on the partial assignment $\{x_i = u\}$. After conditioning on $\{x_i = u\}$, the lower bound, which is a constant term, becomes a linear term and the reduced costs, which are linear, become quadratic terms. □

Algorithm 1, SNC-LAP-GLB, iteratively reformulates the problem P_{alldiff} (lines 7-13) by solving the modified LAP_{GLB}^{iu} (line 5) for all (i, u) and applying the updates (27)–(29). After that, it enforces $F\emptyset IC$ on ALLDIFFERENT (line 16) to get a lower bound, which it improves with EDAC on P_{alldiff} (line 17). EDAC moves costs between unary and binary cost functions, which may invalidate previous optimal LAPs, thus it iterates until a stopping condition is met.

We show next that SNC-LAP-GLB is no weaker than GLB.

Theorem 2. *For any QAP with symmetric flow and distance matrices, the SNC-LAP-GLB bound is no smaller than the GLB bound.*

Proof. We prove first that for all variables i, $f_i(x_i = u) \geq l_{iu}$, where l_{iu} is defined in Equation (9). We show this by induction on the variable i.

For $i = 1$, due to our assumption of symmetric flow and distance matrices, enforcing $F\emptyset IC$ on P_{GLB}^{1u} is equivalent to solving LAP_{GLB}^{1u} for all u. We have $\frac{f_{1j}(x_1 = u, x_j = v)}{2} = \frac{a_{1j}b_{uv} + a_{j1}b_{vu}}{2} = a_{1j}b_{uv}$, thus $l_{1u}' = l_{1u}$.

Assume the inductive hypothesis holds for all $j \leq i$. We show that it holds for $i + 1$. If the reduced costs of LAP_{GLB}^{iu} are all null, then in the ith reformulation of P_{alldiff}, we have $f_{ij}(x_i = u, x_j = v) \leftarrow \frac{f_{ij}(x_i = u, x_j = v)}{2} \forall j > i, v \in D_j$.

---

[4] This is why we call our approach singleton node consistency because we do not want interactions with other binary cost functions that would require introducing new higher-arity (ternary) cost functions to ensure a correct reformulation.

Otherwise, binary costs are higher due to positive reduced costs. Thus, the first term $f_{ji+1}(x_j, x_{i+1} = u) \forall j < i+1$ in Eq. 24 of P_{GLB}^{i+1u} is greater than or equal to $a_{ji+1}b_{vu}$ after the previous i reformulations. The second term $f_{i+1j}(x_{i+1} = u, x_j) \forall j > i+1$ has not been modified by the previous reformulations, and it is also greater than or equal to $a_{i+1j}b_{uv}$.

Since the final step of SNC-LAP-GLB (line 16) solves a LAP with linear costs at least as great as those used in the final step of GLB, we get that the optimum of the former LAP is at least as great as the latter, as required. $\square$

Algorithm 1. SNC-LAP-GLB(P_{alldiff}): Enforcing singleton node consistency with GLB-like LAP on a quadratic assignment problem P_{alldiff}.

1: **repeat**
2:     **for all** $x_i \in X$ **do**
3:         **for all** $u \in D_i$ **do**
4:             */* Temporarily assign $x_i = u$ and enforce F$\emptyset$IC on P_{GLB}^{iu} */*
5:             $(l'_{iu}, rc) = \text{SOLVE-LAP}_{GLB}^{iu}(\{f_{ji} \mid j < i\} \cup \{\frac{f_{ij}}{2} \mid j > i\})$
6:             */* Reformulate P_{alldiff} */*
7:             **for all** $j < i, v \in D_j$ **do**
8:                 $f_{ji}(x_j = v, x_i = u) \leftarrow rc(j, v)$
9:             **end for**
10:           **for all** $j > i, v \in D_j$ **do**
11:              $f_{ij}(x_i = u, x_j = v) \leftarrow \frac{f_{ij}(x_i=u,x_j=v)}{2} + rc(j, v)$
12:           **end for**
13:           $f_i(x_i = u) \leftarrow f_i(x_i = u) + l'_{iu}$
14:         **end for**
15:     **end for**
16:     Enforce F$\emptyset$IC on ALLDIFFERENT(X)
17:     Enforce EDAC and F$\emptyset$IC on P_{alldiff}
18: **until** stopping condition

The time complexity of one pass over all the variables at line 2 is in $O(n^5)$, assuming $O(n^3)$ for solving LAP_{GLB}^{iu}. The space complexity is the same as for representing P_{alldiff}, in $O(n^4)$, using tables for binary cost functions. In our implementation, the stopping condition is that the lower bound $f_\emptyset$ has not increased sufficiently after one pass over all the variables, specifically if $\frac{f_\emptyset - f_\emptyset^{prev}}{f_\emptyset} \leq 1e-4$.

3.1   Improving Further the Iterated Gilmore-Lawler Bound

The intuitive idea of GLB is to distribute the binary costs evenly across the unary costs of all variables before solving the final LAP. We propose another strategy that will try to move all the unary and binary costs towards a single variable in order to increase its own unary cost function as much as possible. We call this strategy GREEDY. We define the following subproblem P_{Greedy}^{iu} where we assign $x_i = u$ and keep only unary and binary cost functions related to $\{x_i = u\}$:

$$P^{iu}_{Greedy} : s_{iu} = \min \sum_{j<i} f_{ji}(x_j, x_i = u)$$

$$+ \sum_{j>i} f_{ij}(x_i = u, x_j) + \sum_{j \neq i} f_j(x_j) \tag{30}$$

$$\textsc{AllDifferent}(X \setminus \{x_i\}) \tag{31}$$

$$x_j \in D_j \setminus \{u\} \qquad \forall j \in \{1, \ldots, n\} \setminus \{i\} \tag{32}$$

Enforcing $F\emptyset IC$ on this problem is equivalent to solving LAP^{iu}_{Greedy}, an LAP with a linear objective function given by Eq. 30.

Algorithm 2. SNC-LAP-GREEDY(P_{alldiff}): Enforcing singleton node consistency with Greedy LAP on a quadratic assignment problem P_{alldiff}.

1: **repeat**
2:     **for all** $x_i \in X$ **do**
3:         /* *Move unary costs into binary cost functions related to x_i* */
4:         **for all** $x_j \in X \setminus \{x_i\}, v \in D_j$ **do**
5:             **for all** $u \in D_i$ **do**
6:                 $f_{ij}(x_i = u, x_j = v) \leftarrow f_{ij}(x_i = u, x_j = v) + f_j(x_j = v)$
7:             **end for**
8:             $f_j(x_j = v) \leftarrow 0$
9:         **end for**
10:        **for all** $u \in D_i$ **do**
11:            /* *Temporarily assign $x_i = u$ and enforce $F\emptyset IC$ on P^{iu}_{Greedy}* */
12:            $(s_{iu}, rc) = \textsc{Solve-LAP}^{iu}_{Greedy}(\{f_{ij} \mid j \neq i\})$
13:            /* *Reformulate P_{alldiff}* */
14:            **for all** $x_j \in X \setminus \{x_i\}, v \in D_j$ **do**
15:                $f_{ij}(x_i = u, x_j = v) \leftarrow rc(j, v)$
16:            **end for**
17:            $f_i(x_i = u) \leftarrow f_i(x_i = u) + s_{iu}$
18:        **end for**
19:        Enforce EDAC and $F\emptyset IC$ on P_{alldiff}
20:    **end for**
21: **until** stopping condition

We define Algorithm 2, SNC-LAP-GREEDY, which reformulates P_{alldiff} by repeatedly solving LAP^{iu}_{Greedy} (line 12) for all (i, u) until a stopping condition is met (the same as for SNC-LAP-GLB). After singleton tests and reformulations for all domain values of the target variable (for-loop line 10), we enforce EDAC on the binary cost functions of P_{alldiff} and $F\emptyset IC$ on ALLDIFFERENT (line 19).

We give the following theorem without proof, as it is largely the same as the proof of Theorem 1. The main difference is that we cannot directly use the unary costs of P_{alldiff} inside LAP^{iu}_{Greedy}, it has to be done through the binary cost functions related to (i, u). Therefore, we move the unary costs to the binary cost functions related to x_i before solving LAP^{iu}_{Greedy} (lines 4-9). Alg. 2 has the same worst-case time and space complexity as Alg. 1. However, the greedy algorithm does not guarantee a better bound than SNC-LAP-GLB or GLB.

Theorem 3. SNC-LAP-GREEDY *preserves equivalence.*

Pruned values $v \in D_j$ found by enforcing $F\emptyset IC$ on P^{iu}_{GLB} in Alg. 1 or P^{iu}_{Greedy} in Alg. 2 become forbidden tuples in P_{alldiff}. If a value v is pruned by all singleton tests ($\forall u \in D_i$), it can be removed from P_{alldiff}, as in SAC and constructive disjunction [6,40]. Because EDAC is equivalent to classical AC on forbidden tuples [10], Alg. 1 (resp. Alg. 2) prunes such values at line 17 (resp. 19).

Example 1. Consider a CFN with 3 variables $\{x_1, x_2, x_3\}$ with 3 values each, an ALLDIFFERENT constraint over them, and 3 binary cost functions $f_{ij}(x_i = u, x_j = v) = 2 \forall i, j, u, v \in \{1,2,3\}^4, i < j, u \neq v$ (we omit unary and binary tuples that have cost 0). Both SNC-LAP-GLB and SNC-LAP-GREEDY produce an optimal lower bound $f_\emptyset$ of 6 and all the unary and binary cost functions are set to zero after reformulation.

4   Exploiting Symmetries in the Flow Matrix of the QAP

For a subset of the QAPLIB instances, it has been shown [19] that is possible to group facilities into equivalence classes. Two facilities i and j are equivalent if the following conditions are met:

$$a_{ij} = a_{ji} \tag{33}$$

$$a_{ik} = a_{jk}, a_{ki} = a_{kj} \qquad \forall k \in \{1, \ldots, n\} \setminus \{i, j\} \tag{34}$$

$$c_{iu} = c_{ju} \qquad \forall u \in \{1, \ldots, n\} \tag{35}$$

$$b_{uv} = b_{vu} \qquad \forall u, v \in \{1, \ldots, n\}^2 \tag{36}$$

In CFNs, such variables are said interchangeable. For any solution of P_{alldiff}, we can permute the values of x_i and x_j to get a new solution with the same cost.

In order to exploit the symmetries, we need to invert the CFN model.[5] In the inverted model, variables are associated to locations and values to facilities. The resulting model is called P^τ_{alldiff}.

Let us assume we have d equivalent classes $\{E_1, \ldots, E_d\}$ of size $m_k = |E_k|, k \in \{1, \ldots, d\}$. We define P^τ_{gcc}, the inverted QAP model of size n, with reduced domains of size d, grouping all facilities/values of the same equivalence class E_k into a single representative value $\varphi(k)$, and replacing ALLDIFFERENT by a Global Cardinality Constraint (GCC) [23]. In GCC, we set the upper bound

---

[5] It has been observed [19] that symmetries can only be exploited in the inverted model and not in the original model.

on the number of occurrences of $\varphi(k)$ to m_k. We define $P^\tau_{gcc} = (X, D, F)$:

$$D_i = \{1, \ldots, d\} \qquad \forall i \in \{1, \ldots, n\} \tag{37}$$

$$f_{ij}(x_i = k, x_j = k') = a_{\varphi(k)\varphi(k')}b_{ij} + a_{\varphi(k')\varphi(k)}b_{ji} \qquad \forall i, j \in \{1, \ldots, n\}^2, i < j, k, k' \in \{1, \ldots, d\}^2 \tag{38}$$

$$f_i(x_i = k) = c_{\varphi(k)i} \qquad \forall i \in \{1, \ldots, n\}, k \in \{1, \ldots, d\} \tag{39}$$

$$f_\varnothing = 0 \tag{40}$$

$$\mathrm{GCC}(X, \{m_1, \ldots, m_d\}) \tag{41}$$

By definition, $d \leq n$ and $\sum_{k=1}^d m_k = n$. The GCC is in *closed form*, *i.e.*, all variables must take a value in $\{1, \ldots, d\}$ and each value k must be used exactly m_k times.

Reformulation with the Global Cardinality Constraint. The combination of the GCC constraint in closed form and the unary cost functions $f_i \forall i \in \{1, \ldots, n\}$ corresponds exactly to the Semi-Assignment Problem (SAP) [8,25]:

$$SAP : o = \min \sum_{i=1}^n \sum_{k=1}^d c_{\varphi(k)i} x_{ik} \tag{42}$$

$$\sum_{k=1}^d x_{ik} = 1 \quad \forall i \in \{1, \ldots, n\} \tag{43}$$

$$\sum_{i=1}^n x_{ik} = m_k \quad \forall k \in \{1, \ldots, d\} \tag{44}$$

$$x_{ik} \in \{0, 1\} \quad \forall i \in \{1, \ldots, n\}, k \in \{1, \ldots, d\} \tag{45}$$

where $0/1$ variable $x_{ik} = 1$ corresponds to $x_i = k$ in P^τ_{gcc}. This problem can be solved efficiently by a modified version of LapJV algorithm in $O(dn^2)$ time [25]. As with the AllDifferent constraint, we can reformulate P^τ_{gcc}. The optimum o of the SAP is added to the problem lower bound $f_\varnothing \leftarrow f_\varnothing + o$ and the reduced costs $rc(i, k)$ of each variable x_{ik} in the SAP become the new unary costs: $f_i(x_i = k) \leftarrow rc(i, k), i \in \{1, \ldots, n\}, k \in \{1, \ldots, d\}$. This reformulation enforces $F\varnothing IC$ on the GCC. This offers the same advantages compared to GCC with costs [35] as $F\varnothing IC$ on AllDifferent does compared to Weighted AllDifferent.

We can apply SNC on P^τ_{gcc} by modifying Algorithm 1 (resp. Alg. 2) to use SAP on GCC instead of LAP on AllDifferent. The resulting algorithms SNC-SAP-GLB (resp. SNC-SAP-Greedy) have a time complexity in $O(d^2 n^3)$.

5   Experimental Results

We implemented in TOULBAR2, an open-source C++ exact WCSP solver, a propagator for enforcing $F\emptyset IC$ on the GCC in closed form, based on the modified LAPJV algorithm [25].[6] TOULBAR2 already includes a propagator for enforcing $F\emptyset IC$ on ALLDIFFERENT [38]. We implemented four methods combining each of the two propagators with each of the two strategies GLB (Alg. 1) and Greedy (Alg. 2). Due to their relatively high cost, we apply them only in preprocessing.

We compared four versions of TOULBAR2 having different preprocessing:

- no preprocessing (TOULBAR2 with default options), called LAP in the results when applied to QAP models with the ALLDIFFERENT constraint (same results as tb2+LAPJV in [38]) and SAP when applied to a model with GCC,
- SNC-LAP-GLB using the GLB strategy (options *-S -glb=1*),
- SNC-LAP-GREEDY using the Greedy strategy (options *-S -glb=0*),
- SNC-LAP-GLB followed by SNC-LAP-GREEDY (options *-S -glb=2*), called SNC-LAP-GLB-GREEDY in the results.

For each version, after preprocessing, TOULBAR2 continues the search, enforcing EDAC [12] on unary and binary cost functions, and $F\emptyset IC$ on ALLDIFFERENT and GCC, at every node of a hybrid best/depth-first branch-and-bound search [2], using binary branching and Maximum Cardinality Search for the DAC ordering.[7] TOULBAR2 used the *dom/wdeg* variable ordering heuristic [7] with last conflict [28] and EAC *support value* ordering heuristic [10,39], combined with solution phase saving [15]. Unless reported otherwise, all tests were run on a 2.5GHz Intel Xeon E5-2680 (2014), using one thread per CPU, and with a CPU-time limit of 1,200 s.

We took the 132 smallest of the 136 instances from the QAPLIB.[8] We compared the four versions of TOULBAR2 against Google OR-Tools CP-SAT, an open-source state-of-the-art constraint programming solver, and IBM CPLEX, a state-of-the-art integer programming solver.[9] Unary and binary cost functions are encoded using the support encoding for CPLEX [22], and to binary and ternary constraints in extension with extra objective variables for CP-SAT. The permutation constraint is encoded for CPLEX as binary constraints enforcing that any pair of two variables cannot take the same value. It is encoded as an ALLDIFFERENT constraint in CP-SAT and TOULBAR2. The GCC constraint is only tested using TOULBAR2 in the inverted model P_{gcc}^{τ}.[10]

---

[6] https://github.com/toulbar2/toulbar2, branch lapjv.

[7] TOULBAR2 options *-d: -o -O=-1*.

[8] http://coral.ise.lehigh.edu/wp-content/uploads/2014/07/qapdata.tar.gz, sizes less than 100 variables.

[9] https://github.com/google/or-tools CP-SAT v9.14 in free-search mode, CPLEX version 22.1.1.0 with non-premature stop parameters *EPAGAP=EPGAP=EPINT=0*. All single-threaded.

[10] CP-SAT uses a decomposition of GCC into linear constraints. Our preliminary tests showed worse results on the inverted model with the decomposed GCC.

Table 1. Initial lower bound gap percent., nb. of iter., and av. time for QAPLIB.

Preprocessing	AvgInitialLbGap	AvgSncIterations	AvgSncTimes
GLB	30.46	N/A	N/A
SNC-LAP-GLB-Greedy	**24.13**	352.53	60.76 sec.
SNC-LAP-Greedy	26.38	368.67	73.08 sec.
SNC-LAP-GLB	26.32	**160.38**	**25.46 sec.**
LAP	83.63	N/A	N/A

Table 2. Optimality gap percentages on the original model for QAPLIB.

Gap	mean	median	std	Score	Nb. instances
CPLEX	16.41	9.35	20.38	25.00	109
CP-SAT	5.12	2.29	11.00	33.00	122
LAP	4.01	1.88	5.27	41.50	**132**
SNC-LAP-Greedy	1.44	**0.00**	**2.29**	83.00	131
SNC-LAP-GLB	1.50	**0.00**	2.56	89.00	**132**
SNC-LAP-GLB-Greedy	**1.42**	**0.00**	2.33	**92.00**	131

5.1 Comparison of Initial Lower Bounds

Table 1 gives average initial lower bound gap $(1 - \frac{lb}{bestsolknown})$ found by the original GLB method and by the four preprocessing methods on P_{alldiff}. It also reports the average number of iterations on variables (for-loop line 2) and CPU-time of SNC methods. Without preprocessing, LAP had a poor lower bound. SNC methods were much better, all being superior on average to original GLB. SNC with the GLB or Greedy strategy got the same quality of lb, but SNC-LAP-GREEDY was 2-3x slower to converge than SNC-LAP-GLB (see also Fig. 2a). Hopefully, SNC-LAP-GLB-GREEDY is a bit faster than SNC-LAP-GREEDY thanks to the reformulations initially made by SNC-LAP-GLB, and it got the best lb gap. Similar results are observed on the other models $P^\tau_{alldiff}$ and P^τ_{gcc}.

For instance *tai100a*, SNC-LAP-GLB (resp. SNC-LAP-GLB-GREEDY) found an initial lower bound of 15,864,722 in 196 s and 300 iterations (resp. 15,931,389 in 405 s, 500 iter.), which are greater than the original GLB bound of 15,824,355 and the best bound of 15,844,731 reported on QAPLIB website using a semidefinite relaxation-matrix splitting bound [34].[11]

5.2 Comparison of Solving Times

On the original model P_{alldiff}, LAP solved 33 instances, whereas all SNC methods solved more than 54. SNC-LAP-GLB-GREEDY got the best results (56

---

[11] https://coral.ise.lehigh.edu/data-sets/qaplib/qaplib-problem-instances-and-solutions/#Ta.

solved). On the inverted model P^τ_{alldiff}, without exploiting symmetries, performances decreased for LAP (23 solved) and SNC-LAP-GREEDY (50 solved), and remained stable for the two other SNCs. By exploiting symmetries in P^τ_{gcc}, we observed a very nice improvement for all the methods, SNC-SAP-GLB being the best (67 solved). The largest solved instance is *lipa90b* with size $n = 90$ in 87 s and 100 nodes by SNC-SAP-GLB (72 s for its preprocessing). As previously reported [38], CFN methods performed much better than CP-SAT (24 solved) and default OR methods (CPLEX, 22 solved) (Fig. 1).

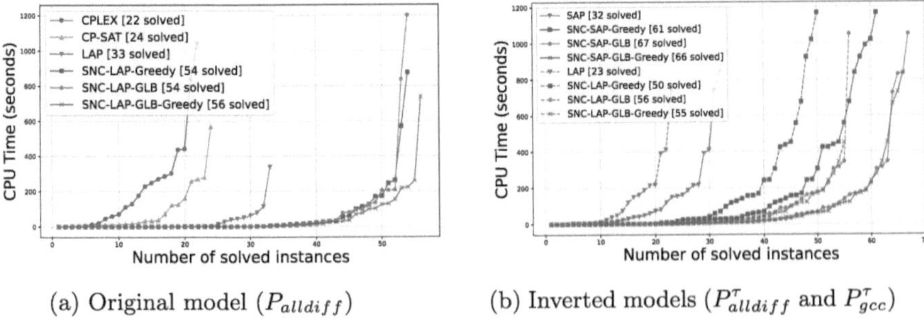

(a) Original model $(P_{alldiff})$ (b) Inverted models $(P^\tau_{alldiff}$ and $P^\tau_{gcc})$

Fig. 1. Cactus plots on the original and inverted models for QAPLIB benchmark.

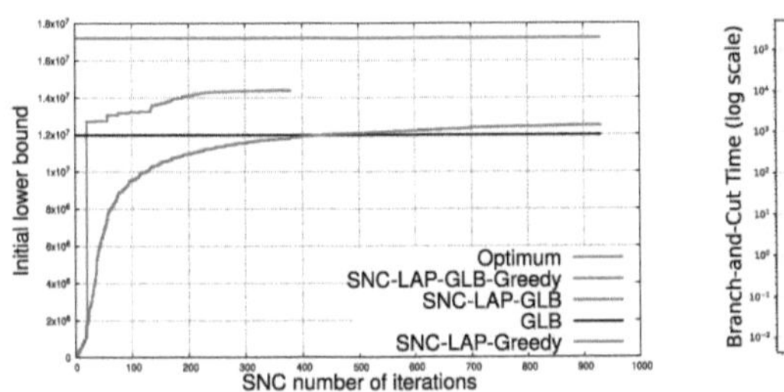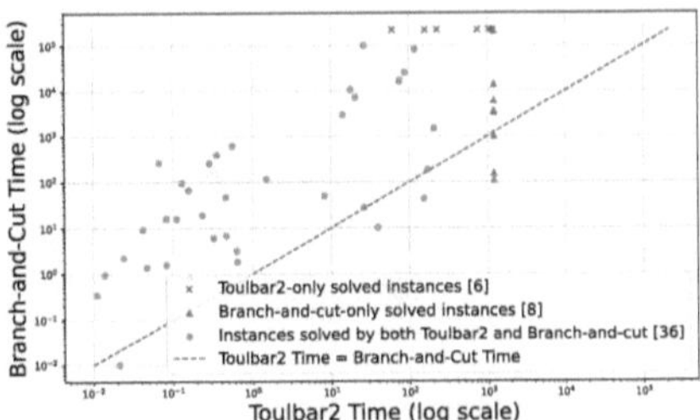

(a) Initial lower bounds obtained on QAPLIB instance *els19* using the original model. SNC-LAP-GLB-GREEDY took 0.69s and 380 iter. in preprocessing and then tb2 solved in 5.82s.

(b) TOULBAR2 vs Branch-and-Cut [18] runtime comparison. Note the drastically different hardware and timeouts (20 minutes for tb2 / 2.5 days for B&C) makes comparison difficult.

Fig. 2. Detailed results for the QAPLIB benchmark.

5.3   Comparison of Solution Quality

In Table 2, we report optimality gap distribution ($\frac{ub-best}{ub}$), XCSP3 competition score [38], and number of instances with at least one solution found on P_{alldiff}

(similar results are observed on the other models). The best average gap of 1.42% is obtained by SNC-LAP-GLB-Greedy. However, it could not finish its preprocessing in less than 20 min for one instance (*tai100b*). SNC-LAP-GLB found a solution for all 132 instances with an average gap of 1.5%. It is clearly better than LAP (4%), CP-SAT (5%), and cplex (16%).

5.4   Comparison with a Dedicated Branch-and-Cut Approach

Figure 2b gives a log-log scatter plot between our best approach using TOULBAR2 (whatever the chosen SNC preprocessing or QAP model) and a branch-and-cut (B&C) method dedicated for QAP [18], using the results reported in that paper. Among the 71 instances TOULBAR2 could solve in less than 20 min, we kept 42 of size less than 30, removing *esc* and *lipa* families as done in B&C results. B&C could solve 44 instances with $n \leq 30$, but often taking much more time than TOULBAR2 (up to 3 orders of magnitude speedups on *bur26* and *chr* families).

Table 3. Comparison on 16 QAPLIB instances with flow matrix symmetry and using the inverted model (P_{gcc}^{τ}). A "-" means CPU-time limit reached.

Instance	n	d	OPT	cplex [19]	SNC-SAP-GLB-Greedy
esc16a	16	9	68	**0.35**	0.65
esc16b	16	7	292	3.07	**2.06**
esc16c	16	12	160	130.98	**73.14**
esc16d	16	12	16	**0.51**	77.22
esc16e	16	8	28	**0.05**	1.11
esc16f	16	1	0	**0.00**	**0.00**
esc16g	16	9	26	**0.04**	0.18
esc16h	16	5	996	0.23	**0.11**
esc16i	16	10	14	**0.18**	7.88
esc16j	16	7	8	**0.03**	0.13
esc32c	32	10	642	**9,643.82**	-
esc32d	32	13	200	**2,973.26**	-
esc32e	32	6	2	**0.04**	106.33
esc32g	32	7	6	**0.06**	5.86
esc64a	64	15	116	**509.87**	-
tai64c	64	2	1,855,928	18,250.40	**995.51**

5.5   Comparison with a Dedicated MILP Approach

In [19], a dedicated MILP formulation is proposed, exploiting symmetry in the flow matrix. In Table 3, we report their results obtained using cplex 12.2 with

8 threads on a quad core Intel Xeon CPU 3.2 GHz with 16 GB RAM. We give also the results obtained by our approach (SNC-SAP-GLB-GREEDY) using TOULBAR2 with the parallel version of its hybrid best/depth-first branch-and-bound search method [4] on a 8-core Intel Xeon E5-2687W v4 3.5 GHz with 256 GB. The CPU-time limit is fixed to 10 h. The MILP approach produced the best results, solving 16 instances and TOULBAR2 only 13. However, on the large *tai64c* instance, our approach was 18 times faster than CPLEX, and developed less search nodes (162,790,320 compared to 1,216,074,081).

6   Conclusion

Taking ideas from CP (singleton consistency, global cardinality constraint) and OR (Gilmore-Lawler bound, linear and semi-assignment problems), we significantly improved a CFN solver on quadratic assignment problems, reaching state-of-the-art OR results in a few cases (mostly on asymmetric instances such as *bur26* and *lipa*, and also *tai64c*). In the future, we will explore other soft local consistency methods [10, 33] and problem formulation [1] to get stronger bounds.

Supplementary Materials. Detailed results (csv files) are available at: https://web-genobioinfo.toulouse.inrae.fr/~degivry/cpaior2026/cpaior2026 supp.zip.

Acknowledgments. This work has benefited from French grants managed by the National Research Agency under the "Investissements d'Avenir" program with the references ANR-18-EURE-0021 and ANR-23-IACL-0002 and ANR-24-CE23-3429-03.

Disclosure of Interests. The authors have no competing interests to declare that are relevant to the content of this article.

References

1. Adams, W.P., Guignard, M., Hahn, P.M., Hightower, W.L.: A level-2 reformulation-linearization technique bound for the quadratic assignment problem. Eur. J. Oper. Res. **180**(3), 983–996 (2007)
2. Allouche, D., de Givry, S., Katsirelos, G., Schiex, T., Zytnicki, M.: Anytime hybrid best-first search with tree decomposition for weighted CSP. In: Proceedings of the CP-15, pp. 12–28. Cork, Ireland (2015)
3. Barták, R., Erben, R.: A new algorithm for singleton arc consistency. FLAIRS, pp. 257–262 (2004)
4. Beldjilali, A., Montalbano, P., Allouche, D., Katsirelos, G., de Givry, S.: Parallel Hybrid Best-First Search. In: Proceedings of the CP-22, vol. 235, pp. 7:1–7:10. Haifa, Israel (2022)
5. Bessiere, C., Cardon, S., Debruyne, R., Lecoutre, C.: Efficient algorithms for singleton arc consistency. Constraints **16**(1), 25–53 (2011)
6. Bessiere, C., Debruyne, R.: Theoretical analysis of singleton arc consistency and its extensions. Artif. Intell. **172**(1), 29–41 (2008)

7. Boussemart, F., Hemery, F., Lecoutre, C., Sais, L.: Boosting systematic search by weighting constraints. In: ECAI, vol. 16, p. 146 (2004)
8. Burkard, R., Dell'Amico, M., Martello, S.: Assignment problems: revised reprint. SIAM (2012)
9. Claus, G., Cambazard, H., Jost, V.: Analysis of reduced costs filtering for alldifferent and minimum weight alldifferent global constraints. In: Proceedings of the 24th European Conference on Artificial Intelligence, vol. 325, pp. 323–330. Santiago de Compostela, Spain (2020)
10. Cooper, M., et al.: Soft arc consistency revisited. Artif. Intell. J. **174**, 449–478 (2010)
11. Cooper, M.C., de Givry, S., Schiex, T.: Valued constraint satisfaction problems. In: A Guided Tour of Artificial Intelligence Research: Volume II: AI Algorithms, pp. 185–207. Springer (2020)
12. De Givry, S., Heras, F., Zytnicki, M., Larrosa, J.: Existential arc consistency: Getting closer to full arc consistency in weighted CSPs. In: IJCAI **5**, 84–89 (2005)
13. Debruyne, R.: Optimal and suboptimal singleton arc consistency algorithms. IJCAI (2005)
14. Debruyne, R., Bessière, C.: Some practicable filtering techniques for the constraint satisfaction problem. In: Proceedings of the Fifteenth International Joint Conference on Artificial Intelligence, IJCAI 97, Nagoya, Japan, August 23–29, pp. 412–417 (1997)
15. Demirović, E., Chu, G., Stuckey, P.J.: Solution-based phase saving for CP: a value-selection heuristic to simulate local search behavior in complete solvers. In: International Conference on Principles and Practice of Constraint Programming, pp. 99–108. Springer (2018)
16. Dlask, T., Werner, T., de Givry, S.: Bounds on weighted CSPs using constraint propagation and Super-Reparametrizations. In: Proceedings of the CP-21. Montpellier, France (2021)
17. Dlask, T., Werner, T., de Givry, S.: Super-Reparametrizations of weighted CSPs: properties and optimization perspective. Constraints **28**, 277–319 (2023)
18. Erdoğan, G., Tansel, B.: A branch-and-cut algorithm for quadratic assignment problems based on linearizations. Comput. Oper. Res. **34**(4), 1085–1106 (2007)
19. Fischetti, M., Monaci, M., Salvagnin, D.: Three ideas for the quadratic assignment problem. Oper. Res. **60**(4), 954–964 (2012)
20. Frieze, A.M., Yadegar, J.: On the quadratic assignment problem. Discret. Appl. Math. **5**(1), 89–98 (1983)
21. Gilmore, P.C.: Optimal and suboptimal algorithms for the quadratic assignment problem. J. Soc. Ind. Appl. Math. **10**(2), 305–313 (1962)
22. Hurley, B., et al.: Multi-language evaluation of exact solvers in graphical model discrete optimization. Constraints **21**(3), 413–434 (2016). https://doi.org/10.1007/s10601-016-9245-y
23. Jean-Charles, R.: Generalized arc consistency for global cardinality constraint. In: American Association for Artificial Intelligence, pp. 209–215. AAAI 1996 (1996)
24. Jonker, R., Volgenant, A.: A shortest augmenting path algorithm for dense and sparse linear assignment problems. Computing **38**, 325–340 (1987)
25. Kennington, J., Wang, Z.: A shortest augmenting path algorithm for the semi-assignment problem. Oper. Res. **40**(1), 178–187 (1992)
26. Koopmans, T.C., Beckmann, M.: Assignment problems and the location of economic activities. Econometrica J. Econometric Soc. 53–76 (1957)
27. Kuhn, H.W.: The Hungarian method for the assignment problem. Naval Res. Logistics Q. **2**(1–2), 83–97 (1955)

28. Lecoutre, C., Saïs, L., Tabary, S., Vidal, V.: Reasoning from last conflict(s) in constraint programming. Artif. Intell. J. **173**, 1592–1614 (2009)
29. Lecoutre, C., Cardon, S.: A greedy approach to establish singleton arc consistency. In: IJCAI, vol. 5, pp. 199–204 (2005)
30. Loiola, E.M., De Abreu, N.M.M., Boaventura-Netto, P.O., Hahn, P., Querido, T.: A survey for the quadratic assignment problem. Eur. J. Oper. Res. **176**(2), 657–690 (2007)
31. Montalbano, P.: Linear Constraints and Conflict-free Learning for Graphical Models, Université de Toulouse (2023). Ph.D. thesis
32. Montalbano, P., de Givry, S., Katsirelos, G.: Multiple-choice knapsack constraint in graphical models. In: International Conference on Integration of Constraint Programming, Artificial Intelligence, and Operations Research, pp. 282–299. Springer (2022)
33. Montalbano, P., de Givry, S., Katsirelos, G.: Virtual arc consistency for linear constraints in cost function networks. In: Proceedings of the ICTAI-25. Athens, Greece (2025)
34. Peng, J., Mittelmann, H., Li, X.: A new relaxation framework for quadratic assignment problems based on matrix splitting. Math. Program. Comput. **2**(1), 59–77 (2010)
35. Régin, J.C.: Cost-based arc consistency for global cardinality constraints. Constraints **7**(3), 387–405 (2002)
36. Rossi, F., Van Beek, P., Walsh, T.: Handbook of constraint programming, Elsevier (2006)
37. Sellmann, M.: An arc-consistency algorithm for the minimum weight all different constraint. In: Van Hentenryck, P. (ed.) CP 2002. LNCS, vol. 2470, pp. 744–749. Springer, Heidelberg (2002). https://doi.org/10.1007/3-540-46135-3_56
38. Sewa, G., Allouche, D., de Givry, S., Katsirelos, G., Montalbano, P., Schiex, T.: Assignment problems in cost function networks. In: Proceedings of the AAAI-26. Singapore (2026)
39. Trösser, F., De Givry, S., Katsirelos, G.: Relaxation-aware heuristics for exact optimization in graphical models. In: International Conference on Integration of Constraint Programming, Artificial Intelligence, and Operations Research, pp. 475–491. Springer (2020)
40. Van Hentenryck, P., Saraswat, V., Deville, Y.: Design, implementation, and evaluation of the constraint language cc (FD). J. Logic Program. **37**(1–3), 139–164 (1998)
41. Xia, Y.: Gilmore-Lawler bound of quadratic assignment problem. Front. Math. China **3**(1), 109–118 (2008)

Exact Certification of Data-Poisoning Attacks Using Mixed-Integer Programming

Philip Sosnin[1] , Jodie Knapp[2], Fraser Kennedy[2], Josh Collyer[2],
and Calvin Tsay[1(✉)]

[1] Department of Computing, Imperial College London, London, UK
{p.sosnin23,c.tsay}@imperial.ac.uk
[2] The Alan Turing Institute, London, UK

Abstract. This work introduces a verification framework that provides both sound and complete guarantees for data poisoning attacks during neural network training. We formulate adversarial data manipulation, model training, and test-time evaluation in a single mixed-integer quadratic programming (MIQCP) problem. Finding the global optimum of the proposed formulation provably yields worst-case poisoning attacks, while simultaneously bounding the effectiveness of all possible attacks on the given training pipeline. Our framework encodes both the gradient-based training dynamics and model evaluation at test time, enabling the first exact certification of training-time robustness. Experimental evaluation on small models confirms that our approach delivers a complete characterization of robustness against data poisoning.

Keywords: Data Poisoning · Verification · Mixed-integer Programming

1 Introduction

Data poisoning attacks pose a fundamental threat to the integrity of machine learning systems. By injecting maliciously crafted samples into training data, adversaries can manipulate model behavior in ways that persist through deployment [7,38]. Recent work has shown that maliciously crafted attacks affecting less than 1% of training samples can severely compromise model predictions [18,49]. These vulnerabilities create an urgent need for formal verification methods that provide provable guarantees about training-time robustness.

While the study of inference-time robustness has led to a mature field of certified defences and verification algorithms [10,13,20,32,45,48], the problem of certifying robustness to training-time perturbations remains comparatively under-explored. In contrast to test-time attacks, where adversarial inputs are evaluated on a fixed model, poisoning attacks alter the learning process itself, potentially influencing every parameter of the trained model and every subsequent prediction. This fundamental coupling between data, optimization, and

© The Author(s), under exclusive license to Springer Nature Switzerland AG 2026
T. Guns (Ed.): CPAIOR 2026, LNCS 16595, pp. 538–556, 2026.
https://doi.org/10.1007/978-3-032-27242-3_32

model behavior makes certification of training-time robustness uniquely challenging. Several recent efforts have introduced alternative approaches to certifying training-time robustness through modified model architectures or inference procedures. For example, aggregation-based defences [22,29,41] provide robustness guarantees for large ensembles of models, and randomized smoothing at inference time can provide guarantees for linear and kernel models [30]. While effective in certain limited settings, these methods rely on strong assumptions about model structure and do not extend naturally to generic training settings.

To overcome these limitations, our previous work introduces a certification framework capable of reasoning about arbitrary training pipelines. In particular, our Abstract Gradient Training (AGT) [33,35,43] framework uses relaxations of the training dynamics that propagate bounds through gradient updates. More recently, MIBP-CERT [24,25] takes an optimization-centered approach to these bounds. These methods formulate training-time certification as a reachability analysis problem, applying interval or optimization-based relaxations at each training iteration to over-approximate the effect of data perturbations on the final learned parameters. Such relaxations allow for computationally scalable certification, and enable general-purpose frameworks for bounding the effect of training-time perturbations without modifying the training algorithm itself.

A key challenge, however, lies in the trade-off between *soundness* and *completeness*. A certification algorithm is *sound* if every guarantee it produces is valid, i.e., no false assurances are made, but may be *incomplete* if it cannot capture all possible safe behaviors. Sound but incomplete methods, such as AGT and MIBP-CERT, rely on relaxations that over-approximate the reachable set of parameters at each training iteration. While this improves the computational scalability to large models, it can lead to vacuous or overly conservative bounds.

In this work, we close this gap by introducing the first sound and complete verification algorithm for robustness to training-time attacks. Our key contribution is a novel formulation of the joint trainingattackevaluation process as a single *mixed-integer quadratic constrained program (MIQCP)*, taking advantage of the sound-and-complete nature of mixed-integer programming solvers. Our formulation exactly encodes the interactions between adversarial data manipulations, gradient-based optimization steps, and test-time objectives, allowing us to reason about a *given training procedure* (i.e., fixed initialization and data ordering) within a single optimization problem. By solving this MIQCP to optimality, we can compute both provably optimal poisoning attacks and exact robustness guarantees for any model trained under the specified threat model.

Contributions. Our main contributions are summarised as follows:

- We introduce the first sound and complete verification framework for data poisoning attacks on gradient-trained models, formulated as a MIQCP.
- We develop several tailored MIQCP approaches, including reformulations, heuristics, and bound tightening approaches for the proposed framework.
- We demonstrate, through empirical evaluation, that our approach achieves exact certification for small, linear models.

While computationally expensive compared to incomplete methods, our proof-of-concept framework can help reveal the structure of optimal poisoning attacks, identify where relaxations in existing certification tools introduce looseness, and inspire new certification and attack algorithms based on MIQCP.

The remainder of this paper is structured as follows. Section 2 reviews related work on poisoning attacks and certification. Section 3 formalizes our problem setting. Sections 4–5 introduce our MIQCP-based verification framework and various improved solution strategies. Finally, Sect. 6 presents empirical results.

2   Related Works

2.1   Inference-Time Adversarial Attacks

Deep neural networks are highly vulnerable to small, carefully designed input perturbations that cause misclassification while remaining nearly imperceptible to humans [15,37]. Such *adversarial attacks* have motivated *formal verification* methods that provide provable robustness guarantees within bounded perturbation regions [16,45,48]. Formally, given a trained machine learning model f_θ and a property $\phi(x, f_\theta(x))$, verification aims to determine whether ϕ holds for all x in some set $\mathcal{X}$. An algorithm is *sound* if it never certifies a false property and *complete* if it can always identify ϕ holds when true.

Incomplete Methods. Incomplete approaches, often referred to as *certification* methods, ensure soundness by over-approximating the possible outputs of a network under the given set of input perturbations. Representative techniques include interval bound propagation [16], convex relaxations [42,48], and abstract interpretation frameworks [32], which can scale to large architectures and be integrated into certified training pipelines [26,44]. However, these over-approximations are typically conservative, often yielding vacuous guarantees even when the model is robust.

Complete Methods. Complete verification algorithms compute *exact* robustness guarantees by exhaustively reasoning over all possible inputs and activation patterns. These methods typically encode the network and property as a constrained optimization problem using mixed-integer programming (MIP) or satisfiability modulo theory (SMT) solvers [6,13,20,21,40]. Representative examples include ReLUplex [21], Planet [12], and NeuralSAT [9]. These approaches are both sound and complete, but can incur high computational costs [2,34,39]. Our work builds directly on this line of complete methods, extending MIP from inference-time robustness to verification over training-time perturbations.

2.2   Adversarial Data Poisoning Attacks

Data poisoning attacks manipulate data *during training* to degrade performance or induce targeted behaviors [3,4,28]. Attacks are typically categorized as *untargeted* (reduce model performance), *targeted* (misclassify specific inputs), or *backdoor* [38]. Backdoor attacks introduce hidden triggers that cause misclassification only when specific patterns are present [8,17,18], and even small fractions of poisoned data can cause severe failures [47,49]. Adversaries may be bounded or unbounded and are often assumed to have full knowledge of the model and training process. Defending against poisoning during training is challenging, as traditional strategies are often attack- or model-specific. These include training classifiers to detect poisoned inputs [23], applying noise or clipping to limit perturbations [19], or combining multiple models via disjoint dataset partitioning [22,29,41]. While effective in specific scenarios, these approaches either fail to generalize to novel attacks or do not apply to general training settings.

On the other hand, certified robustness methods aim to provide guarantees against general poisoning strategies. For linear models, bounds have been established for gradient-based and ℓ_2 perturbation attacks [30,36], and differential privacy can provide statistical guarantees in limited settings [46]. More recent methods extend these guarantees to neural networks using interval bound propagation and MIP [24,25,33,35], though the relaxations used result in incomplete certificates. The work of [31] computes sound and complete certificates for label poisoning in graph neural networks, but with respect to a Neural Tangent Kernel linearisation, which is only exact in the infinite-width limit; for finite-width networks this approximation may yield incomplete guarantees.

Our approach provides the first general framework for sound and complete certification of standard, unmodified training procedures under a broad class of poisoning threats described in Sect. 3.2. We explicitly state the assumptions required for our guarantees and their implications in Sect. 3.3.

3 Background

We first review the necessary background on supervised neural network training and formal methods relevant to poisoning robustness. In particular, we consider a supervised learning setting in which a model $f(x, \theta)$ with parameters θ maps inputs x from a feature space to outputs y in a target space, which may be discrete, continuous, and/or multivariate. The model parameters θ are trained using a labelled dataset $\mathcal{D} = \{(x^{(i)}, y^{(i)})\}_{i=1}^{N}$ of N input-output pairs.

3.1 Neural Networks

While our framework is general, we focus our exposition on feed-forward neural networks with ReLU activation functions. A feed-forward network with K layers is defined by parameters $\theta = \{(W_k, b_k)\}_{k=1}^{K}$ and the layer-wise transformations

$$u_k = W_k z_{k-1} + b_k, \qquad z_k = \sigma(u_k), \quad \forall k \in 1, ..., K, \tag{1}$$

with input $z_0 = x$, output $f(x, \theta) = u_K$, and activation function σ, which we take to be ReLU.

Model parameters are typically learned via gradient-based optimization to minimize a loss function $\mathcal{L}$ that measures discrepancy between predictions and labels. At each iteration t, SGD updates parameters on mini-batches $\mathcal{B}^{(t)} \subset \mathcal{D}$:

$$\theta^{(t)} = \theta^{(t-1)} - \frac{\alpha^{(t)}}{|\mathcal{B}^{(t)}|} \sum_{(x^{(i)}, y^{(i)}) \in \mathcal{B}^{(t)}} \nabla_\theta \mathcal{L}\big(f(x^{(i)}, \theta^{(t-1)}), y^{(i)}\big), \tag{2}$$

where $\alpha^{(t)} > 0$ is the learning rate at iteration t. We denote the parameters obtained by training f on $\mathcal{D}$ (with some fixed training hyperparameters, parameter initialization, and data-ordering) as $\theta = M(\mathcal{D})$.

Although we restrict attention to vanilla SGD for clarity, other gradient-based training algorithms (e.g., momentum, Adam, or RMSProp) can be encoded within the same framework by introducing the corresponding auxiliary state variables and update equations, at the cost of increased formulation size.

3.2 Data Poisoning Threat Models

We formalize poisoning adversaries through a *perturbation model* $\mathcal{T}(\mathcal{D})$, which defines the set of possible modifications to the training dataset. We assume a white-box setting, where the adversary has full knowledge of the training data, model architecture, hyperparameters and parameter values at each training iteration. This white-box setting encompasses dynamic adversaries who can adaptively choose their poisoning objectives during training based on the evolving model state [5]. We consider poisoning attack capabilities in two broad classes:

1) Bounded Perturbations. In the bounded attack setting, an adversary can modify up to n training samples within pre-defined constraints around the features and/or labels of the clean training samples. Formally, the perturbed dataset $\tilde{\mathcal{D}} = \{(\tilde{x}^{(i)}, \tilde{y}^{(i)})\}_{i=1}^N \in \mathcal{T}_{\text{bounded}}^{n,\epsilon,\nu}(\mathcal{D})$ satisfies

$$\|x^{(i)} - \tilde{x}^{(i)}\|_p \leq \epsilon, \quad \|y^{(i)} - \tilde{y}^{(i)}\|_q \leq \nu, \quad \forall i \in \mathcal{I}, \quad |\mathcal{I}| \leq n, \tag{3}$$

where $\mathcal{I}$ indexes the modified samples. We take the feature perturbation norm to be $p = \infty$, and the label perturbation norm to be $q = \infty$ and $q = 0$ for regression and classification tasks, respectively. Bounded perturbations include *clean-label attacks* ($\nu = 0$), where only features are modified, and *label-flipping attacks* ($\epsilon = 0$), where only labels are altered [27,36,49].

2) Arbitrary (Unbounded) Perturbations. In a more general threat model, an adversary may arbitrarily replace data points from the training set. In this setting, the adversary can substitute up to n training samples with entirely arbitrary points unrelated to the removed data, and may optionally modify their associated labels:

$$\tilde{\mathcal{D}} = (\mathcal{D} \setminus \mathcal{S}) \cup \tilde{\mathcal{S}}, \quad |\mathcal{S}| = |\tilde{\mathcal{S}}| \leq n, \tag{4}$$

where $\mathcal{S}$ contains the removed samples and $\tilde{\mathcal{S}}$ contains the injected points. This formulation encompasses fully general poisoning attacks, with clean-label and label-flipping attacks as special cases when the injected points preserve or modify labels accordingly [8,17,27]. To ensure that any resulting verification problem remains bounded, we assume the injected samples to lie within a bounded domain. Unlike the bounded perturbation model above, these bounds are independent of the replaced data samples, and can be chosen to cover the entire input space if the domain is naturally constrained (e.g., physical measurements) or pre-clipped, effectively allowing arbitrary substitutions.

Remark 1. For both threat models, we assume the ordering (batching) of training samples remains fixed before and after poisoning. Specifically, each poisoned sample appears at the same position within mini-batches during training as its corresponding original sample, preserving the sequence in which data are presented across all training iterations.

Adversarial Objectives. We assume the adversary seeks to maximize an objective corresponding to a post-training measure of attack success:

- *Untargeted attacks* aim to degrade overall model accuracy on a chosen dataset. A special case is the *denial of service attack*, which aims to prevent model convergence on the training set.
- *Targeted attacks* aim to force specific outputs on selected inputs while leaving other predictions unaffected.
- *Backdoor attacks* aim to cause misclassification when a trigger is present, while keeping performance on clean inputs unchanged.

We denote the attack objective as a function $J(\theta)$ evaluated on the final trained model. For instance, the test error on a dataset $\{x_{\text{test}}^{(i)}, y_{\text{test}}^{(i)}\}_{i=1}^{N\text{test}}$ can be written as $J(\theta) = \sum_{i=1}^{N_{\text{test}}} \mathbf{I}\left\{ f^\theta(x_{\text{test}}^{(i)}) \neq y_{\text{test}}^{(i)} \right\}$.

3.3   Verification Problem

Our verification framework aims to certify that a machine learning model satisfies specified properties despite adversarial perturbations to its training data. We begin by establishing the assumptions under which our certificates are valid[1]:

- *Assumption 1: Fixed Initialization and Data Ordering.* We assume fixed model parameter initialization and a fixed ordering of training samples. Consequently, our guarantees apply to a specific random seed controlling both initialization and data order during training.
- *Assumption 2: White-Box Adversary with Constrained Influence.* We assume the adversary has complete knowledge of training hyperparameters, model architecture, parameter initialization, data ordering, and the training dataset. However, the adversary's influence is limited exclusively to the allowable perturbations of training data as defined by the perturbation model.

---

[1] We note that these assumptions match those adopted in [24,25,33,35].

Let $\mathcal{D}$ denote the original dataset, M the training procedure (including parameter initialization, data ordering, and hyperparameters), and $\mathcal{T}$ the perturbation model, as described above. The optimal adversarial manipulation can be written:

$$\max_{\tilde{\mathcal{D}}} J(\tilde{\theta}) \quad \text{s.t.} \quad \tilde{\theta} = M(\tilde{\mathcal{D}}), \quad \tilde{\mathcal{D}} \in \mathcal{T}(\mathcal{D}). \tag{5}$$

Computing the global optimum of this problem captures the worst-case effect of allowable perturbations, with respect to $J(\cdot)$. Prior works [24,33,35] compute provably valid upper bounds, rather than this worst case exactly.

Mixed-Integer Programming. We formulate this problem as a mixed-integer quadratically constrained program (MIQCP). Unlike interval or convex relaxation methods, MIQCP explicitly encodes discrete decisions and non-linear dependencies across the entire training procedure, allowing global optimization over all possible perturbations. This comes at a computational cost, as MIP is NP-Hard in general. Formally, a mixed-integer program involves continuous variables $a \in \mathbb{R}^{n_c}$, integer variables $b \in \mathbb{Z}^{n_i}$, and constraints $g_j(a,b)$:

$$\min h(a,b) \quad \text{s.t.} \quad g_j(a,b) \leq 0, \quad j = 1,\ldots,m, \tag{6}$$

where the h and g_j may include linear, quadratic, or bilinear terms.

While [24,35] encode individual training iterations as MIPs, our approach encodes the *entire training and testing procedure* in a single optimization problem, thus incurring no over-approximation.

4 A Mixed-Integer Formulation for Verification of Training-Time Attacks

Our formulation comprises three components: (1) *data-perturbation constraints*, defining the allowed training-set manipulations; (2) *training-dynamics constraints*, encoding gradient computation and parameter updates; and (3) *test-time evaluation constraints*, encoding the adversarial objective over the trained model.

4.1 Encoding Training Data Perturbations

For each sample in $\{x^{(i)}, y^{(i)}\}_{i=1}^{N}$, we introduce perturbed variables $(\tilde{x}^{(i)}, \tilde{y}^{(i)})$ and a binary indicator $s^{(i)} \in \{0,1\}$ denoting whether sample i is modified. Here we provide formulations for both threat models described in Sect. 3.2 assuming $y^{(i)} \in \{0,1\}$, but these may be extended to multi-class and regression settings. The adversarial budget n is enforced by the constraints $\sum_i s^{(i)} \leq n$.

Bounded Feature and Label Perturbations. Perturbations under the bounded threat model can be defined using the following constraints:

$$\left. \begin{aligned} x^{(i)} - \varepsilon\, s^{(i)} &\leq \tilde{x}^{(i)} \leq x^{(i)} + \varepsilon\, s^{(i)}, \\ y^{(i)}(1 - s^{(i)}) &\leq \tilde{y}^{(i)} \leq y^{(i)}(1 - s^{(i)}) + s^{(i)}, \\ \tilde{x}^{(i)} \in \mathbb{R}^d, \quad &\tilde{y}^{(i)}, s^{(i)} \in \{0,1\}, \end{aligned} \right\} \quad i = 1,\ldots,N. \tag{7}$$

Arbitrary Substitutions. For the arbitrary substitution threat model, each sample may be replaced by any point drawn from some MIQCP-representable domain $\mathcal{X}^a$. Using big-M constants $L^x, U^x \in \mathbb{R}^d$, which satisfy $L_j^x \leq x_j \leq U_j^x$ for all $x \in \mathcal{X}^a$, we encode:

$$
\left.
\begin{aligned}
x^{(i)}(1 - s^{(i)}) + L^x s^{(i)} \;\leq\; \tilde{x}^{(i)} \;&\leq\; x^{(i)}(1 - s^{(i)}) + U^x s^{(i)}, \\
y^{(i)}(1 - s^{(i)}) \;\leq\; \tilde{y}^{(i)} \;&\leq\; y^{(i)}(1 - s^{(i)}) + s^{(i)}, \\
\tilde{x}^{(i)} \in \mathcal{X}^a, \quad \tilde{y}^{(i)}, s^{(i)} &\in \{0, 1\},
\end{aligned}
\right\} \quad i = 1, \ldots, N. \quad (8)
$$

4.2 Encoding the Model Training Dynamics

We denote weight matrices $W_k^{(t)}$, biases $b_k^{(t)}$, pre-activations $u_k^{(t)}$, and activations $z_k^{(t)}$ at iteration t for layers $k = 1, \ldots, K$. Then, denoting $\mathcal{I}^{(t)}$ as the index set that specifies samples included in the SGD batch according to the fixed data ordering, $z_0^{(t,i)} = \tilde{x}^{(i)}, i \in \mathcal{I}^{(t)}$ are the inputs considered at iteration t.

Forward Pass. At each layer k of the neural network, the pre-activation vector is given by the bilinear constraint (W_k are variable cf. inference-time verification):

$$
u_k^{(t,i)} = W_k^{(t)} z_{k-1}^{(t,i)} + b_k^{(t)}. \tag{9}
$$

To encode the ReLU activations $z_k^{(t,i)} = \max\{0, u_k^{(t,i)}\}$, we introduce activation binaries $a_k^{(t,i)} \in \{0,1\}^{n_k}$. For each neuron in the network, we require big-M constants $L_k^{(t,i,\mathrm{relu})}, U_k^{(t,i,\mathrm{relu})}$ satisfying $L_k^{(t,i,\mathrm{relu})} \leq u_k^{(t,i)} \leq U_k^{(t,i,\mathrm{relu})}$.[2] Then, the big-$M$ formulation of the ReLU activation function is encoded by the following:

$$
\left.
\begin{aligned}
u_k^{(t,i)} \leq z_k^{(t,i)} &\leq u_k^{(t,i)} - L_k^{(t,i,\mathrm{relu})} \odot \left(1 - a_k^{(t,i)}\right), \\
0 \leq z_k^{(t,i)} &\leq U_k^{(t,i,\mathrm{relu})} \odot a_k^{(t,i)}.
\end{aligned}
\right\}
\quad
\begin{aligned}
&i \in \mathcal{I}^{(t)}, \\
&k = 1, \ldots, K, \\
&t = 0, \ldots, T.
\end{aligned}
\quad (10)
$$

Here we use $\odot$ to denote element-wise multiplication. When $a_{k,j}^{(t,i)} = 1$, the jth neuron in layer k is active and $z_{k,j}^{(t,i)} = u_{k,j}^{(t,i)}$; when $a_{k,j}^{(t,i)} = 0$ the unit is inactive and $z_{k,j}^{(t,i)} = 0$. Overall, (9) is quadratic and (10) is mixed-integer linear.

Training Loss. We select the hinge loss $\mathcal{L}(\hat{y}, y) = \mathrm{ReLU}\left(1 - (2y - 1)\hat{y}\right)$ as our training loss function, since it is readily MIP-representable (we use $2y - 1$ to transform the labels into the domain $\{-1, 1\}$). For training sample i at iteration t with predicted logit $\hat{y}^{(t,i)} = u^{(K,i)} \in \mathbb{R}$, we define the auxiliary variables $r^{(t,i)} = 1 - (2\tilde{y}^{(i)} - 1)\hat{y}^{(t,i)}$. We introduce hinge-activation binary variables $h^{(t,i)} \in \{0,1\}$ and big-M constants $L^{(t,i,\mathrm{hinge})}, U^{(t,i,\mathrm{hinge})}$ that bound $r^{(t,i)}$. Then, the

---

[2] The big-M constants can be obtained using any sound bounding method, such as interval bound propagation. Existing incomplete certifiers (e.g., [33]) naturally provide such bounds, which can be further tightened using bound tightening approaches.

ReLU constraint can be represented using the same big-M formulation described in (10):

$$r^{(t,i)} \leq \mathcal{L}^{(t,i)} \leq r^{(t,i)} - L^{(t,i,\text{hinge})}\left(1 - h^{(t,i)}\right), \quad \left.\begin{array}{l} \\ \\ \end{array}\right\} \quad \begin{array}{l} i \in \mathcal{I}^{(t)}, \\ t = 0,\ldots,T. \end{array} \tag{11}$$
$$0 \leq \mathcal{L}^{(t,i)} \leq U^{(t,i,\text{hinge})} h^{(t,i)}.$$

The derivative of the loss with respect to each predicted logit can be defined using the bilinear constraints

$$\frac{\partial \mathcal{L}^{(t,i)}}{\partial \hat{y}^{(t,i)}} = -2\tilde{y}^{(i)} h^{(t,i)}. \tag{12}$$

The equivalent training loss constraints for regression settings with the squared error loss are simply $\mathcal{L}^{(t,i)} = (\hat{y}^{(t,i)} - \tilde{y}^{(i)})^2$ and $\partial \mathcal{L}^{(t,i)}/\partial \hat{y}^{(t,i)} = 2(\hat{y}^{(t,i)} - \tilde{y}^{(i)})$ for all $i \in \mathcal{I}^{(t)}$ and $t = 0,\ldots,T$. Unlike the classification case, these constraints do not require auxiliary binary variables or the big-M formulation.

Backward Pass. The constraints for the backward pass follow the standard backpropagation rules, which can be written as linear and bilinear constraints. The activation patterns of the ReLU operations are already captured through the binary variables in (10), which can be reused in the backward pass. Again letting $\odot$ denote element-wise multiplication, the gradients for layer k satisfy:

$$\left.\begin{array}{l} \dfrac{\partial \mathcal{L}^{(t,i)}}{\partial u_k^{(t,i)}} = \dfrac{\partial \mathcal{L}^{(t,i)}}{\partial z_k^{(t,i)}} \odot a_k^{(t,i)}, \ \dfrac{\partial \mathcal{L}^{(t,i)}}{\partial z_{k-1}^{(t,i)}} = \left(W_k^{(t)}\right)^{\top} \dfrac{\partial \mathcal{L}^{(t,i)}}{\partial u_k^{(t,i)}}, \\[4mm] \dfrac{\partial \mathcal{L}^{(t,i)}}{\partial W_k^{(t)}} = \dfrac{\partial \mathcal{L}^{(t,i)}}{\partial u_k^{(t,i)}} \left(z_{k-1}^{(t,i)}\right)^{\top}, \ \dfrac{\partial \mathcal{L}^{(t,i)}}{\partial b_k^{(t)}} = \displaystyle\sum_{i \in \mathcal{I}^{(t)}} \dfrac{\partial \mathcal{L}^{(t,i)}}{\partial u_k^{(t,i)}}. \end{array}\right\} \quad k = 1,\ldots,K. \tag{13}$$

Parameter Updates. For learning rates $\alpha^{(t)}$ and mini-batches indexed by $\mathcal{I}^{(t)}$, the parameter updates are encoded by the linear constraint

$$\theta^{(t)} = \theta^{(t-1)} - \frac{\alpha}{|\mathcal{I}^{(t)}|} \sum_{i \in \mathcal{I}^{(t)}} \delta^{(t,i)}, \qquad t = 1,\ldots,T, \tag{14}$$

where $\delta^{(t,i)}$ stacks all layerwise parameter gradients.

4.3   Encoding Post-Training Evaluation

We now formulate the adversarial attack goal as the objective function of our optimization problem. We focus on untargeted attacks, giving examples for attacks that degrade test-time performance or cause denial of service. Extensions to targeted/backdoor objectives follow the same pattern and are discussed below.

Degrading Model Accuracy. First, we present the constraints required to represent an attacker that wishes to maximize the classification error on a particular test set $\{(x_{\text{test}}^{(i)}, y_{\text{test}}^{(i)})\}_{i=1}^{N_{\text{test}}}$. For each test sample i, we encode a forward pass through the network using the final weights $W_k^{(T)}, b_k^{(T)}$. Let $z_K^{(i,\text{test})}$ denote the final-layer logits for test example i obtained via the forward-pass constraints (9)–(10). The model's binary prediction is then

$$p^{(i,\text{test})} = \begin{cases} 1 & \text{if } z_K^{(i,\text{test})} \geq 0, \\ 0, & \text{otherwise.} \end{cases} \tag{15}$$

We introduce one binary variable $p^{(i,\text{test})} \in \{0,1\}$ into our MIQCP formulation per test example to encode this model prediction. Using big-M constants $L^{(i,\text{test})}, U^{(i,\text{test})}$, the predicted label is encoded by:

$$L^{(i,\text{test})} \left(1 - p^{(i,\text{test})}\right) \leq z_K^{(i,\text{test})} \leq U^{(i,\text{test})} p^{(i,\text{test})} - \epsilon \left(1 - p^{(i,\text{test})}\right) \tag{16}$$

where $\epsilon > 0$ is any small constant used to break ties at 0. Thus $p^{(i,\text{test})} = 1$ implies a class 1 prediction, while $p^{(i,\text{test})} = 0$ corresponds to class 0. Finally, we can define the total test error of the model to be

$$J\left(\theta^{(T)}\right) = \sum_{i:y_{\text{test}}^{(i)}=1} p^{(i,\text{test})} + \sum_{i:y_{\text{test}}^{(i)}=0} 1 - p^{(i,\text{test})}. \tag{17}$$

Denial of Service. A second attack objective seeks to degrade training convergence, thereby rendering the trained model ineffective. We encode this goal using the training loss variables $\mathcal{L}^{(t,i)}$ defined earlier. While an attacker could target specific training iterations (e.g., losses in the final epoch), here we formulate the objective as maximizing the cumulative training loss given by $\sum_{t=1}^{T} \sum_{i \in \mathcal{T}^{(t)}} \mathcal{L}^{(t,i)}$. This causes the optimization to choose poisoning samples that impair the learning process across all training steps, rather than just the final iteration.

Alternative Attack Goals. The formulations above extend naturally to targeted and backdoor attack objectives. For a targeted attack, instead of penalizing disagreement with the true label, the binary prediction variables are encouraged to match a specified target label for each test point, and the objective maximizes the number of induced target classifications. For a backdoor attack, certificates can be obtained by additionally permitting adversarial manipulation of the test-time inputs themselves, thereby encoding the possibility of injected trigger patterns when evaluating the model's predictions.

Remark 2. In our complete MIQCP formulation, any feasible primal solution corresponds to a valid poisoning attack, while any dual bound provides a sound certificate on the worst-case poisoning. Both the optimal attack and the exact certificate are obtained by solving the MIQCP. Consequently, existing data poisoning attacks can supply primal solutions to the MIP solver, and, conversely, primal solutions found by the solver directly constitute strong poisoning attacks. Similarly, tighter relaxations improve the resulting certificates given by the dual bound. Section 5.1 demonstrates poisoning heuristics for improving primal solutions, while Sects. 5.2–5.3 describe methods for tightening the formulation.

5   Improved Formulations and Solution Strategies

MIQCPs are generally solved using a branch-and-bound algorithm, where computational efficiency depends on two primary factors: the tightness of the continuous relaxations, and the quality of primal feasible solutions used in pruning and node selection strategies. This section introduces strategies to address these challenges, yielding faster and more reliable certification performance.

5.1   Local Search Primal Heuristic

Standard branch-and-bound solvers treat all decision variables as equally free to vary subject to their constraints, without exploiting any problem-specific structure. In our training-time certification problem, however, the entire objective and all relevant variables are ultimately a deterministic function of the manipulated training data. In other words, significant effort can be spent exploring parameter variables whose values are entirely specified once the label or feature perturbations are fixed. By focusing the search on feasible perturbations, a primal heuristic can more efficiently explore the space of possible solutions.

To exploit this problem structure, we develop a specialized primal heuristic that performs a local search over the poisoning assignment binaries $s^{(i)}$ using fast, batched retraining of the model. The heuristic starts from the current incumbent provided by the MIP solver and explores a sequence of Hamming-ball neighborhoods around this center. At each search radius, it constructs candidate poisoning assignments by swapping a small number of poisoned and clean labels, while preserving the adversary's budget. Each batch of candidates is then evaluated in parallel by retraining under a batched GPU-based SGD procedure, yielding their corresponding test losses. Whenever a candidate improves upon the current incumbent solution, the heuristic immediately returns the updated poisoning assignment to the solver; if no improvement is found, the search radius expands until either a new incumbent arrives from the solver or all feasible neighborhoods have been exhausted.

In certain cases (namely, label-flipping attacks), the primal heuristic can prove optimality by exhaustive search when the search radius exceeds the total poisoning budget. In these instances, the best candidate returned by the heuristic is certifiably optimal. Due to the combinatorial growth of the search space, this situation arises only for very small budgets, e.g., $n \leq 4$. Future work may explore incorporating existing poisoning attacks to further improve primal heuristics.

5.2   Bounds Tightening Within the Test Data Hull

The strength of the continuous relaxation in a Mixed-Integer Quadratic Program (MIQCP) depends heavily on the initial bounds of continuous variables and the resulting big-M constants. To improve these bounds before optimization, *Optimization-Based Bounds Tightening* (OBBT) is commonly applied [14,34]. OBBT refines variable domains by solving auxiliary optimization problems that yield tighter upper and lower bounds.

In classification settings, the constraints with largest effect on the continuous relaxation are those defining the test predictions $p^{(i,\text{test})}$, which rely on big-M constants $L^{(i,\text{test})}$ and $U^{(i,\text{test})}$ that bound the final-layer logits $\hat{y}^{(t,i)}$. These bounds accumulate looseness from all preceding layers, and tightening them is crucial for obtaining a strong dual bound.

Directly tightening the logit bounds for all N^{test} test samples requires solving $2N^{\text{test}}$ MIQCPs, which is often impractical because each has complexity comparable to the original problem. Shorter time limits produce valid, but looser big-M constants. To balance this trade-off, we introduce a single auxiliary input variable $\boldsymbol{x}^{\text{aux}}$ constrained to lie in the integer hull of the test set:

$$\boldsymbol{x}^{\text{aux}} \in \text{IntHull}\Big\{ \boldsymbol{x}^{(1,\text{test})}, \dots, \boldsymbol{x}^{(N^{\text{test}},\text{test})} \Big\}. \tag{18}$$

We then encode its forward pass to obtain $\hat{y}^{\text{aux}}$ and solve two bounding problems (min/max). The resulting values L^{aux} and U^{aux} jointly bound the logits of *all* test samples. This reduces the cost to two sub-problems, allowing longer sub-problem solve times and yielding tighter big-M constants.

This idea extends naturally by clustering test points into P sub-groups (e.g., by class) and constructing two auxiliary sub-problems per group, solving the resulting $2P$ problems trades additional computation for tighter per-group bounds.

5.3   Auxiliary Variable Formulation

The arbitrary substitution threat model is the most expressive, but also the most computationally challenging. When the decision variables $\tilde{x}^{(i)}, \dots, \tilde{x}^{(N)}$ corresponding to each training sample may lie anywhere in the attacker domain $\mathcal{X}^a$, the resulting big-M constants for all forward and backward passes must accommodate worst-case values over $\mathcal{X}^a$. This substantially weakens the continuous relaxation, even for samples that are unmodified in the optimal solution.

To mitigate this effect, we introduce an auxiliary variable formulation that decouples clean and adversarial data at the level of network propagation and gradient computation. The idea is to ensure that only adversarially substituted samples incur loose bounds, while clean training points retain tight bounds derived from the original dataset.

Specifically, we introduce decision variables for two classes of training data:

1. *Clean data.* Each original sample $(x^{(i)}, y^{(i)})$ is propagated through the network using the standard forward and backward constraints described in

Sect. 4.2. Since these samples are not drawn from $\mathcal{X}^a$, their activations admit significantly tighter bounds, yielding a stronger relaxation. Let the resulting gradient decision variables for these samples be denoted $\delta^{(t,i)}$.

2. *Auxiliary data.* We introduce n auxiliary samples $(\tilde{x}^{(j)}, \tilde{y}^{(j)})$, $j = 1, \ldots, n$, representing points that will be substituted into the dataset. These variables satisfy $\tilde{x}^{(j)} \in \mathcal{X}^a, \tilde{y}_{\text{aux}}^{(j)} \in \{0,1\}$, and are propagated independently through the network's forward and backward passes, resulting in (perturbed) gradient decision variables $\tilde{\delta}^{(t,j)}$.

Rather than model manipulations directly in the input space, we now enforce substitutions at the gradient level. We first introduce the following binary decision variables:

- $s^{(i)} \in \{0,1\}$ denotes the removal of the original sample $x^{(i)}$.
- $\tilde{s}^{(i,j)} \in \{0,1\}$ denotes whether perturbed sample $\tilde{x}^{(j)}$ replaces $x^{(i)}$.

The poisoning adversary budget and unique substitutions are enforced via the constraint $\sum_{j=1}^{n} \tilde{s}^{(i,j)} = s^{(i)}$. Then, to enforce the substitution of the data, we modify the parameter update constraint, now given by

$$\theta^{(t)} = \theta^{(t-1)} - \frac{\alpha^{(t)}}{|\mathcal{I}^{(t)}|} \left[\sum_{i \in \mathcal{I}^{(t)}} \left((1 - s^{(i)}) \, \delta^{(t,i)} + \sum_{j=1}^{n} \tilde{s}^{(i,j)}, \tilde{\delta}^{(t,j)} \right) \right]. \tag{19}$$

If an original point is retained, i.e., $s^{(i)} = 0$, its tightly bounded gradient $\delta^{(t,i)}$ contributes to the update. If it is removed, i.e., $s^{(i)} = 1$, its contribution is suppressed and replaced by that of an activated auxiliary sample ($\tilde{s}^{(i,j)} = 1$), with corresponding gradient $\tilde{\delta}^{(t,j)}$ to encode the worst-case substitution.

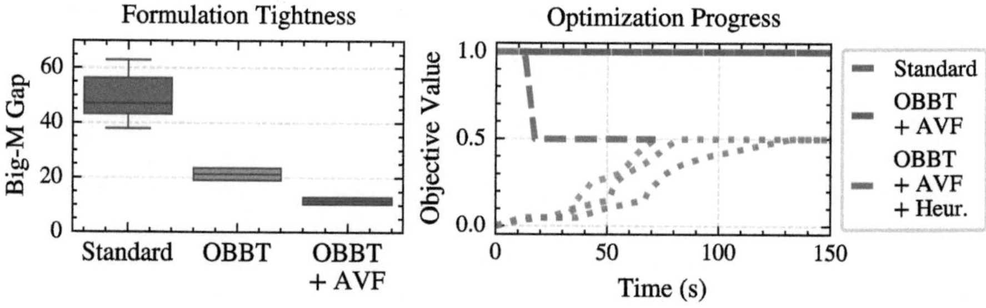

Fig. 1. Comparison of formulation tightness and optimization progress for the Iris dataset under an unbounded attack model ($n = 8$). Left: Tightness of the test-data Big-M constants, defined as $|U^{(i,\text{test})} - L^{(i,\text{test})}|$. Right: Objective value (dashed) and dual bound (dotted) progress over time.

6   Experimental Results

We demonstrate our framework on several moderate-scale learning problems where exact verification is computationally feasible. Note that, exact certification/optimal attacks are often intractable for larger scale problems and impose similar difficulty for MIQCP (see Remark 2). The goal of these experiments is to (i) empirically validate the proposed MIQCP formulation, (ii) qualitatively and quantitatively assess optimal poisoning attacks under different threat models, and (iii) study the effectiveness of the tightening strategies in Sect. 5.

All experiments were run on a server equipped with an AMD EPYC 9334 CPU using Gurobi 13.0. We impose limits of 1 h and 8 threads per instance; unless explicitly stated, we report optimal attacks only where the solver has proven global optimality (the default MIP gap is reached).

Datasets. We evaluate our method on three benchmark datasets:

- Iris ($d = 4$, $N = 80$, $N^{\text{test}} = 20$): The classical Iris dataset [1], restricted to a binary classification task by selecting the first two classes.
- Diabetes ($d = 10$, $N = 320$, $N^{\text{test}} = 89$): The diabetes dataset introduced by [11], in which the regression task is to predict a quantitative measure of disease progression.
- Halfmoons ($d = 2$, $N = 100$, $N^{\text{test}} = 40$): A synthetic two-dimensional binary classification dataset with nonlinearly separable classes.

Models. Across all experiments, we use a single linear layer as the learning model. In classification settings, this corresponds to a linear support vector machine, while in regression settings it reduces to standard linear regression. For the nonlinearly separable Halfmoons dataset, we apply a fixed polynomial feature expansion prior to training, resulting in a total of 9 input features. Note that certification problems remain nonlinear with quadratic terms and loss functions.

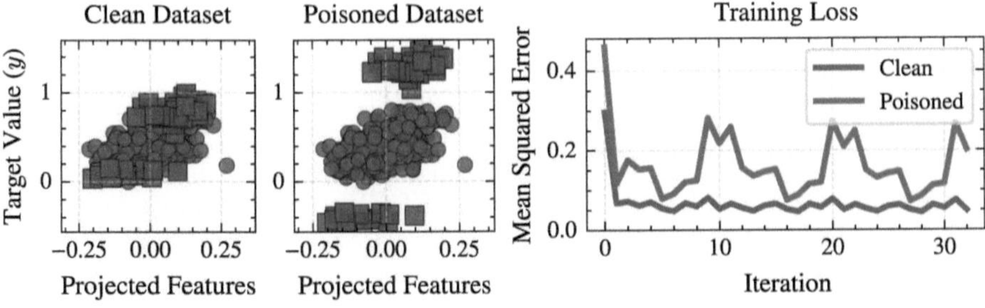

Fig. 2. Optimal denial of service attack on the Diabetes dataset with $n = 50, \epsilon = 0, \nu = 0.5$. The red squares depict the points poisoned by the adversary. The rightmost figure depicts the training loss for the poisoned (green) vs original (blue) datasets. (Color figure online)

Effect of Solution Improvement Strategies. We evaluate the MIQCP solution improvement strategies introduced in Sect. 5 using an unbounded attack on the Iris dataset with $n = 8$. The model is trained for 4 epochs with a batchsize of 20 and a learning rate of $\alpha = 0.03$. Figure 1 illustrates the effect of our improvement strategies on both formulation tightness and solver runtime.

To quantify tightness, we measure the gap between the final Big-M constants on the test data, defined as $|U^{(i,\text{test})} - L^{(i,\text{test})}|$. The left panel of Fig. 1 compares this gap for the standard and tightened formulations. As expected, OBBT significantly tightens the Big-M constants, and the auxiliary variable formulation yields additional tightening. In terms of computational performance (Fig. 1, right), these tightening techniques accelerate convergence of the dual bound but have little effect on the primal solution. Intuitively, the tighter relaxations improve certification, but not identification of the optimal attack. By additionally incorporating our primal heuristic, we can rapidly identify strong attacks, leading to a substantial reduction in the time required to prove optimality.

Optimal Denial-of-Service Attack. We now evaluate a denial-of-service attack on a regression model trained on the Diabetes dataset. We take the adversary's objective to be maximizing the average training loss throughout training, thus preventing model convergence. Figure 2 shows the resulting optimal attack for an adversary capable of modifying $n = 50$ training samples by up to $\nu = 0.5$ in the label space. As seen in the training loss curve, our proposed framework reveals the adversary can effectively prevent stable model convergence by injecting perturbations that induce oscillatory behaviour in the training loss.

Optimal Label Flipping on the Halfmoons Dataset. Finally, we examine a poisoning attack on the Halfmoons dataset. Figure 3 summarizes the attack's effectiveness. On the right, we visually observe the certified worst-case impact of the attack on the model's decision boundary. The model trained on the poisoned dataset exhibits a significantly altered decision boundary, with relatively few label flips required to cause significant mis-classification in the resulting model.

The left side of Fig. 3 shows the resulting optimal attacks and certified accuracies computed for various strengths of the label flipping attack. We observe that optimality is proven in the given time limit only for $n < 6$ perturbed labels. For $n \geq 6$, our formulation still quantifies a sound certificate. This limitation can be attributed to the relatively weaker continuous relaxations induced by larger values of n, resulting in more challenging MIQCPs. Nevertheless, Fig. 3 (left) shows the ability of the proposed framework to *simultaneously* provide users with worst-case attack patterns and certified model performance.

Comparison with Incomplete Certification Baselines. When the MIQCP solver times out before proving optimality, the final dual bound provides a sound but incomplete poisoning certificate, comparable to existing incomplete methods [24,25,33,35]. In all experimental settings considered, existing incomplete methods yield only trivial guarantees (certified test accuracy of 0%), due to the smaller dataset sizes than those considered in these prior works. Additionally,

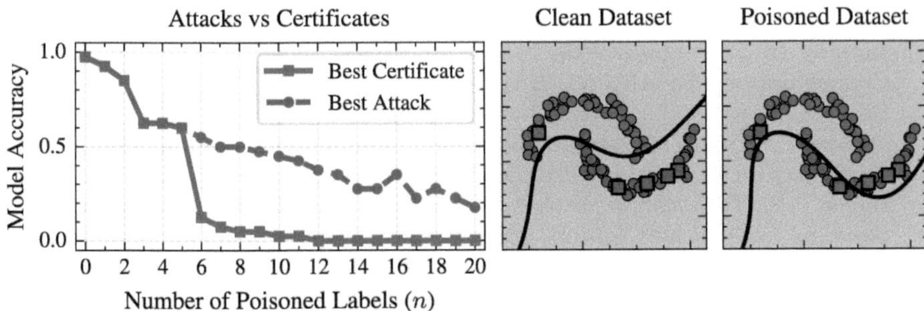

Fig. 3. Right: Label flipping attacks and certificates for the Halfmoons dataset. Left: Example optimal attack for $n = 5$ label flips.

our MIQCP uses big-M constants derived from the same relaxation-based procedures, optionally tightened via OBBT, so the root relaxation is guaranteed to be at least as tight as the corresponding incomplete certificate. Thus, even when optimality is not proven, our bounds are no weaker than those provided by existing incomplete approaches under the same threat model.

7 Conclusion

In this work we present a comprehensive mixed-integer quadratically constrained programming framework for the exact certification of robustness to training-time data perturbations in gradient-based learning. By explicitly encoding adversarial data manipulation, training dynamics, and test-time evaluation, our approach provides sound and complete robustness guarantees under general data poisoning threat models. To improve tractability, we introduce formulation-tightening techniques and primal heuristics to guide the optimization process.

Acknowledgements. This work was supported by the Turing's Defence and Security programme through a partnership with the UK government in accordance with the framework agreement between GCHQ & The Alan Turing Institute. CT was supported by a BASF/Royal Academy of Engineering Senior Research Fellowship.

References

1. Anderson, E.: The species problem in iris. Ann. Mo. Bot. Gard. **23**(3), 457–509 (1936)
2. Anderson, R., Huchette, J., Ma, W., Tjandraatmadja, C., Vielma, J.P.: Strong mixed-integer programming formulations for trained neural networks. Math. Programming **183**(1), 3–39 (2020)
3. Biggio, B., Rieck, K., Ariu, D., Wressnegger, C., Corona, I., Giacinto, G., Roli, F.: Poisoning behavioral malware clustering. In: Proceedings of the 2014 Workshop on Artificial Intelligent and Security Workshop, pp. 27–36 (2014)

4. Biggio, B., Roli, F.: Wild patterns: Ten years after the rise of adversarial machine learning. In: Proceedings of the 2018 ACM SIGSAC Conference on Computer and Communications Security, pp. 2154–2156 (2018)

5. Bose, A., Lessard, L., Fazel, M., Dvijotham, K.D.: Keeping up with dynamic attackers: certifying robustness to adaptive online data poisoning. In: International Conference on Artificial Intelligence and Statistics, pp. 4438–4446. PMLR (2025)

6. Bunel, R.R., Turkaslan, I., Torr, P., Kohli, P., Mudigonda, P.K.: A unified view of piecewise linear neural network verification. In: Advances in Neural Information Processing Systems, vol. 31 (2018)

7. Carlini, N., et al.: Poisoning web-scale training datasets is practical. In: 2024 IEEE Symposium on Security and Privacy (SP), pp. 407–425. IEEE (2024)

8. Chen, X., Liu, C., Li, B., Lu, K., Song, D.: Targeted backdoor attacks on deep learning systems using data poisoning. arXiv preprint arXiv:1712.05526 (2017)

9. Duong, H., Nguyen, T., Dwyer, M.: A DPLL (t) framework for verifying deep neural networks. arXiv preprint arXiv:2307.10266 (2023)

10. Dvijotham, K., Stanforth, R., Gowal, S., Mann, T.A., Kohli, P.: A dual approach to scalable verification of deep networks. In: UAI, vol. 1, p. 3 (2018)

11. Efron, B., Hastie, T., Johnstone, I., Tibshirani, R.: Least angle regression (2004)

12. Ehlers, R.: Formal verification of piece-wise linear feed-forward neural networks. In: D'Souza, D., Narayan Kumar, K. (eds) International Symposium on Automated Technology for Verification and Analysis, pp. 269–286. Springer, Cham (2017). https://doi.org/10.1007/978-3-319-68167-2_19

13. Fischetti, M., Jo, J.: Deep neural networks and mixed integer linear optimization. Constraints **23**(3), 296–309 (2018)

14. Gleixner, A.M., Berthold, T., Müller, B., Weltge, S.: Three enhancements for optimization-based bound tightening. J. Global Optim. **67**(4), 731–757 (2017)

15. Goodfellow, I.J., Shlens, J., Szegedy, C.: Explaining and harnessing adversarial examples. arXiv preprint arXiv:1412.6572 (2014)

16. Gowal, S., et al.: On the effectiveness of interval bound propagation for training verifiably robust models. arXiv preprint arXiv:1810.12715 (2018)

17. Gu, T., Dolan-Gavitt, B., Garg, S.: BadNets: identifying vulnerabilities in the machine learning model supply chain. arXiv preprint arXiv:1708.06733 (2017)

18. Han, X., Xu, G., Zhou, Y., Yang, X., Li, J., Zhang, T.: Physical backdoor attacks to lane detection systems in autonomous driving. In: Proceedings of the 30th ACM International Conference on Multimedia, pp. 2957–2968 (2022)

19. Hong, S., Chandrasekaran, V., Kaya, Y., Dumitraş, T., Papernot, N.: On the effectiveness of mitigating data poisoning attacks with gradient shaping. arXiv preprint arXiv:2002.11497 (2020)

20. Huchette, J., Muñoz, G., Serra, T., Tsay, C.: When deep learning meets polyhedral theory: a survey. arXiv preprint arXiv:2305.00241 (2023)

21. Katz, G., Barrett, C., Dill, D.L., Julian, K., Kochenderfer, M.J.: Reluplex: an efficient SMT solver for verifying deep neural networks. In: Majumdar, R., Kunčak, V. (eds) Computer Aided Verification: 29th International Conference, CAV 2017, July 24-28, 2017, Proceedings, Part I 30, pp. 97–117. Springer, Heidelberg, Germany (2017). https://doi.org/10.1007/978-3-319-63387-9_5

22. Levine, A., Feizi, S.: Deep partition aggregation: provable defense against general poisoning attacks. arXiv preprint arXiv:2006.14768 (2020)

23. Li, S., Cheng, Y., Wang, W., Liu, Y., Chen, T.: Learning to detect malicious clients for robust federated learning. arXiv preprint arXiv:2002.00211 (2020)

24. Lorenz, T., Kwiatkowska, M., Fritz, M.: BiCert: a bilinear mixed integer programming formulation for precise certified bounds against data poisoning attacks. arXiv preprint arXiv:2412.10186 (2024)
25. Lorenz, T., Kwiatkowska, M., Fritz, M.: FullCert: deterministic end-to-end certification for training and inference of neural networks. In: Cremers, D., Lähner, Z., Moeller, M., Nießner, M., Ommer, B., Triebel, R. (eds) DAGM German Conference on Pattern Recognition, pp. 71–85. Springer, Cham (2024). https://doi.org/10.1007/978-3-031-85181-0_5
26. Müller, M.N., Eckert, F., Fischer, M., Vechev, M.: Certified training: small boxes are all you need. arXiv preprint arXiv:2210.04871 (2022)
27. Muñoz-González, L., et al.: Towards poisoning of deep learning algorithms with back-gradient optimization. In: Proceedings of the 10th ACM Workshop on Artificial Intelligence and Security, pp. 27–38 (2017)
28. Newsome, J., Karp, B., Song, D.: Paragraph: Thwarting signature learning by training maliciously. In: Zamboni, D., Kruegel, C. (eds) Recent Advances in Intrusion Detection: 9th International Symposium, RAID 2006 Hamburg, Germany, September 20-22, 2006 Proceedings 9, pp. 81–105. Springer, Berlin, Heidelberg (2006). https://doi.org/10.1007/11856214_5
29. Rezaei, K., Banihashem, K., Chegini, A., Feizi, S.: Run-off Election: improved provable defense against data poisoning attacks. In: International Conference on Machine Learning, pp. 29030–29050. PMLR (2023)
30. Rosenfeld, E., Winston, E., Ravikumar, P., Kolter, Z.: Certified robustness to label-flipping attacks via randomized smoothing. In: International Conference on Machine Learning, pp. 8230–8241. PMLR (2020)
31. Sabanayagam, M., Gosch, L., Günnemann, S., Ghoshdastidar, D.: Exact certification of (graph) neural networks against label poisoning. arXiv preprint arXiv:2412.00537 (2024)
32. Singh, G., Gehr, T., Püschel, M., Vechev, M.: An abstract domain for certifying neural networks. Proc. ACM Program. Lang. **3**(POPL) (2019). https://doi.org/10.1145/3290354
33. Sosnin, P., Müller, M.N., Baader, M., Tsay, C., Wicker, M.: Certified robustness to data poisoning in gradient-based training. Trans. Mach. Learn. Res., 2835-8856 (2025)
34. Sosnin, P., Tsay, C.: Scaling mixed-integer programming for certification of neural network controllers using bounds tightening. In: 2024 IEEE 63rd Conference on Decision and Control (CDC), pp. 1645–1650. IEEE (2024)
35. Sosnin, P., Wicker, M., Collyer, J., Tsay, C.: Abstract gradient training: a unified certification framework for data poisoning, unlearning, and differential privacy. arXiv preprint arXiv:2511.09400 (2025)
36. Steinhardt, J., Koh, P.W.W., Liang, P.S.: Certified defenses for data poisoning attacks. In: Advances in Neural Information Processing Systems, vol. 30 (2017)
37. Szegedy, C., et al.: Intriguing properties of neural networks. arXiv preprint arXiv:1312.6199 (2013)
38. Tian, Z., Cui, L., Liang, J., Yu, S.: A comprehensive survey on poisoning attacks and countermeasures in machine learning. ACM Comput. Surv. **55**(8), 1–35 (2022)
39. Tjeng, V., Xiao, K., Tedrake, R.: Evaluating robustness of neural networks with mixed integer programming. arXiv preprint arXiv:1711.07356 (2017)
40. Tsay, C., Kronqvist, J., Thebelt, A., Misener, R.: Partition-based formulations for mixed-integer optimization of trained ReLU neural networks. In: Advances in Neural Information Processing Systems, vol. 34, pp. 3068–3080 (2021)

41. Wang, W., Levine, A.J., Feizi, S.: Improved certified defenses against data poisoning with (deterministic) finite aggregation. In: International Conference on Machine Learning, pp. 22769–22783. PMLR (2022)
42. Weng, L., et al.: Towards fast computation of certified robustness for ReLU networks. In: International Conference on Machine Learning, pp. 5276–5285. PMLR (2018)
43. Wicker, M., et al.: Certification for differentially private prediction in gradient-based training. In: Forty-second International Conference on Machine Learning (2025)
44. Wong, E., Kolter, Z.: Provable defenses against adversarial examples via the convex outer adversarial polytope. In: International Conference on Machine Learning, pp. 5286–5295. PMLR (2018)
45. Wong, E., Schmidt, F., Metzen, J.H., Kolter, J.Z.: Scaling provable adversarial defenses. In: Advances in Neural Information Processing Systems, vol. 31 (2018)
46. Xie, C., Long, Y., Chen, P.Y., Li, B.: Uncovering the connection between differential privacy and certified robustness of federated learning against poisoning attacks. arXiv preprint arXiv:2209.04030 (2022)
47. Yang, G., Gong, N.Z., Cai, Y.: Fake co-visitation injection attacks to recommender systems. In: NDSS (2017)
48. Zhang, H., Weng, T.W., Chen, P.Y., Hsieh, C.J., Daniel, L.: Efficient neural network robustness certification with general activation functions. In: Advances in Neural Information Processing Systems, vol. 31 (2018)
49. Zhu, C., Huang, W.R., Li, H., Taylor, G., Studer, C., Goldstein, T.: Transferable clean-label poisoning attacks on deep neural nets. In: International Conference on Machine Learning, pp. 7614–7623. PMLR (2019)

From Historical Templates to Hints: Selecting Effective Initializations for the Rack-Loading Problem

Thaïs Souyri[✉][iD], Nadia Lahrichi[iD], and Louis-Martin Rousseau[iD]

Polytechnique Montréal, Montréal, QC H3T 1J4, Canada
{thais.souyri,nadia.lahrichi,louis-martin.rousseau}@polymtl.ca

Abstract. The rack-loading problem consists in placing three-dimensional items in non-stacking, height-adjustable racks under a global height limit while preserving similarity to historical configurations used by operators. We model this setting as a three-dimensional bin-packing variant with fixed rack footprint, variable bin heights, and explicit non-overlap and non-stacking constraints, and solve it with the `CP-SAT` engine of `OR-Tools`. Our main contribution is a systematic study of how to select and construct effective *hints* from historical templates to initialize the solver. We develop a heuristic geometric assignment strategy that chooses the closest template and transfers placements, and a MIP-based assignment framework with Full, Partial, and Critical Assignment policies that trade off guidance and flexibility. A Wasserstein-based similarity measure quantifies how closely solutions mimic historical layouts. Experiments on instances derived from the Hoare and Beasley dataset show that carefully selected *hints* substantially improve feasibility and runtime on difficult cases, while the geometric strategy yields configurations closest to historical templates.

Keywords: Rack-loading problem · Constraint programming · CP-SAT · Hint-based initialization · Operational Logistics · Real-time decision making

1 Introduction

In many industrial and logistics environments, operators must arrange three-dimensional items in storage systems where stacking is prohibited, but space efficiency remains critical. This occurs in pharmacies, where medicines are stored on shelves or in bins without overlap or stacking [8]; in hospitals, where autoclaves load instruments on height-adjustable bins [4]; and in warehouses, where parts are placed on modular bins [5]. In all these settings, bins share the same horizontal footprint, each bin height is the maximum of a fixed minimum and the tallest assigned item, and the sum of bin heights must respect a global rack-height limit.

© The Author(s), under exclusive license to Springer Nature Switzerland AG 2026
T. Guns (Ed.): CPAIOR 2026, LNCS 16595, pp. 557–578, 2026.
https://doi.org/10.1007/978-3-032-27242-3_33

In practice, rack configurations are often chosen manually, based on habits or historical templates. However, the subset of items to be packed together changes from one rack fill to another, quickly making templates inadequate. New combinations may no longer fit into the available space manually, causing efficiency losses, extra equipment, repeated handling, or ad hoc reconfigurations. Our goal is to compute a feasible rack configuration for each item combination, all within real-time limits, while reusing existing templates as much as possible to reduce manual adjustments and operational disruptions.

The problem belongs to the family of three-dimensional orthogonal bin packing problems (3D-BPP), where items are placed in a bounded three-dimensional space under geometric constraints. Classic 3D-BPP formulations are NP-hard [6], typically minimize the number of bins or volume occupied, assume fixed bin dimensions, and allow stacking. Our variant adds: bins with shared horizontal dimensions but variable heights driven by their tallest item; a strict non-stacking requirement, so each item rests on a support surface; a constraint that upper-bin items do not intersect items in lower bins; and an overall rack-height limit.

To better reflect real constraints, several authors have introduced adaptable bin dimensions. Wu et al. study a 3D-BPP with variable bin heights and identical base dimensions [14], while Martin et al. propose a constraint-programming model for an open-dimension packing problem where some bin dimensions can expand [7]. The Multiple Bin-Size Bin Packing Problem (MBSBPP) further generalizes bin packing to multiple bin types or dimensions [2,9,12,13,15]. Our problem can be viewed as an MBSBPP where each admissible bin height defines a bin type, and the total rack height induces a global capacity constraint, yielding homogeneous footprints with variable height capacities.

These contributions move toward more flexible models, but mainly target container and cargo loading, only partially matching our setting. The three-dimensional bin-packing problem of Hoare and Beasley [5] is conceptually closer, since it involves many smaller items, but allows vertical stacking, uses heuristic solution methods, and assumes feasibility as the number of racks is unlimited. In contrast, we prohibit stacking, explicitly determine whether a given item set fits in the rack (allowing infeasibility), and aim to preserve similarity to historical templates to reduce operator adjustment time.

No existing formulation jointly integrates non-stacking constraints, variable bin heights, a global rack-height limit, similarity to historical templates, and real-time solvability within an exact method. We address this gap by formalizing a rack-loading variant of the 3D bin-packing problem and proposing a compact CP-SAT [10] model with explicit bin-height variables, non-overlap, and symmetry-breaking constraints. We treat the problem as a constraint satisfaction problem rather than an optimization model, reflecting practical considerations: industrial settings require rapid solutions, and small deviations from historical templates are acceptable if feasibility is achieved. Defining a formal similarity objective is difficult, as the solver cannot know in advance which template to target, and including multiple templates increases complexity. To tackle this, we design hint-based initialization strategies from historical templates that guide the solver toward familiar layouts while maintaining reasonable solving times. We consider a heuristic geometric assignment strategy that transfers the closest template's placement, and a MIP-based framework with Full, Partial, and Critical Assignment policies that balance guidance and flexibility. We report feasibility and runtime, and measure solution similarity using a Wasserstein-based distance on item positions and dimensions for instances derived from Hoare and Beasley [5].

Section 2 formalizes the problem and presents the CP-SAT model with hint-based initialization. Section 3 reports experimental results comparing the baseline and hint-based strategies.

2   Problem Formulation and Methodology

Our goal is to assign items to bins while satisfying geometric and height constraints and remaining close to historical templates. Each bin has a fixed horizontal footprint and variable height, constrained by the tallest contained item and a total height limit. Items cannot be stacked and can be rotated $90°$ in the horizontal plane, with a set representing all possible orientations for each item [3]. We use the CP-SAT solver from OR-Tools for its efficiency in handling non-overlap and combinatorial constraints [1,11].

The following subsections present the mathematical model, feasibility constraints, initialization strategies based on historical templates, and the similarity measure used to evaluate how closely a configuration matches these templates.

2.1   Parameters and Definitions

Sets	
B	set of bins, $|B| = m$.
I	set of items to be placed, $|I| = n$.
I^*	set of items including their possible $90°$ rotations, $|I^*| = 2n$.
F	set of item families, each grouping items of the same size.
Parameters	
n	total number of items.
w_i, l_i, h_i	width, length, and height of item $i \in I^*$.
W, L, H	total width, length, and height of the rack.
μ_{if}	$= 1$ if i belongs to family $f \in F$ and 0 otherwise.
Decision Variables	
u_{ib}	binary variable indicating if item $i \in I^*$ is placed in bin $b \in B$.
x_{ib}	coordinate along the width axis, of item $i \in I^*$ in bin $b \in B$.
y_{ib}	coordinate along the length axis of item $i \in I^*$ in bin $b \in B$.
q_i	index of the bin containing item $i \in I$.
h_b	height of bin $b \in B$.

2.2 Constraint Programming Model

The problem is formulated as a constraint satisfaction problem (CSP) with the following constraints:

$$u_{ib} + u_{(n+i)b} \leq 1, \quad \forall i \in I, \ \forall b \in B, \tag{1}$$

$$\left(u_{ib} \wedge u_{jb}\right) \Rightarrow \Big((x_{ib} + w_i \leq x_{jb}) \vee (x_{jb} + w_j \leq x_{ib})$$

$$\vee (y_{ib} + l_i \leq y_{jb}) \vee (y_{jb} + l_j \leq y_{ib})\Big),$$

$$\forall b \in B, \ \forall i, j \in I^*, \ i \neq j, \tag{2}$$

$$(u_{ib} = 1) \Rightarrow (h_b \geq h_i), \quad \forall i \in I^*, \ \forall b \in B, \tag{3}$$

$$\sum_{b \in B} h_b \leq H, \tag{4}$$

$$q_i = \sum_{b \in B} b \cdot (u_{ib} + u_{(n+i)b}), \quad \forall i \in I, \tag{5}$$

$$h_{b+1} \leq h_b, \quad \forall b = 1, \ldots, |B| - 1, \tag{6}$$

$$(\mu_{if} \cdot \mu_{(i+1)f} = 1) \Rightarrow q_i \leq q_{i+1}, \quad \forall f \in F, \forall i \in \{1, ..., n-1\}, \tag{7}$$

$$\sum_{b \in B} (u_{ib} + u_{(n+i)b}) = 1, \quad \forall i \in I, \tag{8}$$

$$u_{ib} \in \{0, 1\}, \quad \forall i \in I^*, \forall b \in B, \tag{9}$$

$$x_{ib} \in [0, W - w_i], \quad \forall i \in I^*, \forall b \in B, \tag{10}$$

$$y_{ib} \in [0, L - l_i], \quad \forall i \in I^*, \forall b \in B, \tag{11}$$

$$h_b \in [0, H], \quad \forall b \in B, \tag{12}$$

$$q_i \in [1, |B|], \quad \forall i \in I, \tag{13}$$

$$x_{ib}, y_{ib}, h_b, q_i \in \mathbb{N}, \quad \forall i \in I^*, \forall b \in B. \tag{14}$$

The constraints listed above are intended to ensure the geometric and logical feasibility of placement configurations. Equation (1) prevents an item in both orientations from being simultaneously placed in the same bin. The non-overlap constraint, defined by Eq. (2), requires that two distinct items assigned to the same bin do not overlap in the horizontal plane. The height constraints, represented by Eqs. (3) and (4), ensure that each bin is tall enough to hold the items assigned to it and that the total height of the bins does not exceed the physical limit of the machine. Equation (5) establishes the correspondence between each item and the bin to which it is assigned. Equations (6) and (7) introduce symmetry breaking and ordering constraints: they require that the heights of the bins be decreasing and that items belonging to the same family follow a certain placement order. The introduction of symmetry breaking constraints reduces the search space by eliminating equivalent solutions that differ only in the permutation of bins or items from the same family. This helps the solver to converge more quickly to a feasible solution. Equation (8) ensures that each item is placed exactly once, while (9–14) define the admissible domains of the decision variables, including positions, bin heights, and bin indices.

The non-overlap constraint is implemented using the `AddNoOverlap2D` function from `OR-Tools`. For each item $i \in I^*$ and each bin $b \in B$, we define optional intervals corresponding to horizontal and vertical positions:

$$x_{ib}^{int} = \texttt{NewOptionalIntervalVar}(x_{ib}, w_i, x_{ib} + w_i, u_{ib}), \quad \forall i \in I^*, \forall b \in B, \quad (15)$$

$$y_{ib}^{int} = \texttt{NewOptionalIntervalVar}(y_{ib}, l_i, y_{ib} + l_i, u_{ib}), \quad \forall i \in I^*, \forall b \in B. \quad (16)$$

where x_{ib} and y_{ib} represent the coordinates of item i in bin b, w_i and l_i are its width and length, and $u_{ib} \in \{0, 1\}$ indicates whether the item is actually placed in that bin. A global constraint ensuring non-overlapping is then enforced for all items assigned to a given bin:

$$\texttt{AddNoOverlap2D}(X_b^{int}, Y_b^{int}), \qquad \forall b \in B. \qquad (17)$$

where X_b^{int} (resp. Y_b^{int}) denotes the set of horizontal (resp. vertical) intervals of the items assigned to bin b, i.e.: those for which $u_{ib} = 1$. This formulation ensures that items placed in the same bin are spatially disjoint, while allowing the solver to efficiently handle the different possible positions and rotations.

2.3   Hint-Based Enhancement and Initialization

To improve solver efficiency and generate configurations closer to historical templates–though not explicitly included in the base model's objective–we use the *hint* feature of the `CP-SAT` solver from `OR-Tools`. This allows supplying an initial solution, or *hint*, as a starting point for local search. In our problem, the set of items to place is known in advance, with each occurrence modeled individually, but the combinations of items vary across rack fills. The main idea is to reuse combinations from historical templates, enabling the solver to focus on local adjustments rather than exhaustive search, reducing computation time while maintaining solution quality. Two complementary strategies are implemented to generate effective *hints*.

Heuristic Geometric Assignment Strategy: The Heuristic Geometric Assignment Strategy (HGAS) provides a structured approach for generating *hints* by evaluating geometric similarity between a target configuration and a set of historical templates. For each target item, matching is performed either by exact correspondence when an identical item exists or by greedy geometric association otherwise. The template that minimizes the overall distance to the target configuration is selected as the *hint*. Let:

I_{target} be the set of items in the target configuration, with occurrences $N_{\text{target}}[i]$ for each item $i \in I_{\text{target}}$.

I_{template} be the set of items in a historical template, with occurrences $N_{\text{template}}[j]$ for each item $j \in I_{\text{template}}$.

$d(i,j)$ be the distance between two items i and j is measured as the sum of absolute differences of their dimensions: $d(i,j) = |w_i - w_j| + |l_i - l_j| + |h_i - h_j|$

The distance metric operates at two distinct levels:

- **Direct assignment:** If a target item $i \in I_{\text{target}}$ exists identically in a historical template item $j \in I_{\text{template}}$, i.e., $(w_i, l_i, h_i) = (w_j, l_j, h_j)$, then i is assigned directly to the same bin and position (x, y) as j. The distance contribution is the difference in their occurrences.
- **Indirect assignment:** Otherwise, the target item i is associated with the geometrically closest item $j^* \in I_{\text{template}}$, defined as

$$j^* = \arg \min_{j \in I_{\text{template}}} |w_j - w_i| + |l_j - l_i| + |h_j - h_i|. \tag{18}$$

The item i then inherits the bin and position of j^*. The contribution of this assignment to the total distance is proportional to the dimensional differences, multiplied by the number of matched items and a weight: the weight is 1 if the target item fits entirely within the template item, and 2 otherwise.

Any remaining unassigned target items incur a fixed penalty (e.g., 10 units) in the total distance. The configuration that minimizes the overall distance is selected as the *hint*, ensuring geometric similarity with historical templates. The complete pseudocode for computing this distance is provided in Appendix 1.

MIP-Based Assignment Strategy: This strategy uses a mixed-integer programming (MIP) model to select a single configuration from a set of historical templates and assign target items to corresponding template items or to a placeholder when necessary. Assignments are guided by a cost function that penalizes unassigned items and dimensional deviations, encouraging the solver to maximize assigned items while keeping their placements close to previously validated configurations. The detailed mathematical formulation of the MIP-based assignment model is as follows:

Sets	
I	set of target items to be assigned.
J	set of items contained in the historical template.
$\{j_0\}$	special index representing the placeholder.
Parameters	
w_i, l_i, h_i	dimensions (width, length, height) of item $i \in I \cup J$.
c_{ij}	assignment cost of item $i \in I$ to item $j \in J \cup \{j_0\}$
Decision variables	
x_{ij}	$x_{ij} = \begin{cases} 1 & \text{if item } i \in I \text{ is assigned to } j \in J \cup \{j_0\}, \\ 0 & \text{otherwise.} \end{cases}$

Objective Function

Minimize the total assignment cost:

$$\min \sum_{i \in I} \sum_{j \in J \cup \{j_0\}} c_{ij} x_{ij} \tag{19}$$

where

$$c_{ij} = \begin{cases} w_i + l_i + h_i, & \text{if } j = j_0 \text{ (penalizes unassigned items),} \\ (w_i - w_j)^2 + (l_i - l_j)^2 + (h_i - h_j)^2, & \text{if } j \in J \text{ (dimensional deviation).} \end{cases} \tag{20}$$

The cost function balances two objectives:

1. **Penalization of unassigned items:** Encourages more assignments by penalizing items left unassigned, proportionally to their size.
2. **Dimensional deviation minimization:** Ensures that target items are assigned to template items with similar dimensions to maintain geometric consistency.

Constraints

$$\sum_{j \in J \cup \{j_0\}} x_{ij} = 1, \quad \forall i \in I, \tag{21}$$

$$\sum_{i \in I} x_{ij} \leq 1, \quad \forall j \in J, \tag{22}$$

$$\text{If } |J| < |I|, \quad \sum_{i \in I} x_{ij_0} \leq |I| - |J|, \tag{23}$$

$$\text{If } |I| \leq |J|, \quad x_{ij_0} = 0, \quad \forall i \in I, \tag{24}$$

$$x_{ij} \in \{0, 1\}, \quad \forall i \in I, \forall j \in J \cup \{j_0\}. \tag{25}$$

All model constraints ensure the logical and geometric consistency of item assignments to candidate templates. The bijectivity constraint (21) guarantees that each target item is assigned either to a template item or to the non-assignment placeholder, while the injectivity constraint (22) prevents multiple target items from mapping to the same template item. The non-assignment management constraints (23)-(24) control which items may remain unassigned: only excess items can be unassigned when there are fewer template items than target items, and no unassigned items are allowed otherwise. Finally, domain constraints (25) restrict decision variables to binary values. This formulation supports different assignment strategies–full, partial, or critical-item assignment–allowing control over the level of guidance provided by the *hint*.

Different strategies can control the level of guidance provided by the *hint*:

- **Full Assignment (FA):** All items are assigned for maximum guidance.
- **Partial Assignment (PA):** The solver decides which items to assign, leaving others unassigned.

– **Critical Assignment (CA):** Only the largest items (e.g., top 25% by volume) are assigned first to secure critical placements.

The model is applied to each candidate template in the historical dataset. For each template, the solver minimizes the objective function and selects the template with the lowest cost as a *hint*. The model is solved using the `CBC` solver from the `OR-Tools` library.

Solver initialization with *hints*: Once the closest configuration is identified, item coordinates and bin assignments are provided to the solver via `AddHint()`, guiding the search toward promising regions, although the provided *hints* are not guaranteed to be feasible. `CP-SAT` solver leverages multiple threads, with at least one starting from the *hint*. The number of threads is set empirically based on instance characteristics and available computational resources. Further details on solver parameters are provided in Sect. 3.

2.4   Similarity Measure

To assess the quality of a generated configuration, we compute a geometric similarity measure between the target configuration and a set of historical templates. This measure quantifies how close the positions and dimensions of items in the configuration are to those in historical templates. The smaller the similarity distance, the closer the configuration is to a historical template.

Representation of Configurations: Let a configuration C consist of items $I_C = \{j_1, \ldots, j_n\}$, where each item may appear multiple times and each occurrence is treated separately. Each item $j \in I_C$ is represented by a feature vector $v_j = (x_j, y_j, z_j, w_j, l_j, h_j) \in \mathbb{N}^6$, with (x_j, y_j, z_j) the position of the item and (w_j, l_j, h_j) its dimensions. Given a target configuration T with items $I_T = \{i_1, \ldots, i_m\}$, we compute a distance $d(T, C)$ that reflects the geometric and dimensional differences between T and C.

Normalization of Feature Vectors: To make positions and dimensions comparable across items and configurations, each item vector is normalized using min-max scaling:

$$\tilde{v}_i = \frac{v_i - v_{\min}}{v_{\max} - v_{\min}}, \tag{26}$$

where each component (x, y, z, w, l, h) is scaled to $[0, 1]$ based on the minimum and maximum values over all items in both the target configuration T and the historical template C.

Cost Matrix Between Items: We define a cost function d_{ij} between each target item $i \in I_T$ and each candidate item $j \in I_C$ as a weighted sum of the Euclidean distances of their positions and dimensions:

$$d_{ij} = \lambda_{\text{pos}} \|\mathbf{p}_i - \mathbf{p}_j\|_2 + \lambda_{\text{dim}} \|\mathbf{d}_i - \mathbf{d}_j\|_2, \tag{27}$$

where $\mathbf{p}_i = (x_i, y_i, z_i)$ and $\mathbf{d}_i = (w_i, l_i, h_i)$ denote the position and dimension vectors of item i, respectively. In our experiments, we set $\lambda_{\text{pos}} = \lambda_{\text{dim}} = 1$.

Wasserstein Distance Between Configurations: Let $a \in \mathbb{R}^{|I_T|}$ and $b \in \mathbb{R}^{|I_C|}$ denote discrete probability measures on the items of I_T and I_C, respectively, typically uniform:

$$a_i = \frac{1}{|I_T|}, \quad b_j = \frac{1}{|I_C|}. \tag{28}$$

The Wasserstein distance (or Earth Mover's Distance) between I_T and I_C is defined as the minimum total transport cost:

$$d(T, C) = \min_{\pi \in \Pi(a,b)} \sum_{i=1}^{|I_T|} \sum_{j=1}^{|I_C|} \pi_{ij} \, d_{ij}, \tag{29}$$

subject to

$$\sum_{j \in I_C} \pi_{ij} = a_i, \quad \forall i \in I_T, \tag{30}$$

$$\sum_{i \in I_T} \pi_{ij} = b_j, \quad \forall j \in I_C, \tag{31}$$

$$\pi_{ij} \geq 0, \quad \forall i, j. \tag{32}$$

Selection of the Best Historical Template: Given a set of historical templates $\{C_1, \ldots, C_K\}$, the template that best matches the target configuration T is $C^* = \arg\min_{C_k} d(T, C_k)$.

3   Experimentation

This section presents the experimental evaluation of the proposed CP-SAT model, as well as the impact of the hint-based initialization strategy.

3.1   Experimental Procedure

All experiments were conducted on the Narval supercomputer of the Alliance computing network. Each run used a standard compute node with AMD EPYC 7543 processors (64 cores) and 256 GB of memory. The CP-SAT solver was allocated 16 CPU threads and 8 GB of memory, enabling efficient parallelism. A 1-hour time limit was imposed per instance. The solver was configured with presolve enabled, strengthened linearization (`linearization_level = 2`), portfolio branching (`search_branching = PORTFOLIO_SEARCH`), and maximal symmetry detection (`symmetry_level = 3`); probing was disabled due to its lack of performance benefits. These settings, tuned specifically for the CP model and not for *hints*, support efficient multi-threaded exploration of the solution space and faster convergence, while hint-related CP-SAT parameters were left at their default values.

3.2 Dataset and Instance Generation

We used the publicly available dataset from Hoare and Beasley [5], which contains 11 racks and 17,793 items, each with specified width, length, height, and quantity. The dataset includes diverse boxes and bin arrangements from real operations, enabling realistic and challenging instances while supporting reproducible experiments.

Although the original dataset is intended to load multiple racks simultaneously, we generated instances designed to load a single rack at a time. The instance generation process is detailed in Appendix 2. To reflect differences in item size, we categorized items into three clusters using a K-means algorithm:

Small	many very small items, high combinatorial complexity.
Large	large, flat items with restricted placements.
Tall	medium, tall items, generally easier to pack.

Experiments were conducted on instances composed of a single cluster and on mixed instances combining items from all three clusters. For evaluation, we generated 10 representative instances per cluster covering a wide difficulty range, including infeasible cases. An additional 50 instances per cluster were generated to serve as historical templates.

To ensure comparability of similarity scores across instance types, we normalized the similarity measure so that instances with different geometric characteristics can be evaluated on a common scale. After normalization, all values lie in $[0, 1]$, with values near 0 indicating strong similarity to historical templates. The normalization procedure and its mathematical formulation are provided in Appendix 3.

3.3 Results

Baseline Performance on All Instances: Table 1 reports results of the `CP-SAT` model without hint-based initialization, which serves as a baseline. It summarizes solver performance across all clusters and highlights differences in difficulty and computational effort. The average number of items per instance illustrates the challenges: **Small items** (many and numerous) increase combinatorial complexity, while **Large items** (large and flat) limit placement options.

Notably, **Small items** pose the greatest combinatorial challenge, as evidenced by longer solution times and lower solver success rates–the percentage of instances for which the solver either found a feasible solution or correctly proved infeasibility within the given time limit. In contrast, **Large** and **Tall items** are easier to handle, leading to faster solution times and higher solver success rates within the same time limit.

To focus on instances where hint-based initialization could have a significant impact, we filtered racks based on their estimated packing occupancy:

$$\text{Estimated occupancy} = \frac{\sum_{i \in I}(w_i \cdot l_i) \cdot h_{\text{mean}}}{W \cdot L \cdot H}, \tag{33}$$

Table 1. Baseline performance statistics across clusters.

	Small items	Large items	Tall items	Mix items
Average number of items	717	152	95	285
Solver success rate	85%	100%	100%	98%
Average execution time (s)	216	5	5	151
Average similarity	0.40	0.61	0.70	0.54

Solver success rate: percentage of instances for which a feasible solution was found or infeasibility was proven.

where (W, L, H) are the rack dimensions, (w_i, l_i) the width and length of each item $i \in I$, and h_{mean} the average item height. We considered racks with an estimated occupancy between 50% and 100%, excluding low-density cases that are too easy and high-density cases that are infeasible and efficiently handled by the solver via propagation and presolve. Even after filtering, instances composed of **Tall items** were solved very quickly (mean: 1.7 s, SD: 0.7 s) and were therefore excluded, as they offer little insight into acceleration strategies. Our analysis thus focuses on the remaining difficult instances, where hint-based initialization is most relevant.

The next section compares baseline runs without *hints* to runs using heuristic (HGAS) and MIP-based initialization strategies (FA, PA and CA), evaluating execution time, solver success rates, and the resulting similarity measures to historical templates.

Results with Hint-Based Strategy: To assess whether a strategy improves performance on difficult instances, we consider three criteria. First, feasibility gain: a strategy is superior if it finds a solution for an instance that the baseline solver without *hints* cannot solve. Second, performance gain: when both succeed, we compare solving times. Third, similarity gain: we evaluate how closely each solution resembles historical templates using the normalized similarity measure.

Feasibility gain: Among all methods, the Partial Assignment (PA) strategy achieves the highest success rates across all clusters, highlighting the benefit of incorporating prior structural information into solver initialization. The other methods (HGAS, FA and CA) also outperform the baseline, though with more moderate gains, as shown in Fig. 1.

Since feasibility alone does not reflect solution quality, we further compare execution times. For a fair comparison, this analysis is restricted to instances solved by all five methods (baseline, HGAS, FA, PA, and CA). After filtering, the number of retained instances per category is:

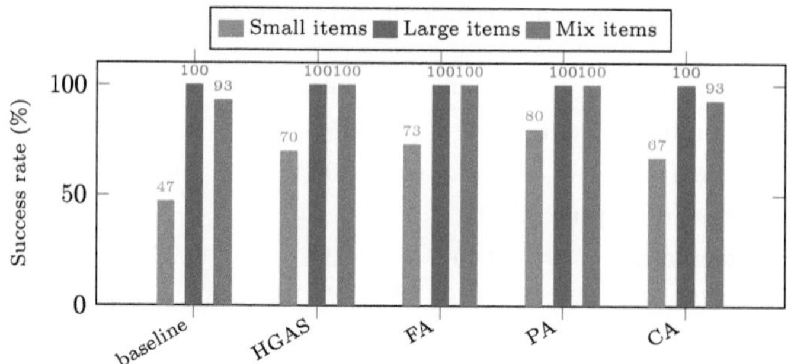

Fig. 1. Success rate (%) on difficult instances per strategy and cluster.

Small items	9 retained instances
Large items	30 retained instances
Mix items	27 retained instances

Performance Gain: For **Small items**, the large number of items creates extreme combinatorial complexity, resulting in long CP-SAT solving times. The PA strategy halves the average execution time from 1060 s to 524 s, but the standard deviation remains high ($\approx$ 600 s), reflecting strong variability across instances. This indicates that hint-based strategies offer limited improvement, suggesting a dedicated heuristic may be more suitable. For **Large items**, the baseline solver is already fast (mean 5.4 s). The best-performing strategy (CA) reduces this to 4.41 s, a measurable but practically minor improvement. The most substantial and consistent gains occur for **Mix items**. As shown in Fig. 2, all hint-based strategies outperform the baseline. Among them, Partial Assignment (PA) and Critical Assignment (CA) achieve the lowest mean solving times and narrowest 95% confidence intervals, demonstrating both efficiency and robustness..

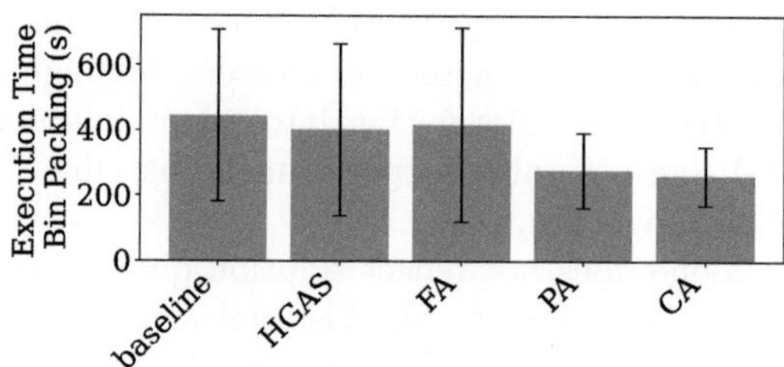

Fig. 2. Mean CP-SAT execution time with 95% confidence intervals for all strategies on the **Mix items** instances.

These results suggest that strategies guiding the solver without overly constraining assignments provide the most reliable acceleration. In contrast, overly forced assignments can generate misleading or low-quality *hints*, introducing structural bias that may slow the search and degrade performance.

For assignment-based strategies, total computation time includes both the MIP assignment phase and the subsequent CP–SAT resolution. For **Large items**, this overhead is unnecessary, as the baseline is already very fast. For **Small items**, the MIP is costly due to the large number of items, making assignment-based strategies less attractive. For **Mix items**, PA and CA remain the most effective strategies in terms of total computation time. As shown in Table 2, they achieve lower average runtimes and greater robustness, indicated by smaller standard deviations, while the baseline, HGAS and FA show high variability across instances.

Table 2. Total computation time for the **Mix items** category.

	baseline	HGAS	FA	PA	CA
Average execution time (s)	445	402	569	414	342
Standard deviation (s)	667	667	754	294	227

Similarity Gain: We now analyze solution similarity with respect to historical templates using the normalized similarity measure. For **Small items**, high item homogeneity leads all methods to similar packings, with values rarely below 0.5, limiting the discriminative power of the metric. For **Large items** and **Mix items**, HGAS achieves the closest similarity, consistently producing lower normalized values than other strategies, as shown in Fig. 3, indicating solutions closer to historical templates. This trend holds across both instance types, demonstrating the robustness of the method.

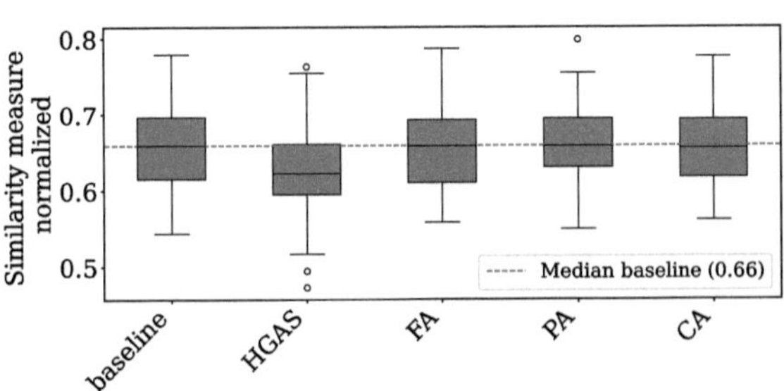

Fig. 3. Normalized Similarity to Historical Templates by Strategy for **Mix items** and **Large items** instances.

In contrast, assignment-based strategies remain close to the baseline distribution and do not significantly improve similarity, even when they improve feasibility or time.

To better understand these results, we analyzed the quality of the generated *hints* by measuring the distance between each *hint* and the historical template closest to the solution. Figure 4 shows a clear positive correlation between this distance and the final solution similarity: the solver tends to reproduce the structural pattern of the *hint* it receives.

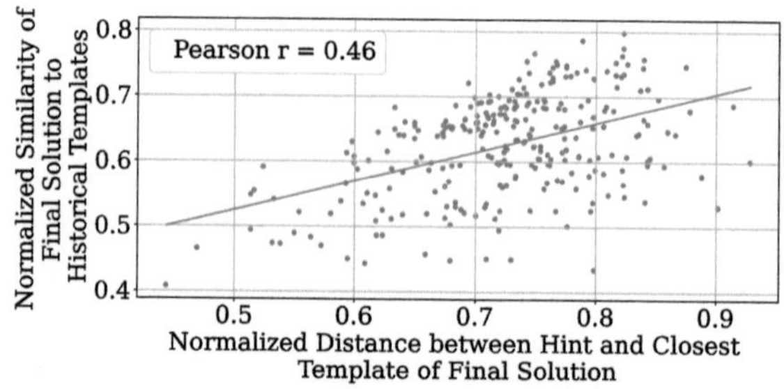

Fig. 4. Correlation Between *hint* Accuracy and Final Solution Similarity for all instances.

This is confirmed by the statistics in Table 3: the heuristic strategy (HGAS) consistently produces *hints* closer to the final historical templates than assignment-based methods, explaining its superior similarity scores on the final solutions.

Table 3. Distance between generated *hints* and the closest historical template of the final solution for **Mix items** and **Large items** instances.

HGAS	FA	PA	CA
0.39 ± 0.38	0.71 ± 0.14	0.69 ± 0.20	0.73 ± 0.15

We also analyzed *hint* feasibility. Feasible *hints* were mainly observed for **Small items**. None of the HGAS *hints* were feasible, while about 30% of MIP-based *hints* satisfied all constraints. Feasible MIP-based *hints* produced solutions closer to historical templates, with a normalized Wasserstein distance of 0.60 versus 0.71 for infeasible ones. *hints* exactly matching the closest template were always feasible, suggesting that feasible *hints* can improve solution quality.

These findings indicate that if the primary objective is to reproduce historical templates, then HGAS is clearly the most appropriate strategy. Conversely, the current formulation of assignment-based methods is not aligned with this objective and would require a redefinition of their objective function.

4   Conclusion

The experiments demonstrate that hint-based initialization markedly enhances solver performance across different item types. For **Small items**, the Partial Assignment (PA) strategy boosted solver success rates from 47% to 80%, although extreme combinatorial complexity limits further improvements. In addition, execution times for **Mix items** were reduced from 445 s to 342 s on average when using the Critical Assignment (CA) strategy. For **Large** and **Mix items**, the heuristic geometric *hints* (HGAS) consistently yielded solutions that were much more aligned with historical templates, with average normalized similarity values dropping to 0.47, compared to values above 0.6 for the other strategies. These results highlight that while HGAS effectively enhances solution similarity and solver convergence for less complex instances, its impact on highly combinatorial cases such as **Small items** remains limited.

However, it is important to note that this study relies on a single benchmark dataset and focuses on static instances, which limits the generalization of the results. Additionally, the number of historical templates was fixed, potentially affecting performance. The influence of instance-specific features on *hint* effectiveness remains largely unexplored.

Despite these limitations, hint-based initialization enables the rapid generation of feasible and operationally consistent rack configurations, reducing manual adjustments and improving workflow predictability. Quantitative improvements in success rates, solution similarity, and execution times demonstrate its practical value, particularly in logistics, warehousing, and healthcare, where maintaining similarity to historical configurations is critical.

Building on these benefits, future research should examine instance-dependent performance to determine which instances gain the most from *hints* and identify the structural features that drive their effectiveness. Additional directions include adaptive, instance-specific *hint* selection, multi-cycle placement, and integration with multi-objective optimization, for example balancing similarity and load distribution. In this setting, creating *hints* that are both feasible and close to historical templates remains a key challenge. Feasible *hints* can guide the solver toward configurations that are easier for operators to adopt. Exploring systematic strategies to better balance feasibility and template similarity may therefore further improve heuristic and hint-based methods when closeness to historical templates is the main objective.

This study shows that hint-based initialization effectively accelerates the CP-SAT solver for the rack-loading problem. By leveraging historical templates, *hints* increase success rates, reduce computation time, and produce solutions closely aligned with past configurations. Their effectiveness varies with item types and instance characteristics, underscoring the importance of strategy choice and instance-specific guidance.

Disclosure of Interests. The authors have no competing interests to declare that are relevant to the content of this article.

Appendix 1   Detailed Algorithm for the Heuristic Geometric Assignment Strategy

This appendix provides the pseudocode for the Heuristic Geometric Assignment Strategy (HGAS) used to generate *hints* from historical templates. The algorithm computes a distance between a target configuration and a candidate template through direct and indirect item assignment.

Algorithm 1. Distance Computation Between Two Configurations

Input: Target configuration, candidate template
Output: Distance $dist$, item correspondences
1. Extract items from target configuration: list I_{target}, occurrences N_{target}
2. Extract items from historical template: list $I_{template}$, occurrences $N_{template}$
3. Initialize $dist \leftarrow 0$ and $correspondence \leftarrow \emptyset$
4. **for each** i in I_{target} **do**
    $n_{target} \leftarrow N_{target}[i]$
    $n_{template} \leftarrow N_{template}[i]$
    if $n_{target} > 0$ and $n_{template} > 0$ **then**
      $a \leftarrow \min(n_{target}, n_{template})$
      $N_{target}[i] \leftarrow N_{target}[i] - a$
      $N_{template}[i] \leftarrow N_{template}[i] - a$
      $dist \leftarrow dist + |n_{target} - n_{template}|$
      $correspondence[i] \leftarrow [i]$ repeated a times
5. **for each** i in I_{target} **do**
    while $N_{target}[i] > 0$ and such that $N_{template}[j] > 0$ **do**
      Compute dimensions of i: (w_i, l_i, h_i)
      Compute dimensions of j: (w_j, l_j, h_j)
      $closest\_j \leftarrow \arg\min_{j \in I_{template}} (|w_i - w_j| + |l_i - l_j| + |h_i - h_j|)$
      $a \leftarrow \min(N_{target}[i], N_{template}[closest\_j])$
      if $w_i \leq w_{closest\_j}$ and $l_i \leq l_{closest\_j}$ and $h_i \leq h_{closest\_j}$ **then**
        $weight \leftarrow \alpha$
      else
        $weight \leftarrow \beta$
      $dist \leftarrow dist + a \times weight \times (|w_i - w_{closest\_j}| + |l_i - l_{closest\_j}| + |h_i - h_{closest\_j}|)$
      $N_{target}[i] \leftarrow N_{target}[i] - a$
      $N_{template}[closest\_j] \leftarrow N_{template}[closest\_j] - a$
      $correspondence[i] \leftarrow [closest\_j]$ repeated a times
    end while
    if $N_{target}[i] > 0$ **then**
      $dist \leftarrow dist + N_{target}[i] \times \delta$
6. **return** $(dist, correspondence)$

The weights $\alpha = 1$ (if the target item fits within the template item) and $\beta = 2$ (otherwise), and the fixed penalty $\delta = 10$ for unassigned items are illustrative constants.

Appendix 2   Instance Generation Procedure

This appendix details the procedure used to construct the test instances employed in the experimental evaluation. All instances are derived from the dataset introduced by Hoare and Beasley [5], which provides a large-scale real-world collection of boxes and rack configurations. The dataset contains 11 racks and a total of 17,793 distinct items, each described by width, length, height, and quantity.

Overview

The objective of the instance-generation process is to create sets of items that:

- vary significantly in size and geometric characteristics,
- include both feasible and infeasible configurations with respect to rack dimensions,
- reflect realistic storage scenarios encountered in industrial settings.

Although the original dataset is designed for multi-rack placement, our problem focuses on filling one rack at a time. We therefore reinterpret the data to generate synthetic, single-rack instances suited to the constraints of our model.

Item Clustering

To structure the item space and create homogeneous groups of objects, we applied a K-means clustering algorithm to all items in the dataset. Based on item dimensions (width, length, height), three clusters were identified:

Cluster 0	small items
Cluster 1	large and flat items
Cluster 2	tall and medium items

Figure 5 illustrates the three item clusters used in our experiments. These clusters allow us to construct instances with controlled dimensional variability and to compare solver performance across differing geometric regimes.

Statistical Preprocessing

For each cluster, we extracted representative geometric values used to guide the construction of instances:

Item Height. The distribution of item heights within each cluster was analyzed. A normal distribution was fitted, and the **third quartile** of this distribution was used as the representative item height for the cluster.

Rack Footprint. We examined the widths, lengths and heights of the 11 racks and selected the **first quartile** of each dimension as the representative rack footprint for generating new instances.

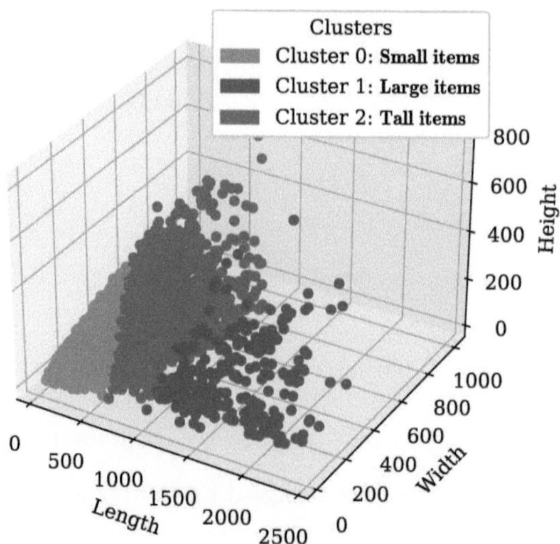

Fig. 5. Clusters of items (K-Means, 3 clusters).

Target Filling Area. Using the representative height $h^\star$, rack dimensions $(W^\star, L^\star, H^\star)$, a target total item area was computed as:

$$A_{\text{target}} = \frac{W^\star \times L^\star \times H^\star}{h^\star}. \tag{34}$$

This formula reflects the volume capacity of the rack while taking into account height constraints.

For each item cluster, we use the third quartile of item heights and the first quartile of rack dimensions as representative values. This choice ensures that generated instances are neither trivially easy nor excessively large: it limits available space.

Item Sampling Procedure

Each instance was generated using the Algorithm 2.

Number of Instances

Using this construction method, we generated:

- 10 instances for each of the three clusters,
- 10 additional instances combining items from all clusters,

leading to a total of 40 test instances.

Additionally, for the hint-based initialization strategy, we created 50 extra instances per cluster using the same methodology. Feasible configurations obtained from these instances were stored to form the historical templates database used during solver initialization. All dimensions provided in millimeters in the original dataset were converted to decimeters for use in our experiments.

Algorithm 2. Instance Generation Procedure

Input: Item dataset $\mathcal{D}$, cluster $C \in \{\text{small}, \text{tall}, \text{large}\}$
Output: Generated instance $\mathcal{I}$
Step 1: Statistical preprocessing
1. Compute representative item height $h^{\star}$ of cluster C (third quartile)
2. Compute representative rack dimensions $(W^{\star}, L^{\star}, H^{\star})$ (first quartiles)
3. Compute average item height in cluster: $h_C^{\star}$
4. Compute target filling area:

$$A_{\text{target}} \leftarrow \frac{W^{\star} \times L^{\star} \times H^{\star}}{h^{\star}}$$

Step 2: Random sampling of items
5. Initialize $A \leftarrow 0$ and $\mathcal{I} \leftarrow \emptyset$
6. **while** $A < A_{\text{target}}$ **do**
   Randomly select an item i from cluster C
   Extract width and length: (w_i, l_i)
   Add item to instance: $\mathcal{I} \leftarrow \mathcal{I} \cup \{i\}$
   Update filled area: $A \leftarrow A + (w_i \times l_i)$
   end while
7. **return** $\mathcal{I}$

Appendix 3   Normalization of the Similarity Measure

The raw similarity values obtained from the Wasserstein distance are not directly comparable across different instance types (clusters), since each cluster exhibits distinct geometric characteristics and distance scales. In particular, instances composed of very small items naturally yield distance values of different magnitudes than instances composed of large or tall items. To enable meaningful cross-cluster comparison, a normalization procedure was applied to the similarity measure.

Empirical Distribution per Cluster

For each cluster type (Small, Large, Tall), we first computed the similarity distance between all pairs of historical templates belonging to the same cluster using the method described in Sect. 2.4. This resulted, for each cluster k, in a set of raw similarity values:

$$\mathcal{D}_k = \{d_1^{(k)}, d_2^{(k)}, \ldots, d_{N_k}^{(k)}\}, \tag{35}$$

where N_k denotes the number of pairwise similarity evaluations within cluster k.

The empirical distributions of these values were visually analyzed and are shown in Fig. 6. They exhibited a strong right-skewness with a long tail, indicating that large distances occur rarely but dominate the scale.

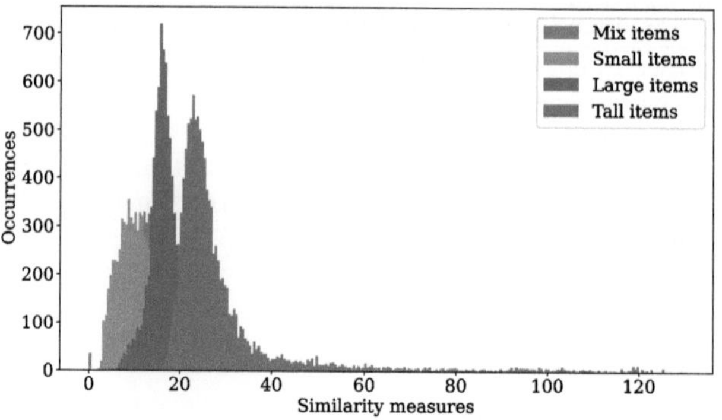

Fig. 6. Similarity measure distributions across instance clusters.

Similarity Measure Distributions Across Instance Clusters

To reduce skewness and obtain a distribution closer to a uniform spread, we applied the following logarithmic transformation to each similarity value:

$$d'_i = \log(1 + d_i) \quad \forall i \in \{1, ..., N_k\}. \tag{36}$$

This logarithmic transformation reduces the skewness of the distribution while preserving the relative ordering of similarity values.

Cluster-Wise Normalization

To map all transformed distances onto a common scale $[0, 1]$, we performed a cluster-wise normalization. Let

$$d^{(k)}_{\max} = \max_{i \in \{1, ..., N_k\}} \log(1 + d^{(k)}_i). \tag{37}$$

denote the maximum transformed similarity value observed within cluster k. Each similarity value is then normalized as:

$$\hat{d}^{(k)}_i = \frac{\log(1 + d^{(k)}_i)}{\log(1 + d^{(k)}_{\max})}. \tag{38}$$

By construction, this ensures:

$$\hat{d}^{(k)}_i \in [0, 1], \quad \forall k. \tag{39}$$

Interpretation

After normalization:

- a value close to 0 indicates a configuration very close to historical templates of the same cluster,

– a value close to 1 indicates a configuration that is among the most dissimilar observed within that cluster.

This procedure allows similarity scores obtained from different instance types to be directly compared on a common and interpretable scale, despite large differences in underlying geometric properties.

References

1. MiniZinc - Challenge. https://www.minizinc.org/challenge/
2. Alonso, M., Alvarez-Valdes, R., Iori, M., Parreño, F.: Mathematical models for multi container loading problems with practical constraints. Comput. Ind. Eng. **127**, 722–733 (2019). https://doi.org/10.1016/j.cie.2018.11.012, https://linkinghub.elsevier.com/retrieve/pii/S0360835218305527
3. Castro, P.M., Oliveira, J.F.: Scheduling inspired models for two-dimensional packing problems. Eur. J. Oper. Res. **215**(1), 45–56 (2011). https://doi.org/10.1016/j.ejor.2011.06.001, https://linkinghub.elsevier.com/retrieve/pii/S0377221711005078
4. Ghiyasinasab, M., Lahrichi, N., Lehoux, N.: A simulation model to analyse automation scenarios in decontamination centers. Health Syst. **12**(2), 181–197 (2023). https://doi.org/10.1080/20476965.2021.2004933, publisher: Taylor & Francis _eprint: https://doi.org/10.1080/20476965.2021.2004933
5. Hoare, N.P., Beasley, J.E.: Placing boxes on shelves: a case study. J. Oper. Res. Soc. **52**(6), 605–614 (2001). https://doi.org/10.1057/palgrave.jors.2601130, publisher: Taylor & Francis _eprint: https://doi.org/10.1057/palgrave.jors.2601130
6. Martello, S., Pisinger, D., Vigo, D.: The Three-dimensional bin packing problem. Oper. Res. **48**(2), 256–267 (Apr 2000). https://doi.org/10.1287/opre.48.2.256.12386, https://pubsonline.informs.org/doi/10.1287/opre.48.2.256.12386
7. Martin, M., Alves De Queiroz, T., Morabito, R.: Solving the three-dimensional open-dimension rectangular packing problem: A constraint programming model. Comput. Oper. Res. **167**, 106651 (2024). https://doi.org/10.1016/j.cor.2024.106651, https://linkinghub.elsevier.com/retrieve/pii/S0305054824001230
8. Oh, H.C., Wong, J.A., Tan, M.C.: Enhancement of patient and staff experience at outpatient pharmacy via optimization of drug–shelf reallocation. Oper. Res. Health Care **3**(1), 15–21 (2014). https://doi.org/10.1016/j.orhc.2013.11.002, https://linkinghub.elsevier.com/retrieve/pii/S2211692313000386
9. Paquay, C., Limbourg, S., Schyns, M.: A tailored two-phase constructive heuristic for the three-dimensional Multiple Bin Size Bin Packing Problem with transportation constraints. Eur. J. Oper. Res. **267**(1), 52–64 (2018). https://doi.org/10.1016/j.ejor.2017.11.010, https://linkinghub.elsevier.com/retrieve/pii/S0377221717310214
10. Perron, L., Didier, F.: Cp-sat. https://developers.google.com/optimization/cp/cp-solver/
11. Stuckey, P.J., Feydy, T., Schutt, A., Tack, G., Fischer, J.: The MiniZinc Challenge 2008–2013. AI Mag. **35**(2), 55–60 (2014). https://doi.org/10.1609/aimag.v35i2.2539, https://onlinelibrary.wiley.com/doi/10.1609/aimag.v35i2.2539
12. Tian, T., Zhu, W., Lim, A., Wei, L.: The multiple container loading problem with preference. Eur. J. Oper. Res. **248**(1), 84–94 (2016). https://doi.org/10.1016/j.ejor.2015.07.002, https://linkinghub.elsevier.com/retrieve/pii/S0377221715006232

13. Tsao, Y.C., Tai, J.Y., Vu, T.L., Chen, T.H.: Multiple bin-size bin packing problem considering incompatible product categories. Expert Syst. Appl. **247**, 123340 (2024). https://doi.org/10.1016/j.eswa.2024.123340, https://linkinghub.elsevier.com/retrieve/pii/S0957417424002057
14. Wu, Y., Li, W., Goh, M., De Souza, R.: Three-dimensional bin packing problem with variable bin height. Eur. J. Oper. Res. **202**(2), 347–355 (2010). https://doi.org/10.1016/j.ejor.2009.05.040, https://linkinghub.elsevier.com/retrieve/pii/S0377221709003919
15. Zhao, X., Bennell, J.A., Bektaş, T., Dowsland, K.: A comparative review of 3D container loading algorithms. Int. Trans. Oper. Res. **23**(1-2), 287–320 (2016). https://doi.org/10.1111/itor.12094, https://onlinelibrary.wiley.com/doi/abs/10.1111/itor.12094

Complete Anytime Decision Diagram Search with GPU-Accelerated State Expansion

Fabio Tardivo[1]([✉]) [iD], Laurent Michel[1] [iD], and Willem-Jan van Hoeve[2] [iD]

[1] School of Computing, University of Connecticut, Storrs, CT, USA
`{fabio.tardivo,laurent.michel}@uconn.edu`
[2] Tepper School, Carnegie Mellon University, Pittsburgh, PA, USA
`vanhoeve@andrew.cmu.edu`

Abstract. We present a GPU-accelerated method for compiling multi-valued decision diagrams (MDDs) for dynamic programming models. MDD construction expands states in layers defined by their distance from the root, a structure that enables independent parallel expansion of all states in a layer. We exploit this by partitioning each layer into limited-size fragments and executing the full expansion of each fragment on the GPU, including successor generation, dominance filtering, duplicate elimination, and pruning of suboptimal states. By systematically traversing fragments in a depth-first manner, we obtain a parallel Complete Anytime Decision Diagram Search that preserves exactness and anytime behavior. Applied to the Traveling Salesman Problem with Time Windows, our implementation yields speedups of up to two orders of magnitude, and outperforms other state-based methods as well as a state-of-the-art dedicated optimization method for this application.

Keywords: Decision Diagrams · Dynamic Programming · GPU Computing · TSPTW

1 Introduction

Dynamic programming (DP) models provide a general and expressive formalism for representing combinatorial optimization problems through a state space and transitions between states. Two generic solver frameworks have recently been developed around this paradigm. The DIDP system [11] follows the heuristic-search tradition from AI and implements methods such as A* search and Complete Anytime Beam Search (CABS). Decision diagram-based solvers [9,15,20] instead construct relaxed and restricted diagrams that provide dual and primal bounds in a branch-and-bound search [2].

The effectiveness of decision-diagram methods depends on the strength of the relaxed or restricted diagrams. Relaxed diagrams yield dual bounds through state merging, but effective merging rules can be problem-specific and are often

T. Guns (Ed.): CPAIOR 2026, LNCS 16595, pp. 579–595, 2026.
https://doi.org/10.1007/978-3-032-27242-3_34

weak for applications such as the Traveling Salesman Problem with Time Windows (TSPTW). In these cases, restricted diagrams, that are constructed simply by retaining the best states per layer and that require no merging rule, become the primary practical mechanism for search and anytime performance.

A key structural property of decision diagrams is that they are compiled layer by layer, while the expansion of a layer is inherently parallel: each state can be processed independently to generate successors, evaluate feasibility, and apply dominance checks. This structure aligns well with modern GPU architectures, which are designed to apply uniform operations to large, contiguous sets of data. Prior work on parallelizing decision-diagram methods, first proposed in [3] and implemented in Ddo, distributes branch-and-bound subproblems across CPU workers, but does not parallelize the construction of a single diagram. Consequently, the core MDD compilation loop remains sequential in existing DD-based solvers, despite its highly parallelizable structure. DIDP offers a parallel version of CABS as well, which is executed on a CPU [12]. Prior work combining GPUs and MDDs has been conducted in the context of Large Neighborhood Search [21].

In this paper, we investigate GPU acceleration of restricted MDD construction for solving DP models. Our method is implemented within the CODD architecture but does not construct relaxed diagrams, and therefore does not require a merging operator. Instead, we solve pure DP models using restricted diagrams only. This brings the approach closer in spirit to heuristic-search methods such as CABS that exhibit strong anytime performance. At the same time, the layer-structured compilation of MDDs offers the opportunity for massively parallel GPU execution. To this end, we partition each layer into bounded fragments and offload their expansion –including successor generation, dominance filtering, and pruning– to the GPU. A depth-first traversal over fragments preserves the anytime and completeness properties of the search. This improves upon sequential Beam-Stack-Search [25], which requires re-expanding nodes during backtracking and lacks equality and dominance filtering.

We evaluate our approach on the Traveling Salesman Problem with Time Windows since it is a challenging benchmark with hard instances even for modern DD-based approaches [19]. Nonetheless, our approach is generic and applicable to any problem that can be modeled in DIDP [11]. The main purpose of the experiments is to demonstrate the value of using GPUs –generically– for solving optimization problems with CADDS.

We experimentally show that the GPU acceleration achieves an order of magnitude speedup compared to the sequential CPU version. We also obtain an order of magnitude speedup compared to the parallel CABS implementation of DIDP, and up to two orders of magnitude to its sequential implementation in Rust as well as the state-of-the-art optimization solver RouteOpt. These results indicate that the structured, layer-by-layer construction of MDDs is particularly amenable to GPU parallelism, and that accelerating the state-expansion step can significantly enhance the scalability of DP-based solution methods.

2   Decision Diagram-Based Optimization

We first introduce the dynamic programming modeling formalism and the associated decision diagram compilation methodology that we will use in this work. We follow the terminology and notation from CODD [15]. We assume optimization problems for which the feasible set of solutions is a finite set P of decision sequences. The sequences are assumed to be bounded but need not be of the same length. Each sequence $x \in P$ has an associated cost $f(x)$ and the objective is to find a feasible sequence with minimum cost.

2.1   Dynamic Programming Model

Let $\mathcal{U}$ be the set of decision labels for the problem. The dynamic program represents solutions as sequences of labels from $\mathcal{U}$, defined by a labeled transition system with an associated cost function. Let $\mathcal{S}$ denote the set of states of the DP model, with initial state $r \in \mathcal{S}$ and goal state $g \in \mathcal{S}$. For each state $s \in \mathcal{S}$, let $\lambda(s) \subseteq \mathcal{U}$ denote the set of applicable labels. The transition function $\tau(s, \ell)$ maps state $s \in \mathcal{S}$ and label $\ell \in \lambda(s)$ to a successor state. Each transition has an associated cost function $c(s, \ell) \in \mathbb{R}$ for state $s \in \mathcal{S}$ and label $\ell \in \lambda(s)$. Each state $s \in \mathcal{S}$ has an associate value function $v(s)$ that is recursively defined as

$$v(s_{i+1}) = \min_{\substack{\ell \in \lambda(s_i) \\ \tau(s_i, \ell) = s_{i+1}}} v(s_i) + c(s_i, \ell) \quad \forall s_i \in \mathcal{S} \setminus \{g\}$$

where we assume the objective is to minimize the value function $v(g)$ of the goal state. We can optionally initialize $v(r)$ with a non-zero constant. The model is correct when the set of label sequences of all transitions from r to g equals P and the sum of their associated costs equals f. The optimal solution can be obtained by backwards induction using the arguments that minimize the value function.

Example 1. Let $G = (V, E)$ be a complete directed graph with distances d_{ij} for all $(i, j) \in E$, representing time. We identify $i_0 \in V$ as the depot. Let each location $j \in V$ have a time window $[a_j, b_j]$. The time window $[a_{i_0}, b_{i_0}]$ applies to the final return to the depot. The traveling salesman problem with time windows (TSPTW) is to find a closed tour that starts at the depot, then visits all locations in V exactly once and within their time window, and ends at the depot, with minimum total distance. We allow waiting at location j to meet the lower time window a_j. To model this problem as a dynamic program, we let each state be a tuple (U, i, t) where $U \subseteq V$ is the set of unvisited locations, $i \in V$ is the current location, and $t \geq 0$ is the accumulated time. The available labels for state $s = (U, i, t)$ are $\lambda(s) = \{j \in U : t + d_{ij} \leq b_j\}$. For $j \in \lambda(s)$, the transition function is $\tau\big((U, i, t), j\big) = (U \setminus \{j\}, j, \max\{t + d_{ij}, a_j\})$. The associated transition cost function is $c\big((U, i, t), j\big) = d_{ij}$. The initial state is $(V, i_0, 0)$ and the target state is $(\varnothing, i_0, -)$. A complete tour is obtained by applying a feasible sequence of transitions from the initial state to the goal state. $\qquad\square$

2.2 Decision Diagram Representation

In the context of optimizing DP models, a decision diagram is a representation of the state space $\mathcal{S}$ as a weighted layered acyclic directed graph. The first layer is defined as $\mathcal{L}_0 = \{r\}$ and contains the initial state. Each subsequent layer is the set of states that are the successors of the previous layer, i.e., $\mathcal{L}_i = \cup_{s \in \mathcal{L}_{i-1}} \cup_{\ell \in \lambda(s)} \tau(s, \ell)$ for $1 \leq i \leq n$, assuming that each solution sequence has maximum length n. This definition recursively defines the expansion of feasible successor states (via the label function λ) and merges equivalent successor states in each layer. Thus, layer $\mathcal{L}_i$ represents all states that are of distance i from the root state. The directed acyclic graph $D = (N, A)$ is now defined by the state set partitioned as $N = (\mathcal{L}_0, \mathcal{L}_1, \ldots, \mathcal{L}_n)$ and arc set $A = \{(u, v) : \exists \ell \in \lambda(u) \text{ s.t. } v = \tau(u, \ell)\}$. Each arc $(u, v) \in A$ represents the transition $\tau(u, \ell) = v$ for some $\ell \in \lambda(u)$ and has associated weight $c(u, \ell)$. Observe that in the above definition, each layer may contain a goal state g. In standard decision diagram literature, these are usually merged together as a single state in the last layer of the diagram [2], but this is not required in an actual implementation. Each r-g path represents a solution, and the optimal solution is obtained by finding a shortest path from the root to the goal state in D.

A decision diagram is *exact* if all reachable states appear in their corresponding layers. Because the state space can grow exponentially large, [2] proposed using restricted and relaxed diagrams of polynomial size. Let $\mathrm{Sol}(D)$ denote the set of solutions represented by decision diagram D. For each arc-specified r-g path p in D, let $w(p)$ denote its total weight, and let x^p denote its sequence of labels. Recall that P is the set of feasible sequences of the problem and $f(x)$ is the cost of sequence $x \in P$.

Definition 1. *A decision diagram D is* restricted *if* $\mathrm{Sol}(D) \subseteq P$ *and* $w(p) \geq f(x^p)$ *for all r-g paths p in D. A decision diagram D is* relaxed *if* $\mathrm{Sol}(D) \supseteq P$ *and* $w(p) \leq f(x^p)$ *for all r-g paths p in D such that $p \in P$.*

The general approach of decision diagram-based optimization is to compile restricted and relaxed diagrams with a given maximum *width*, the number of states in any layer. Building restricted decision diagrams can be done relatively easily by a top-down compilation method. Starting at the root state, we iteratively expand each layer with its successor state, keeping only the best states (according to some scoring function) within the maximum allotted width. Computationally, this requires sorting the states by their score. Building relaxed decision diagrams is done by merging non-equivalent states whenever the size of layer is more than the maximum width. This requires a merging rule that maps two states to a new merged state without excluding any solution as a result. The merging rule is iteratively applied until the size of the diagram meets the width. Using relaxed diagrams is more involved than using restricted diagrams, because the DP model needs to be extended with a merging rule, and applying the rule is computationally more expensive than a simple sorting of states.

Restricted and relaxed decision diagrams respectively provide a primal and dual bound for the problem. An exact branch-and-bound method can be defined by identifying a cutset of exact states in the relaxed diagram (for example, the last exact layer), and recursively applying the process to each state in the cutset [2]. This is particularly effective when the relaxed decision diagrams provide strong bounds, justifying the additional computational work of building them. An alternative approach is to iteratively expand the restricted diagram by systematically expanding unexplored states in each layer, until no more states are unexplored [16]. This approach does not necessarily require relaxed decision diagrams, and makes it more suitable for problems for which strong relaxations are difficult to build. In this work, we will adopt the latter approach.

2.3   State Filtering Rules

State-based search methods for DP models can take advantage of several rules to discard states that are provably sub-optimal or infeasible.

Feasibility Rules. The feasibility rules are encoded in the function $\lambda(s)$ that provides the set of feasible labels for state s. By default, this function has a myopic view, i.e., a one-step look-ahead, but it can be enhanced by more involved constraint reasoning to discard labels that lead to infeasibility.

Dominance Rules. Let s_1 and s_2 be two states. A *dominance rule* $\mathrm{dom}(s_1, s_2)$, specifying that s_1 dominates s_2, is a sufficient condition under which the best possible completion from s_1 is at least as good as the best possible completion from s_2. As a result, s_2 can be discarded without compromising exactness of the search.

Local Bound Rules. A state-based local bound function $\mathrm{cost}(s)$ for a state s provides a lower bound on the objective value of any feasible solution that extends s. By default, this is the shortest path value from the root to s, but it can be enhanced by computing the cost-to-go estimate from s to the goal, also known as an admissible heuristic. If $\mathrm{cost}(s)$ exceeds a given primal bound, then s can be discarded without compromising exactness of the search.

Example 2. Consider the DP model for the TSPTW from Example 1. We can define a dominance rule as follows. Let $s_1 = (U, i, t_1)$ and $s_2 = (U, i, t_2)$ be two states with the same set of unvisited locations U and the same current location i, but possibly different arrival times t_1 and t_2. Then s_1 dominates s_2 if $t_1 \leq t_2$. Indeed, any feasible continuation of s_2 is also feasible from s_1, and the remaining schedule can only benefit from the earlier arrival time t_1. The default local bound for a state s is the value of the shortest r-s path. A simple admissible heuristic can be defined by sorting the distances non-decreasingly in a list L once. For state $s = (U, i, t)$ we can add to $\mathrm{cost}(s)$ the lower bound $\sum_{j=1}^{|U|} L[j]$. A more involved option is to compute a combinatorial bound, for example the value of a minimum spanning tree on the graph induced by U, as is done in [15]. This paper uses the same enhanced heuristic as DIDP. Namely, for each vertex j it precomputes the shortest incoming arc d_j^{in} and outgoing arc d_j^{out}. Then, it establishes $\mathrm{cost}(s)$ by adding $\max\left(\sum_{j \in U \cup i_0} d_j^{in}, \sum_{j \in U \cup i} d_j^{out}\right)$ where i is the last vertex. □

3   Complete Anytime Decision Diagram Search

We next introduce *Complete Anytime Decision Diagram Search* (CADDS), an exact search method that operates on restricted decision diagrams. It formalizes the restricted-diagram expansion strategy introduced in Sect. 2.2 by systematically exploring the remaining unexplored states in the diagram, using a width-bounded selection scheme analogous to that employed by Complete Anytime Beam Search (CABS) [24] and Beam-Stack-Search [25]. CADDS follows the structure of the decision diagram: rather than maintaining a global beam over the frontier, it maintains and expands bounded sets of states within individual layers and explores these sets in a depth-first manner.

3.1   Motivation and Relation to Complete Beam Search Methods

CABS iteratively constructs restricted search frontiers of bounded size, guided by a scoring function, and restarts the search with increasing beam widths until optimality can be proven. This strategy can find good solutions and provides strong anytime behavior, but it repeatedly regenerates large parts of the search space [24]. Beam-Stack-Search is an upgrade on breadth-first branch-and-bound search in which no layer of the breadth-first search tree is allowed to grow beyond the beam width [25]. Several other variants exist, including Limited Discrepancy Beam Search [7] and Incremental Beam Search [22].

These complete beam search variants are developed for a generic state space without explicitly exploiting the layered structure induced by the DP model. In contrast, CADDS works directly on the layered decision diagram. Each layer groups states by their distance from the root, and the DP transitions define all outgoing arcs from one layer to the next. CADDS preserves the key ideas of beam search, i.e., retaining a subset of states of bounded width, but it does so within each layer of the decision diagram. And like Beam-Stack-Search, CADDS explores the limited-width layers in a depth-first traversal.

3.2   Fragments and Search Structure

The basic search unit in CADDS is a *fragment*. For a given layer $\mathcal{L}_i$ and width parameter w, a fragment is defined as a subset $F \subseteq \mathcal{L}_i$ containing at most w states. In our implementation, F consists of the w best states in $\mathcal{L}_i$ according to a scoring function f_s; if such a function is not provided we select the first w states in F. Intuitively, a fragment represents the portion of the layer that will be expanded next.

CADDS maintains a sequence of layers $(\mathcal{L}_0, \mathcal{L}_1, \dots)$ as in Sect. 2.2 and a current best solution s^*. At each iteration, the algorithm selects the deepest nonempty layer $\mathcal{L}_i$, extracts a fragment F from $\mathcal{L}_i$, and expands all states in F to generate their successors in $\mathcal{L}_{i+1}$. The remaining states in $\mathcal{L}_i \setminus F$ are left in place and may be expanded later. This induces a depth-first traversal over fragments: the search tends to go deeper whenever possible, but it eventually returns to shallower layers to process the fragments that were left unexplored.

Function: $\mathrm{cadds}(r\colon \text{state},\ w\colon \mathbb{N},\ f_s\colon \text{score function}) \rightarrow s^*\colon \text{best solution}$

```
 1  s* ← ⊥                                          // No solution found yet
 2  L ← ({r}, ∅, …, ∅)
 3  while ∪ᵢ Lᵢ ≠ ∅ do
 4      i ← max{i : Lᵢ ≠ ∅}                          // Deepest nonempty layer
 5      k ← min{w, |Lᵢ|}
 6      F ← top-k states of Lᵢ according to fₛ
 7      S ← calcSuccessors(F, s*)                     // Successors after filtering
 8      Lᵢ ← Lᵢ \ F
 9      Lᵢ₊₁ ← Lᵢ₊₁ ∪ S
10      forall s ∈ S do
11          if isSolution(s) ∧ cost(s) < cost(s*) then
12              s* ← s
13  return s*
```

Algorithm 1: Sequential CADDS (high-level specification)

The width parameter w controls both the memory usage and the amount of work that can be exposed to potential parallel execution. For small w, CADDS behaves similarly to a depth-first search, with limited memory but also limited potential for parallelism. For large w, it approaches a breadth-first expansion at each layer, which increases the potential for parallelism but also increases memory usage and delays the discovery of complete solutions.

3.3 Sequential CADDS Algorithm

Algorithm 1 gives a high-level specification of CADDS in its sequential form. The input consists of the initial state r, a width parameter w, and a scoring function f_s on states. The algorithm maintains a layered representation $\mathcal{L} = (\mathcal{L}_0, \mathcal{L}_1, \dots)$ of the current restricted diagram and an incumbent solution s^*. Initially, $\mathcal{L}_0$ contains only the root state and all other layers are empty.

At each iteration, the deepest nonempty layer is selected (line 4), and a fragment F of size at most w is extracted (lines 5–6). The function CALCSUCCESSORS expands all states in F and applies the available pruning rules, returning the surviving successors S in the next layer (line 7). The fragment F is then removed from the current layer, and its successors are inserted into $\mathcal{L}_{i+1}$ (lines 8–9). Any complete solutions among the successors update the incumbent solution (lines 10–13).

The algorithm terminates when all layers are empty, which means that every reachable state that was not discarded by feasibility, dominance, or local bound rules has been expanded. Under the assumption that these rules are sound, the final incumbent s^* is therefore optimal.

The memory complexity involves δ, i.e., an upper bound on the branching factor for states in F ($\delta \geq \max\{\, |\lambda(s)| : s \in F \,\}$). Since each iteration generates at most $w \cdot \delta$ successors, and the algorithm expands the deepest nonempty layer, the memory complexity is $O(|\mathcal{L}| \cdot w \cdot \delta)$.

Function: calcSuccessors(F: set of states, s^*: best solution)$\rightarrow$$S$: successors
1 $T \leftarrow \{\tau(s,\ell) : s \in F,\ \ell \in \lambda(s)\}$ // `Successor states`
2 $S \leftarrow \{s : s \in T, \text{cost}(s) < \text{cost}(s^*), \nexists s' \in T : \text{dom}(s',s)\}$ // `Apply filters`
3 **return** S

Algorithm 2: Sequential successor generation and pruning

3.4   Successor Generation and Pruning

Algorithm 2 specifies the CALCSUCCESSORS procedure in abstract form. Given a fragment F and incumbent solution s^*, it generates all successors of states in F and applies the pruning rules introduced in Sect. 2.3.

The set T contains all feasible successors obtained by applying the transition function τ to each state in F and each feasible label in $\lambda(s)$. The local bound function $\text{cost}(\cdot)$ is used to discard states whose bound exceeds the cost of the incumbent solution (line 2). Dominance rules, as defined in Sect. 2.3, are used to discard states that are dominated by other successors in T. The surviving states form the set S of successors that will be inserted into the next layer of the decision diagram. While described as two separate phases, the abstract specification does not assume any particular implementation of the dominance test or heuristic bound. The next section refines this procedure to obtain a GPU-accelerated implementation that expands all states in a fragment in parallel.

3.5   Anytime Behavior and Exactness

CADDS is an anytime algorithm because it maintains a valid incumbent solution throughout the search. Whenever a complete solution is generated in CALCSUCCESSORS, it is compared to the current incumbent, and the better of the two is retained. The incumbent therefore improves monotonically as the search proceeds, and the current best solution can be reported at any time.

Proposition 1. *The complete anytime decision diagram search defined in Algorithm 1 is an exact search.*

Proof. Each iteration removes a fragment F from some layer $\mathcal{L}_i$ and inserts its (filtered) successors into $\mathcal{L}_{i+1}$. States that are not included in F remain in $\mathcal{L}_i$ and must eventually be selected as part of a fragment in a later iteration, unless they are discarded by sound pruning rules. Since we always choose the deepest nonempty layer, every reachable, nondominated, and non-suboptimal state is eventually expanded. Therefore, when all layers are empty, the incumbent solution is optimal. When no solution is returned, the problem is infeasible. $\square$

In summary, CADDS provides an exact search procedure over layered decision diagrams with anytime behavior, and its structure naturally exposes potential for fragment-level parallelism. In the next section, we show how to exploit this parallelism on GPUs by offloading the expansion and pruning of fragments to the device.

4   GPU-Accelerated State Expansion

In this section we describe how to accelerate the state expansion step of CADDS on a GPU. The goal is to offload the most expensive part of the search, namely the generation and filtering of successors for a fragment, while keeping the overall control of the search on the CPU. Although GPUs often accelerate matrix operations, they are capable of much more [10]. A useful abstraction model of a GPU is a collection of multi-core processors (each with its own dedicated cache) that excels at parallelizing independent loop iterations. The key ideas are to represent states in a compact, contiguous layout, to partition fragments into batches that fit on the device, and to run successor generation, local bound filtering, and dominance checks on the GPU.

We note that a parallel beam search method was proposed in [6] while a parallel CABS implementation has been implemented in DIDP [12]. Both methods utilize CPU workers, whereas we focus on massive parallelization on a GPU.

4.1   Design Objectives

The CALCSUCCESSORS procedure in Algorithm 1 is the natural target for GPU acceleration. Given a fragment F of states from some layer $\mathcal{L}_i$ and an incumbent solution s^*, it generates all feasible successors, discards suboptimal and dominated states, and returns the remaining successors S for layer $\mathcal{L}_{i+1}$. For problems such as the TSPTW, fragments can contain tens of thousands of states, and the number of raw successors can be much larger. In this regime, the cost of state expansion dominates the total running time.

The design of the GPU implementation is guided by the following objectives: First, we want to exploit data parallelism by applying the same operations to large sets of states. Second, we want to keep the memory layout simple and contiguous to favor coalesced memory access. Third, we aim to limit device memory usage by processing large fragments in batches of bounded size. Fourth, we want to preserve the pruning behavior of the sequential specification, within the efficient context of each batch. The CPU remains responsible for managing the layers, selecting fragments, updating the incumbent solution, and orchestrating calls to the GPU.

4.2   Fragments, Batches, and Data Layout

Recall that a fragment F is a subset of a layer $\mathcal{L}_i$ of size at most w. To offload a fragment, the CPU packs the states in F into a contiguous array and transfers it to the GPU. Device memory is limited, and the number of successors per state may be large, so it is not always possible to expand the entire fragment at once. We therefore partition F into disjoint *batches* $B_0, \ldots, B_k$ and process them sequentially on the device. The batch size is chosen as large as possible under an upper bound on the memory required for storing the batch, its successors, and temporary working arrays. Let M be the the maximum number of states that can be stored in the GPU memory, δ the upper bound on the branching factor,

Function: calcSuccessorsGPU(F: fragment, s^*: best solution, δ: branching factor)$\rightarrow$(S: successors, δ': successor branching factor)

```
 1  S ← []                                          // Initialize successor array
 2  δ' ← 0                                  // Initialize successor branching factor
 3  b ← ⌊M/(1 + δ + δt)⌋                                       // Largest batch size
 4  {B₀, B₁, …, B_k} ← Partition(F, b)                       // Divide F in batches
 5  forall B ∈ {B₀, B₁, …, B_k} do
 6      T₀ ← []                                        // Feasible successors of B
 7      parall s ∈ B do
 8          parall ℓ ∈ λ(s) do
 9              T₀ ← T₀ ∥ [τ(s, ℓ)]          // Array concatenation (thread-safe)
10      T₁ ← []                                       // Filter sub-optimal states
11      parall s ∈ T₀ do
12          if cost(s) < cost(s*) then
13              T₁ ← T₁ ∥ [s]
14      H ← parSortBy(T₁, h ∘ π)              // Array of states sorted by hash
15      R ← [⊥, …, ⊥]                                   // State dominance flags
16      parall i ∈ {0, …, |H| − 1} do
17          forall j > i s.t. h(π(H[i])) = h(π(H[j])) do
18              if H[i] = H[j] ∨ dom(H[i], H[j]) then
19                  R[j] ← ⊤                           // State H[j] is dominated
20              else if dom(H[j], H[i]) then
21                  R[i] ← ⊤                           // State H[i] is dominated
22      T₂ ← []                                     // Filter dominated states
23      parall i ∈ {0, …, |H| − 1} do
24          if R[i] = ⊥ then
25              T₂ ← T₂ ∥ [H[i]]
26      parall s ∈ T₂ do
27          δ' ← max(δ', |λ(s)|)                    // Update branching factor
28      S ← S ∥ T₂
29  return (S, δ')
```

Algorithm 3: GPU-based successor generation and pruning

and t a constant accounting for temporary buffers, we choose the largest integer b, representing the batch size, such that

$$b + b \cdot \delta + b \cdot \delta \cdot t \leq M. \tag{1}$$

The batch $B_j \subseteq F$ of size at most b form a partition of F. The b states in B_j produce up to $b \cdot \delta$ successor states, each of which require auxiliary space $b \cdot \delta \cdot t$.

4.3 GPU Successor Generation and Pruning

We next refine the CALCSUCCESSORS procedure to its GPU-based counterpart. Algorithm 3 shows the high-level structure of the procedure. It starts by initializing the successor array S and branching factor δ'. It then computes the largest batch size b for fragment F, and partitions F into batches $B_0, B_1, \ldots, B_k$ using the function PARTITION(F, b) (the details are omitted for brevity). It then processes each batch B in sequence, by generating its feasible successors (lines 6–9), discarding suboptimal states based on the local bounding rule (lines 10–13), and discarding dominated states (lines 14–25). The resulting array of remaining successors states is added to S, and the next batch is considered. Each stage uses a temporary array (denoted T_0, T_1, T_2) to store the intermediate set of successor states. Observe the use of PARALL in the algorithm, denoting the parallel execution of a set of instructions on the GPU.

The computationally most demanding part of the algorithm is identifying dominated states, because this requires a pairwise comparison of states. To check dominance efficiently in our setting, we first identify the components of the states that take the same value in the dominance rule. Let $s = (a_1, a_2, \ldots, a_k)$ be the state definition as a list of attributes and let $\mathrm{dom}(s, s')$ be the dominance rule under consideration. Let $J = \{j : a_j = a'_j \wedge \mathrm{dom}(s, s')\}$ be the subset of state attributes a_j that are identical in the application of $\mathrm{dom}(s, s')$. We let $\pi(s) = (a_j : j \in J)$ be the projection of state s under the dominance rule.

Example 3. Consider the TSPTW for which the dominance rule $\mathrm{dom}(s_1, s_2)$ is defined for states $s_1 = (U, i, t_1)$ and $s_2 = (U, i, t_2)$ with the same value for U and for i. The associated projection of state s is $\pi(s) = (U, i)$. $\square$

We use the projection of a state relative to the dominance rule as follows. First, we sort the successor states by their hash function value that is defined on the projection π (function PARSORTBY$(T_1, h \circ \pi)$ in line 14) in a temporary array H. We then execute parallel threads for each state in H to flag dominance (line 16); this can be done quickly for contiguous elements with the same hash value (lines 17–21), because these are the only states to which the dominance rule may apply. We flag all dominated states in a temporary array R, which is subsequently used to discard dominated states in parallel (lines 22–25), storing the remaining states in the temporary array T_2. An important feature of this approach is that dominance is only applied within each batch. This improves the computational performance on the GPU, as it avoids an explicit quadratic number of comparisons and maintains a flat, contiguous memory layout. The cost of sorting T_1 by the hash value is offset by the simplicity of the subsequent dominance checks. The potential downside, however, is that fewer dominated states can be identified across a layer compared to CABS.

Algorithm 3 concludes by appending the outcome T_2 of each batch to the successor array S, and updating the successor branching factor. It then returns the resulting complete set of successors S and the associated branching factor δ'.

Note: Algorithm 3 defines the successor computation at the abstract level of states. In the actual GPU implementation, F is represented as an array of nodes of the form (s, c, p), where s is the state, c is its cost (the current value of the local bound $\mathrm{cost}(s)$), and p is a compact representation of the path from the root to s. This provides access to the information needed for the transition rule and the pruning rules. We keep the entire path in memory because each fragment is discarded after processing it. This design choice improves the performance, even if it decreases the maximum batch size b for partitioning the fragment.

4.4   Integration with CADDS

The procedure CALCSUCCESSORSGPU integrates into CADDS by replacing the sequential call to CALCSUCCESSORS in Algorithm 1 (line 7). The CPU provides the fragment F, the current incumbent solution s^*, and an estimate δ of the branching factor for states in F. The GPU returns the successors S and an updated branching factor δ', which is used to estimate the batch sizes for the next layer. From the point of view of the CADDS algorithm, the semantics of the state expansion step remain unchanged: a fragment is removed from the current layer, its surviving successors are inserted into the next layer, and any complete solutions among the successors are used to update the incumbent. The only difference is that successor generation and pruning are now carried out in parallel on the GPU.

The combination of fragment-based search and GPU-accelerated state expansion yields two complementary benefits. First, it exposes large amounts of parallel work whenever layers contain many states, which is precisely the regime where sequential solvers can have scalability issues. Second, it preserves the anytime and exactness properties of CADDS, since the pruning rules applied on the GPU are the same as in the sequential specification and dominance is applied in a sound, batch-local manner.

5   Experimental Evaluation

We next present an experimental evaluation using the TSPTW as a case study, using five TSPTW benchmark sets from the literature [1,8,17–19] for a total of 637 instances. We compare the following methods:

- *CADDS-GPU:* GPU implementation of CADDS, without scoring function.
- *CADDS-GPU-Greedy:* GPU implementation with cost($\cdot$) as scoring function.
- *CADDS-Sequential:* Sequential version of CADDS, without scoring function.
- *DIDP-Parallel:* Multi-threaded implementation of CABS in DIDP [12].
- *RPID:* Sequential Rust implementation of CABS in DIDP [13].
- *RouteOpt:* State-of-the-art exact solver for VRPTW that implements a branch-price-and-cut algorithm [23].

We do not include an extension of the relaxed-DD-based approach from CODD [15] to the TSPTW, as preliminary experiments showed it to be non-competitive for this problem. All CADDS[1], DIDP, and RPID models use the same dynamic programming model, dominance rules, and cost heuristics following the literature for TSPTW [4,5,14].

---

[1] Code available at https://github.com/ldmbouge/CODD/tree/CPAIOR26

The hardware used is an AMD Ryzen Threadripper PRO 7995WX[2], 768 GB of RAM, and a NVIDIA RTX 6000 Ada Generation[3]. The system operates on Ubuntu Linux 24.04 LTS and uses CUDA 12.5, GCC 13.3, Rust 1.90, and Gurobi 13.0. We bound CADDS-Sequential to 48 GB of working memory to process fragments and mimic the GPU memory size. The parallel CABS in DIDP uses 96 threads, 1 per CPU core with no other restrictions. We apply a 10-minute timeout per instance.

The main results are presented in Table 1 over a set of 506 instances that were solved within the timeout by the CADDS and DIDP solvers. The reason is that all these solvers *use the same DP model specification* which makes it more meaningful to compare their performance in terms of time and states explored. For each benchmark and each solution method, the table reports the total solving time over all solved instances, the maximum solving time, the geometric mean of the solving time, the total number of search states, and the maximum memory. Unsolved instances never exhausted memory. All CADDS methods adjust $|F|$ at each iteration to yield a single batch of maximal size according to eq. (1). We note that RouteOpt is not designed for asymmetric or infeasible instances, which are present in four benchmarks, and we indicate this with a dash $(-)$.

GPU Acceleration. We first inspect the impact of the GPU acceleration, for which we compare the total time of CADDS-GPU and CADDS-GPU-Greedy with the baseline CADDS-Sequential. The GPU acceleration achieves a speedup of 10–20x compared to the sequential version, while having a lower memory footprint. This confirms our expectation that state expansion can indeed be offloaded very efficiently on a GPU to be handled in parallel. Observe that the scoring function can have a (marginal) computational benefit, but not always.

Solver Comparison. We next compare the performance of CADDS with CABS in DIDP. The sequential versions CADDS-Sequential and RPID are competitive, although CADDS-Sequential is on average about three times faster while using more memory. Naturally the multi-threaded implementation of DIDP is more competitive. Yet, the GPU still derives speed-ups from 2.6x (Solomon-Pesant) to 16.7x (Solnon Infeasible) compared to DIDP-Parallel. The total number of nodes explored gives a more nuanced story. Indeed, for some benchmarks, CADDS requires significantly fewer nodes than DIDP variants, while for others (e.g., Solomon-Pesant) the opposite holds. The latter happens exactly when the speedups are the weakest (2.6x). An analysis of the behavior indicates that it should be attributed to the weakening of the dominance check caused by the reliance on fragments and batches (recall that dominance is local to fragments and batches in CADDS). That also manifests itself in the increased memory usage (fewer dominated nodes are discarded).

The performance of RouteOpt (on the symmetric and feasible benchmark sets) is weaker than the DP solvers, as it generally solves fewer instances within the time limit. Because it uses strong relaxations during search, it explores fewer

---

[2] 96 Cores at 5.1 GHz, 384 MB of L3 Cache.
[3] 18176 CUDA Cores at 2.5 GHz, 96 MB of L2 Cache, 48 GB of VRAM.

Table 1. Solving performance on instances solved by all DP solvers.

Benchmark	Solver	Solved	T_{sum} [s]	$T_{\max}$ [s]	T_{GM} [s]	$N_{\text{sum}}[\times 10^6]$	$M_{\max}$[GB]
AFG	CADDS-GPU	45	9.95	1.6	0.06	66.7	1.8
	CADDS-GPU-Greedy	45	10.78	1.9	0.06	66.7	1.8
	CADDS-Sequential	45	120.55	30.6	0.00	66.7	7.8
	DIDP-Parallel	45	109.52	22.6	0.37	202.3	2.1
	RPID	45	642.28	179.7	0.09	201.6	1.7
	RouteOpt	–	–	–	–	–	–
Gendreau Dumas Extended	CADDS-GPU	84	98.83	21.6	0.19	740.5	7.0
	CADDS-GPU-Greedy	84	112.12	31.5	0.19	823.9	7.3
	CADDS-Sequential	84	1278.87	301.2	0.00	726.5	24.4
	DIDP-Parallel	84	824.35	104.9	1.53	1453.8	2.8
	RPID	84	3720.75	469.8	1.67	1449.0	1.9
	RouteOpt	60	2442.90	368.3	7.17	0.0	2.9
Solnon (Feasible)	CADDS-GPU	235	60.53	7.2	0.05	683.7	10.2
	CADDS-GPU-Greedy	235	62.13	7.1	0.05	673.3	10.0
	CADDS-Sequential	235	643.14	102.6	0.00	690.7	28.4
	DIDP-Parallel	235	505.86	41.9	0.28	1452.7	4.0
	RPID	235	2513.05	306.3	0.05	1441.7	2.9
	RouteOpt	219	3064.80	341.8	3.46	0.0	5.3
Solnon (Infeasible)	CADDS-GPU	95	18.89	3.8	0.00	254.6	4.2
	CADDS-GPU-Greedy	95	20.11	4.5	0.00	260.3	5.2
	CADDS-Sequential	95	221.45	52.7	0.00	256.1	14.3
	DIDP-Parallel	95	317.96	72.2	0.13	772.3	7.1
	RPID	95	1508.56	447.9	0.01	771.5	5.1
	RouteOpt	–	–	–	–	–	–
Solomon Pesant	CADDS-GPU	21	29.17	16.3	0.07	404.8	24.3
	CADDS-GPU-Greedy	21	29.17	16.3	0.07	404.8	24.3
	CADDS-Sequential	21	607.36	326.1	0.00	698.4	47.7
	DIDP-Parallel	21	75.33	35.4	0.34	220.0	3.8
	RPID	21	435.48	213.7	0.10	219.9	2.6
	RouteOpt	–	–	–	–	–	–
Solomon Potvin Bengio	CADDS-GPU	26	46.63	26.8	0.07	764.1	18.7
	CADDS-GPU-Greedy	26	36.36	14.0	0.08	425.3	19.8
	CADDS-Sequential	26	731.85	365.1	0.00	696.5	44.6
	DIDP-Parallel	26	167.05	69.2	0.39	489.4	4.5
	RPID	26	1128.60	523.8	0.07	485.0	3.0
	RouteOpt	–	–	–	–	–	–

search nodes than the CADDS and DIDP variants. The total number of nodes appears as zero in the table as it is too small for the scale. However, CADDS-GPU outperforms RouteOpt by about two orders of magnitude.

Performance Plots. The plots in Fig. 1 show the percentage of solved instances as a function of time for each method, grouped by benchmark. For this analysis, all instances were included. On the AFG and Gendreau-Dumas-Extented, DIDP-Parallel closes one more instance than CADDS-GPU. Once again, it appears that this is a consequence of the dominance weakening. On the feasible Solnon instances, CADDS-GPU solves six more instances than any other solver. Yet, for all the plots, CADDS-GPU dominates in terms of anytime performance, climbing

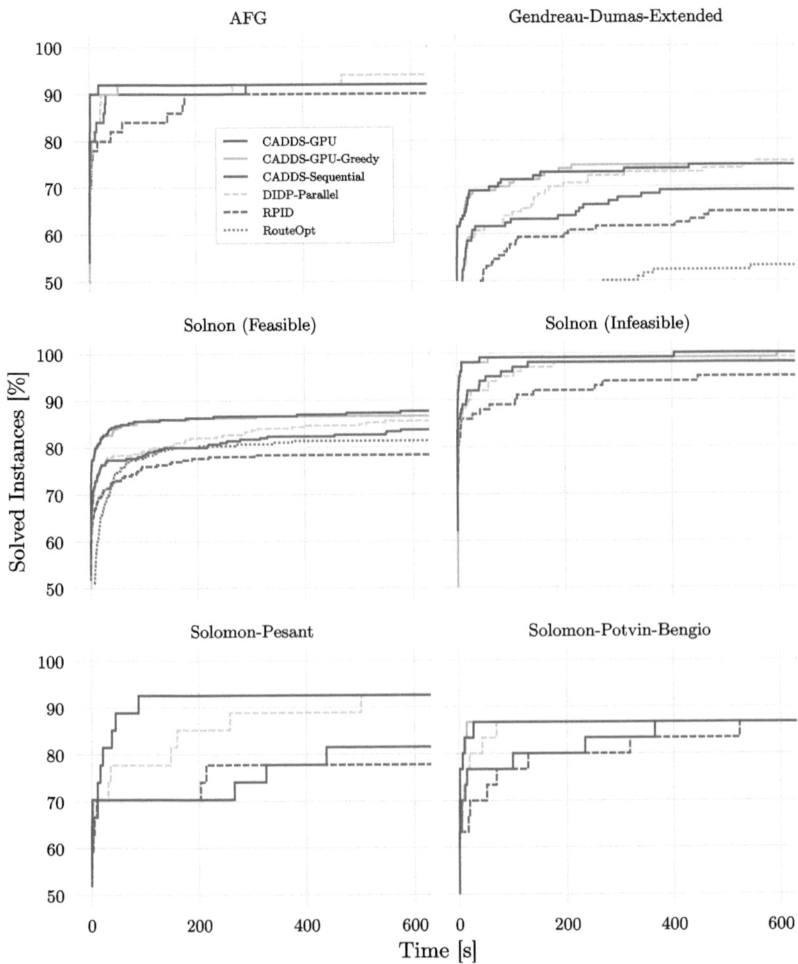

Fig. 1. Empirical cumulative distributions of performance across all instances (truncated at 50% for readability).

to a high success rates early on, and retaining that advantage throughout in almost all classes.

6   Conclusions

We introduced a GPU implementation of the state expansion process for a complete anytime decision diagram search for dynamic programming models. We rely on the structure of restricted decision diagrams of limited width to create fragments of each node layer, which are processed by the GPU using a massively parallel implementation. By traversing the fragments of the restricted decision diagram in a depth-first manner, we obtain an exact search method with anytime behavior. We empirically demonstrate on the Traveling Salesman Problem with Time Windows that GPU acceleration obtains an order of magnitude speedup compared to the sequential version, and up to two orders of magnitude speedup compared to the CABS solver in DIDP and the dedicated solver RouteOpt. We conclude that GPU acceleration of state-based search methods provides substantial computational improvements, at no cost to the modeler who states their problem as a dynamic programming model.

Acknowledgments. The authors would like to thank Prof. John D. Owens for his thoughtful insights on GPU-based hash tables and duplicate detection.

References

1. Ascheuer, N.: Hamiltonian Path Problems in the On-line Optimization of Flexible Manufacturing Systems. Ph.D. thesis, Konrad-Zuse-Zentrum für Informationstechnik, Berlin (1996). https://nbn-resolving.org/urn:nbn:de:0297-zib-5328
2. Bergman, D., Cire, A.A., van Hoeve, W.J., Hooker, J.N.: Decision Diagrams for Optimization. Springer (2016). https://doi.org/10.1007/978-3-319-42849-9
3. Bergman, D., Cire, A.A., Sabharwal, A., Samulowitz, H., Saraswat, V., van Hoeve, W.-J.: Parallel combinatorial optimization with decision diagrams. In: Simonis, H. (ed.) CPAIOR 2014. LNCS, vol. 8451, pp. 351–367. Springer, Cham (2014). https://doi.org/10.1007/978-3-319-07046-9_25
4. Coppé, V., Gillard, X., Schaus, P.: Modeling and exploiting dominance rules for discrete optimization with decision diagrams. In: Dilkina, B. (ed.) Proceedings of CPAIOR. Lecture Notes in Computer Science, vol. 14742, pp. 226–242. Springer (2024). https://doi.org/10.1007/978-3-031-60597-0_15
5. Dumas, Y., Desrosiers, J., Gélinas, É., Solomon, M.M.: An optimal algorithm for the traveling salesman problem with time windows. Oper. Res. **43**(2), 367–371 (1995). https://doi.org/10.1287/OPRE.43.2.367
6. Frohner, N., Gmys, J., Melab, N., Raidl, G.R., Talbi, E.: Parallel beam search for combinatorial optimization. In: Workshop Processing of ICCP, pp. 21:1–21:8. ACM (2022). https://doi.org/10.1145/3547276.3548633
7. Furcy, D., Koenig, S.: Limited discrepancy beam search. In: Kaelbling, L.P., Saffiotti, A. (eds.) Proceedings of IJCAI, pp. 125–131. Professional Book Center (2005). http://ijcai.org/Proceedings/05/Papers/0596.pdf
8. Gendreau, M., Hertz, A., Laporte, G., Stan, M.: A generalized insertion heuristic for the traveling salesman problem with time windows. Oper. Res. **46**(3), 330–335 (1998). https://doi.org/10.1287/OPRE.46.3.330
9. Gillard, X., Schaus, P., Coppé, V.: Ddo, a generic and efficient framework for MDD-based optimization. In: Bessiere, C. (ed.) Proceedings of IJCAI, pp. 5243–5245. ijcai.org (2020). https://doi.org/10.24963/IJCAI.2020/757
10. Kirk, D.B., Wen-Mei, W.H.: Programming Massively Parallel Processors: A Hands-on Approach, 4 edn. Morgan Kaufmann (2022). https://doi.org/10.1016/C2020-0-02969-5
11. Kuroiwa, R., Beck, J.C.: Domain-independent dynamic programming: generic state space search for combinatorial optimization. In: Koenig, S., Stern, R., Vallati, M. (eds.) Proceedings of ICAPS, pp. 236–244. AAAI Press (2023). https://doi.org/10.1609/ICAPS.V33I1.27200

12. Kuroiwa, R., Beck, J.C.: Parallel beam search algorithms for domain-independent dynamic programming. In: Wooldridge, M.J., Dy, J.G., Natarajan, S. (eds.) Proceedings of AAAI, pp. 20743–20750. AAAI Press (2024). https://doi.org/10.1609/AAAI.V38I18.30062

13. Kuroiwa, R., Beck, J.C.: RPID: rust programmable interface for domain-independent dynamic programming. In: de la Banda, M.G. (ed.) Proceedings of CP. LIPIcs, vol. 340, pp. 23:1–23:21. Schloss Dagstuhl - Leibniz-Zentrum für Informatik (2025). https://doi.org/10.4230/LIPICS.CP.2025.23

14. Libralesso, L., Bouhassoun, A., Cambazard, H., Jost, V.: Tree search for the sequential ordering problem. In: Giacomo, G.D., et al. (eds.) Proceedings of ECAI. Frontiers in Artificial Intelligence and Applications, vol. 325, pp. 459–465. IOS Press (2020). https://doi.org/10.3233/FAIA200126

15. Michel, L., van Hoeve, W.: CODD: a decision diagram-based solver for combinatorial optimization. In: Proceedings of ECAI, vol. 392, pp. 4240–4247. IOS Press (2024). https://doi.org/10.3233/FAIA240997

16. O'Neil, R.J., Hoffman, K.: Decision diagrams for solving traveling salesman problems with pickup and delivery in real time. Oper. Res. Lett. 47(3), 197–201 (2019). https://doi.org/10.1016/J.ORL.2019.03.008

17. Pesant, G., Gendreau, M., Potvin, J., Rousseau, J.: An exact constraint logic programming algorithm for the traveling salesman problem with time windows. Transp. Sci. 32(1), 12–29 (1998). https://doi.org/10.1287/TRSC.32.1.12

18. Potvin, J., Bengio, S.: The vehicle routing problem with time windows part II: genetic search. INFORMS J. Comput. 8(2), 165–172 (1996). https://doi.org/10.1287/IJOC.8.2.165

19. Rifki, O., Solnon, C.: On the phase transition of the euclidean travelling salesman problem with time windows. J. Artif. Intell. Res. 82, 2167–2188 (2025). https://doi.org/10.1613/JAIR.1.18334

20. Rudich, I., Cappart, Q., Rousseau, L.: Improved peel-and-bound: methods for generating dual bounds with multivalued decision diagrams. J. Artif. Intell. Res. 77, 1489–1538 (2023). https://doi.org/10.1613/JAIR.1.14607

21. Tardivo, F., Pontelli, E.: Parallel declarative solutions of sequencing problems using multi-valued decision diagrams and gpus. In: Cheney, J., Perri, S. (eds.) PADL 2022. LNCS, vol. 13165, pp. 191–207. Springer, Cham (2022). https://doi.org/10.1007/978-3-030-94479-7_13

22. Vadlamudi, S.G., Aine, S., Chakrabarti, P.P.: Incremental beam search. Inf. Process. Lett. 113(22–24), 888–893 (2013). https://doi.org/10.1016/J.IPL.2013.08.010

23. You, Z., Yang, Y.: RouteOpt: an open-source modular exact solver for vehicle routing problems. SSRN (2025). https://doi.org/10.2139/ssrn.5314242

24. Zhang, W.: Complete anytime beam search. In: Mostow, J., Rich, C. (eds.) Proceedings AAAI, pp. 425–430. AAAI Press/The MIT Press (1998). http://www.aaai.org/Library/AAAI/1998/aaai98-060.php

25. Zhou, R., Hansen, E.A.: Beam-stack search: integrating backtracking with beam search. In: Biundo, S., Myers, K.L., Rajan, K. (eds.) Proceedings of ICAPS, pp. 90–98. AAAI (2005). https://aaai.org/papers/icaps-05-010-beam-stack-search-integrating-backtracking-with-beam-search

Optimization over Trained Neural Networks: Going Large with Gradient-Based Algorithms

Jiatai Tong[1], Yilin Zhu[2], Thiago Serra[2(✉)], and Samuel Burer[2]

[1] Northwestern University, Evanston, IL, USA
jiataitong2026@u.northwestern.edu
[2] University of Iowa, Iowa City, IA, USA
{yilin-zhu,thiago-serra,samuel-burer}@uiowa.edu

Abstract. When optimizing a nonlinear objective, one can employ a neural network as a surrogate for the nonlinear function. However, the resulting optimization model can be time-consuming to solve globally with exact methods. As a result, local search that exploits the neural-network structure has been employed to find good solutions within a reasonable time limit. For such methods, a lower per-iteration cost is advantageous when solving larger models. The contribution of this paper is two-fold. First, we propose a gradient-based algorithm with lower per-iteration cost than existing methods. Second, we further adapt this algorithm to exploit the piecewise-linear structure of neural networks that use Rectified Linear Units (ReLUs). In line with prior research, our methods become competitive with—and then dominant over—other local search methods as the optimization models become larger.

Keywords: Constraining learning · Gradient ascent · Linear regions · Piecewise-linear functions · Rectified linear units

1 Introduction

In the field of mathematical programming, piecewise-linear functions play an important role in modeling nonlinear functions [25,34,36,37,60]. In deep learning, a popular model that provides a piecewise-linear approximation of a nonlinear function is the neural network with the ReLU activation function [2,24]. Researchers have long known that some neural network architectures are universal function approximators [11,15,22], and in particular, this is also true of the ReLU activation function if the architecture is sufficiently wide [62] or deep [20]. When neural-network approximations are used as surrogates for solving nonlinear optimization problems, algorithms that exploit the piecewise-linear structure of the neural networks are of particular interest. In this paper, we propose gradient-based algorithms for this setting.

Optimizing a piecewise-linear function over a polyhedron can be modeled using a Mixed-Integer Linear Programming (MILP) formulation. In the specific

T. Guns (Ed.): CPAIOR 2026, LNCS 16595, pp. 596–613, 2026.
https://doi.org/10.1007/978-3-032-27242-3_35

case of ReLU networks, existing MILP formulations either have a weak linear relaxation due to big M coefficients [14] or become prohibitively large when using a disjunctive formulation [23]. Researchers have found success by improving the big M coefficients [3,10,16,21,30,51,63], strengthening formulations using valid inequalities [1], and using a hybrid of both formulations [56]. Other improvements include reformulation [23,32,45], parameter rescaling [40], pruning the search space by inference [6,46,54,61], and working with sparser neural networks [7,39, 44,61]. Those improvements can also help using MILP formulations in neural networks for verification [1,10,43,52], compression [48,49], evaluating output variation [8,29,31,46,47], and counterfactual explanations [28,57].

There are also local search methods, which do not solve MILPs exactly, but instead are designed to find good solutions in limited time. Indeed, both Perakis & Tsiourvas [38] and Tong et al. [55] are closely related to this paper for taking a geometric view of the input space of the ReLU network, focusing on the linear pieces of the function approximation. Known as *linear regions* in machine learning, each piece corresponds to a polyhedron associated with a distinct set of active neurons. Within a linear region, changes to the input have a linear impact on the output. In Fig. 1, we illustrate these concepts on a neural network having inputs x_1 and x_2, neurons with outputs h_1 to h_5, and output y.

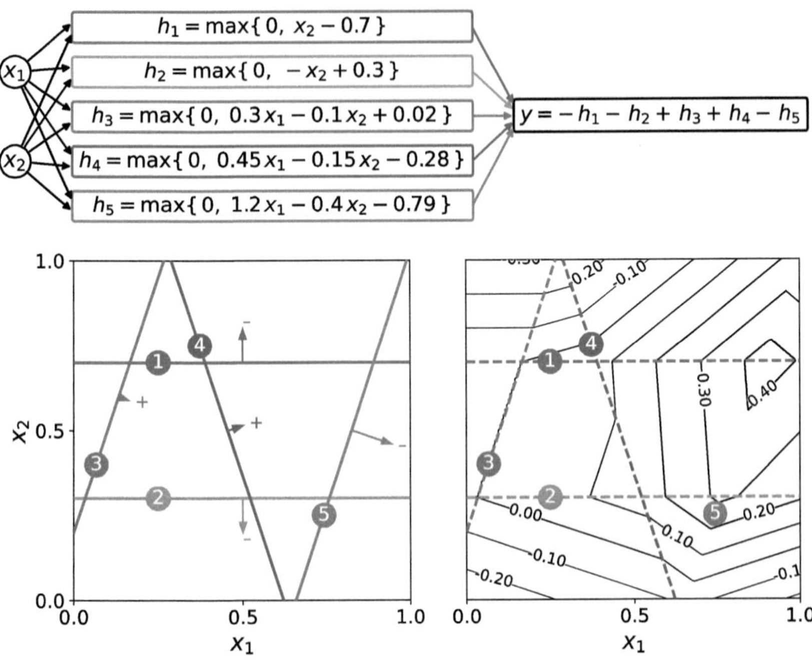

Fig. 1. Top: Visual description of neural network used as example. **Bottom left:** Lines partitioning the space based on what inputs produce a positive output for each neuron, with the arrow pointing to the positive side, the length of the arrow proportional to the magnitude of the parameters, and the arrow label denoting the influence on y. **Bottom right:** Contour plots of y over the lines associated with neural activations.

We summarize the works [38,55] just mentioned, taking the liberty to name them *MILP Walk* and *LP Walk*, respectively, in order to draw parallels between these methods and our methods introduced in Sect. 3:

- **MILP Walk:** Perakis & Tsiourvas [38] solve a sequence of restricted MILP models. Each MILP model finds the best solution across all linear regions containing the current solution. If a better solution is found, the same process is repeated from the new solution. In Fig. 2 Left, solution A lies only in the linear region in darkest gray, in which the best solution is B. In turn, solution B lies in the four linear regions with the three darker tones of gray, where the best solution is C. Finally, solution C lies in the four linear regions with the two lighter tones of gray, where the best solution is C again. Once no improvement is found, the algorithm stops.
- **LP Walk:** Tong et al. [55] solve a sequence of LP models. Each LP model finds the best solution in a linear region containing the current solution. If a better solution is found, they repeat the process by moving slightly past the new solution along the line from the last solution. Moving slightly past avoids using a solution lying in multiple linear regions. In Fig. 2 Right, the linear region of solution A is in darker gray, and its best solution is along the line from A to B. In turn, the linear region of solution B is in slightly lighter gray, and its the best solution is along the line from B to C. From C we find solution E and move towards D. From D we find solution E again. At this point, the algorithm stops.

For a broader discussion about linear regions and their nexus with mathematical optimization, we recommend the survey by Huchette et al. [24].

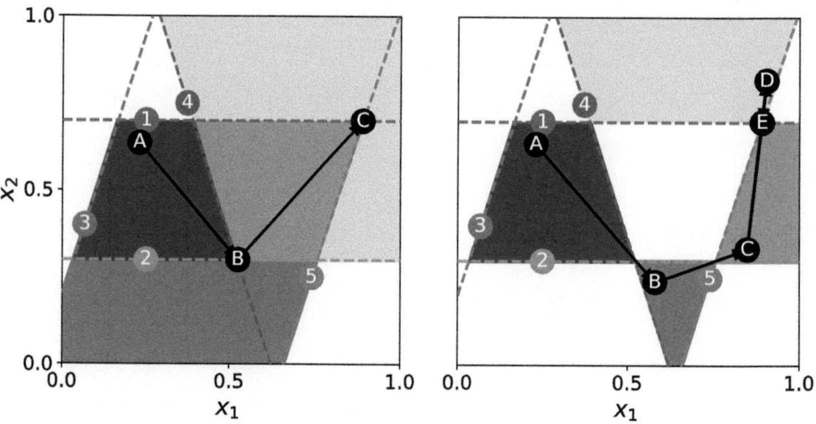

Fig. 2. Left: Solutions found by MILP Walk from the initial solution $A = (0.23, 0.636)$ until convergence. **Right:** Solutions found by LP Walk from A until convergence.

Which algorithm, MILP Walk or LP Walk, is best suited for a particular instance often depends on the size of that instance. Starting from the same solution, it is easy to see that MILP Walk will move next to a solution that is at least as good as the one found by LP Walk. On the other hand, each iteration of MILP Walk solves the same MILP model used to optimize over the entire neural network, albeit restricted to the neighborhood of the current solution. Hence, the per-iteration cost of MILP Walk is higher than the per-iteration cost of LP Walk. Consequently, LP Walk can perform more iterations in the same amount of time. Indeed, an empirical comparison of both methods shows that LP Walk performs better for neural networks with more inputs, layers, and neurons [55], all of which imply a larger number of linear regions [47]. Thus, LP Walk conducts a style of search that is more akin to sampling than to enumeration [46].

Of course, solving an LP model for each linear region, as in LP Walk, may eventually become too costly in ever larger neural networks. Hence, in this paper, we propose a new local search approach with an even smaller per-iteration cost:

- **Gradient Walk:** We compute a sequence of gradient steps. Each step may find a better solution within the current linear region, or a solution in another linear region that is better or worse. We keep track of the best solution found. If the improvement is too small for a predefined number of steps, we restart from a perturbation of the best solution found thus far. In Fig. 3 Left, we move from A to B through nine intermediary steps in the same linear region as A, all in the same direction, and then from B to C with five intermediary steps. Note that the steps are orthogonal to the contour plots within the linear regions. In Fig. 3 Right, we continue from C until H by zig-zagging among linear regions and improving in all steps except G. The steps following H would not find a much better solution. Hence, the algorithm stops.

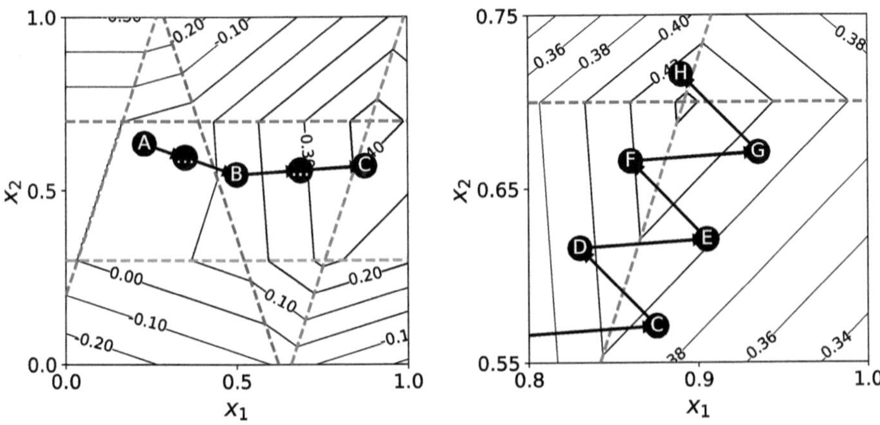

Fig. 3. Left: First solutions found by Gradient Walk from the initial solution $A = (0.23, 0.636)$. **Right:** Next solutions found over a narrower region of the input space.

In what follows, we define our problem of interest and its conventional MILP model in Sect. 2. Then we present an algorithm for Gradient Walk as well as a variant that further exploits knowledge of linear regions in Sect. 3. We evaluate those algorithms in Sect. 4. Conclusions are given in Sect. 5.

2   Preliminaries

We optimize the function $f : \mathbb{R}^{n_0} \to \mathbb{R}$ associated with a neural network over a polytope $\mathbb{X} \subset \mathbb{R}^{n_0}$:

$$\text{maximize}_x \quad f(\boldsymbol{x}) \tag{1}$$

$$\text{s.t.} \quad \boldsymbol{x} \in \mathbb{X} \tag{2}$$

i.e., we want an input $\boldsymbol{x} = [x_1\ x_2\ \ldots\ x_{n_0}]^\top \in \mathbb{X}$ maximizing the prediction $f(\boldsymbol{x})$. In the experiments of Sect. 4, $\mathbb{X}$ equals a box, but our theoretical development requires only that $\mathbb{X}$ be polyhedral.

Let the neural network have L hidden layers, each hidden layer $l \in \mathbb{L} := \{1, \ldots, L\}$ having preactivation values $\boldsymbol{g}^l = [g_1^l\ g_2^l \ldots g_{n_l}^l]^\top$ and outputs $\boldsymbol{h}^l = [h_1^l\ h_2^l \ldots h_{n_l}^l]^\top$ from neurons indexed by $i \in \mathbb{N}_l = \{1, 2, \ldots, n_l\}$. The output layer $L + 1$ has a single preactivation value g_1^{L+1} as the network output. Let $\boldsymbol{W}^l$ and $\boldsymbol{b}^l$ denote the weight matrix and bias vector associated with the l-th layer, for which we assume that all elements are within $[-1, 1]$. The preactivation value g_i^l of neuron i in layer l is given by $g_i^l = \boldsymbol{W}_i^l \boldsymbol{h}^{l-1} + b_i^l$, and the output h_i^l of neuron i in hidden layer l follows by the ReLU activation $h_i^l = \max\{0, g_i^l\}$. In this setting, $\boldsymbol{x} = \boldsymbol{h}^0$ and $f(\boldsymbol{x}) = g_1^{L+1}$.

For an input $\boldsymbol{x} \in \mathbb{X}$, let $\boldsymbol{z}^l(\boldsymbol{x}) = [z_1^l\ z_2^l \ldots z_{n_l}^l]^\top$ be the *layer activation pattern* produced in layer l of the neural network when given $\boldsymbol{x}$ as input, where:

$$z_i^l = \begin{cases} 1, & \text{if } h_i^l = g_i^l \geq 0, \\ 0, & \text{if } g_i^l \leq 0. \end{cases} \tag{3}$$

A neuron is *binding* if $g_i^l = h_i^l = 0$. In this case, z_i^l can be either 0 or 1.

Let $\boldsymbol{z}(\boldsymbol{x}) = \{\boldsymbol{z}^1(\boldsymbol{x}), \boldsymbol{z}^2(\boldsymbol{x}), \ldots, \boldsymbol{z}^L(\boldsymbol{x})\}$ be the *activation pattern* produced across all layers of the neural network when given $\boldsymbol{x}$ as input. A *linear region* $\mathcal{R}_{z'} \in \mathbb{X}$ is a set where every input $\boldsymbol{x}$ has the same activation pattern $\boldsymbol{z}'$ as all the other inputs, i.e., $\boldsymbol{z}(\boldsymbol{x}) = \boldsymbol{z}' \ \forall \boldsymbol{x} \in \mathcal{R}_{z'}$. The output of f varies linearly within each linear region. Inputs lie in multiple linear regions if a neuron is binding.

We can formulate the optimization problem as the following MILP model:

$$\text{maximize} \quad f(\boldsymbol{x}) = g_1^{L+1} \tag{4}$$

$$\text{s.t.} \quad \boldsymbol{h}^0 = \boldsymbol{x}, \qquad \boldsymbol{x} \in \mathbb{X} \tag{5}$$

$$\boldsymbol{W}_i^l \boldsymbol{h}^{l-1} + b_i^l = g_i^l, \qquad\qquad \forall l \in \mathbb{L} \cup \{L+1\}, i \in \mathbb{N}_l \tag{6}$$

$$(z_i^l = 1) \to (h_i^l = g_i^l), \qquad\qquad \forall l \in \mathbb{L}, i \in \mathbb{N}_l \tag{7}$$

$$(z_i^l = 0) \to (g_i^l \leq 0 \wedge h_i^l = 0), \qquad\qquad \forall l \in \mathbb{L}, i \in \mathbb{N}_l \tag{8}$$

$$h_i^l \geq 0, \qquad z_i^l \in \{0, 1\}, \qquad\qquad \forall l \in \mathbb{L}, i \in \mathbb{N}_l \tag{9}$$

The indicator constraints (7)–(8) can be modeled with big M constraints [5].

We can formulate an LP model optimizing over a single linear region $\mathcal{R}_{z'}$ by fixing $z = z'$. For example, this is what LP Walk does in each iteration.

3   Our Proposed Algorithms

We propose two gradient-based algorithms for solving (1)–(2). We assume throughout that the polyhedral domain $\mathbb{X}$ is bounded, i.e., $\mathbb{X}$ is a polytope. Indeed, neural networks are often used within a confined domain. There is a growing body of work on preventing a neural network from extrapolating outside the nominal or implied domain defined by a training set [50,57,64]. Moreover, the possibility of embedding a neural network as part of an MILP model depends on the input set being bounded [47], which otherwise would also prevent us from benchmarking against existing algorithms [38,55].

Our first algorithm, the Perturbed Projected Gradient Ascent (PPGA) algorithm, is described in Sect. 3.1. PPGA makes use of both the piecewise-constant nature of ∇f and projection onto $\mathbb{X}$. Our second algorithm, PPGA with Linear Region Valve (PPGA$_{\mathrm{LR}}$), is described in Sect. 3.2. PPGA$_{\mathrm{LR}}$ enhances PPGA by further exploiting the structure of the linear regions, being particularly beneficial if the number of linear regions is large.

3.1   Perturbed Projected Gradient Ascent (Algorithm 1)

We first mention the standard approach, called the Projected Gradient Ascent (PGA) algorithm, of projecting the gradient steps over the feasible set to produce iterates of the form

$$x^{t+1} = P_{\mathbb{X}}(x^t + \gamma \nabla f(x^t)), \tag{10}$$

where γ is the learning rate and L_2 projection [41,42] solves the convex quadratic program (QP)

$$P_{\mathbb{X}}(\dot{x}) = \arg \min_{x} \quad \|x - \dot{x}\|^2 \tag{11}$$

$$\text{s.t.} \quad x \in \mathbb{X}, \tag{12}$$

which takes only $O(n_0)$ time when $\mathbb{X}$ is a box.

Based on PGA, we now introduce our method PPGA. For any point $x' \in \mathcal{R}_{z(x)}$, i.e., any point x' in the same linear region as x, we can calculate $f(x')$ with the affine transformation

$$f|_{\mathcal{R}_z(x)}(x') = T(z(x))x' + t(z(x)), \tag{13}$$

where

$$T(z(x)) = w^{L+1} \left(\prod_{l=1}^{L} (z^l(x)I)W^l \right) \tag{14}$$

and $t(z(x))$ is constant [24]. Hence, it follows that

$$\nabla f(x) = w^{L+1} \left(\prod_{l=1}^{L} (z^l I) W^l \right) \tag{15}$$

for any point x at the interior of its linear region, i.e., not binding for any neuron. Hence, f has a piecewise affine landscape, which may contain local maxima, saddle points, and local minima.

For smooth maximization problems, one may generally escape from (interior feasible) saddle points and local minima by introducing a perturbation when $\nabla f(x)$ is small enough [17,26,27,59]. Our case is quite different, however, since all differentiable regions have constant gradients and because saddle points and local minima occur at the boundary of those regions, where the function is

---

Algorithm 1. Perturbed Projected Gradient Ascent

Input: Model f with input size n_0, learning rate γ, time limit T, restart noise coefficient Ξ, error threshold ϵ, tolerance window k; input domain $\mathbb{X}$

Output: Best solution x^* and objective value $f(x^*)$

1: $\delta \leftarrow \dfrac{\Xi}{\sqrt{n_0}}$
2: Sample initial $x \sim \text{Uniform}(\mathbb{X})$
3: $x^* \leftarrow x$ ▷ Initialized best solution so far
4: $x' \leftarrow x$ ▷ Initialized best solution since reset
5: $r \leftarrow 0$ ▷ Initialized reset counter
6: **while** Time $< T$ **do**
7:     $x \leftarrow P_{\mathbb{X}}(x + \gamma \nabla f(x))$ ▷ Gradient step; replaced with Algorithm 2 in PPGA$_{\text{LR}}$
8:     **if** $f(x) > f(x')$ **then**   ▷ If found best solution since reset
9:         $\Delta \leftarrow f(x) - f(x')$ ▷ Calculate local improvement before update
10:        $x' \leftarrow x$ ▷ Update best solution since reset
11:        **if** $f(x) > f(x^*)$ **then**   ▷ If found best overall solution
12:            $x^* \leftarrow x$ ▷ Update best overall solution
13:        **end if**
14:        **if** $\Delta < f(x) \cdot \epsilon$ **then**   ▷ If local improvement is below threshold
15:            $r \leftarrow r + 1$ ▷ Increment reset counter
16:            **if** $r = k$ **then**   ▷ If reached reset trigger
17:                $x \leftarrow P_{\mathbb{X}}(x^* + \xi), \quad \xi \sim \mathcal{N}(0, \delta)$ ▷ Reset event
18:                $x' \leftarrow x$
19:                $r \leftarrow 0$
20:            **end if**
21:        **else**   ▷ If the improvement is above the threshold
22:            **if** $f(x) = f(x^*)$ **then** ▷ If current solution matches best overall solution
23:                $r \leftarrow 0$ ▷ Reset the counter
24:            **end if**
25:        **end if**
26:    **end if**
27: **end while**
28: **return** x^*, $f(x^*)$

---

nondifferentiable. Hence, the size of $\nabla f(x)$ is not a reliable measure of local optimality in our case.

We hence use a different mechanism to measure progress. If we accumulate k improvements that are relatively small in comparison to the current objective value $f(x^t)$, while not producing a better overall solution, then we continue the next iteration from a perturbation $\xi \sim \mathcal{N}(0, \frac{\Xi}{\sqrt{n_0}})$ around the current best solution x^*. If that happens at step t, then the next update is

$$x^{t+1} = P_\mathbb{X}(x^* + \xi + \gamma \nabla f(x^* + \xi)) \tag{16}$$

The resultant algorithm, our Perturbed Projected Gradient Ascent (PPGA) algorithm, is described as Algorithm 1.

3.2 PPGA With Linear Region Valve (Algorithms 1 and 2)

When the ReLU network gets deeper, the gradient calculated with Equation (15) may either explode or diminish, which makes it important to find an appropriate learning rate. As an alternative to calibrate the learning rate, we propose using linear region information for making local decisions about the size of the gradient step. Because we assume in this paper that all weights are within $[-1, 1]$, we expect the gradient to diminish when the network gets deeper. Notably, a similar approach can also be applied to resolve gradient explosion.

Suppose that we are in the linear region $\mathcal{R}_z$ with activation pattern z. We can calculate how far we may move to reach the next linear region in the direction of the gradient by a ratio test:

$$u = \min_{i \in \mathcal{I}}\left(-\frac{g_i}{\Delta g_i(x)}\right) \tag{17}$$

where $\mathcal{I} = \left\{ i \mid i \in \mathbb{N}_l, l \in \mathbb{L}, -\frac{g_i}{\Delta g_i(x)} \geq 0 \right\}$ is a subset of all neurons, and

$$\Delta g_i(x) = g_i(x + \nabla f(x)) - g_i(x). \tag{18}$$

Here, u is the relative step size to the next linear region, while the actual step size is $u \cdot \|\nabla f(x^t)\|$. We can use u as an estimate for the size of a linear region $\mathcal{R}_{z'}$ near $\mathcal{R}_z$, i.e., $\|z - z'\|_1 \leq \zeta$ with a relatively small ζ. Given also the relative step size γ, then we estimate the gradient step to stretch over

$$v = \left\lceil \frac{\gamma}{u} \right\rceil \tag{19}$$

linear regions around $\mathcal{R}_z$.

Let $V > 1$ be a predetermined valve value, which corresponds to the number of linear regions that we would like to stretch over at each gradient step. If γ is such that $v < V$ at the current linear region, then we use a scale factor c over the magnitude of the gradient:

$$x^{t+1} = \begin{cases} P_\mathbb{X}\left(x^t + c\,\dfrac{\nabla f(x^t)}{\|\nabla f(x^t)\|}\right), & \text{if } \gamma \leq V \cdot u, \\ P_\mathbb{X}\left(x^t + \gamma\,\nabla f(x^t)\right), & \text{otherwise.} \end{cases} \tag{20}$$

Algorithm 2. Adaptive Gradient Step with Linear Region Valve

Input: Current solution x, learning rate γ, valve value V, scale factor c
Output: New iterate x' ▷ Replaces iterate calculated in Line 7 of Algorithm 1

1: $\nabla f(x) \leftarrow w^{L+1}\left(\prod_{l=1}^{L}(z^l I)W^l\right)$
2: $g \leftarrow f(x)$
3: $g' \leftarrow f\big(x + \nabla f(x)\big)$
4: $\Delta g \leftarrow g' - g$
5: $\rho \leftarrow -g/\Delta g$
6: Mask all negative entries in ρ with $+\infty$
7: $u \leftarrow \min(\rho)$
8: **if** $V \cdot u \geq \gamma$ **then**
9:     $x' \leftarrow P_{\mathbb{X}}\left(x + c \cdot \dfrac{\nabla f(x)}{\|\nabla f(x)\|}\right)$
10: **else**
11:     $x' \leftarrow P_{\mathbb{X}}\big(x + \gamma \cdot \nabla f(x)\big)$
12: **end if**
13: **return** x'

The adaptive gradient step replaces Line 7 in Algorithm 1 with Algorithm 2. In our implementation, we simply chose $V = \frac{1}{\|\nabla f(x^t)\|}$ as an adaptive valve value, so that a small gradient will more likely trigger the rescaled gradient update. We also set $c = u$, so that we force the step size stretching more than V linear regions, since each linear region is estimated to have length $u \cdot \|\nabla f(x^t)\| = \frac{u}{V}$. The main reason using adaptive hyperparameters is that applying grid search to extra hyperparameters is very costly, and we want to be fair to those algorithms with fewer hyperparameters, such as PPGA and [55].

We denote this variant of PPGA using Algorithm 2 and adaptive (V, c) design as PPGA$_{\mathrm{LR}}$.

4 Numerical Experiments

We devised numerical experiments to evaluate algorithms PPGA and PPGA$_{\mathrm{LR}}$ on standard benchmarks and compare them with other methods. In the following subsections, we define the concept of a basic experiment, then describe our method for generating multiple experiments, and finally detail the optimization results. All numerical experiments were implemented in Python 3.10.8 using Gurobi 11.0 and evaluated on a single Xeon E5-2680v4 core running at 2.4 GHz with 16 GB of memory under the CentOS Linux operating system. The source code is publicly shared at https://github.com/yillzhu/nn_opt.

4.1 Definition of an Experiment

In our study, an *experiment* $\mathcal{P}$ refers to a complete specification of five design options that determine the structure of a neural network and the algorithm used:

- *Input size*, i.e., the input dimension n_0 of the neural network.
- *Depth*, i.e., the number of hidden layers d of the neural network.
- *Width*, i.e., the number of neurons m in each hidden layer.
- *Seed*, i.e., the seed s used to instantiate the network parameters.
- *Algorithm*, i.e., the local search method $\mathcal{M}$ used for optimization.

Given a particular combination of these specifications, the experiment proceeds as follows. With seed s fixed, we generate a neural network having input dimension n_0, depth d, and width m. For each optimization algorithm $\mathcal{M}$, we then conduct a grid search to determine the optimal hyperparameters: shaking noise σ, error threshold ϵ, and tolerance window k. Once these parameters are fixed, the algorithm $\mathcal{M}$ is used to optimize the network within a specified time limit.

In line with prior work [38,55] and to properly benchmark with it, we optimize over neural networks with their weights as defined at initialization. Hence, instead of optimizing over neural networks approximating specific functions based on their training, we work with neural networks representing distinct and random functions. We believe that this makes the results more representative.

4.2   Generating Multiple Experiments

In addition to PPGA and PPGA$_{\mathrm{LR}}$, we use the PGA algorithm from Sect. 3.1 as a baseline and benchmark against the algorithm proposed for LP Walk in [55], which we denote as SimplexWalk. SimplexWalk has shown better scalability than solving directly with Gurobi [19] or with MILP Walk [38]. Hence,

$$\mathcal{M} \in \{\mathrm{PPGA}_{\mathrm{LR}}, \mathrm{PPGA}, \mathrm{PGA}, \mathrm{SimplexWalk}\}.$$

We use input size $n_0 \in \{10, 100, 1000\}$, depth $d \in \{2, 4, 6, 8\}$, and width $m \in \{100, 1000, 10000\}$, resulting in $3 \times 3 \times 4 = 36$ distinct network configurations.

For gradient-based algorithms (PPGA$_{\mathrm{LR}}$, PPGA, PGA), performance is very susceptible to hyperparameters. Therefore, we use a preprocessing phase to search for better hyperparameters for each gradient-based algorithm through grid search and voting. For each network configuration (n_0, d, m), we generate five random networks using seeds $s \in \{5, 6, 7, 8, 9\}$. These networks, denoted as *grid search instances*, are used for choosing hyperparameters over a grid with

$$\gamma \in \{0.001, 0.01, 0.1, 1, 5\}, \quad \sigma \in \{0.2, 2, 20\}, \quad k \in \{100, 500, 1000\}.$$

For each grid search instance, we evaluate all $5 \times 3 \times 3 = 45$ parameter combinations, running the optimization for 5 min per grid point. The five combinations achieving the best objective values are recorded for each instance. After processing all five grid search instances, we obtain five sets of top-performing parameter combinations. The most frequently occurring combination across these sets is selected as the final grid search result for that algorithm and network configuration. This process runs independently for each gradient-based algorithm.

Once the hyperparameters are calibrated, we generate 20 additional networks for each (n_0, d, m) combination using seeds $s \in \{10, 11, \ldots, 29\}$. These are referred to as *optimization instances*. Each algorithm $\mathcal{M}$ is then applied to these instances, and the gradient-based algorithms use the hyperparameters obtained from the grid search. Every run is executed for 7200 s, during which we record, at every second, the best objective value and the number of iterations.

4.3   Computational Complexity of Walk Steps

The complexity of the problem can be affected by all three setup options (n_0, d, m). For gradient-based algorithms (PPGA$_{\text{LR}}$, PPGA, PGA), the asymptotic complexity of both calculating the gradient and making a prediction is

$$O(dm^2 + n_0 m).$$

which can be deduced from a sequence of vector-matrix multiplications. Through Eq. (18), each step of the linear region algorithm (PPGA$_{\text{LR}}$) may cost twice as much as PGA and PPGA steps, which we may consider still acceptable. On the other hand, SimplexWalk carries out the solution of one LP model per step. Notably, the computational cost of a SimplexWalk step, if the LP model is solved using the simplex algorithm, is exponential on n_0 in worst-case.

4.4   Results

We use the Dolan–More performance profile [12,53] to compare the performance of different algorithms on multiple problems. A performance profile shows, for each algorithm, the fraction of test instances on which it performs within a given factor of the best observed result. The horizontal axis represents the performance factor $\tau \geq 1$ in log scale, and the vertical axis $\rho_{\mathcal{M}}(\tau) \in [0, 1]$ represents the fraction of experiments for which the algorithm $\mathcal{M}$ attains a performance ratio of at most τ. The vertical intercept $\rho_{\mathcal{M}}(1)$ indicates how often an algorithm achieves the best result among all competitors, and a curve approaching $\rho = 1$ more rapidly reflects an algorithm whose performance is consistently close to the best algorithm across all experiments.

Figure 4 shows the overall performance profiles across all setups after the methods have been run for 30, 60, and 120 min. The other two plots focuses on the results after 120 min. Figure 5 shows the performance profiles from partitioning the instances according to the input size of the neural network. Figure 6 partitions the instances according to the depth and the width of the neural network.

4.5   Analysis

In general, the relationship between the three gradient algorithms is similar across different time budgets, with PPGA and PPGA$_{\text{LR}}$ outperforming the vanilla PGA. However, the gap between the first PPGA and vanilla PGA shrinks for the largest width (10000). Since width dominates the complexity of gradient steps, it is possible that PPGA algorithms do not gain much in performance advantage

Performance Profiles by Time Limit

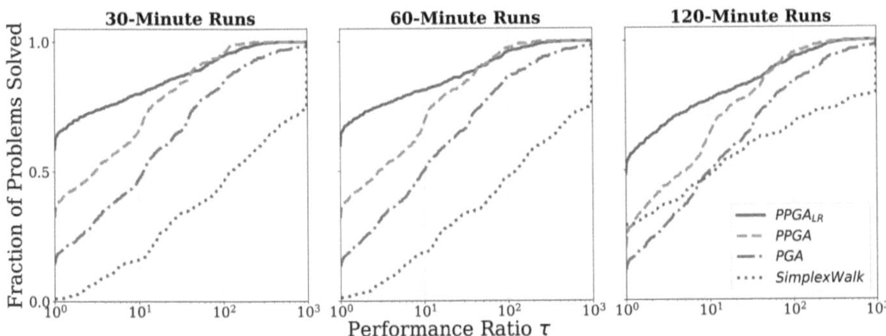

Fig. 4. Comparison of algorithm performance over all instances by varying time limit.

Performance Profile for 120-Minute Runs: Input Size

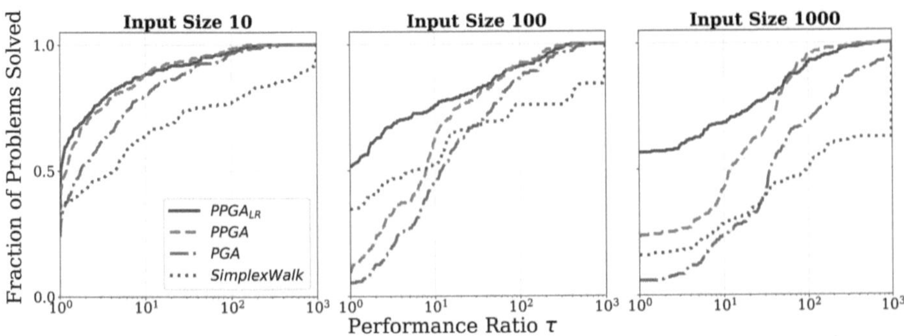

Fig. 5. Comparison of algorithm performance by varying input size in 120-minute runs.

in the early steps compared to the vanilla algorithm. Therefore, the advantage that we observe is likely gained in the later steps, when the vanilla algorithm is captured in some local optimum while PPGA escapes through perturbation.

While PPGA$_{\mathrm{LR}}$ is overall better than PPGA in the aggregate of instances from Fig. 4, the advantage of PPGA$_{\mathrm{LR}}$ is due to the instances with larger dimensions:

- From Fig. 5, we see that PPGA$_{\mathrm{LR}}$ matches PPGA for the smallest input (10) but does significantly better than PPGA for larger input sizes (100 and 1000).
- From the rows of Fig. 6, PPGA$_{\mathrm{LR}}$ is generally better for larger widths.
- From the columns of Fig. 6, we see that PPGA$_{\mathrm{LR}}$ is worse for the smallest depth (2), but that it dominates the results for deeper networks (6 and 8).

Hence, larger input size and depth seem to be the most determinant factors for PPGA$_{\mathrm{LR}}$ performing better than PPGA. Larger width also contributes in some cases. Those conditions conform with the cases in which neural networks tend to have more linear regions, which has inspired the design of PPGA_{LR} in the first place. Moreover, as indicated by Eq. 15 and given the weight distribution, neural networks with greater depth are bound to have smaller gradients.

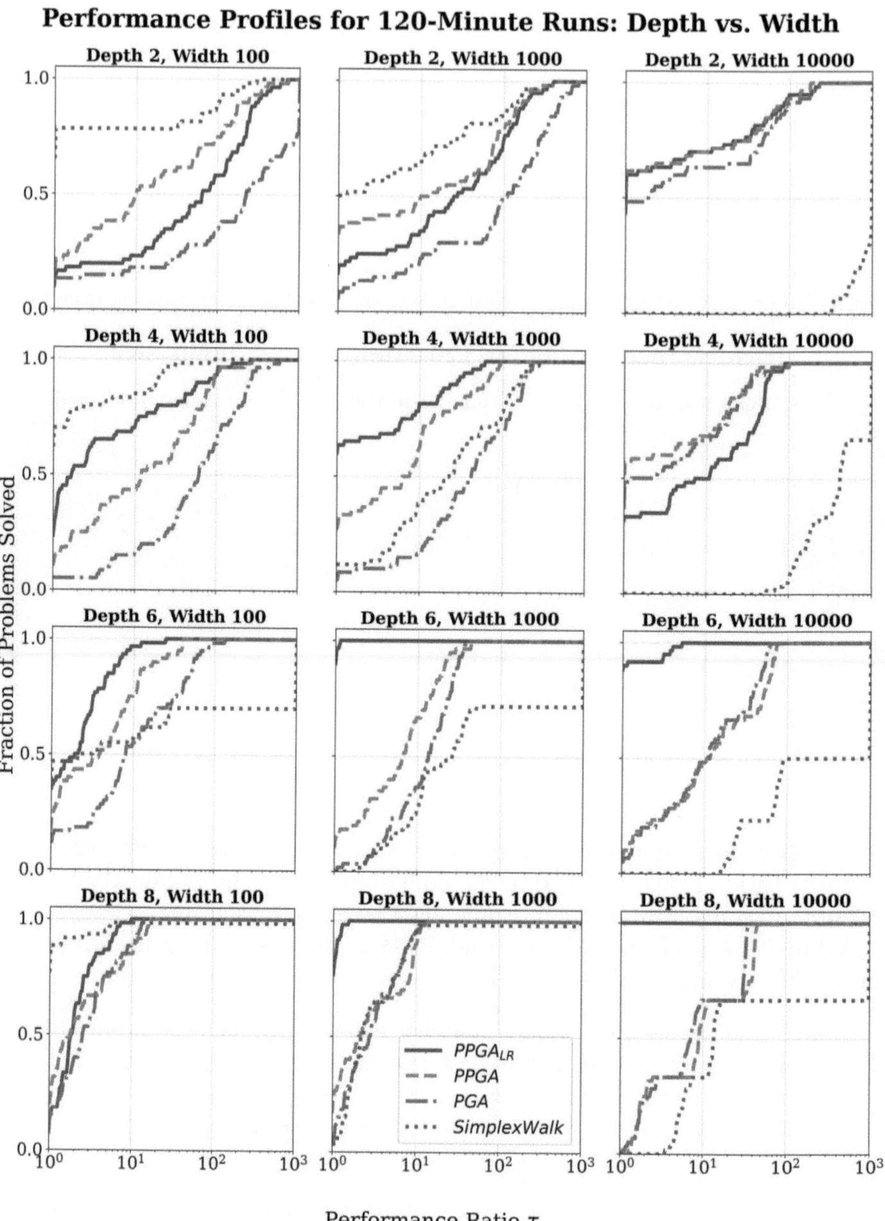

Fig. 6. Comparison of performance by varying depth and width in 120-minute runs.

Meanwhile, notably, `SimplexWalk` gets comparatively better with longer runtimes, as shown in Fig. 4. That conforms with the intuition that it takes longer to converge due to the more costly steps, but that each step tends to provide greater improvements. Indeed, we observe `SimplexWalk` dominating in the three scenarios with smallest depth and width in Fig. 6.

4.6   A Case Study on Adaptive Stepsize for PPGA$_{\mathrm{LR}}$

To better understand why PPGA$_{\mathrm{LR}}$ dominates in deeper instances, we present a case study on a randomly generated instance with $(n_0 = 1000, d = 6, w = 1000)$ and seed $s = 30$. We ran 1000 iterations of PPGA with five learning rates, $\gamma \in \{5, 50, 500, 5000, 50000\}$. We also ran 1000 iterations of PPGA$_{\mathrm{LR}}$, but with only three different learning rates, $\gamma \in \{5, 500, 50000\}$. We use fewer learning rates with PPGA$_{\mathrm{LR}}$ because there is a greater overlap in step sizes and objective values across learning rates in this case, making it harder to visually distinguish them. Figure 7 shows the influence of the learning rate γ on the step size over time. Figure 8 shows the influence of the learning rate γ on the solution value.

Our three main observations are the following:

– In PPGA, when we increase the learning rate γ, the step size changes proportionally. In PPGA$_{\mathrm{LR}}$, on the other hand, the step size is less affected by the learning rate and only the size of the smallest steps are clipped.
– In PPGA, the range of step sizes for a specific learning rate is small. In PPGA$_{\mathrm{LR}}$, on the other hand, the valve mechanism defined by Eq. (20) allows for a larger range of step sizes regardless of learning rate. That seems to lead to similar convergence for step sizes in different orders of magnitude.
– Since there is no mechanism designed for avoiding excessively large step sizes, both algorithms perform similarly worse when $\gamma = 50000$. We could potentially address this by designing another valve and truncating large steps.

Hence, PPGA is more sensitive to the learning rate and does not converge well with lower learning rates ($\gamma \in \{5, 50\}$). In turn, PPGA$_{\mathrm{LR}}$ is more robust and converges well with lower learning rates in deep neural networks, as observed in Fig. 6.

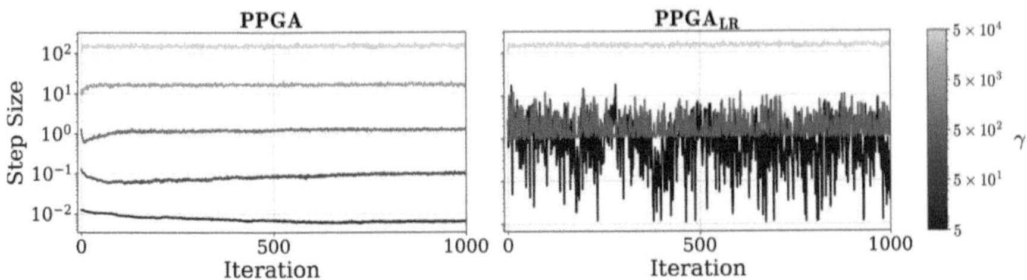

Fig. 7. Step size in log scale over iterations for PPGA (Left) and PPGA$_{\mathrm{LR}}$ (Right) for each learning rate.

Objective Value per Iteration

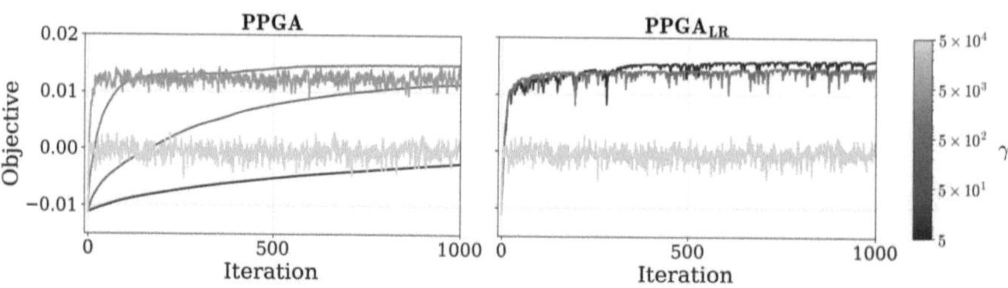

Fig. 8. Solution value over iterations for PPGA (Left) and PPGA$_{\mathrm{LR}}$ (Right) for each learning rate.

5   Conclusion

In this paper, we proposed Gradient Walk: a new approach for local search based on gradient steps to optimize over the piecewise affine landscape of a trained ReLU network function within a polytope. We also proposed two algorithms for this approach, PPGA and PPGA$_{\mathrm{LR}}$. PPGA is based on projected gradient steps and empirically-crafted perturbations. PPGA is designed to search for a feasible solution while avoiding entrapment in arbitrary local optima constructed by the special landscape of ReLU networks. PPGA$_{\mathrm{LR}}$ extends PPGA by using adaptive gradient steps that stretch further when the step size is too small relative to the local affine landscape. PPGA$_{\mathrm{LR}}$ is designed to circumvent the issue of diminishing step sizes caused by the combination of a misaligned learning rate with small gradients observed in deep ReLU networks. We report that PPGA$_{\mathrm{LR}}$ has a better performance than PPGA in neural networks with larger dimensions, whereas both outperform the baseline projected gradient ascent algorithm PGA and compare favorably against the LP Walk algorithm SimplexWalk, which has previously shown better results than other alternatives in the benchmark setting [55].

In a nutshell, we double down on the trend of using specialized local search with lower per-iteration cost to find better solutions in constraint learning. We achieve that with gradient steps leveraging the local structure in ReLU networks.

Given the emergence of constraint learning frameworks [4,9,13,18,33,35,58] for applications in operations research, the development of new algorithms to timely find better solutions is of particular importance.

In future work, we intend to devise a variant of PPGA that is also robust against large learning rates, explore how other characteristics of the landscape of ReLU networks can be leveraged to produce better algorithms, apply the Gradient Walk approach to solve other types of constraint learning models, and explore its performance on specific families of functions approximated by neural networks.

Disclosure of Interests. The authors have no competing interests to declare that are relevant to the content of this article.

References

1. Anderson, R., Huchette, J., Ma, W., Tjandraatmadja, C., Vielma, J.P.: Strong mixed-integer programming formulations for trained neural networks. Math. Program. (2020)
2. Arora, R., Basu, A., Mianjy, P., Mukherjee, A.: Understanding deep neural networks with rectified linear units. In: ICLR (2018)
3. Badilla, F., Goycoolea, M., Muñoz, G., Serra, T.: Computational tradeoffs of optimization-based bound tightening in ReLU networks. arXiv:2312.16699 (2023)
4. Bergman, D., Huang, T., Brooks, P., Lodi, A., Raghunathan, A.U.: JANOS: an integrated predictive and prescriptive modeling framework. INFORMS J. Comput. (2022)
5. Bonami, P., Lodi, A., Tramontani, A., Wiese, S.: On mathematical programming with indicator constraints. Math. Program. **151**(1), 191–223 (2015). https://doi.org/10.1007/s10107-015-0891-4
6. Botoeva, E., Kouvaros, P., Kronqvist, J., Lomuscio, A., Misener, R.: Efficient verification of (ReLU)-based neural networks via dependency analysis. In: AAAI (2020)
7. Cacciola, M., Frangioni, A., Lodi, A.: Structured pruning of neural networks for constraints learning. Oper. Res. Lett. (2024)
8. Cai, J., et al.: Getting away with more network pruning: from sparsity to geometry and linear regions. In: CPAIOR (2023)
9. Ceccon, F., et al.: OMLT: optimization & machine learning toolkit. J. Mach. Learn. Res. **23**(349), 1–8 (2022)
10. Cheng, C., Nührenberg, G., Ruess, H.: Maximum resilience of artificial neural networks. In: ATVA (2017)
11. Cybenko, G.: Approximation by superpositions of a sigmoidal function. Math. Control, Signals Syst. (1989)
12. Dolan, E.D., Moré, J.J.: Benchmarking optimization software with performance profiles. Math. Program. **91**(2), 201–213 (2002). https://doi.org/10.1007/s101070100263
13. Fajemisin, A., Maragno, D., den Hertog, D.: Optimization with constraint learning: a framework and survey. Eur. J. Oper. Res. (2023)
14. Fischetti, M., Jo, J.: Deep neural networks and mixed integer linear optimization. Constraints **23**(3), 296–309 (2018). https://doi.org/10.1007/s10601-018-9285-6
15. Funahashi, K.I.: On the approximate realization of continuous mappings by neural networks. Neural Networks (1989)
16. Grimstad, B., Andersson, H.: ReLU networks as surrogate models in mixed-integer linear programs. Comput. Chem. Eng. (2019)
17. Guo, X., Han, J., Tajrobehkar, M., Tang, W.: Escaping saddle points efficiently with occupation-time-adapted perturbations. arXiv preprint arXiv:2005.04507 (2020)
18. Gurobi Optimization: Gurobi Machine Learning. https://github.com/Gurobi/gurobi-machinelearning (2025). Accessed 09 Feb 2025
19. Gurobi Optimization, LLC: Gurobi Optimizer Reference Manual (2024). https://www.gurobi.com
20. Hanin, B., Sellke, M.: Approximating continuous functions by ReLU nets of minimal width. arXiv:1710.11278 (2017)
21. Hojny, C., Zhang, S., Campos, J.S., Misener, R.: Verifying message-passing neural networks via topology-based bounds tightening. In: ICML (2024)

22. Hornik, K., Stinchcombe, M., White, H.: Multilayer feedforward networks are universal approximators. Neural Networks (1989)
23. Huchette, J.: Advanced mixed-integer programming formulations: methodology, computation, and application. Ph.D. thesis, Massachusetts Institute of Technology (2018)
24. Huchette, J., Muñoz, G., Serra, T., Tsay, C.: When deep learning meets polyhedral theory: a survey. INFORMS J. Comput. (2026)
25. Huchette, J., Vielma, J.P.: Nonconvex piecewise linear functions: advanced formulations and simple modeling tools. Oper. Res. (2023)
26. Jin, C., Ge, R., Netrapalli, P., Kakade, S.M., Jordan, M.I.: How to escape saddle points efficiently. In: International Conference on Machine Learning, pp. 1724–1732. PMLR (2017)
27. Jin, C., Netrapalli, P., Jordan, M.I.: Accelerated gradient descent escapes saddle points faster than gradient descent. In: Conference On Learning Theory, pp. 1042–1085. PMLR (2018)
28. Kanamori, K., Takagi, T., Kobayashi, K., Ike, Y., Uemura, K., Arimura, H.: Ordered counterfactual explanation by mixed-integer linear optimization. In: AAAI (2021)
29. Kumar, A., Serra, T., Ramalingam, S.: Equivalent and approximate transformations of deep neural networks. arXiv:1905.11428 (2019)
30. Liu, C., Arnon, T., Lazarus, C., Strong, C., Barrett, C., Kochenderfer, M.J., et al.: Algorithms for verifying deep neural networks. Found. Trends® Optim. (2021)
31. Liu, X., Han, X., Zhang, N., Liu, Q.: Certified monotonic neural networks. In: NeurIPS, vol. 33 (2020)
32. Liu, X., Dvorkin, V.: Optimization over trained neural networks: difference-of-convex algorithm and application to data center scheduling. IEEE Control Syst. Lett. (2025)
33. Lueg, L., Grimstad, B., Mitsos, A., Schweidtmann, A.M.: reluMIP: open source tool for MILP optimization of ReLU neural networks (2021). https://doi.org/10.5281/zenodo.5601907, https://github.com/ChemEngAI/ReLU_ANN_MILP
34. Mangasarian, O.L., Rosen, J.B., Thompson, M.E.: Global minimization via piecewise-linear underestimation. J. Global Optim. (2005)
35. Maragno, D., Wiberg, H., Bertsimas, D., Birbil, S.I., Hertog, D.D., Fajemisin, A.: Mixed-integer optimization with constraint learning. Oper. Res. (2023)
36. McCormick, G.P.: Computability of global solutions to factorable nonconvex programs: part I — convex underestimating problems. Math. Program. (1976)
37. Misener, R., Floudas, C.A.: Piecewise-linear approximations of multidimensional functions. J. Optim. Theory Appl. (2010)
38. Perakis, G., Tsiourvas, A.: Optimizing objective functions from trained ReLU neural networks via sampling. arXiv:2205.14189 (2022)
39. Pham, H., Ren, A., Tahir, I., Tong, J., Serra, T.: Optimization over trained (and sparse) neural networks: a surrogate within a surrogate. In: CPAIOR (2026)
40. Plate, C., Hahn, M., Klimek, A., Ganzer, C., Sundmacher, K., Sager, S.: An analysis of optimization problems involving ReLU neural networks. arXiv:2502.03016 (2025)
41. Polyak, B.T., Levitin, E.: Constrained minimization methods. USSR Comput. Math. Math. Phys. **6**(5), 1–50 (1966)
42. Rosen, J.B.: The gradient projection method for nonlinear programming. part i. linear constraints. J. Soc. Ind. Appl. Math. **8**(1), 181–217 (1960)
43. Rössig, A., Petkovic, M.: Advances in verification of ReLU neural networks. J. Global Optim. (2021)

44. Say, B., Wu, G., Zhou, Y.Q., Sanner, S.: Nonlinear hybrid planning with deep net learned transition models and mixed-integer linear programming. In: IJCAI (2017)
45. Schweidtmann, A.M., Mitsos, A.: Deterministic global optimization with artificial neural networks embedded. J. Optim. Theory Appl. (2019)
46. Serra, T., Ramalingam, S.: Empirical bounds on linear regions of deep rectifier networks. In: AAAI (2020)
47. Serra, T., Tjandraatmadja, C., Ramalingam, S.: Bounding and counting linear regions of deep neural networks. In: ICML (2018)
48. Serra, T., Yu, X., Kumar, A., Ramalingam, S.: Scaling up exact neural network compression by ReLU stability. In: NeurIPS (2021)
49. Serra, T., Kumar, A., Ramalingam, S.: Lossless compression of deep neural networks. In: CPAIOR (2020)
50. Shi, C., Emadikhiav, M., Lozano, L., Bergman, D.: Constraint learning to define trust regions in optimization over pre-trained predictive models. INFORMS J. Comput. (2024)
51. Sosnin, P., Tsay, C.: Scaling mixed-integer programming for certification of neural network controllers using bounds tightening. In: CDC (2024)
52. Strong, C.A., et al.: Global optimization of objective functions represented by ReLU networks. Mach. Learn. (2021)
53. Tits, A.L., Yang, Y.: Globally convergent algorithms for robust pole assignment by state feedback. IEEE Trans. Automat. Control $41(10)$, 1432–1452 (1996). https://doi.org/10.1109/9.539425
54. Tjeng, V., Xiao, K., Tedrake, R.: Evaluating robustness of neural networks with mixed integer programming. In: ICLR (2019)
55. Tong, J., Cai, J., Serra, T.: Optimization over trained neural networks: taking a relaxing walk. In: International Conference on the Integration of Constraint Programming, Artificial Intelligence, and Operations Research, pp. 221–233. Springer (2024). https://doi.org/10.1007/978-3-031-60599-4_14
56. Tsay, C., Kronqvist, J., Thebelt, A., Misener, R.: Partition-based formulations for mixed-integer optimization of trained ReLU neural networks. In: NeurIPS (2021)
57. Tsiourvas, A., Sun, W., Perakis, G.: Manifold-aligned counterfactual explanations for neural networks. In: AISTATS (2024)
58. Turner, M., Chmiela, A., Koch, T., Winkler, M.: PySCIPOpt-ML: embedding trained machine learning models into mixed-integer programs. In: CPAIOR (2020)
59. Vahedi, A.M., Ilies, H.T.: SPGD: steepest perturbed gradient descent optimization. arXiv preprint arXiv:2411.04946 (2024)
60. Vielma, J.P., Ahmed, S., Nemhauser, G.: Mixed-integer models for nonseparable piecewise-linear optimization: Unifying framework and extensions. Oper. Res. (2010)
61. Xiao, K.Y., Tjeng, V., Shafiullah, N.M., Madry, A.: Training for faster adversarial robustness verification via inducing ReLU stability. In: ICLR (2019)
62. Yarotsky, D.: Error bounds for approximations with deep ReLU networks. Neural Networks (2017)
63. Zhao, H., Hijazi, H., Jones, H., Moore, J., Tanneau, M., Hentenryck, P.V.: Bound tightening using rolling-horizon decomposition for neural network verification. In: CPAIOR (2024)
64. Zhu, Y., Burer, S.: An extended validity domain for constraint learning. arXiv:2406.10065 (2024)

Graph Isomorphism: Mixed-Integer Convex Optimization from First-Order Methods

Wenjie Xiao[1,2]($\boxtimes$) (iD), Mathieu Besançon[1,4] (iD), Patrick Gelß[2] (iD),
Deborah Hendrych[1,2] (iD), Stefan Klus[3] (iD), and Sebastian Pokutta[1,2] (iD)

[1] Zuse Institute Berlin, Berlin, Germany
[2] Technische Universität Berlin, Berlin, Germany
xiao@zib.de
[3] Heriot–Watt University, Edinburgh, UK
[4] Université Grenoble Alpes, Inria, CNRS, LIG, Grenoble, France

Abstract. The *graph isomorphism (GI) problem*, which asks whether two graphs are structurally identical, occupies a unique position in computational complexity—it is neither known to be solvable in polynomial time, nor proven to be NP-complete. We propose a convex mixed-integer formulation of the problem and leverage first-order convex optimization to tackle it, following a stream of recent work on optimization-driven graph isomorphism detection. We strengthen our formulation with variable fixing techniques that prove highly effective while preserving the polyhedral structure. We perform extensive computations evaluating the performance of different families of methods including a mixed-integer convex formulation, mixed-integer linear optimization, local search and spectral heuristics over a collection of challenging GI instances. We find that a high level of symmetry is beneficial for optimization-based methods. On the other hand, presolving techniques that detect local substructures to fix variables are crucial for asymmetric instances. The proposed method outperforms the second best approach, the integer feasibility approach, on 6 of the 12 graphs families and is on par with it on symmetric families.

Keywords: Graph Isomorphism · Frank–Wolfe Method · Combinatorial Optimization

1   Introduction

Graphs are versatile representations of complex systems and appear in many applications ranging from social networks, biology to engineering. A question of interest in many of these fields is to determine whether two graphs are structurally identical (identical up to vertex ordering), i.e. whether they are *isomorphic*. This is called the *Graph Isomorphism problem* (GI) [4,36,39]. While it

T. Guns (Ed.): CPAIOR 2026, LNCS 16595, pp. 614–630, 2026.
https://doi.org/10.1007/978-3-032-27242-3_36

is not known whether the problem can be solved in polynomial time, for certain classes of graphs, there are polynomial-time algorithms [5,21,27]. For an overview and the historic background, we refer the reader to [17].

A powerful off-the-shelf solver for GI is **nauty** [30], originally developed in the 1980s [31]. The solver computes the automorphism group of a graph which can be utilized to efficiently check if two graphs are isomorphic. **nauty** is the state of the art and very fast by, in particular, detecting different graph structures and employing tailored preprocessing algorithms.

In this paper, we investigate optimization formulations and algorithms for the GI problem which can then be embedded within a larger problem. This work proposes a new optimization-based method to compute isomorphisms or certify that none exists, leveraging in particular first-order methods similar to [23] to solve an optimization formulation of GI, but instead build an exact method rather than a heuristic. We utilize the Boscia framework [20], a mixed-integer convex optimization framework based on Frank–Wolfe methods, and adapt the solving process to the structure of the GI problem. In particular, we exploit the formulation of the problem as seeking the minima of a function on the Birkhoff polytope, for which we have access to specialized algorithms for linear assignment. Additionally, we utilize the detection of local substructures to fix variables as a preprocessing step. Note that Boscia does not include any preprocessing out-of-the-box as it assumes only oracles access to the objective, its gradient and the feasible region. We compare this approach to the method proposed in [23], a difference-of-convex algorithm (DCA) approach for which FW methods have recently been studied, integer feasibility and minimization problems and a heuristic spectral method proposed in [22] on a benchmark of selected graph families from the **nauty** library. The goal is to computationally assess graph properties that render the problem easy or challenging for the different methods. High symmetric is beneficial for optimization-based methods as it increases the number of possible solutions. The proposed variable fixing techniques are crucial for asymmetric graphs as they greatly reduce the size of the optimization problem.

The optimization-based approaches are not expected to compete with specialized GI solvers. However, constrained and inexact variants of the problem appear in many applications, akin to the Quadratic Assignment Problem (QAP). These can be handled by optimization methods but not by the specialized solvers. Furthermore, the ideas and formulations we investigate can tackle graphs of up to 500 nodes (250000 variables) despite using generic optimization techniques, showing a high potential for their integration within specialized GI solvers. While focusing on the GI problem for unweighted undirected graphs, the proposed methods can be naturally extended to weighted and/or directed graphs and to the graph matching problem in which one seeks the permutation that minimizes the distance between two graphs. In contrast, **nauty** does only supports directed graphs through graph expansions and weighted graphs have to presented by in an unweighted fashion, greatly increasing problem size.

Notation

Throughout this work, we use the following notation: A function f is L-smooth if its gradient ∇f is L-Lipschitz continuous over its domain. We denote the standard inner product as $\langle \cdot, \cdot \rangle$, the Euclidean norm of a vector as $\|\cdot\|$ and the Frobenius norm of a matrix as $\|\cdot\|_F$. Matrices are denoted by uppercase letters, vectors are lowercase bold, scalars are lowercase letters. The convex hull of a set $\mathcal{X}$ is denoted by $\mathrm{conv}(\mathcal{X})$. The $n \times n$ identity matrix is denoted as I_n, the all-one vector as $\mathbf{1}$. The set of permutation matrices is denoted by $\mathcal{P}_n$ and its convex hull, the Birkhoff polytope or set of doubly stochastic matrices, is denoted by $\mathcal{D}_n$.

2   Graph Isomorphism Problem

Given two graphs $G_1 = (V, E_1)$ and $G_2 = (V, E_2)$ on a common vertex set V with $|V| = n$, the graphs are said to be *isomorphic* if there exists a bijection $\pi : V \to V$ such that two vertices $u, v \in V$ are adjacent in G_1 if and only if $\pi(u)$ and $\pi(v)$ are adjacent in G_2. In other words, an isomorphism is a relabeling of the vertices that preserves adjacency.

Let A and B be the adjacency matrices of G_1 and G_2, respectively. A vertex bijection corresponds to a permutation matrix $P \in \mathcal{P}_n$, and the graphs are isomorphic if and only if $PAP^\top = B$, which can be rewritten as $PA = BP$. The GI problem thus consists in finding a permutation matrix that satisfies this equation. This viewpoint naturally casts the GI problem as a feasibility problem over the discrete set $\mathcal{P}_n$.

To place this feasibility problem within an optimization framework, one may instead measure the violation of the matching constraint and search for a permutation that minimizes it. This leads to the quadratic formulation[1]

$$\min_{P \in \mathcal{P}_n} \|PA - BP\|_F^2,$$

whose optimal value is zero precisely when the matching condition is satisfied. The objective is convex in P but the feasible set $\mathcal{P}_n$ is combinatorial.

Additionally, the presence of graph symmetries may give rise to many distinct permutations that achieve the same optimal value. These challenges motivate the use of continuous relaxations, which can provide certificates of non-isomorphism or serve as effective initializations for discrete methods.

A widely used relaxation replaces $\mathcal{P}_n$ by its convex hull, the set of doubly stochastic matrices, by the Birkhoff–von Neumann theorem [35]:

$$\mathcal{D}_n := \left\{ X \in \mathbb{R}^{n \times n} \mid X\mathbf{1} = \mathbf{1}, \ X^\top \mathbf{1} = \mathbf{1}, \ X \geq 0 \right\}.$$

The relaxed problem

$$\min_{X \in \mathcal{D}_n} \|XA - BX\|_F^2$$

---

[1] We note that the l_1-vector-norm-based formulation is equivalent but leads to a non-smooth objective rendering our proposed method not applicable.

is therefore a continuous relaxation of the GI problem and can be approached using convex constrained optimization techniques.

This relaxation is tight for certain classes of graphs, such as asymmetric graphs or graphs satisfying additional structural conditions [17], but may admit fractional optimal solutions for other graph families, which leaves the existence of a permutation with a zero objective undecided.

3 Integer Optimization with Frank–Wolfe

In this section, we present the method developed for the GI problem based on a first-order method and a branch-and-bound algorithm tailored around it.

3.1 Frank–Wolfe Methods

The Frank–Wolfe (FW) [13] or conditional gradient method [26] is a first-order algorithm for constrained convex optimization. Given a convex, L-smooth function f and a compact convex feasible region $\mathcal{X}$, the method exploits a *linear minimization oracle* (LMO) over the feasible region, which solves

$$\mathbf{v}_t \in \arg\min_{\mathbf{y} \in \mathcal{X}} \langle \mathbf{y}, \nabla f(\mathbf{x}_t) \rangle.$$

In particular, FW algorithms do not require projections nor algebraic representations of the constraints. The oracle returns an extreme point of $\mathcal{X}$ that minimizes the linearization of f at the current iterate $\mathbf{x}_t$. The standard FW algorithm is detailed in Algorithm 1.

Algorithm 1 Frank–Wolfe Algorithm

1: **Input:** Initial feasible point $\mathbf{x}_0 \in \mathcal{X}$, step sizes $\{\gamma_t\}_{t \geq 0}$
2: **for** $t = 0, \ldots, T$ **do**
3:     $\mathbf{v}_t := \arg\min_{\mathbf{y} \in \mathcal{X}} \langle \mathbf{y}, \nabla f(\mathbf{x}_t) \rangle$
4:     $\mathbf{x}_{t+1} := \mathbf{x}_t - \gamma_t(\mathbf{x}_t - \mathbf{v}_t)$
5: **end for**

The sequence produces feasible iterates without requiring orthogonal projections onto $\mathcal{X}$, which can be computationally expensive. In contrast, the LMO is often significantly cheaper, especially for polytopes where linear optimization reduces to selecting an appropriate extreme point. In particular, the Birkhoff polytope admits a specialized implementation of the LMO using the Hungarian algorithm [24,32]. Several methods have enhanced FW in important classes of problems; we refer readers to the survey [10] for a detailed overview. We highlight in particular the *Blended Pairwise Conditional Gradient* (BPCG) method [37], which produces sparse iterates by maintaining the current iterate as a sparse convex decomposition of vertices, and the *Decomposition Invariant Conditional Gradient* (DICG) [15] which removes the need to maintain the active set, by requiring instead an oracle for minimizing functions on particular faces of the feasible region. BPCG can benefit from so-called *lazification*: one can replace the exact LMO at many iterations by a heuristic providing a vertex with sufficient decrease, see [10,11], while DICG performs linear optimization on specific faces of the feasible region. In particular for the Birkhoff polytope, the Hungarian algorithm can be adapted to efficiently compute the LMO restricted to a face, which for the Birkhoff polytope corresponds to the same changes performed when branching. If an entry can be fixed to one, the corresponding row and column can be removed since all other entries have to be zero. One can also set an entry to zero by removing the edge in the bipartite graph of the linear assignment problem. Since the dimensions of the problem grow quadratically with the number of nodes, this is a significant reduction in terms of computational cost compared to utilizing BPCG. While the worst-case complexity at the root-node remains identical, we highlight that the Hungarian algorithm becomes more tractable when fixed to faces when using specific data structures [14].

BPCG has linear convergence under certain conditions, for example under a $1/2$–sharp objective and polyhedral constraints. The constants of the convergence rate can however result in a dependency on the ambient dimension which could result in a slow rate in practice; we therefore specify these constants to compute an explicit convergence rate for the convex relaxations we solve depending only on the problem data (i.e., the adjacency matrices). This rate can then be linked with the cost per iteration dominated by the $\mathcal{O}(n^3)$ Hungarian algorithm.

The original definition of the *pyramidal width* of a polytope is given in [25], several equivalent definitions were derived in [33] and are typically easier to manipulate and derive for specific polytopes. We use the facial distance shown to be equal to the pyramidal width [33].

Definition 1 (Pyramidal Width [33]). *For any polytope $\mathcal{X}$ with vertex set $\mathcal{V}$, its pyramidal width δ is equal to the minimum distance of any of its proper faces F to the convex hull of the vertices of $\mathcal{X}$ not lying on F. That is:*

$$\delta = \min_{F \in \text{faces}(\mathcal{X}):\emptyset \subset F \subset \mathcal{X}} \text{dist}(F, \text{conv}(\mathcal{V} \setminus F)). \tag{1}$$

Computing the pyramidal width is non-trivial. Thus, we use the observation from [38] that the pyramidal width of the Birkhoff polytope is lower bounded by the pyramidal width of the n^2-dimensional unit cube. Therefore, we have $\delta \geq 1/n$, see [25] for the proof. We can now state the linear convergence rate of BPCG.

Proposition 1. *Define $f(X) = \|XA - BX\|_F^2$, $f^* = \min_{X \in \mathcal{D}_n} f(X)$ Then the iterations $\{X_t\}_{t=0}^T$ of the BPCG algorithm guarantee the following:*

$$f(X_t) - f^* \leq 2n(\|A\|_F + \|B\|_F)^2 \left(1 - \frac{1}{4n^3(\|A\|_F + \|B\|_F)^2}\right)^{\lceil (t-1)/2 \rceil}. \tag{2}$$

In particular, if the graphs are undirected, unweighted, and have the same number of edges, $\|A\|_F = \|B\|_F = \sqrt{m}$, where $m = |E|$ if the graph is directed and $2|E|$ if the graph is undirected, leading to:

$$f(X_t) - f^* \leq 8mn \left(1 - \frac{1}{16n^3\,m^2}\right)^{\lceil (t-1)/2 \rceil}.$$

Proof. By [19, Theorem 6] and [10, Corollary 3.33], we have that

$$f(X_t) - f^* \le \left(1 - \frac{\delta^2}{2M^2 LD^2}\right)^{\lceil (t-1)/2 \rceil} \frac{LD^2}{2}$$

for a $(M, 1/2)$-sharp function. Our objective function f is, in particular, $(1, 1/2)$-sharp. We now show for self-containment of our proof that the diameter of the Birkhoff polytope is $D = \sqrt{2n}$. This can be verified by considering the maximal distance between two permutation matrices, attained when the sets of n nonzero entries do not intersect.

$$\|X - Y\|_F = \sqrt{\sum_{i=1}^{n}\sum_{j=1}^{n}(X_{ij} - Y_{ij})^2} = \sqrt{2n}.$$

From the previous discussion, we can set the pyramidal width $\delta = 1/n$. The Lipschitz smoothness constant is $L = 2(\|A\|_F + \|B\|_F)^2$. Combining the different constants into the generic convergence rate, we obtain the required result. $\square$

In the computational experiments, we utilize DICG as it has better empirical performance and memory efficiency. The different elements of the above proof can be adapted to derive the linear convergence of DICG.

3.2   The Boscia Framework

Boscia [20] is a branch-and-bound framework for convex mixed-integer optimization that employs Frank–Wolfe (FW) algorithms to solve convex relaxations. The novel aspect of the method is the propagation of the integer constraints to the LMO. Thus, FW optimizes the objective over the convex hull of integer feasible points, called the *integer hull*, intersected with the node-specific bound constraints. The LMO always returns an integer feasible point and the B&B tree obtains an upper bound (incumbent) from the root node. The *Frank–Wolfe gap* is an upper bound for the primal gap, from the convexity of the objective and the optimality of $\mathbf{v}$ for the linear subproblem:

$$f(\mathbf{x}) - f^* \le \max_{\mathbf{v} \in \mathcal{D}_n} \langle \nabla f(\mathbf{x}), \mathbf{x} - \mathbf{v} \rangle =: g(\mathbf{x}).$$

Hence, FW provides a valid lower bound on the optimal value at any iteration.

The Birkhoff polytope is an uni-modular polytope, thus propagating the integer constraints to the LMO does not incur any additional computational cost.

Boscia features a *callback* mechanism, which allows problem-specific logic to be executed after each node is processed and branched on. Callbacks can inspect intermediate FW iterates, prune nodes prematurely, terminate the entire algorithm when a desired condition is reached, or inject structural information back into the solver. This makes Boscia highly adaptable and particularly effective for problems where feasibility, symmetry, or combinatorial structure can be exploited during the tree search.

Within each node, the choice of FW variant plays an important role. BPCG achieves fast iteration progress at the cost of maintaining an active set. DICG avoids storing active sets altogether and is well-suited for structured polytopes like the Birkhoff polytope.

We assess two node selection strategies: the best-lower-bound search and the depth-first search. The former is the most common in branch-and-bound algorithms and selects the node with the smallest lower bound as the next node to process; the latter explores the tree by following a branch until a leaf node is reached, after which it restarts the exploration from the best-bound node. We also leverage the asymmetry of branching decisions on the Birkhoff polytope, always branching up (i.e., rounding to one) the branching variable during the depth-first search, thus building search trees of depth at most n.

3.3   Variable Fixing Through Local Graph Structures and Perturbations

Problem presolving is an essential driver of optimization methods efficiency. Unlike specialized GI software or mixed-integer programming (MIP) solvers, Boscia does not perform presolving techniques because it only assumes an oracle access to both the objective and constraints. In our GI formulation optimizing over the Birkhoff polytope through the Hungarian algorithm, adding general cutting planes would break the constraint structure and require a generic solver for the LMO. Fixing variables of the permutation matrix to zero or one can however be performed, and results in an LMO consisting of a modified and less expensive call to the Hungarian algorithm. We propose two presolving techniques detecting local graph structures and one optimization-based fixing technique.

The local graph structure consists in two steps: quantifying a property of each vertex that is independent of the vertex ordering, and fixing to zero the entry X_{ij} if i and j do not agree on that property. The two properties we identify are the number of cliques of size $3 \ldots k$ that contain the current vertex. This can be performed in polynomial time for a fixed k. When the graph has a bounded degree, we can even use the Bron-Kerbosch algorithm [12] to compute all maximal cliques and thus derive the number of maximal cliques of each size which each given vertex belongs to. A pair of vertices from the first and second graphs can correspond to one another only if they have the same clique count for each clique size. As a second substructure, we count maximal stars on the subgraphs induced by the neighborhood of each vertex, which correspond to independent sets on that graph. Using a similar reasoning, if the graph has bounded degree, each induced subgraph has a fixed maximum size and we can compute all maximal independent sets. Otherwise, we can compute independent sets up to a fixed size in these induced subgraphs.

As a last presolving step, we propose an optimization-based technique that tentatively fixes entries in the permutation matrix. It exploits the non-negativity of the objective and only requires calls to the Hungarian algorithm and gradient computations. For a given entry (i, j) of the permutation matrix, we run a FW method with a limited iteration budget on the modified problem

$$\min_{X \in \mathcal{D}_n} \|XA - BX\|^2 + X_{ij}.$$

If at an iteration t, the dual bound given by the primal value minus the FW gap is above zero, the optimal value cannot attain zero. From nonnegativity of the two objective parts, this implies that $X_{ij} > 0$ in any solution, and thus that X_{ij} can be fixed to one, eliminating a row and column. We can equivalently optimize with a term $1 - X_{ij}$ to attempt to fix X_{ij} at zero. This technique can be viewed as a modified optimization-based bound tightening (OBBT) using an augmented Lagrangian formulation, relating to earlier work on generic mixed-integer optimization [16]. Furthermore, a set of variable fixings that is infeasible on the Birkhoff polytope provides a proof of non-isomorphism of the graphs. This technique has a polynomial runtime but is much costlier in practice than the substructure presolve.

4   Solution Approaches

First, we consider the exact integer formulation

$$\min_{X \in \mathcal{P}_n} \|XA - BX\|_F^2,$$

which we solve using **Boscia**. At each node, Boscia optimizes the objective to a given tolerance over the corresponding integer hull using DICG. This variant consistently exhibited superior performance both in terms of progress per iteration and wall-clock time compared to standard FW and BPCG. The LMO is computed with the Hungarian algorithm which, especially for large dimensions, is computationally more efficient than calling an LP solver. The callback mechanism provides effective control over the search. After a node is processed, its lower bound is inspected; if the bound is strictly positive, the subtree cannot contain a feasible solution and is pruned. The callback also checks the incumbent and the lower bound of the tree; if the incumbent reaches zero, a permutation matrix satisfying $XA = BX$ has been found, and the algorithm terminates, certifying isomorphism. On the other hand, if the lower bound is greater than zero, it certifies non-isomorphism.

As discussed earlier, the convex relaxation of the GI problem may admit fractional solutions that do not correspond to valid isomorphisms. One approach to address this issue is to augment the formulation with a penalty term that promotes solutions closer to permutation matrices, as proposed in [23]. The resulting optimization problem is

$$\min_{\substack{X \in \mathcal{D}_n \\ XA = BX}} -\|X\|_F^2,$$

where the constraint $XA = BX$ enforces feasibility with respect to the isomorphism condition, and the objective $-\|X\|_F^2$ biases the solution toward the extreme points of $\mathcal{D}_n$, i.e., the permutation matrices. We denote it by **Penalty**. The above problem is concave, hence, there is no guarantee that Frank–Wolfe will converge to the global minimum. This may be mitigated by generating a good initial point, see [23] for more details. Note that [23] uses an LP solver to generate the initial point from the convex relaxation. The relaxation is naturally suited to Frank–Wolfe methods, which can produce a feasible fractional point far more efficiently. Nevertheless, this method can only serve as a heuristic.

A drawback of the previous formulation is the constraint $XA = BX$ as it leads to expensive LP subproblems. The difference-of-convex (DC) formulation

$$\min_{X \in \mathcal{D}_n} \ \|XA - BX\|_F^2 - \lambda\|X\|_F^2,$$

moves the constraint into the objective. Thus, the LMO can be computed with the Hungarian algorithm. Like the aforementioned penalty formulation, the problem is nonconvex but its structure allows the use of FW-type algorithm for DC, proposed in [28] and extended in [34]. It will be further identified by **DC-FW**. This approach cannot guarantee finding an optimal permutation, but it may lead to faster convergence compared to running Frank–Wolfe directly. While DCA guarantees convergence to a stationary point, it offers no guarantee of global optimality. The quality of the solution is sensitive to the choice of the parameter λ; for instance, a large λ may steer the iterates away from matrices satisfying $\|XA - BX\|_F = 0$. In our experiments, we have set $\lambda = 10^{-2}$.

The GI problem can also be formulated as a pure binary feasibility problem. Given adjacency matrices A and B, the graphs are isomorphic if and only if there exists a permutation matrix X such that

$$XA = BX, \ X \in \mathcal{P}_n.$$

where $\mathcal{P}$ is the set of all permutation matrices. This formulation can be handed directly to a **MIP** solver, in this case SCIP, by enforcing the linear and integrality constraints defining $\mathcal{P}_n$ together with the equation $XA = BX$. Although potentially expensive for large instances, this provides an exact and simple baseline against which we compare our other approaches. Furthermore, to the best of our knowledge, there are no assessment of the practical performance of a recent MIP solver on the GI problem.

A **Spectral** assignment approach for detecting isomorphisms was proposed in [22]. The method relies on iteratively perturbing the adjacency matrices of the two graphs to break symmetries and assigning vertices of G_1 to vertices of G_2 by solving linear assignment problems that are based on the eigenvalues and eigenvectors of the perturbed matrices. While it can be shown that the graph isomorphism problem can be solved in polynomial time if the graphs are friendly (see, e.g., [2]), the method might require backtracking for graphs with repeated eigenvalues and therefore does not have a polynomial runtime.

5   Computational Experiments

Our benchmark consists of 12 graph families from the **nauty** benchmark library [29], ranging from 10 to 500 nodes. The **exact** set is from the PACE 2023 challenge [1]. All graphs except the **exact** set are regular. For each graph, we generate three isomorphic instances by applying random permutation matrices. An instance is counted as solved if the method correctly identifies isomorphism within the time limit of 1 h. We compare methods by the number of solved instances and their solving times. The methods described in Sect. 4 are implemented in `Julia` v1.10.3. We use SCIP 9.2.4 [8,9] as the MIP solver through its MATHOPTINTERFACE.JL wrapper. The integer Frank–Wolfe algorithms rely on BOSCIA.JL v0.2.3 [20] and FRANKWOLFE.JL v0.6.1 [6,7]. The linear assignment oracle uses HUNGARIAN.JL v0.7. All experiments are conducted on a 32-core node (Intel(R) Xeon(R) Gold 6338, 2.00GHz, 1024GB RAM). We utilize the *secant* line search in FW variants, which amounts to exact line search with a single additional gradient call on quadratic functions [18].

The families are split by symmetry in the tables where we consider families with normalized orbits of 0.0–0.3 as highly symmetric (top), 0.3–0.7 as varying degrees of symmetry (middle) and 0.7–1.0 as low symmetry (bottom). Table 1 shows the performance of the selected methods on the benchmark. Note that both penalty heuristics as well as the spectral method were always dominated by either one of the Boscia versions and/or the MIP approach, so we do not report the results in the table. Their performance is shown for selected graph families in later figures. As highlighted in the Sect. 4, the spectral method potentially has to backtrack to find an isomorphism. Due to the nonconvexity of the penalty-based heuristics, they may converge to a local minimum. The first four columns of the table denote different versions of Boscia. The first version is Boscia without any preprocessing and the depth-first-search (DFS) traverse strategy. The next three utilize different preprocessing steps in Boscia: star detection, clique and star detection and the OBBT fixings. All of these versions use the DFS traverse strategy. Note that the presolving methods of SCIP are enabled and it calls **nauty** as part of its preprocessing. We also conducted additional experiments with the ℓ_1-vector-norm formulation solved via MIP solver, which consistently underperformed the feasibility formulation proposed in the paper, thus it is not reported separately.

Table 1 also shows the results for **nauty** for completeness. Yet we highlight that we do not compare the proposed methods to **nauty** as it is specialized for GI and not an optimization method.

High symmetry seems to have an overall positive effect. The reason is that the number of possible solution increases with the symmetry. The relative weakness of the effect is likely due to the graphs in the symmetric families being rather dense.

The multitude of solutions in the symmetric case also explains why the preprocessing steps do not improve the performance of Boscia. On the other hand, it has a big positive impact on the asymmetric instances. Especially on the **usr** set, Boscia with preprocessing outperforms the MIP approach. We further highlight that even **nauty** is struggling on this set compared to its performance on the other sets.

Table 1. Performance summary by graph family. Best performance for each family is highlighted in bold. The reported times are the geometric mean of the total solving times shifted by 1 s for all instances. This includes timed out instances.

Family	Inst.	Boscia DFS			Boscia Star			Boscia Clique & Star			Boscia Fixings			MIP			Nauty		
		#	%	Time (s)	#	%	Time (s)	#	%	Time (s)	#	%	Time (s)	#	%	Time (s)	#	%	Time (s)
latin	63	34	**54%**	289.71	34	**54%**	298.79	34	**54%**	303.30	15	24%	1530.52	34	**54%**	257.23	63	100%	0.38
Lattice	21	15	71%	201.26	15	71%	202.80	13	62%	218.84	6	29%	1621.49	18	**86%**	96.07	21	100%	0.37
paley_power	24	19	**79%**	109.66	19	**79%**	124.95	19	**79%**	122.61	6	25%	1282.27	10	42%	365.73	24	100%	0.37
paley_prime	21	18	**86%**	46.26	18	**86%**	51.45	18	**86%**	67.53	6	29%	1316.20	10	48%	234.53	21	100%	0.35
Triangular	21	15	71%	241.13	15	71%	256.89	15	71%	254.69	6	29%	2189.35	17	**81%**	122.98	21	100%	0.35
CHH_cc	24	8	33%	740.13	8	33%	870.83	8	33%	842.36	6	25%	2150.97	22	**92%**	80.67	24	100%	0.40
tnn	42	13	31%	1299.91	14	33%	887.11	14	33%	919.52	6	14%	2261.95	30	**71%**	106.28	35	83%	14.55
cfi	12	0	0%	3600.00	0	0%	3600.00	0	0%	3600.00	0	0%	3600.00	11	**92%**	383.75	12	100%	0.57
exact	21	21	**100%**	35.57	21	**100%**	5.21	18	86%	22.75	6	29%	1302.23	12	57%	79.76	21	100%	0.36
iso_r01N	21	18	**86%**	35.14	18	**86%**	5.91	18	**86%**	6.26	9	43%	1025.71	18	**86%**	32.33	21	100%	0.35
sts	21	0	0%	3600.00	12	57%	444.28	18	**86%**	411.44	0	0%	3600.00	0	0%	3600.00	21	100%	1.10
usr	36	5	14%	2538.99	30	**83%**	58.27	18	50%	200.90	4	11%	3164.43	6	17%	1867.12	30	83%	15.94

For the middle level symmetric families, all of the Boscia variants are outperformed by the MIP approach. They seem to exhibit enough symmetry that the preprocessing is not very effective but do not admit too many possible solutions. Nonetheless, the MIP feasibility approach is beating our method on only four families out of twelve.

Other exact approaches for GI are based on the NP-hard Quadratic Assignment Problem (QAP) [3], using in particular a lifted formulation resulting in $\mathcal{O}(n^4)$ variables. We did not test it given its high computational cost for moderate graphs compared to the compact MIP feasibility problem, also reported by the authors.

The OBBT fixing, while effective for low symmetric graphs, is in its current implementation far too costly, see Table 2. Thus, we solve hardly more instances than the baseline DFS version. For symmetric graphs in particular, OBBT hardly fixes any entries. Despite this, it will be worthwhile to improve the implementation as it can be a powerful tool.

Clique and star detection likewise do not have much impact for symmetric graphs. In contrast to the fixings, they are very cheap and thus also do not negatively effect the performance. On the asymmetric instances, the proposed method benefits a lot from clique and star detection. Notable exception is the `exact` set. The reason is that it is the only set which does, in fact, include non-regular graphs. The assumption of regularity is crucial for the clique and star detection to be effective.

In Figs. 1, 2, 3, 4 and 5, we showcase the number of solved instances over time for the different methods on a selection of families. On the left side, we compare the different methods. Boscia with DFS and the star detection has overall the best performance of all variants and it is thus the chosen method for the comparison plots, denoted just as Boscia. On right, we compare the different versions of Boscia. The baseline is standard Boscia with best-lower-bound search and without any preprocessing. By Boscia, we denote Boscia with DFS and the star detection enabled like on the left side. The DFS_left variant is the depth-first-search but favoring the left child, so fixing an entry to zero. Given

Table 2. Fixings analysis by graph family. # Graphs: number of instances; % Solved: percentage solved within time limit; Time (s): geometric mean of total solving time shifted by 1 s; Fixing Time (s) geometric mean of time spent on optimization-based bound tightening (OBBT) fixings shifted by 1 s; Fixings/n^2: percentage of permutation matrix entries fixed relative to n^2; Avg Iters: average number of Frank–Wolfe iterations for the fixing procedure.

Family	# Graphs	% Solved	Time (s)	Fixing Time (s)	Fixings/n^2	Avg Iters
latin	63	23.8%	1530.52	1523.19	0.16%	23
Lattice	21	28.6%	1621.49	1614.70	0.00%	0
paley_power	24	25.0%	1282.27	1275.75	0.00%	0
paley_prime	21	28.6%	1316.20	1310.75	0.00%	0
Triangular	21	28.6%	2189.35	2182.52	0.00%	0
CHH_cc	24	25.0%	2150.97	2145.75	64.25%	33
tnn	42	14.3%	2261.95	2259.73	77.44%	13
cfi	12	0.0%	3600.00	3600.00	0.00%	0
exact	21	28.6%	1302.23	1299.37	59.93%	11
iso_r01N	21	42.9%	1025.71	1022.23	30.60%	16
sts	21	0.0%	3600.00	3600.00	0.00%	0
usr	36	11.1%	3164.43	2818.93	0.00%	0

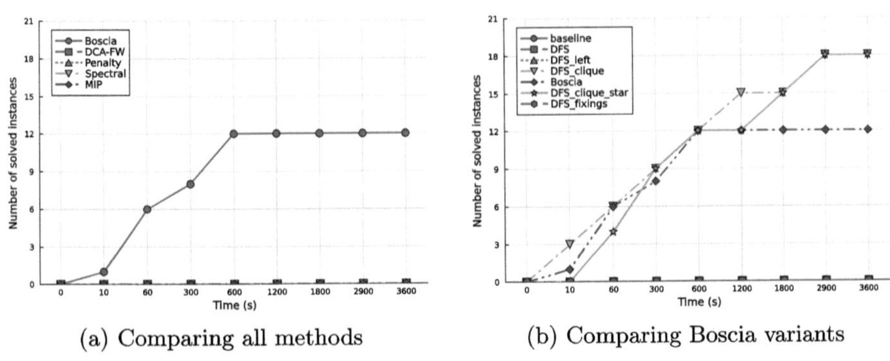

(a) Comparing all methods   (b) Comparing Boscia variants

Fig. 1. Solved instances over time for `sts` graph family.

the structure of the Birkhoff polytope, this is predictably not as performant as favouring the right child.

On the `sts` family, shown in Fig. 1, only our method is able to solve a significant amount of instances, neither of the other approaches solved any instance. The graphs in this family are sparse but exhibit low symmetry. The penalty and spectral methods generally benefit from symmetry and struggle on more asymmetric instances. We also observe from Fig. 1b that clique detection makes a significant performance difference.

The same behavior can be observed for the `usr` family on Fig. 2. The MIP performs better here than on the `sts` family, but is not competitive. This graph family is the densest one in our benchmark, so this could account for performance of the MIP approach.

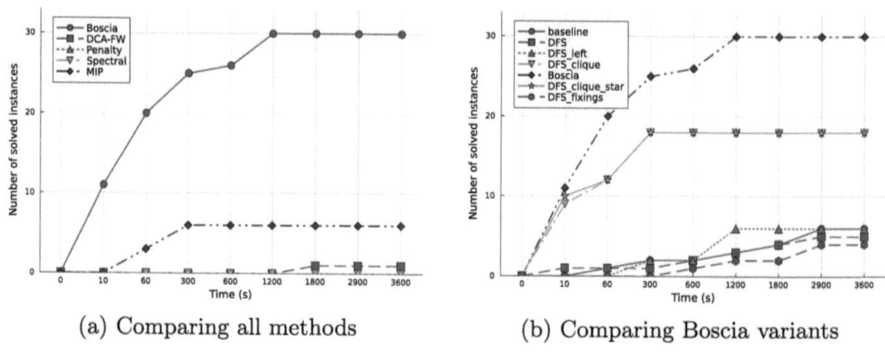

(a) Comparing all methods   (b) Comparing Boscia variants

Fig. 2. Solved instances over time for `usr` graph family.

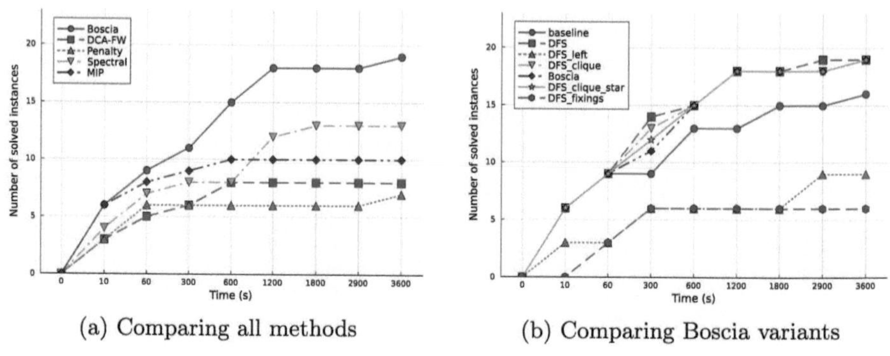

(a) Comparing all methods   (b) Comparing Boscia variants

Fig. 3. Solved instances over time for `paley_power` graph family.

For both `Paley` graph families, Figs. 3 and 4, the clique and star detection are nearly equally effective. Our method outperforms all other methods on these sets as well. Interestingly, the spectral method outperforms the feasibility approach on these sets. It, in particular, fares well on the `paley_prime` family.

The `iso_r01N` family, Fig. 5, is relatively easy to solve for most methods. This is somewhat counterintuitive as the graphs exhibit very low symmetry. On the other hand, this family includes smaller graphs which might contribute to the ease of solving.

In addition to the isomorphic instances, we also created non-isomorphic instances by flipping random edges of the graphs. The results are summarized in Table 3. Our proposed method based outperforms the MIP feasibility approach. In particular, we observe that the presolving method based on substructure detection determines non-isomorphism at the root node for all instances. We highlight however that having identical spectra is a necessary condition for isomorphism and is not respected by these graphs. Spectral analysis would therefore also discard them as non-isomorphic. Investigating the performance of optimization-based methods on isospectral but non-isomorphic graphs is a promising direction for future work.

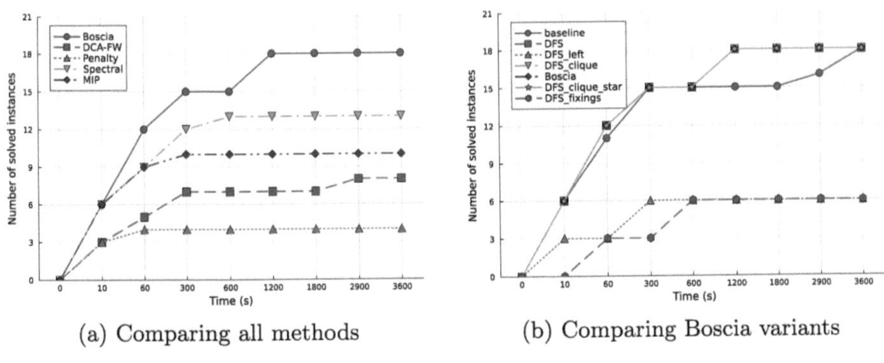

(a) Comparing all methods   (b) Comparing Boscia variants

Fig. 4. Solved instances over time for `paley_prime` graph family.

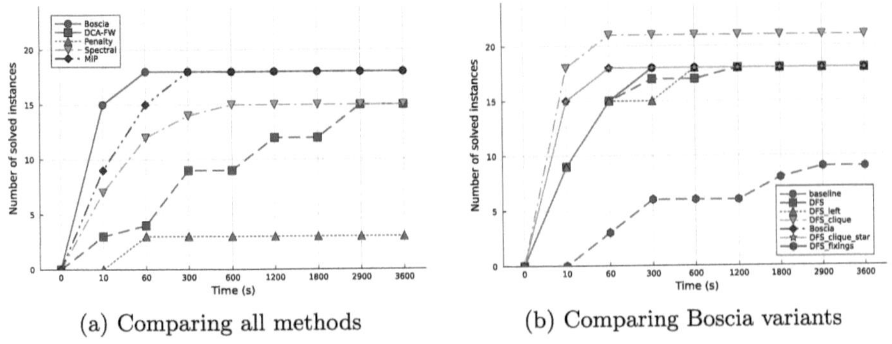

(a) Comparing all methods   (b) Comparing Boscia variants

Fig. 5. Solved instances over time for `iso_r01N` graph family.

Table 3. Performance summary on the non-isomorphic instances by graph family. Best performance for each family is highlighted in bold. The reported times are the geometric mean of the total solving times shifted by 1 s for all instances. This includes timed out instances.

Family	Inst.	Boscia DFS			Boscia Clique			Boscia Star			Boscia Fixings			MIP		
		#	%	Time (s)	#	%	Time (s)	#	%	Time (s)	#	%	Time (s)	#	%	Time (s)
latin	63	13	21%	719.69	63	**100%**	0.03	63	**100%**	0.51	17	27%	981.22	22	35%	772.60
Lattice	21	4	19%	789.79	21	**100%**	0.00	21	**100%**	0.04	6	29%	965.80	10	48%	307.76
paley_power	24	6	25%	494.10	24	**100%**	0.23	24	**100%**	2.93	9	38%	879.39	9	38%	526.51
paley_prime	21	6	29%	374.47	21	**100%**	0.13	21	**100%**	0.96	7	33%	702.10	12	57%	307.43
Triangular	21	3	14%	1170.75	21	**100%**	0.01	21	**100%**	0.09	5	24%	1343.66	12	57%	360.24
CHH_cc	24	7	29%	582.42	24	**100%**	0.00	24	**100%**	0.01	6	25%	1366.70	24	**100%**	30.55
tnn	42	12	29%	867.90	42	**100%**	0.01	42	**100%**	0.04	12	29%	1559.77	34	81%	72.93
cfi	12	0	0%	3600.00	12	**100%**	0.01	12	**100%**	0.02	0	0%	3600.00	12	**100%**	135.69
exact	21	12	57%	108.25	18	86%	0.11	21	**100%**	0.18	9	43%	636.68	12	57%	60.77
iso_r01N	21	12	57%	250.89	21	**100%**	0.01	18	86%	0.28	10	48%	660.03	15	71%	41.00
sts	21	0	0%	3600.00	21	**100%**	0.16	21	**100%**	4.20	0	0%	3600.00	3	14%	2124.66
usr	36	6	17%	1559.75	18	50%	130.22	36	**100%**	2.55	10	28%	1402.56	15	42%	853.38

Finally, we performed preliminary experiments for the graph matching problem, in which one seeks permutation that minimizes the distance between two graphs. Most of our preprocessing and early-stopping steps are not directly applicable in this setting and are deactivated. Thus, the integer linear formulation minimizing the ℓ_1-norm of the mismatch as the objective exhibits a better performance than Boscia. This warrants further investigation, in particular with regards to preprocessing techniques.

6   Conclusion

In this paper, we develop and evaluate mixed-integer convex formulations and algorithms for the graph isomorphism problem. Despite the problem being well-studied on theoretical and algorithmic aspects, we showed that a mixed-integer convex approach leveraging first-order methods and graph-specific presolving techniques could achieve great performance.

From the experiments, we can conclude that optimization-based methods benefit from a high degree of symmetry as it increases the number of solutions. To effectively solve GI for asymmetric graphs, presolving techniques that detect substructures and infer variable fixings from them are crucial. We hope that this insight will provide guidance for the design of specialized GI detection solvers.

The fundamental operations performed in Boscia, namely gradient computation and calls to the Hungarian algorithm, could both be ported to GPU for increased performance, unlike solving methods based on constraint programming or linear optimization, offering a promising avenue for future research.

Considering the encouraging performance of our approach, we plan to extend our work to the graph matching problem, with a particular focus on preprocessing techniques.

Acknowledgments. Research reported in this paper was partially supported through the Research Campus Modal funded by the German Federal Ministry of Education and Research (fund numbers 05M14ZAM,05M20ZBM), the Deutsche Forschungsgemeinschaft (DFG) through the DFG Cluster of Excellence MATH+ (EXC-2046/1, project ID 390685689, project AA3-15), and the ANR through MIAI Cluster (reference ANR-23-IACL-0006). The authors also thank Pascal Schweitzer for insightful discussions on GI at SEA 2024 motivating parts of this work.

References

1. PACE challenge 2023: Twinwidth. PACE 2023 challenge on computing low-width contraction sequences; exact and heuristic tracks (2023). https://pacechallenge.org/2023/
2. Aflalo, Y., Bronstein, A., Kimmel, R.: On convex relaxation of graph isomorphism. Proc. Natl. Acad. Sci. **112**(10), 2942–2947 (2015)
3. Aurora, P., Mehta, S.K.: The QAP-polytope and the graph isomorphism problem. J. Comb. Optim. **36**(3), 965–1006 (2018). https://doi.org/10.1007/s10878-018-0266-x

4. Babai, L.: Group, graphs, algorithms: the graph isomorphism problem. In: Proceedings of the International Congress of Mathematicians: Rio de Janeiro 2018, pp. 3319–3336. World Scientific (2018)

5. Babai, L., Grigoryev, D.Y., Mount, D.M.: Isomorphism of graphs with bounded eigenvalue multiplicity. In: Proceedings of the Fourteenth Annual ACM Symposium on Theory of Computing, pp. 310–324 (1982)

6. Besançon, M., Carderera, A., Pokutta, S.: FrankWolfe.jl: a high-performance and flexible toolbox for Frank-Wolfe algorithms and conditional gradients. INFORMS J. Comput. (2022)

7. Besançon, M., et al.: Improved algorithms and novel applications of the FrankWolfe.jl library. ACM Trans. Math. Software (2025)

8. Bestuzheva, K., et al.: Enabling research through the SCIP Optimization Suite 8.0. ACM Trans. Math. Software $49(2)$, 1–21 (2023)

9. Bolusani, S., et al.: The SCIP Optimization Suite 9.0. Technical report, Optimization Online (2024). https://optimization-online.org/2024/02/the-scip-optimization-suite-9-0/

10. Braun, G., et al.: Conditional Gradient Methods. MOS-SIAM Series on Optimization (2025)

11. Braun, G., Pokutta, S., Zink, D.: Lazifying conditional gradient algorithms. In: International Conference on Machine Learning, pp. 566–575. PMLR (2017)

12. Eppstein, D., Löffler, M., Strash, D.: Listing all maximal cliques in large sparse real-world graphs. J. Exper. Algorithmics (JEA) $\mathbf{18}$, 3:1–3:21 (2013)

13. Frank, M., Wolfe, P.: An algorithm for quadratic programming. Naval Res. Logistics Q. $\mathbf{3}$, 95–110 (1956). https://onlinelibrary.wiley.com/doi/abs/10.1002/nav.3800030109

14. Fredman, M.L., Tarjan, R.E.: Fibonacci heaps and their uses in improved network optimization algorithms. J. ACM (JACM) $\mathbf{34}(3)$, 596–615 (1987)

15. Garber, D., Meshi, O.: Linear-memory and decomposition-invariant linearly convergent conditional gradient algorithm for structured polytopes. In: Advances in Neural Information Processing Systems, vol. 29 (2016)

16. Gleixner, A.M., Berthold, T., Müller, B., Weltge, S.: Three enhancements for optimization-based bound tightening. J. Global Optim. $\mathbf{67}(4)$, 731–757 (2017)

17. Grohe, M., Schweitzer, P.: The graph isomorphism problem. Commun. ACM $\mathbf{63}(10)$, 128–134 (2020). https://doi.org/10.1145/3372123

18. Hendrych, D., Besançon, M., Martínez-Rubio, D., Pokutta, S.: Secant line search for Frank-Wolfe algorithms. In: Forty-second International Conference on Machine Learning (2025)

19. Hendrych, D., Besançon, M., Pokutta, S.: Solving the optimal experiment design problem with mixed-integer convex methods. In: Proceedings of the Symposium on Experimental Algorithms (2024). https://doi.org/10.4230/LIPIcs.SEA.2024.16

20. Hendrych, D., Troppens, H., Besançon, M., Pokutta, S.: Convex mixed-integer optimization with Frank-Wolfe methods. Math. Program. Comput. $\mathbf{17}$, 731–757 (2025)

21. Hopcroft, J.E., Tarjan, R.E.: Isomorphism of planar graphs. In: Complexity of Computer Computations: Proceedings of a Symposium on the Complexity of Computer Computations, held March 20–22, 1972, at the IBM Thomas J. Watson Research Center, Yorktown Heights, New York, and sponsored by the Office of Naval Research, Mathematics Program, IBM World Trade Corporation, and the IBM Research Mathematical Sciences Department, pp. 131–152. Springer (1972)

22. Klus, S., Sahai, T.: A spectral assignment approach for the graph isomorphism problem. Inf. Infer. A J. IMA **7**, 689–706 (2018). https://doi.org/10.1093/imaiai/iay001
23. Klus, S., Gelß, P.: Continuous optimization methods for the graph isomorphism problem. Inf. Inference: A J. IMA **14**(2), iaaf011 (2025)
24. Kuhn, H.W.: The Hungarian method for the assignment problem. Naval Res. Logistics Q. **2**(1–2), 83–97 (1955)
25. Lacoste-Julien, S., Jaggi, M.: On the global linear convergence of Frank-Wolfe optimization variants. In: Advances in Neural Information Processing Systems, vol. 28 (2015)
26. Levitin, E., Polyak, B.: Constrained minimization methods. USSR Comput. Math. Math. Phys. **6**, 1–50 (1966). https://www.sciencedirect.com/science/article/pii/0041555366901145
27. Luks, E.M.: Isomorphism of graphs of bounded valence can be tested in polynomial time. J. Comput. Syst. Sci. **25**(1), 42–65 (1982)
28. Maskan, H., Hou, Y., Sra, S., Yurtsever, A.: Revisiting Frank-Wolfe for structured nonconvex optimization. arXiv preprint arXiv:2503.08921 (2025)
29. McKay, B., Piperno, A.: Nauty and traces benchmark graph libraries. https://pallini.di.uniroma1.it/Graphs.html (2025), collections of benchmark graphs (DIMACS and dreadnaut formats); current version 2_9_1 (2025)
30. McKay, B.D., Piperno, A.: Practical graph isomorphism, II. J. Symb. Comput. **60**, 94–112 (2014)
31. McKay, B.D., et al.: Practical graph isomorphism (1981)
32. Munkres, J.: Algorithms for the assignment and transportation problems. J. Soc. Ind. Appl. Math. **5**(1), 32–38 (1957)
33. Pena, J., Rodriguez, D.: Polytope conditioning and linear convergence of the Frank-Wolfe algorithm. Math. Oper. Res. **44**(1), 1–18 (2019)
34. Pokutta, S.: Scalable DC optimization via adaptive Frank-Wolfe algorithms. arXiv preprint arXiv:2507.17545 (2025). https://arxiv.org/abs/2507.17545
35. Read, R., Corneil, D.: The graph isomorphism disease. J. Graph Theory **1**, 339–363 (1977). https://onlinelibrary.wiley.com/doi/abs/10.1002/jgt.3190010410
36. Read, R.C., Corneil, D.G.: The graph isomorphism disease. J. Graph Theory **1**(4), 339–363 (1977)
37. Tsuji, K., Tanaka, K., Pokutta, S.: Sparser kernel herding with pairwise conditional gradients without swap steps. arXiv preprint arXiv:2110.12650 (2021)
38. Valls, V., Iosifidis, G., Tassiulas, L.: Birkhoff's decomposition revisited: sparse scheduling for high-speed circuit switches. IEEE/ACM Trans. Networking **29**(6), 2399–2412 (2021)
39. Zemlyachenko, V.N., Korneenko, N.M., Tyshkevich, R.I.: Graph isomorphism problem. J. Sov. Math. **29**(4), 1426–1481 (1985)

Large Neighborhood Search Meets Iterative Neural Constraint Heuristics

Yudong W. Xu[1(✉)], Wenhao Li[1], Scott Sanner[1,2], and Elias B. Khalil[1]

[1] Department of Mechanical & Industrial Engineering, University of Toronto, Toronto, Canada
{wil.xu,chriswenhao.li}@mail.utoronto.ca, ssanner@mie.utoronto.ca, elias.khalil@utoronto.ca
[2] Vector Institute, Toronto, Canada

Abstract. Neural networks are being increasingly used as heuristics for constraint satisfaction. These neural methods are often recurrent, learning to iteratively refine candidate assignments. In this work, we make explicit the connection between such iterative neural heuristics and Large Neighborhood Search (LNS), and adapt an existing neural constraint satisfaction method—ConsFormer—into an LNS procedure. We decompose the resulting neural LNS into two standard components: the destroy and repair operators. On the destroy side, we instantiate several classical heuristics and introduce novel prediction-guided operators that exploit the model's internal scores to select neighborhoods. On the repair side, we utilize ConsFormer as a neural repair operator and compare the original sampling-based decoder to a greedy decoder that selects the most likely assignments. Through an empirical study on Sudoku, Graph Coloring, and MaxCut, we find that adapting the neural heuristic to an LNS procedure yields substantial gains over its vanilla settings and improves its competitiveness with classical and neural baselines. We further observe consistent design patterns across tasks: stochastic destroy operators outperform greedy ones, while greedy repair is more effective than sampling-based repair for finding a single high-quality feasible assignment. These findings highlight LNS as a useful lens and design framework for structuring and improving iterative neural approaches.

Keywords: Recurrent Transformer · Large Neighborhood Search · Constraint Satisfaction Problems

1 Introduction

Constraint satisfaction is ubiquitous in automated decision-making applications that involve planning, scheduling, and resource management, among others. By combining search, inference, and heuristics, constraint solvers have exhibited continual improvement over the past few decades, leading to reliable performance on a diverse range of challenging problems. That being said, there is also emerging interest in a complementary direction, namely the development of fast heuristics based on end-to-end machine learning (ML) [4,25].

© The Author(s), under exclusive license to Springer Nature Switzerland AG 2026
T. Guns (Ed.): CPAIOR 2026, LNCS 16595, pp. 631–648, 2026.
https://doi.org/10.1007/978-3-032-27242-3_37

By design, ML heuristics exploit statistical patterns within a homogeneous distribution of training instances, e.g., coloring problems on graphs with a similar structure. While an appropriately trained heuristic can perform well on test instances similar to those seen in training, it is unlikely to perform well out of distribution, e.g., when the number of nodes or the density of the graph changes in a coloring problem. One-shot solution prediction methods are particularly prone to this failure mode as they have a fixed per-instance computational budget that cannot adapt to instance size or complexity [26]. Classical solvers, on the other hand, simply search further, as desired.

To address this limitation and adapt to out-of-distribution instances, *recurrence* has been introduced to many existing neural architectures. This includes masked diffusion models [30], recurrent Transformers [11,40], and single-step iterative Transformers [39]. In these approaches, the model is applied repeatedly over multiple iterations. The number of iterations can scale with the difficulty of the instance, partially bypassing the fixed-compute constraint of one-shot solution prediction and often improving generalization to harder or larger problems [11,39]. A common thread across these iterative neural methods is that they implicitly or explicitly refine a solution while optimizing some objective that captures the degree of constraint satisfaction. Viewed at a higher level, this is closely related to classical local search techniques and, in particular, Large Neighborhood Search (LNS) [24,29].

In this paper, we make this connection explicit by viewing a recent neural constraint heuristic, ConsFormer [39], as an implicit neural LNS procedure. We formalize the deployment of ConsFormer within a generic LNS loop and decompose the resulting neural LNS into two standard components: the destroy operator and the repair operator. We refer to the resulting procedure as ConsFormer-LNS. This perspective allows us to import standard LNS design dimensions and to ask: which classical LNS heuristics carry over to the neural setting, and what new heuristics become possible when the repair operator is a learned model with access to rich internal signals? To this end, we make the following contributions:

- We reinterpret the ConsFormer [39] as an implicit neural LNS solver in which the Transformer acts as part of the repair operator and embed it in an explicit LNS procedure.
- We propose prediction-guided destroy operators that exploit the neural network's logits as signals for variable selection, alongside classical heuristics.
- We conduct a systematic empirical study of classical and neural-specific LNS heuristics on several CSP benchmarks. Adapting the neural heuristic to an LNS procedure yields substantial gains over its vanilla iterative deployment.

2   Related Work and Background

2.1   Constraint Satisfaction Problems

A *constraint satisfaction problem* (CSP) is given by a tuple $(\mathbf{X}, \mathcal{D}, \mathcal{C})$, where $\mathbf{X} = \{X_1, \ldots, X_n\}$ is a set of variables, each X_i takes values in a finite discrete domain $\mathcal{D}_i \in \mathcal{D} = \{\mathcal{D}_1, \ldots, \mathcal{D}_n\}$, and $\mathcal{C} = \{c_1, \ldots, c_m\}$ is a set of constraints. Each constraint $c_j \in \mathcal{C}$ is defined over a subset of variables $\mathbf{X}_j \subseteq \mathbf{X}$ and restricts the assignments to $\mathbf{X}_j$. Let $\mathcal{X} \triangleq \prod_{i=1}^n \mathcal{D}_i$ denote the Cartesian product over variable domains, and an assignment by $\mathbf{x} = (x_1, \ldots, x_n) \in \mathcal{X}$. The objective of a CSP is to find $\mathbf{x}$ that satisfies every constraint: $c_j(\mathbf{x}) = \texttt{true}, \quad \forall j \in [m]$.

2.2 Large Neighborhood Search

Large Neighborhood Search is a popular approach for solving CSPs [24,29]. LNS iteratively destroys part of an existing solution and then rebuilds it. A destroy operator removes specific components of a solution. A repair operator then re-instantiates the removed elements to construct a new, complete feasible solution.

Neural LNS has been an active area of recent research [5,12]. Existing work includes both utilizing neural methods as the destroy operator as well as the repair operator. For example, various works have studied learning a neural destroy operator to select the neighborhood in integer programs using reinforcement learning [31,38], imitation learning [32], contrastive learning [18], graph convolutional networks [42], and hindsight relabeling [13]. Hottung et al. use reinforcement learning to learn destroy and repair operators for vehicle routing problems [15–17]. Falkner et al. [10] attack similar problems with graph neural networks as repair operators.

2.3 Iterative Neural Heuristics

Masked Generative Transformers [7], Recurrent Transformers [8], and Masked Diffusion Models [30] have gained popularity for handling discrete data, and have often been applied to solve simple constraint satisfaction problems such as Sudoku [3,19,40] and harder combinatorial optimization problems in general [28, 33,39]. Within these iterative neural heuristics, techniques like remasking and subset improvement have been used to enable iterative solution improvement where the framework forgets or refines parts of the solution [36,39], which closely mirrors the LNS destroy-repair pattern.

We view an iterative neural heuristic as an update function u that takes a complete but potentially infeasible assignment $\mathbf{x}$ together with a subset of variables $S \subseteq \mathbf{X}$ to be revised. We denote by $I := \{\, i \in \{1, \ldots, n\} : X_i \in S \,\}$ the corresponding set of indices and write

$$\mathbf{x}' \sim u(\, \cdot \mid \mathbf{x}, I\,), \qquad x'_i = x_i \ \forall i \notin I.$$

ConsFormer. In this work, we adapt the ConsFormer [39], a recent model utilizing the Transformer architecture to solve constraint satisfaction problems. ConsFormer instantiates the update function u by taking in a complete, potentially infeasible variable assignment $\mathbf{x} = (x_1, \ldots, x_n)$ as a sequence of tokens.

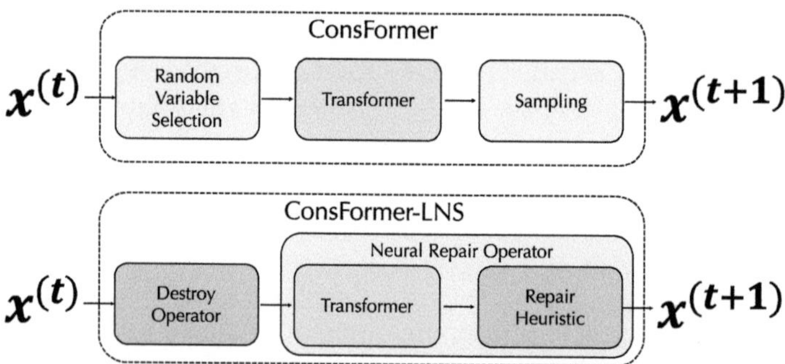

Fig. 1. Comparison between the original ConsFormer update and our ConsFormer-LNS. **Top:** ConsFormer takes the current assignment $\mathbf{x}^{(t)}$, selects a random subset of variables, applies the Transformer, and samples a new assignment $\mathbf{x}^{(t+1)}$. **Bottom:** In ConsFormer-LNS, a destroy operator selects variables to modify, and the Transformer acts as part of a neural repair operator that proposes the next assignment $\mathbf{x}^{(t+1)}$.

Each variable's input representation combines a learned value embedding as well as positional information, i.e., information about the index i of variable X_i.

The model randomly selects a subset of variables $S \subset \mathbf{X}$ to update during a forward pass. A specialized learnable embedding $\mathbf{e}_s$ is added to the tokens corresponding to variables in S to signal their eligibility for modification. The Transformer processes this sequence to produce updated assignments; for variables in S, new values are sampled via the Gumbel-Softmax operator applied to the output logits to maintain differentiability, while variables not in S remain unchanged.

ConsFormer is trained in a self-supervised way using differentiable constraint penalty functions p_k, such that $p_k(\mathbf{x}) = 0$ if $\mathbf{x}$ satisfies the k-th constraint and $p_k(\mathbf{x}) > 0$ otherwise. Given the model's soft outputs for an instance, the training loss is a weighted sum of these penalties, $\mathcal{L}(\mathbf{x}; \Theta) = \sum_k \lambda_k f(p_k(\mathbf{x}))$, where f is typically a quadratic transformation, weights λ_k are hyperparameters, and Θ are the learnable model parameters. This loss directly measures constraint violation under the model's predictions and provides gradients with respect to the prediction scores. In our work, we reuse ConsFormer's architecture and self-supervised loss, and reinterpret it as part of a neural repair operator inside an LNS framework.

3   Methodology

We reinterpret ConsFormer as a neural Large Neighborhood Search (LNS) procedure, allowing us to modify its components along the destroy and repair axes.

3.1   Neural LNS View of ConsFormer

We consider a CSP instance as defined in Sect. 2.2. A complete assignment is denoted by $\mathbf{x} = (x_1, \ldots, x_n)$, with a cost function $\mathrm{cost}(\mathbf{x})$ measuring constraint violation. As described in Sect. 2.3, ConsFormer iteratively refines an assignment $\mathbf{x}^{(t)}$. We map this iterative process to the standard LNS components as follows:

- **Destroy (The Subset Selection):** The random selection of the subset $S^{(t)} \subseteq \mathbf{X}$ (and the application of the embedding $\mathbf{e}_s$) functions as a stochastic destroy operator. By flagging these variables for update, the model effectively "unassigns" them, rendering them eligible for modification while keeping $\mathbf{X} \setminus S^{(t)}$ fixed.
- **Repair (The Transformer Pass):** The Transformer's forward pass acts as a learned repair operator. Given the current assignment to the variables and the "destroyed" subset $S^{(t)}$, the model proposes a new assignment $x_i^{(t+1)}$ for $X_i \in S^{(t)}$ by sampling from their corresponding output logits $z_i^{(t)}$ via Gumbel-Softmax.

Figure 1 illustrates this reinterpretation: the original ConsFormer loop can be seen as an LNS procedure with a random destroy operator and a neural repair operator that samples for the next solution. We build on this view to study alternative design choices for each of the components.

3.2 Destroy Operator

At each iteration t, the destroy operator defines a binary mask $m^{(t)} \in \{0,1\}^n$, where $m_i^{(t)} = 1$ means variable X_i is selected. Following ConsFormer, the size of the selected set $S^{(t)} = \{X_i : m_i^{(t)} = 1\}$ adheres to a specified degree of destruction $\rho \in (0,1]$. We adapt several classical destroy heuristics from the LNS literature [21] and introduce neural prediction-guided methods that exploit internal signals from the ConsFormer model.

Classical Destroy Operators

Random removal. Let $\pi_i^{(t)} \in [0,1]$ denote the probability of destroying variable X_i at iteration t. The simplest strategy is to select the subset at random by setting all $\pi_i^{(t)}$ to the rate of destruction ρ, which corresponds to the original ConsFormer. We sample the binary mask by drawing a Bernoulli random variable for each variable:

$$\pi_i^{(t)} = \rho, \quad m_i^{(t)} \sim \text{Bernoulli}(\pi_i^{(t)}).$$

Greedy worst removal. A more targeted strategy destroys variables that contribute the most to constraint violations. Using ConsFormer's relaxed constraint penalty loss $\mathcal{L}$, we define violation scores per variable to be $v_i(\mathbf{x}^{(t)})$ by evaluating the loss on the discretized variable assignments and extracting the per-variable contributions. In practice, we compute

$$v_i(\mathbf{x}^{(t)}) = \left\| \frac{\partial \mathcal{L}(\text{OneHot}(\mathbf{x}^{(t)}))}{\partial (\text{OneHot}(\mathbf{x}_i^{(t)}))} \right\|_1 .$$

We can then set $m_i^{(t)} = 1$ for the top-ranked variables with the largest scores (ties broken arbitrarily), yielding a greedy worst variable removal operator.

Stochastic worst removal. To improve diversification, we can randomize the selection while biasing towards highly violating variables [24]. We compute a normalized score from $v_i(\mathbf{x}^{(t)})$ such that the average probability matches ρ:

$$\pi_i^{(t)} \propto v_i(\mathbf{x}^{(t)}), \quad \text{s.t.} \quad \tfrac{1}{n}\sum_{i=1}^{n} \pi_i^{(t)} \approx \rho,$$
$$m_i^{(t)} \sim \text{Bernoulli}(\pi_i^{(t)})$$

Variables with higher violation scores thus have a higher selection probability.

Stochastic related removal. Introduced by Shaw [29], related removal destroys a group of "related" variables at once. In our setting, a natural notion of relatedness is participation in a common constraint. At each iteration we sample a random subset of constraints and then destroy all variables that belong to any selected constraint. Concretely, we draw a Bernoulli mask over constraints and define the destroy set accordingly:

$$m_k^{\text{constr}} \sim \text{Bernoulli}(\rho), \quad S^{(t)} = \bigcup_{k:\, m_k^{\text{constr}}=1} \text{vars}(c_k).$$

For graph-based problems where the number of constraints significantly exceeds the number of variables, we rescale the per-constraint Bernoulli parameter so that the expected number of selected constraints remains proportional to the desired variable-level degree of destruction ρ.

Greedy related removal. We can build on random related removal by greedily selecting the constraints with the largest per constraint penalty p_k introduced in Sect. 2.3. Given the current assignment $\mathbf{x}^{(t)}$, each constraint has a violation score $p_k(\mathbf{x}^{(t)}) \geq 0$, we select the subset of constraints with the largest penalties and destroy all variables that participate in them:

$$\mathcal{K}^{(t)} = \text{top}_L\left(\left\{p_k(\mathbf{x}^{(t)})\right\}_{c_k \in \mathcal{C}}\right), \qquad S^{(t)} = \bigcup_{k \in \mathcal{K}^{(t)}} \text{vars}(c_k).$$

where top_L returns the indices of the constraints with the largest penalties, with ties broken arbitrarily.

Prediction-Guided Destroy Operators We can design heuristics unique to the neural setting by utilizing the model's internal latent embeddings.

Gradient-guided removal. After the final hidden layer, ConsFormer produces a vector of logits $z_i^{(t)} \in \mathbb{R}^{|\mathcal{D}_i|}$ corresponding to the variables X_i, which we can be interpreted as the model's current belief over the values of X_i. Our first neural strategy uses the gradient of the loss function with respect to the model's current belief to drive the destroy operator.

To do this, we evaluate the penalty loss used during training on the current belief. We then compute the gradient of $\mathcal{L}$ with respect to the logits

$$g_i^{(t)} = \frac{\partial \mathcal{L}(\mathrm{softmax}(z^{(t)}))}{\partial z_i^{(t)}} \quad \text{for each variable } X_i,$$

and use the gradients as a per-variable score. We note that this is different from the gradient used for worst removal, since we rely on the model's internal signals instead of a discretized variable assignment. We include a greedy and a stochastic variant similar to classical approaches, where the greedy variant selects the top-ranked variables with the largest gradients, and the random variant defines

$$\pi_i^{(t)} \propto \left\| g_i^{(t)} \right\|_1, \quad \text{s.t.} \quad \tfrac{1}{n} \sum_{i=1}^{n} \pi_i^{(t)} \approx \rho,$$
$$m_i^{(t)} \sim \mathrm{Bernoulli}(\pi_i^{(t)}).$$

Variables with larger $\pi_i^{(t)}$ are those for which small changes in the prediction would most strongly affect the loss. In contrast to purely violation-based heuristics, this gradient-based removal explicitly leverages the neural model and the training loss to identify variables that are most "responsible" for the current soft constraint violations.

Confidence-margin removal. Inspired by recent work in Masked Diffusion Models [3,20], we introduce a strategy using solely the model's current belief. Intuitively, we target variables for which the model is highly uncertain, regardless of their current violation.

We first transform the logits into probabilities $q_i^{(t)} = \mathrm{softmax}(z_i^{(t)})$, we then quantify confidence using the gap between the top two probabilities in $q_i^{(t)}$. Let v_1, v_2 be the assignment values with the largest and second-largest entries in q_i, we define a confidence margin

$$\mathrm{margin}_i^{(t)} = q_i^{(t)}(v_1) - q_i^{(t)}(v_2).$$

This margin can be used as the score for selecting the variables. Intuitively, small margins indicate that the model is unsure between multiple values. We again introduce a greedy and a stochastic variant where $\pi_i^{(t)} \propto (-margin_i^{(t)})$. This encourages the search to repeatedly revisit parts of the assignment where the model has low confidence and may benefit from additional refinement.

3.3 Repair Operator

The repair operator takes the current assignment $\mathbf{x}^{(t)}$ and destroy set $S^{(t)}$, and proposes a candidate assignment $\mathbf{x}^{(t+1)}$ by modifying only variables in $S^{(t)}$. In our framework, ConsFormer is utilized as a neural repair operator. Given $\mathbf{x}^{(t)}$ and mask $m^{(t)}$, the model produces logits and corresponding probabilities $q_i^{(t)}(v)$ for each variable $x_i^{(t)}$ and value assignment v. We consider two decoding strategies: sampling and greedy decoding.

Stochastic sampling. This is the original ConsFormer implementation. For each variable with $m_i^{(t)} = 1$, we sample $x_i^{(t+1)} \sim q_i^{(t)}(v)$ using the Gumbel–Softmax sampler; for $m_i^{(t)} = 0$, we keep $x_i^{(t+1)} = x_i^{(t)}$. This yields a stochastic repair operator that explores the neighborhood induced by the destroy mask.

Greedy decoding. A deterministic alternative selects the most likely value for each variable with $m_i^{(t)} = 1$:

$$x_i^{(t+1)} \in \arg \max_{v \in \mathcal{D}_i} q_i^{(t)}(v)$$

This corresponds to a greedy repair step based on the model's current beliefs.

4   Experiments

This section aims to empirically answer the following research questions:

- **RQ1 (ConsFormer vs. ConsFormer-LNS):** Does adapting ConsFormer to a neural LNS procedure improve its performance?
- **RQ2 (Classical vs. Prediction-guided Destroy Operators):** How do classical and prediction-guided destroy operators contribute to performance gains?
- **RQ3 (Greedy vs. Stochastic Destroy Operators):** How important is randomness in the destroy operator?
- **RQ4 (Greedy vs. Stochastic Repair Operators):** Given a fixed destroy strategy, does greedy repair generally outperform the original sampling-based repair, and how does its per-instance behavior differ from sampling?
- **RQ5 (ConsFormer-LNS vs. Other Solvers):** How does the LNS-enhanced ConsFormer perform compared to other neural or classical solvers?

4.1   Experimental Setup

Problem Selection. We evaluate on the same datasets as Xu et al. [39], excluding the nurse rostering problem for which ConsFormer already solves 100% of the instances.

Sudoku is a simple CSP that involves filling a 9×9 grid with digits from 1 to 9 such that each row, column, and 3×3 sub-grid contains all 9 numbers. A single Sudoku instance is defined by a partially filled board and its difficulty is determined by the number of initial values: fewer initially filled cells in the board involve a larger space of possible assignments to the unfilled cells and is therefore harder. Following Xu et al. [39], we use the dataset from SATNet [37] for training and in-distribution testing, and the dataset from RRN [22] for harder out-of-distribution testing.

Graph Coloring seeks an assignment of colors to vertices in a graph such that no two neighboring nodes share the same color. The problem is defined by the graph's structure and the number of available colors k. We study the same 5-coloring and 10-coloring datasets, where training graphs have 50 vertices for $k = 5$ and 100 vertices for $k = 10$ whereas OOD graphs have 100 for $k = 5$ and 200 for $k = 10$.

MaxCut aims to identify a cut in a graph such that the number of edges crossing the partition is maximized. MaxCut can be alternatively viewed as a 2-coloring Max-CSP which seeks an assignment that maximizes the number of satisfied constraints. Since our framework works by reducing the number of violated constraints, it can be applied directly for MaxCut. The model is trained on small generated graphs and tested on benchmark instances from the GSET dataset [41], which includes graphs with sizes ranging from 800 to 10000 vertices.

Baselines. For RQ5, we compare against a robust set of neural and classical baselines established in the literature following Xu et al. [39]. OR-Tools [23][1], a state-of-the-art constraint programming solver, is the primary non-learning baseline; note that many of its internal sub-solving routines utilize Large Neighborhood Search (LNS) [23]. We omit OR-Tools from the Sudoku experiments, as these instances are computationally trivial for exact solvers and are primarily used to benchmark the reasoning capabilities of neural approaches.

Implementation Details. We train our models on single H100 GPU nodes. We adopt the hyperparameter configurations from Xu et al. [39] and retrain a separate model for each destroy/repair operator pairing[2].

4.2 Results and Analysis

Experimental results comparing the different destroy and repair operators are presented in Tables 1, 2 and 3. Italic numbers correspond to the original Cons-Former configuration (random destroy + sampling-based repair); bold and underlined numbers mark the best and second best result within each dataset, respectively.

RQ1: ConsFormer Vs. ConsFormer-LNS. Across all benchmarks, adapting ConsFormer into an explicit LNS procedure enhances model performance. On Sudoku (Table 1), ConsFormer-LNS improves the out-of-distribution (OOD) percentage of instances solved from 85.8% to 91.8%. On MaxCut (Table 2), ConsFormer-LNS reduces the gap to the best known cut sizes from 16.33 to 4.44 for $|V| = 800$, from 12.44 to 8.00 for $|V| = 1K$, from 52.11 to 30.56 for

---

[1] All OR-Tools results are reported using version 9.10 and were run on the same hardware as the ConsFormer-LNS models.

[2] Our code is available at https://github.com/khalil-research/ConsFormer.

Table 1. Sudoku instances solved (%) for different combinations of destroy and repair operators. Columns correspond to the destroy operators specified in Sect. 3.2 where the Greedy and Stochastic variants are labeled (Gr.) and (St.) respectively. Rows vary the repair operator (sampling-based vs. greedy). Test instances contain $1,000$ instances from the SATNet dataset, OOD refers to Out-of-Distribution evaluation on the RRN test dataset which contains 18K instances. All configurations are evaluated for 2K iterations.

Dataset	Repair	Random	Worst		Related		Gradient		Confidence	
			Gr.	St.	Gr.	St.	Gr.	St.	Gr.	St.
Test	Sampled	*100*	94.3	**100**	99.7	**100**	20.6	**100**	0.0	**100**
	Greedy	**100**	85.6	**100**	<u>99.9</u>	**100**	0.0	**100**	0.0	**100**
OOD	Sampled	*85.8*	16.7	71.7	32.1	48.7	0.1	90.6	0.0	90.8
	Greedy	81.2	11.8	**91.8**	30.9	63.9	0.1	84.1	0.0	<u>91.5</u>

Table 2. MaxCut performance on GSET: average gap to the best known cut (lower is better) for different graph sizes. Columns and Rows correspond to destroy and repair operators as in Table 1. All configurations are evaluated with a 180 s time limit.

Size	Repair	Random	Worst		Related		Gradient		Confidence	
			Gr.	**St.**	**Gr.**	**St.**	**Gr.**	**St.**	**Gr.**	**St.**
$|V|=800$	Sampled	*16.33*	430.11	473.00	923.67	472.33	167.78	45.11	489.67	<u>9.78</u>
	Greedy	31.67	503.11	38.56	915.00	71.33	137.00	**4.44**	397.00	21.00
$|V|=1K$	Sampled	*12.44*	382.89	417.67	781.00	418.56	154.89	30.89	469.22	<u>9.78</u>
	Greedy	25.00	414.11	33.22	714.33	57.00	116.56	**8.00**	364.11	18.78
$|V|=2K$	Sampled	*52.11*	832.89	944.78	1646.11	944.67	297.00	69.89	1014.00	<u>37.11</u>
	Greedy	74.67	842.89	97.00	1472.33	157.33	207.44	**30.56**	674.00	62.78
$|V| \geq 3K$	Sampled	*115.25*	1268.75	1618.00	2311.50	1467.38	530.63	122.75	1921.13	<u>106.63</u>
	Greedy	142.50	1286.50	188.00	2085.50	229.88	354.50	**63.63**	1230.13	140.25

$|V| = 2K$, and from 115.25 to 63.63 for $|V| \geq 3K$. On graph coloring (Table 3), OOD percentage of instances solved increased from 46.3% to 54.2% for $k = 5$ and from 10.2% to 18.4% for $k = 10$.

RQ2: Classical Vs. Prediction-Guided Destroy. Overall, the best performing classical and prediction-guided destroy operators achieve similar performance gain from the baseline, with the exception of MaxCut, where the gradient-guided removal achieves much stronger performance. We further compare the best performing methods by examining the instance solved percentage across iterations in Fig. 2. We observe that the confidence-based removal rapidly improves in early iterations, but is overtaken by the random worst removal operator towards the end.

Table 3. Graph-Coloring instances solved (%). Top: Coloring-5, Bottom: Coloring-10. Columns and Rows correspond to destroy and repair operators as in Table 1. OOD refers to Out-of-Distribution evaluation for ANYCSP and ConsFormer where the number of vertices n in the graph is larger than that of the training instances. All datasets have 1200 instances. All configurations are evaluated with a 10 s time limit.

Dataset	Repair	Random	Worst		Related		Gradient		Confidence	
			Gr.	St.	Gr.	St.	Gr.	St.	Gr.	St.
Graph-Coloring-5 ($n = 50 \rightarrow n = 100$)										
Test	Sampled	_81.6_	33.4	77.8	32.0	81.8	41.7	79.9	25.7	82.3
	Greedy	**82.9**	32.9	82.5	22.1	<u>82.8</u>	41.2	**82.9**	32.7	81.8
OOD	Sampled	_46.3_	0.0	41.0	0.0	47.6	0.7	44.8	0.0	49.0
	Greedy	**54.2**	0.1	52.5	0.0	53.7	2.6	<u>54.0</u>	0.0	49.4
Graph-Coloring-10 ($n = 100 \rightarrow n = 200$)										
Test	Sampled	_53.6_	0.0	<u>53.8</u>	0.0	53.5	11.9	**53.9**	0.0	<u>53.8</u>
	Greedy	**53.9**	0.4	**53.9**	0.0	**53.9**	14.8	**53.9**	5.5	**53.9**
OOD	Sampled	_10.2_	0.0	14.7	0.0	14.2	2.6	14.5	0.0	14.8
	Greedy	<u>17.4</u>	0.0	16.8	0.0	**18.4**	5.1	16.6	1.0	16.3

The strong early performance of the confidence-margin model is consistent with the intuition that the model is prioritizing the variable assignments it has the least confidence in. However, this destroy operator receives no explicit signal from the current solution's quality, therefore, as the model becomes more certain in the later iterations, its improvement slows down and eventually gets overtaken by methods with richer penalty-derived signals. This observation aligns with the efficiency-accuracy trade-off reported in the Masked Diffusion literature [3].

The lack of a clear winner among destroy operators, across all benchmarks, is in line with the findings of classical LNS literature: problem structure determines the most suitable Destroy operator [24]. Further augmentation of ConsFormer-LNS with an adaptive mechanism [21] is therefore a promising direction for future work.

RQ3: Greedy Vs. Stochastic Destroy Operators. The failure of purely greedy destroy operators is observed across all benchmarks. Greedy variants of the different strategies often collapse to near-zero solved percentages on OOD Sudoku and coloring and exhibit very poor cut values on MaxCut. This echoes the classical LNS literature, where randomization is often needed to escape local optima. In contrast, the randomized variants tend to perform well, highlighting the importance of controlled randomness in the destroy step. Figure 3 shows the per-cell accuracy scores of the best performing destroy operators, i.e., the fraction of variables that are assigned to the correct value in the groundtruth unique feasible Sudoku solution. It can be seen that the greedy variant plateaus much earlier while the stochastic variants are able to achieve much higher scores.

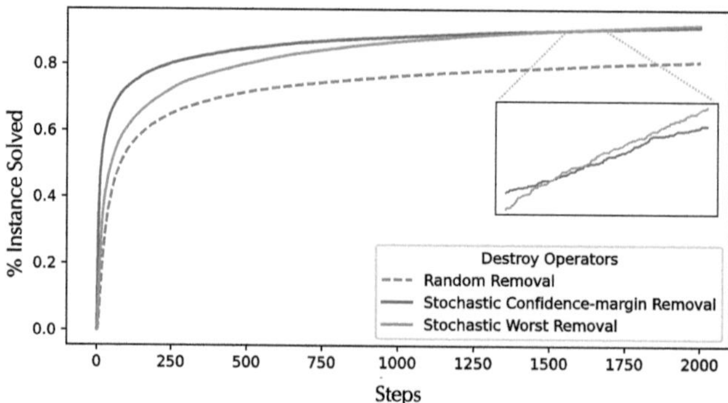

Fig. 2. Sudoku instances solved (%) over LNS steps for the baseline random destroy and the best-performing classical (stochastic worst removal) and prediction-guided (stochastic confidence-margin removal) destroy operators. Both heuristics outperform random removal, with confidence-margin removal improving fastest early on and stochastic worst removal eventually achieving the highest solved percentage.

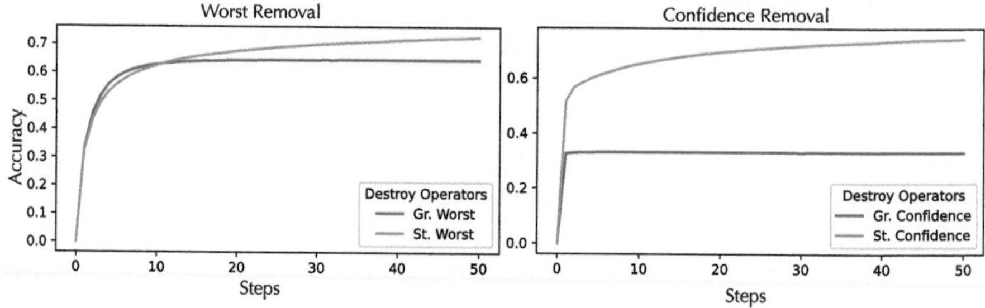

Fig. 3. Cell accuracy for Sudoku over iterations for greedy (blue) and stochastic (orange) variants of Worst and Confidence destroy operators. Greedy variants plateau early, while stochastic variants continue to improve and reach substantially higher accuracies. (Color figure online)

RQ4: Greedy Vs. Stochastic Repair Operators. The greedy repair operator almost always outperforms stochastic sampling-based repair across all benchmarks. To better understand this result, we run ConsFormer-LNS with the baseline and best-performing destroy operators for a small number of iterations for Sudoku. We plot the distribution of the constraint-satisfaction rate as well as the cumulative percentage of instance solved in Fig. 4. Across all three configurations, we observe a few clear patterns.

The Greedy repair operator has a better improvement in the first step, but gets overtaken by the stochastic repair operator quickly. At later iterations, the stochastic repair operator has a better overall distribution with a mean closer to 100% accuracy, while greedy repair has a long tail of unsolved instances and a spike at 100% accuracy. Despite the worse performance per step, the

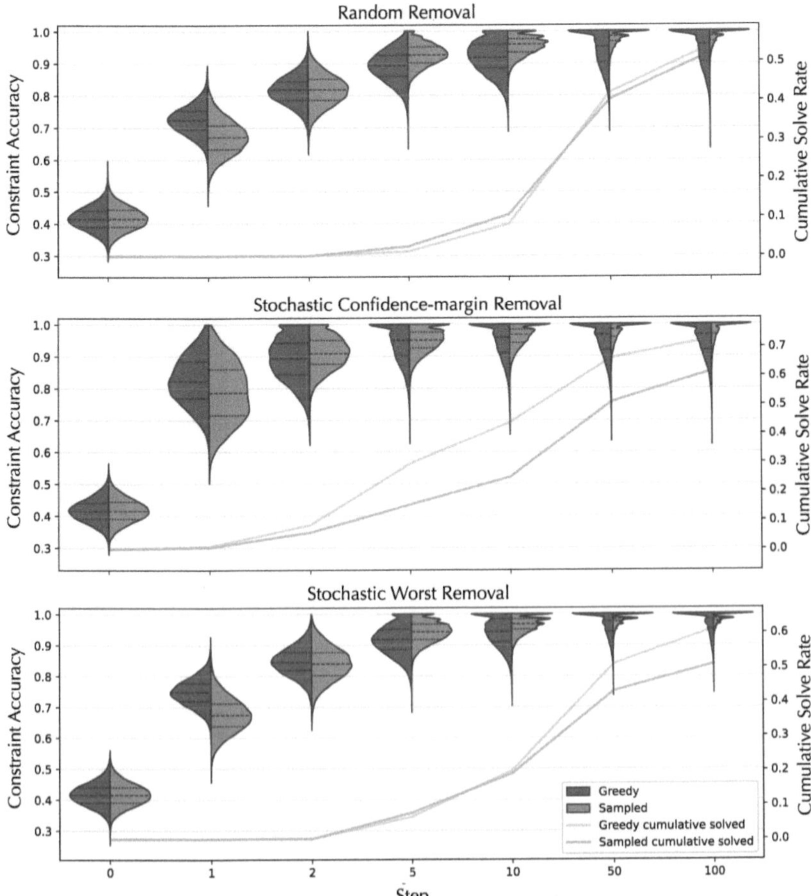

Fig. 4. Sudoku constraint accuracy on OOD Test dataset over LNS steps for greedy (blue) and sampling-based (orange) repair under Random and two best-performing destroy operators (Stochastic Confidence-margin, Stochastic Worst). Split violin plots show the distribution of per-instance constraint satisfaction at selected steps, while the lines on the right axis show the cumulative fraction of instances solved. Greedy repair achieves higher final solved percentages but exhibits a wider tail of unsolved instances, whereas sampling attains higher average constraint accuracy at intermediate steps. (Color figure online)

greedy repair operator consistently achieves a higher cumulative instance solved percentage in the later steps.

This suggests that while the stochastic sampling-based repair operator learns the solution distribution better through exploration, the greedy repair operator can find the single best solution more often. Intuitively, the superior performance of the greedy operator in our setting is due to our objective of finding a single best solution minimizing our constraint penalty. Our evaluation metrics for the tasks only depend on the best solution and not on the quality or diversity of the underlying solution distribution. Given that the neural model is trained to minimize a constraint violation penalty for the assignments, greedy repair is effectively a *maximum a posteriori*—or MAP—decoding step that directly targets a single high-quality assignment, whereas sampling is better aligned with exploring multiple diverse assignments (Table 5).

Table 4. Performance comparison for Sudoku. In-distribution test instances contain $1{,}000$ instances; OOD refers to RRN test dataset (18K instances). For all but ConsFormer and ConsFormer-LNS, the reported instance solved (%) are based on the results from Xu et al. [39].

Dataset	SATNet [37]	RRN [22]	Recurrent Trans. [40]	IRED [9]	ConsFormer	ConsFormer -LNS
Test	98.3	99.8	**100**	99.4	**100**	**100**
OOD	3.2	28.6	32.9	62.1	85.8	**91.8**

Table 5. Performance comparison for Graph-Coloring tasks. All methods are evaluated with a 10 s time limit. OOD refers to Out-of-Distribution evaluation where n is larger than training instances. For all but ConsFormer and ConsFormer-LNS, the reported instance solved (%) are based on the results from Xu et al. [39].

Dataset	Greedy	OR-Tools [23]	ANYCSP [35]	ConsFormer	ConsFormer -LNS
Graph-Coloring-5 ($n = 50 \rightarrow n = 100$)					
Test	32.42	**83.08**	79.17	81.60	82.90
OOD	0.00	**57.16**	34.83	47.33	54.20
Graph-Coloring-10 ($n = 100 \rightarrow n = 200$)					
Test	0.75	52.41	0.00	53.60	**53.90**
OOD	0.00	10.25	0.00	10.20	**18.40**

RQ5 ConsFormer-LNS vs. Other Solvers. We compare the best ConsFormer-LNS configuration on each benchmark against classical and neural baselines (Tables 4 to 6). Overall, enhancing ConsFormer with LNS substantially improves its position among existing methods. On Sudoku, ConsFormer-LNS is compared to other neural approaches and attains the highest OOD percentage of instance solved (91.8% vs. 62.1% for IRED [9]), while matching the best methods on the in-distribution test set. For graph coloring with $k = 5$, ConsFormer-LNS improves on ConsFormer but is still slightly behind OR-Tools

Table 6. Performance comparison for MaxCut tasks on GSET. Numbers reported are the average gap to the best known cut size, the lower the better. Values are as reported in Xu et al. [39]. We similarly set a time limit of 180 s.

Size	Greedy	SDP [14]	RUNCSP [34]	ECO-DQN [1]	ECORD [2]	ANYCSP [35]	OR-Tools [23]	ConsFormer	ConsFormer-LNS
$\|V\|$=800	411.44	245.44	185.89	65.11	8.67	**1.22**	143.89	16.33	<u>4.44</u>
$\|V\|$=1K	359.11	229.22	156.56	54.67	8.78	**2.44**	112.78	12.44	<u>8.0</u>
$\|V\|$=2K	737.00	-	357.33	157.00	39.22	**13.11**	365.89	52.11	<u>30.56</u>
$|V|\geq$3K	774.25	-	401.00	428.25	187.75	**51.63**	378.62	115.25	<u>63.63</u>

on the smaller instances. On the harder graph coloring tasks with $k = 10$, it becomes strongest overall. It marginally surpasses OR-Tools on the test set and achieves a substantially higher OOD solved percentage (18.4% vs.10.25%). For MaxCut, ConsFormer-LNS consistently improves upon ConsFormer, achieving second-best performance across all graph sizes.

5   Conclusion

In this work, we have made the connection between iterative neural constraint solvers and Large Neighborhood Search (LNS) explicit. We reinterpreted Cons-Former, a recent neural heuristic, as a neural repair operator and adapted it into an LNS procedure. We implemented classical as well as novel prediction-guided destroy operators that leverage the model's internal confidence and gradient signals. We systematically evaluated combinations of destroy and repair strategies.

Our empirical evaluation on Sudoku, Graph Coloring, and MaxCut demonstrated a few key findings:

- Porting the ConsFormer into the LNS framework yielded substantial performance gains across all benchmarks.
- Echoing classical LNS findings, purely greedy destroy operators frequently collapsed to poor local minima, whereas their stochastic counterparts continued to improve over iterations and reached much better final performance.
- Our novel destroy operators showed strong performance in early iterations and specific tasks like MaxCut, but no single destroy operator dominated across all tasks. This suggests that an Adaptive LNS [21,27] that utilizes multiple destroy operators is a promising direction for future work.
- Although sampling-based repair attained better average constraint satisfaction at intermediate steps, greedy decoding almost always achieved higher overall performance, consistent with a MAP-style decoding optimizing for a single best assignment rather than exploring a diverse solution distribution.
- Compared to other approaches, ConsFormer-LNS attains the strongest OOD performance among neural Sudoku solvers, becomes competitive with OR-Tools on graph coloring and strongest on the hardest k=10 OOD setting, and closes the gap to ANYCSP on MaxCut instances.

While neural methods offer promising heuristics for constraint satisfaction, the overlay of an LNS framework has clearly provided many avenues for improvement. To this end, future work is needed to fully cross-pollinate from the LNS literature to iterative neural approaches in order to reliably outperform both historical LNS and neural methods across a variety of problems. Finally, there is further potential to extend the LNS paradigm to neural Diffusion models [3, 6, 30] that bear a striking resemblance to LNS methods and may offer more constrained and controllable diffusion in a range of generative AI applications.

Acknowledgments. We thank the anonymous reviewers for their insightful feedback. This work was supported by the Institute of Information & Communications Technology Planning & Evaluation (IITP) grant funded by the Korean Government (MSIT) (No. RS-2024-00457882, National AI Research Lab Project).

References

1. Barrett, T., Clements, W., Foerster, J., Lvovsky, A.: Exploratory combinatorial optimization with reinforcement learning. In: Proceedings of the AAAI Conference on Artificial Intelligence, vol. 34, pp. 3243–3250 (2020)
2. Barrett, T.D., Parsonson, C.W., Laterre, A.: Learning to solve combinatorial graph partitioning problems via efficient exploration. arXiv preprint arXiv:2205.14105 (2022)
3. Ben-Hamu, H., Gat, I., Severo, D., Nolte, N., Karrer, B.: Accelerated sampling from masked diffusion models via entropy bounded unmasking. In: The Thirty-ninth Annual Conference on Neural Information Processing Systems (2025). https://openreview.net/forum?id=WBcBhT1NKO
4. Bengio, Y., Lodi, A., Prouvost, A.: Machine learning for combinatorial optimization: a methodological tour d'horizon. Eur. J. Oper. Res. **290**(2), 405–421 (2021)
5. Cappart, Q., Guns, T., Lombardi, M., Pesant, G., Tsouros, D.: Combining constraint programming and machine learning: from current progress to future opportunities. J. Artif. Intell. Res. **84** (2025)
6. Cardei, M., Christopher, J.K., Hartvigsen, T., Bartoldson, B.R., Kailkhura, B., Fioretto, F.: Constrained discrete diffusion. arXiv preprint arXiv:2503.09790 (2025)
7. Chang, H., Zhang, H., Jiang, L., Liu, C., Freeman, W.T.: MaskGIT: masked generative image transformer. In: Proceedings of the IEEE/CVF Conference on Computer Vision and Pattern Recognition (CVPR), pp. 11315–11325 (2022)
8. Dehghani, M., Gouws, S., Vinyals, O., Uszkoreit, J., Kaiser, L.: Universal transformers. In: International Conference on Learning Representations (2019). https://openreview.net/forum?id=HyzdRiR9Y7
9. Du, Y., Mao, J., Tenenbaum, J.B.: Learning iterative reasoning through energy diffusion. In: International Conference on Machine Learning (ICML) (2024)
10. Falkner, J.K., Thyssens, D., Schmidt-Thieme, L.: Large neighborhood search based on neural construction heuristics. arXiv preprint arXiv:2205.00772 (2022)
11. Fan, Y., Du, Y., Ramchandran, K., Lee, K.: Looped transformers for length generalization. In: The Thirteenth International Conference on Learning Representations (2025). https://openreview.net/forum?id=2edigk8yoU

12. Feng, S., Sun, W., Li, S., Talwalkar, A., Yang, Y.: FrontierCO: Real-world and large-scale evaluation of machine learning solvers for combinatorial optimization. In: The Fourteenth International Conference on Learning Representations (2026). https://openreview.net/forum?id=BVprkacwFY

13. Feng, S., Sun, Z., Yang, Y.: SPL-LNS: sampling-enhanced large neighborhood search for solving integer linear programs. arXiv preprint arXiv:2508.16171 (2025)

14. Goemans, M.X., Williamson, D.P.: Improved approximation algorithms for maximum cut and satisfiability problems using semidefinite programming. J. ACM (JACM) **42**(6), 1115–1145 (1995)

15. Hottung, A., Tierney, K.: Neural large neighborhood search for the capacitated vehicle routing problem. In: ECAI 2020, pp. 443–450. IOS Press (2020)

16. Hottung, A., Wong-Chung, P., Tierney, K.: Neural deconstruction search for vehicle routing problems. Trans. Mach. Learn. Res. (2025). https://openreview.net/forum?id=bCmEP1Ltwq

17. Hottung, A., Tierney, K.: Neural large neighborhood search for routing problems. Artif. Intell. **313**, 103786 (2022)

18. Huang, T., Ferber, A.M., Tian, Y., Dilkina, B., Steiner, B.: Searching large neighborhoods for integer linear programs with contrastive learning. In: International Conference on Machine Learning, pp. 13869–13890. PMLR (2023)

19. Jolicoeur-Martineau, A.: Less is More: recursive reasoning with tiny networks. arXiv preprint arXiv:2510.04871 (2025)

20. Kim, J., Shah, K., Kontonis, V., Kakade, S.M., Chen, S.: Train for the worst, plan for the best: understanding token ordering in masked diffusions. In: Forty-second International Conference on Machine Learning (2025). https://openreview.net/forum?id=DjJmre5IkP

21. Mara, S.T.W., Norcahyo, R., Jodiawan, P., Lusiantoro, L., Rifai, A.P.: A survey of adaptive large neighborhood search algorithms and applications. Comput. Oper. Res. **146**, 105903 (2022)

22. Palm, R., Paquet, U., Winther, O.: Recurrent relational networks. In: Advances in Neural Information Processing Systems, vol. 31 (2018)

23. Perron, L., Didier, F.: CP-SAT. https://developers.google.com/optimization/cp/cp_solver/

24. Pisinger, D., Ropke, S.: Large neighborhood search. In: Handbook of Metaheuristics, pp. 99–127. Springer (2018). https://doi.org/10.1007/978-3-319-91086-4_4

25. Popescu, A., et al.: An overview of machine learning techniques in constraint solving. J. Intell. Inf. Syst. **58**(1), 91–118 (2022)

26. Qiu, R., Sun, Z., Yang, Y.: Dimes: a differentiable meta solver for combinatorial optimization problems. Adv. Neural. Inf. Process. Syst. **35**, 25531–25546 (2022)

27. Ropke, S., Pisinger, D.: An adaptive large neighborhood search heuristic for the pickup and delivery problem with time windows. Transp. Sci. **40**(4), 455–472 (2006)

28. Sanokowski, S., Hochreiter, S., Lehner, S.: A diffusion model framework for unsupervised neural combinatorial optimization. In: International Conference on Machine Learning, pp. 43346–43367. PMLR (2024)

29. Shaw, P.: Using constraint programming and local search methods to solve vehicle routing problems. In: Maher, M., Puget, JF. (eds.) International Conference on Principles and Practice of Constraint Programming, pp. 417–431. Springer, Berlin, Heidelberg (1998). https://doi.org/10.1007/3-540-49481-2_30

30. Shi, J., Han, K., Wang, Z., Doucet, A., Titsias, M.: Simplified and generalized masked diffusion for discrete data. Adv. Neural. Inf. Process. Syst. **37**, 103131–103167 (2024)

31. Song, J., Yue, Y., Dilkina, B., et al.: A general large neighborhood search framework for solving integer linear programs. Adv. Neural. Inf. Process. Syst. **33**, 20012–20023 (2020)
32. Sonnerat, N., Wang, P., Ktena, I., Bartunov, S., Nair, V.: Learning a large neighborhood search algorithm for mixed integer programs. arXiv preprint arXiv:2107.10201 (2021)
33. Sun, Z., Yang, Y.: DIFUSCO: graph-based diffusion solvers for combinatorial optimization. Adv. Neural. Inf. Process. Syst. **36**, 3706–3731 (2023)
34. Toenshoff, J., Ritzert, M., Wolf, H., Grohe, M.: Graph neural networks for maximum constraint satisfaction. Front. Artif. Intell. **3**, 580607 (2021)
35. Tönshoff, J., Kisin, B., Lindner, J., Grohe, M.: One model, any CSP: graph neural networks as fast global search heuristics for constraint satisfaction. In: Proceedings of the Thirty-Second International Joint Conference on Artificial Intelligence. IJCAI '23 (2023). https://doi.org/10.24963/ijcai.2023/476
36. Wang, G., Schiff, Y., Sahoo, S.S., Kuleshov, V.: Remasking discrete diffusion models with inference-time scaling. In: The Thirty-ninth Annual Conference on Neural Information Processing Systems (2025). https://openreview.net/forum?id=IJryQAOy0p
37. Wang, P.W., Donti, P., Wilder, B., Kolter, Z.: SATNet: bridging deep learning and logical reasoning using a differentiable satisfiability solver. In: International Conference on Machine Learning, pp. 6545–6554. PMLR (2019)
38. Wu, Y., Song, W., Cao, Z., Zhang, J.: Learning large neighborhood search policy for integer programming. In: Beygelzimer, A., Dauphin, Y., Liang, P., Vaughan, J.W. (eds.) Advances in Neural Information Processing Systems (2021). https://openreview.net/forum?id=IaM7U4J-w3c
39. Xu, Y., Li, W., Sanner, S., Khalil, E.B.: Self-supervised transformers as iterative solution improvers for constraint satisfaction. In: Forty-second International Conference on Machine Learning (2025). https://openreview.net/forum?id=IQN6ID0snT
40. Yang, Z., Ishay, A., Lee, J.: Learning to solve constraint satisfaction problems with recurrent transformer. In: The Eleventh International Conference on Learning Representations (2023). https://openreview.net/forum?id=udNhDCr2KQe
41. Ye, Y.: The GSet dataset (2003)
42. Zhou, J., Wu, Y., Cao, Z., Song, W., Zhang, J., Chen, Z.: Learning large neighborhood search for vehicle routing in airport ground handling. IEEE Trans. Knowl. Data Eng. **35**(9), 9769–9782 (2023)

Author Index